Feature Papers in Communications Section 2022

Feature Papers in Communications Section 2022

Editor

Peter Chong

MDPI • Basel • Beijing • Wuhan • Barcelona • Belgrade • Manchester • Tokyo • Cluj • Tianjin

Editor
Peter Chong
Auckland University of Technology
New Zealand

Editorial Office
MDPI
St. Alban-Anlage 66
4052 Basel, Switzerland

This is a reprint of articles from the Special Issue published online in the open access journal *Sensors* (ISSN 1424-8220) (available at: https://www.mdpi.com/journal/sensors/special_issues/FP_CS).

For citation purposes, cite each article independently as indicated on the article page online and as indicated below:

LastName, A.A.; LastName, B.B.; LastName, C.C. Article Title. *Journal Name* **Year**, *Volume Number*, Page Range.

ISBN 978-3-0365-7120-1 (Hbk)
ISBN 978-3-0365-7121-8 (PDF)

Contents

About the Editor

Peter Chong

Professor Peter Chong is the Associate Head of School (Research) at the School of Engineering, Computer and Mathematical Sciences, Auckland University of Technology, New Zealand. Between 2016 and 2021, he was the Head of Department of Electrical and Electronic Engineering at AUT. He received his Ph.D. degree in Electrical and Computer Engineering from the University of British Columbia, Canada, in 2000. He is currently an Adjunct Professor at the Department of Information Engineering, Chinese University of Hong Kong, Hong Kong. He is an Honorary Professor at Amity University, India. He is a Fellow of the Institution of Engineering and Technology (FIET), UK. Prof. Chong is listed in the World's Top 2% Scientists published by Stanford University in 2022. Before joining AUT in 2016, Professor Chong was an Associate Professor (tenured) from July 2009 to April 2016 and Assistant Professor from May 2002 to June 2009 at the School of Electrical and Electronic Engineering, Nanyang Technological University (NTU), Singapore. Between 2013 and 2016, he was the Director of Infinitus, Centre for Infocomm Technology. He was the recipient of the 'EEE Teaching Excellence Award' and 'Nanyang Award Excellence in Teaching' in 2010, and 'Nanyang Education Award (College)' in 2015. From February 2001 to May 2002, he was with the Radio Communications Laboratory at the Nokia Research Center, Finland. Between July 2000 and January 2001, he worked in the Advanced Networks Division at Agilent Technologies Canada Inc., Canada. He co-founded P2 Wireless Technology in Hong Kong in 2009 and Zyetric Technologies in Hong Kong, New Zealand, and the US in 2017. His current research projects focus on machine learning techniques applied to software-defined vehicular networks. He has been developing techniques of deep reinforcement learning (DRL)-based resource management for future 5G-V2X networks. His research interests are in the areas of wireless/mobile communications systems, including radio resource management, multiple access, MANETs/VANETs, green radio networks, and 5G-V2X networks. He has published over 300 journal and conference papers, 1 edited book,13 book chapters, and 4 US patents in the relevant areas.

sensors

MDPI

Article

Non-Intrusive Privacy-Preserving Approach for Presence Monitoring Based on WiFi Probe Requests

Aleš Simončič [1,2], Miha Mohorčič [1], Mihael Mohorčič [1,2,*,†] and Andrej Hrovat [1,2,†]

1 Jožef Stefan Institute, Jamova 39, 1000 Ljubljana, Slovenia; ales.simoncic@ijs.si (A.S.); mmohorcic@ijs.si (M.M.); andrej.hrovat@ijs.si (A.H.)
2 Jozef Stefan International Postgraduate School, Jamova 39, 1000 Ljubljana, Slovenia
* Correspondence: mihael.mohorcic@ijs.si
† These authors contributed equally to this work.

Abstract: Monitoring the presence and movements of individuals or crowds in a given area can provide valuable insight into actual behavior patterns and hidden trends. Therefore, it is crucial in areas such as public safety, transportation, urban planning, disaster and crisis management, and mass events organization, both for the adoption of appropriate policies and measures and for the development of advanced services and applications. In this paper, we propose a non-intrusive privacy-preserving detection of people's presence and movement patterns by tracking their carried WiFi-enabled personal devices, using the network management messages transmitted by these devices for their association with the available networks. However, due to privacy regulations, various randomization schemes have been implemented in network management messages to prevent easy discrimination between devices based on their addresses, sequence numbers of messages, data fields, and the amount of data contained in the messages. To this end, we proposed a novel de-randomization method that detects individual devices by grouping similar network management messages and corresponding radio channel characteristics using a novel clustering and matching procedure. The proposed method was first calibrated using a labeled publicly available dataset, which was validated by measurements in a controlled rural and a semi-controlled indoor environment, and finally tested in terms of scalability and accuracy in an uncontrolled crowded urban environment. The results show that the proposed de-randomization method is able to correctly detect more than 96% of the devices from the rural and indoor datasets when validated separately for each device. When the devices are grouped, the accuracy of the method decreases but is still above 70% for rural environments and 80% for indoor environments. The final verification of the non-intrusive, low-cost solution for analyzing the presence and movement patterns of people, which also provides information on clustered data that can be used to analyze the movements of individuals, in an urban environment confirmed the accuracy, scalability and robustness of the method. However, it also revealed some drawbacks in terms of exponential computational complexity and determination and fine-tuning of method parameters, which require further optimization and automation.

Keywords: clustering; information element; MAC de-randomization; OPTICS; probe request; WiFi-enabled device detection; wireless sensing device

Citation: Simončič, A.; Mohorčič, M.; Mohorčič, M.; Hrovat, A. Non-Intrusive Privacy-Preserving Approach for Presence Monitoring Based on WiFi Probe Requests. *Sensors* **2023**, *23*, 2588. https://doi.org/10.3390/s23052588

Academic Editor: Peter Chong

Received: 23 January 2023
Revised: 21 February 2023
Accepted: 24 February 2023
Published: 26 February 2023

1. Introduction

Although presence and movement monitoring can have a negative connotation when applied to an individual, it also has the potential to improve and even save lives, but it must be conducted in accordance with some social consensus on preserving privacy. This challenge is being addressed with an increasing number of existing and emerging services and applications, for instance in the fields of public safety, transportation, urban planning, disaster and crisis management, mass events organization, etc. that depend on the information about people's presence, proximity, occupancy, crowdedness, crowd dynamics, and movement patterns. Many of these services and applications relied on vision-based solutions until recently, when many camera-based monitoring systems have been shut down because they violated the new General Data Protection Regulation laws [1]. These systems are now being replaced by the development of new, less intrusive methods based on the indirect monitoring of devices equipped with various sensors and radio interfaces, e.g., WiFi, Bluetooth, 3G/4G/5G, etc.

As per development indicators collected by the World Bank, the number of cellular subscriptions in the world per 100 people grew from 87 in 2012 to 110 in 2021, reaching more than 100 in most developed countries, indicating that the mobile phone is the most omni-present personal device. In recent years, practically all smart mobile phones come with a number of wireless technologies including WiFi and Bluetooth in addition to supporting present and past generations of mobile network technologies (i.e., 2G–5G). For the association with available networks and provision of connectivity, all these technologies rely on exchanging some network management messages with network devices such as base stations and access points. Data in these messages can be used for non-intrusive detection of presence and movement patterns of individuals by way of tracking their carry-on devices. While there are strict legally enforced rules in place regarding the access to such data collected by mobile operators from mobile network technologies, wireless technologies have been more prone for potential misuse for instance due to the use of globally unique medium access control (MAC) addresses. Bluetooth relies on the short-range and point-to-point nature of its protocols to avoid tracking. Personal Bluetooth devices (e.g., mobile phones) also by default do not publicly advertise their presence. Wi-FI devices constantly scan and advertise their capabilities for better power management and quicker connection; therefore, they can be easily detected, recognized and tracked. This potential privacy breach led to the introduction of WiFi MAC address randomization, making the MAC address time-varying and random rather than globally unique. Recent randomization procedures in WiFi make it even more difficult to identify individual devices, as the sequence numbers of messages can also be randomized and the amount of data contained in messages can be reduced to only a few mandatory fields.

The recent COVID-19 pandemic with the introduction of preventive measures such as social distancing intensified investigations in non-intrusive privacy-preserving monitoring of individual's social interactions largely by means of smart phones and various mobile apps (e.g., [2–4]). On the other hand, many applications still require only data on monitoring the presence and movement of anonymous individuals or crowds to provide an insight into actual patterns of behavior and enable the identification of hidden trends. In our study as part of the RESILOC project (https://www.resilocproject.eu/ (accessed on 22 February 2023)), we were focusing on the public safety domain, where knowing crowd densities, their dynamics and patterns in specific locations or strategic areas enables the planning and implementation of appropriate preventive measures and management strategies to improve the resilience of local communities. In particular, our goal was to develop a reliable system for the non-intrusive collection of anonymized information about the presence or movement patterns in terms of statistical counts without identifying and saving any privacy-sensitive information that would enable backtracking a device or an individual person, while efficiently avoiding possible double or multiple counts.

In this paper, we address the MAC address randomization problem for non-intrusive and privacy-preserving detection of WiFi-enabled devices. We propose a new MAC derandomization method, which is able to recognize unique devices with high accuracy by grouping similar network management messages (called Probe Requests, PRs) based on the data in the message and the corresponding radio channel characteristics. The method uses a novel clustering and matching procedure that was calibrated with a publicly available labeled dataset. It was validated by the measurements performed using a custom-designed, low-cost system for capturing, transferring and storing WiFi PRs. The validation measurements took place in a fully controlled rural environment and a semi-controlled indoor environment at the Jozef Stefan Institute (JSI). An additional verification of the method in terms of scalability and accuracy was performed in a real, crowded urban environment in the city of Catania with high-density pedestrian traffic and consequently a high number of devices transmitting a large number of PRs. The main contributions of this work are:

1. The design and implementation of a low-cost system for capturing, transferring and storing WiFi PRs and corresponding radio channel characteristics.
2. Open datasets of the captured WiFi PRs and corresponding radio channel characteristics in a controlled rural outdoor, semi-controlled indoor and uncontrolled urban outdoor environments.
3. A novel MAC de-randomization method for distinguishing individual WiFi-capable devices including new clustering and matching procedures based on PRs and corresponding radio channel characteristics.
4. Validation of the proposed method by the measurements in controlled, semi-controlled and completely uncontrolled environments.

The rest of the paper is organized as follows. Section 2 provides the necessary background on the structure of WiFi PRs and randomization of MAC addresses, and it outlines the related work on WiFi based monitoring of population behavior. Section 3 describes the proposed system architecture and its implementation using wireless sensing devices (WSDs) and a remote server with database for storing raw data. Section 4 gives a detailed description of the proposed MAC de-randomization method including the procedures for the collection and pre-processing of data as well as for the clustering and matching of PRs for identifying unique WiFi-enabled devices. Section 5 defines the validation scenarios considering different numbers and groupings of devices and different operating environments before it provides performance evaluation results. Finally, Section 6 concludes the paper and outlines some ideas for future work.

2. Background and Related Work

Exploiting the information acquired from the WiFi network management messages has been a widely used approach for user tracking, crowd monitoring, presence detection, etc. This approach has become quite challenging in recent years, which is mainly due to the privacy guarantees that manufacturers achieve by randomizing the MAC addresses of WiFi interfaces. Therefore, device-based passive tracking approaches that detect a device carried by a user had to address the issue by new methods that consider several additional parameters/data extracted from the WiFi management frames. In the following, we briefly present the basic types of WiFi management frames that contain data useful for methods to determine the number of devices or to distinguish between individual detected WiFi-enabled devices and MAC addressing randomization followed by a brief overview of related research and applications.

2.1. Probe Requests and MAC Randomization

The basic idea of non-intrusive presence monitoring is to exploit the standard operation of WiFi-enabled devices, i.e., their activity when not connected to an operational WiFi network. In general, WiFi technology is based on the 802.11 standard that defines several frame types which are categorized in three major groups, namely, (i) data frames for data transmission, (ii) control frames for controlling the access to the wireless medium and for validation of received frames, and (iii) management frames used by supervisory functions (e.g., associating the device with the network, roaming between access points, etc.). When a device with an enabled WiFi interface is in an unassociated state, i.e., not connected to a network, it carries out passive and/or active scanning for available WiFi networks that it could connect to. Both approaches rely on WiFi management frames which are not encrypted as they do not contain any user data. Passive scanning refers to devices that are waiting and listening for announcements from access points sent in beacon frames, successively moving through the entire set of channels. Active scanning relies on WiFi-enabled devices sending PR frames on a selected channel and waiting for a probe response from nearby access points, and if no response is received for a certain time, repeating the procedure on the next channel.

Figure 1 shows a generic WiFi management frame. Management frames are denoted by the *Subtype* field of the Frame Control field. Frame body of management frames uses fixed-length fields called *fixed fields* and variable-length fields called *Information Elements (IE)*. While fixed fields do not have a header as their length and order are predefined, the first octet in IE fields defines the element ID and the second octet defines its length. The mandatory IEs in the PRs' body frame are Service Set Identifiers (SSIDs), i.e., the unique names of WiFi networks that the device was already associated with in the past, and supported data rates. Based on IE data in the received PR, the access point determines whether the device fulfills the conditions to join the network or not.

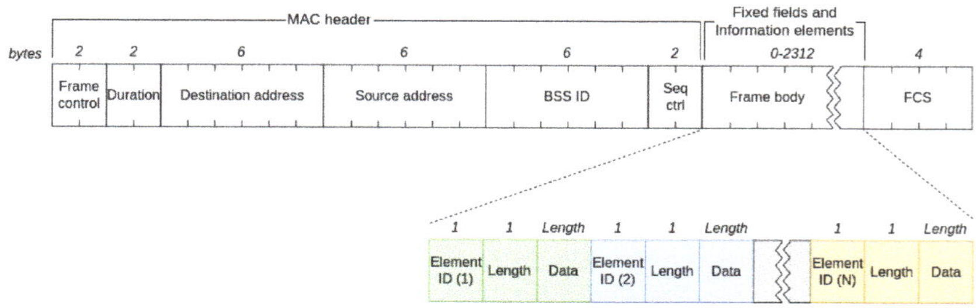

Figure 1. Generic 802.11 management frame with IE fields in the frame body.

By actively scanning for available WiFi networks, a WiFi-enabled device becomes discoverable to nearby listeners since it needs to include its MAC address in the source address field of PR. In case of having a globally unique MAC address, this makes it distinguishable from other devices in the network. MAC addresses have a standardized length of six octets. In a globally unique MAC address, the first three octets, called an Organization Unique Identifier (OUI), are unique to a manufacturer and are defined by IEEE. The last three octets, called the Network Interface Controller (NIC), are assigned by a device manufacturer to make each device uniquely distinguishable.

Since the MAC address is tied to a device and consequently to the person carrying it, manufacturers started to implement different schemes to randomize MAC adressess and thus protect user privacy. The seventh bit in the first octet of the MAC address indicates whether a device is using a globally unique address that is constant over time (bit set to '0') or a randomized address (bit set to '1'). Apple with iOS version 8.0 was the first to release its devices with MAC randomization for mass sale in 2014. Although MAC randomization is now implemented by almost every device for pre-association state and for post-association state with an access point [5], the process of randomization is not standardized. Some manufacturers randomize the entire MAC address, while others use a fixed Company Identifier (CID) for first three octets and randomize the remaining three octets. To prevent tracking a particular MAC address, devices are changing their MAC address over time. The frequency or specific time events of generating new MAC addresses are also not standardized. Some devices change their MAC address for each burst that PRs are sent, while others change it less frequently. The entire process of generating and changing MAC addresses is managed by the operating system.

Undocumented and proprietary source codes make it difficult to analyze vulnerabilities in the randomization process. However, previous works show that tracking WiFi-enabled devices is possible using various techniques that differ in terms of the frame type, state of communication, physical characteristics of the radio waves, radio channel and transceiver characteristics of the particular device [5]. Some techniques, such as using the WPS field in PR to infer the actual MAC address of the device or using the sequence number to distinguish mobile devices, are already outdated, which shows the efforts of manufacturers to quickly fix the potential privacy vulnerabilities. To make it even more challenging to identify a unique WiFi device, some manufacturers have introduced sending subsequent PRs that contain different data, which makes it difficult for the observer to associate them to a single device, or they send PRs that contain only the mandatory IEs. In this way, the data from different mobile devices look very similar and are therefore harder to distinguish.

2.2. Related Work

Numerous solutions have been proposed in the literature for monitoring population behavior in terms of crowdedness, density, presence, proximity, etc., based on WiFi traffic analysis. Early solutions were based on simply tracking MAC addresses, assuming that MAC addresses are unique for each WiFi-enabled device [6]. These solutions are now outdated due to MAC address randomization. The privacy became important in any system that collects data and in particular if it processes user-related information, as data protection rights under the GDPR (e.g., Regulation EU 2016/679) must be fully met. In recent years, MAC address randomization has attracted a lot of attention from vendors, who have developed and implemented various solutions [7]. An extensive test of different solutions was carried out [5] to determine the usage of randomization, under what conditions MAC address randomization is performed, and if the tracking vulnerabilities are suitably mitigated.

To distinguish between different WiFi-enabled devices while protecting user privacy, different approaches and techniques have been applied. Initially, differentiating between unique devices was based on the timing analysis of PR frames. In [8], the authors measured the time between received PRs on different channels. Since the time delay is not specified by the standard and depends on the configuration of a device, it is a suitable feature for fingerprinting. A similar approach based on timing analyses of the PR frames is reported in [9]. The de-randomization of MAC addresses can be based on timing attacks, where the inter-frame arrival times of PRs are used to group frames coming from the same device, although they use distinct MAC addresses as proposed in [10]. Frames are grouped by several distance metrics based on the timing and incremental learning algorithm.

Timing as a distinction feature is unreliable in real-world environments due to scattering and multi-path phenomena which introduce some random delays between probes and bursts [11]. Thus, more reliable solutions to fingerprint the devices based on IEs in PR have been widely addressed in the literature. In [12], a study of IEs is presented, and new fields and techniques to track users are identified. It was demonstrated that scrambler seeds of commodity WiFi radios are predictable and can be used for device identification. In addition, two attacks that reveal the real MAC address of devices were presented.

Many applications for counting people at a specific location based on PRs were developed as in [13], where the goal was to count the passengers in public transport. It is worth noting that the majority of Android handsets did not use randomized techniques in that study. A solution for counting participants in public demonstrations proposed in [14] was based on a WiFi PR broadcast by the phones. The basic signal behavior was investigated by applying a distance filter based on RSSI, which is impractical due to applying a common threshold, and time-based filters, which have extra requirements regarding the scanner setup and increase the likelihood that counted devices actually belong to participants. Results showed that the count from analyzing PRs represented only a small fraction of the actual attendance. A method for pedestrian counting, classifying them as moving or static and locating them in a road intersection, is exploited in [15] and compared with machine learning (ML) techniques in [16]. It is based on power measurements and PRs; however, it does not provide a description of the MAC randomization procedure used. The solution for understanding the behavior of visitors in [17] addresses the issue of the device randomization by using more than 1.7 million PR frames. historical transition probability and a Hidden Markov Model (HMM)-based trajectory inference algorithm. The group behavior detection system introduced in [18] is also based on capturing PRs and tackling MAC address randomization by the method described in [7] based on collective matrix factorization, which reveals the hidden associations by factorizing mobility information and usage patterns simultaneously.

In [19], an approach for estimating the presence of mobile devices at a certain place in time is proposed, which is immune to MAC address randomization. The approach is based on the state machine to detect the arrival, presence and departure of devices in the sensor proximity. A novel architecture for people mobility monitoring and analyzing a solution based on PRs is presented in [20]. The main features include the preservation of user privacy, extraction of key metrics on user return and permanence, and computation of mobility heat maps. Another solution for studying the mobility patterns proposed in [21] is based on PRs which correlate the multi-dimensional statistical properties of the captured PRs split by brands with the actual ground truth, which is manually labeled for creating a regression model. A low-cost, high-reliability and low-complexity real-time passenger counting system is reported in [22] with a highly accurate mathematical model in conjunction with the MAC data provided by the developed system and applied Kalman filter. The authors in [23] have developed a crowd estimation scheme that addresses MAC randomization by exploiting the fact that the number of PRs in a defined time interval changes proportionally to the number of devices present. The conversion factor between the number of devices and the number of PRs was determined by measuring the bursts per minute of PRs from 75 devices, whereas the authors in [24] used a statistical approach to estimate the client population from PR counts without requiring an additional ground truth technique.

Vision and TrueSight crowd monitoring algorithms that estimate the number of devices were exploited in [25] to prove that despite MAC address randomization, MAC address-based crowd monitoring is still a viable solution. Both approaches mitigate the influence of the randomization by PRs. While Vision uses data and beacon packets, TrueSight uses PR sequence numbers and hierarchical clustering. A similar approach of tracking the sequence number of the PR, assuming that the number increases with newly received PR, is given in [26], which also incorporates the timestamp of the received PR. Based on these parameters, a de-randomization algorithm calculates the score for each couple of random MAC addresses and identifies which MAC addresses belong to the same device based on the defined threshold. An algorithm incorporated in the system for analyzing urban mobility is applied in [27] for the privacy-preserving detection of people's flow by an ML approach [27] and in [28] where two methodologies for people counting and mobility detection are proposed. However, new privacy-preserving approaches tend to completely randomize the sequence number of PR which makes the previously mentioned methods unreliable [5]. Despite the enhanced randomization, the authors in [29] developed an efficient crowd monitoring system based on the passive detection of PRs. The algorithm counts the devices from a measured rate of PR burst (PRs received in 10 ms) transmissions using a statistical estimator. Another solution for crowd monitoring based on the information in PR is proposed in [30], where the SSID information of the preferred WiFi access points included in PR is exploited in post-processing together with the information of the existing WiFi access points to determine the daily number of visitors at different locations.

A solution for estimating the number of people proposed in [31] is based on a footprint mechanism to overcome MAC address randomization, generating an identifier for probes based on MAC addresses and other IE data. The test in real scenarios showed that the solution can achieve the accuracy close to 95%. Based on the analyses of the influence that different MAC address randomization schemes have on statistical counts of the WiFi-based monitoring systems, another approach was proposed and verified in [1], which is based on the DBSCAN clustering algorithm and two fingerprinting features. The first feature consists of an ordered list of IEs and certain bitmasks available in PR, while the second feature is based on the burst length and the arrival time difference between PRs within that burst, or the Inter-Frame Arrival Time (IFAT).

To estimate the number of WiFi-enabled devices in a given area, the authors in [11] also used clustering algorithms, namely DBSCAN and OPTICS. They extracted the most relevant IEs from PRs and used their data lengths as features to distinguish between different devices. Additionally, they wrote an algorithm to dynamically detect which IEs are changing between different PRs and use only the most relevant ones. In laboratory testing, they achieved 91.3% accuracy with the OPTICS algorithm. The authors extended their work in [32], where they also considered pseudo-random MAC address detection, HDBSCAN as a clustering algorithm, and additional features for burst identification: namely, sequence number, RSSI, and time of arrival. In a controlled environment, they achieved 96% accuracy, while in the real scenario, they achieved an average accuracy of 75%. Since PRs from two different devices can contain IEs of the same lengths, this method is prone to undercounting the number of unique WiFi devices. The authors in [33] also used the DBSCAN algorithm and a publicly available dataset [34] for clustering PRs. They determined the impact of IE as features for the clustering algorithm by computing Gini importance with Random Forests, and by using the most important features, they achieved a clustering accuracy of 92%. Since the approaches applying the DBSCAN and OPTICS algorithms gave the most promising results in the related literature, they were also used as the basis for developing the algorithm described in this paper.

3. System Design, Implementation and Deployment

Leveraging the existing off-the-shelf technologies, a low-cost system shown in Figure 2 was designed and built to monitor and capture WiFi network management messages. These messages are captured passively (without user or device interaction) by listening for PR packets on a single channel of the 2.4 GHz WiFi band.

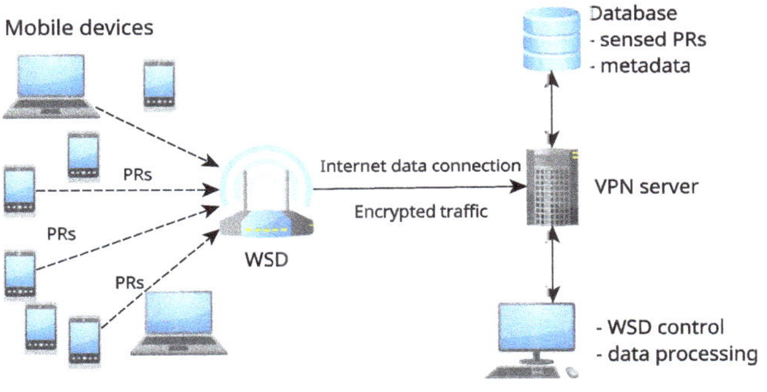

Figure 2. Schematic of system design and flow of data.

3.1. System Architecture

The system shown in Figure 2 is composed of a WSD for capturing WiFi network management traffic which is connected either via the wireless access point or with an Ethernet cable through a secure VPN connection to the remote database. The database stores raw data for further processing, together with additional metadata about the deployment and optional pre-processed data. Once the data are stored in the remote database, it can be further analyzed either in real time or by using more complex algorithms for post-processing the collected data.

3.2. Capturing the WiFi Network Management Traffic

Passive monitoring of the network management traffic is performed by a WSD listening to the WiFi traffic in a 2.4 GHz band and filtering out the PR packets. On a WSD, certain minimum computing power is required for filtering, pre-processing, and transmitting the collected data. This dictates the minimum capabilities of CPU and connectivity options and limits the software stack used.

Due to its relatively low cost, good support and known capabilities, the Raspberry Pi (rPi) device (version 4 with 2 GB of internal memory) was selected to be used as the WSD. The rPi's built-in WiFi adapter and drivers do not support the operation in a so-called monitor mode required for capturing the WiFi traffic. Therefore, the built-in adapter is used as a default connection to the Internet for the transfer of captured data to the database, whereas an additional USB dongle using a Realtek RTL88212au chipset with open-source drivers was used to enable the monitor mode of operation. In addition to capturing WiFi PR packets, the USB dongle on the WSD is also capable of capturing Bluetooth beacon advertisements from Bluetoth tags or devices if enabled.

The operating system used for the WSD is a minimal installation of 64 bit Raspberry Pi OS. A custom service is run upon the device startup to automatically start wireless data collection. The traffic monitoring data are collected using the Tshark packet capture program, filtered and parsed in Python using the PyShark package. The collected data are transferred in a JSON format to the PostgreSQL database via HTTPS REST (Representational State Transfer) protocol calls to the remote server through a secure VPN connection, ensuring that the sensitive data are not exposed at any point in the communication pipeline. The described data flow is depicted in Figure 2.

The database used to store raw data collected from WSDs is an instance of the PostgreSQL database running on remote infrastructure. The collected raw data are stored together with the corresponding metadata. Periodically, every ten seconds, the WSD transmits collected data to the database. Each PR contains all of the information from the captured packet in a verbose, human readable JSON format.

3.3. System Deployment

Before the system is deployed in the targeted operating environment, WSDs are pre-programmed, pre-installed and tested in a controlled laboratory environment. When a device is set up and turned on in the field, it automatically creates an access point for the user who connects to the access point and sets the metadata (e.g., location and ID of the device). The connection to the Internet can be established either via another access point or via the Ethernet cable. The system was deployed at multiple locations as described in Section 5.1, each with specific observable variations in WiFi traffic.

4. Detecting Unique WiFi Interfaces

To determine the actual number of WiFi-enabled devices at a certain location covered by WSD in a completely anonymous and unobtrusive way, we developed and implemented a method for the de-randomization of MAC addresses of the WiFi interfaces integrated in devices. In this respect, we exploit the data collected by WSDs which can be, after the processing phase, further used for different use cases, namely (i) defining the number of people at a location of interest in given time or in different time frames; (ii) analyzing population movement patterns in streets, shopping malls, etc. in different timeframes; (iii) counting crossings or passages over bridges, entrances, streets, etc.; (iv) real-time adaptations of emergency exits/directions; etc.

4.1. Data Collection

The authors in [11,32] used only the length of the data from the IE fields in PR to de-randomize MAC addresses of WiFi-enabled devices and identify their number in a given area. In this study, we adopted a similar approach as in [34] and also considered the data itself, along with time of arrival (ToA) and RSSI, to reduce the problem of undercounting. With the collected information, two different WiFi-enabled devices can be distinguished, or a WiFi-enabled device sending PRs with different MAC addresses can be detected. The proposed procedure takes into account that data in PRs are strongly influenced by the chipset, the device driver and the WiFi software stack.

The first step in collecting the information regarding PRs was to identify which IEs and other PR information are more specifically characterizing a given device. In [12,35], the authors calculated the entropy and stability of IEs in datasets and investigated which IEs are changing between devices and which IEs are stable for a particular device. The goal was to choose IEs with high entropy, meaning that the data from IEs for different devices are considerably different, and at the same time highly stable, so that the data from IEs for a particular device are stable over time and do not estimate a single device as multiple devices.

Based on this, we decided to collect from each received PR the information about MAC address, *Supported Data Rates*, *Extended Supported Rates*, *HT Capabilities*, *Extended Capabilities*, *Interworking*, *VHT Capabilities*, data under *Extended Tag* and *Vendor-Specific Tag*, *RSSI*, *SSID* and the timestamp when PR was received. Note that not every PR includes all the parameters specified above, as all IE fields except *Supported Data Rates* and *SSID* are optional, so the information about which IEs a given WiFi-enabled device is transmitting is also relevant in device characterization. Selected IEs with information about the stored data type and data length are listed in Table 1. *Supported Data Rates* and *Extended Supported Rates* are represented as arrays of values that encode information about the rates supported by a mobile device. The rest of the data from IEs is represented in hexadecimal format. The *Vendor-Specific Tag* is structured differently than the other IEs. This field can contain multiple vendor IDs with multiple data IDs and corresponding data. Similarly, the *Extended Tag* can contain multiple data IDs with corresponding data.

Table 1. Selected IEs, their data type and data length.

IE Name	Data Type	Data Length in Octets
SSID	UTF-8 encoded	Variable (max 32)
Supported Data Rates	Each data rateencoded as one octet	Variable (max 8)
Extended Supported Rates	Each data rateencoded as one octet	Variable (max 255)
HT Capabilities	Hex	26
Extended Capabilities	Hex	Variable
Interworking	Hex	1–9
VHT Capabilities	Hex	12
Vendor Specific Tag	Hex	Variable
Extended Tag	Hex	Variable

4.2. Data Pre-Processing and Storing

WSDs scan for PRs for a predefined scan time, and during this time, the data from IE fields are pre-processed and saved in a predefined JSON structure before being transferred to the database. For more information about the structure of saved data, see Appendix A. The pre-processing procedure is depicted graphically in Figure 3.

For each new PR, the procedure first checks if its MAC address has already been detected and saved in the current scan time. If not, a new data structure is created under the new MAC address for storing PR's IE data and SSIDs. If the new PR contains one of the already existing MAC addresses from the current scan time, the procedure compares new IE data with already recorded IE data for the same MAC address. If identical PR's IE data from the same MAC address are already stored, then only data for *ToA*, *RSSI* and *SSID* are appended to the existing data structure. Thus, the database is reduced and PRs can be compared more efficiently. However, if no identical PR's IE data have yet been recorded with the same MAC address, then a new data structure under the same MAC address with new PR's IE data and possible new SSIDs is appended.

At the end of each scan time, all processed data are sent to the database along with additional metadata about the collected data such as WSD serial number and scan *start time* and *stop time*.

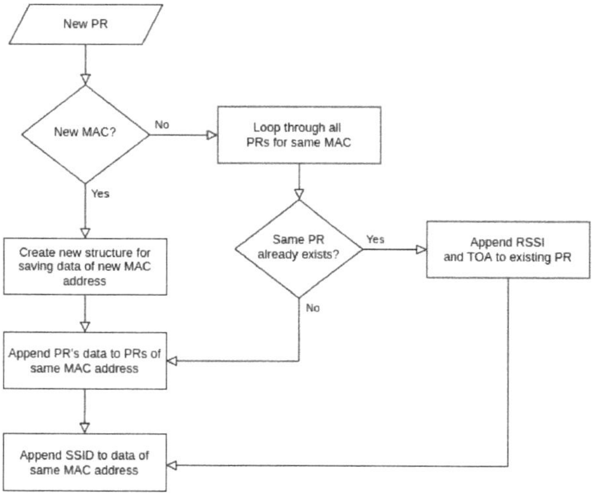

Figure 3. Algorithm for saving data from received new PRs.

4.3. De-Randomization Method

To estimate the number of unique WiFi devices during a selected time interval in the area covered by a specific WSD, the de-randomization method of gathered PRs has to be performed. This can be accomplished by clustering PRs with respect to their similarity. The data from different MAC addresses are compared by calculating their *distance* whereby MAC addresses with very similar data have small distance and vice versa.

The proposed method for matching MAC addresses comprises the following steps:

1. MAC addresses are first divided into two groups: global and random addresses. Additionally, random MAC addresses are also subgrouped with respect to the CID part of the MAC address.
2. The clustering of random MAC addresses is applied to all groups with random MAC addresses to obtain clusters from individual WiFi-enabled devices.
3. The clustering of global addresses with clusters of random addresses is applied to match global MAC addresses with clusters of random MAC addresses obtained in the previous step.
4. The number of individual WiFi-enabled devices is estimated by counting the number of clusters.

4.3.1. Initial Grouping of MAC Addresses

Since the data for each scan interval are sent to the database in packets, these packets are merged to match the desired time interval in which the de-randomization method is performed. In addition, the packets are filtered based on the serial number of the WSD that sent data to the database so that only data from one WSD is processed at any given time. Next, the algorithm for merging PRs from the same MAC address is applied, similar as used in the first pre-processing phase. In addition, each MAC address and its data are placed into a global or random group based on the *local* bit in the MAC address.

Furthermore, random MAC addresses are mapped according to their CID value. If no *CID group* is found for a particular MAC address, it is assigned to a *Random group* which corresponds to devices with random MAC address and no manufacturer identifier. This additional subgrouping allows for the finer matching of similar MAC addresses, as it is assumed that a WiFi-enabled device that sends PRs with a completely random MAC address will not send subsequent PRs with a known CID as part of its MAC address. The described initial grouping procedure is shown in Figure 4.

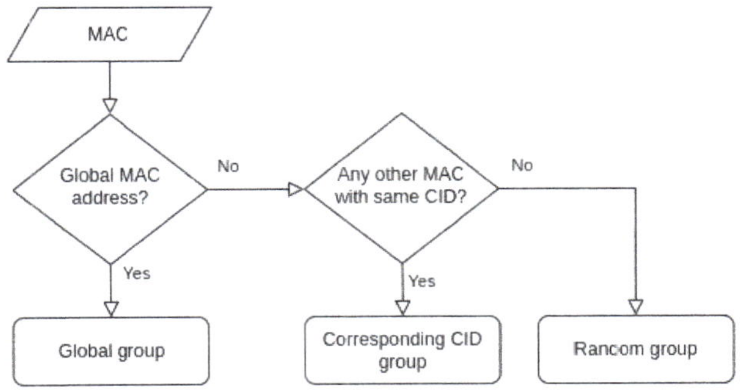

Figure 4. Algorithm for the initial grouping of MAC addresses and corresponding data.

4.3.2. Clustering of Random MAC Addresses

The clustering of random MAC addresses starts with the calculation of the distance matrix for each *CID group* and for the *Random group*. In particular, the distance between all pairs of MAC addresses is calculated and stored in a 2D array. The distance depends on the similarity of the PR's IE fields. Each IE is assigned one or more coefficients that reflect its weight. IEs that vary considerably between different WiFi-enabled devices and also have high stability for the same WiFi-enabled device have a higher weight and vice versa. For example, the distance decreases for each *SSID* that the two PRs have in common, while it increases for each supported data rate that they do not have in common. The distance is also affected by *RSSI* (increased when the absolute difference exists) and *ToA* (decreased when the absolute difference is less than a certain threshold). For other fields considered (i.e., *HT Capabilities*, *Extended Capabilities*, *Extended Tag*, *Vendor-Specific Tag*, *Interworking* and *VHT Capabilities*), the distance is increased proportionally to the number of different bits. The pseudocode for calculating the distance between two PRs can be found in Appendix B as Algorithm A1.

The distance for each PR from one MAC address (N) to each PR of another MAC address (M) is calculated. Then, the distance for these two MAC addresses is defined as an average of one-third of the shortest distances, which is inserted in an NxM distance matrix.

The distance matrix of random MAC addresses represents an input to a density-based clustering OPTICS (Ordering Points to Identify the Clustering Structure) algorithm [36]. In the proposed approach, we used a specific implementation of the OPTICS algorithm from the scikit-learn library (https://scikit-learn.org/stable/ (accessed on 22 February 2023)). The algorithm orders the data points so that the spatially closest points become neighbors. A point is classified as a core point if at least *MinPts* points are found within its neighborhood with a predefined radius. To detect a change in the density of points, the OPTICS algorithm defines two additional parameters for each point: the core distance and the reachability distance. The core distance is undefined if a point is not a core point; otherwise, it is equal to the minimum value of the neighborhood radius required to classify a given point as a core point. The reachability distance is defined for a selected point in relation to another point. It is the maximum value of the distance between these two points or the core distance of a point. If the selected point is not a core point, the reachability distance is set to undefined.

The OPTICS algorithm does not explicitly cluster the data into groups. Its output is a visualization of the reachability distances of points in the same order as processed by the algorithm. The order in which the algorithm selects points is based on the reachability distances. The point with the smallest reachability distance is selected first. The resulting 2D representation is called a reachability graph and is shown in Figure 5. The points belonging to the same cluster have low reachability distance, so they appear as valleys on the reachability diagram. The valleys are separated by spikes corresponding to the distances between clusters or between a cluster and a noise point. In other words, the peaks on the reachability plot indicate the beginning of a new cluster, as also indicated in Figure 5 by different colors.

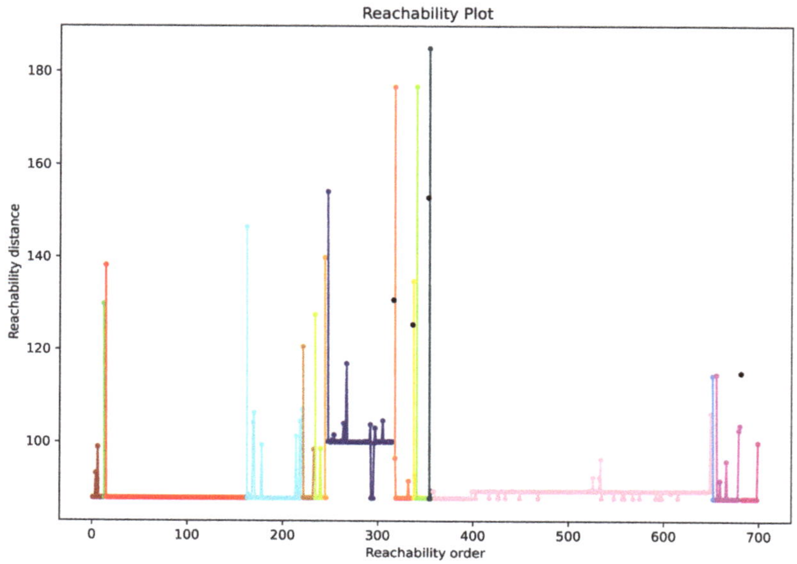

Figure 5. Reachability plot for data collected at the city square Piazza Università in Catania in an early morning 15-min interval. Points of the same cluster have the same color.

In the last step, the reachability distances obtained by the OPTICS algorithm are used in a new optimized algorithm for clustering random MAC addresses, as provided in Appendix B as Algorithm A2. The main principle of the algorithm is to detect the point in the reachability distances at which the curve starts to drop. If the drop is larger than the predefined threshold, the subsequent points are grouped in a new cluster until the values start to increase and exceed the predefined threshold, indicating the end of the cluster.

The described algorithms are applied to each *CID group* and to the *Random group*. As a result, clusters of MAC addresses that likely correspond to the same WiFi-enabled device within each *CID group* and the *Random group* are obtained. The data from MAC addresses clustered together are merged and used later for matching with global MAC addresses.

4.3.3. Matching of Global MAC Addresses with Clusters of Random MAC Addresses

The next step of the de-randomization method matches the global MAC addresses with already clustered random MAC addresses. The distance matrix is first calculated between the data of each global MAC address and the data of each cluster of random MAC addresses, whereby the distances between the global MAC addresses and the distances between the clusters of random MAC addresses are set to infinity to prevent matching.

Then, the matching algorithm iterates through the global MAC addresses and checks the distances to the clusters of random MAC addresses. If the smallest distance is less than the specified threshold and the next smallest distance is larger for a predefined factor, then the cluster of random MAC addresses with the smallest distance is a good candidate for matching with the global MAC address. In addition, all the distances for the selected cluster of random MAC addresses are checked. If the smallest distance corresponds to the same global MAC address and the next smallest distance to other global MAC address is larger by a certain factor, the global MAC address is matched to the cluster with random addresses, and their data are merged.

The code for the data pre-processing and the de-randomization method was written in the Pyhton programming language and is freely available to allow replication and continuation of the work (https://gitlab.com/e62Lab/resiloc_project/wireless-data-analysis (accessed on 22 February 2023)).

5. Performance Evaluation and Discussion

The proposed MAC de-randomization method has been tested and validated on datasets collected in different operating environments with different WiFi-enabled devices and with different test scenarios, including labeled datasets, datasets from a controlled and a semi-controlled environment, and a dataset from a challenging, completely uncontrolled environment to test robustness and scalability.

5.1. Testing Scenarios and Methodology

The datasets for performance evaluation of the proposed MAC de-randomization approach include the publicly available labeled dataset [34] as well as datasets from the measurements in three different environments, namely (i) a controlled rural environment, (ii) a semi-controlled indoor environment, and (iii) an uncontrolled urban environment, as described in the following. Measurements in rural and indoor environments were carried out under different testing scenarios, while no specific scenarios could be implemented in the uncontrolled urban environment, where only the robustness and scalability of the proposed approach could be tested under high density of WiFi-enabled devices.

For the initial testing and setting the parameters of the MAC de-randomization method, we used the publicly available labeled dataset [34] obtained with 22 devices among which 18 used MAC randomization. It contains 20-min captures of PRs in three non-overlapping WiFi channels (1, 6, and 11) for each individual device in six different operating modes (i.e., combinations of settings based on display status, Wi-Fi connectivity, and power saving) saved in 315 .pcap files. Data collection was performed in an anechoic chamber and in pseudo-isolated environments. In pseudo-isolated environments, additional filtering was performed based on the MAC address of known nearby devices and based on the power threshold of unknown nearby devices.

For validation in a controlled rural environment, we deployed three WSDs for the acquisition of PRs, which were sending data to the remote database using a cellular network. As shown in Figure 6a, they were placed in three different locations with non-overlapping WiFi coverage. The three WSDs recorded no PRs when test devices were turned off; thus, the deployment can be characterized as completely controlled without external interference. Data gathered at this location are taken as a ground truth for each device individually and as group behavior when multiple devices were active near WSD.

A dataset for validation in a semi-controlled indoor environment was also obtained with three WSDs placed in the corridors of the Jozef Stefan Insitute, as depicted in Figure 6b. The WiFi coverage areas of the devices at locations 2 and 3 were partly overlapped, while the device at location 1 had no overlapping with other devices. In this case, data were sent to the collocated database via WiFi with a known global MAC address, which was subsequently filtered out of the dataset.

The last dataset was captured in an uncontrolled real-world environment with high-density pedestrian traffic in the city center of Catania, Italy. In this case, we deployed four WSDs on two main squares, Piazza del Duomo (locations 1 and 2) and Piazza Università (locations 3 and 4), which were connected with a busy pedestrian street Via Etnea, as shown in Figure 6c, and were collecting data over a period of several months. For sending data to the remote database, WSDs were connected to the Internet via ethernet or via WiFi access points with known global MAC addresses.

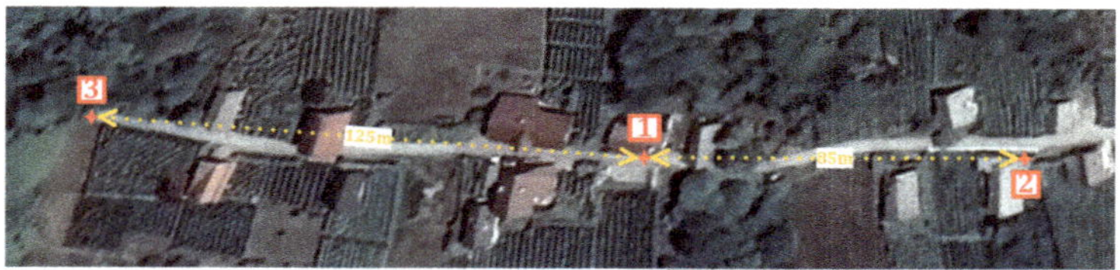

(a) outdoor rural deployment (controlled)

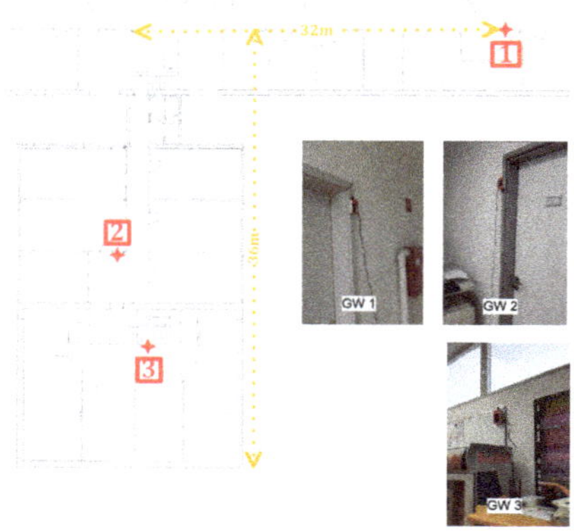

(b) indoor deployment at JSI (semi-controlled)

(c) outdoor urban deployment (uncontrolled)

Figure 6. Locations of probe requests acquisitions.

During the development and initial testing, the parameters of Algorithms A1 and A2 were first determined by observing PRs collected in an office environment and then fine-tuned using a publicly available labeled dataset [34]. Subsequently, the algorithms and the entire MAC de-randomization method were further validated in a controlled rural environment, where the results of the method were compared with ground truth, and in a semi-controlled indoor environment. Before starting the measurement campaign, we turned off all measurement devices and checked the environments for any remaining active WiFi devices. In the rural outdoor environment, we did not detect any unknown active device sending WiFi packets within the coverage areas of the deployed WSDs. In the indoor environment, the deployed WSDs detected several devices with active WiFi interfaces. These devices were excluded from the database used for validating the de-randomization method. In addition, to minimize the potential impact of uncontrolled WiFi-enabled devices carried by random people passing by, the measurement campaign was performed during the weekend.

After the verification of the environment, we started with data collection for individual devices. A brief summary of the devices used in measurement campaigns is given in Table 2. For each device, PRs generated within one minute were collected with the screen on, which was followed by PRs collected for one minute with the screen off. During these measurements, the other devices involved in the campaign were turned off in rural environment and had disabled WiFi interfaces in an indoor environment, so the recorded data for each device could later be used to determine the ground truth.

Table 2. Summary of devices used for the acquisition of PRs.

Device Name	OS	MAC Type	Assigned Group
Apple iPhone 12 Pro	iOS 16	Random MAC only	1
Nokia 7 Plus	Android 10	Random MAC only(CID: da:a1:19)	1
Samsung S10E	Android 12	Random MAC only	1
Samsung J3 2016	Android 5.1.1	Global MAC only (d0:b1:28:d2:de:e5)	2
Samsung S3	Android 4.4.4	Global MAC only (34:23:ba:d5:34:1b)	2
Samung Galaxy Nexus	Android 4.3	Global MAC only (a0:0b:ba:da:64:7e)	2
Samsung S10E	Android 12	Random MAC only	2
Samsung S7 Edge	Android 8	Random MAC only	2
Samsung J5	Android 6	Global MAC only (20:55:31:fc:4c:86)	2
Samsung S7	Android 8	Random MAC only	2
Samsung S7	Android 8	Random MAC only	3
Samsung Tab S8	Android 12	Random MAC only	2
Huawei Nexus 6P	Android 8.1.0	Global MAC (dc:ee:06:fd:8c:9e) + Random MAC (CID: da:a1:19)	3
Huawei P20	Android 10	Global MAC (e4:34:93:b5:f0:74) + Random MAC (CID: da:a1:19)	3
Huawei P20	Android 10	Global MAC (e4:0e:ee:3e:3e:44) + Random MAC (CID: da:a1:19)	3
Huawei P30 Lite	Android 10	Random MAC only (CID: da:a1:19)	1
Huawei P20 Lite	Android 9	Random MAC only (CID: da:a1:19)	3
Asus Tab 8″	Android 5.0	Global MAC only (54:a0:50:0e:8f:ee)	1
Asus Tab 7″	Android 4.2.2	Global MAC only (08:62:66:72:ac:1f)	3
OnePlus 3	Android 9	Random MAC only (CID: da:a1:19)	3
OnePlus 6	Android 11	Global MAC only (64:a2:f9:28:98:6c)	1
Lenovo VIBE A7020	Android 6	Global MAC only (54:27:58:30:ac:5a)	1
Xiaomi Poco F1	Android 10	Random MAC only	1

To test the MAC de-randomization method with three different levels of difficulty, the devices were divided into three groups. The first group comprised only devices from different manufacturers, so larger differences were expected between the PRs received from different devices. The second group contained only devices from one manufacturer (Samsung), which may aggravate the de-randomization process and thus distinguish unique devices. For the third group, a medium degree of difficulty in MAC de-randomization was expected, since half of the devices were from one manufacturer (Huawei) and the other half were from other different manufacturers. The distribution of devices among the groups is denoted in the last column in Table 2.

The same data collection procedure was applied for all three groups of devices in the rural and indoor environments at three different locations of WSDs indicated in Figure 6a,b. At each WSD location, PRs were collected from each group of devices for 10 min. Then, all three groups switched locations between WSDs, and the process was repeated. Thus, the database contains PR measurements from all three WSD locations at both measurement sites for all three groups of devices.

In the last testing scenario, all devices listed in Table 2 were grouped together. In the rural environment, data from received PRs were collected at one location for 10 min with screens on and for 10 min with screens off, whereas in the indoor environment, screens were on for 10 min. In the next step, all devices were moved to the location of the next WSD and PRs were measured for 10 min with screens off. In the rural outdoor scenario, the final step was to move the devices to location 2, where PRs were recorded for 10 min with the screens off.

The final testing and validation of the proposed MAC de-randomization method was conducted in a completely uncontrolled environment characterized by a much larger amount of PRs collected than the rural environment and indoor environments but with no ground truth. Thus, these tests did not focus on the accuracy of the proposed approach but rather on the robustness and scalability of the WSDs used to capture WiFi network management traffic, the remote database used to store the data, the PR pre-processing procedures and the MAC de-randomization method. Due to the large number of PRs collected, the statistical behavior of the proposed method can be better observed. In addition, such long-term data can also be exploited to identify daily/weekly/monthly patterns in user behavior, and combinations of WSDs can also be used to identify movement directions of identified WiFi-enabled devices.

The datasets used in this paper and the corresponding description are freely available to allow replication and continuation of the work. The dataset for a rural indoor environment is published in [37], while the dataset with all PRs collected by four WSDs for one week in the uncontrolled urban environment is published in [38].

5.2. MAC De-Randomization and Results Analysis

The results of the MAC de-randomization method for the labeled dataset are provided along with the list of devices in Table 3. In the first step, the proposed method was applied to the PRs of each device separately to determine the maximum differences between the PRs transmitted by the same device. In most cases, the method correctly identified a device. There were only a few cases where two devices were identified instead of one because (i) the PRs sent by the device were very different, (ii) the algorithm for matching global addresses with clusters of random addresses could not match a random MAC address with a global MAC address, and (iii) the device used two different types of addresses in PRs: one completely random and another with a fixed CID part. Since the devices sending two different addresses (random and with CID part) are rare (only one device in the labeled dataset used), this type of clustering is not considered as a separate case by the proposed method.

Table 3. MAC de-randomization results for individual devices from the labeled dataset.

Device	Global Addresses Detected	Random Addresses Detected	Devices Identified
Samsung Galaxy M31	0	15	1
Xiaomi Redmi 4	0	531	2
Samsung Galaxy S4	1	0	1
Huawei ALE-L21	1	0	1
Xiaomi Mi A2 Lite	0	435	2
Huawei CLT-L09 (P20)	1	0	1
Samsung Galaxy S6 edge (SM-G928F)	1	0	1
Samsung Galaxy S7	0	38	1
Xiaomi Redmi 5 Plus	0	253	2
Samsung Galaxy J6	1	26	2
Google Pixel 3A	0	46	2
Apple XS max	0	103	1
Apple iPhone 6	0	57	1
One Plus Nord	0	35	1
Huawei VTR-L09 (P10)	1	0	1
Huawei STF-L09 (Honor 9)	1	88	1
Xiaomi Redmi Note 7	1	153	1
Xiaomi Redmi Note 9S	0	138	1
Apple iPhone XR	0	36	1
Google Pixel 3A	0	23	1
Apple iPhone 12	0	1206	1
Apple iPhone 7	0	19	1
All devices combined	8	3201	21/22 (95.5%)

To further analyze the performance of the proposed method, the PRs of all 22 devices were aggregated. In this case, the proposed method correctly identified 21 devices (95.5%). This estimate of the total number of devices is a good approximation of the actual value, but we made a more detailed analysis of the clustered devices as allowed by the labeled dataset. It turned out that some of the 21 clusters formed by the MAC de-randomization contained PRs from more than one device. The method uniquely identified 11 devices (i.e., 11 clusters contained only one device with all its PRs). Two devices were fully identified by the random MAC addresses clustering algorithm, but they did not match or could not be matched with a global MAC address. The remaining clusters contained combinations of PRs from other devices due to some very similar PR's IE data. Since the core of the MAC de-randomization method is the clustering of PRs with random MAC addresses, analyzing the clustering of only random addresses gives us a deeper insight into how the method works. When only random addresses were considered, the proposed method uniquely identified 10 out of 17 devices.

The final estimate of the total number of devices is still a good approximation of the actual value. If multiple devices send two different PRs, one of which contains only mandatory data that is very similar among the devices, then a cluster of multiple devices is formed based on these PRs, resulting only in one overcount. An undercount is caused by the similarity of PRs sent by similar devices (considering manufacturer and OS), so the proposed method has difficulty distinguishing between them. If the method is not able to match a global and a random MAC address of the same device, an undercount could be sometimes compensated by an overcount.

The validation results for the MAC de-randomization method in the rural and indoor environments are presented in Tables 4–7. Similar as with the labeled dataset, the proposed method was first validated for data from each individual device separately for time intervals when the screen was on and off, which also revealed the differences in the behavior of the devices. In the rural environment, where no interfering devices were present, the proposed method successfully identified only one device for each of the devices tested. In the indoor semi-controlled environment, additional devices with global and random MAC addresses were also present during the data collection of individual tested devices. As explained in Section 5.1, these devices were identified and considered in the de-randomization. Thus, the final estimate of the de-randomization method includes the tested device and the additional uncontrolled devices whose PRs were detected during the corresponding time interval. The average ratio between the number of estimated devices and the actual number of devices was 96.7% . The main identified cause of errors was a mismatch between clustered devices with random MAC addresses and devices with global MAC addresses, while clustering only devices with random MAC addresses was error-free.

For further validation of the proposed method, three groups were formed according to Table 2. Table 5 shows the estimated values in the rural environment for each group for 10-min intervals at three different locations with the devices' screen on. The mean values for identified devices in all scenarios were 91.7%, 87.5%, and 100% of the actual number of devices for the groups 1, 2, and 3, respectively, confirming the expectation that the distinction between devices of the same manufacturer is the most challenging. If considering only PRs with random MAC addresses detected during the data collection in the corresponding time interval, the proposed method detected 100%, 75%, and 100% of the actual devices. The source of error for the group 1 devices was an incorrect match between the device with the global MAC address and the cluster of devices with random MAC addresses. Since the group 2 devices consisted only of devices from the same manufacturer, the main error was due to the similar PRs of the devices, which caused the method to undercount the devices by one.

The results for the last scenario, where all devices are grouped together at three different time intervals, are summarized in Table 6. The proposed method detected on average 71.2% of actual devices. The percentage of the correctly identified devices decreases when the devices' screens were turned off, since some devices did not send any PRs in this measurement period. If considering only PRs with random MAC addresses detected during the corresponding time interval, the method identified on average 67.5% of devices. This performance deterioration was caused by two sources of error, namely (i) false matches between a device with a global MAC address and a cluster of devices with random MAC addresses, and (ii) the similarity of PRs received from the same or similar devices that the proposed method could not distinguish.

Table 4. MAC de-randomization results for individual devices for the rural and indoor environments.

Device \ Location	Global Addresses Detected		Random Addresses Detected		Devices Identified/Devices Present	
	Rural	**Indoor**	**Rural**	**Indoor**	**Rural**	**Indoor**
Apple iPhone 12 Pro	0	6	31	19	1/1	8/8
Nokia 7 Plus	0	6	6	20	1/1	8/8
Samsung S10e	0	5	6	15	1/1	8/9
Samsung J3 2016	1	7	0	7	1/1	8/8
Samsung S3	1	7	0	6	1/1	9/9
Samsung Galaxy Nexus	1	6	0	10	1/1	8/9
Samsung S7	0	5	4	22	1/1	9/9
Huawei Nexus 6P	1	6	2	14	1/1	8/8
Asus Tab 8″	1	6	0	10	1/1	7/7
Asus Tab 7″	1	6	0	11	1/1	7/7
OnePlus 3	0	5	1	16	1/1	7/9
Samsung S10e	0	8	4	4	1/1	9/9
Samsung S7 Edge	0	8	4	4	1/1	9/9
Samsung J5	1	8	0	0	1/1	8/8
Samsung S7	0	8	3	1	1/1	9/9
Samsung Tab S8	0	9	4	12	1/1	12/12
Huawei P20	1	8	3	0	1/1	8/8
Huawei P20		8		3		8/8
Huawei P30 Lite	0	8	1	1	1/1	8/9
Huawei P20 Lite	0	8	1	1	1/1	8/9
OnePlus 6	1	10	0	0	1/1	10/10
Lenovo VIBE A7020	1	8	0	7	1/1	9/10
Xiaomi Poco F1	0	7	2-5	16	1/1	10/10
				Mean	**100%**	**96.7 %**

Table 5. MAC de-randomization results for groups of devices for the rural environment.

	Global Addresses Detected			Random Addresses Detected			Devices Identified			Devices Identified (Only Random MACs)		
Loc.	**Group 1**	**Group 2**	**Group 3**	**Group 1**	**Group 2**	**Group 3**	**Group 1**	**Group 2**	**Group 3**	**Group 1**	**Group 2**	**Group 3**
Rural 1	3	4	3	50	16	6	7/8	7/8	6/6	5/5	3/4	4/4
Rural 3	3	4	3	92	17	7	8/8	7/8	6/6	5/5	3/4	3/3
Rural 2	3	4	3	54	25	5	7/8	7/8	6/6	5/5	3/4	3/3
						Mean	**91.7%**	**87.5%**	**100%**	**100%**	**75%**	**100%**

Table 6. MAC de-randomization results for all devices in a single group for the rural environment.

Loc./Scenario	Global Addresses Detected	Random Addresses Detected	Devices Identified	Devices Identified (Only Random MACs)
Rural 2/screen on	10	92	17/22	8/12
Rural 2/screen off	8	97	14/22	8/12
Rural 2/screen on + screen off	10	188	16/22	9/13
		Mean	**71.2 %**	**67.5 %**

In the indoor environment, the coverage areas of two WSDs were overlapping. Therefore, the proposed method was tested by each group of devices at the non-overlapping location and with the combinations of the two groups at the overlapping locations. Finally, the proposed method was validated with PRs sent by all devices merged into a single group. The results in Table 7 show that on average, the proposed algorithm identified 91.3% of the actual present devices. When only PRs with random MAC addresses were clustered, it recognized 83% of the devices. A more in-depth analysis of the results showed that the sources of errors were the same as in the rural environment.

Table 7. MAC de-randomization results for groups of devices for the indoor environment.

Group/Loc	Global Addresses Detected	Random Addresses Detected	Devices Identified	Devices Identified (Only Random MACs)
Group 1/1	8	63	13/13	6/6
Group 2/1	9	11	12/13	3/4
Group 3/1	9	10	12/12	4/5
Groups 1,3/2,3	18	54	24/26	7/8
Groups 2,3/2,3	19	47	25/26	6/7
Groups 1,2/2,3	16	57	22/25	7/9
All devices/3	20	105	27/32	9/12
		Mean	**91.3 %**	**83 %**

The robustness and scalability of the proposed method was tested with the dataset collected by the WSD installed at Piazza del Duomo in Catania (location 1 in Figure 6c). The numbers of received PRs and identified WiFi-enabled devices are listed in Table 8. The proposed method was executed at hourly intervals over a 24 h period from midnight to midnight on 23 September 2022. The starting hour in the table corresponds to the local time of the beginning of the hourly interval. The number of PRs recorded in an hour and the PRs with actual unique data are shown to illustrate the amount of computation required by the proposed method. It should be noted that the number of PRs in such urban environment increases significantly compared to less populated environments, up to several hundreds per minute.

Table 8. MAC de-randomization results for one day at Piazza del Duomo, Catania.

Start of Hourly Interval	All PRs/ Unique PRs	Global Addresses Detected	Random Addresses Detected	Devices Identified
00:00:00	18,980 /5637	82	3112	124
01:00:00	16,531/4448	59	2055	91
02:00:00	14,895/3985	37	1657	61
03:00:00	13,973/3115	28	802	50
04:00:00	13,174/2655	24	598	50
05:00:00	13,700/2716	32	609	57
06:00:00	17,736/3902	50	1638	86
07:00:00	21,342/6141	124	3044	181
08:00:00	29,980/10,918	166	8121	232
09:00:00	48,273/21,079	237	17,606	385
10:00:00	46,029/21,150	347	17,225	485
11:00:00	76,586/38,601	520	32,369	793
12:00:00	58,161/26,319	401	22,234	572
13:00:00	72,632/35,171	305	30,214	544
14:00:00	42,608/20,848	257	17,938	370
15:00:00	33,156/14,899	160	12,573	231
16:00:00	54,233/25,086	405	20,521	556
17:00:00	59,599/28,587	356	23,353	547
18:00:00	74,070/33,673	508	27,763	745
19:00:00	67,298/31,440	592	25,321	777
20:00:00	43,776/20,274	254	16,735	366
21:00:00	50,731/23,810	409	19,529	518
22:00:00	32,345/15,197	228	12,855	325
23:00:00	44,087/21,058	259	17,792	372

The number of PRs and the number of devices identified by the proposed method in an hourly time slot for the 24 h period are also shown in Figure 7. It can be observed that the number of PRs coincides very well with the number of identified devices. During night-time hours, a higher number of devices than expected are identified due to nearby static devices not being filtered out. Since the method compares each PR with all the remaining PRs, the computational complexity of the method increases exponentially with the number of recorded PRs. Therefore, in busy urban environments with a high volume of pedestrians and consequently a huge number of recorded PRs, the identification of unique devices may not be possible in real time, indicating one possible direction for further optimization of the algorithm.

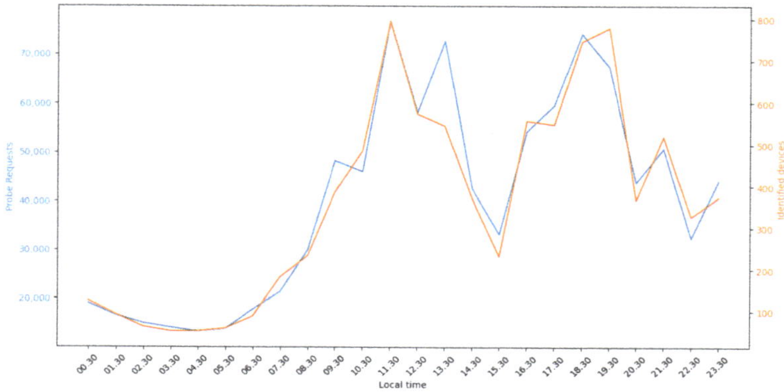

Figure 7. Hourly number of collected PRs and identified devices for one day at Piazza del Duomo, Catania.

5.3. Discussion

Extensive validation of the proposed method using datasets from controlled rural, semi-controlled indoor, and uncontrolled urban environments confirms its ability to accurately detect unique active WiFi interfaces.

Comparison of the results with existing work applying similar approaches is difficult since their validation was performed in public places by counting the number of people and without the information about the number of WiFi-enabled devices carried by the individual (e.g., tickets sold [31]), so no ground truth can be determined. An approximate comparison can be made with the solutions where validations were performed with measurements that includes the information on the number of WiFi-enabled devices as a ground truth. In [11], the accuracy of 91.3% was achieved in the laboratory environment, while in [32], 96% accuracy in the laboratory environment and 75% in a real-world scenario is reported. With the publicly available dataset [34], we obtained similar results of 95.5%, while in the controlled environment, we achieved accuracy of at least 87.5% for smaller groups of devices. In rural environments the achieved accuracy was over 70%, and in indoor environments for larger groups, it was 84%.

A more accurate comparison can be made with the findings reported in [32,33], where the authors used the same labeled, publicly available dataset as in this work [34]. Although the results in [32] are not explicit, since only the behavior of the proposed method was tested under different parameters, they claim to achieve an accuracy of up to 95%, which is similar to the accuracy achieved with our proposed method. In [33], the authors achieved an accuracy of 92% and also performed a more thorough analysis of the homogeneity of the formed clusters, achieving values of up to 0.97. High homogeneity values can be obtained in the case where devices grouped in clusters send different number of PRs. From the paper, it is not clear whether they filtered PRs with the same data and thus considered only unique PRs of devices. Related works focused only on the de-randomization of random MAC addresses, while we considered the real-world scenario that includes some special cases, such as devices transmitting both random and global MAC addresses that need to be matched to avoid overcounting the devices.

In general, the results show that the proposed method tends to undercount the devices, which is mainly due to the similarity of PRs of the same or similar devices. The method performs well for small groups of devices, while for larger groups, the differences between PRs become less significant, resulting in an underestimate of the number of identified devices. This effect could be reduced by fine-tuning the model parameters depending on the scenario. In addition, the scaling test with an extremely large urban dataset revealed some drawbacks in terms of exponential computational complexity. Thus, the two main drawbacks of the proposed de-randomization method based on PR's IE values are (i) the need to configure multiple parameters, which also can be seen as an advantage, as it allows adapting the method to the specificites of the environment or application, and (ii) the computational effort required to compare the PRs of a single device with all the PRs of other devices. On the other hand, the proposed approach has several advantages. It is a low-cost solution that requires only WSDs operating in the monitor mode and remote data storage capabilities, the system is easy to set up, and it is completely unobtrusive to people, since no particular actions are required from the owners of the WiFi-enabled devices. The system not only estimates the number of people in close proximity to the WSD but can also provide information on clustered data that can be used to analyze the movements of the crowd and individuals.

Validation of the proposed method showed that many of the tested devices which use random MAC addresses are changing them very rarely. Therefore, the approach could be further improved by identifying these devices based on the number of PRs detected. When comparing PRs sent by devices that change the MAC address at every burst and PRs sent by devices that change it infrequently, a large difference in the number of PRs per address was observed, which can be used to set a higher threshold for matching these two types of devices. This would also reduce the errors caused by incorrectly matching a global MAC address with a clusters of random MAC addresses.

Similarly, the observation of a high correlation between the number of collected PRs and the number of identified devices, as seen in Figure 7, could lead to a substantial simplification of the method in which the number of devices could be estimated from the number of PRs recorded. Further data collection would be required to determine whether the correlation is indeed strong enough. Such a simplification would only be applicable in areas where there is a large number of devices so that the statistics becomes sufficiently robust. As manufacturers change the frequency of sending PRs, the simplified approach would also need to be re-evaluated over time. However, simplifying the method in this way would also preclude the ability to use the same data obtained from PRs for monitoring the movement between the coverage areas of neighboring WSDs.

Fine-tuning the parameters of the proposed method proved to be time-consuming and not optimized. Therefore, a procedure to automatically determine and optimize the coefficients would have to be implemented. A more detailed analysis of the importance of each bit of the IE fields could also be performed to include only the most important bits to distinguish between devices. In addition, other features could be extracted from ToA, such as the inter-frame and inter-burst times.

The exponential complexity of the proposed method is another challenge for future optimization if it is expected to run for longer periods of time or for multiple locations. The results presented in this work were calculated on a consumer-level hardware, but further method optimization should be considered if more data are to be processed or it should be adapted for the use on high-performance computing infrastructure.

6. Conclusions

This paper presented an approach for monitoring the presence of individuals at specific locations based on collected PRs, taking into account the increasing adoption of MAC address randomization due to privacy concerns. The main contribution of this work is the development of a method for MAC de-randomization based on the similarity of PRs, more specifically IE data. We described the designed system and deployment for capturing PRs sent by WiFi-enabled devices. The detection of unique WiFi interfaces is implemented in two stages. After grouping PRs based on random or global MAC addresses, clustering is performed on PRs from devices with random MAC addresses. The generated clusters are then matched with PRs of devices with global MAC addresses.

The proposed approach was validated in a controlled rural, semi-controlled indoor, and uncontrolled urban environment, first with PRs from only individual WiFi-enabled devices and later with all devices and three formed groups of devices to account for three different levels of difficulty for de-randomization. Validation on individual devices showed that in some cases, the de-randomization method detected two devices instead of one. For the formed groups, the results show the disadvantage of the proposed method when multiple devices of the same manufacturer and version of OS were present. Although the performance of the proposed method decreases in this case, it is still above 70% for rural and 80% for indoor environments.

In the paper, we identified some potential improvements and shortcomings of the proposed methods that could be addressed as part of future work such as the automatic fine-tuning of parameters, reduction of computational complexity, optimization of the threshold for matching devices with large differences in the number of detected PRs, and extraction of additional features from ToA. Another challenge for future work is to extend the proposed approach of detecting unique WiFi-enabled devices to a system that analyzes the collected data from multiple locations in the urban area to determine the movement patterns of users and use these data for traffic management, route optimization, resilience planning, etc.

Author Contributions: Conceptualization, M.M. (Mihael Mohorčič) and A.H.; methodology, M.M. (Mihael Mohorčič) and A.H.; hardware, M.M. (Miha Mohorčič); software, A.S. and M.M. (Miha Mohorčič); validation, M.M. (Miha Mohorčič), A.S. and A.H.; data analysis and curation, A.S. and M.M. (Miha Mohorčič); writing—original draft preparation, A.S., M.M. (Miha Mohorčič) and A.H.; writing—review and editing, M.M. (Mihael Mohorčič); visualization, A.S.; supervision, A.H.; funding acquisition, M.M. (Mihael Mohorčič). All authors have read and agreed to the published version of the manuscript.

Funding: This research was partly funded by the Slovenian Research Agency (Grant no. P2-0016, J2-3048, and J2-2507) and the European Community under the H2020 RESILOC project (Grant no. 833671).

Institutional Review Board Statement: Not applicable.

Informed Consent Statement: Not applicable.

Data Availability Statement: The data presented in this study are openly available in repositories Zenodo [37,38] and Mendeley Data at https://doi.org/10.17632/j64btzdsdy.1.

Conflicts of Interest: The author declare no conflict of interest. The funders had no role in the design of the study; in the collection, analyses, or interpretation of data; in the writing of the manuscript, or in the decision to publish the results.

Abbreviations

The following abbreviations are used in this manuscript:

CID	Company Identifier
GDPR	General Data Protection Regulation
HMM	Hidden Markov Models
HT	High Throughput
IE	Information Element
IFAT	Inter-Frame Arrival Time
IoT	Internet-of-Things
JSI	Jozef Stefan Institute
JSON	JavaScript Object Notation

MAC	medium access control
ML	Machine learning
NIC	Network Interface Controller
OPTICS	Ordering Points to Identify the Clustering Structure
OS	Operating system
OUI	Organization Unique Identifier
PR	Probe Request
REST	Representational State Transfer
rPi	Raspberry Pi
RSSI	Received Signal Strength Indicator
SSID	Service Set Identifier
ToA	Time of Arrival
VHT	Very High Throughput
WSD	Wireless Sensor Device

Appendix A. Structure of Stored Data from PR

Data from PRs of each MAC address are stored in the following JSON format:

```
{'MAC': MAC_address, 'SSIDs': [ SSID ], 'PROBE_REQs': [PR_data] },
```

where *PR_data* is structured as:

```
PR_data =
{
    'TIME': [ DATA_time ],
    'RSSI': [ DATA_rssi ],
    'DATA': pr_IE_data
}
```

and *PR_IE_data* as:

```
PR_IE_data =
{
    'DATA_RTS': {'SUPP': DATA_supp , 'EXT': DATA_ext},
    'HT_CAP': DATA_htcap,
    'EXT_CAP': {'length': DATA_len, 'data': DATA_extcap},
    'VHT_CAP': DATA_vhtcap,
    'INTERWORKING': DATA_inter,
    'EXT_TAG': {'ID_1': DATA_1_ext, 'ID_2': DATA_2_ext ...},
    'VENDOR_SPEC': {VENDOR_1:{
                              'ID_1': DATA_1_vendor1,
                              'ID_2': DATA_2_vendor1
                              ...},
                    VENDOR_2:{
                                'ID_1': DATA_1_vendor2,
                                'ID_2': DATA_2_vendor2
                                ...}
                    ...}
}
```

Appendix B. Algorithms

Algorithm A1 Algorithm for calculating distance between two probe requests

procedure DIST_BTWN_TWO_PROBE_REQ($PR\_IE\_data_1, SSIDs_1 PR\_IE\_data_2, SSIDs_2$)

$distance \leftarrow STARTING\_DISTANCE$

for each IE_key $PR\_IE\_data_1$ and $PR\_IE\_data_2$ have in common **do**
  if IE_key **equal** DATA_RTS **then**
    merge Supported Data Rates and Extended Data Rates for both PRs
    increase *distance* for each data rate of symetric difference multiplied by IE's specific scale

Algorithm A1 *Cont.*

if IE_key **equal** HT_CAP **Or** VHT_CAP **then**
$distance \leftarrow distance + \frac{number\_of\_distinguishing\_bits}{number\_of\_all\_bits} * specific\_IE\_scale$

if IE_key **equal** EXT_CAP **Or** INTERWORKING **then**
if $length(probe\_req\_data_1[IE\_key] \mathrel{!}= length(probe\_req\_data_2[IE\_key]$ **then**
$distance \leftarrow distance + specific\_IE\_value$
else
$distance \leftarrow distance + \frac{number\_of\_distinguishing\_bits}{number\_of\_all\_bits} * specific\_IE\_scale$

if IE_key **equal** VENDOR_SPEC **then**
$distance \leftarrow distance - special\_IE\_koef$
increase *distance* for vendor ID of symetric difference multiplied by IE's specific scale
decrease *distance* for each intersecting vendor ID multiplied by IE's specific scale
for each intersecting vendor ID **do**
decrease *distance* for each intersecting data ID multiplied by IE's specific scale
for each intersecting data ID **do**
$distance \leftarrow distance + \frac{number\_of\_distinguishing\_bits}{number\_of\_all\_bits} * specific\_IE\_scale$
if IE_key **equal** EXT_TAG **then**
$distance \leftarrow distance - special\_IE\_koef$
increase *distance* for each data ID of symetric difference multiplied by IE's specific scale
decrease *distance* for each intersecting data ID multiplied by IE's specific scale
for each intersecting data ID **do**
$distance \leftarrow distance + \frac{number\_of\_distinguishing\_bits}{number\_of\_all\_bits} * specific\_IE\_scale$
decrease *distance* for intersecting SSID multiplied by IE's specific scale

increase *distance* by number of IEs of symetric difference relative to number of intersecting IEs multiplied by specific scale
$tmp\_dst \leftarrow Inf$
for each $RSSI_1, TOA_1$ in $probe\_req\_data_1$ **do**
for each $RSSI_2, TOA_2$ in $probe\_req\_data_2$ **do**
$dst\_RSSI \leftarrow RSSI\_KOEF^{|RSSI_2 - RSSI_1|} - 1$
$dst\_TOA \leftarrow 0$
if $|TOA_2 - TOA_1| < TIMING\_THRESHOLD$ **then**
$dst\_TOA \leftarrow TIMING\_KOEF$
if $(dst\_RSSI + dst\_TOA) < tmp\_dst$ **then**
$tmp\_dst \leftarrow dst\_RSSI + dst\_TOA$
$distance \leftarrow distance + tmp\_dst$

if $distance < MINIMUM\_DISTANCE$ **then**
$distance \leftarrow MINIMUM\_DISTANCE$
return *distance*

Algorithm A2 Reachability distance-based clustering

```
procedure CLUSTERING(reach_dists, xi_up, xi_down, epsMax)
    in_cluster ← False
    labels[:] ← −1
    current_label ← 0
    climbing ← False
    falling ← False
    for i ← 1 to length(reach_distss) − 1 do
        if in_cluster then
            if reach_dists[i + 1] > reach_dists[i] then
                if not climbing then
                    climbing ← True
                    value_start_climbing ← reach_dists[i]
                if reach_dists[i+1] / value_start_climbing > xi_up Or reach_dists[i + 1] > epsMax then
                    in_cluster ← False
                    current_label ← current_label + 1
                    climbing ← False
                else
                    labels[i + 1] ← current_label
            else
                climbing ← False
                labels[i + 1] ← curr_label
        else if reach_dists[i + 1] < reach_dists[i] then
            if not falling then
                value_start_falling ← reach_dists[i]
                index_start_falling ← i
                falling ← True
            if reach_dists[i + 1] < epsMax And reach_dists[i+1] / value_start_falling < xi_down then
                labels[index_start_falling − 1 : i + 1] ← curr_label
                in_cluster ← True
                falling ← False
        else
            falling ← False
        i ← i + 1
    return labels
```

References

1. Covaci, A.I. Wi-Fi MAC Address Randomization vs. Crowd Monitoring. Bachelor's Thesis, University of Twente, Enschede, The Netherlands, 2022.
2. Ahmed, N.; Michelin, R.A.; Xue, W.; Ruj, S.; Malaney, R.; Kanhere, S.S.; Seneviratne, A.; Hu, W.; Janicke, H.; Jha, S.K. A survey of COVID-19 contact tracing apps. *IEEE Access* **2020**, *8*, 134577–134601. [CrossRef]
3. Su, Z.; Pahlavan, K.; Agu, E. Performance evaluation of COVID-19 proximity detection using bluetooth LE signal. *IEEE Access* **2021**, *9*, 38891–38906. [CrossRef] [PubMed]
4. Švigelj, A.; Hrovat, A.; Javornik, T. User-Centric Proximity Estimation Using Smartphone Radio Fingerprinting. *Sensors* **2022**, *22*, 5609. [CrossRef] [PubMed]
5. Fenske, E.; Brown, D.; Martin, J.; Mayberry, T.; Ryan, P.; Rye, E. Three Years Later: A Study of MAC Address Randomization in Mobile Devices and When It Succeeds. *Proc. Priv. Enhancing Technol.* **2021**, *2021*, 164–181. [CrossRef]
6. Bonne, B.; Barzan, A.; Quax, P.; Lamotte, W. WiFiPi: Involuntary tracking of visitors at mass events. In Proceedings of the 2013 IEEE 14th International Symposium on "A World of Wireless, Mobile and Multimedia Networks" (WoWMoM), Madrid, Spain, 4–7 June 2013; pp. 1–6. [CrossRef]
7. Martin, J.; Mayberry, T.; Donahue, C.; Foppe, L.; Brown, L.; Riggins, C.; Rye, E.C.; Brown, D. A Study of MAC Address Randomization in Mobile Devices and When it Fails. *arXiv* **2017**, arXiv:1703.02874.
8. Franklin, J.; McCoy, D. Passive Data Link Layer 802.11 Wireless Device Driver Fingerprinting. In Proceedings of the 15th USENIX Security Symposium (USENIX Security 06), Vancouver, BC, Canada, 31 July–4 August 2006.

9. Desmond, L.C.C.; Yuan, C.C.; Pheng, T.C.; Lee, R.S. Identifying Unique Devices through Wireless Fingerprinting. In Proceedings of the First ACM Conference on Wireless Network Security, Alexandria, VA, USA, 31 March–2 April 2008; Association for Computing Machinery: New York, NY, USA, 2008; pp. 46–55. [CrossRef]
10. Matte, C.; Cunche, M.; Rousseau, F.; Vanhoef, M. Defeating MAC Address Randomization Through Timing Attacks. In Proceedings of the 9th ACM Conference on Security & Privacy in Wireless and Mobile Networks, Darmstadt, Germany, 18–20 July 2016; pp. 15–20. [CrossRef]
11. Uras, M.; Cossu, R.; Ferrara, E.; Bagdasar, O.; Liotta, A.; Atzori, L. WiFi Probes sniffing: An Artificial Intelligence based approach for MAC addresses de-randomization. In Proceedings of the 2020 IEEE 25th International Workshop on Computer Aided Modeling and Design of Communication Links and Networks (CAMAD), Pisa, Italy, 14–16 September 2020; pp. 1–6. [CrossRef]
12. Vanhoef, M.; Matte, C.; Cunche, M.; Cardoso, L.S.; Piessens, F. Why MAC Address Randomization is not Enough: An Analysis of Wi-Fi Network Discovery Mechanisms. In Proceedings of the 11th ACM on Asia Conference on Computer and Communications Security, Xi'an, China, 30 May–3 June 2016; pp. 413–424. [CrossRef]
13. Myrvoll, T.A.; Hakegard, J.E.; Matsui, T.; Septier, F. Counting public transport passenger using WiFi signatures of mobile devices. In Proceedings of the 2017 IEEE 20th International Conference on Intelligent Transportation Systems (ITSC), Yokohama, Japan, 16–19 October 2017; pp. 1–6. [CrossRef]
14. Groba, C. Demonstrations and people-counting based on Wifi probe requests. In Proceedings of the 2019 IEEE 5th World Forum on Internet of Things (WF-IoT), Limerick, Ireland, 15–18 April 2019; pp. 596–600. [CrossRef]
15. Guillen-Perez, A.; Cano Banos, M.D. A WiFi-based method to count and locate pedestrians in urban traffic scenarios. In Proceedings of the 2018 14th International Conference on Wireless and Mobile Computing, Networking and Communications (WiMob), Limassol, Cyprus, 15–17 October 2018; pp. 123–130. [CrossRef]
16. Guillen-Perez, A.; Cano, M.D. Counting and locating people in outdoor environments: A comparative experimental study using WiFi-based passive methods. *ITM Web Conf.* **2019**, *24*, 01010. [CrossRef]
17. Hong, H.; De Silva, G.D.; Chan, M.C. CrowdProbe: Non-invasive Crowd Monitoring with Wi-Fi Probe. In Proceedings of the ACM on Interactive, Mobile, Wearable and Ubiquitous Technologies, Singapore, 8–12 October 2018; Volume 2, pp. 1–23. [CrossRef]
18. Shen, J.; Cao, J.; Liu, X. BaG: Behavior-Aware Group Detection in Crowded Urban Spaces Using WiFi Probes. *IEEE Trans. Mob. Comput.* **2021**, *20*, 3298–3310. [CrossRef]
19. Oliveira, L.; Schneider, D.; De Souza, J.; Shen, W. Mobile Device Detection Through WiFi Probe Request Analysis. *IEEE Access* **2019**, *7*, 98579–98588. [CrossRef]
20. Uras, M.; Cossu, R.; Atzori, L. PmA: A solution for people mobility monitoring and analysis based on WiFi probes. In Proceedings of the 2019 4th International Conference on Smart and Sustainable Technologies (SpliTech), Split, Croatia, 18–21 June 2019; pp. 1–6. [CrossRef]
21. Wu, F.J.; Huang, Y.; Doring, L.; Althoff, S.; Bitterschulte, K.; Chai, K.Y.; Mao, L.; Grabarczyk, D.; Kovacs, E. PassengerFlows: A Correlation-Based Passenger Estimator in Automated Public Transport. *IEEE Trans. Netw. Sci. Eng.* **2020**, *7*, 2167–2181. [CrossRef]
22. Vieira, T.; Almeida, P.; Meireles, M.; Ribeiro, R. Public Transport Occupancy Estimation using WLAN Probing and Mathematical Modeling. *Transp. Res. Procedia* **2020**, *48*, 3299–3309. [CrossRef]
23. Furuya, Y.; Asahina, H.; Yoshida, M.; Sasase, I. Indoor Crowd Estimation Scheme Using the Number of Wi-Fi Probe Requests under MAC Address Randomization. *IEICE Trans. Inf. Syst.* **2021**, *E104.D*, 1420–1426. [CrossRef]
24. Yang, F.; Ahriz, I.; Denby, B. Statistical Approach to Estimating Audience from MAC-Randomized WiFi Probe Requests. *Sensors* **2022**, *22*, 8679. [CrossRef] [PubMed]
25. Cai, Y.; Tsukada, M.; Ochiai, H.; Esaki, H. MAC address randomization tolerant crowd monitoring system using Wi-Fi packets. In Proceedings of the 16th Asian Internet Engineering Conference, Virtual Event, Japan, 14–16 December 2021; pp. 27–33. [CrossRef]
26. Nitti, M.; Pinna, F.; Pintor, L.; Pilloni, V.; Barabino, B. iABACUS: A Wi-Fi-Based Automatic Bus Passenger Counting System. *Energies* **2020**, *13*, 1446. [CrossRef]
27. Gebru, K. A Privacy-preserving Scheme for Passive Monitoring of People's Flows through WiFi Beacons. In Proceedings of the 2022 IEEE 19th Annual Consumer Communications & Networking Conference (CCNC), Las Vegas, NV, USA, 8–11 January 2022; pp. 421–424. [CrossRef]
28. Gebru, K.; Rapelli, M.; Rusca, R.; Casetti, C.; Chiasserini, C.F.; Giaccone, P. Edge-based passive crowd monitoring through WiFi Beacons. *Comput. Commun.* **2022**, *192*, 163–170. [CrossRef]
29. Determe, J.F.; Azzagnuni, S.; Singh, U.; Horlin, F.; De Doncker, P. Monitoring Large Crowds With WiFi: A Privacy-Preserving Approach. *IEEE Syst. J.* **2022**, *16*, 2148–2159. [CrossRef]
30. Berenguer, A.; Ros, D.F.; Gómez-Oliva, A.; Ivars-Baidal, J.A.; Jara, A.J.; Laborda, J.; Mazón, J.N.; Perles, A. Crowd Monitoring in Smart Destinations Based on GDPR-Ready Opportunistic RF Scanning and Classification of WiFi Devices to Identify and Classify Visitors' Origins. *Electronics* **2022**, *11*, 835. [CrossRef]
31. Vega-Barbas, M.; Álvarez Campana, M.; Rivera, D.; Sanz, M.; Berrocal, J. AFOROS: A Low-Cost Wi-Fi-Based Monitoring System for Estimating Occupancy of Public Spaces. *Sensors* **2021**, *21*, 3863. [CrossRef]
32. Uras, M.; Ferrara, E.; Cossu, R.; Liotta, A.; Atzori, L. MAC Address De-Randomization for WiFi Device Counting: Combining Temporal- and Content-Based Fingerprints. *Comput. Netw.* **2022**, *218*, 109393. [CrossRef]

33. Pintor, L.; Atzori, L. Analysis of Wi-Fi Probe Requests Towards Information Element Fingerprinting. In Proceedings of the 2022 IEEE Global Communications Conference GLOBECOM 2022, Rio de Janeiro, Brazil, 4–8 December 2022; pp. 3857–3862. [CrossRef]
34. Pintor, L.; Atzori, L. A dataset of labelled device Wi-Fi probe requests for MAC address de-randomization. *Comput. Netw.* **2022**, *205*, 108783. [CrossRef]
35. Robyns, P.; Bonné, B.; Quax, P.; Lamotte, W. Noncooperative 802.11 MAC Layer Fingerprinting and Tracking of Mobile Devices. *Secur. Commun. Netw.* **2017**, *2017*, 6235484. [CrossRef]
36. Ankerst, M.; Breunig, M.M.; Kriegel, H.P.; Sander, J. OPTICS: Ordering points to identify the clustering structure. *ACM Sigmod Rec.* **1999**, *28*, 49–60 [CrossRef]
37. Simončič, A.; Mohorčič, M.; Mohorčič, M.; Hrovat, A. *Labeled Dataset of IEEE 802.11 Probe Requests;* Zenodo: Geneva, Switzerland, 2023. [CrossRef]
38. Mohorčič, M.; Simončič, A.; Mohorčič, M.; Hrovat, A. *Dataset of IEEE 802.11 Probe Requests from an Uncontrolled Urban Environment;* Zenodo: Geneva, Switzerland, 2023. [CrossRef]

sensors

MDPI

Article

An Adaptive Traffic-Flow Management System with a Cooperative Transitional Maneuver for Vehicular Platoons

Lopamudra Hota [1], Biraja Prasad Nayak [1], Bibhudatta Sahoo [1], Peter H. J. Chong [2,*] and Arun Kumar [1,*]

1 Department of Computer Science and Engineering, National Institute of Technology, Rourkela 769008, India
2 Department of Electrical and Electronic Engineering, Auckland University of Technology, Auckland 1010, New Zealand
* Correspondence: peter.chong@aut.ac.nz (P.H.J.C.); kumararun@nitrkl.ac.in (A.K.)

Abstract: Globally, the increases in vehicle numbers, traffic congestion, and road accidents are serious issues. Autonomous vehicles (AVs) traveling in platoons provide innovative solutions for efficient traffic flow management, especially for congestion mitigation, thus reducing accidents. In recent years, platoon-based driving, also known as vehicle platoon, has emerged as an extensive research area. Vehicle platooning reduces travel time and increases road capacity by reducing the safety distance between vehicles. For connected and automated vehicles, cooperative adaptive cruise control (CACC) systems and platoon management systems play a significant role. Platoon vehicles can maintain a closer safety distance due to CACC systems, which are based on vehicle status data obtained through vehicular communications. This paper proposes an adaptive traffic flow and collision avoidance approach for vehicular platoons based on CACC. The proposed approach considers the creation and evolution of platoons to govern the traffic flow during congestion and avoid collisions in uncertain situations. Different obstructing scenarios are identified during travel, and solutions to these challenging situations are proposed. The merge and join maneuvers are performed to help the platoon's steady movement. The simulation results show a significant improvement in traffic flow due to the mitigation of congestion using platooning, minimizing travel time, and avoiding collisions.

Keywords: platoon; traffic congestion; collision avoidance; CACC; merge maneuver; join maneuver; lane change

Citation: Hota, L.; Nayak, B.P.; Sahoo, B.; Chong, P.H.J.; Kumar, A. An Adaptive Traffic-Flow Management System with a Cooperative Transitional Maneuver for Vehicular Platoons. *Sensors* **2023**, *23*, 2481. https://doi.org/10.3390/s23052481

Academic Editor: Jose F. Monserrat

Received: 30 December 2022
Revised: 9 February 2023
Accepted: 13 February 2023
Published: 23 February 2023

1. Introduction

The explosive growth in the number of vehicles has resulted in many critical social problems worldwide, such as road safety, traffic congestion, and fuel consumption. The high cost of construction and the lack of available land make road development unsustainable; even though it can somewhat decrease traffic congestion, it is not an efficient strategy. To deal with these issues, a platoon-based driving pattern, also known as a vehicle platoon, has received much attention in the past few years. A platoon consists of a vehicle which follows another vehicle and keeps a close and relatively constant safe distance from the one in front of it, termed cooperative driving [1]. Vehicle platooning is a technique where highway traffic is organized into groups of close-following vehicles called platoons [2]. Platooning enables vehicles to drive closer to others than regular vehicles at the same speed, which improves traffic throughput [3]. The communication can be V2V, i.e., inter-platoon, intra-platoon, and platoon to the non-platoon vehicle, and V2I, via RSU and base stations. Dedicated short range communication (DSRC) is an effective mode for short-range communication, whereas LTE/5G is found to be more efficient and reliable for long-range communication. Vehicle platooning is primarily used to alleviate traffic congestion and increase traffic flow in even dense scenarios. Figure 1 depicts a generalized platoon architecture consisting of the platoon and non-platoon vehicles with V2V and V2I communication.

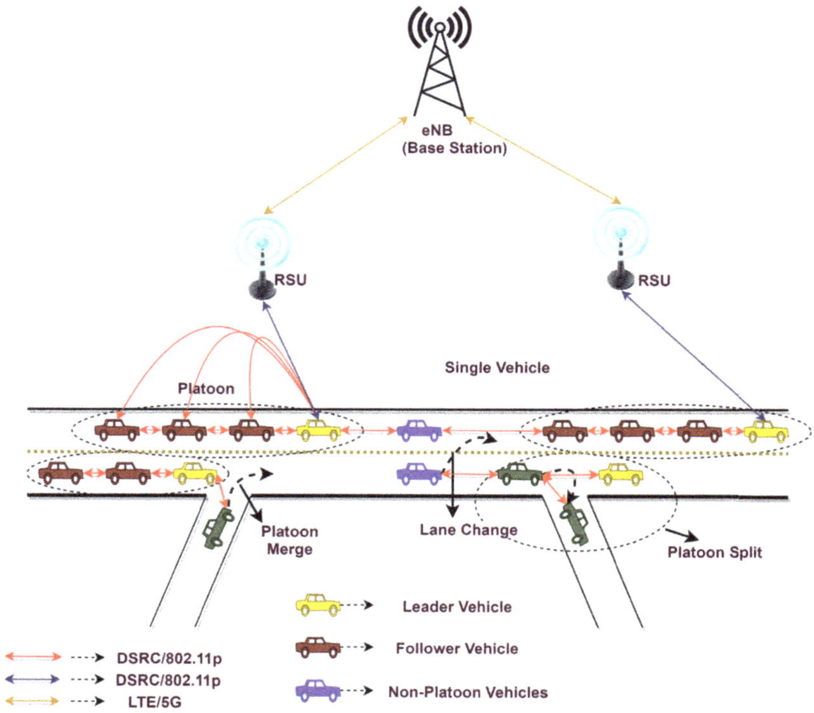

Figure 1. Platoon architecture.

The vehicles in the platoon coordinate among themselves by frequently exchanging periodic broadcast cooperative awareness messages (CAMs). The ETSI EN 302 637-2 standard specifies CAM-triggering conditions which depend on the dynamics of an originating vehicle [4]. The CAM is broadcast with a multi-channel access mechanism with a congestion control approach. The use of IEEE 802.11p in platooning is the subject of numerous studies and proposals, including message prioritization based on their type, dissemination, rate adaptation, and re-transmission [5]. An analysis of connectivity probability within a platoon is presented in [6]. Still, there is a scope for the design of protocols for the delay-tolerant delivery of CAM in a platoon scenario and decentralized congestion control to avoid collision of CAM.

Cooperative driving can improve fuel economy, enhance infrastructure efficiency, and improve road safety [7,8]. Platooning implements the mutual assistance driving mechanism that aims to achieve semi-autonomous platooning by having a leader vehicle handled by either humans or an automation that drives a group of vehicles as followers. The onboard system uses the data from the leader and nearby vehicles via inter-vehicular communication (IVC) to manage the engine, brakes, and steering of followers, eliminating the need for the driver to steer manually, accelerate, or brake. Platooning involves lateral steering, coordinated acceleration, and braking via longitudinal collision mitigation [9], management protocols that monitor the creation of the platoon, driving maneuvers, and lane changing, thereby confirming that vehicle control is not the sole responsibility of either the human driver or automation to provide safety measures [10].

For performing basic maneuvers in the platoon, a lane is reserved for the vehicles traveling in the platoon. Each can perform several maneuvers to maintain the optimal size of the platoon and increase its efficiency. It becomes essential to differentiate between the platoon and non-platoon vehicles. Platoons in vehicular ad-hoc networks (VANet) depend very highly on effective communication between platoon members to carry out the basic platoon maneuvers such as platoon merge and join maneuvers. Failing to acquire efficient communication among the vehicles lead to the failure of the platoon. A platoon-enabled vehicle meets all of the functional requirements implemented for platoons. A non-platooned vehicle is driven manually or with the assistance of an automated cruise control (ACC) system [11]. In contrast, a platooned vehicle is a platoon member and can be either a leader or a follower.

Platooning is a viable solution as it eliminates frequent stopping and starting, which wastes time and fuel. Globally, the paradigm is shifting toward AVs, and the necessity arises to look for advanced long-term solutions to reduce congestion, travel time, and fuel consumption. The AVs' penetration rate, defined as the ratio of AVs to total vehicles in a specified network, is predicted to increase from 24% to 87% by 2045 [12]. The massive increment of AVs in the future is the reason for platooning receiving serious attention in the past years. Because AVs are free of restrictions connected to driver behavior, such as reaction time and coordination, they could be one of the best possible platoon implementations. AVs with inter-vehicle communication (IVC) avoid traffic congestion, provide efficient coordination, and require a minimum response time by drivers, thereby avoiding accidents. Furthermore, analyzing safety features in scenarios such as crashes or accidents [13] is a critical consideration for AVs and human-driven vehicles. Current collision avoidance systems [14], such as ACC, adapt themselves using radar and lidar data, or communicate via breaking signals when delays or system failures could result in catastrophic damage. The proper routing of vehicles in the platoon is necessary to reduce traffic congestion. Therefore, the proposed approach's primary objective is to minimize vehicle collision and travel time, leading to effective and reliable traffic-flow management.

1.1. Motivation

Traffic congestion and road safety are important issues of public safety worldwide. Annually, on average, approximately 150,000 people are killed, and 500,000 are injured in road accidents in India due to human errors such as distracted driving, over-speeding, and not complying with safety rules.Global statistics state that approximately 1.3 million people die yearly due to road traffic crashes. AVs moving in platoons are one of the best ways to reduce the accidents caused by human error and save passengers' time by reducing traffic congestion. The World Health Organization (WHO), in November 2018, stated that the number of road traffic collisions has outreached the mark of 1.5 million per year. According to the 2008 World Health Statistics, road accidents were the 9th most significant cause of death, and at current rates, they will be the 5th leading cause of death by 2030 [15].

1.2. Contribution

The contributions of the proposed work are as follows:

1. This work proposes a mechanism to efficiently manage traffic during high congestion, thus reducing road fatalities.
2. The proposed work focuses on merging platoons into one platoon, improvising the traffic flow and reducing travel time.
3. Finally, traffic performance is enhanced by joining a single non-platooned vehicle into a vehicle platoon, and collision is reduced by lane-changing mechanisms.

The novelty of the proposed methodology lies in enhancing the overall management of traffic flow for AVs in a platoon. A structured traffic flow is achieved by configuring various algorithms such as maneuvering, re-routing, and lane-changing. In a real-time scenario, the platooning approach reduces the time to reach the destination and reduces the number of accidents as vehicles move synchronously.

1.3. Paper Organisation

The rest of the paper is organised as follows. Section 2 presents the related work, the proposed approach is demonstrated in Section 3, followed by simulations and results in Section 4. Finally, the conclusion and future work are presented in Section 5.

2. Related Works

Numerous research studies have been conducted to maintain traffic flow for vehicles in platoons in various aspects, such as leader vehicle failure [16], obstructions within the platoon, etc. [12]. Many researchers and automobile companies have proposed various techniques to increase AVs' safety and road capacity. ACC, including no-communication functions, and CACC [17], including communication functions, are basic controller strategies followed for platoons. Platoons using CACC are fascinating because of their ability to increase road capacity in a very precise way [18].

A platoon-management protocol based on VANET and CACC vehicles which incorporate basic platooning maneuvers, such as platoon split, merge, and join maneuvers, is presented in [2]. However, this method has several disadvantages, such as a high delay during the merging of the platoon. An integrated ACC and CACC model with string stability in various traffic scenarios is implemented in [19]. Ploeg et al. [20] developed a hybrid controller for platoon merge and join maneuvers. The longitudinal control is handled by a continuous time system, while a discrete event supervisor decides the platoon merge-and-join maneuvers. However, where vehicle density is high, this technique suffers from frequent connection loss, resulting in a high packet-loss ratio. Huang et al. [21] demonstrated a cooperative platoon maneuver switching paradigm for merging and split maneuvers. However, this protocol causes significant latency and frequent connection loss. The authors in [22] proposed a distributed coordinated brake-control mechanism for longitudinal collision avoidance for multiple connected AVs in realistic scenarios. A brief statistical comparative study is conducted considering the initial velocity of the vehicle, inter-vehicular distance, communication topology, braking process, and different control mechanisms such as direct brake control, coordinated brake control, and driver reaction-based brake control.

The authors in [23] proposed an algorithm for platoon merge in cooperate driving that ensures effective platoon merging in all possible conditions. Wu et al. [24] proposed an adaptive velocity-based V2I fair access scheme based on IEEE 802.11, a distributive coordinate function for platoon vehicles. Roy et al. [25] proposed a model based on headway distribution of two-lane roads under different traffic situations with varying distributions such as Poisson, log-logistic and Pearson. In [26], authors investigated an event-based control and scheduling co-design technique for a platoon with packet disorder and communication delay, reducing congestion and stabilizing the system. Nevigato et al. [27] provide a collision-avoidance solution in a mobile edge computing-based environment to avoid accidents.

Hu et al. [28] proposed a reliable, trustworthy platoon-based service recommendation technique to help the user eliminate malicious platoons using V2V communication. Zhang et al. [29] present a trust-based and privacy-preserving platoon approach to enable the user vehicle to avoid the malicious leader vehicle. However, the method suffers from a problem in accurately identifying the trust value. The author in [30] proposed a method to distribute urban platooning towards high flexibility, adaptability and stability to solve the problem of vehicle density in urban areas, traffic lights etc. This method suffers from communication failure in dense vehicle scenarios.

The authors in [31] have argued that platoon-based driving offers a plethora of benefits. Firstly, since vehicles in the same platoon are substantially closer to one another, the traffic congestion may be reduced, and the road capacity is increased appropriately. Secondly, the platoon structure can significantly decrease energy use and exhaust emissions, as a platoon's vehicles can be streamlined to reduce air resistance. Third, driving has become safer due to contemporary technologies; it becomes more secure and comfortable in a platoon.

Lastly, platoon-based data dissemination mechanisms are efficient due to their position and synchronization, significantly enhancing vehicular network performance.

Segata et al. [32] developed an integrated simulator as a novel contribution towards research in platooning techniques. This is the first attempt to describe a high-level platoon management protocol in VANET-enabled vehicles that employ wireless V2V and V2I communication with IEEE 802.11p. Moreover, there is scope to explore the topic in-depth, and more simulation situations can be tested utilizing this simulator for vehicular platooning.

In this paper, three scenarios (join maneuver, merge maneuver, and lane change) are implemented, and the behavior of their speed, acceleration, and distance are analyzed. Taking different aspects from the literature, a meticulous solution is designed for AVs moving in platoons. Further, the total time to reach the destination for a platoon, non-platoon, or obstruction within the platoon (i.e., lane change or failure of leader vehicle) and multiple platoon scenarios (i.e., merge maneuvers) are explored.

3. Proposed Work

This section describes the system model, assumptions, and the methodology used for the proposed work to manage traffic flow using platooning.

3.1. System Model

Each vehicle is equipped with sensors and a GPS, and uses wireless access in vehicular environment (WAVE) as the inter-vehicle communication protocol based on DSRC. The onboard unit is capable of localization and time synchronization due to the equipped GPS. To exchange information about vehicle dynamics and emergency data, each vehicle in the platoon periodically transmits beacons. The vehicle movement is based on CACC, and thus, the cooperative movement considerably reduces the inter-vehicle spacing within a platoon. The intra-platoon communication must be relayed when the platoon is too large. To ensure safety, the inter-platoon separation is greater than the intra-platoon spacing. The size of the platoon is constrained to allow direct communication between platoon members inside the same platoon via one-hop communication. During the evaluation, the platoon's topology is considered to be static. This leads to proper channel utilization with a reduced synchronization overhead. In a real-world situation, there is a possibility of more than one platoon driving in the same lane, here, we have considered N-platoon. The jth platoon is designated as P_j. The ith vehicle within j^{th} platoon is denoted as $V_{j,i}$, where $i = 1 \ldots N$ and $j = 1 \ldots M$. The communication only takes place between the last vehicle of leading platoon and first vehicle of the following platoon. This reduces the interference between non-neighboring platoons.

3.2. Assumptions

1. All vehicles on the road are AVs to make communication reliable and compatible; there are no human-driven vehicles .
2. Let d_1, d_2, d_3 be the current density, threshold density, and normal density, respectively. The density of the AV represents the number of AVs per unit length-segment of the lane. AVs are generated by Poisson distribution with arrival rate λ as V_i, where $i = 1, 2, 3, \ldots, N$.
3. The initial route and alternate routes are generated against each source destination.The source and destination of each AV are assumed to be known, creating platoons P having a minimum of four AVs, where $P = V_1, V_2, V_3, \ldots, V_N$. The set of AVs fetched in each platoon can be stated as P_jV_i where $j = 1, 2, 3, \ldots, M$, for example, P_1V_4, P_1V_5, P_2V_6, P_2V_4, P_3V_4, P_3V_8, and so on.
4. The speed of the AV, acceleration, minimum gap, and distance to the leader are assumed to be known. The AVs in the platoons are induced to proceed from the source towards the destination, following the leader AV. The platoon vehicles move in a dedicated lane of the four-way highway. This mechanism minimizes the hindrance of human-driven vehicles in the other lanes.

5. The AVs broadcast CAM via a dedicated channel (CCH) at a frequency of 10 Hz, as per 802.11p specification. As an example, standard single-radio transceivers for platooned AVs are considered which are continually modulated to the CCH to broadcast and receive CAM [33]. The information about the density of AVs, speed, acceleration, and flow of AVs individually and in the platoon are utilized for congestion detection and avoidance during rerouting.
6. The car-following mobility model is similar to the one used in the PLEXE simulator i.e., the CACC approach. The CACC approach exploits the communication among vehicles via IVC. The control law for the CACC model considered for our implementation is based on the theory of consensus [32].

3.3. *Proposed Methodology*

The proposed algorithm (Algorithm 1) assumes that all the vehicles on the lanes are autonomous, and that there are no cooperation or communications glitches among the AVs. The AVs are assumed to know each other's location and speed details. Platoons of different sizes are considered to travel in highway lanes. Initially, the source and destination of the AVs and the current and alternate routes are assumed to be known. The AVs with the same destination are in one platoon according to the proposed algorithm. After the formation of all the platoons, the densities of the platoons are analyzed. Based on the platoon density, it is decided whether to keep the platoon on the same route or reroute. Let d be the total density of AVs in a road length stated as AVs per unit road length. The current density d_1, which is the maximum density for free flow traffic in the platoon at a particular instance of time, and the threshold density d_2, which is the maximum density for a road length, are the two basic types of densities considered in traffic-flow management. The total number of AVs generated by the free flow mobility model per unit length of the road is denoted by d_3. The distance between two AVs is inversely proportional to the density of the AVs in that particular lane. The free-flow model is designed with reference to [34], taking d_1 to be 20 and d_2 to be 50 for a single lane.

Apart from density, the time delay of platoons is checked and compared with the threshold value. If the delay time is greater than the threshold value, then the route of the platoon is changed (re-routing). If the delay is less than or equal to the threshold value, the same path is followed. Sometimes, it may be possible that the platoon is not at its best efficiency due to the presence of many smaller platoons on the road, which leads to many leaders. Therefore, to improve the algorithm's efficiency, two smaller platoons are merged into one larger platoon, resulting in one leader. In addition, it may be possible that instead of merging platoons, a single AV is added to the platoon. Therefore, the joining maneuver is performed. During this maneuver, the AVs in a platoon increase or decrease their speed synchronously. When an obstruction occurs between the platoon to eliminate these situations, a collision-avoidance mechanism is also implemented. The mechanism will reroute the platoons in the lane to another lane if it is blocked/obstructed. At the end of the journey, if all platoon AVs reach the destination safely, the platoon is marked as "Successfully Reached". Figure 2 presents the gist of the approach for analyzing platooning maneuvers. Table 1 shows the summary of the notation used in the algorithm.

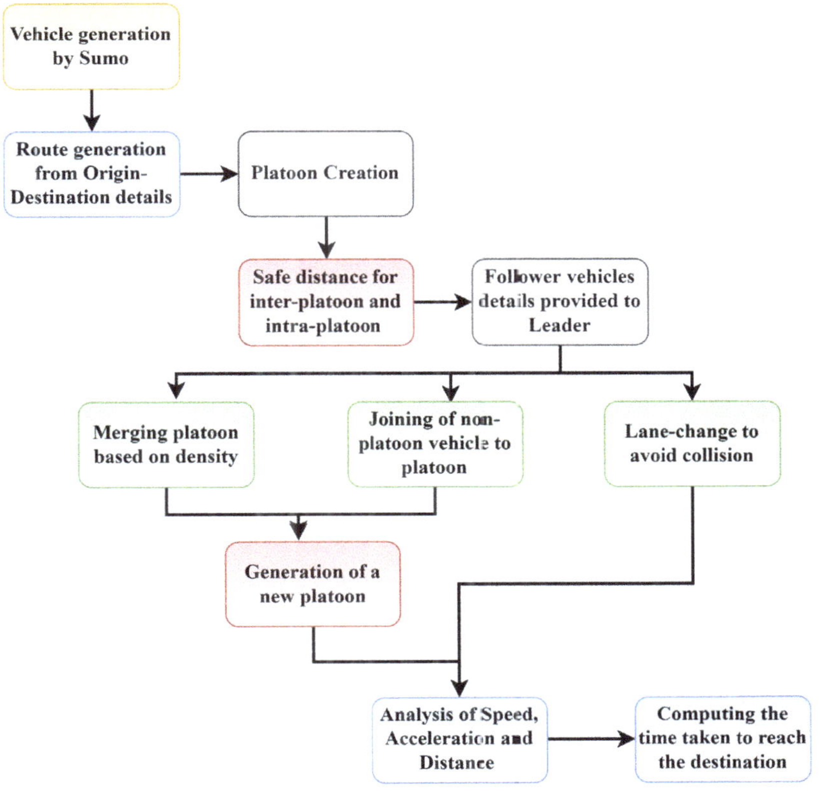

Figure 2. Flow diagram of proposed model.

Table 1. Summary of notations.

Notations	Description
d_1	Current Density of AVs
d_2	Threshold Density of AVs
d_3	Normal Density of As
T_d	Time Delay
Th_d	Threshold Value of Delay
Th_{den}	Threshold Density of Platoon
P_j	Identity of the jth platoon
V_i	Identity of the ith AV
L	Lane number

Algorithm 1 An algorithm for traffic management using platooning

Start
Set d_1, d_2, d_3 ;
for each P_j do
if (d_1 >= d_2) **then**
    Platoon Rerouting
else if (d_1 <= d_3) **then**
    No Rerouting
end if
for each P_j following P_1 do
if ($T_d > Th_d$) **then**
    Platoon Rerouting
else if ($T_d < Th_d$) **then**
    No Rerouting
end if
if (P_1 receives merge request from P_2) **then**
    if (P_1 , P_2 ϵ *Same Lane)* **then**
        if ((size of P_1 + size of P_2) $<= Th_{den}$) **then**
            Merge P_1 and P_2
        end if
    end if
end if
if (P_j receives join request from V_i) **then**
    if (P_j , V_i ϵ *Same Lane)* **then**
        if ((size of P_j) $< Th_{den}$) **then**
            Join V_i into P_j
        end if
    end if
end if
if (P_jL==inactive) **then**
  Assign new Leader to P_j
end if
if (Obstruction between P_j) **then**
  Split P_j in P_j and P_j' and assign L$'$ to P_j'
end if
for each P_j do
if (collision) **then**
    lane change
end if
End

4. Simulation and Results

This section presents the various scenarios and an analysis of results in detail.

4.1. Simulation Tool

This section describes the simulation tool used for the proposed approach. Simulation is performed on PLEXE, which supports vehicle platooning, based on network simulation platforms OMNeT++, VEINs, and simulation of urban mobility (SUMO). SUMO is an open-source traffic simulator used to optimize traffic signals, investigate route choice and forecast traffic. VEINS and OMNeT++ handle V2V and V2I communications. VEINs are used as a vehicle communication technology. Its broad class of libraries improve the realism and efficiency of traffic simulations. VEINs and OMNeT++ provide several capabilities that allow vehicles to communicate with one another. The VEINs framework communicates vehicles via the IEEE 802.11p and 1609.4 DSRC/wireless access.

4.2. Simulation Parameters

This section presents the simulation parameters. Table 2 shows the different parameters used in the simulation.

Table 2. Simulation Parameters.

Parameter	Values
AV's length	4 m
Optimal platoon size	8
Controller	ACC, CACC
Leader headway	1.2 s
Maximum speed (leader)	33.34 m/s
Maximum acceleration	3 m/s^2
Maximum deceleration	3 m/s^2
Lanes	4
Platoon size	4,6,8
Simulation rime	Merge-and-join maneuver (120 s), lane change (300 s)
PHY/MAC Model	IEEE 802.11p
MAC Model	1609.4

4.3. Merge Maneuver (Scenario 1)

Several maneuvers are performed for merging, joining, and collision avoidance. This section presents the results and discussions. In this section, the AVs are represented as cars in the figures.

The scenario taken for merging platoons is depicted in Figure 3. Two or more platoons moving in the same lane and joining to form a single large platoon is a merging maneuver [35]. Whenever the platoon size is smaller than the optimal size, a merge maneuver is initiated. We assume two platoons Y and X, where Y is the rear and X is the front platoon, with one platoon leader and three followers. The optimal size of the platoon is taken to be 8. Initially, the leader of platoon X receives the merge request from the leader of platoon Y. The X-platoon leader can either accept or decline the merge request based on several factors, such as the capacity of the platoon.

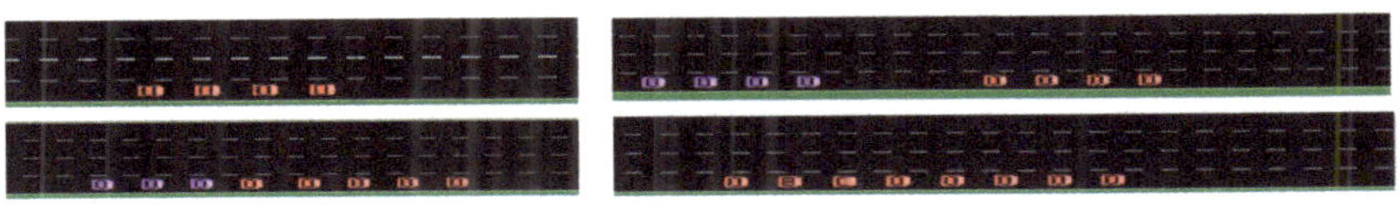

Figure 3. Merging of platoons.

On receiving approval for merging from the leader of X, the leader of Y reduces its intra-platoon-spacing [36] by increasing its speed to catch up with the front platoon. After the leader of platoon Y is in tune with the front platoon X, Y's leader sends a request to all of the platoon's members for leader change. After changing the leader, all the follower AVs start communicating with the platoon leader, and finally, the rear platoon leader becomes a follower of platoon X.

Figure 4 represents the speed versus time of merge maneuver of the platoon. The P_1 (including Car1–Car4) and the P_2 (including Car5–Car8) line shows the speed of the front and rear platoons, respectively. Initially, both the platoons are moving at the same speed, and after approximately 12 s, the rear platoon increases its speed and then moves constantly for a few seconds. After reaching a time period of nearly 45 s, the rear platoon starts reducing its speed to merge into the first platoon. After merging, the platoon's AVs move at a constant speed agaiuntil the destination.

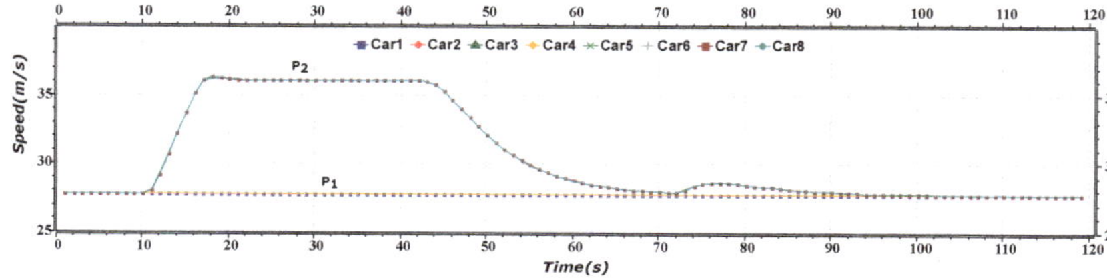

Figure 4. Speed versus time graph of merge maneuver.

Figure 5 represents the distance versus time of the merge maneuver of the platoon. The graph shows the distance between both the platoons. Initially, both the platoons have an inter-platoon distance of approximately 330 m. After approximately 12 s, as the rear platoon leader increases its speed, and the distance between the platoon starts decreasing. After nearly 80 s, all the AVs have a constant distance between them as both platoons merge to form one platoon.

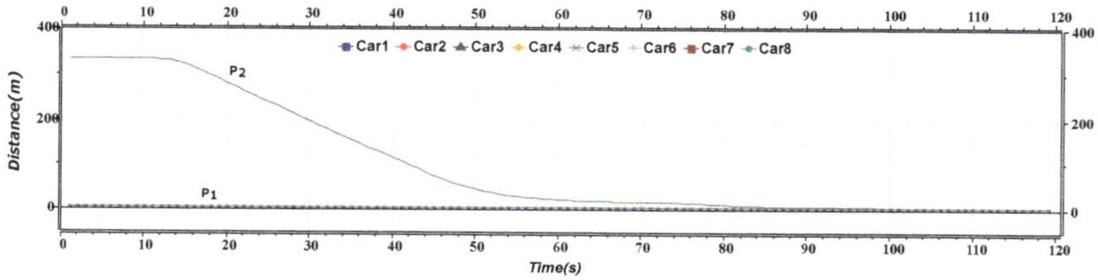

Figure 5. Distance versus time graph of merge maneuver.

Figure 6 represents the acceleration versus the time of the merge maneuver of the platoon. The P_1 (including Car1–Car4) line shows the acceleration of the front platoon, whereas the P_2 (including Car5–Car8) shows the acceleration of the rear platoon. Initially, both platoons are moving with the same acceleration. After approximately 12 s, as the rear platoon increases its speed, its acceleration changes from 0 to 1.5 m/s^2. After nearly 43 s, acceleration starts reducing as the rear leader AV decreases its speed to match the speed of the front platoon AV.

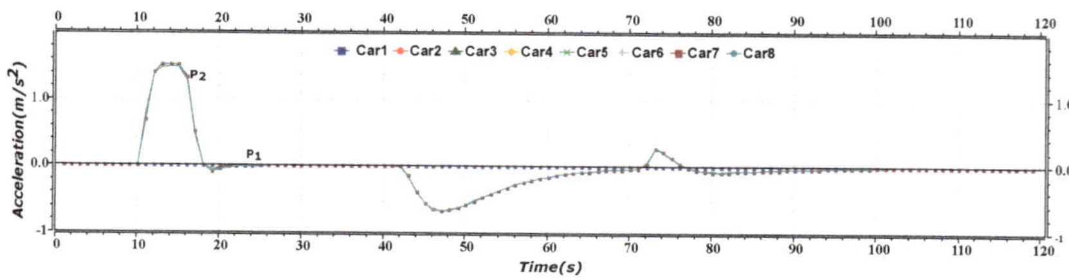

Figure 6. Acceleration versus time graph of merge maneuver.

4.4. Join Maneuver (Scenario 2)

The scenario of joining platoons is depicted in Figure 7.

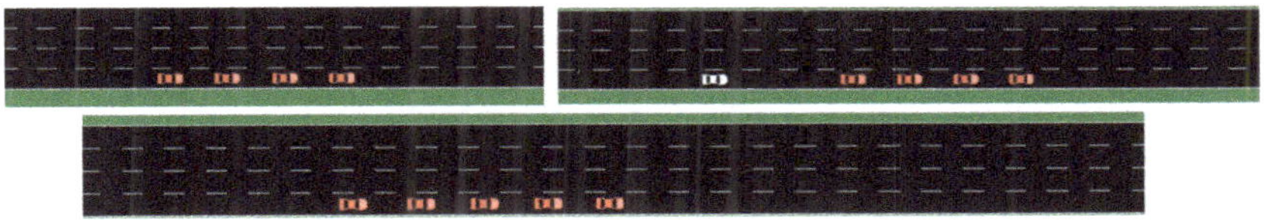

Figure 7. Joining maneuver.

In the joining maneuver scenario, a single platoon of four AVs traveling on the freeway is taken, while a fifth AV requests the platoon leader to join at the tail of the platoon. The non-platooned AV asks the platoon's leader to join the platoon and waits for a response. The leader accepts the request of the fifth AV (Car5) and responds with a message that includes essential information about the platoon, such as lane number and joining position, and waits for the joiner to become closer. The joiner AV uses that information to come close to the platoon's last AV, increasing its speed and managing the distance. When a joiner AV reaches a predetermined distance from the platoon's last AV, it signals to the leader that it is ready to join.

Figure 8 represents the speed versus time of the joining maneuver of the platoon and AV. P_1 (including Car1–Car4) shows the speed of the front platoon, whereas the Car5 line indicates the speed of the non-platooned AV. Initially, both are moving at the same speed. At 17 s, the non-platooned AV increases speed and constantly moves until the time reaches 23 s. After reaching 32 s, the AV starts reducing its speed to merge into the platoon. After joining, the AV becomes part of the platoon and starts moving at a constant speed.

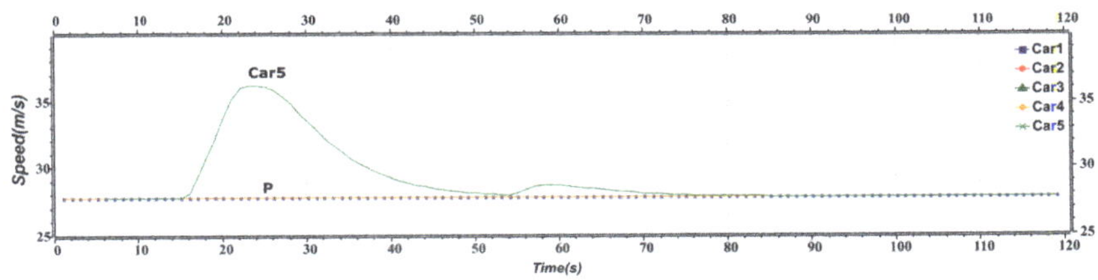

Figure 8. Speed versus time graph of joining maneuver.

Figure 9 represents the distance versus time of the joining maneuver of the platoon and AV. The graph shows the distance between the platoon and the AV. After approximately 16 s, the distance between the platoon and the AVs decreases as the AV increases its speed. After reaching a time of nearly 50 s, all the AVs have a constant distance between them as AVs merge into the platoon merge to form one platoon.

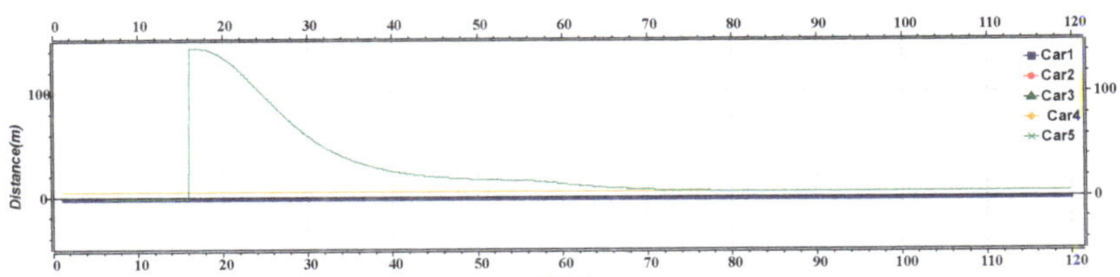

Figure 9. Distance versus time graph of joining maneuver.

Figure 10 represents the acceleration versus time of joining maneuver of the platoon and AV. The P_1 shows the speed of the front platoon, whereas the Car5 line shows the AV's speed. Initially, both are moving at the same speed. After approximately 16 s, the AV increases speed and then reduces at 22 s. After reaching a time of nearly 32 s, the AV begins reducing its speed to match the speed of the first platoon. As the AV is entirely ready to merge into the platoon, it needs to adjust its speed, so at approximately 45 s, its acceleration shows some fluctuations. After merging, the AVs of the platoon move at a constant speed.

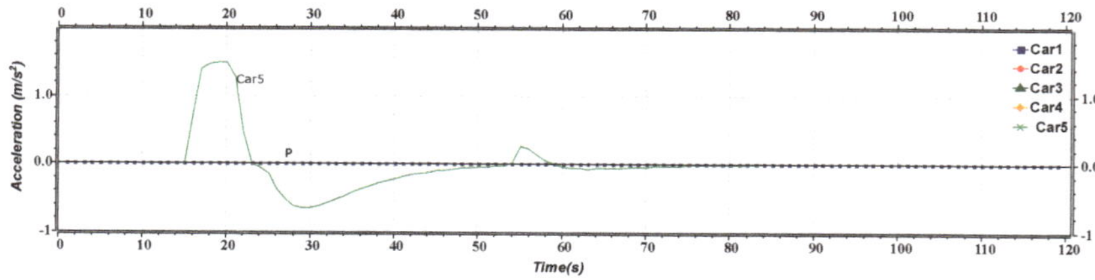

Figure 10. Acceleration versus time graph of joining maneuver.

4.5. Collision-Avoidance (Scenario 3)

This section focuses on avoiding collision by changing the lane of the whole platoon whenever any obstruction occurs during the journey. Figure 11 shows two platoons with 6 and 8 AVs moving in different lanes. The figure depicts collision avoidance due to obstruction by lane change, where the platoon of AVs with 6 AVs moves to the alternate lane.

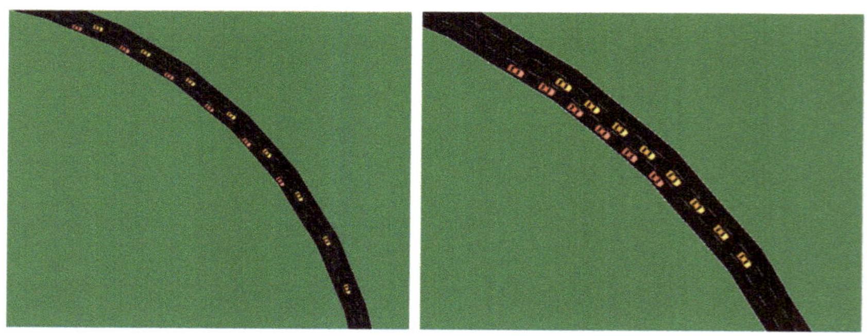

Figure 11. Collision avoidance by lane change.

Figure 12 represents the speed versus the time of lane change of the platoon leaders. Initially, both the platoons are moving at the same speed. After approximately 50 s and 90 s, the second platoon needs to change the lane to avoid the collision, decreasing its speed to move to another lane and then moving with the same speed as earlier. After approximately 90 s, as the first platoon also needs to change lanes to avoid the collision, it also decreases its speed to move to another lane and then moves with the same speed as earlier.

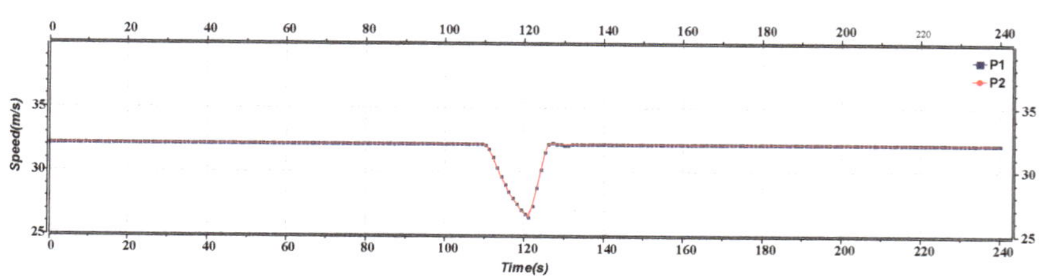

Figure 12. Speed versus time graph of lane change.

Figure 13 represents the platoon leaders' acceleration versus the time of lane change. Initially, both the platoons are moving with the same acceleration. After approximately 50 s and 90 s, as the second platoon reduces its speed for a lane change, its acceleration varies from 0 to $-1.5\ m/s^2$ every time. The first platoon also needs to change lane to avoid a collision; it adjusts its speed to approximately 90 s, and its acceleration shows some fluctuations due to the lane change.

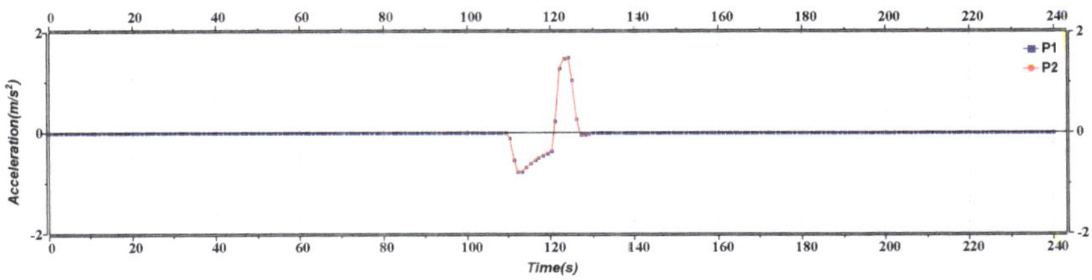

Figure 13. Acceleration versus time graph of lane change.

Figure 14 represents the distance versus the time of inter-platoon lane change, specifically between the leader of both platoons. Initially, the distance between the platoon leaders is constant. As the lane change is about to occur, the inter-platoon distance deviates, with the first deviation occurring at around 30 s. The graph predicts that it is challenging to maintain a constant distance between platoons during the lane change, but a safe distance is still maintained. This provides a research scope to formulate a critical safe distance model during the lane change, thus avoiding crashes. The proposed algorithm has focused on the safe distance between inter and intra-platoon to some extent which can be considered in future work.

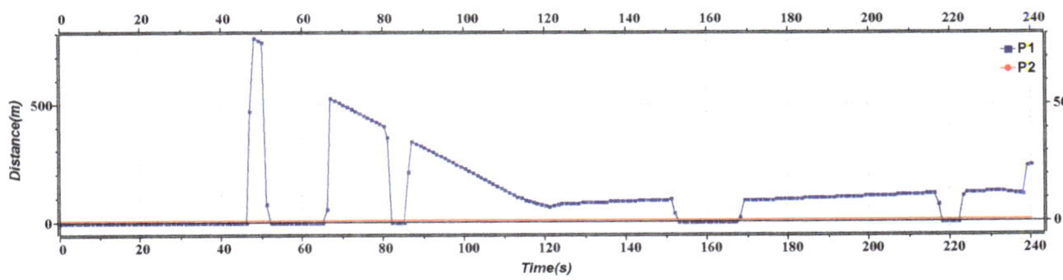

Figure 14. Distance versus time graph of lane change.

4.6. Comparison of All Scenarios

This section describes a comparative analysis of the number of AVs versus the total time to reach the destination in different scenarios with and without platoons, lane change, leader failure, and merge maneuver. For a comparative analysis of the total time taken to reach the destination for 500 AVs, these AVs were randomly generated by the free flow model in veh/hr, and their travelling time is computed.

The time an AV takes to get from its starting point to its destination is known as the travel time. AVs on highways move at a reasonable speed (90 to 140 km/h) to consume the least amount of fuel. A successful platooning mechanism also reduces fuel consumption due to the reduced travel time. In our simulation environment, the traveling time is defined as the interval between the AV's generation (entry of lane) and the moment it reaches its destination point. Each AV keeps track of its generation time. When it reaches its destination, it uses the time difference to calculate the distance traveled and time taken, based on relative speed.

Every AV predicts the time it will take to reach its destination when it merges onto the freeway, assuming a constant relative speed with the leader. AVs calculate the journey time by recording their actual travel time after arriving. This statistic allows us to demonstrate how platooning affects travel time. In this study, the platoon leader determines the speed of platoon members. The platoon maneuvers reduce the travel time to a great extent, contributing towards fuel-consumption minimization.

Figure 15 compares AV's movements with and without platooning, computing their travel time. The simulated results show that the traveling time of 500 AVs is 1360 s (nearly 23 min) without platooning. For a platoon scenario where the AVs moved in a platoon with cooperative speed and time-headway and rerouted to an alternate path in case of collision, the traveling time is 396 s (nearly 7 min). Therefore, we can conclude that the travel time is minimized by approximately 69% in a platoon scenario.

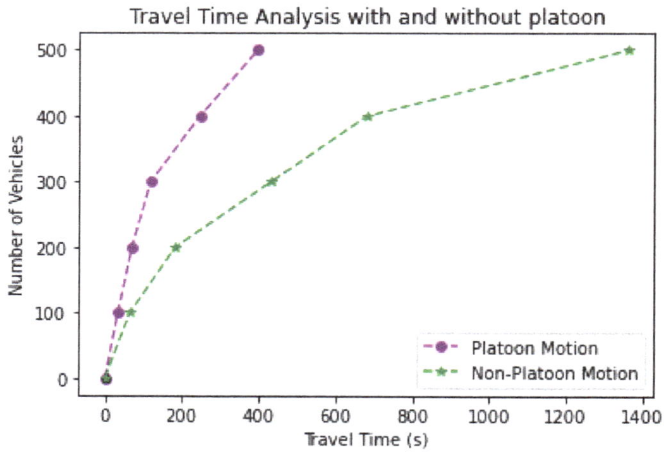

Figure 15. Comparison of platoon and non-platoon scenarios in terms of total time taken to reach destination.

In Figure 16, three scenarios are analyzed. The graph depicts the total traveling time taken by 500 AVs to reach their destinations. The three analyses include the scenario of lane change due to obstruction or chance of collision, the scenario when the leader AV fails, and the scenario with multiple platoons, i.e., merge maneuver. The results show that it takes 1008s (nearly 16 min) for 500 AVs to reach their destination when the lane change is performed. Identifying and choosing a new leader if the leader fails is time-consuming in case of any platoon interruptions. When the leader fails, it takes almost 1040 s (nearly 17 min) for 500 AVs to reach their destination. The travel time in case of leader AV failure is more than the travel time in a lane change. Finally, when small platoons merge into one large platoon, as in the merging maneuver, it significantly improves the AV platooning performance. It takes almost 365 s (nearly 6 min) for 500 AVs to arrive at the destinations. Thus, based on these results, it can be observed that merging platoons into one platoon could improve the traffic flow by reducing the travel time by 64%. However, in a real-time scenario, there will be a trade-off between travel time and platoon size, as a very long platoon can tend to congestion of road traffic. This analysis provides scope for designing an optimized algorithm considering the number of AVs, platoon size, and travel time, thereby monitoring fuel consumption.

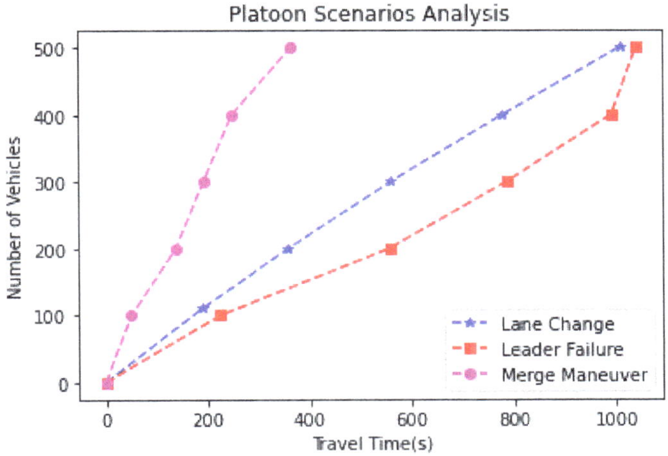

Figure 16. Number of AVs vs travel time to reach destination in different scenarios.

5. Conclusions and Future Works

This paper has presented an adaptive traffic-flow management mechanism with collision avoidance for vehicular platoons. The proposed model has contributed toward traffic-flow management of AVs using platooning by rerouting platoons in a collision situation, thereby avoiding traffic congestion. The merge and join maneuvers are performed to reduce the total travel time and road overhead by merging two small platoons into a single platoon. The lane-change mechanism of the platoon has been implemented to avoid vehicular collisions, thereby reducing the time to reach the destination. The comparative study shows that merging small platoons can minimize traveling time in a platoon scenario. Thus, this study focuses on optimizing and analyzing travelling delay for platoon scenarios.

In future, leaving and splitting maneuvers will be implemented for different sections of the platoon. A headway control will be designed for platoon modeling to minimize rear-end and side crashes.

Author Contributions: Conceptualization, L.H. and A.K.; Methodology, L.H., B.P.N. and P.H.J.C.; Software, B.S.; Formal analysis, L.H., B.P.N. and A.K.; Investigation, A.K.; Resources, B.S. and P.H.J.C.; Data curation, B.P.N.; Writing—original draft, L.H.; Writing—review & editing, B.P.N., P.H.J.C. and A.K.; Supervision, B.S. and A.K.; Funding acquisition, P.H.J.C. All authors have read and agreed to the published version of the manuscript.

Funding: This research received no external funding.

Data Availability Statement: Not applicable.

Conflicts of Interest: The authors declare no conflict of interest.

References

1. Caveney, D. Cooperative Vehicular Safety Applications. *IEEE Control Syst. Mag.* **2010**, *30*, 38–53.
2. Amoozadeh, M.; Deng, H.; Chuah, C.N.; Zhang, H.M.; Ghosal, D. Platoon Management with Cooperative Adaptive Cruise Control enabled by VANET. *Veh. Commun.* **2015**, *2*, 110–123. [CrossRef]
3. Qiao, L.; Shi, Y.; Chen, S.; Gao, W. Modeling and Analysis of Safety Messages Propagation in Platoon-based Vehicular Cyber-Physical Systems. *Wirel. Commun. Mob. Comput.* **2018**, *2018*, 12. [CrossRef]
4. *ETSI EN 302 637-2 V1.3.0 (2013-08)*; Intelligent Transport Systems (ITS); Vehicular Communications; Basic Set of Applications; Part 2: Specification of Cooperative Awareness Basic Service. Europeean Comission: Belgium, Brussels, 2013.
5. Böhm, A.; Jonsson, M.; Kunert, K.; Vinel, A. Context-aware retransmission scheme for increased reliability in platooning applications. In Proceedings of the International Workshop on Communication Technologies for Vehicles, Offenburg, Germany, 6–7 May 2014; Springer: Cham, Switzerland, 2014.

6. Shao, C.; Leng, S.; Zhang, Y.; Vinel, A.; Jonsson, M. Analysis of connectivity probability in platoon-based vehicular ad hoc networks. In Proceedings of the 2014 International Wireless Communications and Mobile Computing Conference (IWCMC), Nicosia, Cyprus, 4–8 August 2014; pp. 706–711. [CrossRef]
7. Se, G.U.O.Q.L.; Xu, S. Performance Enhanced Predictive Control for Adaptive Cruise Control System Considering Road Elevation Information. *IEEE Trans. Intell. Veh.* **2017**, *2*, 150–160. [CrossRef]
8. Faber, T.; Sharma, S.; Snelder, M.; Klunder, G.; Tavasszy, L.; van Lint, H. Evaluating Traffic Efficiency and Safety by Varying Truck Platoon Characteristics in a Critical Traffic Situation. *Transp. Res. Rec.* **2020**, *2674*, 525–547. [CrossRef]
9. Wang, J.; Li, S.; Zheng, Y.; Lu, X.-Y. Longitudinal collision mitigation via coordinated braking of multiple vehicles using model predictive control. *Integr. Comput. Aided Eng.* **2015**, *22*, 171–185. [CrossRef]
10. Santini, S.; Salvi, A.; Valente, A.S.; A. Segata, P.M.; Cigno, R.L. Platooning Maneuvers in Vehicular Networks: A Distributed and Consensus-Based Approach. *IEEE Trans. Intell. Veh.* **2019**, *4*, 59–72. [CrossRef]
11. Singh, P.K.; Sharma, S.; Nandi, S.K.; Singh, R.; Nandi, S. *Leader Election in Cooperative Adaptive Cruise Control Based Platooning*; Association for Computing Machinery: New York, NY, USA, 2018; pp. 8–14, ISBN 9781450359252.
12. Mushtaq, A.; Haq, I.U.; Nabi, W.U.; Khan, A; Shafiq, O. Traffic Flow Management of Autonomous Vehicles using Platooning and Collision Avoidance Strategies. *Electronics* **2021**, *10*, 1221. [CrossRef]
13. Kirthima, A.M.; Verma, R.; Hegde, C.R.; Shanbhag, A.S. Intelligent Accident Prevention in VANETs. *Int. J. Recent Technol. Eng. (IJRTE)* **2019**, *8*, 2401–2405. [CrossRef]
14. Wang, Z.; Xu, G.; Zhang, M.; Guo, Y. Collision avoidance models and algorithms in the era of internet of vehicles. In Proceedings of the IEEE 3rd International Conference of Safe Production and Informatization (IICSPI), Chongqing, China 28–30 November 2020; pp. 123–126.
15. 2008 World Health Statistics. Available online: https://morth.nic.in/ (accessed on 22 June 2022).
16. Jia, D.; Lu, K.; Wang, J. A Disturbance-Adaptive Design for VANET-enabled Vehicle Platoon. *IEEE Trans. Veh. Technol.* **2014**, *63*, 527–539. [CrossRef]
17. Segata, M.; Cigno, R.L.; Hardes, T.; Heinovski, J.; Schettler, M.; Bloessl, B.; Sommer, C.; Dressler, F. Multi-Technology Cooperative Driving: An Analysis Based on PLEXE. *IEEE Trans. Mob. Comput.* **2022**, *early access*. [CrossRef]
18. Wang, C.; Gong, S.; Zhou, A.; Li, T.; Peeta, S. Cooperative Adaptive Cruise Control for Connected Autonomous Vehicles by Factoring Communication-related Constraints. *Transp. Res. Procedia* **2020**, *113*, 124–145. [CrossRef]
19. Lu, X.Y.; Shladover, S. Integrated ACC and CACC development for heavy-duty truck partial automation. In Proceedings of the 2017 American Control Conference (ACC), Seattle, WA, USA, 24–26 May 2017; pp. 4938–4945. [CrossRef]
20. Ploeg, J.; Semsar-Kazerooni, E.; Medina, A.I.M.; de Jongh, J.F.; van de Sluis, J.; Voronov, A.; Englung, C.; Bril, R.J.; Salunkhe, H.; Arrue, A.; et al. Cooperative Automated Maneuvering at the 2016 Grand Cooperative Driving Challenge. *IEEE Trans. Intell. Transp. Syst.* **2018**, *19*, 1213–1226. [CrossRef]
21. Huang, Z.; Chu, D.; Wu, C.; He, Y. Path Planning and Cooperative Control for Automated Vehicle Platoon Using Hybrid Automata. *IEEE Trans. Intell. Transp. Syst.* **2019**, *20*, 959–974. [CrossRef]
22. Hu, M.; Li, J.; Bian, Y.; Wang, J.; Xu, B.; Zhu, Y. Distributed Coordinated Brake Control for Longitudinal Collision Avoidance of Multiple Connected Automated Vehicles. *IEEE Trans. Intell. Veh.* **2023**, *8*, 745–755. [CrossRef]
23. Paranjothi, A.; Atiquzzaman, M.; Khan, M.S. Pmcd: Platoon-Merging Approach for Cooperative Driving. *Internet Technol. Lett.* **2020**, *3*, e139. [CrossRef]
24. Qiong, W.; Xia, S.; Fan, P.; Fan, Q.; Li, Z. Velocity-Adaptive V2I Fair-Access Scheme based on IEEE 802.11 DCf for Platooning Vehicles. *Sensors* **2018**, *18*, 4198.
25. Roy, R.; Saha, P. Headway Distribution Models of Two-Lane Roads under Mixed Traffic Conditions: A Case Study from India. *Eur. Transp. Res. Rev.* **2018**, *10*, 1–12. [CrossRef]
26. Wu, L.; Zhang, L.; Zhou, Q. Event-based Control and Scheduling of a Platoon of Vehicles in VANETs. *IEEE Access* **2021**, *9*, 166223. [CrossRef]
27. Nevigato, N.; Tropea, M.; De Rango, F. Collision Avoidance Proposal in a MEC based VANET Environment. In Proceedings of the 2020 IEEE/ACM 24th International Symposium on Distributed Simulation and Real Time Applications (DS-RT), Prague, Czech Republic, 14–16 September 2020; pp. 1–7.
28. Hu, H.; Lu, R.; Zhang, Z.; Shao, J. Replace: A Reliable Trust-based Platoon Service Recommendation Scheme in VANET. *IEEE Trans. Veh. Technol.* **2017**, *66*, 1786–1797. [CrossRef]
29. Zhang, C.; Zhu, L.; Xu, C.; Sharif, K.; Ding, K.; Liu, X.; Du, X.; Guizani, M. Tppr: A Trust-based and Privacy-Preserving Platoon Recommendation Scheme in VANET. *IEEE Trans. Serv. Comput.* **2022**, *15*, 806–818. [CrossRef]
30. Jeong, S.; Baek, Y.; Son, S.H. Distributed Urban Platooning towards High Flexibility, Adaptability, and Stability. *Sensors* **2021**, *21*, 2684. [CrossRef] [PubMed]
31. Jia, D.; Lu, K.; Wang, J.; Zhang, X.; Shen, X. A Survey on Platoon-Based Vehicular Cyber-Physical Systems. *IEEE Commun. Surv. Tutorials* **2016**, *18*, 263–284. [CrossRef]
32. Segata, M.; Joerer, S.; Bloessl, B.; Sommer, C.; Dressler F.; Cigno, R.L. Plexe: A Platooning Extension for Veins. In Proceedings of the IEEE Vehicular Networking Conference (VNC), Paderborn, Germany, 3–5 December 2014; pp. 53–60.
33. Santa, J.; Pereniguez-Garcia, F.; Moragón, A.; Skarmeta, A. Experimental evaluation of CAM and DENM messaging services in vehicular communications. *Transp. Res. Part C Emerg. Technol.* **2014**, *46*, 98–120. [CrossRef]

34. Sala, M.; Soriguera, F. Capacity of a freeway lane with platoons of autonomous vehicles mixed with regular traffic. *Transp. Res. Part B Methodol.* **2021**, *147*, 116–131. [CrossRef]
35. Fida, N.A.; Ahmad, N.; Cao, Y.; Jan, M.A.; Ali, G. An Improved Multiple Manoeuver Management Protocol for Platoon Mobility in Vehicular Ad hoc Networks. *IET Intell. Transp. Syst.* **2021**, *15*, 886–901. [CrossRef]
36. Boubakri, A.; Gammar, S.M. Intra-platoon communication in autonomous vehicle: A survey. In Proceedings of the 9th IEEE International Conference on rPerformance Evaluation and Modeling in Wireless Networks (PEMWN), Berlin, Germany, 1–3 December 2020; pp. 1–6.

sensors

MDPI

Article

Enhancing Circular Polarization Performance of Low-Profile Patch Antennas for Wearables Using Characteristic Mode Analysis

Zhensheng Chen [1], Xuezhi Zheng [1,*], Chaoyun Song [2], Jiahao Zhang [3], Vladimir Volskiy [1], Yifan Li [1] and Guy A. E. Vandenbosch [1,*]

1 Division ESAT-WaveCoRE, KU Leuven, 3001 Leuven, Belgium
2 Department of Engineering, Strand Campus, King's College London, London WC2R 2LS, UK
3 National Key Laboratory of Science and Technology on Vessel Integrated Power System, Naval University of Engineering, Wuhan 430000, China
* Correspondence: xuezhi.zheng@kuleuven.be (X.Z.); guy.vandenbosch@kuleuven.be (G.A.E.V.)

Abstract: A wearable antenna functioning in the 2.4 GHz band for health monitoring and sensing is proposed. It is a circularly polarized (CP) patch antenna made from textiles. Despite its low profile (3.34 mm thickness, 0.027 λ_0), an enhanced 3-dB axial ratio (AR) bandwidth is achieved by introducing slit-loaded parasitic elements on top of analysis and observations within the framework of Characteristic Mode Analysis (CMA). In detail, the parasitic elements introduce higher-order modes at high frequencies that may contribute to the 3-dB AR bandwidth enhancement. More importantly, additional slit loading is investigated to preserve the higher-order modes while relaxing strong capacitive coupling invoked by the low-profile structure and the parasitic elements. As a result, unlike conventional multilayer designs, a simple single-substrate, low-profile, and low-cost structure is achieved. While compared to traditional low-profile antennas, a significantly widened CP bandwidth is realized. These merits are important for the future massive application. The realized CP bandwidth is 2.2–2.54 GHz (14.3%), which is 3–5 times that of traditional low-profile designs (thickness < 4 mm, 0.04 λ_0). A prototype was fabricated and measured with good results.

Keywords: characteristic mode; parasitic loading; slit loading; low-profile; wearable antenna

Citation: Chen, Z.; Zheng, X.; Song, C.; Zhang, J.; Volskiy, V.; Li, Y.; Vandenbosch, G.A.E. Enhancing Circular Polarization Performance of Low-Profile Patch Antennas for Wearables Using Characteristic Mode Analysis. *Sensors* **2023**, *23*, 2474. https://doi.org/10.3390/s23052474

Academic Editor: Peter Chong

Received: 13 January 2023
Revised: 8 February 2023
Accepted: 20 February 2023
Published: 23 February 2023

1. Introduction

Due to the restrictions on cost, reliability, productivity, safety, and robustness, as well as the requirement for a good user experience in advanced communications and wireless body area networks (WBAN) for health care, health monitoring, sensing, and athlete training, etc., a tremendous effort has been devoted to minimizing traditional rigid antennas [1,2]. However, as a compromise, the miniaturization sacrifices the antenna performance since the antenna gain is essentially related to its dimensions [3]. Alternatively, flexible wearable antennas based on textiles [4–6], polyimide [7], PDMS [8–10], etc., are supposed to be good candidates for maintaining good performance and user experience while relieving the dimension restriction. For instance, [6] reports a wearable antenna for energy harvesting applications; however, the reported antenna has a low front-to-back ratio (FBR) without ground shielding. It is well known that the unshielded design results in a significant gain decrease (e.g., it decreases by a factor of 2 or more in [6]) due to the coupling and power absorption caused by the vicinity loading of the lossy human body. In contrast, patch antennas are reported to be insensitive to the human body and can obtain high FBR due to their shielding ground, which can benefit wearable applications. Their merits of low profile and low cost are also favorable for massive usage. Additionally, one needs to address the issue of polarization mismatch, which inevitably happens due to unpredictable movements in realistic circumstances when linearly polarized (LP) antennas are used. Hence, CP antennas are a more appropriate choice [11–16]. Ref. [11] reports an S-shape slot CP antenna with a narrow 3-dB AR bandwidth (1.23%) for Iridium and GPS communications. Ref. [12] presents an omnidirectional button antenna using conceptual magnetic dipoles for 5 GHz applications. There are some other wearable CP antennas designed using a low-profile approach since clothes mostly have a thin thickness [17–22]. However, according to the above-mentioned literature, most of these wearable antennas have a limited CP bandwidth and a degraded performance in harsh environments. There are some antennas with bandwidth-enhancing techniques reported for traditional rigid antenna designs, for example, using aperture coupling [23], metamaterials/metasurfaces [24–28], parasitic elements [14,23], thick-air substrates [14,24], multilayer stacking [13], pin-loading [29], reflectors (e.g., AMC, PRS) [30,31], AI-based approaches [25,32], etc. However, most of these popular techniques are not suitable for wearable applications because of the thickness restrictions.

In this paper, we present a low-profile slit-loaded CP antenna with enhanced 3-dB AR bandwidth using Characteristic Mode Analysis (CMA). Its CP bandwidth is 2.2–2.54 GHz (14.3%). This design features two essential merits over the literature: (1) a superior CP bandwidth and high gain are achieved for low-profile wearable CP antennas in the 2.4 GHz band; (2) increased bandwidth with only single layer substrate, which is easy to fabricate and low-cost. It is worth mentioning that not many designs are capable of obtaining a decent 3-dB AR bandwidth for low-profile CP antennas (LPCPA), as depicted in Table 1, whereas our work draws a clear distinction. In detail, a multilayer structure with stacked patches and parasitic elements is reported in [13], which leads to 20.7% CP bandwidth at 6 GHz. Ref. [14] reports a strip-loaded antenna with 24% CP bandwidth. However, they are only suitable for traditional rigid substrates. Ref. [17] presents a wearable antenna array fed by a sequential feeding network. The feeding is embedded in the inner layer of the antenna, which makes the fabrication complicated. A modified metasurface is also applied to the antenna design in [18], which results in a wideband property with a peak gain of 8.5 dBic. In [12,19], shorting pins are used to design flexible antennas for off-body or on-body communication. However, these antennas work in the 5 GHz band [12,13,17–19]. If such antennas are scaled to the 2.4 GHz band, the physical thickness of the antennas increases. The scaled designs may be too bulky for wearable applications. As for the 2.4 GHz band low-profile wearable designs [16,20–22], they all have a limited CP bandwidth. This is due to the fact that with a physically similar low-profile thickness, their electrical thickness in the 2.4 GHz band is smaller than that one in the 5 GHz band, and thus, leads to a high Q, in other words, a narrow bandwidth in the 2.4 GHz band. Comparing to these reported wearable designs with a low profile, the proposed antenna achieves the highest CP bandwidth of 14.3%, around 3–5 times that of the counterparts. This is accomplished by adopting parasitic elements, as in [14], and a novel slit-loading technique. Notably, except for the parasitic loadings, Ref. [14] also utilizes a widely used bandwidth boosting method, i.e., air gap loading [14,23], and two blind disk-loaded pins to realize a single-feed design. However, the approach in [14] is unfavorable for wearables because of the thick air gap, and the blind disk and pins make the antenna fabrication complicated and costly (which are not easy to be addressed for wearables). Moreover, it is worth emphasizing that unlike traditional parasitic loadings [14,29], the bandwidth enhancement in this work is also attributed to the slit-loading approach, which introduces higher-order modes and also unblocks the requirement of inductive matching compensation in traditional designs. Overall, both the parasitic elements and the slit-loading lead to a decent CP bandwidth with a low-profile single-layer substrate.

This work is organized as follows. Section 2 illustrates the antenna configuration. Section 3 introduces the CP bandwidth-enhancing procedure from the perspective of CMA. Further, under the guidance of the obtained CMA results, CST microwave studio 2020 is employed to validate the antenna design with full-wave simulations. Then the antenna is fabricated, and the measurement results are discussed in Section 4. Finally, conclusions are given in Section 5.

Table 1. Comparison with counterpart CP antennas in the literature.

Ref.	f_0 (GHz)	Size (mm^3, $\lambda_0 \times \lambda_0 \times \lambda_0$	Impedance BW (GHz)	3-dB AR BW(GHz)	Peak Gain	Efficiency	FBR (dB)	Wearable Material
[12]	5.8	14.1 × 14.1 × 10.29, 0.273 × 0.273 × **0.199**	5.68–5.91 3.97%	5.72–5.89 2.93%	2.1 dBic	72.6%	NA	Yes
[13]	6	40 × 40 × 4.5, 0.8 × 0.8 × **0.09**	5.0–7.0 33%	5.38–6.62 20.7%	8.6 dBic	NA	20 *	No, rigid
[14]	2.49	130 × 130 × 15, 1.079 × 1.079 × **0.125**	1.87–3.11 49.8%	2.32–2.95 **24%**	8.7 dBic	NA	25 *	No, rigid
[17]	5.44	40 × 40 × 4, 0.725 × 0.725 × **0.072**	4.25–6.63 43%	4.71–6.67 34%	7.5 dBic	NA	30 *	Yes
[18]	5.47	54 × 54 × 4.4, 0.985 × 0.985 × **0.080**	4.51–6.43 35.1%	5.02–5.98 17.5%	8.5 dBic	77%	35 *	Yes
[19]	5.86	35 × 35 × 2.24, 0.684 × 0.684 × **0.044**	5.67–6.05 6.6%	5.73–5.955 3.85%	7.2 dBic	NA	28 *	Yes
[20]	2.43	65.6 × 58.9 × 3.94, 0.531 × 0.477 × **0.032**	2.26–2.6 * 14.0%	2.365–2.499 **5.5%**	6.5 dBic	73%	25 *	Yes
[21]	2.45	100 × 100 × 3.94, 0.818 × 0.818 × **0.032**	2.34–2.57 9.4%	2.4–2.45 * **2.1%**	6 dBic	62%	30 *	Yes
[22]	2.5	60 × 60 × 3, 0.5 × 0.5 × **0.025**	2.36–2.64 11%	2.42–2.49 **2.86%**	1.8 dBic	30.7%	18 *	Yes
[16]	2.45	60 × 60 × 3.4, 0.48 × 0.48 × **0.027**	2.26–2.64 * 15.5%	2.3–2.4 **4.08%**	2.5 dBic	42.26%	25 *	Yes
This work	**2.4**	**130 × 130 × 3.34, 1.04 × 1.04 × 0.027**	**2.18–2.61 18%**	**2.2–2.54 14.3%**	**8.9 dBic**	**52.96%**	**35**	**Yes**

NA: not available. *: evaluated from figures.

2. Antenna Configuration

The geometry of the antenna is depicted in Figure 1. The antenna is composed of a corner-truncated patch surrounded by parasitic elements. The patch is with a cross-slot at the center. The antenna is deposited on top of a grounded single substrate layer made of felt (ε_r = 1.35, tan δ = 0.03). Both the patch and the ground are made of conductive textiles with a conductivity of $\sigma = 1.18 \times 10^5$ S/m (Shieldit Super from LessEMF Inc, Latham NY, USA.). The materials are flexible so that the antenna is friendly to be worn on the body. A ground shielding topology rather than unshielded ones [6,33] is adopted to reduce the human body loading effect on the antenna performance. Since a large area is available on the body, the antenna ground can be kept large to have higher gain and less radiation exposure issues. The thickness of the felt substrate and the conductive textiles are 3 mm and 0.17 mm, respectively. This results in a low-profile structure whose thickness is only 0.027 λ_0. An antenna prototype is fabricated, as shown in Figure 9a. All the materials were cut manually. The textile radiator, the ground layer, and the felt substrate were ironed together with thermal melting glue. An SMA connector was soldered to the ground and to the patch to feed the antenna for measurements using a temperature-tunable soldering iron to avoid burning the textile.

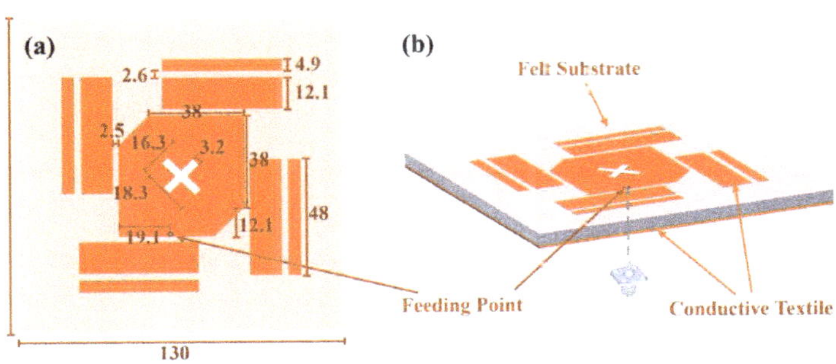

Figure 1. Structure of the antenna. (**a**) Front view. (**b**) 3D view.

3. Working Mechanism

3.1. Characteristic Mode (CM) Theory

To ease the illustration of the working mechanism of the CP bandwidth enhancement, here we review the main theoretical elements of CMA first.

According to the CM theory [34], an impedance matrix $\boldsymbol{Z}$ can be written as

$$\boldsymbol{Z} = \boldsymbol{R} + j\boldsymbol{X},$$

where $\boldsymbol{R}$ and $\boldsymbol{X}$ are the real and imaginary parts of the impedance matrix, respectively. Accordingly, a generalized eigenvalue problem is defined as:

$$\boldsymbol{X}\boldsymbol{J}_{\mathbf{n}} = \lambda_n \boldsymbol{R}\boldsymbol{J}_{\mathbf{n}},$$

where $\boldsymbol{J}_{\mathbf{n}}$ is the nth characteristic current mode of a conductive structure and λ_n is the corresponding eigenvalue of the nth mode. For each mode, modal significance MS is defined as [35]:

$$\mathrm{MS} = |1/(1 + j\lambda_n)|.$$

It is an intrinsic property of the structure and is independent of external excitations. Thus, by investigating the MSs of a certain structure, the properties of the structure can be envisioned and modified. Especially for $\lambda_n = 0$, i.e., MS = 1, the nth mode is in resonance. If we take an antenna as an example, with proper feeding excitation, the antenna may effectively radiate.

Another important factor for antennas design in CM theory is the characteristic angle (CA), which is defined as [35]:

$$\alpha_n = 180^{\mathrm{o}} - \tan^{-1}\lambda_n.$$

It presents the constant phase lag between the current $\boldsymbol{J}_{\mathbf{n}}$ and the tangential component of characteristic fields $\boldsymbol{E}_{\mathbf{n}}$, where $\boldsymbol{E}_{\mathbf{n}}$ is the field produced by current $\boldsymbol{J}_{\mathbf{n}}$.

To design a CP antenna, in practice, it is required that two orthogonal modes be effectively excited, and their MSs be close. Meanwhile, the characteristic angles should have a phase difference of nearly 90° (70°–110°).

3.2. Characteristic Mode Analysis (CMA) of the Low-Profile CP Antenna

CMA Multilayer Solver in CST studio 2020 is applied to carry out the predesign and analysis of the proposed antenna. Figure 2 shows the design in 3 steps. The design starts from a traditional corner-truncated patch, and then a cross-slot and four parasitic elements are adopted to introduce extra resonant modes in the frequency band of interest. Afterward, slits are cut on the parasitic elements to further broaden the bandwidth. Figure 3 illustrates the MSs and CAs of the first six modes of the three-step design, and Figures 4–6 illustrate the corresponding modal current distributions. As we can see in Figure 3a, mode 1 and mode 2 have large MSs in the 2.5–2.8 GHz band, while the MSs of mode 3–6 are almost 0, which means mode 1 and mode 2 could be excited. However, the MS and CA conditions to generate CP waves are effective in a limited bandwidth, i.e., 2.62–2.72 GHz, indicated in the colored zone 1 in Figure 3b. This also explains the natural narrow CP bandwidth of traditional corner-truncated patch antennas [21,22,36].

To increase the bandwidth, four parasitic elements are utilized to introduce extra modes in the interested band in the second step. The MSs and CAs for the interested mode 1–6 are shown in Figure 3c,d. It can be seen that mode 1 and mode 2 are the dominant ones and satisfy the CA requirements for generating CP waves around 2.34 GHz (colored zone 1, 2.3–2.39 GHz), while other modes around 2.34 GHZ have low MSs so that they can be ignored. Similarly, mode 2 and mode 3/4 contribute to a potential CP around 2.45 GHz (colored zone 2), mode 3/4 and mode 5 contribute to the frequency band of 2.47–2.51 GHz (colored zone 3), and mode 5 and mode 6 contribute to 2.51–2.55 GHz (colored zone 4). To sum up, the antenna may theoretically have a CP bandwidth in the frequency band of 2.3–2.55 GHz as long as all the modes can be effectively excited. Notably, the resonant frequencies of mode 1 and mode 2 are lower than the ones in step 1. This is due to the introduction of the cross-slot at the patch center and due to the coupling between the parasitic elements and the patch. This can be seen from the modal current distributions of mode 1 and mode 2, as shown in Figure 4a,b and Figure 5a,b. Such coupling can be modeled using LC model methods [5,28]. Moreover, it is worth mentioning that the parasitic patches and the traditional center patch without slot-loading in step 1 induce a strong capacitive effect with the antenna ground. This makes a traditional high-profile design preferable since a long feed probe can be used to compensate for the capacitive effect by its intrinsic inductance. However, for a wearable design with low-profile restrictions, the feed probe is too short to fulfill this requirement. Hence, a cross-slot is cut at the center patch to reduce the capacitive effect and to release the need for a long probe.

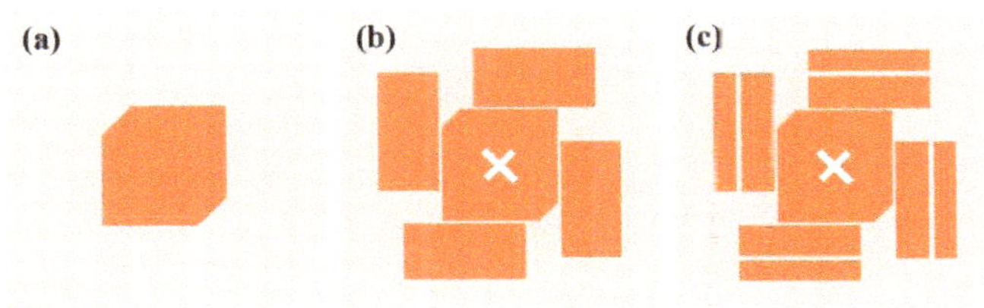

Figure 2. Evolution of the LPCPA design. (**a**) Step 1, a traditional corner truncated patch antenna. (**b**) Step 2, a corner truncated patch with cross-slot and parasitic elements. (**c**) Step 3, the proposed LPCPA with slit-loading.

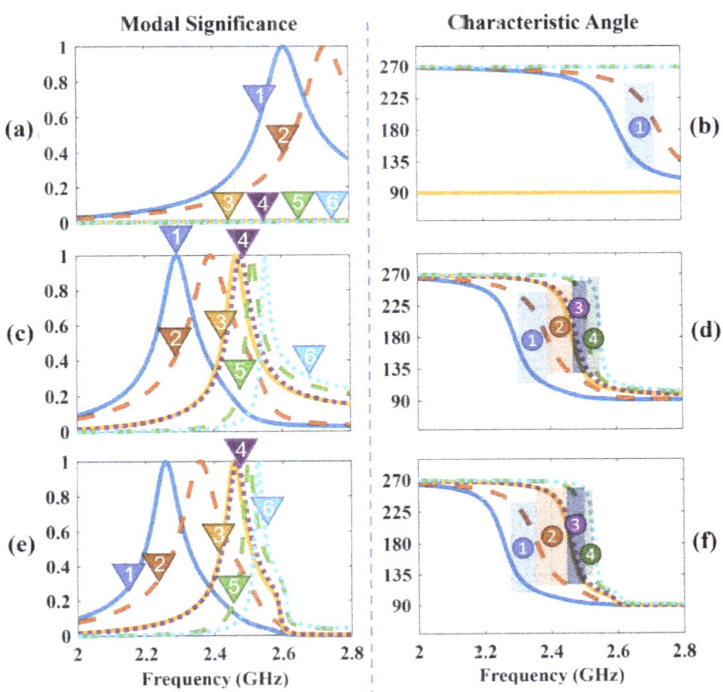

Figure 3. MSs and CAs evolution of the LPCPA. (**a**,**b**) MSs and CAs of Step 1. (**c**,**d**) MSs and CAs of Step 2. (**e**,**f**) MSs and CAs of Step 3. Colored zones present where the CA phase differences are in the range of 70°–110°.

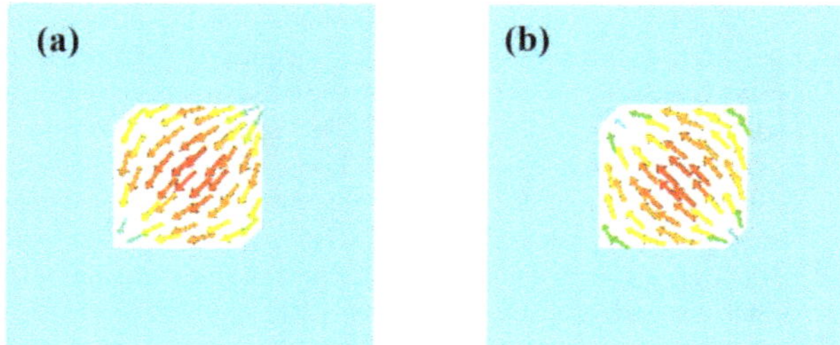

Figure 4. Modal current distributions of basic corner-truncated structure. (**a**) Mode 1. (**b**) Mode 2.

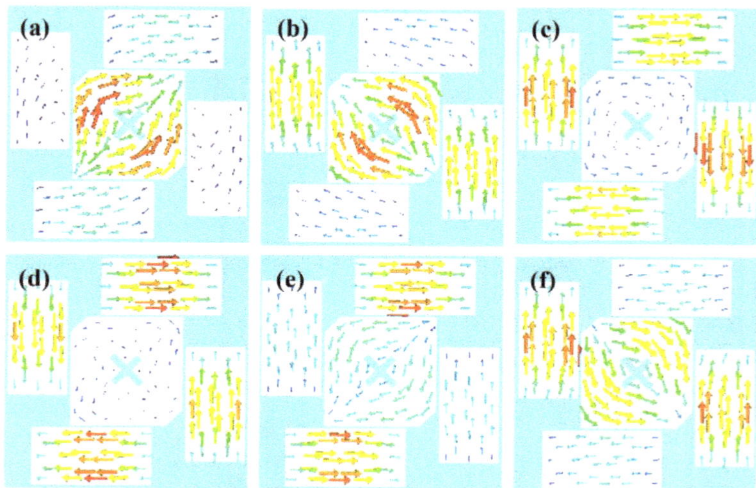

Figure 5. Modal current distributions of the structure without slit-loading. (**a**–**f**) are Mode 1–6.

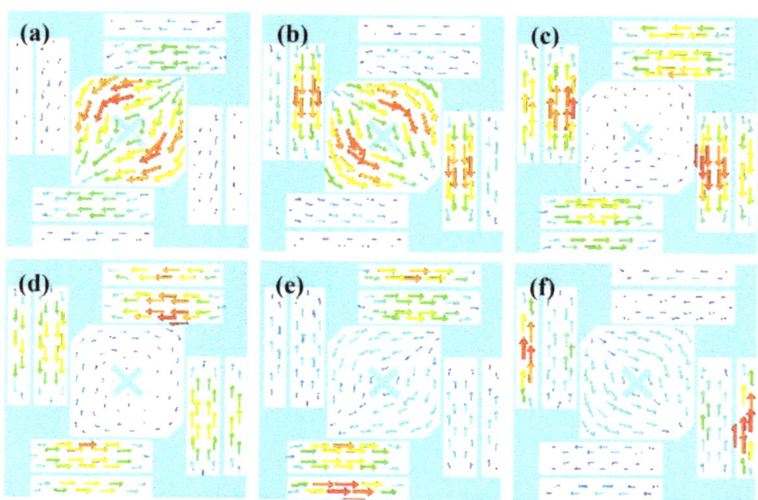

Figure 6. Modal current distributions of the structure with slit-loading. (**a**)–(**f**) are Mode 1–6.

Last, we find that the topology with slit-loadings on the parasitic elements proposed in step 3 has similar MSs, CAs, and modal currents as in step 2 (see Figure 3e,f and Figure 6). The potential CP bandwidth is 2.26–2.52 GHz. This is due to the fact that all the modal currents of modes 1–6 on the parasitic elements flow along the x or y direction so that slits along x or y directions have limited impact on the modal currents and the characteristic properties. Thanks to this, narrowing down the size of parasitic patches in step 2 by means of additional slits increases the parasitic inductance and decreases the capacitance while keeping the characteristic properties, meaning that a short, less inductive feeding structure may be sufficient to feed and excite the antenna. Overall, this approach with slits and the cross-slot is significantly different from the thick multilayer designs in [14,29], which need extra disks and pins for impedance matching.

3.3. Full-Wave Simulation

Summarized, topology 3 has the potential to obtain a wide CP bandwidth if the desired modes 1–6 are effectively excited with proper feeding. In the following, a coaxial probe (inner-pin diameter is 1 mm, outer-shell diameter is 4.1 mm) is used as feeding, and full-wave simulations are carried out to validate the proposed design. The final optimized parameters are given in Figure 1. The simulated reflection coefficient and AR results of the three topologies are shown in Figure 7. The full-wave simulation results agree with the above-analyzed results very well. It is clear that the proposed LPCPA with slit-loadings has significantly improved the performance of the low-profile antennas of topology 1 and topology 2. Its impedance bandwidth ($S_{11} < -10$ dB) and CP bandwidth (AR < 3 dB) are 2.2–2.64 GHz and 2.29–2.53 GHz, respectively.

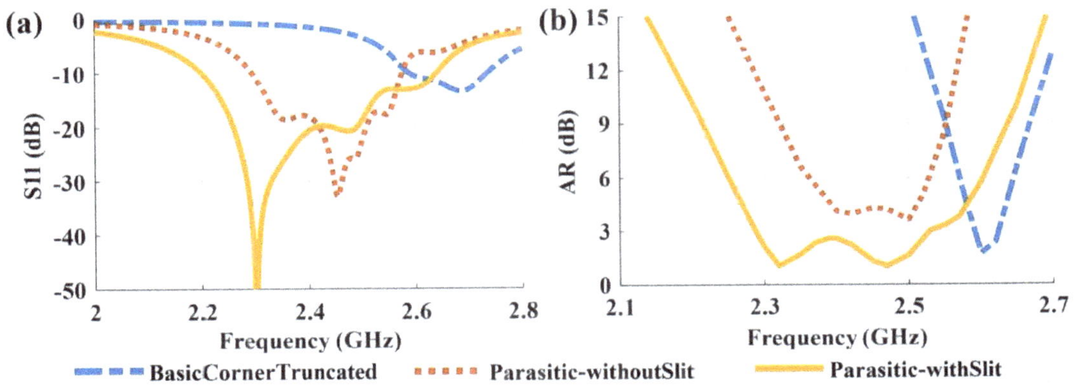

Figure 7. Simulated (**a**) S11 and (**b**) AR evaluation of the antennas.

The Specific Absorption Rate (SAR) over 10 g of tissue is simulated by CST Studio 2020 according to the IEEE C95.3 standard. For realistic wearing applications, the antenna is placed about 5 mm above the body most of the time due to the insertion of air gaps and mid-layers to prevent scratching, potential allergic reactions, etc. Figure 8 shows the simulation model, which has been validated in [37,38]. The permittivity of the fat, skin, and muscle is 5.28, 38.01, and 52.73, respectively [39]. The simulated SAR result is 0.0159 W/kg at 2.4 GHz. The simulated SAR value is significantly lower than the European standard threshold of 2 W/kg. This is a benefit of the ground shielding of the proposed design.

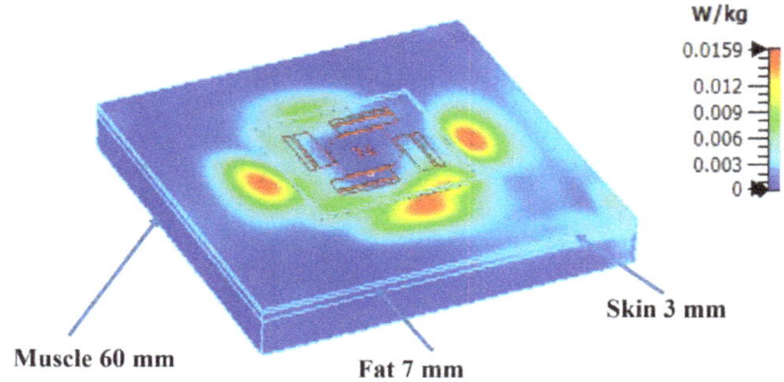

Figure 8. SAR simulation result at 2.4 GHz.

4. Measurement and Discussion

The fabricated antenna was measured in an anechoic chamber with a Vector Network Analyzer 8510 system. Three scenarios were considered: free-space, on-body, and on-phantom. The measurement setups are shown in Figure 9. Since a wearable antenna is inevitably bent when being worn on the body in practice, bent antennas with curvature radii of 5 cm and 8.5 cm (i.e., conformal situations) were measured for the free space scenario. Further, the reflection coefficient of on-body scenarios, for example, on-arm and on-chest, was measured. As for the AR measurements, body tissue simulating liquid MSL2450v2 [40] was used to mimic the body loading due to the fact that a human may move during the measurement and cause unstable transmissions and reflections. A cylinder with a 10 cm diameter and a liquid bag packed with a 10-mm-thick foam were used as phantoms to mimic the on-arm and on-chest situations, respectively.

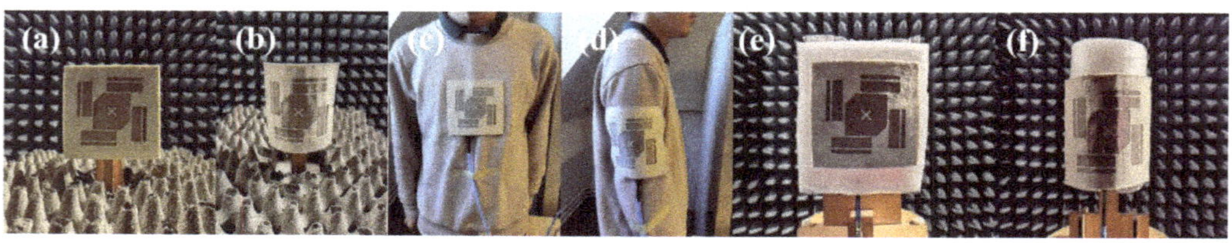

Figure 9. Measurement setups. (**a**,**b**) Free space. (**c**,**d**) On-body. (**e**,**f**) On-phantom.

4.1. Reflection Coefficient

As shown in Figure 10a, in free space, the impedance bandwidth ($S_{11} < -10$ dB) is 2.2–2.64 GHz in simulation and 2.18–2.61 GHz in measurement when the antenna is flat. The slight discrepancy between simulation and measurement may be caused by manual fabrication errors. For the conformal cases, a slight frequency shift is observed. Nevertheless, the antenna still shows a good performance. Further, the on-body results are shown in Figure 10b. The simulated impedance bandwidth is 2.21–2.64 GHz, and the measured bandwidth of on-arm and on-chest is 2.18–2.59 GHz. It is evident that, due to the benefit of ground shielding, the proximity of the body results in a limited influence when the antenna is worn on the arm and on the chest. Overall, we can conclude that the simulation and the measurement results match well. The antenna shows a robust impedance performance in free space and on-body scenarios.

4.2. Axial Ratio

The simulated and measured AR results of the antenna in free-space and on-phantom scenarios are shown in Figure 11. They are evaluated on the normal axis ($\theta = 0°$) of the antenna. For the free-space scenario (see Figure 11a), the simulated flat antenna has a

CP bandwidth (AR < 3 dB) of 2.29–2.53 GHz, and, accordingly, the measurement result is 2.2–2.54 GHz. The measurement bandwidth is wider than that of the simulation. This may be caused by fabrication errors during the manual fabrication process. The measured results when the antenna is bent show a frequency shift (depression) compared to the flat scenario. This is due to the symmetry of the structure that is broken. In detail, the modal currents of the dominant modes at high frequencies, i.e., modes 3–6 in Figure 6c–f, mainly flow on the rotationally symmetric parasitic elements. So, when the symmetric is broken, the CP waves cannot be effectively generated at high frequencies, and a frequency shift (depression) is observed. As for the on-phantom scenarios (see Figure 11b), a similar phenomenon can be observed.

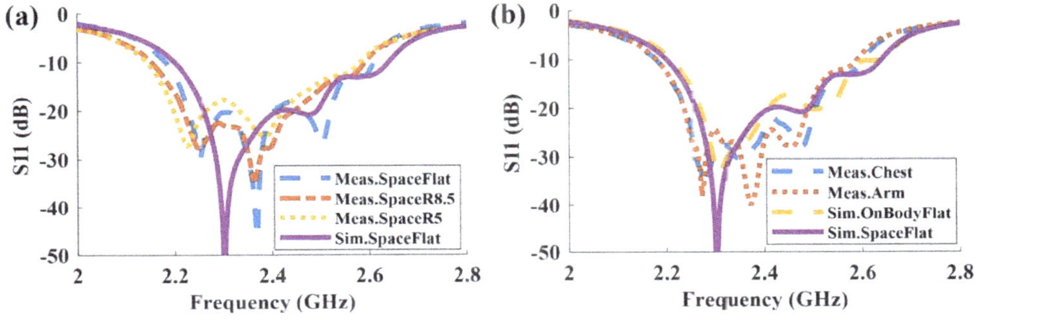

Figure 10. Simulated and measured reflection coefficient in (**a**) free-space and (**b**) on-body scenarios.

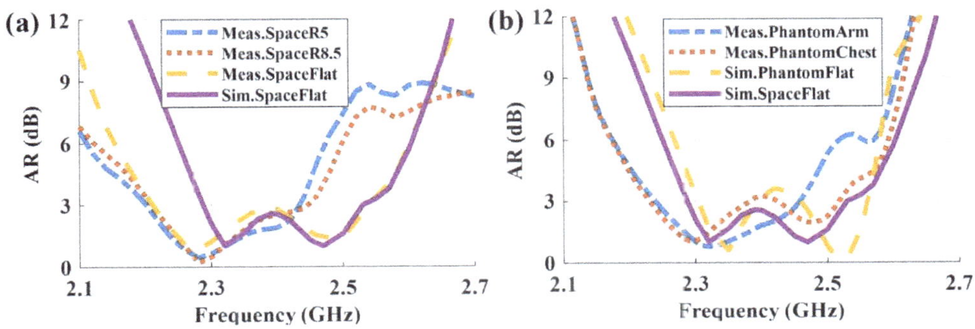

Figure 11. Simulated and measured AR in (**a**) free-space and (**b**) on-phantom scenarios.

4.3. Radiation Pattern and FBR

Figure 12 shows the LHCP and RHCP realized the gain pattern of the LPCPA at 2.3 GHz and 2.4 GHz for both the free space and the on-phantom scenarios. The LHCPs are 10 dB larger than the RHCPs, which means the main polarization of the antenna is left-handed circular polarization. In free space (Figure 12a,b), the flat antenna has a measured realized gain of 8.9 dBic and 7.1 dBic at 2.3 GHz and 2.4 GHz, respectively. Correspondingly, the estimated proximate efficiencies are 52.96% and 44.12% based on the half-power beamwidth [41]. Compared to the literature in Table 1, the designed low-profile antenna has achieved the highest peak gain even with lossy textile materials rather than the less-lossy PCB substrates that are used in [13,14]. Due to the bending, a decrease in the LHCP gain is observed. The measured realized gain is 8.7 dBic (2.3 GHz) and 5.4 dBic (2.4 GHz) when the antenna is bent with a curvature radius of 5 cm. As for the on-phantom scenario, see Figure 12c,d, the measured realized gains at 2.3 GHz are 8.6 dBic for on-chest and 8.4 dBic for on-arm, respectively, and at 2.4 GHz the corresponding ones are 7.1 dBic and 5.9 dBic. According to the patterns, we can also obtain the FBR values. The evaluated FBR values are all larger than 18 dB, and the peak one is 35 dB for the flat antenna in free space, which shows good radiation shielding due to the presence of the ground. As for positions having limited space or with drastic bending, for example, the wrist, the ground should be decreased accordingly for good wearing comfortability. Alternatively, the materials could be changed to more flexible and stretchable ones.

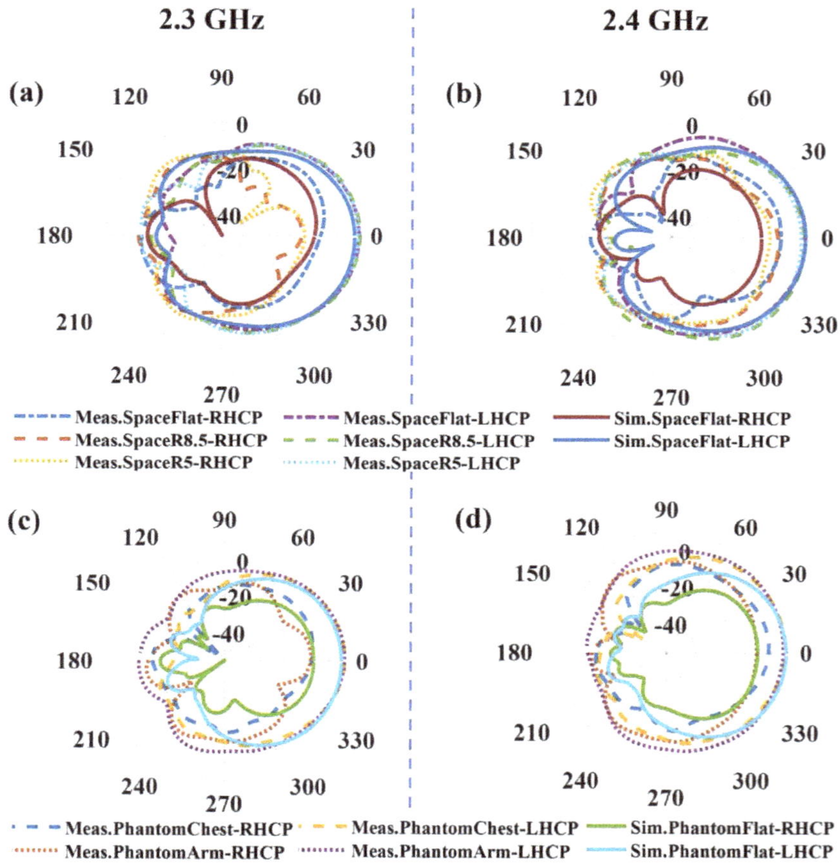

Figure 12. Simulated and measured realized gain pattern in (**a**,**b**) free-space and (**c**,**d**) on-phantom scenarios at 2.3 GHz and 2.4 GHz.

5. Conclusions

A wearable low-profile antenna with circular polarization is designed. Slit-loaded parasitic elements are used to improve the CP bandwidth in the 2.4 GHz band. CM theory is used to reveal the modal current distributions of the antenna. According to these current distributions, we find that the antenna can achieve a wide CP bandwidth by loading slits on the parasitic elements whilst the antenna profile remains low. Full wave simulations and measurements validate our design. The size of the prototype antenna is $1.04 \times 1.04\ \lambda_0^2$ while the profile is only 0.027 λ_0. The antenna has an impedance bandwidth of 2.18–2.61 GHz (18%) and a CP bandwidth of 2.2–2.54 GHz (14.3%) and realized a peak gain of 8.9 dBic. Compared to other low-profile wearable designs in the 2.4 GHz band, this design achieves a bandwidth that is 3–5 times larger.

Author Contributions: Z.C., Conceptualization, methodology, software, validation, writing; X.Z., validation, writing—review and editing, supervision; C.S., validation, writing—review and editing; J.Z., conceptualization, methodology; V.V., software, measurement; Y.L., measurement; G.A.E.V., validation, writing—review and editing, supervision. All authors have read and agreed to the published version of the manuscript.

Funding: Z. Chen is grateful for the scholarship from China Scholarship Council (CSC). G.A.E. Vandenbosch and X. Zheng are grateful for KU Leuven Internal Funds: the C1 project (C14/19/083).

Institutional Review Board Statement: Not applicable.

Informed Consent Statement: Not applicable.

Data Availability Statement: Not applicable.

Conflicts of Interest: The authors declare no conflict of interest.

References

1. Sharma, M.; Parini, C.G. A miniature wideband antenna for wearable systems. In Proceedings of the 2013 Loughborough Antennas and Propagation Conference, Loughborough, UK, 11–12 November 2013; pp. 619–623. [CrossRef]
2. Rao, S.; Llombart, N.; Moradi, E.; Koski, K.; Bjorninen, T.; Sydanheimo, L.; Rabaey, J.M.; Carmena, J.M.; Rahmat-Samii, Y.; Ukkonen, L. Miniature implantable and wearable on-body antennas: Towards the new era of wireless body-centric systems [antenna applications corner]. *IEEE Antennas Propag. Mag.* **2014**, *56*, 271–291. [CrossRef]
3. Harrington, R.F. Effect of antenna size on gain, bandwidth, and efficiency. *J. Res. Natl. Bur. Stand. Sect. D Radio Propag.* **1960**, *64*, 1–12. [CrossRef]
4. Yan, S.; Soh, P.J.; Vandenbosch, G.A.E. Low-Profile Dual-Band Textile Antenna with Artificial Magnetic Conductor Plane. *IEEE Trans. Antennas Propag.* **2014**, *62*, 6487–6490. [CrossRef]
5. Zhang, J.; Yan, S.; Vandenbosch, G.A.E. A Miniature Feeding Network for Aperture-Coupled Wearable Antennas. *IEEE Trans. Antennas Propag.* **2017**, *65*, 2650–2654. [CrossRef]
6. Wagih, M.; Weddell, A.S.; Beeby, S. Omnidirectional Dual-Polarized Low-Profile Textile Rectenna with Over 50% Efficiency for Sub-μW/cm^2 Wearable Power Harvesting. *IEEE Trans. Antennas Propag.* **2021**, *69*, 2522–2536. [CrossRef]
7. Khaleel, H.R.; Al-Rizzo, H.M.; Rucker, D.G.; Mohan, S. A Compact Polyimide-Based UWB Antenna for Flexible Electronics. *IEEE Antennas Wirel. Propag. Lett.* **2012**, *11*, 564–567. [CrossRef]
8. Jiang, Z.H.; Cui, Z.; Yue, T.; Zhu, Y.; Werner, D.H. Compact, Highly Efficient, and Fully Flexible Circularly Polarized Antenna Enabled by Silver Nanowires for Wireless Body-Area Networks. *IEEE Trans. Biomed. Circuits Syst.* **2017**, *11*, 920–932. [CrossRef]
9. Zhang, S.; Zhu, J.; Zhang, Y.; Chen, Z.; Song, C.; Li, J.; Yi, N.; Qiu, D.; Guo, K.; Zhang, C.; et al. Standalone stretchable RF systems based on asymmetric 3D microstrip antennas with on-body wireless communication and energy harvesting. *Nano Energy* **2022**, *96*, 107069. [CrossRef]
10. Sayem, A.S.M.; Lalbakhsh, A.; Esselle, K.P.; Buckley, J.L.; O'Flynn, B.; Simorangkir, R.B.V.B. Flexible Transparent Antennas: Advancements, Challenges, and Prospects. *IEEE Open J. Antennas Propag.* **2022**, *3*, 1109–1133. [CrossRef]
11. Kaivanto, E.K.; Berg, M.; Salonen, E.; de Maagt, P. Wearable Circularly Polarized Antenna for Personal Satellite Communication and Navigation. *IEEE Trans. Antennas Propag.* **2011**, *59*, 4490–4496. [CrossRef]
12. Hu, X.; Yan, S.; Zhang, J.; Volskiy, V.; Vandenbosch, G.A.E. Omni-Directional Circularly Polarized Button Antenna for 5 GHz WBAN Applications. *IEEE Trans. Antennas Propag.* **2021**, *69*, 5054–5059. [CrossRef]
13. Yang, W.; Zhou, J.; Yu, Z.; Li, L. Single-Fed Low Profile Broadband Circularly Polarized Stacked Patch Antenna. *IEEE Trans. Antennas Propag.* **2014**, *62*, 5406–5410. [CrossRef]
14. Hu, W.; Tang, Z.-Y.; Fei, P.; Yin, Y.-Z. Broadband circularly polarized Z-shaped dipole antenna with parasitic strips. *Int. J. RF Microw. Comput. Eng.* **2015**, *27*, e21052. [CrossRef]
15. Ray, M.K.; Mandal, K.; Nasimuddin, N.; Lalbakhsh, A.; Raad, R.; Tubbal, F. Two-Pair Slots Inserted CP Patch Antenna for Wide Axial Ratio Beamwidth. *IEEE Access* **2020**, *8*, 223316–223324. [CrossRef]
16. Sayem, A.S.M.; Simorangkir, R.B.V.B.; Esselle, K.P.; Lalbakhsh, A.; Gawade, D.R.; O'Flynn, B.; Buckley, J.L. Flexible and Transparent Circularly Polarized Patch Antenna for Reliable Unobtrusive Wearable Wireless Communications. *Sensors* **2022**, *22*, 1276. [CrossRef]
17. Chen, Y.; Liu, X.; Fan, Y.; Yang, H. Wearable Wideband Circularly Polarized Array Antenna for Off-Body Applications. *IEEE Antennas Wirel. Propag. Lett.* **2022**, *21*, 1051–1055. [CrossRef]
18. Yang, H.; Liu, X.; Fan, Y. Design of Broadband Circularly Polarized All-Textile Antenna and Its Conformal Array for Wearable Devices. *IEEE Trans. Antennas Propag.* **2022**, *70*, 209–220. [CrossRef]
19. Yang, H.C.; Liu, X.Y.; Fan, Y.; Tentzeris, M.M. Flexible circularly polarized antenna with axial ratio bandwidth enhancement for off-body communications. *IET Microw. Antennas Propag.* **2021**, *15*, 754–767. [CrossRef]
20. Moro, R.; Agneessens, S.; Rogier, H.; Bozzi, M. Circularly-polarised cavity-backed wearable antenna in SIW technology. *IET Microw. Antennas Propag.* **2018**, *12*, 127–131. [CrossRef]

21. Hertleer, C.; Rogier, H.; Vallozzi, L.; Van Langenhove, L. A Textile Antenna for Off-Body Communication Integrated Into Protective Clothing for Firefighters. *IEEE Trans. Antennas Propag.* **2009**, *57*, 919–925. [CrossRef]
22. Li, J.; Jiang, Y.; Zhao, X. Circularly Polarized Wearable Antenna Based on NinjaFlex-Embedded Conductive Fabric. *Int. J. Antennas Propag.* **2019**, *2019*, 3059480. [CrossRef]
23. Row, J.-S.; Wu, S.-W. Circularly-Polarized Wide Slot Antenna Loaded With a Parasitic Patch. *IEEE Trans. Antennas Propag.* **2008**, *56*, 2826–2832. [CrossRef]
24. Zhu, H.L.; Cheung, S.W.; Chung, K.L.; Yuk, T.I. Linear-to-Circular Polarization Conversion Using Metasurface. *IEEE Trans. Antennas Propag.* **2013**, *61*, 4615–4623. [CrossRef]
25. Zheng, Q.; Guo, C.; Ding, J.; Akinsolu, M.O.; Liu, B.; Vandenbosch, G.A.E. A Wideband Low-RCS Metasurface-Inspired Circularly Polarized Slot Array Based on AI-Driven Antenna Design Optimization Algorithm. *IEEE Trans. Antennas Propag.* **2022**, *70*, 8584–8589. [CrossRef]
26. Lalbakhsh, A.; Afzal, M.U.; Hayat, T.; Esselle, K.P.; Mandal, K. All-metal wideband metasurface for near-field transformation of medium-to-high gain electromagnetic sources. *Sci. Rep.* **2021**, *11*, 9421. [CrossRef]
27. Esfandiari, M.; Lalbakhsh, A.; Shehni, P.N.; Jarchi, S.; Ghaffari-Miab, M.; Mahtaj, H.N.; Reisenfeld, S.; Alibakhshikenari, M.; Koziel, S.; Szczepanski, S. Recent and emerging applications of Graphene-based metamaterials in electromagnetics. *Mater. Des.* **2022**, *221*, 110920. [CrossRef]
28. Lalbakhsh, A.; Afzal, M.U.; Esselle, K.P.; Smith, S.L. All-Metal Wideband Frequency-Selective Surface Bandpass Filter for TE and TM Polarizations. *IEEE Trans. Antennas Propag.* **2022**, *70*, 2790–2800. [CrossRef]
29. Fu, S.; Kong, Q.; Fang, S.; Wang, Z. Broadband Circularly Polarized Microstrip Antenna with Coplanar Parasitic Ring Slot Patch for L-Band Satellite System Application. *IEEE Antennas Wirel. Propag. Lett.* **2014**, *13*, 943–946. [CrossRef]
30. Feng, D.; Zhai, H.; Xi, L.; Yang, S.; Zhang, K.; Yang, D. A Broadband Low-profile Circular Polarized Antenna on an AMC Reflector. *IEEE Antennas Wirel. Propag. Lett.* **2017**, *16*, 2840–2843. [CrossRef]
31. Lalbakhsh, A.; Afzal, M.; Esselle, K.; Smith, S. A High-gain Wideband EBG Resonator Antenna for 60 GHz Unlicenced Frequency Band. In Proceedings of the 12th European Conference on Antennas and Propagation (EuCAP 2018), London, UK, 9–13 April 2018. [CrossRef]
32. Lalbakhsh, A.; Simorangkir, R.B.; Bayat-Makou, N.; Kishk, A.A.; Esselle, K.P. *Artificial Intelligence and Data Science in Environmental Sensing*; Academic Press: Cambridge, MA, USA, 2022. [CrossRef]
33. Chaudhary, P.; Kumar, A. Compact ultra-wideband circularly polarized CPW-fed monopole antenna. *AEU Int. J. Electron. Commun.* **2019**, *107*, 137–145. [CrossRef]
34. Harrington, R.; Mautz, J. Theory of characteristic modes for conducting bodies. *IEEE Trans. Antennas Propag.* **1971**, *19*, 622–628. [CrossRef]
35. Yikai, C.; Chao-Fu, W. *Characteristic Modes: Theory and Applications in Antenna Engineering*; Wiley: Hoboken, NJ, USA, 2015.
36. Sharma, P.; Gupta, K. Analysis and optimized design of single feed circularly polarized microstrip antennas. *IEEE Trans. Antennas Propag.* **1983**, *31*, 949–955. [CrossRef]
37. Gemio, J.; Parron, J.; Soler, J. Human body effects on implantable antennas for ism bands applications: Models comparison and propagation losses study. *Prog. Electromagn. Res.* **2010**, *110*, 437–452. [CrossRef]
38. Soh, P.J.; Vandenbosch, G.A.E.; Wee, F.H.; van den Bosch, A.; Martinez-Vazquez, M.; Schreurs, D.M.M.P. Specific Absorption Rate (SAR) evaluation of biomedical telemetry textile antennas. In Proceedings of the 2013 IEEE MTT-S International Microwave Symposium Digest (MTT), Seattle, WA, USA, 2–7 June 2013. [CrossRef]
39. Microwave Studio. Computer Simulation Technology (CST). 2020. Available online: https://www.3ds.com/products-services/simulia/products/cst-studio-suite/ (accessed on 21 February 2023).
40. 7 SAR Tissue Ingredients. Available online: https://fcc.report/FCC-ID/ACJ9TGWL16B/3754300.pdf (accessed on 19 December 2022).
41. Balanis, C.A. *Antenna Theory: Analysis and Design*; John Wiley & Sons: Hoboken, NJ, USA, 2016.

sensors

MDPI

Article

Non-Data-Aided SNR Estimation for Bandlimited Optical Intensity Channels

Wilfried Gappmair

Institute of Communication Networks and Satellite Communications, Graz University of Technology, Inffeldgasse 12, 8010 Graz, Austria; gappmair@tugraz.at

Abstract: Powerful and reliable estimation of transmission parameters is an indispensable task in each receiver unit—not only for radio frequency, but also for optical wireless communication systems. In this context, the signal-to-noise ratio (SNR) plays an eminent role, especially for adaptive scenarios. Assuming a bandlimited optical intensity channel, which requires a unipolar waveform design, an algorithm for SNR estimation is developed in this paper, which requires no knowledge of the transmitted data. This non-data-aided approach benefits to a great extent from the fact that very long observation windows of payload symbols might be used for the estimation process to increase the accuracy of the result; this is in striking contrast to a data-aided approach based on pilot symbols reducing the spectral efficiency of a communication link. Since maximum likelihood, moment-based or decision-directed algorithms are not considered for complexity and performance reasons, an expectation-maximization solution is introduced whose error performance is close to the Cramer-Rao lower bound as the theoretical limit, which has been derived as well.

Keywords: SNR estimation; optical wireless communications; intensity modulation

1. Introduction

When comparing radio frequency (RF) to optical wireless communication (OWC) techniques, the advantages of the latter are well known: there are no regulatory and license issues, they are rather inexpensive and easy to deploy, have extremely high throughput and cause no problems with data security, just to mention the most significant aspects in this context [1–4].

However, not only for RF but also for OWC solutions, the relevant transmission parameters have to be recovered by powerful algorithms, because otherwise subsequent receiver stages, e.g., detectors or error correction algorithms cannot be reliably operated [5,6]. Of course, in case that bandlimited optical intensity solutions are envisaged, a unipolar waveform design is indispensable with respect to pulse shaping and symbol constellation. Investigated in [7,8] for a PAM scheme and root-raised cosines as typical pulse shapes used for RF solutions, this is simply achieved by a suitably selected bias or offset signal. Unfortunately, such concepts are not very efficient in terms of power and energy if no harvesting is implemented. Hence, squared raised cosine and double jump functions have been suggested in [9] as viable alternatives.

Nevertheless, focusing on the pulse shapes proposed in [9] also means that recovery methods developed in the RF context are not automatically applicable. Of course, synchronization of carrier frequency and phase need not be considered in case of intensity modulation, whereas the recovery of the symbol timing is still of paramount importance since this is a pre-requisite for many other estimation and detection procedures. This problem has been tackled in a couple of papers recently published by the author in [10–12].

Apart from symbol timing and clock recovery, some knowledge about the signal-to-noise ratio (SNR) is equally important for the reliable transmission of data, e.g., for adaptive communication systems to select modulation and coding schemes according to the given

Citation: Gappmair, W. Non-Data-Aided SNR Estimation for Bandlimited Optical Intensity Channels. *Sensors* **2023**, *23*, 802. https://doi.org/10.3390/s23020802

Academic Editor: Yang Yue

Received: 7 December 2022
Revised: 27 December 2022
Accepted: 4 January 2023
Published: 10 January 2023

channel conditions so that the link might be operated close to the Shannon bound [13], but also powerful error correction methods—like turbo or LDPC algorithms—need this sort of information [14]. Scanning the open literature, numerous papers are available about SNR estimation in RF channels, e.g., the frequently cited overview by Pauluzzi and Beaulieu [15], but little or no information is published for OWC systems. One of the rare examples is the paper in [16], but the authors discuss on-off keying (OOK), i.e., a binary concept with rectangular pulse shapes and no bandwidth limitation.

This background was the incentive for the author to study in [17] data-aided SNR estimation for a bandlimited optical intensity channel, i.e., data are known to the receiver unit in form of pilot symbols. However, it could be shown that the accuracy of SNR estimates depends on the length of the pilot sequence used for this purpose, although this reduces the spectral efficiency of the communication link as such. Therefore, if SNR estimation could be organized in a non-data-aided (NDA) way by employing payload or user symbols, the error performance of the estimates might be increased by much longer observation windows with no impact on the spectral efficiency. This is the main motivation of the current paper, which is structured as follows:

The signal and channel models used for analytical and simulation work are introduced in Section 2. In Section 3, the Cramer-Rao lower bound (CRLB) is derived as the theoretical limit of the jitter variance for an SNR estimator developed in this respect. Based on the expectation-maximization (EM) principle, an estimator algorithm is introduced in Section 4 and verified in Section 5 by numerical means. Finally, Section 6 concludes the paper.

2. Signal and Channel Model

Of course, for the NDA scenario to be investigated in this contribution, we are working with the same signal and channel model used in the companion paper addressing a DA situation [17]. For clarity and readability reasons, this model is briefly recapitulated in the sequel.

Due to the unipolar waveform design mentioned previously, it is assumed that the data symbols a_k, $k \in \mathbb{Z}$, are independent and identically distributed (i.i.d.) elements of an M-ary PAM alphabet $\mathcal{A}$. In this context, it makes sense to normalize them to unit energy, i.e., $\mathbb{E}[a_k^2] = 1$, where $\mathbb{E}[\cdot]$ denotes the expectation operator. Therefore, by definition of $\eta_M = \frac{1}{6}(M-1)(2M-1)$, we obtain $a_k \in \mathcal{A} = \frac{1}{\sqrt{\eta_M}}\{0, 1, \ldots, M-1\}$. As a consequence, the average value is given by

$$\mu_a = \mathbb{E}[a_k] = \frac{1}{\sqrt{\eta_M}} \frac{M-1}{2} = \sqrt{\frac{3(M-1)}{2(2M-1)}}. \tag{1}$$

If $h(t)$ describes the pulse shape satisfying the non-negativity as well as the Nyquist constraint, the signal at the output of the opto-electrical receiver unit can be expressed by

$$r(t) = A \sum_k a_k h(t - kT - \tau) + w(t), \tag{2}$$

where T is the symbol period and τ indicates the propagation delay between transmitter and receiver. The channel gain $A > 0$ is assumed to be a constant, which is justified by the fact that the coherence time of fading events is usually much larger than the observation window needed for estimation purposes. In line with the previous publication about this topic in [17], the noise component in (2) is assumed to be a zero-mean white Gaussian process with variance σ_w^2.

In addition, if we introduce the average optical power as $P_0 = \mu_a \bar{h}$, where

$$\bar{h} = \frac{1}{\sqrt{T}} \int_{-\infty}^{\infty} h(t)\, dt, \tag{3}$$

the average electrical SNR at the receiver can be defined as

$$\gamma_s = \frac{A^2 P_0^2}{\sigma_w^2}. \quad (4)$$

Nevertheless, before $r(t)$ is processed in further receiver stages, e.g., in the SNR estimator to be developed in the sequel, it must be filtered appropriately. Assuming an impulse response $q(t)$, the filter output is determined by $z(t) = q(t) \otimes r(t)$, where $\otimes$ denotes the convolutional operator. For convenient reasons, the signal model used for SNR estimation is illustrated in Figure 1.

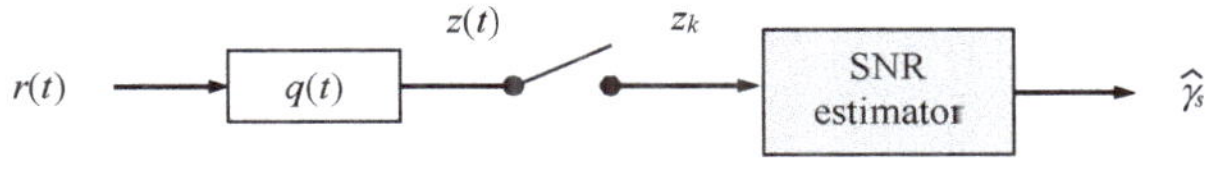

Figure 1. Signal model for SNR estimation.

It has been proved in [9] that there exists no simple solution for a matched receiver filter in case of a bandlimited optical intensity link. Hence, it is suggested to focus on a solution performing a rectangular shape over the complete spectrum occupied by the user component in (2). By application of the Fourier-transform [18], we have that $Q(f) = \mathcal{F}[q(t)] = \sqrt{T}$ for $|f| \leq (1+\alpha)/T$ and $Q(f) = 0$ elsewhere, with α as the roll-off factor (excess bandwidth) of the selected pulse shape. In this case, the signal parts of $r(t)$ and $z(t)$ are the same, whereas the noise component is determined by $n(t) = w(t) \otimes q(t)$ representing a zero-mean non-white Gaussian process. Under the assumption that the symbol timing has been reliably recovered and corrected, e.g., by one of the algorithms proposed in [10–12], the T-spaced samples at the output of the receiver filter are obtained as

$$z_k = z(kT) = A \cdot a_k + n_k, \quad (5)$$

where $\mathbb{E}[n_k] = 0$, $\mathbb{E}[n_i n_k] = 2(1+\alpha)\,\sigma_w^2 \operatorname{sinc}[2(1+\alpha)(i-k)]$, and $\operatorname{sinc}(x) = \sin(\pi x)/(\pi x)$.

3. Cramer-Rao Lower Bound

3.1. Log-Likelihood Function and Fisher Information Matrix

For parameter estimation, in general, the Cramer-Rao lower bound (CRLB) is a major figure of merit [19]. It turns out to be most helpful for comparison purposes, since the bound represents the theoretical limit of the error performance of any algorithm developed in this context.

By detailed inspection of (4), it is obvious that the average electrical SNR is a function of the channel gain and the standard deviation of the noise component in (2), A and σ_w, respectively, whereas P_0 might be considered as a constant factor depending on the PAM scheme and the selected pulse shape. Therefore, it makes sense to focus on the SNR normalized by P_0^2, henceforth denoted by $\rho_s = \gamma_s / P_0^2$, and to employ ρ_s and $P_n = \sigma_w^2$, instead of A and σ_w, as elements constituting the parameter vector $\mathbf{u}$ to be estimated in the sequel, i.e., $\mathbf{u} = (u_1, u_2) = (\rho_s, P_n)$. On top of that, it is assumed that L observables given by (5) are available for the estimation procedure, which might be elegantly expressed in vector form:

$$\mathbf{z} = A \cdot \mathbf{a} + \mathbf{n}. \quad (6)$$

It is to be recalled that the noise samples in $\mathbf{n}$ are not independent. The related covariance matrix is given by $\mathbb{E}[\mathbf{n} \cdot \mathbf{n}^T] = 2(1+\alpha)\,\sigma_w^2\,\mathbf{\Omega}$, where $\mathbf{\Omega}$ describes a symmetric Toeplitz matrix [20] with entries $\omega_{ik} = \operatorname{sinc}[2(1+\alpha)(i-k)]$ for line i and column k.

Conditioned on the knowledge of the data sequence $\mathbf{a}$ and the unknown but deterministic parameter vector $\mathbf{u}$, the log-likelihood function (LLF) characterizing the estimation problem has been derived in [17] as

$$\Lambda(\mathbf{z}|\mathbf{a};\mathbf{u}) = -\frac{L}{2}\log P_n - \frac{\mathbf{z}^T\mathbf{\Psi}\,\mathbf{z} - 2\sqrt{\rho_s P_n}\,\mathbf{z}^T\mathbf{\Psi}\,\mathbf{a} + \rho_s P_n\,\mathbf{a}^T\mathbf{\Psi}\,\mathbf{a}}{4(1+\alpha)P_n}, \tag{7}$$

where $\mathbf{\Psi} = \mathbf{\Omega}^{-1}$. However, the computation of the CRLB for NDA estimation requires that the LLF does not depend on $\mathbf{a}$, which is achieved by averaging the related likelihood function, i.e., $\Pr(\mathbf{z}|\mathbf{a};\mathbf{u}) = e^{\Lambda(\mathbf{z}|\mathbf{a};\mathbf{u})}$, with respect to $\mathbf{a} \in \mathcal{A}^L$, where $\mathcal{A}^L$ denotes the L-dimensional symbol space spanned by L i.i.d. elements of $\mathcal{A}$. Therefore, we have that

$$\Lambda(\mathbf{z};\mathbf{u}) = \log\Pr(\mathbf{z};\mathbf{u}) = \log\left(\frac{1}{M^L}\sum_{\mathbf{a}\in\mathcal{A}^L} e^{\Lambda(\mathbf{z}|\mathbf{a};\mathbf{u})}\right). \tag{8}$$

As a consequence, the entries of the Fisher information matrix (FIM) are obtained as

$$J_{i,k} = \mathbb{E}_{\mathbf{z}}\left[\frac{\partial\Lambda(\mathbf{z};\mathbf{u})}{\partial u_i}\frac{\partial\Lambda(\mathbf{z};\mathbf{u})}{\partial u_k}\right], \tag{9}$$

where $\mathbb{E}_{\mathbf{z}}[\cdot]$ denotes expectation with respect to $\mathbf{z}$, i.e., an averaging procedure with respect to $\mathbf{a}$ and $\mathbf{n}$, which is only possible by numerical means.

According to theory, the CRLB for parameter u_i is formally given by the i-th diagonal entry of the inverted FIM. Since we are only interested in the estimation of ρ_s, representing the first element of $\mathbf{u}$ in our definition introduced before, the corresponding CRLB develops as

$$\mathrm{CRLB}(\rho_s) = \frac{J_{22}}{J_{11}J_{22} - J_{12}J_{21}} = \frac{J_{22}}{J_{11}J_{22} - J_{12}^2}. \tag{10}$$

3.2. Low-Complexity Solution

From the complexity point of view, it is clear that the LLF in (8) is the most demanding ingredient for the computation of the CRLB in (10). This is mainly due to the averaging procedure, which is in the order of M^L operations. Even smaller values of M and L are challenging in this respect, but for values of L between 100 and 1000, which are typical for an NDA scenario, the computational load would be intractable. Luckily, for some values of the excess bandwidth, in particular for $\alpha \in \left\{0,\, \frac{1}{2},\, 1\right\}$, it turns out that $\mathbf{\Omega}$ boils down to an L-dimensional identity matrix $\mathbf{I}_L$, which means that $\mathbf{\Psi} = \mathbf{\Omega}^{-1} = \mathbf{I}_L$. In consequence, the likelihood function in (8) can be re-organized as

$$\Pr(\mathbf{z};\mathbf{u}) = \frac{P_n^{-L/2}}{M^L}\sum_{\mathbf{a}\in\mathcal{A}^L}\exp\left(-\frac{\mathbf{z}^T\mathbf{z} - 2\sqrt{\rho_s P_n}\,\mathbf{z}^T\mathbf{a} + \rho_s P_n\,\mathbf{a}^T\mathbf{a}}{4(1+\alpha)P_n}\right). \tag{11}$$

Because of $\mathbf{\Psi} = \mathbf{I}_L$ the elements of $\mathbf{z}$ might be considered as statistically independent entries. By taking into account that the symbols $a_i \in \mathcal{A}$ are i.i.d., we just obtain

$$\Pr(\mathbf{z};\mathbf{u}) = \frac{P_n^{-L/2}}{M^L}\sum_{\mathbf{a}\in\mathcal{A}^L}\prod_{k=0}^{L-1} e^{\Lambda(z_k|\,a_i;\mathbf{u})}, \tag{12}$$

where

$$\Lambda(z_k|a_i;\mathbf{u}) = -\frac{\left(z_k - \sqrt{\rho_s P_n}\,a_i\right)^2}{4(1+\alpha)P_n}. \tag{13}$$

Finally, the averaging of $\Pr(\mathbf{z}|\mathbf{a};\mathbf{u})$ with respect to $\mathbf{a} \in \mathcal{A}^L$ is simply achieved, if we exchange in (12) the order of sum and product, i.e.,

$$\Pr(\mathbf{z};\mathbf{u}) = \frac{P_n^{-L/2}}{M^L} \prod_{k=0}^{L-1} \sum_{a_i \in \mathcal{A}} e^{\Lambda(z_k|a_i;\mathbf{u})}, \tag{14}$$

resulting in a computational complexity in the order of $M \times L$, which is much less compared to M^L.

In the next step, with $\Lambda(\mathbf{z};\mathbf{u}) = \log \Pr(\mathbf{z};\mathbf{u})$, the first-order derivatives in (9) are expressed by

$$\frac{\partial \Lambda(\mathbf{z};\mathbf{u})}{\partial \rho_s} = \sum_{k=0}^{L-1} \frac{\sum_{a_i \in \mathcal{A}} \Lambda_s(z_k|a_i;\mathbf{u})\, e^{\Lambda(z_k|a_i;\mathbf{u})}}{\sum_{a_i \in \mathcal{A}} e^{\Lambda(z_k|a_i;\mathbf{u})}} \tag{15}$$

and

$$\frac{\partial \Lambda(\mathbf{z};\mathbf{u})}{\partial P_n} = -\frac{L}{2P_n} + \sum_{k=0}^{L-1} \frac{\sum_{a_i \in \mathcal{A}} \Lambda_n(z_k|a_i;\mathbf{u})\, e^{\Lambda(z_k|a_i;\mathbf{u})}}{\sum_{a_i \in \mathcal{A}} e^{\Lambda(z_k|a_i;\mathbf{u})}}, \tag{16}$$

where

$$\Lambda_s(z_k|a_i;\mathbf{u}) = \frac{\partial \Lambda(z_k|a_i;\mathbf{u})}{\partial \rho_s} = \frac{1}{4(1+\alpha)} \left(\frac{a_i\, z_k}{\sqrt{\rho_s\, P_n}} - a_i^2 \right) \tag{17}$$

and

$$\Lambda_n(z_k|a_i;\mathbf{u}) = \frac{\partial \Lambda(z_k|a_i;\mathbf{u})}{\partial P_n} = \frac{1}{4(1+\alpha)} \left(\frac{z_k^2}{P_n^2} - \sqrt{\frac{\rho_s}{P_n^3}}\, a_i\, z_k \right). \tag{18}$$

For the simplified scenario, the computation of FIM entries and CRLB does not differ from the general case such that the relationships in (9) and (10) might be used in the same way.

3.3. Asymptotic Scenario

For increasing values of M, the density of the PAM symbols a_i will increase accordingly, when we assume that their average energy is normalized to unity, i.e., $\mathbb{E}[a_k^2] = 1$. Hence, in case that $M \to \infty$, the symbol alphabet $\mathcal{A}$ turns out to be equally distributed between 0 and $\sqrt{3}$. Applying the framework developed previously to compute the CRLB for such a scenario, it is clear that the sums over $a_i \in \mathcal{A}$ in (15) and (16) have to be replaced by the related integrals, i.e.,

$$\sum_{a_i \in \mathcal{A}} a_i^m\, e^{\Lambda(z_k|a_i;\mathbf{u})} \to \mathcal{I}_m(z_k;\mathbf{u}) = \int_{a_i=0}^{\sqrt{3}} a_i^m e^{\Lambda(z_k|a_i;\mathbf{u})} da_i, \tag{19}$$

where $m \in \{0, 1, 2\}$. By taking into account the solutions for (19) computed in Appendix A, the first-order derivatives for the FIM entries in (9) are after some lengthy but straightforward manipulations given by

$$\frac{\partial \Lambda(\mathbf{z};\mathbf{u})}{\partial \rho_s} = \frac{1}{4(1+\alpha)\sqrt{\rho_s\, P_n}} \sum_{k=0}^{L-1} \frac{z_k \mathcal{I}_1(z_k;\mathbf{u}) - \sqrt{\rho_s\, P_n}\, \mathcal{I}_2(z_k;\mathbf{u})}{\mathcal{I}_0(z_k;\mathbf{u})} \tag{20}$$

and

$$\frac{\partial \Lambda(\mathbf{z};\mathbf{u})}{\partial P_n} = -\frac{L}{2P_n} + \frac{1}{4(1+\alpha)P_n^2}\sum_{k=0}^{L-1} z_k^2 - \frac{1}{4(1+\alpha)}\sqrt{\frac{\rho_s}{P_n^3}}\sum_{k=0}^{L-1}\frac{z_k \mathcal{I}_1(z_k;\mathbf{u})}{\mathcal{I}_0(z_k;\mathbf{u})}. \tag{21}$$

It is to be recalled that the simplified computation of the CRLB applies in the strict sense only to values of $\alpha \in \left\{0,\ \frac{1}{2},\ 1\right\}$, where $\mathbf{\Psi} = \mathbf{\Omega}^{-1} = \mathbf{I}_L$. However, since the entries of $\mathbf{\Omega}$ are given by $\omega_{ik} = \mathrm{sinc}[2(1+\alpha)(i-k)]$, it is clear that the off-diagonal elements of the matrix are rapidly decaying for $\alpha \notin \left\{0,\ \frac{1}{2},\ 1\right\}$, which means that $\mathbf{\Omega}$ and $\mathbf{\Psi}$ might be approximated by an identity matrix of the same dimension such that the simplification would be applicable. This results in a tight approximation of the true bound confirmed by numerical results in Section 5.

4. Expectation-Maximization Estimator

Since the normalized SNR value is given by $\rho_s = A^2/P_n$, it is clear that any estimator algorithm must provide the estimates of channel gain as well as noise power, which means that we are focusing in the sequel on a parameter vector $\mathbf{u} = (A, P_n)$. Formally, a maximum likelihood solution for $\mathbf{u}$ is easily obtained by using the LLF in (8), i.e., $\hat{\mathbf{u}} = \mathrm{argmax}_{\tilde{\mathbf{u}}} \Lambda(\mathbf{z};\tilde{\mathbf{u}})$. However, this problem cannot be solved in closed form so that we must resort to numerical methods, e.g., the iterative Newton-Raphson procedure [21]. Apart from troubles in terms of initialization, convergence and stability, it is the computational complexity which prohibits this approach, even if the simplified variant with $\mathbf{\Psi} = \mathbf{I}_L$ would be envisaged.

Alternatively, a rather simple solution based on first- and second-order moments given by $\mathbb{E}[z_k]$ and $\mathbb{E}[z_k^2]$, respectively, delivered reliable estimates only in the very low SNR range. On the other hand, for an algorithm based on symbol decisions [15], useful results were solely achievable for very large SNRs. As a consequence, an expectation-maximization (EM) estimator [22–24] is proposed, whose error variance turned out to be close to the CRLB over a wide SNR range as will be demonstrated in Section 5.

The EM algorithm is an iterative procedure using in step l the parameter estimates calculated in step $l-1$. The conditional LLF for this approach is for $\mathbf{\Psi} = \mathbf{I}_L$ expressed by

$$\Lambda(\mathbf{z}|\hat{\mathbf{u}}^{(l-1)};\tilde{\mathbf{u}}) = -\frac{L}{2}\log \tilde{P}_n - \frac{\mathbf{z}^T\mathbf{z} - 2\tilde{A}\,\mathbf{z}^T\dot{\mathbf{a}}^{(l-1)} + \tilde{A}^2\ddot{a}^{(l-1)}}{4(1+\alpha)\tilde{P}_n}. \tag{22}$$

This is very similar to (7), but $\mathbf{u}$ is replaced by the trial value $\tilde{\mathbf{u}}$ used for optimization purposes; $\mathbf{a}$ as well as $\mathbf{a}^T\,\mathbf{a}$ are substituted by the corresponding soft decisions in step $l-1$, henceforth denoted by $\dot{\mathbf{a}}^{(l-1)}$ and $\ddot{a}^{(l-1)}$. The k-th element of vector $\dot{\mathbf{a}}^{(l-1)}$ develops as

$$\dot{a}_k^{(l-1)} = \sum_{a_i \in \mathcal{A}} a_i P_{i,k}^{(l-1)}, \tag{23}$$

whereas $\ddot{a}^{(l-1)}$, representing a scalar, is a finite double sum given by

$$\ddot{a}^{(l-1)} = \sum_{k=0}^{L-1}\sum_{a_i \in \mathcal{A}} a_i^2 P_{i,k}^{(l-1)}. \tag{24}$$

By inspection of (23) and (24), we observe that the soft decisions are characterized by an averaging of the symbols $a_i \in \mathcal{A}$ via the *a posteriori* probabilities [25]:

$$P_{i,k}^{(l-1)} = \Pr(a_i \in \mathcal{A}|z_k, \hat{\mathbf{u}}^{(l-1)}) = \frac{\Pr[z_k|a_i \in \mathcal{A}, \hat{\mathbf{u}}^{(l-1)}]\Pr[a_i \in \mathcal{A}|\hat{\mathbf{u}}^{(l-1)}]}{\Pr[z_k|\hat{\mathbf{u}}^{(l-1)}]}. \tag{25}$$

Because the symbols a_i are i.i.d., we have that $\Pr[a_i \in \mathcal{A}|\hat{\mathbf{u}}^{(l-1)}] = \Pr[a_i \in \mathcal{A}] = \frac{1}{M}$. Furthermore, by considering the relationship in (13), the probability $\Pr[z_k|a_i \in \mathcal{A}, \hat{\mathbf{u}}^{(l-1)}]$ develops as

$$\Pr[z_k|a_i \in \mathcal{A}, \hat{\mathbf{u}}^{(l-1)}] = \frac{1}{\sqrt{4\pi(1+\alpha)\hat{P}_n^{(l-1)}}} \exp\left(-\frac{(z_k - \hat{A}^{(l-1)}a_i)^2}{4(1+\alpha)\hat{P}_n^{(l-1)}}\right). \tag{26}$$

Since the denominator in (25) does not depend on a_i, it might be replaced by a constant including also the *a priori* probability $\Pr[a_i \in \mathcal{A}|\hat{\mathbf{u}}^{(l-1)}] = \frac{1}{M}$, which is determined by the fact that $\sum_{i=0}^{M-1} P_{i,k}^{(l-1)} = 1$.

Finally, computing the first-order derivatives of (22) with respect to $\widetilde{A}$ as well as $\widetilde{P}_n$ and equating them to zero for $\widetilde{A} = \hat{A}^{(l)}$ and $\widetilde{P}_n = \hat{P}_n^{(l)}$, the parameter estimates for step l are achieved in closed form by

$$\hat{A}^{(l)} = \frac{\mathbf{z}^T \dot{\mathbf{a}}^{(l-1)}}{\ddot{a}^{(l-1)}} \tag{27}$$

and

$$\hat{P}_n^{(l)} = \frac{\mathbf{z}^T\mathbf{z} - 2\hat{A}^{(l)}\,\mathbf{z}^T\dot{\mathbf{a}}^{(l-1)} + (\hat{A}^{(l)})^2\ddot{a}^{(l-1)}}{2L(1+\alpha)}. \tag{28}$$

Nevertheless, the algorithm has to be initialized by appropriately selected values for the probabilities in (25). Since the symbols a_i are assumed to be i.i.d., it makes sense to start the EM algorithm with $P_{i,k}^{(0)} = \frac{1}{M}$ for all values of i and k. For convenient reasons, the iterative procedure is summarized as follows:

1. Initialization

   $i = 0 \dots M-1,\ k = 0 \dots L-1:\ P_{i,k}^{(0)} = \frac{1}{M}$
2. Iteration: $l = 1 \dots l_{\max}$
   - Compute $\dot{\mathbf{a}}^{(l-1)}$, $\ddot{a}^{(l-1)}$
   - Compute $\hat{A}^{(l)}$, $\hat{P}_n^{(l)}$
   - Compute $P_{i,k}^{(l)}$
3. Final estimate

   $\hat{\rho}_s = \frac{(\hat{A}^{(l_{\max})})^2}{P_n^{(l_{\max})}}$

It is not difficult to see that the iterative step of the EM algorithm has a computational complexity in the order of $M \times L$ additive and multiplicative operations, only relationship (26) requires the evaluation of square root and exponential functions, which might be elegantly handled via look-up tables. For the numerical results in Section 5, the iterative procedure is organized such that it stops as soon as the relative error between two successive estimates achieves a predefined value of 10^{-3} or when a maximum number of 10^3 iterations is achieved; by extensive tests it turned out that this would be a good compromise between complexity and accuracy. Furthermore, it is to be remembered that the algorithm applies in the strict sense only to $\mathbf{\Psi} = \mathbf{I}_L$, i.e., $\alpha \in \left\{0,\ \frac{1}{2},\ 1\right\}$. However, since good results are obtained for other values of α as well, it makes sense to employ the EM algorithm in these cases as well.

5. Numerical Results

In the following, the EM algorithm developed previously will be verified by numerical means in terms of jitter (error) performance and bias. For comparison purposes, the CRLB is included to the related diagrams, which show also the modified Cramer-Rao lower bound (MCRLB). Derived in closed form in [17], the MCRLB is much simpler than the CRLB, but it is in general less tight [26–29], i.e., MCRLB(ρ_s) $\leq$ CRLB(ρ_s), in particular at lower SNRs as demonstrated subsequently.

For 2-PAM and 4-PAM signals operated with $\alpha = 0$ and two different observation lengths, $L = 100$ and 1000, Figure 3 illustrates the evolution of the error performance as a function of the SNR value; for convenient reasons, error performance and theoretical limits are normalized by ρ_s^2. By detailed inspection, we observe that

- For larger SNRs, the CRLB (dashed line) approaches the corresponding MCRLB (solid line) irrespective of the selected modulation scheme and the value of L.
- For very low SNRs, the ratio between CRLB and MCRLB seems to approach a small but non-negligible constant, which decreases somewhat by increasing values of M.
- In the medium SNR range, we see a significant difference between MCRLB and CRLB whose maximum grows with increasing values of M and which moves to larger SNR values.
- For medium-to-low SNRs and $L = 100$, the error performance of the EM estimator, indicated by markers in different style, is characterized by a considerable difference to the CRLB, which shrinks more and more with increasing values of the SNR. This degradation is basically explained by the fact that the algorithm performs a bias effect evolving in the same way, which is depicted in Figure 2 (in this case, the dashed lines do not correspond to an analytical relationship; they are due to an interpolation procedure in order to achieve a better readability of these numerical results). This drawback might be circumvented with larger observation windows, in Figures 2 and 3 exemplified by $L = 1000$.

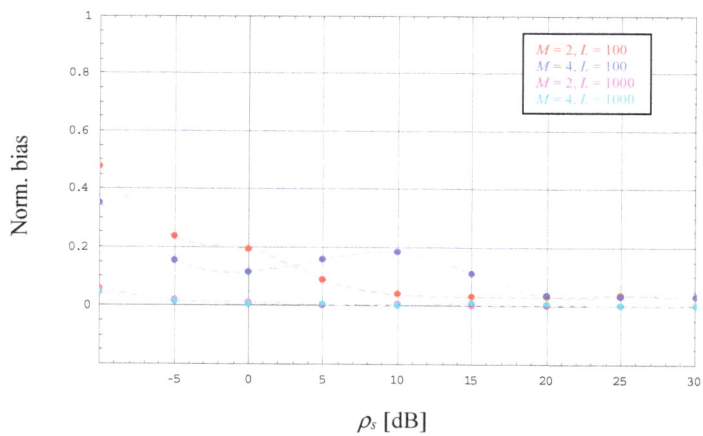

Figure 2. Evolution of the normalized bias (2/4-PAM, $\alpha = 0$).

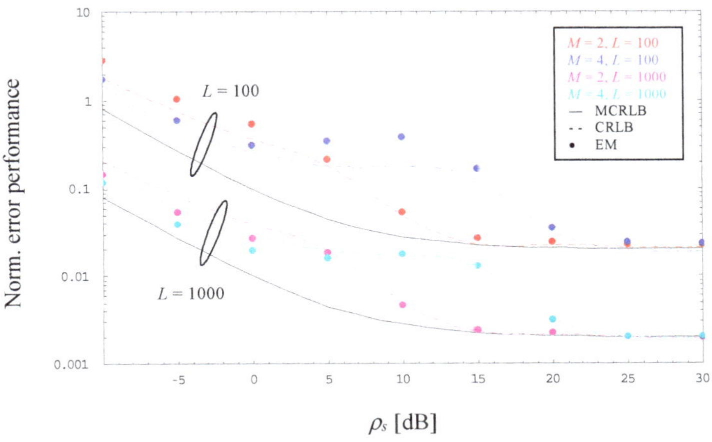

Figure 3. Evolution of the normalized error performance (2/4-PAM, $\alpha = 0$).

The observations made with $M = 2$ and 4 hold also true for higher orders of M, in Figures 4 and 5 verified by a 16-PAM scheme operated with $\alpha = 0$ and $L = 100$ or 1000. However, Figure 4 includes also the evolution of the CRLB for $M \to \infty$ as it has been derived in Section 4. One can see that the theoretical limit for $M \to \infty$ is close to that computed for $M = 16$ as long as the latter does not start to approach the MCRLB, whereas for $M \to \infty$ it continues to increase with increasing SNRs. This property suggests that NDA estimation of the SNR becomes more and more problematic in case we increase the order of the selected PAM scheme.

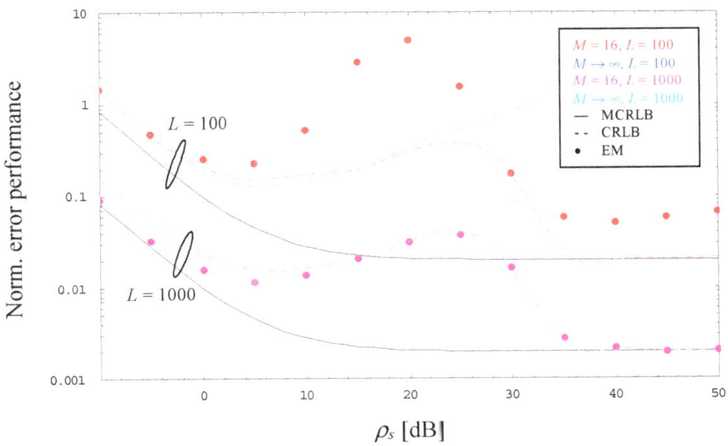

Figure 4. Evolution of the normalized error performance (16-PAM, $\alpha = 0$).

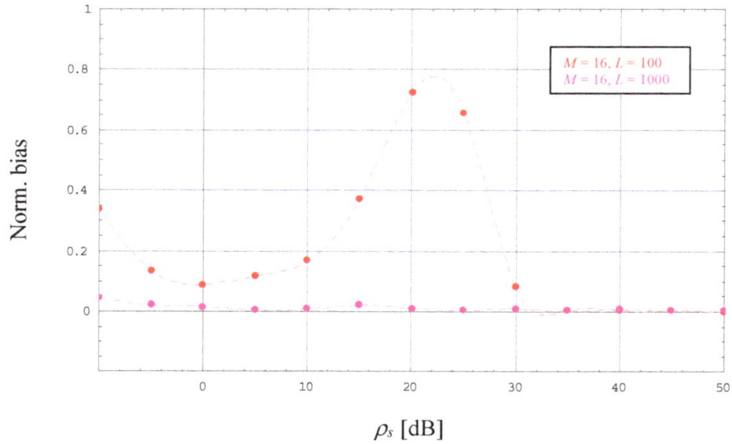

Figure 5. Evolution of the normalized bias (16-PAM, $\alpha = 0$).

The diagrams above visualize the error and bias performance for different PAM constellations operated with $\alpha = 0.0$, which is perhaps most interesting in practice, since it represents the scenario with minimum bandwidth. Nevertheless, in order to complete the portrait, Figure 6 shows the normalized error performance for a 4-PAM signal operated with $L = 1000$ and $\alpha \in \{0.0, 0.3, 1.0\}$. According to the exact relationship derived in [17], the MCRLB is proportional to $2(1+\alpha)/L$ for very low SNRs, whereas for $\rho_s \to \infty$ it approaches $2/L$ regardless of the selected α, or in other words: with respect to $\alpha = 0$, the MCRLB for $\alpha > 0$ appears as shifted to the right depending on the chosen value.

Exactly the same behavior is reflected by the CRLB, although the results are in the very low SNR range somewhat higher than those achieved with the MCRLB, but for rather large values, say $\rho_s > 25$ dB, the CRLB approaches more and more the horizontal floor characterizing the MCRLB performance. Of course, in the medium SNR range, the CRLB deviates significantly from the MCRLB, which applies also to the simplified computation for $\alpha = 0.3$.

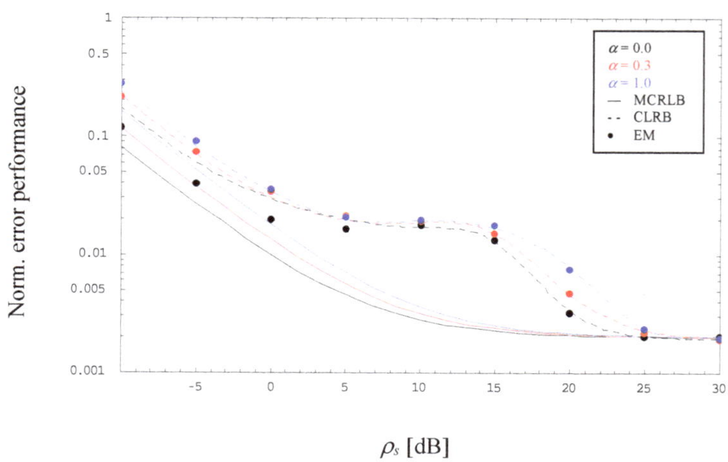

Figure 6. Evolution of the normalized error performance (4-PAM, $L = 1000$).

Figure 6 includes also the normalized jitter variance of the EM algorithm developed in the previous section. By detailed inspection, we observe that the performance differs for $\alpha = 0$ in the lower SNR domain somewhat from the CRLB, which is mainly due to a residual bias effect, whereas for medium-to-high SNR values the jitter variances are very close to the corresponding CRLBs regardless of the selected excess bandwidth.

6. Concluding Remarks

The availability of reliable SNR estimates is most helpful in many communication systems, particularly when adaptive solutions have to be considered in terms of modulation and coding schemes. This is not only true for radio frequency, but also for optical wireless links. Assuming a bandlimited optical intensity channel, an algorithm for SNR estimation has been developed, which does not require any knowledge about the transmitted data symbols. Such an NDA approach is very appreciated, because the larger observation lengths do not adversely affect the spectral efficiency as it would happen with a DA solution. Maximum likelihood, moment-based and decision-directed methods were out of scope because of complexity and/or performance reasons, but it turned out that the developed expectation-maximization algorithm exhibits an error performance close to the CRLB as the theoretical limit, which is mainly true for longer observation windows where bias effects are negligible.

Funding: Part of the results have been obtained in the context of the Austrian Space Application Programme (ASAP), funded by the Austrian Research Promotion Agency (FFG), for a project on "Optical Pulse Shaping for Power Efficient Transmission", Contract No. 885376.

Data Availability Statement: Data are available from the author upon mail request.

Acknowledgments: Open Access Funding by Graz University of Technology.

Conflicts of Interest: The author declares no conflict of interest.

Appendix A

In the following, a closed form solution is provided for the integrals specified in (19). Regarding that $\Lambda(z_k|a_i;\mathbf{u})$ is expressed by (13), it is obvious that the integral for $m = 0$ is given by an instance of the error function [30] (3.321/2). Applying the basic rules of integration, we simply obtain

$$\begin{aligned} \mathcal{I}_0(z_k;\mathbf{u}) &= \int_{a_i=0}^{\sqrt{3}} e^{\Lambda(z_k|a_i;\mathbf{u})} da_i \\ &= \sqrt{\frac{\pi(1+\alpha)}{\rho_s}} \left[\operatorname{erf}\left(\frac{z_k}{2\sqrt{(1+\alpha)P_n}} \right) - \operatorname{erf}\left(\frac{z_k - \sqrt{3\rho_s P_n}}{2\sqrt{(1+\alpha)P_n}} \right) \right]. \end{aligned} \tag{A1}$$

Doing the same for $m = 1$ by taking into account the result in (A1), the corresponding integral is solved as

$$\begin{aligned} \mathcal{I}_1(z_k;\mathbf{u}) &= \int_{a_i=0}^{\sqrt{3}} a_i\, e^{\Lambda(z_k|a_i;\mathbf{u})} da_i \\ &= \frac{z_k\, \mathcal{I}_0(z_k;\mathbf{u})}{\sqrt{\rho_s P_n}} + \frac{2(1+\alpha)}{\rho_s} \left[\exp\left(-\frac{z_k^2}{4(1+\alpha)P_n} \right) - \exp\left(-\frac{\left(z_k - \sqrt{3\rho_s P_n}\right)^2}{4(1+\alpha)P_n} \right) \right]. \end{aligned} \tag{A2}$$

Finally, for $m = 2$ and considering (A1) as well as (A2) together with [30] (3.361/1), we get after some algebra

$$
\begin{aligned}
\mathcal{I}_2(z_k;\mathbf{u}) &= \int_{a_i=0}^{\sqrt{3}} a_i^2 e^{\Lambda(z_k|a_i;\mathbf{u})} da_i \\
&= \frac{[z_k^2+2(1+\alpha)P_n]\,\mathcal{I}_0(z_k;\mathbf{u})}{\rho_s P_n} \\
&\quad - \frac{2(1+\alpha)}{\rho_s^2 P_n}\Bigg[(\sqrt{3}\rho_s P_n + \sqrt{\rho_s P_n} z_k)\exp\left(-\frac{(z_k-\sqrt{3\rho_s P_n})^2}{4(1+\alpha)P_n}\right) \\
&\qquad - \sqrt{\rho_s P_n}\, z_k \exp\left(-\frac{z_k^2}{4(1+\alpha)P_n}\right)\Bigg].
\end{aligned}
\tag{A3}
$$

References

1. Hranilovic, S. *Wireless Optical Communication Systems*; Springer: New York, NY, USA, 2004.
2. Arnon, S.; Barry, J.; Karagiannidis, G.; Schober, R.; Uysal, M. *Advanced Optical Wireless Communication Systems*; Cambridge University Press: New York, NY, USA, 2012.
3. Khalighi, M.A.; Uysal, M. Survey on free space optical communication: A communication theory perspective. *IEEE Commun. Surv. Tutor.* **2014**, *16*, 2231–2258. [CrossRef]
4. Ghassemlooy, Z.; Arnon, S.; Uysal, M.; Xu, Z.; Cheng, J. Emerging optical wireless communications—Advances and challenges. *IEEE J. Select. Areas Commun.* **2015**, *33*, 1738–1749. [CrossRef]
5. Mengali, U.; D'Andrea, A.N. *Synchronization Techniques for Digital Receivers*; Plenum Press: New York, NY, USA, 1997.
6. Meyr, H.; Moeneclaey, M.; Fechtel, S.A. *Digital Communication Receivers: Synchronization, Channel Estimation, and Signal Processing*; Wiley: New York, NY, USA, 1998.
7. Tavan, M.; Agrell, E.; Karout, J. Bandlimited intensity modulation. *IEEE Trans. Commun.* **2012**, *60*, 3429–3439. [CrossRef]
8. Czegledi, C.; Khanzadi, M.R.; Agrell, E. Bandlimited power-efficient signaling and pulse design for intensity modulation. *IEEE Trans. Commun.* **2014**, *62*, 3274–3284. [CrossRef]
9. Hranilovic, S. Minimum-bandwidth optical intensity Nyquist pulses. *IEEE Trans. Commun.* **2007**, *55*, 574–583. [CrossRef]
10. Gappmair, W. On parameter estimation for bandlimited optical intensity channels. *Computation* **2019**, *7*, 11. [CrossRef]
11. Gappmair, W.; Nistazakis, H.E. Blind symbol timing estimation for bandlimited optical intensity channels. In Proceedings of the 12th IEEE/IET International Symposium on Communication Systems, Networks and Digital Signal Processing (CSNDSP), Porto, Portugal, 20–22 July 2020.
12. Gappmair, W.; Schlemmer, H. Feedback solution for symbol timing recovery in bandlimited optical intensity channels. In Proceedings of the IEEE 4th International Conference Broadband Communications for Next Generation Networks and Multimedia Applications (CoBCom), Graz, Austria, 12–14 July 2022.
13. Chung, T.S.; Goldsmith, A.J. Degrees of freedom in adaptive modulation: A unified view. *IEEE Trans. Commun.* **2001**, *49*, 1561–1571. [CrossRef]
14. Summers, T.A.; Wilson, S.G. SNR mismatch and online estimation in turbo decoding. *IEEE Trans. Commun.* **1998**, *46*, 421–423. [CrossRef]
15. Pauluzzi, D.R.; Beaulieu, N.C. A comparison of SNR estimation techniques for the AWGN channel. *IEEE Trans. Commun.* **2000**, *48*, 1681–1691. [CrossRef]
16. D'Amico, A.A.; Colavolpe, G.; Foggi, T.; Morelli, M. Timing synchronization and channel estimation in free-space optical OOK communication systems. *IEEE Trans. Commun.* **2022**, *70*, 1901–1912. [CrossRef]
17. Gappmair, W. Data-aided SNR estimation for bandlimited optical intensity channels. *Sensors* **2022**, *22*, 8660. [CrossRef] [PubMed]
18. Proakis, J.G.; Manolakis, D.G. *Digital Signal Processing: Principles, Algorithms, and Applications*; Prentice Hall: Upper Saddle River, NJ, USA, 1996.
19. Kay, S.M. *Fundamentals of Statistical Signal Processing: Estimation Theory*; Prentice Hall: Upper Saddle River, NJ, USA, 1993.
20. Gray, R.M. *Toeplitz and Circulant Matrices: A Review*; Now Publishers: Hanover, MA, USA, 2006.
21. Press, W.H.; Teukolsky, S.A.; Vetterling, W.T.; Flannery, B.P. *Numerical Recipes in C: The Art of Scientific Computing*; Cambridge University Press: New York, NY, USA, 1992.
22. Dempster, A.P.; Laird, N.M.; Rubin, D.B. Maximum likelihood from incomplete data via the EM algorithm. *J. Roy. Stat. Soc. B* **1977**, *39*, 1–38.
23. Moon, T.K. The expectation-maximization algorithm. *IEEE Signal Process. Mag.* **1996**, *13*, 47–60. [CrossRef]
24. Gappmair, W.; Lopez-Valcarce, R.; Mosquera, C. Cramer-Rao lower bound and EM algorithm for envelope-based SNR estimation of nonconstant modulus constellations. *IEEE Trans. Commun.* **2009**, *57*, 1622–1627. [CrossRef]
25. Papoulis, A. *Probability, Random Variables, and Stochastic Processes*; McGraw-Hill: New York, NY, USA, 1991.
26. D'Andrea, A.N.; Mengali, U.; Reggiannini, R. The modified Cramer-Rao bound and its application to synchronization problems. *IEEE Trans. Commun.* **1994**, *42*, 1391–1399. [CrossRef]
27. Gini, F.; Reggiannini, R.; Mengali, U. The modified Cramer-Rao bound in vector parameter estimation. *IEEE Trans. Commun.* **1998**, *46*, 52–60. [CrossRef]

28. Moeneclaey, M. On the true and the modified Cramer-Rao bounds for the estimation of a scalar parameter in the presence of nuisance parameters. *IEEE Trans. Commun.* **1998**, *46*, 1536–1544. [CrossRef]
29. Gappmair, W. Cramer-Rao lower bound for non-data-aided SNR estimation of linear modulation schemes. *IEEE Trans. Commun.* **2008**, *56*, 689–693. [CrossRef]
30. Gradshteyn, I.S.; Ryzhik, I.M. *Table of Integrals, Series, and Products*; Academic Press: New York, NY, USA, 1994.

sensors

MDPI

Article

Secure Vehicular Platoon Management against Sybil Attacks

Danial Ritzuan Junaidi [1], Maode Ma [2] and Rong Su [1,*]

[1] School of Electrical and Electronic Engineering, Nanyang Technological University, Singapore 639798, Singapore
[2] College of Engineering, Qatar University, Doha P.O. Box 2713, Qatar
* Correspondence: rsu@ntu.edu.sg

Abstract: The capacity of highways has been an ever-present constraint in the 21st century, bringing about the issue of safety with greater likelihoods of traffic accidents occurring. Furthermore, recent global oil prices have inflated to record levels. A potential solution lies in vehicular platooning, which has been garnering attention, but its deployment is uncommon due to cyber security concerns. One particular concern is a Sybil attack, by which the admission of fake virtual vehicles into the platoon allows malicious actors to wreak havoc on the platoon itself. In this paper, we propose a secure management scheme for platoons that can protect major events that occur in the platoon operations against Sybil attacks. Both vehicle identity and message exchanged are authenticated by adopting key exchange, digital signature and encryption schemes based on elliptic curve cryptography (ECC). Noteworthy features of the scheme include providing perfect forward secrecy and both group forward and backward secrecy to preserve the privacy of vehicles and platoons. Typical malicious attacks such as replay and man-in-the-middle attacks for example can also be resisted. A formal evaluation of the security functionality of the scheme by the Canetti–Krawczyk (CK) adversary and the random oracle model as well as a brief computational verification by CryptoVerif were conducted. Finally, the performance of the proposed scheme was evaluated to show its time and space efficiency.

Keywords: authentication; digital signature; elliptic curve cryptography; key exchange; platoons; Sybil attack

Citation: Junaidi, D.R.; Ma, M.; Su, R. Secure Vehicular Platoon Management against Sybil Attacks. *Sensors* **2022**, *22*, 9000. https://doi.org/10.3390/s22229000

Academic Editor: Unai Hernandez

Received: 11 October 2022
Accepted: 16 November 2022
Published: 21 November 2022

1. Introduction

The spontaneous formation of a network of mobile devices interconnected wirelessly has introduced the concept of mobile ad hoc networks (MANETs). In recent years, the iteration of MANETs utilizing vehicles as mobile devices has evolved the research field to what it is known as today: vehicular ad hoc networks (VANETs). Throughout their evolution, VANETs have seen much progress in terms of different applications. One notable application that has been gaining much interest is vehicle platooning. Platooning, in brief, involves having a manually driven vehicle, known as a platoon leader, spearheading a convoy of semi-automated vehicles in a single lane on the road. These vehicles can be referred to as platoon members, and among this train of vehicles, they are all spaced closely and uniformly apart from each other [1].

The benefits of platooning include decreasing the headway time (gap) between platoon members, thereby providing better traffic management. A platoon can also help to increase the capacity on highways as vehicles in the platoon take up less space than when vehicles are independently and individually controlled [2]. Other known benefits of platooning include cutting down on fuel consumption due to the reduced aerodynamic drag from the slipstream effect of travelling in a close convoy [3], increased driving comfort and the removal of human errors in traffic accidents since trailing vehicles can be semi-autonomously driven. The smoother cruise also reduces engine wear [4].

Due to the automated nature of platooning, vehicles need to periodically broadcast messages containing crucial information such as vehicle identity, position and speed [2].

Hence, vehicle-to-vehicle (V2V) and vehicle-to-infrastructure (V2I) communications are needed to facilitate the broadcast of such messages among the vehicles as well as to base stations. These base stations are situated along the sides of roads and are known as roadside units (RSUs). Interactions with RSUs and other platoon members provide crucial information such as knowing whom to interact with and how to behave while inside the platoon, thereby allowing the vehicles to move in tandem smoothly [5]. However, if this network and its supporting actors are compromised, it could potentially cause the platoon to be unsafe.

Such a scenario could occur if the communication channel of the network is disrupted. For instance, vehicle nodes could experience a loss of signal for technical reasons when multiple simultaneous transmissions among vehicles interfere with one another. Vehicles would share the same channel and transmit at the same time, causing transmission packets to collide and subsequently be dropped. Researchers such as in [6] have addressed this issue in the platoon context by devising a channel allocation algorithm based on the orientation of vehicles in a platoon. Since platoon vehicles are ordered in a single file and separated uniformly from one another, vehicles can be allocated specific channels depending on their distances from the platoon leader. In short, only platoon vehicles that are out of the interference range can use the same channel. Their real-world experiment using Android-based mobile devices showed that their scheme outperformed typical platooning systems with regards to packet drop rates and delays.

On the other hand, an alternative hazardous scenario could arise whereby nodes that are intentionally malicious worm their way into the network to influence the behaviors of the vehicles in the platoon. One case where these malicious nodes can inhibit the operation of platoons is through a Sybil attack. The definition of such an attack is when a malicious actor forges fake virtual vehicles via pseudonyms [1] so that it can mask itself as multiple simultaneous vehicles that seem to be physically present on the roads [1,2,7–11]. These illegitimate vehicles, known as Sybil vehicles, can be preloaded with any number of false credentials that are not of the original malicious node [7]. They can also steal the credentials of legitimate vehicles and use the stolen information to impersonate them. If these Sybil vehicles are admitted into platoons, the malicious node can send false messages to the other platoon members through these virtual vehicles. These false messages are shared within the platoon to fabricate fake traffic scenarios. The platoon and its members would then have to consider the given traffic scenario and act accordingly, altering the platoon's original intentions and affecting its overall performance as a result. Furthermore, within the platoon itself, the sheer presence of these Sybil vehicles causes undesirable gaps to appear between the platoon members as they have to accommodate for the "physical presence" of these virtual Sybil vehicles. In turn, it directly impedes the platooning benefits such as fuel economy and road capacity [8]. Thus, it is imperative that only honest vehicles be authenticated and admitted into platoons.

Motivation and Contributions

We have reviewed existing research work utilizing technologies of our interest. We then focused on cryptographic-based works and discovered that they exhibited incomplete authentication of vehicle identity and messages. Those that achieve complete authentication only do so at high computational costs.

To overcome the shortcomings of these existing solutions, we designed a hybrid security scheme that authenticated both vehicle identity and messages exchanged into a platoon management scheme to prevent threats by Sybil vehicles. The proposed solution is called secure platoon management against Sybil attacks (SPMSA) and employs the ECC to provide a secure yet lightweight solution against the Sybil attacks hampering platoon operations. Other typical attacks that SPMSA resists include replay, man-in-the-middle and distributed denial-of-service attacks. Lastly, to preserve the privacy of the vehicles and the platoon, our scheme offers perfect forward secrecy as well as group forward and backward secrecy.

The remainder of the paper is as follows: Section 2 discusses existing research work related to platoon management against Sybil attacks. Section 3 discloses the system model and preliminaries. Section 4 describes the SPMSA scheme in detail. The security and performance of SPMSA are evaluated in Sections 5 and 6, respectively. Section 7 concludes the paper.

2. Related Works

In this section, we provide a holistic overview of existing research work that has addressed the problem of Sybil attacks on vehicular platoons. Although various technologies have been used to address this research problem, we only focus on three main technologies: blockchain, machine learning and cryptography. However, investigators found that there is a scarcity of work that explicitly addressed securing against Sybil attacks in the context of vehicle platoons [2]. To the best of our knowledge, a general approach was taken by most researchers, who addressed the problem of Sybil attacks in VANETs instead. Nevertheless, we review those works that we believe can be adapted for platooning.

2.1. Blockchain

Bochem et al. [12] proposed a fully decentralized blockchain that monetarily disincentivizes the creation of Sybil identities in MANETs. It works with any proof-of-work (PoW)-based blockchains (e.g., Ethereum) and binds vehicular node identities to cryptographic public–private key pairs known as blockchain wallet addresses. Whenever a vehicle wants to join the network, an identity proof is created for it. In this process, the vehicle sends a cryptocurrency deposit to the deposit wallet, where this transaction is subsequently mined into a block in the blockchain. Through the mined block, an identity proof is created for the joining vehicle that contains the details of the transaction and the wallet address (i.e., vehicle's identity in the form of its public key) responsible for it. Before the vehicle nodes can start to communicate, they exchange their uniquely generated identity proofs as part of a two-way handshake to verify their peers' identities and prevent Sybil nodes from entering the network for free. The vehicle it exchanges its identity proof with (i.e., identifier) is selected at random to prevent a scenario where a malicious vehicle colludes with one of its Sybil vehicles to disingenuously validate itself. The scheme was initially used for the offline verification of identity proofs, but Bochem and Leiding [13] later adapted the scheme for Internet-of-Things (IoT) environments as well and included the functionality of revoking false identities: If the identity proof of a vehicle is rejected, the identity is revoked. The identifier deems the vehicle a possible Sybil vehicle, and no further communication takes place between the vehicles. Moreover, since generating new identities is expensive and there are likely to be a considerable number of honest nodes in the network, it is infeasible for a malicious entity to perform attacks utilizing Sybil vehicles. However, disregarding the costs for malicious users, the costs to honest users in this scheme could be cause for concern. As the scheme leverages PoW, the costs of transactions are higher since more blocks are mined, and hence, more energy is consumed. As we stated, the cost of a revocation transaction increased by more than tenfold in just one year.

On the other hand, Liu et al. [14] proposed a dual cyber-physical blockchain instead for building a secure and trusted communication for connected vehicle (CV) applications. Their approach involved using the proof-of-stake (PoS) as its consensus mechanism and adopting sharding to partition vehicles into regions and decrease computation, communication, and storage costs. Two blockchains were used in their scheme: the trust points blockchain and the proof-of-travel blockchain. The former quickly identifies and records malicious misbehavior and can be regarded as a misbehavior detector. Intuitively, telling the truth and being acknowledged by most neighbor vehicles earn trust points, while telling a lie loses trust points. Meanwhile, the second blockchain records each vehicle's historical contribution to the CV community and can be regarded as a reputation management system. This contribution comes in the form of a vehicle's travel activities. The more traffic information a vehicle shares with other vehicles around it, the greater its contribution to the CV community, and the more proof-of-travel credits it receives. The outputs of the two blockchains are then added up to form the stake of a vehicle. The higher the vehicle stake, the more trustworthy it is within the network. Hence, Sybil vehicles in the network are detected if they have a small stake from the two blockchains. However, breaking the area into multiple regions does intuitively mean more transfers of records between blockchains in different regions when a vehicle travels frequently between regions. This could accumulate considerable latency as more blockchain mining should occur.

Didouh et al. [15] proposed a novel cyber-physical blockchain architecture to prevent position spoofing attackers such as Sybil nodes from becoming validated nodes for the highway ETC application. The scheme is a witness-based approach using proof-of-location (PoL) as its consensus mechanism. Smart contracts were used as an authentication method to determine the legitimacy of a node. A consortium blockchain was maintained by the RSUs in the network, granting these RSUs access and authority over the network. Thus, only RSUs can mine new blocks and permit nodes into the blockchain. When a vehicle (prover) enters the network, RSUs collect the prover's PoLs provided by other vehicles (witnesses) and calculate its overall score. Since the witnesses are likely to be honest, the prover's PoL should prove that the prover is in the position where it claims to be. If the prover's PoL score is below a specified threshold, it is determined to be a Sybil vehicle and denied entry into the network. To ensure that the PoL scores are transparent for any node to view and verify the information without the need for an external party, smart contracts were published in the blockchain as their nonrepudiation nature guarantees that no vehicle can deny the authenticity of its information on the blockchain.

2.2. Machine Learning

A form of machine learning is the support vector machine (SVM), which Gu et al. [16] used to detect Sybil attacks in VANETs. The SVM was used to classify the driving patterns of vehicles such that Sybil vehicles could be distinguished from legitimate ones. The flexible nature and good generalization performance of the SVM allow it to reduce overfitting of the data and solve various problems without requiring major tuning. This makes SVM a suitable learning algorithm for the dynamic environment of VANETs. The driving patterns of vehicles are thus defined as aggregations of the quantifiable vehicle data, which include time, location, velocity, acceleration and acceleration variation. Crucially, the scheme works on the hypothesis that when traffic is dense, vehicles drive in a similar manner, and hence, any obvious outliers with relatively high variance in their driving patterns can be detected as Sybil vehicles.

Similarly, the authors in [17] used the mobility information of vehicles sent to the RSUs to form the input matrix that represents the driving patterns of the vehicles. Rather than use an SVM, their scheme used extreme learning machine (ELM) to determine the similarity of the mobility patterns and detect the Sybil nodes. ELM was used over traditional artificial neural networks due to its greater advantage with regards to stopping criteria, learning rate and minimum local and over-tuning, contributing to a faster and lower complexity feed-forward learning algorithm. In both schemes [16,17], however, there are doubts as to how effective they are when traffic density is low because then there is greater variation in the benign vehicles' driving patterns.

The concept of radio frequency (RF) fingerprinting was explored by [18], who inserted signatures into transmitted I (in-phase) and Q (quadrature) samples so that a transmitter could be passively identified as a friendly or authorized party. This was carried out by the detection of these signatures through a deep convolutional neural network (CNN) at the physical layer. Reus-Muns et al. [19] adapted this concept for use in 5G and open radio access networks (open RANs) to identify trustworthy base stations instead. Doing so could prevent Sybil attacks whereby base stations attempt to spoof as other base stations. Using real-world datasets, they were able to demonstrate an accuracy of 99.86% irrespective of the training and testing time gap. This outlines the potential benefit of building trust for future open RAN networks and then for use in platoon networks to identify spoofing-based attacks such as Sybil attacks. Similarly, Comert et al. [20] used deep learning-based RF fingerprinting methods to identify the malicious transmitters in cyber-physical systems. However, the authors found that using real-world datasets that considered all environmental scenarios was impractical. Hence, they used unobserved data instead and obtained a peak accuracy of 87.94%, a modest drop from the results obtained by [19].

2.3. Cryptography

The work in [8] is one such cryptography-based work that has explicitly addressed securing platoons against Sybil attacks. The authors did so by integrating a hybrid key management with a witness-based mechanism as a defense protocol. Evaluations performed using OMNeT++, SUMO and Veins framework showed that Sybil attacks could be significantly deterred with minimal overhead effect on network throughput and delay. The minimal overhead is realized by bootstrapping the credentials of the public key infrastructure to establish pairwise symmetric keys, making the proposed protocol lightweight as a result. The identities of the vehicles are encrypted during transmission and decrypted upon reception with this symmetric key, ensuring the safe transfer of vehicle identities from one vehicle to another. A vehicle in the network is then able to verify whether a neighboring vehicle is a Sybil node or not by inspecting its witness table. The witness table holds the information of the other vehicles (witnesses) it passed along the way to reach the verifier node, i.e., the route to the verifier vehicle. If there are missing witnesses, the verifier deems the route invalid, and the node is detected as a Sybil vehicle. Hence, by symmetric key cryptography and the witness-based algorithm, vehicle identities are authenticated. However, the scheme is unable to verify the authenticity of messages exchanged among the vehicles, which brings about the possibility of supposed trustworthy vehicles sending false messages.

Authors in [9] proposed privacy-aware Sybil attack detection (PASAD) that aimed to detect Sybil attacks in both V2V and V2I communications while preserving vehicle privacy. The proposed scheme adopted the Boneh–Shacham (BS) short group signature and safe physical authentication to set up a secure and privacy-aware communication channel for the vehicles. The PASAD scheme is able to detect Sybil attacks on two levels. The first level detects Sybil attacks from an outsider attacker by checking unique registrations of vehicles for any double registrations. This level of detection utilizes the safe physical authentication in trusted RSUs (TRSU) and the BS group signature in RSUs. Meanwhile, the second level prevents insider Sybil attacks by checking for the retransmission of warning messages in V2V communication and double-registrations of vehicles. At this level, the BS group signature is used to ensure the uniqueness of warning messages and the vehicles that sent them. In short, PASAD is able to prevent both outsider and insider attacks in a decentralized manner, even when vehicles are out of range of the RSUs. Thus, the scheme can achieve authentication by verifying both a vehicle's physical presence and the short group signatures attached to the broadcasted warning messages. Although additional overhead is not imposed by the scheme, its increase in computational delay when there are many invalid signatures is a cause for concern when VANETs have to be time efficient.

An alternative approach to detecting Sybil vehicles is seen in [10], who used timestamps in conjunction with a hybrid public key infrastructure. Specifically, a chain timestamp was utilized to provide secure communication in the public channel. The proposed scheme concatenated an encrypted message with a timestamp before it was transmitted to the receiver. Another timestamp was generated upon the receipt of the message, and each timestamp was subsequently recorded by the RSU. Under the chain timestamp concept, if the RSU records consecutive timestamps that are the same, then a Sybil attack has been detected and the session concludes. Otherwise, the message will be sent to the receiver as normal. However, it is not shown how the identities of the vehicles are secured and hence validated. Although a public key architecture is in place to ensure that a trusted authority is in control of the private–public key pairs of the vehicles, the message exchange between a sender and receiver only requires the receiver's set of private and public keys. As it is a one-way communication system, the receiver is unaware of the identity of the sender, who could very well be a malicious node. In such a scenario, the message received might also not be safe. Thus, expanding the proposed scheme to authenticate the identity of the sender and the messages received could help to provide greater security.

One work that adopted message authentication is [11], where elliptic curve cryptography (ECC)-based digital signatures were attached to messages to verify their origin and authenticity, thereby strengthening the vehicle's privacy against Sybil attacks. The elliptic curve digital signature algorithm (ECDSA) is a secure hash algorithm that generates and verifies these signatures. When a vehicle sends a message, it signs the message using its private key. The receiver is then able to verify the digital signature attached to the message by using the sender's public key. If the verification fails, the signature is rejected, and the message is suspected of coming from a Sybil vehicle instead. Moreover, even if an attacker were to steal this signature and use it as its own when communicating with other vehicles, the verification process would still fail because the signature had been generated using the private key and message of the original vehicle. The minimal delay in signature generation and verification implies that the efficiency of the proposed scheme is dependent on the traffic and message size instead. The proposed ECDSA can provide greater security and safety for the vehicles during message transmission while being more time and memory space efficient than other key cryptographic algorithms. In a future plan, one suggestion was to extend the proposed scheme by assigning and registering unique identification numbers (IDs) for vehicles to authenticate their identities. Although this would ensure greater message security, it would also inherently increase the processing time.

3. System Model and Preliminaries

3.1. System Model

A typical highway scenario is considered wherein only vehicles equipped with the appropriate on-board unit (OBU) can communicate wirelessly with other similarly equipped vehicles and trusted RSUs within the VANET. Hence, these OBUs facilitate the vehicles' communications. The fifth-generation cellular technology (5G), which is protected by the security scheme specified as part of the Third Generation Partnership Project (3GPP) standard, is used as the means for the V2I communication between the vehicles and the RSU. Meanwhile, V2V communication that allowed vehicles to talk to one another utilized dedicated short-range communication (DSRC). Each vehicle's OBU was equipped with two separate interfaces to allow for parallel V2I and V2V communications via 5G and DSRC, respectively. A 5G core network database that was used to securely store the vehicles' confidential data has a wired connection to each RSU. At least one platoon led by a platoon leader is assumed to be present in the VANET at all times of the platoon operation. The role of the platoon leader is assumed to remain with the same vehicle while the proposed scheme is being run. In addition, no platoon members can communicate with any entities outside of the platoon except for the platoon leader. A simple illustration of the architecture of the system under the study can be found in Figure 1. In this paper, we focus on the management of a platoon when a vehicle intends to join the platoon to the moment it eventually leaves the same platoon.

3.2. Threat Model

The threat model we consider is the Canetti–Krawczyk (CK) adversary model [21–23], which is defined as follows:

1. Participants: Let JV be the vehicle attempting to join a platoon, PL be the leader of the specific platoon, RSU be a trusted RSU and P be any of the participants. All the participants are considered oracles.
2. Partners: If two oracles. e.g., JV and PL, share the same session key, then they are known as partners.
3. Adversary: $\mathcal{A}$ represents a Sybil vehicle adversary running in polynomial bounded time that can attack by eavesdropping, modifying, injecting messages, etc.

4. Queries: Various actions $\mathcal{A}$ can make are defined in the following queries:
   a. Send(P, m): $\mathcal{A}$ modifies and sends the message m to P, then receiving a response from P.
   b. Execute(JV, PL): $\mathcal{A}$ passively eavesdrops on the communication between JV and PL, returning a copy of the information exchanged between the two participants.
   c. Corrupt(P): $\mathcal{A}$ obtains a long-term private key of P.
   d. ESReveal(P): $\mathcal{A}$ obtains the ephemeral private key of P.
   e. SKReveal(P): $\mathcal{A}$ obtains the session key of P.
   f. Expire(P): $\mathcal{A}$ deletes a completed session key of P.
   g. Hash(m): $\mathcal{A}$ obtains random hash r due to the hashing of message m, i.e., Hash$(m) = r$. Any subsequent Hash(m) of the same m produces the same r.
   h. Test(P): Used to test a session key's semantic security. An unbiased coin c is flipped. If $c = 1$, session key of P is sent to $\mathcal{A}$. Otherwise, a random value is sent to $\mathcal{A}$.
5. Semantic Security: A semantically secure system is one in which within a reasonable amount of time, it is infeasible for $\mathcal{A}$ to obtain significant information about a plaintext message given only its ciphertext. Given that $\mathcal{A}$'s objective is to predict the result of a test query correctly, let $Pr[S]$ denote the probability that $\mathcal{A}$ succeeds in its prediction. Subsequently, the advantage of $\mathcal{A}$ in breaking the semantic security is generally defined as $Adv(A) = |2Pr[S] - 1|$. Hence, if $Adv(A) \leq \varepsilon$ is satisfied for any sufficiently small value $\varepsilon > 0$, the scheme is safe by the CK adversary model.

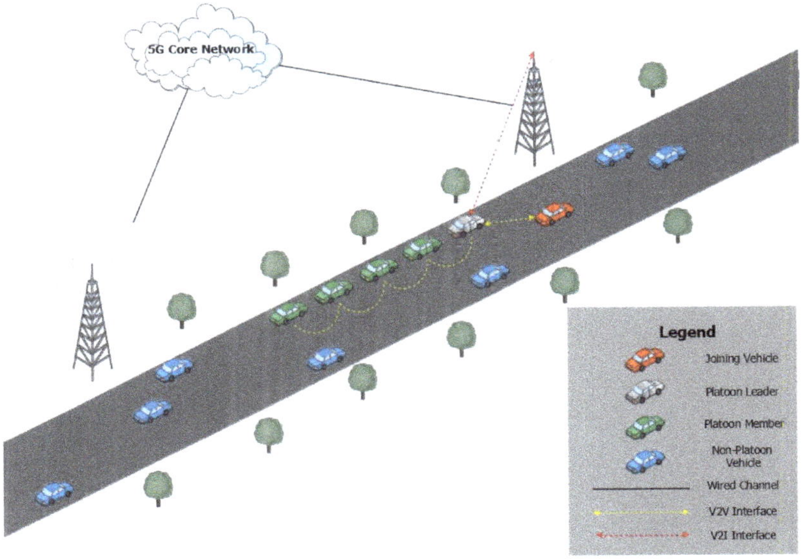

Figure 1. System architecture.

3.3. Elliptic Curve Cryptography

ECC is a type of public key cryptography that offers security equivalent to more widely used cryptosystems such as the Rivest–Shamir–Adleman (RSA) algorithm while requiring fewer bits for computation [11,22]. Hence, the ECC is ideal for systems that are limited in terms of storage, bandwidth and power [11].

An elliptic curve in its simplest form satisfies the equation $y^2 = x^3 + Ax + B$, where $A, B \in F_P$ are constants with $4A^3 + 27B^2 \neq 0 \ mod \ p$ and $p \geq 5$ is a prime number. Hence, the constants A and B control the elliptic curve that is produced. The set of (x, y) 2-tuple that satisfies the elliptic curve equation lies on the curve itself and is referred to as $E(F_P)$. One pair that fundamentally exists for an elliptic curve used in a system deploying ECC is the generator point G, i.e., G is a point on the curve satisfying the equation and has the coordinates of (x, y). Let d represent the private key, which is chosen randomly in the interval $[1, n-1]$, where n is the order of G greater than $2^{Bit\ Size}$. This scalar multiplication of d and G would then produce the public key D i.e., $D = dG$. It forms the basis of the ECC, which works on the elliptic curve private–public key pair (d, D). Lastly, the security of the ECC is guaranteed if the following computations hold [22]:

1. Elliptic curve discrete logarithm problem (ECDLP): Given two points $G \in E(F_P)$ and $aG \in E(F_P)$, it is computationally hard for a polynomial time bound algorithm to compute $a \in F_P$.
2. Elliptic curve computational Diffie-Hellman (ECCDH) problem: Given three points $G \in E(F_P)$, $aG \in E(F_P)$ and $bG \in E(F_P)$, it is computationally hard for a polynomial time-bound algorithm to calculate abG where $a, b \in F_P$ are unknown parameters.
3. Elliptic curve decisional Diffie–Hellman (ECDDH) problem: Given four points G, $A = aG$, $B = bG$ and $C = cG$ in $E(F_P)$, where a, b and c are unknown parameters and $a, b \in F_P$, it is difficult to determine if $C = abG$.

4. Proposed SPMSA

We proposed the SPMSA with the main motivation of establishing a platoon management scheme that would be secure against Sybil attacks. As mentioned, Sybil attacks against platoons are defined as instances where fake virtual vehicles use forged identities to admit themselves legitimately into platoons so as to disrupt the platoons' operation. The proposed scheme was designed with the intention of addressing most of the drawbacks of current cryptographic implementations discussed in Section 2.3. These drawbacks include incomplete identity and message authentication as well as high computation costs.

As it concerns platoon management, the proposed scheme operated over three main events: platoon entry, platoon communication and platoon exit. In short, a vehicle's entire journey from joining the platoon to when it eventually leaves the platoon was covered.

4.1. Platoon Entry Event

Platoon entry is defined as when a vehicle is attempting to join a platoon by interacting with its leader who has the authorization of admitting vehicles into its platoon. The major security issue in this event is ascertaining the identity of the joining vehicle. The legitimacy of this identity should be ensured by verifying that it truly is a registered vehicle. Another security issue is verifying that the messages exchanged in the platoon were not altered. Messages could be intercepted, altered and/or repeated by the Sybil vehicles to falsely admit them into the platoon or deny entry to legitimate vehicles. Hence, the hybrid authentication of both identity and message for this event is required and subsequently provided by the SPMSA.

This event is further broken into four individual phases: initialization, identity authentication, message authentication and platoon key update.

4.1.1. Initialization Phase

In the initialization phase, vehicles equipped with OBUs in the VANET first register their unique vehicle IDs to the 5G core network database through an RSU to obtain the common generator point G used throughout the network. The vehicles also obtain their respective unique long-term private–public key pair (d, D) in return, which is used as a pseudonym to preserve identity privacy [1]. Since a malicious node holds multiple identities when carrying out a Sybil attack, each vehicle is only allowed one identity and as such, can only possess one key pair (d, D) at one time. The initial platoon key qG is generated so that all platoon members can use it to communicate with each other. Thus, this platoon key is only shared among the platoon members. Timestamps were added to each message throughout the operation of the platoon to ensure the freshness of messages, where T is the timestamp when a message is sent, while T' is the timestamp when the same message is received. Whenever a message is intercepted and relayed to the intended recipient, additional time is taken up for the message transmission. Should the time delta between reception and transmission exceed a specified threshold, that is, the estimated time taken for the message to be in transit, i.e., if $T' - T > \sigma$, the message might have been intercepted, so the message and the associated session are discarded. Note that the threshold adjusts independently for each message transmission to accommodate any additional computational operations that could take place before a message is sent out as well as after a message is received. Finally, it is assumed that the ECDLP, ECCDH and ECDDH problems are hard to solve. The initialization phase happens before the start of platoon operation.

4.1.2. Identity Authentication Phase

The identity authentication phase is visualized in Figure 2, where a Sybil vehicle is detected whenever an if' statement is not fulfilled. It is important to note that a Sybil vehicle is detected in this manner throughout the entire operation of the platoon and not just in this specific phase. The main objective of this phase is for the joining vehicle and platoon leader to generate and agree upon a secret key that only they will share. This secret key is commonly referred to as a session key and is used to encrypt and decrypt messages so that unintended vehicles cannot decipher and read them. It can be seen as an attempt to prevent external attacks, namely Sybil vehicles spoofing as authentic identities.

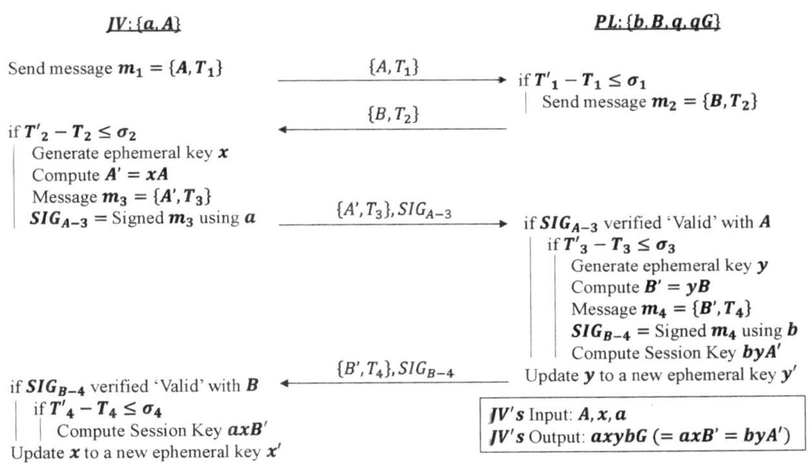

Figure 2. Identity Authentication Phase of the SPMSA.

The ECDSA is used to secure the message. It is an elliptic curve variant of the DSA and boasts a shorter key length than the RSA. The ECDSA allows a sender to sign a message with its own private key and attach the generated digital signature to the message. By verifying the attached signature with the sender's public key, the recipient can check whether the signature and hence the message it is attached to are authentic and came from the sender instead of an adversary. The ECDSA works as follows [11]:

Key Generation:

1. Elliptic Curve Parameters: A, $B \in F_P$: Domain Parameters; $G \in E(F_P)$: Generator Point; n: Order of G greater than 2^{256}; $d \in [1,\ n-1]$: Randomly selected Private Key of Sender; $D = d{\cdot}G$: Generated Public Key of Sender

Signature Generation [Input: message m, d]:

2. $t{\cdot}G = (A_1, B_1)$: A point on elliptic curve of randomly selected number $t \in [1,\ n-1]$

3. $r = A_1 \ mod \ n$: Go back to Step 1 if $r = 0$
4. $s = (t^{-1}(SHA(m) + d * r) \ mod \ n$: Go back to Step 1 if $s = 0$, where SHA is the hash function
3. (r, s): Output ECDSA Signature

Signature Verification [Input: message *m*, (*r*,*s*), *D*]:

6. $r, s \notin [1, n-1]$: Signature is invalid if this condition occurs
7. $w = s^{-1}(mod \ n)$: Compute w
8. $u_1 = [(SHA(m) \cdot w \ mod \ n]$: Compute u_1
9. $u_2 = (r \cdot w) \ mod \ n$: Compute u_2
10. $V = (u_1 \cdot G + u_2 \cdot D) \ mod \ n$: Signature is valid if $V = r$

As a result of the initialization phase, the joining vehicle JV and platoon leader PL own the private–public key pairs (a, A) and (b, B), respectively. To realize this exchange, the existing elliptic curve Diffie–Hellman (ECDH) key exchange protocol is modified, which itself is an ECC-based variant of the original Diffie–Hellman protocol [21]. The identity authentication phase starts with the standard ECDH, where JV and PL exchange their public keys A and B so that they can each compute the same session key thereafter. Upon receiving A, PL can then compute $bA = baG$. Similarly, after receiving B, JV computes $aB = abG$, which is equivalent, and thus, a common key is set up between the two vehicles.

However, to secure this secret key further, another round of key exchange between the two vehicles occurs. In this round, both vehicles first generate a random temporary private key known as an ephemeral private key, which expires and is updated to a new random value once the identity authentication phase ends. A corresponding ephemeral public key is then generated such that JV, for example, would have an ephemeral private–public key pair $(x, A') = (x, xA)$ where $x \in [1, n-1]$. JV would send to PL its ephemeral public key $A' = xA$ with an ECDSA signature SIG_{A-3} attached to it that has been signed using its initial private key a. Upon reception, PL first verifies the signature SIG_{A-3} and then the timestamp T_3. If either fails to be validated, the message and session are discarded as a potential Sybil vehicle has been detected. Otherwise, PL carries out similar actions as JV and sends back its own signed ephemeral public key as acknowledgment. If all the timestamps and signatures are valid, a session key $axB' = byA' = axybG$ can be derived and secretly agreed upon on both sides. Finally, the ephemeral private keys of both vehicles are refreshed.

4.1.3. Message Authentication Phase

Figure 3 demonstrates the procedure of the message authentication phase. The main purpose of this phase is to hand the platoon key over to the joining vehicle in a secure manner. Hence, to establish a secure handover of the platoon key, message authentication is used to ensure the data integrity, origin and authenticity of the messages being transmitted. In contrast to identity authentication, message authentication can prevent internal attacks, particularly, which is an attack by a Sybil vehicle that has been authenticated as a legal user within the VANET. A modified version of the elliptic curve variant of integrated encryption scheme (ECIES) is used to provide semantic security in this phase. Briefly, the standard ECIES encrypts a plaintext message and attaches a message authentication code (MAC) to this encrypted message [24]. The MAC is also known as a keyed hash function as it takes as input a secret key (i.e., MAC key) and the encrypted message to produce a hash i.e., MAC as the output. The MAC guarantees the message's integrity and authenticity because only an actor who has the knowledge of the secret key can generate the MAC [25]. Thus, the receiver can check the MAC for authenticity of the message. Since the receiver shares the same secret key with the sender, if the MAC is deemed invalid by the receiver, then the message might have been tampered and is not safe to be decrypted. The modified ECIES works as follows:

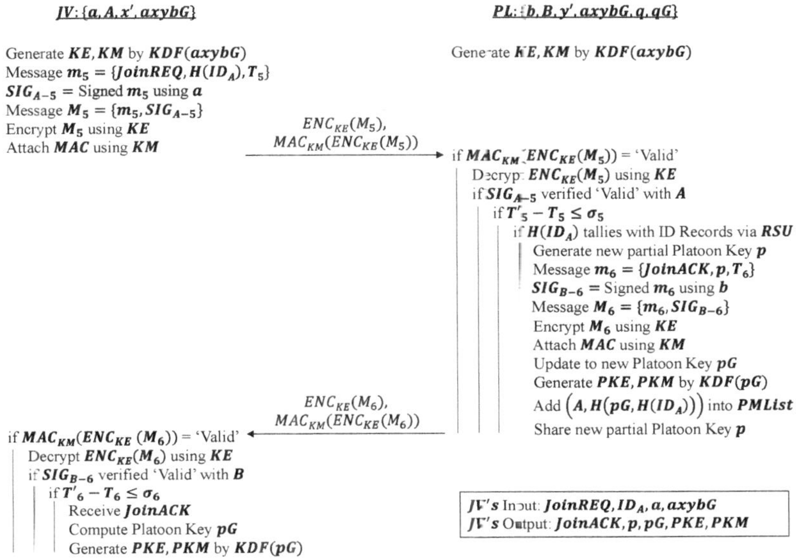

Figure 3. Message Authentication Phase of the SPMSA Scheme.

Encryption of message m [Input: $axybG$]:

1. $KDF(axybG) = KE||KM$: A Key Derivation Function (KDF) is used to derive Symmetric Encryption Key KE and MAC Key KM from the shared Session Key $axybG$
2. $ENC(KE; m) = c$: Encrypt message m using Symmetric Encryption Key KE
3. $MAC(KM; c) = d_A$: Generate the MAC tag d_A of encrypted message c using MAC Key KM
4. $c||d_A$: Ciphertext output of joining vehicle JV

Decryption of ciphertext $c||d_A$ [Input: $c||d_A$, $axybG$]:

5. $KDF(axybG) = KE||KM$: Symmetric Encryption Key KE and MAC Key KM is derived by PL in the same manner as JV did
6. $MAC(KM; c) = d_B$: MAC is Valid' and encrypted message c has not been tampered with in transit if $d_B = d_A$
7. $ENC^{-1}(KE; c) = m$: Decrypt c using Symmetric Encryption Key KM to obtain message m

This phase is initiated by JV sending PL a request to join PL's platoon with a hash of its ID, and the accompanying ECDSA signature SIG_{A-5} belonging to JV. The session key $axybG$ computed at the end of the identity authentication phase is used as the input to the KDF to derive the symmetric encryption key KE and MAC key KM. The benefit of generating the two keys from the KDF is that key diversification can be achieved. The session key is essentially hashed because a master key (i.e., session key) is being separated into two children keys (i.e., KE and KM). Thus, even if an attacker wants to obtain one of the derived keys, it is unable to reverse engineer the stolen key to get the entire session key or the other derived keys. Thus, JV uses KE to encrypt the concatenated message M_5, and KM to generate and attach a MAC to the encrypted form of M_5 before sending them over to PL. It ensures that any tamper attempt on the encrypted message (i.e., joining request, ID, timestamp, and signature) can be checked and detected by the PL through this MAC upon message reception.

Only if the following conditions occur in order is a new partial platoon key p generated and JV's identity added into the platoon members list $PMList$ as a 2-tuple $(A, H(pG, H(ID_A)))$:

1. MAC from JV is valid
2. JV's digital signature SIG_{A-5} is valid
3. Timestamp delta is within threshold range σ_5
4. Hashed ID_A tallies with the ID records in the 5G core network database after the PL sends a database check query through the RSU

If the above conditions are satisfied, a new partial platoon key p is generated and sent to JV using the same message composition procedures conducted by JV at the start of the message authentication phase. It allows JV to compute the updated platoon key pG, thereby authenticating it as an official platoon member PM that can communicate with others from now on. At the same time, PL shares this new partial platoon key p with other current PMs, who will update their platoon key according to the new key pG to preserve the privacy of the previous platoon key qG from the new platoon member, JV. This can ensure group backward secrecy on the old platoon key qG achieved.

Platoon-wide symmetric encryption and MAC key are then generated by a KDF of the common platoon key pG upon JV entering the platoon, and these keys are referred to as PKE and PKM, respectively. Consequently, all authenticated platoon members PMs share the same set of PKE and PKM keys. As can be seen, the use of digital signatures and MACs ensures the origin, data integrity and authenticity of messages exchanged in this phase.

4.1.4. Platoon Key Update Phase (Entry)

This phase occurs whenever a vehicle is authenticated to enter the platoon. As mentioned in the message authentication phase, all existing platoon members will have the new platoon key pG shared with them by the PL. Hence, the primary objective is to allow the PL to distribute the new partial platoon key p to the existing members in its platoon. In this phase, the PL will send out an update request $UpdateREQ$ and the partial platoon key p it generated to all current platoon members by attaching a timestamp and old platoon signature SIG_{qG-7} to it. The platoon signature is similarly signed using the old partial platoon key q.

Subsequently, the message is encrypted with a MAC attached to it. Once again, the previous set of platoon encryption and MAC keys PKE' and PKM' are used to carry these actions out. When the validity of the MAC, platoon signature and timestamp attached to the message sent are verified by the PM, the PM is able to derive the new platoon key pG and thereby generate a new set of platoon encryption and MAC keys PKE and PKM. A platoon key sends acknowledgment $UpdateACK$ back to the PL to indicate the correct reception of the new platoon keys. This acknowledgment is accompanied by a timestamp and new platoon signature SIG_{pG-8} before being encrypted and tagged with a MAC using the new keys PKE and PKM instead of the previous keys PKE' and PKM'. The algorithm works as shown in Figure 4.

4.2. Platoon Communication Event

The platoon communication event refers to a scenario where successfully authenticated platoon members PMs intend to transmit payload information to other members during platooning. Message authentication is once again used, but this time its purpose is to conduct payload communication rather than transfer a platoon key. Let JV be the vehicle that has just joined the platoon and is the latest authenticated PM. Additionally, let an example scenario for this event be a PM informing JV to close the physical distance between them. The message exchange algorithm for this event is similar to that of the message authentication phase found in Section 4.1.3.

However, to protect the privacy of the vehicles in the platoon, the only information being sent over the communication channel of the platoon in this event is action requests and acknowledgements. In this instance, only $CloseUpREQ$ and $CloseUpACK$ messages are exchanged between the two vehicles. Each message is assigned with a timestamp and signed afterward using the partial platoon key p. The messages are then encrypted using the PKE key, and a MAC generated by the PKM key is attached to the resulting ciphertext.

The receiving party of the request message, i.e., JV, then verifies the attached MAC and decrypts the ciphertext using the same set of PKE and PKM keys used by PM. If the platoon signature can be verified as valid using platoon key pG and the timestamp delta is less than the threshold σ_9, JV is able to acknowledge PM's request and execute it. JV then replies to PM in an equivalent manner by attaching a timestamp and platoon signature to the message before encrypting it and tagging it with a MAC using the same PKE and PKM keys. At the end of the message exchange, PM receives an acknowledgement $CloseUpACK$ informing it that JV is executing the $CloseUp$ action. Figure 5 details the algorithm for this event.

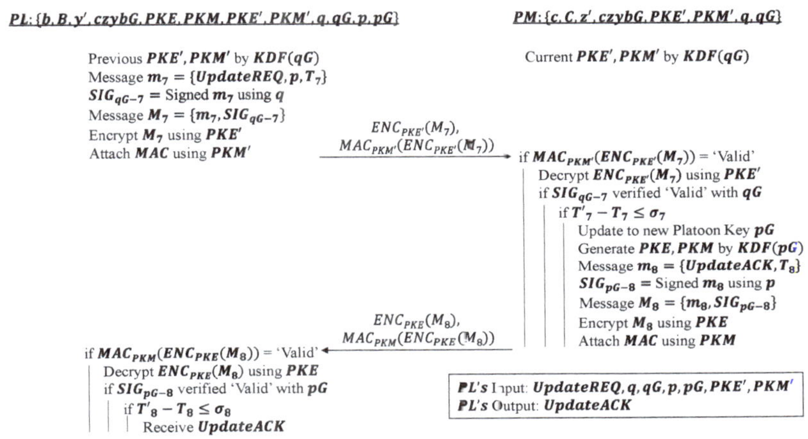

Figure 4. Platoon Key Update Phase of the SPMSA Scheme when a Vehicle Enters the Platoon.

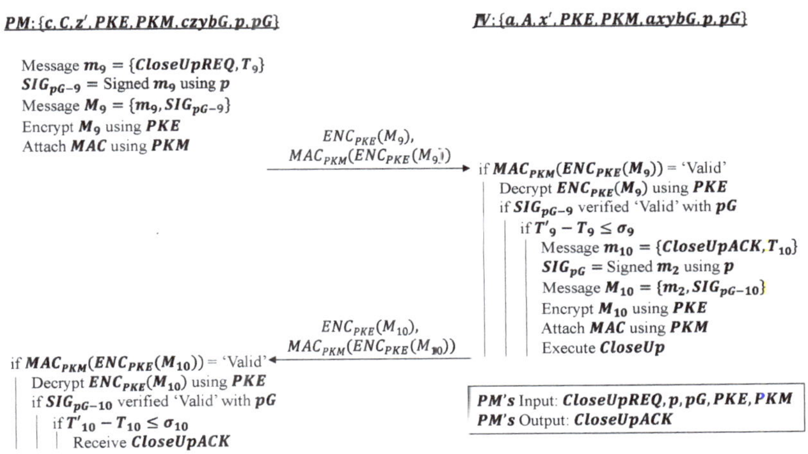

Figure 5. Platoon Communication Event of the SPMSA Scheme.

4.3. Platoon Exit Event

A platoon exit occurs when a current platoon member informs its platoon leader that it wishes to leave the platoon. Once again, since it is an interaction between authenticated platoon members, the message comes from the leaving vehicle, and it should be confirmed that the message has not been tampered with by a malicious platoon member. Hence, only message authentication is used throughout the two phases of this event.

4.3.1. Exit Request Phase

The aim of this phase is to allow a vehicle to leave the platoon without compromising the privacy of the platoon afterwards. For simplicity's sake, it is assumed that the leaving vehicle is JV, which holds the same set of keys and maintains information after the platoon entry and communication events. The JV is denoted as LV to indicate it is a leaving vehicle. Once again, the algorithm for this phase is similar to that in Section 4.1.3 as it ultimately is the inverse operation of the message authentication phase. The contents of the message sent by LV contains a platoon leaving request $LeaveREQ$ and a hashed 2-tuple of the current platoon key pG and LV's hashed identity $H(ID_A)$ i.e., $H(pG, H(ID_A))$. The composition of the message is the same as that of Section 4.2, where a timestamp and platoon signature are attached to the message before PKE and PKM keys are used to encrypt the message and tag it with a MAC.

On the reception of the encrypted message, the PL first checks the validity of the MAC, platoon signature and timestamp. If they are all valid, the PL then verifies if LV's double identity is in the platoon members list $PMList$, i.e., if LV is an authenticated member of the platoon. If LV's record is in the $PMList$, its entry is removed from the list and a platoon exit acknowledgment $LeaveACK$ is sent back to the LV. Similar to the first message sent by LV in this phase, the same procedure is used by PL to prepare the message for transmission to the LV. At the same time, a partial platoon key r is generated by PL so that it can update its current platoon key pG to the latest key rG. This updated partial platoon key is then shared only with the other platoon members PMs that will be staying in the platoon.

Upon receipt of $LeaveACK$, the LV can leave the platoon as it has been deemed safe to exit. This is because the LV does not retain any significant information pertaining to the platoon and its members. For example, to ensure group forward secrecy, the platoon key rG being used by the platoon after LV's exit is different from the one that LV still possesses, i.e., pG. The only information that LV retains with regards to the platoon is the PL's long-term public key B. The algorithm for this phase can be found in Figure 6.

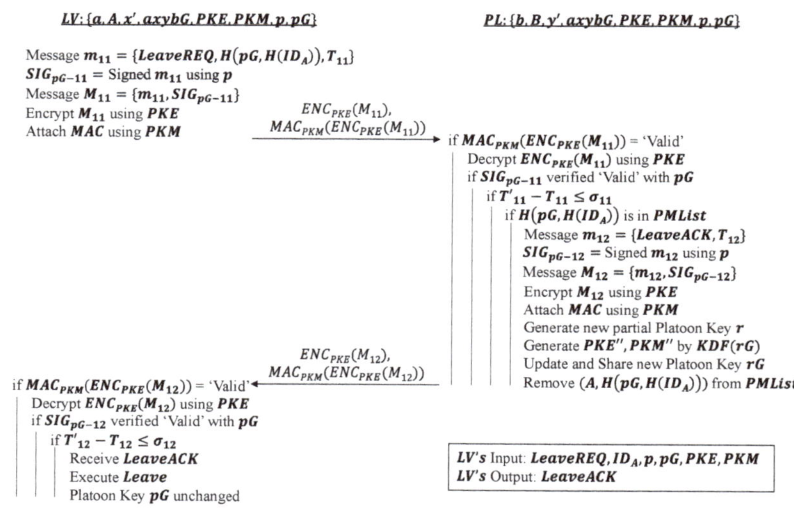

Figure 6. Exit Request Phase of the SPMSA Scheme.

4.3.2. Platoon Key Update Phase (Exit)

This phase is initiated when a vehicle is allowed to exit the platoon, and its primary goal is to allow the PL to distribute a new platoon key r to the PMs that remain in the platoon. Its operation is similar to that in the platoon key update phase in Section 4.1.4. The differences merely lie in the keys being used. From the exit request phase in Section 4.3.2, it is clear that the new platoon key to be used in the platoon is now rG instead of pG. Hence, the partial platoon key r needs to be shared with all PMs so that they can derive the new platoon key rG and preserve the privacy of the platoon from outgoing vehicle LV. The old platoon signature SIG_{pG} is attached to the update request and new partial platoon key r before being sent over by using the old platoon encryption and MAC keys PKE and PKM. The usual verification of the received message is performed by the PM before it can safely generate the new set of platoon keys: rG, PKE'', PKM'' The new platoon signature SIG_{rG} and timestamp are attached to the update acknowledgment $UpdateACK$ and encapsulated as an encrypted message using the new PKE'' key. The new MAC key PKM'' is duly used to tag the encrypted message where the PL is able to verify its validity. The algorithm can be found in Figure 7 and concludes the SPMSA scheme.

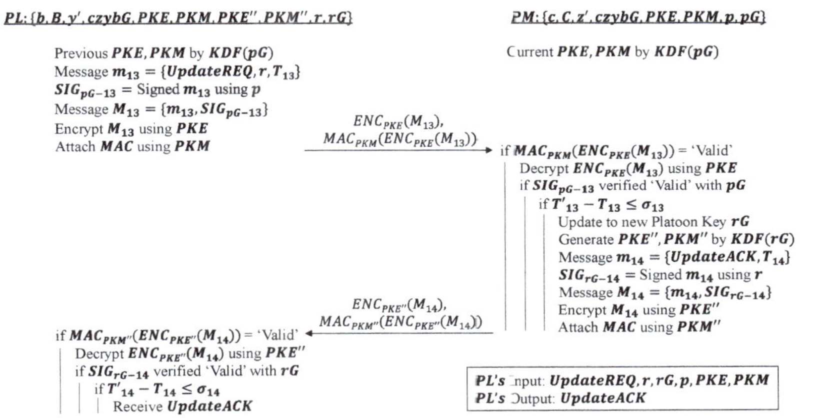

Figure 7. Platoon Key Update Phase of the SPMSA Scheme when a Vehicle Leaves the Platoon.

5. Security Evaluation

5.1. Formal Proof of Security by Random Oracle Model

We first evaluate the SPMSA scheme formally by proving its semantic security under the CK adversary with random oracle model. With the CK adversary model, a probabilistic polynomial–time adversary $\mathcal{A}$ can eavesdrop, modify and inject information into the message exchange process by interacting with the participants involved. In the random oracle model [23], there exists a random oracle that models cryptographic hash functions as ideally random functions. With this model, all participants can interact with one another including $\mathcal{A}$. The queries covered in Section 3.2 that detail the various actions $\mathcal{A}$ can take are assumed to be sent to this random oracle for execution [22].

5.1.1. Formal Proof of Platoon Entry Event

Ultimately, the goal of adversary $\mathcal{A}$ is to determine the real platoon key pG from a random number that occurs in the test query. It requires $\mathcal{A}$ to break the semantic security of the SPMSA for the entry event. To evaluate whether the SPMSA can withstand $\mathcal{A}$'s attempt, it is run through a series of games outlined by the random oracle model. In this section, we omit the initialization phase as it involves the preparation of the keying materials for the system rather than the running of the SPMSA. Meanwhile, the platoon key update phase will be verified in Section 5.2.1. Thus, the proof will only cover the identity and message authentication phases of the platoon entry event. Let $Pr[S_i]$ be the probability that $\mathcal{A}$ succeeds in predicting the results of the test query for Game i. Let the joining vehicle and platoon leader be denoted by JV and PL. q_h, q_s, q_e and q represent the number of hash, send, execute and total random oracle queries sent by $\mathcal{A}$, respectively, while H denotes the hash space size such that $H = 2^{Hash\ Length\ (in\ Bits)}$. As mentioned in Section 3.2, the entry event of the SPMSA offers semantic security under the CK adversary with random oracle model if the advantage for $\mathcal{A}$ winning all the games is $Adv_{Entry}(\mathcal{A}) \leq \varepsilon$ for any sufficiently small value $\varepsilon > 0$.

Lemma 1 (Difference Lemma). *Let E_1, E_2, F denote events in a certain probability distribution where F is known as the failure event. The two events E_1 and E_2 will execute in a similar manner as long as the failure event F does not happen, i.e., $E_1 \wedge \neg F \Leftrightarrow E_2 \wedge \neg F$. As both $Pr[E_1]$ and $Pr[E_2]$ are between 0 and $Pr[F]$. The subsequent difference between the probabilities of the two events is $|Pr[E_1] - Pr[E_2]| \leq Pr[F]$.*

Game 0: This game is the initial attacking game set out in Section 3.2. It is a real attack by $\mathcal{A}$ in the semantic security framework under a random model. Hence, the advantage to $\mathcal{A}$ is:

$$Adv_{Entry}(\mathcal{A}) = |2Pr[S_0] - 1| \tag{1}$$

Game 1: $\mathcal{A}$ launches a passive attack on both parties in the authentication agreement in this game. $\mathcal{A}$ sends an Execute(JV, PL) query to acquire the information exchanged between both parties, which includes $\{A, B, A', B', SIG_{A-3}, SIG_{B-4}, ENC_{KE}(M_5), ENC_{KE}(M_6), MAC_{KM}(ENC_{KE}(M_5)), MAC_{KM}(ENC_{KE}(M_6)), T_1, T_2, T_3, T_4\}$. $\mathcal{A}$ is then unable to compute session key $axybG$ and thus is unable to derive the symmetric decryption key KE. As such, $\mathcal{A}$ cannot acquire partial platoon key p, which is encrypted in $ENC_{KE}(M_6)$. Thus, the probability that $\mathcal{A}$ succeeds is:

$$Pr[S_1] = Pr[S_0] \tag{2}$$

Game 2: This game follows Game 1, with $\mathcal{A}$ now using send queries to initiate active attacks. The following events are omitted, however, as either event would cause the game to be over instantly:

Event E_1: The collision of the hash query outputs in different sessions. The birthday paradox states that E_1 happens with probability $|Pr[E_1]| \leq \frac{q_h^2}{2H}$.

Event E_2: The collision of the random numbers generated in different sessions. As the random numbers are only generated in send and execute queries, $|Pr[E_2]| \leq \frac{(q_s+q_e)^2}{2q}$.

As long as the ECDLP and ECCDH assumptions hold, $\mathcal{A}$ does not have sufficient information to reconstruct the previous session key $axybG$ to decrypt $ENC_{KE}(M_2)$ and obtain partial platoon key p. $\mathcal{A}$ is also unable to establish a new ephemeral key x (or y) using a send query as it needs either JV's or PL's long-term private key to sign the message so that its identity can be verified by the other party. Therefore, according to the difference lemma, (3) is obtained as follows:

$$|Pr[S_2] - Pr[S_1]| \leq \frac{q_h^2}{2H} + \frac{(q_s + q_e)^2}{2q} + q_h \cdot max\{Adv_{ECDLP}(\mathcal{A}),\ Adv_{ECCDH}(\mathcal{A})\} \tag{3}$$

Game 3: Game 3 involves running Game 2 while $\mathcal{A}$ then tries to guess the hash values $KDF(axybG) = KE||KM$ and $H(ID_A)$ without querying the random oracle. If the guess is correct, the game is over. Thus, the resulting polynomial is:

$$|Pr[S_3] - Pr[S_2]| \leq \frac{q_s^2}{2H} \tag{4}$$

Game 4: This game continues from Game 3 but with the consideration of semantic security. $\mathcal{A}$ obtains any two of $\{a,\ b, x, y\}$ by making ESReveal and corrupt queries to the random oracle. However, according to the CK adversary model, $\mathcal{A}$ is unable to obtain both the long-term and ephemeral private keys of the same vehicle at the same time (e.g., acquiring JV's a and x). As a result, $\mathcal{A}$ is unable to recompute the session key $axybG$ and subsequently obtain platoon key pG because it needs both private keys of the same vehicle.

The only way it can do so is to solve the ECDLP and find a or x from A and A', respectively, or to solve the ECCDH problem. Thus, since Game 4 is similar to Game 3,

$$\Pr[S_4] = Pr[S_3] \tag{5}$$

Subsequently, $\mathcal{A}$ initiates a test query where an unbiased coin c is flipped. Since the probability of such an event is $\frac{1}{2}$,

$$Pr[S_4] = \frac{1}{2} \tag{6}$$

Combining all the advantages from Game 0 to Game 4 i.e., Equations (1) to (6) through back substitution, we can obtain (7):

$$Adv_{Entry}(\mathcal{A}) \leq \frac{q_h^2}{H} + \frac{(q_s + q_e)^2}{q} + \frac{q_s^2}{H} + 2q_h \cdot max\{Adv_{ECDLP}(\mathcal{A}),\ Adv_{ECCDH}(\mathcal{A})\} \tag{7}$$

Since $Adv_{Entry}(\mathcal{A}) \leq \varepsilon$, where $\varepsilon > 0$, the entry event of the SPMSA is safe under the CK adversary with random oracle model.

5.1.2. Formal Proof of Platoon Exit Event

For the formal proof of platoon exit, the platoon key update phase will also be verified in Section 5.2.1. Thus, we only evaluate Section 4.3.1 of the SPMSA, i.e., the exit request phase. We evaluate a scenario where JV in the platoon is now looking to exit the platoon. Following the notation in Figure 6, JV is synonymous with LV. It is assumed that no vehicle has joined or left the platoon since LV's entry into the platoon. In other words, the platoon key that is established throughout all platoon members is pG.

A goal of adversary $\mathcal{A}$ is to determine the current platoon key pG from a random number that occurs in the test query. Since a vehicle leaving the platoon causes the platoon key to be updated, $\mathcal{A}$ must intercept the $LeaveREQ$ from LV to prevent it from reaching PL. It ensures that the platoon key $\mathcal{A}$ obtained will remain valid for use in the platoon and not be outdated. The notations used for the formal proof of the platoon entry phase are reused here. There is no change to the lemma difference or the games conducted. Similarly, we can say that the exit phase of our scheme offers semantic security under the CK adversary with the random oracle model if the advantage for $\mathcal{A}$ winning all of the games is $Adv_{Exit}(\mathcal{A}) \leq \varepsilon$ for any sufficiently small $\varepsilon > 0$.

Game 0: This is the initial attacking game set out in Section 3.2, which outlines a real attack by $\mathcal{A}$ in the semantic security framework under a random model. Hence, the advantage of $\mathcal{A}$ is the same as that of (1).

Game 1: A passive attack on both parties is first launched by $\mathcal{A}$ in this game. $\mathcal{A}$ sends an Execute(LV, PL) query to steal the information exchanged between the parties, which includes $\{ENC_{PKE}(M_{11}), ENC_{PKE}(M_{12}), MAC_{PKM}(ENC_{PKE}(M_{11})), MAC_{PKM}(ENC_{PKE}(M_{12}))\}$. Since $\mathcal{A}$ does not hold platoon key pG, it is unable to decrypt the information it has stolen as it cannot generate the necessary platoon encryption and MAC keys PKE and PKM. Hence, $\mathcal{A}$ cannot obtain partial platoon key r, and the probability that $\mathcal{A}$ succeeds is found in (2).

Game 2: Once again, this game follows Game 1 with $\mathcal{A}$ using send queries thereafter to initiate active attacks. Similarly, the following events are omitted:

Event E_1: The collision of the hash query outputs in different sessions. The birthday paradox states that E_1 happens with probability $|Pr[E_1]| \leq \frac{q_h^2}{2H}$

Event E_2: The collision of the random numbers generated in different sessions. As the random numbers are only generated in send and execute queries, $|Pr[E_2]| \leq \frac{(q_s+q_e)^2}{2q}$

Only if the two events above occur will $\mathcal{A}$ have a plausible amount of information to potentially forge a legitimate message to intercept the communication between LV and PL. Therefore, according to the Difference Lemma, (8) is obtained as follows:

$$|Pr[S_2] - Pr[S_1]| \leq \frac{q_h^2}{2H} + \frac{(q_s + q_e)^2}{2q} \tag{8}$$

Game 3: After the conclusion of Game 2, $\mathcal{A}$ tries to guess the hash values $KDF(pG) = PKE||PKM$ and $H(pG, H(ID_A))$ without querying the random oracle. Recall that the platoon encryption and MAC keys PKE and PKM are generated from $KDF(pG)$. If the guess is correct, the game is over, and the resulting polynomial is (4).

Game 4: Following Game 3, semantic security is taken into consideration. $\mathcal{A}$ obtains any two of $\{a,\ b, x', y'\}$ by making ESReveal and corrupt queries to the random oracle. However, obtaining any of them will not enable $\mathcal{A}$ to procure platoon key pG. This is because the queries only uncover the components of session key $axybG$. As shown in Figure 6, this key is not involved in this phase. Hence, with no additional advantage for $\mathcal{A}$ to obtain platoon key pG, Game 4 is no different from Game 3 which can be seen in (5).

$\mathcal{A}$ then initiates a test query in which an unbiased coin c is flipped and the probability of such an event is $\frac{1}{2}$, as shown in (6).

Finally, we combine the advantages from Game 0 to Game 4, i.e., Equations (1), (2), (4), (5), (6) and (8) through back substitution to get the advantage of $\mathcal{A}$, as seen in (9).

$$Adv_{Exit}(\mathcal{A}) \leq \frac{q_h^2}{H} + \frac{(q_s + q_e)^2}{q} + \frac{q_s^2}{H} \tag{9}$$

Since $Adv_{Exit}(\mathcal{A}) \leq \varepsilon$, where $\varepsilon > 0$, the platoon exit event of the SPMSA is also safe under the CK adversary with random oracle model.

5.2. Formal Verification of Security Functionality by CryptoVerif

In this section, we verify the security functionality of the platoon key update phases of the SPMSA that we covered in Sections 4.1.4 and 4.3.2. Although these phases occur in two different events, the algorithms are the same, with the only difference the names of the keys. Hence, verifying the security of one of the platoon key update phases could verify the security of another.

First, we briefly review CryptoVerif. CryptoVerif is an automatic protocol verifier on security that is sound in the computational model. It can verify secrecy and correspondences such as authentication. It provides formal verifications as a sequence of games, similar to the CK adversary model that we used to prove the other parts of the SPMSA. However, instead of being manually implemented, CryptoVerif can be automatically executed via a programming model. The generated verifications are valid for any number of sessions of the protocol. Hence, it can provide an upper bound on the probability of the success of an adversary against the protocol as a function of the likelihood of breaking each cryptographic primitive and of the number of sessions it takes to do so [25].

The input script for CryptoVerif to run contains the cryptographic assumptions and properties to verify. CryptoVerif uses the technique of game hopping where the first game models the actual protocol that we wrote in the input script to verify. From the second game onwards, CryptoVerif applies syntactic transformations on the game until the game satisfies the security properties realized. Note that an adversary is unable to distinguish one game from another after transformation as the difference of probability between consecutive games is negligible, i.e., $|Pr[Si] - Pr[Sj]| \approx 0$, where $j = i + 1$. Consequently, the advantage of the adversary for the final game is $Adv(\mathcal{A}) = 0$. Figure 8 shows the game-hopping procedure of the CryptoVerif. After CryptoVerif finishes execution, it will output the sequence of games that occurred, a brief explanation of the transformations that took place between the games and finally, the upper bound of probability of an adversary being successful against the protocol [26]. For our formal verification results, we show the first and last games and the upper bound probability of the adversary of breaking the security properties of the SPMSA.

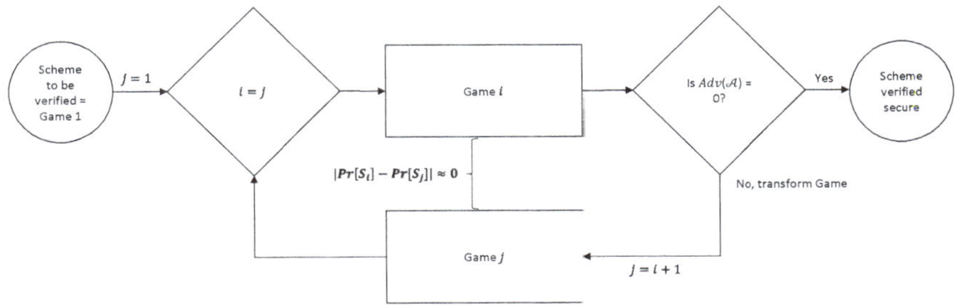

Figure 8. Security Verification by the Game-Hopping Process of CryptoVerif.

5.2.1. Formal Verification of Platoon Key Update Phases

As mentioned beforehand, the platoon key update phases for entry and exit events have the same algorithm, with the name of the keys being the only difference. The actors involved in both phases are the same, i.e., a platoon leader and its members, and the messages exchanged are of the same structure. Specifically, the messages being exchanged exist in the structure of an encrypted message $ENC_{PKE}(M)$, and a MAC $MAC_{PKM}(ENC_{PKE}(M))$, where M is the plaintext encapsulated message, while PKE and PKM are the platoon encryption and MAC keys, respectively. Since the algorithm encrypts the plaintext message before attaching a MAC of the encrypted message, we can therefore deem it an Encrypt-then-MAC cryptographic scheme.

CryptoVerif has a library of predefined cryptographic primitives that can be used to model the SPMSA scheme. For the platoon key update phases of the SPMSA scheme, the core principle is Encrypt-then-MAC. We use the following primitives that have already been specified in CryptoVerif's library [26]:

- Expand IND_CPA_sym_enc(key, cleartext, ciphertext, enc, dec, injbot, Z, Penc). This primitive defines an indistinguishable under a chosen plaintext attack (IND-CPA) probabilistic symmetric encryption scheme. In other words, given the encryption of two messages of the same length, an adversary has a negligible probability of telling the two encryptions apart. We denote this probability as Penc.
- Expand SUF_CMA_det_mac(mkey, macinput, macres, mac, check, Pmac). This primitive defines a strongly unforgeable under chosen message attacks (SUF-CMA) deterministic MAC. This means that for an adversary that is given access to the MAC and verification oracles, it has a negligible probability of forging a MAC. This probability is denoted as Pmac.

We use the oracle front-end of CryptoVerif, which is more suitable in our case because its syntax of games resembles manual cryptographic verification better. This falls in line with the previous proofs of the SPMSA scheme in Section 5.1 that were performed manually. We adopt the input scripts written by the author in [25] to verify two security properties of the platoon key update phases of the SPMSA scheme: that the encryption of plaintext message is indistinguishable (IND-CPA) and that the integrity of the ciphertext generated by the encryption is hard to break (INT_CTXT). By verifying these two properties, we can safely say that the partial platoon key, update request and timestamp are transferred securely to the PMs with the SPMSA.

For the IND-CPA property verification, two oracles called L and R are required. CryptoVerif uses equivalences to transform the processes that call the L oracles into processes that call the R oracles. If the oracles on the two sides return different results, the event is deemed unreachable, and CryptoVerif declares that the two sides, i.e., the encryption of messages, are indistinguishable.

To verify the IND-CPA property in a discernible manner, a query secret Boolean b is used where if $b = 1$, then message = $m1$, while if $b = 0$, message = $m2$. After a specific game transformation, if b has no influence on which message is encrypted, then we can confirm the IND-CPA property [26]. Figure 9 shows the process of this verification.

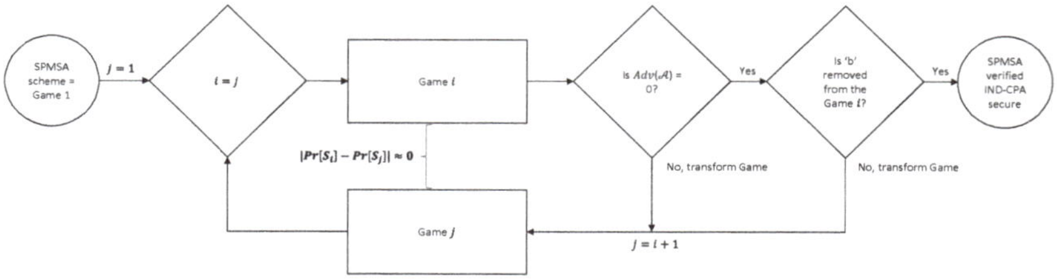

Figure 9. Game-Hopping to Verify IND-CPA Property of Platoon Key Update Phases.

Figure 10 shows the initial game of the verification, while Figure 11 shows it takes eight games (seven game transformations) for the query secret b to not be used in the games anymore because the line of code containing b is missing and the game goes straight into encrypting the message. This is because CryptoVerif merges the two "if" branches of the test "$m0$: bitstring <- (if b then $m1$ else $m2$);" as the same code to be executed in either branch. In short, $m1$ and $m2$ are indistinguishable because the two use the same code. Finally, the RESULT header shows the upper bound probability of the adversary to be successful in telling the encryption of messages apart. This upper bound is shown to be double that of Penc, which is the probability of breaking the IND-CPA property of the underlying encryption scheme, as previously discussed.

```
======================== Proof starts ========================
Initial state
Game 1 is
     Ostart() :=
     b <-R bool;
     k_3 <-R key;
     mk_2 <-R mkey;
     return();
     foreach i <= qEnc do
     Oenc(m1: bitstring, m2: bitstring) :=
     if Z(m1) = Z(m2) then
     m0: bitstring <- (if b then m1 else m2);
     return((m: bitstring <- m0; c1: bitstring <- (m_1: bitstring <- m; k_2: key
 <- k_3; r <-R enc_seed; enc_r(m_1, k_2, r)); concat(c1, mac(c1, mk_2))))
```

Figure 10. First Game of Verifying the IND-CPA Property of Platoon Key Update Phases.

```
Game 8 is
     Ostart() :=
     b <-R bool;
     k_4 <-R key;
     mk_2 <-R mkey;
     return();
     foreach i <= qEnc do
     Oenc(m1: bitstring, m2: bitstring) :=
     if Z(m1) = Z(m2) then
     r_3 <-R enc_seed;
     c1: bitstring <- enc_r'(Z(m2), k_4, r_3);
     return(concat(c1, mac(c1, mk_2)))


Proved secrecy of b in game 8



RESULT Proved secrecy of b up to probability 2 * Penc(time_1, qEnc, max(maxlengt
h(game 4: m1), maxlength(game 4: m2)))



All queries proved.
```

Figure 11. Final Output of IND-CPA Verification of Platoon Key Update Phases.

For the INT_CTXT property verification, encryption and decryption test oracles are required. A query event "bad" is used to show whether the adversary has successfully broken the INT_CTXT property. If event bad occurs, the adversary has managed to produce a ciphertext that decrypted successfully and has not been produced by the encryption oracles. Hence, the verification is only successful when event bad does not happen, i.e., when the occurrence of event bad is false [26]. Figure 12 shows the process of the INT_CTXT property verification.

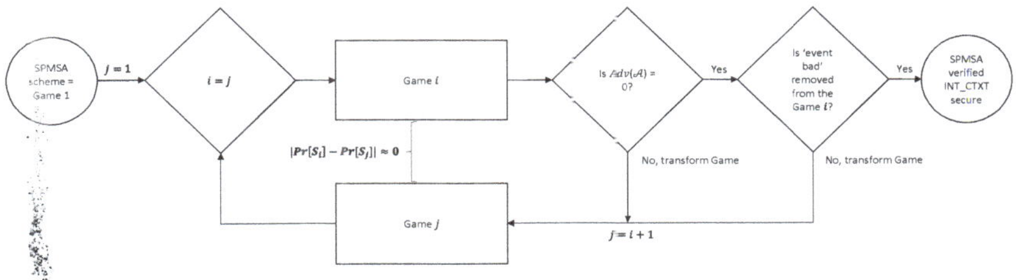

Figure 12. Game-Hopping to Verify INT_CTXT Property of Platoon Key Update Phases.

Figure 13 portrays the initial game of the INT_CTXT verification where event bad can be seen in the fifth and last line. The final result of the INT_CTXT security verification in Figure 14 shows that nine games (eight game transformations) are required for event bad to no longer occur in the game. Hence, CryptoVerif has verified that the adversary will not be able to forge a ciphertext that can be decrypted successfully and has not been produced by the encryption oracles. Finally, the RESULT header shows the upper bound probability that the adversary will be successful in breaking the ciphertext integrity to be equivalent to Pmac, which is the probability of breaking the SUF-CMA property of the MAC.

To conclude, through the use of a computational verifier tool CryptoVerif, we showed that the platoon key update phases of the SPMSA resist an adversary $\mathcal{A}$ distinguishing between encrypted messages. It is also resistant to $\mathcal{A}$ forging ciphertexts that can be decrypted to obtain the original plaintext message, which in our case crucially includes the partial platoon key p for the entry event and r for the exit event.

```
======================== Proof starts ========================
Initial state
Game 1 is
    Ostart() :=
    k_3 <-R key;
    mk_2 <-R mkey;
    return();
    ((
      foreach ienc <= qEnc do
      OEnc(m0: bitstring) :=
      c0: bitstring <- (m: bitstring <- m0; c1: bitstring <- (m_1: bitstring <-
 m; k_1: key <- k_3; r <-R enc_seed; enc_r(m_1, k_1, r)); concat(c1, mac(c1, mk_
2)));
      insert ciphertexts(c0);
      return(c0)
    ) | (
      foreach idec <= qDec do
      ODec(c_1: bitstring) :=
      get ciphertexts(=c_1) in
        return(true)
      else
        if (let concat(c1_1: bitstring, mac1: macs) = c_1 in if verify(c1_1, mk
_2, mac1) then dec(c1_1, k_3) else bottom else bottom) <> bottom then
          event bad;
          return(true)
        else
          return(false)
    ))
```

Figure 13. First Game of Verifying the INT_CTXT Property of Platoon Key Update Phases.

```
Game 9 is
    Ostart() :=
    k_3 <-R key;
    mk_2 <-R mkey;
    return();
    ((
      foreach ienc <= qEnc do
      OEnc(m0: bitstring) :=
      r <-R enc_seed;
      c1: bitstring <- enc_r(m0, k_3, r);
      ma2: macs <- mac'(c1, mk_2);
      c0: bitstring <- concat(c1, ma2);
      return(c0)
    ) | (
      foreach idec <= qDec do
      ODec(c_1: bitstring) :=
      find u = u_1 <= qEnc suchthat defined(c0[u_1]) && (c0[u_1] = c_1) then
        return(true)
      else
        return(false)
    ))

Proved event(bad) ==> false in game 9


RESULT Proved event(bad) ==> false up to probability Pmac(time_1, qEnc, qDec, 0,
 max(maxlength(game 4: c1), maxlength(game 4: c1_1)))


All queries proved.
```

Figure 14. Output of INT_CTXT Verification of Platoon Key Update Phases.

5.2.2. Formal Verification of Security Functionality for Communication Event

In fact, by nature, the communication event is a much more simplified version of the key update phases. They both have a pair of request and acknowledgment messages exchanged using Encrypt-then-MAC. However, the platoon communication event involves purely payload communication. In comparison, the key update phases require an additional partial platoon key to be transmitted over the channel to generate the platoon key. Transferring these additional data does not make the algorithms more complex; rather, it introduces additional potential vulnerability.

Thus, since we have verified that the key update phases are secure by CryptoVerif, we can then deduce that the algorithm of the platoon communication event that has fewer potential data vulnerabilities is secure as a result. To conclude, the platoon communication event resists an adversary $\mathcal{A}$ distinguishing between its encrypted messages as well as $\mathcal{A}$ forging ciphertexts of valid plaintext messages. To reiterate, these messages only include platoon requests/acknowledgments and the accompanying timestamps.

5.3. *Security Analysis*

In this section, a qualitative analysis of the security properties and the abilities against some of the typical malicious attacks of the SPMSA scheme is presented.

Mutual authentication: Mutual authentication can be achieved by both identity and message authentication as discussed in Sections 4.1.2 and 4.1.3. Digital signatures are used to ensure the identity of the vehicle, while MACs are used to confirm the message's origin and integrity.

As stated in Section 3.2, the Canetti–Krawczyk (CK) adversary model was used against the SPMSA to test it for any vulnerabilities. Participants, partners and the adversary are the parties involved based on the model. The adversary represents a Sybil vehicle that can make queries to disrupt and obtain information to authorize itself as a legitimate platoon member. The adversary's main goal is to obtain a valid and working platoon key by tethering the communication between any pair of partners, including JV/LV, PL and RSU. With the help of the CK threat model, the SPMSA is secure against Sybil attackers.

Key agreement: The session key $axybG$ and platoon key pG can be computed by both JV and PL after the mutual authentication. As long as the long-term private key is inaccessible and the ECDLP and the ECCDH assumptions hold, an attacker cannot compute the session key.

Perfect forward secrecy: The previously established session key $axybG$ will still be secure even if the long-term private keys a and b are compromised. This is due to an attacker's inability to obtain the previous ephemeral private keys x or y, which have expired and have been erased from the OBU's memory.

Group backward secrecy: Whenever a vehicle joins a platoon, even if the platoon is one it has joined before, it cannot compute or possess the previously used platoon key qG.

Group forward secrecy: Whenever a vehicle leaves a platoon, it is unable to compute or possess the new platoon key rG.

Ability against replay attacks: A replay attack is launched so that $\mathcal{A}$ can spoof a legitimate vehicle by sending previous data to the vehicles. Adding short-term keys generated by random numbers and timestamps can ensure the freshness of messages.

Ability against man-in-the-middle (MitM) Attacks: With a MitM attack, $\mathcal{A}$ tries to establish connections with vehicles individually to make them mistakenly believe that they are connected to each other. In the entry event of the SPMSA, where JV and PL try to set up a session key and new platoon key, we mentioned in Game 4 of Section 5.1.1 that an attacker cannot acquire both the long-term and the ephemeral private keys of the same vehicle at the same time. Hence, even if the messages are intercepted and modified by $\mathcal{A}$, it cannot have the generator point G forge a signature that will be validated by JV and PL. The other phases of the scheme involve authenticated communication between platoon members PMs, so $\mathcal{A}$ can only establish a connection with vehicles individually if it has the platoon key. Otherwise, a MitM attack will fail.

Ability against distributed denial-of-service (DDoS) attacks: A DDoS attack is an attack in which multiple malicious attackers overwhelm a single server/node to deny other nodes access to it. Identity authentication should be able to detect and disregard these malicious vehicles in time before the platoon leader is overwhelmed.

Ability against Sybil attacks: We discussed in Section 4 that Sybil vehicles can be detected by the SPMSA with a combination of identity and message authentications. This detection can be achieved by invalidating timestamps, digital signatures, MACs and hashed vehicle identities that accompany transmitted messages. When detected, the message and session pertaining to that Sybil vehicle are discarded afterwards, and the vehicle is denied entry into the platoon.

6. Performance Evaluation

The performance of the SPMSA was evaluated by estimating computation and communication overheads as well as simulation experiments to determine the average elapsed time. We primarily contrasted the performance of the SPMSA with that of PASAD in [9]. We deem it a comparable and appropriate scheme to use for platooning purposes, in large part due to its group membership phase.

Since the platoon entry event is where the major performance issues lie, only Section 4.1 of the SPMSA is considered for the entirety of the performance evaluation. For a fair comparison of the two schemes, we only apply PASAD when a vehicle joins a new RSU group. This is because this event can plausibly resemble a platoon entry event where a vehicle tries to join a platoon. Hence, only Algorithms 4–7 of PASAD are considered. Since this instance of PASAD only involves the initial authentication of vehicles for entry into the new RSU group, we also exclude the platoon key update phase in Section 4.1.4 of the SPMSA from this point onwards.

For reference, the PASAD scheme in [9] is executed through seven algorithms. We shall provide a brief description of each algorithm as follows:

Algorithm 1: System initialization that is conducted by the Center of Authority (CA) to generate the common parameters for the TRSUs and RSUs to register the vehicles.

Algorithm 2: Generation of the first private key for a vehicle by a TRSU.

Algorithm 3: Generation of a signature by a vehicle to prove its unique existence.

Algorithm 4: RSU ensures no double-registrations of vehicles entering its group by verifying the signature provided by the entering vehicle.

Algorithm 5: Group initialization by TRSU or RSU to generate the local group parameters as well as the vehicles' secondary private key.

Algorithm 6: A vehicle that has joined a group generates a signature for issuing event-reporting messages.

Algorithm 7: Verification of signatures by the vehicles which have received a specific event from another vehicle within the group.

6.1. Computational Overheads

The various cryptographic operations that comprise the SPMSA and PASAD are estimated to find the computational delay of these schemes. We adopt the evaluation method in [27] using the following parameters in our experiments including an Intel Core i3 2.4-GHz processor with MIRACL and Crypto++ libraries. Table 1 summarizes the execution times of the cryptographic operations.

Table 1. Execution Time of Cryptographic Operations.

Notation	Description	Execution Time (ms)
$Pair$	Bilinear Pairing Operation	23.625
Exp	Exponentiation Operation	3.3421
Mul	Scalar Multiplication	1.258
$Hash$	SHA256 Hash Function	0.005
$ECIES$	Operation of ECIES	4.35
$ECDSA-S$	Signing Operation of ECDSA	3.01
$ECDSA-V$	Verifying Operation of ECDSA	8.89

To calculate the computational delay of the two schemes, we only consider time-consuming operations that are not involved in the initialization stages. Hence, Section 4.1.1 of the SPMSA and Algorithm 5 of PASAD were excluded. Low computational modular arithmetic operations such as addition and subtraction were also excluded. The computational delay for the SPMSA scheme is calculated in (10):

$$T_{SPMSA} = 4T_{ECIES} + 4T_{ECDSA-S} + 4T_{ECDSA-V} + 6T_{Mul} + 2T_{Hash} = 72.558 \text{ ms} \tag{10}$$

We assume a best-case scenario where there are no invalid signatures i.e., no Sybil nodes. The computational costs for Algorithms 4, 6 and 7 of PASAD are as follows:

$$T_{PASAD-Alg4} = 2T_{Mul} + 4T_{Pair} \tag{11}$$

$$T_{PASAD-Alg6} = 8T_{Exp} + 7T_{Mul} + T_{Pair} + T_{Hash} \tag{12}$$

$$T_{PASAD-Alg7-Best} = 9nT_{Exp} + (9n+2)T_{Mul} + (2n+4)T_{Pair} \tag{13}$$

To create a similar environment to compare with the SPMSA, we assume a minimalist VANET system for PASAD where two vehicles communicate in isolation as in the SPMSA. Hence, $n = 1$, and the total computation overhead of PASAD is

$$T_{PASAD} = T_{PASAD-Alg4} + T_{PASAD-Alg6} + T_{PASAD-Alg7-Best} = 323.1189 \text{ ms} \tag{14}$$

Comparing (10) and (14) shows that the SPMSA has an estimated lower time complexity than that of PASAD.

6.2. Communication Overheads

We estimate the communication delay of the two schemes by calculating the total sum of transmission and propagation delays of each. The transmission delay is the amount of time to transmit the packets of data onto the transmission medium. It can be determined by L/R, where L is the size of a data packet in bits and R is the data transmission rate in bits per second (bps). We can assume that the data rates of V2V and V2I communication, i.e., DSRC and 5G, to be 6 and 50 Mbps, respectively.

To determine the transmission delay, we first approximate the amount of memory required to run the two schemes. The keys and hashes used in the SPMSA scheme require 256 bits of data each as the secp256r1 and SHA256 protocols have been used as the initial parameters of the elliptic curve and hash function, respectively. In addition, the MACs attached to the encrypted messages also take up 256 bits each. Meanwhile, plaintext messages and timestamps require 32 bits each, while each ECDSA signature uses up 512 bits. Lastly, an additional parity bit is needed for each key and signature. It culminates in a data size of 5642 bits being transmitted by the SPMSA scheme. Meanwhile, assuming that PASAD also uses the ECIES for its symmetric encryption, it requires an estimated 5324 bits to be transmitted [9].

On the other hand, propagation delay is the amount of time for the packets of data to reach the destination over the physical medium. The formula for the propagation delay is thus D/S, where D is the physical distance between the two vehicles and S is the propagation speed of the communication link. On average, we assume that a single V2V link covers a distance of 50 m while a V2I link covers about 200 m. In both cases, the communication is wireless, so a common propagation speed of the speed of light (3×10^8 m/s) can be assumed. For each scheme, the number of transmissions that occur using DRSC and 5G technology are summed up to determine their respective propagation delays. We ignore queuing and processing delays as they are dependent on several factors that we cannot predict well. The communication delays of the two schemes as well as that of another relevant scheme by Santhosh et al. [8] can be found in Table 2.

Table 2. Comparison of the Communication Delays of Schemes.

	PASAD	SPMSA	Santhosh [8]
Transmitted Data (bits)	5324 (DSRC)	5354 (DSRC) 288 (5G)	8192 (DSRC)
Transmission Delay (ms)	0.88733	0.89809	1.36533
Number of Transmissions	3 V2I (DSRC)	6 V2V (DSRC) 2 V2I (5G)	6 V2V (DSRC) 2 V2I (DSRC)
Propagation Delay (ms)	0.00200	0.00233	0.00233
Total Communication Delay (ms)	0.88933	0.90042	1.36767

It can be seen that the propagation delay makes a tiny contribution to the overall communication delay for all schemes. Instead, the communication overhead is predominantly determined by the transmission delay. Both PASAD and the SPMSA performed considerably better than the scheme by Santhosh et al.

Focusing only on PASAD and the SPMSA, they portrayed comparable communication delays despite the SPMSA requiring a couple hundred more bits to be transmitted than PASAD. However, since their respective communication delays are relatively insignificant when compared with the computational overheads for both schemes, the computational overhead will be the dominant factor in the execution times of the schemes.

6.3. Performance Comparison by Simulations

In this section, we evaluate the performance of the SPMSA using simulations. First, we observe its performance under some unknown attacks and compare it with that of PASAD. The simulation is built on MATLAB software. A known attack is an attack that should have been picked up by the SPMSA or by PASAD. In contrast, an unknown attack is one that was not analyzed or claimed previously. Intuitively, numerous unknown attacks could be launched at any time. They could break the execution of a protocol, and if they did so, they would likely break it at different points of execution. The probability that they will do either is difficult to predict with full certainty. Thus, we model these two processes as independent random processes with uniform probability. In turn, the objective of this simulation is to predict the negative effects to the performance of the system of unknown attacks.

In the simulation, one million attacks were launched for each specified ratio of the unknown attacks to the total attacks, with ratio ranging from 0 to 0.8 in increments of 0.1. An unknown attack has a uniform probability of breaking the scheme at a random step in the execution of the protocol. Note that the execution time of a protocol is defined as the sum of the computational overhead and communication delay that has been calculated beforehand in Sections 6.1 and 6.2. If a protocol can resist an attack, we consider the scheme successful. However, when an unknown attack breaks the scheme, only the execution time up till the point the scheme stops running is recorded. Subsequently, it is not deemed a successful run. For a given ratio of unknown attack:

$$Average\ Execution\ Time = \frac{Total\ Execution\ Time\ After\ 1\ Million\ Attacks}{Number\ of\ Successful\ Runs\ of\ Scheme} \tag{15}$$

The average execution times of the SPMSA and PASAD at each ratio of unknown attacks is plotted in Figure 15. When the ratio of unknown attacks is 0, it represents both the execution time as well as the communication reconnection time of the schemes in the event of disturbances such as channel interference that was discussed in Section 1. There is an exponential increase in the execution/reconnection times of both schemes that is to be expected when there are more unknown attacks obstructing the schemes from completion. More importantly, our SPMSA is less time-consuming than PASAD regardless of the ratio of the unknown attacks that appear, peaking at 155.4 ms while achieving a similar major goal of preventing Sybil attacks.

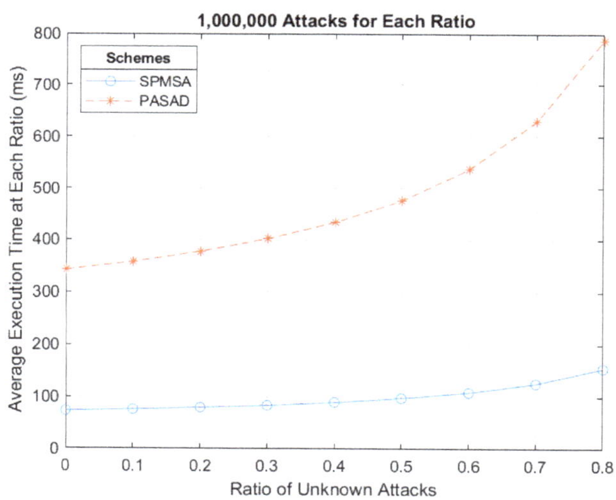

Figure 15. Performance of the Two Schemes against Unknown Attacks.

We conducted a second simulation experiment to evaluate the performance of the SPMSA when Sybil nodes were present in the VANETs. We compared the performance with regards to the time it takes to authenticate all honest vehicles with that of PASAD. Let n represent the number of vehicles intending to join the platoon, while x represents the number of Sybil vehicles among these n vehicles. We assume an average-case scenario of PASAD this time around. Hence, the equation in (13) becomes (16), and the total computation cost of PASAD is seen in (17):

$$T_{PASAD-Alg7-Avg} = 9nT_{Exp} + \left(3n\log x + x\log\left(\tfrac{n}{x}\right) + 11n - x + \tfrac{n(n+1)}{2}\right)T_{Mul} + \left(2x\log\left(\tfrac{n}{x}\right) + 4x + n(n+1)\right)T_{Pair} \quad (16)$$

$$T_{PASAD-Avg} = T_{PASAD-Alg4} + T_{PASAD-Alg6} + T_{PASAD-Alg7-Avg} \quad (17)$$

Since the SPMSA discards the message and session of a detected Sybil vehicle, only the time taken to reach such an occurrence was recorded by the scheme as the authentication time of a Sybil vehicle. In contrast, in the case of the authentication of an honest vehicle, the time taken for a full run of the SPMSA was recorded. For both schemes, we assume that at least one honest node is present in the VANET such that $n - x \geq 1$. In this simulation, we set n to be 9 throughout and vary the number of Sybil vehicles x from 1 to 8. Thus, there will be eight independent runs of the simulation to outline the authentication time cost of each scheme when the number of Sybil vehicles varies. For example, in the first run, there is $x = 1$ Sybil vehicle among the $n = 9$ vehicles that would like to join the platoon. The total time taken to authenticate all 9 vehicles is cumulatively summed up and denoted by T_{Auth}. To calculate the average authentication time $\overline{T_{Auth}}$, the cumulative authentication time is divided by the number of honest vehicles, as seen in (18). This process will then be repeated for $x = 2, 3, \ldots, 8$ for both schemes.

$$\overline{T_{Auth}} = \frac{T_{Auth}}{n - x} \quad (18)$$

The average authentication times for the two schemes for a varying number of Sybil vehicles is plotted in Figure 16. As more Sybil vehicles are added, the difference in performance of the two schemes in terms of authentication time cost grows greater. The performance achieved by PASAD is not a surprise as its computational delay increases considerably when there are many invalid signatures, as mentioned in Section 2.3. More notably, the number of Sybil vehicles in the VANET is not a factor in determining which scheme is faster in authenticating honest vehicles as our scheme has been shown to be consistently faster. At its peak, PASAD takes roughly 3.6 s. In comparison, the SPMSA only consumes approximately 0.34 s.

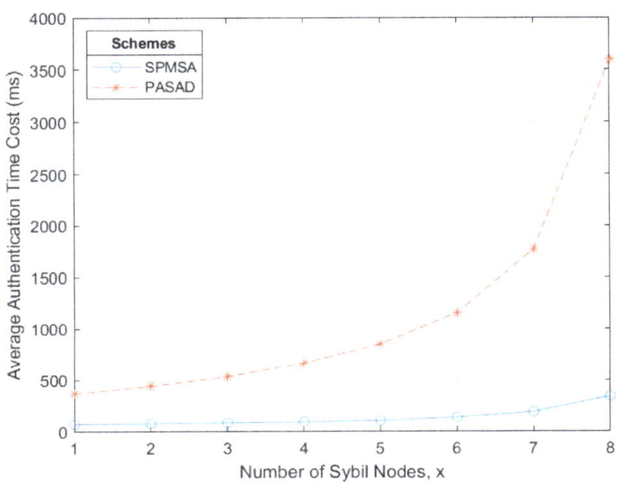

Figure 16. Performance of the Two Schemes in the Presence of Sybil Nodes.

7. Conclusions

In this paper, we have proposed a secure management scheme for platoon access that is resistant to Sybil attacks using elliptic curves. The SPMSA can achieve both identity and message authentication between a platoon leader and a vehicle intending to join the platoon and maintain message authentication throughout the vehicle's tenure in the platoon. The security functionality of the proposed SPMSA was then proven in the CK adversarial model with the random oracle model as well as with the CryptoVerif protocol verifier. We also conducted a qualitative analysis of the scheme's security to show its security features including perfect forward secrecy and both group forward and backward secrecy. Finally, we evaluated the performance of the proposed scheme with numerical analysis and simulation experiments to show its time efficiency in the face of unknown attacks and the minimal resource costs when Sybil vehicles are present. Future work is expected to explore the authentication of a new platoon leader when the existing leader intends to leave the platoon. As evidenced by the proposed scheme, the platoon leader carries important information that needs to be handed over to the right vehicle in a secure manner.

Author Contributions: Conceptualization, D.R.J. and M.M.; methodology, D.R.J.; software, D.R.J.; validation, M.M.; formal analysis, D.R.J.; investigation, D.R.J.; resources, M.M.; data curation, D.R.J.; writing—original draft preparation, D.R.J.; writing—review and editing, D.R.J. and M.M.; visualization, D.R.J.; supervision, M.M.; project administration, R.S.; funding acquisition, M.M. and R.S. All authors have read and agreed to the published version of the manuscript.

Funding: This research is funded by A*STAR under its RIE2020 Advanced Manufacturing and Engineering (AME) Industry Alignment Fund–Pre Positioning (IAF-PP) (Award A19D6a0053). Any opinions, findings and conclusions or recommendations expressed in this material are those of the author(s) and do not reflect the views of A*STAR. The APC is funded by the Guest Editor, Peter Chong.

Institutional Review Board Statement: Not applicable.

Informed Consent Statement: Not applicable.

Data Availability Statement: The data presented in this study are available in the article.

Conflicts of Interest: The authors declare no conflict of interest.

References

1. Boeira, F.; Barcellos, M.P.; de Freitas, E.P.; Vinel, A.; Asplund, M. On the Impact of Sybil Attacks in Cooperative Driving Scenarios. In Proceedings of the 2017 IFIP Networking Conference and Workshops, Stockholm, Sweden, 12–16 June 2017. [CrossRef]
2. Boeira, F.; Barcellos, M.P.; de Freitas, E.P.; Vinel, A.; Asplund, M. Effects of colluding Sybil nodes in message falsification attacks for vehicular platooning. In Proceedings of the 2017 IEEE Vehicular Networking Conference (VNC), Torino, Italy, 27–29 November 2017. [CrossRef]
3. Solyom, S.; Coelingh, E. Performance Limitations in Vehicle Platoon Control. *IEEE Intell. Transp. Syst. Mag.* **2013**, *5*, 112–120. [CrossRef]
4. Vahidi, A.; Eskandarian, A. Research advances in intelligent collision avoidance and adaptive cruise control. *IEEE Trans. Intell. Transp. Syst.* **2003**, *4*, 143–153. [CrossRef]
5. Samara, G.; Al-Raba'nah, Y. Security Issues in Vehicular Ad Hoc Networks (VANET): A survey. *Int. J. Sci. Appl. Res.* **2015**, *2*, 50–55. [CrossRef]
6. Sarker, A.; Qiu, C.; Shen, H. Connectivity Maintenance for Next-Generation Decentralized Vehicle Platoon Networks. *IEEE ACM Trans. Netw.* **2020**, *28*, 1449–1462. [CrossRef]
7. Rabieh, K.; Mahmoud, M.M.; Guo, T.N.; Younis, M. Cross-Layer Scheme for Detecting Large-scale Colluding Sybil attack in VANETs. In Proceedings of the 2015 IEEE International Conference on Communications (ICC), London, UK, 8–12 June 2015. [CrossRef]
8. Santhosh, J.; Sankaran, S. Defending against Sybil Attacks in Vehicular Platoons. In Proceedings of the 2019 IEEE International Conference on Advanced Networks and Telecommunications Systems (ANTS), Goa, India, 16–19 December 2019. [CrossRef]
9. Parham, M.; Pouyan, A.A. An Effective Privacy-Aware Sybil Attack Detection Scheme for Secure Communication in Vehicular Ad Hoc Network. *Wirel. Pers. Commun.* **2020**, *113*, 1149–1182. [CrossRef]
10. Soni, M.; Jain, A. Secure Communication and Implementation Technique for Sybil Attack in Vehicular Ad-Hoc Networks. In Proceedings of the 2018 Second International Conference on Computing Methodologies and Communication (ICCMC), Erode, India, 15–16 February 2018. [CrossRef]

11. Kushwah, R.; Kulshreshtha, A.; Singh, K.; Sharma, S. ECDSA for Data Origin Authentication and Vehicle Security in VANET. In Proceedings of the 2019 Twelfth International Conference on Contemporary Computing (IC3), Noida, India, 8–10 August 2019. [CrossRef]
12. Bochem, A.; Leiding, B.; Hogrefe, D. Unchained identities: Putting a price on sybil nodes in mobile ad hoc networks. In Proceedings of the International Conference on Security and Privacy in Communication Systems, Singapore, 8–10 August 2018; pp. 358–374. [CrossRef]
13. Bochem, A.; Leiding, B. Rechained: Sybil-resistant distributed identities for the Internet of Things and mobile ad hoc networks. *Sensors* **2021**, *21*, 3257. [CrossRef] [PubMed]
14. Liu, X.; Luo, B.; Abdo, A.; Abu-Ghazaleh, N.; Zhu, Q. Securing Connected Vehicle Applications with an Efficient Dual Cyber-Physical Blockchain Framework. In Proceedings of the 2021 IEEE Intelligent Vehicles Symposium (IV), Nagoya, Japan, 11–17 July 2021. [CrossRef]
15. Didouh, A.; Lopez, A.B.; Hillali, Y.E.; Rivenq, A.; Faruque, M.A.A. Eve, You Shall Not Get Access! A Cyber-Physical Blockchain Architecture for Electronic Toll Collection Security. In Proceedings of the 2020 IEEE 23rd International Conference on Intelligent Transportation Systems (ITSC), Rhodes, Greece, 20–23 September 2020. [CrossRef]
16. Gu, P.; Khatoun, R.; Begriche, Y.; Serhrouchni, A. Support Vector Machine (SVM) Based Sybil Attack Detection in Vehicular Networks. In Proceedings of the 2017 IEEE Wireless Communications and Networking Conference (WCNC), San Francisco, CA, USA, 19–22 March 2017. [CrossRef]
17. Quevedo, C.H.O.O.; Quevedo, A.M.B.C.; Campos, G.A.; Gomes, R.L.; Celestino, J.; Serhrouchni, A. An Intelligent Mechanism for Sybil Attacks Detection in VANETs. In Proceedings of the ICC 2020—2020 IEEE International Conference on Communications (ICC), Dublin, Ireland, 7–11 June 2020. [CrossRef]
18. Mohanti, S.; Soltani, N.; Sankhe, K.; Jaisinghani, D.; Di Felice, M.; Chowdhury, K. AirID: Injecting a custom RF fingerprint for enhanced UAV identification using deep learning. In Proceedings of the GLOBECOM 2020—2020 IEEE Global Communications Conference, Taipei, Taiwan, 7–11 December 2020. [CrossRef]
19. Reus-Muns, G.; Jaisinghani, D.; Sankhe, K.; Chowdhury, K.R. Trust in 5G Open RANs through Machine Learning: RF Fingerprinting on the POWDER PAWR Platform. In Proceedings of the GLOBECOM 2020—2020 IEEE Global Communications Conference, Taipei, Taiwan, 7–11 December 2020. [CrossRef]
20. Comert, C.; Kulhandjian, M.; Gul, O.M.; Touazi, A.; Ellement, C.; Kantarci, B.; D'Amours, C. Analysis of Augmentation Methods for RF Fingerprinting under Impaired Channels. In Proceedings of the 2022 ACM Workshop on Wireless Security and Machine Learning (WiseML'22), San Antonio, TX, USA, 19 May 2022; Association for Computing Machinery: New York, NY, USA, 2022; pp. 3–8. [CrossRef]
21. Canetti, R.; Krawczyk, H. Analysis of Key-Exchange Schemes and Their Use for Building Secure Channels. In Proceedings of the International Conference on the Theory & Application of Cryptographic Techniques, Innsbruck, Austria, 6–10 May 2001; Pfitzmann, B., Ed.; Springer: Berlin/Heidelberg, Germany, 2000; Volume 2045, pp. 453–474. [CrossRef]
22. Chen, W.-C.; Huang, Y.-T.; Wang, S.-D. Provable Secure Group Key Establishment Scheme for Fog Computing. *IEEE Access.* **2021**, *9*, 158682–158694. [CrossRef]
23. Bellare, M.; Pointcheval, D.; Rogaway, P. Authenticated Key Exchange Secure against Dictionary Attacks. In Proceedings of the International Conference on the Theory and Applications of Cryptographic Techniques, Bruges, Belgium, 14–18 May 2000; Preneel, B., Ed.; Springer: Berlin/Heidelberg, Germany, 2000; Volume 1807, pp. 139–155. [CrossRef]
24. Martínez, V.G.; Encinas, L.H.; Dios, A.Q. Security and Practical Considerations When Implementing the Elliptic Curve Integrated Encryption Scheme. *Cryptologia* **2015**, *39*, 244–269. [CrossRef]
25. Blanchet, B. *CryptoVerif: A Computationally-Sound Security Protocol Verifier*; Techical Report; Centre Inria de Paris: Paris, France, 2017.
26. Blanchet, B.; Cadé, D. *CryptoVerif Computationally Sound, Automatic Cryptographic Protocol Verifier User Manual*; User Manual; Centre Inria de Paris: Paris, France, 2021.
27. Pan, J.; Cui, J.; Wei, L.; Xu, Y.; Zhong, H. Secure data sharing scheme for VANETs based on edge computing. *J. Wirel. Com. Netw.* **2019**, *2019*, 169. [CrossRef]

sensors

MDPI

Article

Data-Aided SNR Estimation for Bandlimited Optical Intensity Channels

Wilfried Gappmair

Institute of Communication Networks and Satellite Communications, Graz University of Technology, Inffeldgasse 12, 8010 Graz, Austria; gappmair@tugraz.at

Abstract: Not only for radio frequency but also for optical communication systems, knowledge of the signal-to-noise ratio (SNR) is essential, e.g., for an adaptive network, where modulation schemes and/or error correction methods should be selected according to the varying channel states. In the current paper, this topic is discussed for a bandlimited optical intensity link under the assumption that the data symbols are known to the receiver unit in form of pilot sequences. This requires a unipolar signal design regarding the symbol constellation, but also a non-negative pulse shape satisfying the Nyquist criterion is necessary. Focusing on this kind of scenario, the modified Cramer–Rao lower bound is derived, representing the theoretical limit of the error performance of the data-aided SNR estimator developed in this context. Furthermore, we derive and analyze a maximum likelihood algorithm for SNR estimation, which turns out to be particularly simple for specific values of the excess bandwidth, among them the most attractive case of minimum bandwidth occupation. Numerical results confirming the analytical work conclude the paper.

Keywords: SNR estimation; optical wireless communications; intensity modulation

Citation: Gappmair, W. Data-Aided SNR Estimation for Bandlimited Optical Intensity Channels. *Sensors* **2022**, *22*, 8660. https://doi.org/10.3390/s22228660

Academic Editor: Peter Han Joo Chong

Received: 20 September 2022
Accepted: 4 November 2022
Published: 9 November 2022

1. Introduction

In a series of papers recently published by the author [1–3], parameter estimation and synchronization for a bandlimited optical intensity link have been discussed. In this context, a unipolar waveform design is indispensable with respect to pulse shaping and symbol constellation [4,5]. Furthermore, it is most helpful that pulse shapes satisfy the Nyquist criterion, which allows for a simple detection process in the receiver unit [6].

However, not only for radio frequency (RF) but also for optical wireless communication (OWC) solutions [7–10], the relevant transmission parameters have to be recovered reliably, because otherwise subsequent receiver stages, such as error correction algorithms, cannot be operated in an efficient way [11,12]. In particular, recovery of the symbol timing is of paramount importance in this respect, since this is a prerequisite for many other estimation and detection procedures. In [1–3], it has been shown how this might be achieved for a bandlimited optical intensity link under different conditions, e.g., whether data are known to the receiver unit or not in the form of pilot sequences, or if the estimator or synchronizer module is to be implemented in a feedforward or feedback manner.

Usually, the estimation of the signal-to-noise ratio (SNR) requires that the symbol timing has been established before by a properly selected algorithm. It is to be recalled that knowledge of the SNR is normally needed for adaptive communication systems to select modulation and coding schemes according to the given channel conditions [13], but also powerful error correction methods—such as turbo or LDPC algorithms—need this sort of information [14]. Scanning the open literature, numerous papers are available about SNR estimation in RF channels, e.g., the frequently cited overview by Pauluzzi and Beaulieu [15], but little or no information is published for OWC systems. This has been the main motivation of the current contribution addressing data-aided SNR estimation for a bandlimited optical intensity channel. Finally, it is to be noticed that the article was prepared for a Special Issue of Feature Papers 2022 in the Communications Section of MDPI Sensors.

The rest of the paper is organized as follows: the signal and channel model for analytical and simulation work is introduced in Section 2, whereas Section 3 focuses on the derivation of the Cramer–Rao lower bound (CRLB) as the theoretical limit of the jitter performance for any algorithm discussed in the context of SNR estimation. In Section 4, we derive a maximum likelihood (ML) algorithm and analyze it in terms of mean value and variance. Numerical results are shown in Section 5, and Section 6 concludes the paper.

2. Signal and Channel Model

As already mentioned in the introductory section, properly selected pulse shapes and modulation schemes are necessary to satisfy the non-negativity as well as the Nyquist constraints required for a bandlimited optical intensity channel. In this respect, it is assumed that the real-valued data symbols a_k, $k \in \mathbb{Z}$, are independent and identically distributed (i.i.d.) elements of an M-ary PAM alphabet $\mathcal{A}$. It makes sense to organize the alphabet such that the symbols are normalized to unit energy, i.e., $\mathbb{E}[a_k^2] = 1$, where $\mathbb{E}[\cdot]$ denotes the expectation operator. Then, with $\eta_M = \frac{1}{6}(M-1)(2M-1)$, we have $a_k \in \mathcal{A} = \frac{1}{\sqrt{\eta_M}}\{0, 1, \ldots, M-1\}$. This means that the average value is given by

$$\mu_a = \mathbb{E}[a_k] = \frac{1}{\sqrt{\eta_M}} \frac{M-1}{2} = \sqrt{\frac{3\,(M-1)}{2\,(2M-1)}} \tag{1}$$

On the other hand, the signal at the output of the opto-electrical receiver module is obtained as

$$r(t) = A \sum_k a_k\, h(t - kT - \tau) + w(t) \tag{2}$$

where $A > 0$ is the channel gain, $h(t)$ describes the pulse shape, T and τ symbolize the symbol period and the propagation delay between receiver and transmitter station, respectively. We assume that A is a constant regarding the observation interval used for estimation purposes, because variations of the channel state are normally slow enough so that fading effects need not be taken into account. As already required previously, $h(t)$ must satisfy the non-negativity as well as the Nyquist criterion, e.g., achieved by a squared raised cosine function [1,6]. In line with the investigations carried out in [1–6], the receiver signal in (2) is also assumed to be distorted by additive white Gaussian noise (AWGN), in the following expressed by $w(t)$, with zero mean and variance σ_w^2.

In addition, we introduce the average optical power as $P_0 = \mu_a\, \overline{h}$, where

$$\overline{h} = \frac{1}{\sqrt{T}} \int_{-\infty}^{\infty} h(t)\, dt \tag{3}$$

so that the average electrical SNR at the receiver can be defined as

$$\gamma_s = \frac{A^2 P_0^2}{\sigma_w^2} \tag{4}$$

However, before being treated in further stages of operation, the signal in (2) has to pass the receiver filter $q(t)$, whose output is given by $z(t) = q(t) \otimes r(t)$, where $\otimes$ denotes the convolutional operator. For convenient reasons, this is summarized in Figure 1.

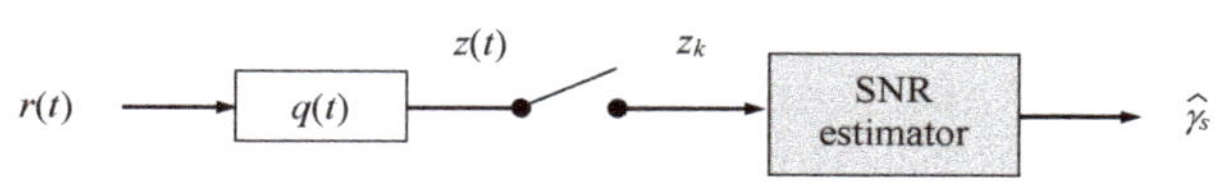

Figure 1. Signal model for SNR estimation.

Since there exists no simple solution for a matched filter structure in the context of a bandlimited optical intensity link [6], it is suggested that $q(t)$ exhibits a flat behavior over the spectrum occupied by the user component in (2). This straightforward approach guarantees that the waveform will not be distorted, but the price to be paid is an increased amount of noise the subsequent receiver stages have to cope with. In particular, this means that the transfer function of the filter performs a rectangular shape in the frequency domain, i.e., $Q(f) = \mathcal{F}[q(t)] = \sqrt{T}$ for $|f| \leq (1+\alpha)/T$ and $Q(f) = 0$ elsewhere, with α as the roll-off factor (excess bandwidth) of the selected pulse shape; recall that $\alpha = 0$ indicates the minimum bandwidth scenario. The related impulse response is then furnished by application of the inverse Fourier transform [16], i.e.,

$$q(t) = \mathcal{F}^{-1}[Q(f)] = \frac{2(1+\alpha)}{\sqrt{T}} \text{sinc}[2(1+\alpha)t/T] \tag{5}$$

with $\text{sinc}(x) = \sin(\pi x)/(\pi x)$. Of course, the noise signal at this flat filter output develops as $n(t) = w(t) \otimes q(t)$ representing a zero-mean non-white Gaussian process. Assuming in the next step that the symbol timing has been reliably recovered and corrected, e.g., by the algorithm proposed in [1], the T-spaced samples at the output of the receiver filter are obtained as

$$z_k = z(kT) = A \cdot a_k - n_k \tag{6}$$

where $\mathbb{E}[n_k] = 0$ and $\mathbb{E}[n_i\, n_k] = 2(1+\alpha)\, \sigma_w^2\, \text{sinc}[2(1+\alpha)(i-k)]$.

3. Cramer–Rao Lower Bound

3.1. Derivation of the Log-Likelihood Function

The Cramer–Rao lower bound (CRLB) is a major figure of merit when it comes to the estimation of a parameter [17]. The reason behind this is the fact that the bound represents the theoretical limit of the jitter (error) variance of any estimator developed in this context.

According to the signal model specified previously, we have to consider the parameter vector $\mathbf{u} = (u_1, u_2) = (A, \sigma_w)$. The CRLB for u_i is determined by

$$\text{CRLB}(u_i) = [\mathbf{J}^{-1}(\mathbf{u})]_i \tag{7}$$

where $[\cdot]_i$ indicates the i-th diagonal entry of the inverted Fisher information matrix (FIM) expressed by $\mathbf{J}(\mathbf{u})$. In the case that no nuisance parameter needs to be taken into account, the FIM entries are computed as

$$J_{i,k} \equiv [\mathbf{J}(\mathbf{u})]_{i,k} = -\mathbb{E}\left[\frac{\partial^2 \Lambda(\mathbf{z};\mathbf{u})}{\partial u_i\, \partial u_k}\right] \tag{8}$$

with $\mathbf{z}$ as the given vector of observables, $\Lambda(\mathbf{z};\mathbf{u})$ denotes the log-likelihood function (LLF) characterizing the communication link, and $\mathbb{E}[\cdot]$ symbolizes expectation with respect to the noise model.

By inspection of (8), it is clear that the computation of the CRLB requires the knowledge of the LLF describing the subject of investigation. To this end, we assume that a sequence of L receiver samples (6) forms the vector $\mathbf{z}$ expressed by

$$\mathbf{z} = A \cdot \mathbf{a} - \mathbf{n} \tag{9}$$

The vector $\mathbf{a}$ of known data symbols specifies the pilot sequence, which is to be used for estimation purposes in the sequel, and $\mathbf{n}$ denotes the noise vector with covariance matrix

$$\mathbf{R} = \mathbb{E}[\mathbf{n} \cdot \mathbf{n}^T] = 2(1+\alpha)\, \sigma_w^2\, \boldsymbol{\Omega} \tag{10}$$

where the entries for line i and column k of $\mathbf{\Omega}$ are given by $\omega_{ik} = \text{sinc}[2(1+\alpha)(i-k)] = \omega_{ki}$ forming this way a symmetric Toeplitz matrix [18]. As a result, the likelihood function for our estimation problem can be written as [19,20],

$$\Pr(\mathbf{z};\mathbf{u}) = \frac{1}{\sqrt{(2\pi)^L \det(\mathbf{R})}} e^{-\frac{1}{2}\,(\mathbf{z}-A\,\mathbf{a})^T\mathbf{R}^{-1}(\mathbf{z}-A\,\mathbf{a})} \tag{11}$$

However, instead of using $\mathbf{u} = (A, \sigma_w)$, it is easier to concentrate in the following on the average electrical SNR normalized by P_0^2, i.e., $\rho_s = \gamma_s / P_0^2 = A^2/\sigma_w^2$. Then, by introduction of $P_n = \sigma_w^2$, we have that $\mathbf{u} = (\rho_s, P_n)$ and the related LLF is furnished by

$$\Lambda(\mathbf{z};\mathbf{u}) = \log \Pr(\mathbf{z};\mathbf{u}) \sim -\frac{L}{2}\log P_n - \frac{\mathbf{z}^T\mathbf{\Psi}\,\mathbf{z} - 2\sqrt{\rho_s P_n}\,\mathbf{z}^T\mathbf{\Psi}\,\mathbf{a} + \rho_s P_n\,\mathbf{a}^T\mathbf{\Psi}\,\mathbf{a}}{4(1+\alpha)P_n} \tag{12}$$

which has been achieved by $\mathbf{\Psi} = \mathbf{\Omega}^{-1}$ as well as omitting all immaterial constants and factors not depending on $\mathbf{u}$.

3.2. Modified Cramer–Rao Lower Bound

In the next step, the FIM entries are obtained by computing the second-order derivatives according to (8), the results of which have then to be averaged with respect to the noise vector $\mathbf{n}$. However, this approach means that the CRLB will be a function of the selected pilot sequence $\mathbf{a}$. Therefore, it is suggested to extend the averaging procedure to $\mathbf{a}$ as well, which creates the so-called modified Cramer–Rao lower bound (MCRLB) [21–23]. Doing so, we get after some algebra:

$$J_{11} = -\mathbb{E}\left[\frac{\partial^2\Lambda(\mathbf{z};\mathbf{u})}{\partial\rho_s^2}\right] = \frac{1}{8(1+\alpha)\rho_s}\mathbb{E}_{\mathbf{a}}[\mathbf{a}^T\mathbf{\Psi}\,\mathbf{a}] \tag{13}$$

$$J_{22} = -\mathbb{E}\left[\frac{\partial^2\Lambda(\mathbf{z};\mathbf{u})}{\partial P_n^2}\right] = -\frac{L}{2P_n^2} + \frac{\rho_s}{8(1+\alpha)P_n^2}\mathbb{E}_{\mathbf{a}}[\mathbf{a}^T\mathbf{\Psi}\,\mathbf{a}] + \frac{1}{2(1+\alpha)P_n^3}\mathbb{E}_{\mathbf{n}}[\mathbf{n}^T\mathbf{\Psi}\,\mathbf{n}] \tag{14}$$

$$J_{12} = J_{21} = -\mathbb{E}\left[\frac{\partial^2\Lambda(\mathbf{z};\mathbf{u})}{\partial\rho_s\partial P_n}\right] = \frac{1}{8(1+\alpha)P_n}\mathbb{E}_{\mathbf{a}}[\mathbf{a}^T\mathbf{\Psi}\,\mathbf{a}] \tag{15}$$

Evaluating (7) for $u_1 = \rho_s$, the corresponding MCRLB is given by

$$\text{MCRLB}(\rho_s) = \frac{J_{22}}{J_{11}J_{22} - J_{12}^2} \tag{16}$$

Substituting (13)–(15) into (16) and scaling the result with respect to ρ_s^2, we obtain the normalized MCRLB expressed as

$$\begin{aligned}\text{NMCRLB}(\rho_s) &= \frac{\text{MCRLB}(\rho_s)}{\rho_s^2} \\ &= 2(1+\alpha)\left(\frac{P_n}{\mathbb{E}_{\mathbf{n}}[\mathbf{n}^T\mathbf{\Psi}\,\mathbf{n}] - L(1+\alpha)P_n} + \frac{4}{\rho_s\,\mathbb{E}_{\mathbf{a}}[\mathbf{a}^T\mathbf{\Psi}\,\mathbf{a}]}\right)\end{aligned} \tag{17}$$

Introducing the auxiliary terms

$$\overline{\Psi}_0 = \frac{1}{L}\sum_{i=0}^{L-1}\psi_{ii},\ \overline{\Psi}_1 = \frac{1}{L}\sum_{i=0}^{L-1}\sum_{k=i+1}^{L-1}\psi_{ik},\ \overline{\Psi}_2 = \frac{1}{L}\sum_{i=0}^{L-1}\sum_{k=0}^{L-1}\omega_{ik}\,\psi_{ik} \tag{18}$$

where ψ_{ik} is the entry of $\boldsymbol{\Psi}$ indicating line i and column k, the expected operations in (17) can be written as

$$\begin{aligned}\mathbb{E}_{\mathbf{a}}[\mathbf{a}^T\boldsymbol{\Psi}\,\mathbf{a}] &= \sum_{i=0}^{L-1}\sum_{k=0}^{L-1}\mathbb{E}[a_i\,a_k]\,\psi_{ik} \\ &= \sum_{i=0}^{L-1}\mathbb{E}[a_i^2]\,\psi_{ii} + \sum_{i=0}^{L-1}\sum_{k=0,k\neq i}^{L-1}\mathbb{E}[a_i a_k]\,\psi_{ik} \\ &= \eta_a\sum_{i=0}^{L-1}\psi_{ii} + 2\mu_a^2\sum_{i=0}^{L-1}\sum_{k=i+1}^{L-1}\psi_{ik} = L(\overline{\Psi}_0 + 2\mu_a^2\overline{\Psi}_1)\end{aligned} \tag{19}$$

and

$$\begin{aligned}\mathbb{E}_{\mathbf{n}}[\mathbf{n}^T\boldsymbol{\Psi}\,\mathbf{n}] &= \sum_{i=0}^{L-1}\sum_{k=0}^{L-1}\mathbb{E}[n_i n_k]\,\psi_{ik} \\ &= \sigma_n^2\sum_{i=0}^{L-1}\sum_{k=0}^{L-1}\omega_{ik}\,\psi_{ik} = 2L(1+\alpha)P_n\,\overline{\Psi}_2\end{aligned} \tag{20}$$

Finally, by plugging (19) and (20) into (17), we have that

$$\mathrm{NMCRLB}(\rho_s) = \frac{2}{L}\left(\frac{1}{2\overline{\Psi}_2 - 1} + \frac{4(1+\alpha)}{\rho_s(\overline{\Psi}_0 + 2\mu_a^2\overline{\Psi}_1)}\right) \tag{21}$$

Nevertheless, the relationship might be simplified for $\alpha \in \left\{0,\ \frac{1}{2},\ 1\right\}$, because in this case $\omega_{ik} = \ \mathrm{sin\,c}[2(1+\alpha)(i-k)] = 1$ for $i = k$ and zero elsewhere. This means that $\boldsymbol{\Omega} = \boldsymbol{\Omega}^{-1} = \boldsymbol{\Psi} = \mathbf{I}_L$, with $\mathbf{I}_L$ as the L-dimensional identity matrix, which means also that $\overline{\Psi}_0 = \overline{\Psi}_2 = 1$ and $\overline{\Psi}_1 = 0$. Hence, the normalized bound boils down to

$$\mathrm{NMCRLB}(\rho_s) = \frac{2}{L}\left(1 + \frac{4\,(1+\alpha)}{\rho_s}\right) \tag{22}$$

4. Maximum Likelihood Estimation

4.1. Derivation of the Estimator Algorithm

By means of the LLF in (12), we are basically in the position to derive a maximum likelihood (ML) algorithm for SNR estimation. However, the SNR parameter is composed of two ingredients—channel gain A and the noise power P_n—the estimates of which are needed to compute the SNR estimate. This is simply achieved by substituting $\rho_s = A^2/P_n$ into (12), deriving the resulting LLF with respect to A and P_n, equating both relationships to zero and solving them analytically. Doing this, we obtain for $\mathbf{u} = (A, P_n)$

$$\left.\frac{\partial\Lambda(\mathbf{z};\mathbf{u})}{\partial A}\right|_{\mathbf{u}=\hat{\mathbf{u}}} = \frac{\mathbf{z}^T\boldsymbol{\Psi}\,\mathbf{a} - \hat{A}\,\mathbf{a}^T\boldsymbol{\Psi}\,\mathbf{a}}{2(1+\alpha)\hat{P}_n} = 0 \tag{23}$$

and

$$\left.\frac{\partial\Lambda(\mathbf{z};\mathbf{u})}{\partial P_n}\right|_{\mathbf{u}=\hat{\mathbf{u}}} = -\frac{L}{2\hat{P}_n} + \frac{\mathbf{z}^T\boldsymbol{\Psi}\,\mathbf{z} - 2\hat{A}\,\mathbf{z}^T\boldsymbol{\Psi}\,\mathbf{a} + \hat{A}^2\mathbf{a}^T\boldsymbol{\Psi}\,\mathbf{a}}{4(1+\alpha)\hat{P}_n^2} = 0 \tag{24}$$

Then, by introduction of $M_{aa} = \mathbf{a}^T\boldsymbol{\Psi}\,\mathbf{a}$, $M_{az} = \mathbf{z}^T\boldsymbol{\Psi}\,\mathbf{a}$, and $M_{zz} = \mathbf{z}^T\boldsymbol{\Psi}\,\mathbf{z}$, we find the estimates for channel gain and noise power in closed form:

$$\hat{A} = \frac{\mathbf{z}^T\boldsymbol{\Psi}\,\mathbf{a}}{\mathbf{a}^T\boldsymbol{\Psi}\,\mathbf{a}} = \frac{M_{az}}{M_{aa}} \tag{25}$$

$$\hat{P}_n = \frac{\mathbf{z}^T\boldsymbol{\Psi}\,\mathbf{z} - 2\hat{A}\,\mathbf{z}^T\boldsymbol{\Psi}\,\mathbf{a} + \hat{A}^2\mathbf{a}^T\boldsymbol{\Psi}\,\mathbf{a}}{2(1+\alpha)L} = \frac{1}{2(1+\alpha)L}\left(M_{zz} - \frac{M_{az}^2}{M_{aa}}\right) \tag{26}$$

By inspection of (25) and (26), it is clear that a viable solution is only achievable for $M_{aa} > 0$, i.e., a pilot sequence consisting of zero symbols only would not work. Finally, according to the invariance principle for ML estimates [24], the SNR solution is given by

$$\hat{\rho}_s = \frac{\hat{A}^2}{\hat{P}_n} \tag{27}$$

4.2. Probability Analysis

In this subsection, we want to derive the probability density function (PDF) of the SNR estimate in (27) and analyze it in terms of bias and variance. It turns out that this is possible in closed form for $\mathbf{\Psi} = \mathbf{I}_L$, i.e., $\alpha \in \left\{0,\ \frac{1}{2},\ 1\right\}$, whereas for other values of α, it is verified in Section 5 that the analytical results achieved with $\mathbf{\Psi} = \mathbf{I}_L$ are very close to the true ones obtained by numerical means.

By plugging $\mathbf{\Psi} = \mathbf{I}_L$ into (25), the estimate for the channel gain develops as

$$\hat{A} = \frac{\mathbf{z}^T\mathbf{a}}{\mathbf{a}^T\mathbf{a}} = \frac{(A\ \mathbf{a} + \mathbf{n})^T\mathbf{a}}{\mathbf{a}^T\mathbf{a}} = A + \frac{\mathbf{n}^T\mathbf{a}}{\mathbf{a}^T\mathbf{a}} = A + \frac{1}{M_{aa}}\sum_{k=0}^{L-1} a_k n_k \tag{28}$$

which means that $\hat{A}$ is a zero-mean Gaussian variate. Computing the variance of the latter, we have to consider that the noise samples n_k are zero-mean and i.i.d. in case that $\mathbf{\Psi} = \mathbf{I}_L$. Therefore,

$$\sigma_A^2 = \mathbb{E}[(\hat{A} - A)^2] = \frac{1}{M_{aa}^2}\mathbb{E}\left[\left(\sum_{k=0}^{L-1} a_k n_k\right)^2\right] = \frac{1}{M_{aa}^2}\mathbb{E}\left[\sum_{k=0}^{L-1} a_k^2 n_k^2\right] = \frac{\sigma_n^2}{M_{aa}} \tag{29}$$

where $\sigma_n^2 = \mathbb{E}[n_k^2] = 2(1+\alpha)\ \sigma_w^2$. The related PDF is then straightforwardly given by

$$f_A(\hat{A}) = \frac{1}{\sqrt{2\pi}\sigma_A} e^{-(\hat{A}-A)^2/2\sigma_A^2} \tag{30}$$

On the other hand, $Y = \hat{A}^2$ corresponds to a non-central Gamma variate [19] characterized by the distribution

$$f_Y(y) = \frac{1}{\sqrt{2\pi y}\,\sigma_A} e^{-(y+A^2)/2\sigma_A^2} \cosh\left(\frac{A\sqrt{y}}{\sigma_A^2}\right),\ y > 0 \tag{31}$$

If we consider in the next step the estimate of the noise power for $\mathbf{\Psi} = \mathbf{I}_L$, we have

$$\hat{P}_n = \frac{\mathbf{z}^T\mathbf{z} - 2\hat{A}\ \mathbf{z}^T\mathbf{a} + \hat{A}^2\mathbf{a}^T\mathbf{a}}{2(1+\alpha)L} \tag{32}$$

With $\mathbf{z} = A \cdot \mathbf{a} + \mathbf{n}$ and $\hat{A}$ determined by (28), the numerator in (32) simplifies to

$$(A\ \mathbf{a} + \mathbf{n})^T(A\ \mathbf{a} + \mathbf{n}) - 2\hat{A}\ (A\ \mathbf{a} + \mathbf{n})^T\mathbf{a} + \hat{A}^2\mathbf{a}^T\mathbf{a} = \mathbf{n}^T\mathbf{n} - \frac{(\mathbf{n}^T\mathbf{a})^2}{\mathbf{a}^T\mathbf{a}} \tag{33}$$

which represents a central Gamma variate with variance σ_n^2 and L—1 degrees of freedom [20,25]. Hence, by introduction of $X = \hat{P}_n$, the PDF of (32) can be written as

$$f_X(x) = \frac{1}{(2\sigma_x^2)^{\frac{L-1}{2}}\Gamma(\frac{L-1}{2})} x^{\frac{L-1}{2}-1} e^{-x/2\sigma_x^2},\ x \geq 0 \tag{34}$$

where $\sigma_x^2 = \frac{\sigma_n^2}{2(1+\alpha)L} = \frac{\sigma_w^2}{L}$. Employing in the following the PDFs in (31) and (34), the distribution of the SNR estimate is determined by (A2) derived in the Appendix A, i.e.,

$$f(\hat{\rho}_s) = \frac{K_0}{\sqrt{\hat{\rho}_s(\hat{\rho}_s+\beta)^L}} e^{-\lambda} {}_1F_1\left(\frac{L}{2}, \frac{1}{2}; \frac{\lambda\,\hat{\rho}_s}{\hat{\rho}_s+\beta}\right), \hat{\rho}_s > 0 \tag{35}$$

Regarding (A3) and (A4), the parameters β, λ, and K_0 are functions of the true SNR value denoted by ρ_s, the roll-off factor α, the observation length L, as well as the selected pilot sequence $\mathbf{a}$.

By means of the relationships (A9) and (A10) detailed in the Appendix A, we can specify the first- and second-order moments of (35) as follows:

$$\mathbb{E}[\hat{\rho}_s] = \int_0^\infty \hat{\rho}_s\, f(\hat{\rho}_s)\, d\hat{\rho}_s = \frac{L}{L-3}\left(\rho_s + \frac{2(1+\alpha)}{M_{aa}}\right) \tag{36}$$

$$\mathbb{E}[\hat{\rho}_s^2] = \int_0^\infty \hat{\rho}_s^2\, f(\hat{\rho}_s)\, d\rho_s = \frac{L^2}{(L-3)(L-5)}\left(\rho_s^2 + \frac{12(1+\alpha)\rho_s}{M_{aa}} + \frac{12(1+\alpha)^2}{M_{aa}^2}\right) \tag{37}$$

Therefore, bias and variance of $\hat{\rho}_s$, normalized by ρ_s and ρ_s^2, respectively, are given by

$$\mathrm{NBias}(\hat{\rho}_s) = \frac{\mathbb{E}[\hat{\rho}_s] - \rho_s}{\rho_s} = \frac{L}{L-3}\left(1 + \frac{2(1+\alpha)}{M_{aa}\rho_s}\right) - 1 \tag{38}$$

and

$$\begin{aligned} \mathrm{NVar}(\hat{\rho}_s) &= \frac{\mathbb{E}[\hat{\rho}_s^2] - \mathbb{E}^2[\hat{\rho}_s]}{\rho_s^2} \\ &= \frac{2L^2}{(L-3)^2(L-5)}\left(1 + \frac{4(1+\alpha)(L-2)}{M_{aa}\rho_s} + \frac{4(1+\alpha)^2(L-2)}{M_{aa}^2\rho_s^2}\right) \end{aligned} \tag{39}$$

Via M_{aa}, it is obvious that (38) and (39) depend on the selected pilot sequence. In order to avoid this drawback, we could average the relationships with respect to $\mathbf{a}$. The problem in this context is that there exists no closed form solution. A way out of this dilemma is Jensen's inequality [26] (Appendix 1B), which provides us with

$$\mathbb{E}[\frac{1}{M_{aa}}] \geq \frac{1}{\mathbb{E}[M_{aa}]} = \frac{1}{L\mathbb{E}[a_k^2]} = \frac{1}{L} \tag{40}$$

and

$$\mathbb{E}[\frac{1}{M_{aa}^2}] \geq \frac{1}{\mathbb{E}[M_{aa}^2]} = \frac{1}{L\mathbb{E}[a_k^4] + L(L-1)\mathbb{E}^2[a_k^2]} = \frac{1}{L\kappa_a + L(L-1)} \tag{41}$$

where κ_a denotes the symbol kurtosis of the PAM alphabet, which is given by

$$\kappa_a = \mathbb{E}[a_k^4] = \frac{6}{5} \cdot \frac{3M(M-1)-1}{(2M-1)(M-1)} \tag{42}$$

By taking into account the auxiliary results in (40) and (41), we finally obtain

$$\overline{\mathrm{NBias}}(\hat{\rho}_s) = \frac{L}{L-3}\left(1 + \frac{2(1+\alpha)}{L\rho_s}\right) - 1 \tag{43}$$

and

$$\overline{\mathrm{NVar}}(\hat{\rho}_s) = \frac{2L^2}{(L-3)^2(L-5)}\left(1 + \frac{4(1+\alpha)(L-2)}{L\rho_s} + \frac{4(1+\alpha)^2(L-2)}{[L\,\kappa_a + L(L-1)]\,\rho_s^2}\right) \tag{44}$$

as lower bounds for the relationships in (38) and (39), respectively.

5. Numerical Results

The analytical results achieved for SNR estimation in Sections 3 and 4 will be verified by Monte Carlo (MC) simulations. In the following, the former are indicated by lines, whereas the latter are shown by markers. Each point in the diagrams below has been obtained by averaging a number of 10^5 estimates, which turned out to be large enough to verify the analytical results with sufficient accuracy.

Assuming a 4-PAM constellation operated with $\rho_S = 0$ dB and a roll-off factor $\alpha \in \{0.0, 1.0\}$, Figure 2 illustrates the evolution of the normalized bias as a function of the observation length L. It is to be noticed that the lines in different colors represent the lower bound given by (43); verified by simulation results, we observe that the lower limit is very tight over the full range of L. We observe that the bias decreases rapidly with increasing values of L, which is also confirmed by (43). In addition, the diagram depicts the results obtained for 16-PAM, $\rho_S = 10$ dB, and $\alpha \in \{0.2, 0.8\}$. In the strict sense, the relationship in (43) applies only to values of $\alpha \in \{0.0, 0.5, 1.0\}$, but the 16-PAM scenario in Figure 2 demonstrates that it represents also a very good approximation for other values of the excess bandwidth.

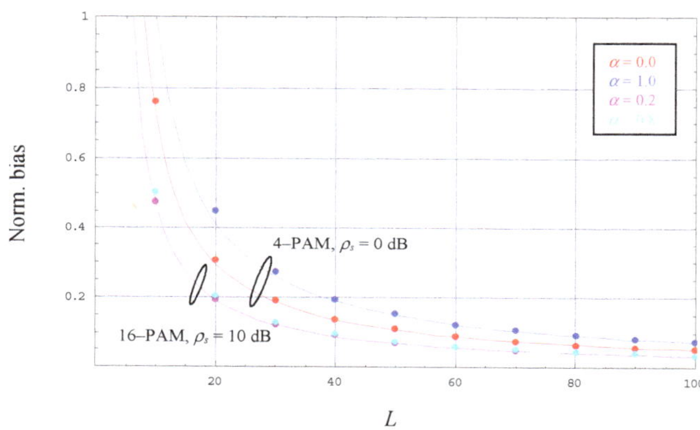

Figure 2. Evolution of the normalized bias.

The evolution of the normalized bias has been simulated and verified for modulation schemes other than 4-PAM and 16-PAM, e.g., 2-PAM and 8-PAM, as well as for roll-off factors different to those exemplified in Figure 2. It turned out that the bias of the estimator algorithm is reflected accurately enough by the formula in (43), disappearing for very large values of L irrespective of the selected values of M or α.

Using a 4-PAM scheme with $L = 10$ and the same roll-off factors as before, Figure 3 illustrates the evolution of the normalized variance as a function of the true SNR value in dB. For comparison purposes, the normalized MCRLB expressed by (22) is shown in dashed style. We observe that the latter is fairly loose for such small observation windows, whereas the lower bound of the variance in (44) appears to be very tight as confirmed by simulation results, in particular at larger SNR values. However, the diagram illustrates also that the MCRLB is more and more approximated by the jitter variance of the related estimator algorithm, when we increase the observation length in Figure 3, verified for 16-PAM, $L = 100$, and $\alpha \in \{0.2, 0.8\}$; it is to be recalled that for $\alpha \notin \{0.0, 0.5, 1.0\}$, the NMCRLB is furnished by (21). Finally, we see that the MC output is very close to (44) over the full SNR range, although the relationship is, in a strict sense, only applicable to roll-off factors $\alpha \in \{0.0, 0.5, 1.0\}$. These observations also hold true for modulation schemes and roll-off factors other than those used in Figure 3; especially, one can see that for $L \gg 1$ and $\rho_S \gg 1$, the normalized variance is simply given by $2/L$.

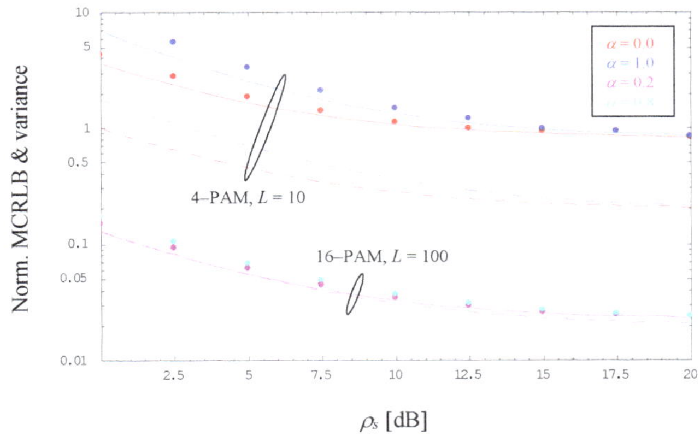

Figure 3. Evolution of the normalized MCRLB and variance.

6. Concluding Remarks

Assuming a data-aided situation, i.e., data symbols are known to the receiver in the form of a pilot sequence, SNR estimation for a bandlimited optical intensity link has been investigated in the current paper. This requires a signal design achieved by an M-ary PAM scheme and a non-negative pulse shape also satisfying the Nyquist criterion. By means of a flat receiver filter, it is avoided that the waveforms of the user signal are distorted, but the price to be paid is an additional amount of noise which the subsequent receiver stages are suffering from.

Conditioned on reliable recovery and correction of the symbol timing, the modified CRLB could be derived as the theoretical limit of the jitter variance produced by the SNR estimator developed in the context of this paper. With respect to the latter, an ML solution has been obtained in closed form, which turned out to be particularly simple from a computational point of view for specific values of excess bandwidth, among them being the minimum bandwidth scenario. For these values, the analytical relationships for bias and jitter variance have been obtained in closed form as well.

Verified by simulation results, it could be shown that—irrespective of the chosen PAM constellation and the value of the excess bandwidth—the bias effect vanishes more and more with increasing values of the true SNR value and the observation length L the link is operated with. This is also confirmed in view of jitter performance insofar as the CRLB is successively approached by increasing values of L.

Funding: Part of the results have been obtained in the context of the Austrian Space Application Programme (ASAP), funded by the Austrian Research Promotion Agency (FFG), for a project on "Optical Pulse Shaping for Power Efficient Transmission". Contract No. 885376.

Data Availability Statement: Data are available from the author upon mail request.

Conflicts of Interest: The author declares no conflict of interest.

Appendix A

Using the identities introduced in Section 4, i.e., $X = \hat{P}_n$ as well as $Y = \hat{A}^2$, together with $Z = \hat{\rho}_s = Y/X$, the PDF of the SNR estimate can be expressed as [27]

$$f_Z(z) = \int_0^\infty f_{Z|X}(z|x) f_X(x)\, dx = \int_0^\infty x\, f_Y(x\, z) f_X(x)\, dx \quad \text{(A1)}$$

Substituting in the sequel the PDFs given by (31) and (34), we obtain by means of [28] (3.462/1) and (9.240) after some lengthy but straightforward manipulations,

$$f_Z(z) = \frac{K_0}{\sqrt{z(z+\beta)^L}} e^{-\lambda} {}_1F_1\left(\frac{L}{2}, \frac{1}{2}; \frac{\lambda z}{z+\beta}\right) \tag{A2}$$

where ${}_1F_1(\cdot)$ denotes the confluent hypergeometric (Kummer) function [29] and the parameters β, λ, and K_0 are specified as follows:

$$\lambda = \frac{A^2}{2\sigma_A^2} = \frac{M_{aa}\rho_s}{4(1+\alpha)}, \; \beta = \frac{\sigma_A^2}{\sigma_x^2} = \frac{2(1+\alpha)L}{M_{aa}} \tag{A3}$$

$$K_0 = \frac{\Gamma(\frac{L}{2})}{\sqrt{\pi}\,\Gamma(\frac{L-1}{2})}\left(\frac{\sigma_A^2}{\sigma_x^2}\right)^{(L-1)/2} = \frac{\Gamma(\frac{L}{2})}{\sqrt{\pi}\,\Gamma(\frac{L-1}{2})}\beta^{(L-1)/2} \tag{A4}$$

Deriving the m-th order moment of (A2), we first express the confluent hypergeometric function by its Meijer G-equivalent [30] (8.4.45/2), i.e.,

$${}_1F_1(a,b;\,z) = \frac{\Gamma(b)}{\Gamma(a)} G_{2,1}^{1,1}\left(-\frac{1}{z} \,\middle|\, \begin{matrix} 1,\, b \\ a \end{matrix}\right) \tag{A5}$$

Applying then the integration rules for Meijer G-functions [30] (2.24.2/6), we get

$$\begin{aligned} M_m &= \int_0^\infty z^m f_Z(z)\,dz\,,\; m \in \mathbb{N}_0 \\ &= \tfrac{K_0\,\Gamma(\frac{1}{2})}{\Gamma(\frac{L}{2})} e^{-\lambda} \int_0^\infty \tfrac{z^{m-1/2}}{(z+\beta)^{L/2}} G_{2,1}^{1,1}\left(-\tfrac{z+\beta}{\lambda z} \,\middle|\, \begin{matrix} 1,\, \frac{1}{2} \\ \frac{L}{2} \end{matrix}\right) dz \\ &= K_0 \tfrac{\Gamma(\frac{1}{2})\,\Gamma(\frac{L-2m-1}{2})}{\Gamma(\frac{L}{2})} \beta^{(2m+1-L)/2} e^{-\lambda} G_{3,2}^{2,1}\left(-\tfrac{1}{\lambda} \,\middle|\, \begin{matrix} 1,\, \frac{1}{2},\, \frac{L}{2} \\ m+\frac{1}{2},\, \frac{L}{2} \end{matrix}\right) \end{aligned} \tag{A6}$$

Employing the integral definition for Meijer G-functions [30] (8.2.1/1) as well as the identity in (A5), which means that

$$G_{3,2}^{2,1}\left(-\frac{1}{\lambda} \,\middle|\, \begin{matrix} 1,\, \frac{1}{2},\, \frac{L}{2} \\ m+\frac{1}{2},\, \frac{L}{2} \end{matrix}\right) = G_{2,1}^{1,1}\left(-\frac{1}{\lambda} \,\middle|\, \begin{matrix} 1,\, \frac{1}{2} \\ m+\frac{1}{2} \end{matrix}\right) = \frac{\Gamma(m+\frac{1}{2})}{\Gamma(\frac{1}{2})} {}_1F_1\left(m+\frac{1}{2}, \frac{1}{2}; \lambda\right) \tag{A7}$$

the relationship in (A6) might be simplified to

$$M_m = K_0 \frac{\Gamma(m+\frac{1}{2})\,\Gamma(\frac{L-2m-1}{2})}{\Gamma(\frac{L}{2})} \beta^{(2m+1-L)/2} e^{-\lambda} {}_1F_1\left(m+\frac{1}{2}, \frac{1}{2}; \lambda\right) \tag{A8}$$

Finally, by taking into account the properties of confluent hypergeometric functions, the first- and second-order moments are provided as

$$M_1 = \frac{\sqrt{\pi}K_0}{2} \frac{\Gamma(\frac{L-3}{2})}{\Gamma(\frac{L}{2})} \beta^{-(L-3)/2}(2\lambda+1) \tag{A9}$$

and

$$M_2 = \frac{\sqrt{\pi}K_0}{4} \frac{\Gamma(\frac{L-5}{2})}{\Gamma(\frac{L}{2})} \beta^{-(L-5)/2}(4\lambda^2+12\lambda+3) \tag{A10}$$

References

1. Gappmair, W. On parameter estimation for bandlimited optical intensity channels. *Comput. Spec. Issue Opt. Wirel. Commun. Syst.* **2019**, *7*, 11. [CrossRef]

2. Gappmair, W.; Nistazakis, H.E. Blind symbol timing estimation for bandlimited optical intensity channels. In Proceedings of the 12th IEEE/IET International Symposium on Communication Systems, Networks and Digital Signal Processing (CSNDSP), Porto, Portugal, 20–22 July 2020.
3. Gappmair, W.; Schlemmer, H. Feedback solution for symbol timing recovery in bandlimited optical intensity channels. In Proceedings of the IEEE 4th International Conference on Broadband Communications (CoBCom), Graz, Austria, 12–14 July 2022.
4. Tavan, M.; Agrell, E.; Karout, J. Bandlimited intensity modulation. *IEEE Trans. Commun.* **2012**, *60*, 3429–3439. [CrossRef]
5. Czegledi, C.; Khanzadi, M.R.; Agrell, E. Bandlimited power-efficient signaling and pulse design for intensity modulation. *IEEE Trans. Commun.* **2014**, *62*, 3274–3284. [CrossRef]
6. Hranilovic, S. Minimum-bandwidth optical intensity Nyquist pulses. *IEEE Trans. Commun.* **2007**, *55*, 574–583. [CrossRef]
7. Hranilovic, S. *Wireless Optical Communication Systems*; Springer: New York, NY, USA, 2004.
8. Arnon, S.; Barry, J.; Karagiannidis, G.; Schober, R.; Uysal, M. *Advanced Optical Wireless Communication Systems*; Cambridge University Press: New York, NY, USA, 2012.
9. Khalighi, M.A.; Uysal, M. Survey on Free Space Optical Communication: A Communication Theory Perspective. *IEEE Commun. Surv. Tutor.* **2014**, *16*, 2231–2258. [CrossRef]
10. Ghassemlooy, Z.; Arnon, S.; Uysal, M.; Xu, Z.; Cheng, J. Emerging Optical Wireless Communications—Advances and Challenges. *IEEE J. Select. Areas Commun.* **2015**, *33*, 1738–1749. [CrossRef]
11. Mengali, U.; D'Andrea, A.N. *Synchronization Techniques for Digital Receivers*; Plenum Press: New York, NY, USA, 1997.
12. Meyr, H.; Moeneclaey, M.; Fechtel, S.A. *Digital Communication Receivers: Synchronization, Channel Estimation, and Signal Processing*; Wiley: New York, NY, USA, 1998.
13. Chung, T.S.; Goldsmith, A.J. Degrees of freedom in adaptive modulation: A unified view. *IEEE Trans. Commun.* **2001**, *49*, 1561–1571. [CrossRef]
14. Summers, T.A.; Wilson, S.G. SNR mismatch and online estimation in turbo decoding. *IEEE Trans. Commun.* **1998**, *46*, 421–423. [CrossRef]
15. Pauluzzi, D.R.; Beaulieu, N.C. A comparison of SNR estimation techniques for the AWGN channel. *IEEE Trans. Commun.* **2000**, *48*, 1681–1691. [CrossRef]
16. Proakis, J.G.; Manolakis, D.G. *Digital Signal Processing: Principles, Algorithms, and Applications*; Prentice Hall: Upper Saddle River, NJ, USA, 1996.
17. Kay, S.M. *Fundamentals of Statistical Signal Processing: Estimation Theory*; Prentice Hall: Upper Saddle River, NJ, USA, 1993.
18. Gray, R.M. *Toeplitz and Circulant Matrices: A Review*; Now Publishers: Hanover, MA, USA, 2006.
19. Proakis, J.G. *Digital Communications*; McGraw-Hill: New York, NY, USA, 1989.
20. Papoulis, A. *Probability, Random Variables, and Stochastic Processes*; McGraw-Hill: New York, NY, USA, 1991.
21. D'Andrea, A.N.; Mengali, U.; Reggiannini, R. The modified Cramer-Rao bound and its application to synchronization problems. *IEEE Trans. Commun.* **1994**, *42*, 1391–1399. [CrossRef]
22. Gini, F.; Reggiannini, R.; Mengali, U. The modified Cramer-Rao bound in vector parameter estimation. *IEEE Trans. Commun.* **1998**, *46*, 52–60. [CrossRef]
23. Moeneclaey, M. On the true and the modified Cramer-Rao bounds for the estimation of a scalar parameter in the presence of nuisance parameters. *IEEE Trans. Commun.* **1998**, *46*, 1536–1544. [CrossRef]
24. Scharf, L.L. *Statistical Signal Processing*; Prentice-Hall: Upper Saddle River, NJ, USA, 1990.
25. Evans, M.; Hastings, N.; Peacock, B. *Statistical Distributions*; John Wiley & Sons: New York, NY, USA, 2000.
26. Viterbi, A.J.; Omura, J.K. *Principles of Digital Communication and Coding*; McGraw-Hill: New York, NY, USA, 1979.
27. Li, Y.; He, Q. On the ratio of two correlated complex Gaussian random variables. *IEEE Commun. Lett.* **2019**, *23*, 2172–2176. [CrossRef]
28. Gradshteyn, I.S.; Ryzhik, I.M. *Table of Integrals, Series, and Products*; Academic Press: New York, NY, USA, 1994.
29. Abramowitz, M.; Stegun, I.A. *Handbook of Mathematical Functions*; Dover Publications: New York, NY, USA, 1970.
30. Prudnikov, A.P.; Brychkov, Y.A.; Marichev, O.I. *Integrals and Series, Volume 3: More Special Functions*; Gordon & Breach: New York, NY, USA, 1990.

sensors

MDPI

Article

Multi-Objective Routing Optimization for 6G Communication Networks Using a Quantum Approximate Optimization Algorithm

Helen Urgelles, Pablo Picazo-Martinez, David Garcia-Roger and Jose F. Monserrat *

iTEAM Research Institute, Universitat Politècnica de València, 46022 València, Spain; heurpe@iteam.upv.es (H.U.); picazopau@gmail.com (P.P.-M.); dagarro@iteam.upv.es (D.G.-R.)
* Correspondence: jomondel@iteam.upv.es

Abstract: Sixth-generation wireless (6G) technology has been focused on in the wireless research community. Global coverage, massive spectrum usage, complex new applications, and strong security are among the new paradigms introduced by 6G. However, realizing such features may require computation capabilities transcending those of present (classical) computers. Large technology companies are already exploring quantum computers, which could be adopted as potential technological enablers for 6G. This is a promising avenue to explore because quantum computers exploit the properties of quantum states to perform certain computations significantly faster than classical computers. This paper focuses on routing optimization in wireless mesh networks using quantum computers, explicitly applying the quantum approximate optimization algorithm (QAOA). Single-objective and multi-objective examples are presented as robust candidates for the application of quantum machine learning. Moreover, a discussion about quantum supremacy estimation for this problem is provided.

Keywords: multi-objective; quantum computing; quantum optimization algorithms; quantum routing optimization; 6G communication networks

Citation: Urgelles, H.; Picazo-Martinez, P.; Garcia-Roger, D.; Monserrat, J.F. Multi-Objective Routing Optimization for 6G Communication Networks Using a Quantum Approximate Optimization Algorithm. *Sensors* **2022**, *22*, 7570. https://doi.org/10.3390/s22197570

Academic Editor: Peter Han Joo Chong

Received: 2 September 2022
Accepted: 3 October 2022
Published: 6 October 2022

1. Introduction

Quantum computing has boosted worldwide interest in different research areas, including telecommunications. The advent of complex sixth-generation (6G) technologies suggests that a quantum computing approach may better serve some use cases for high-performance computing in wireless communications. Optimization in communication networks has been a hot research topic throughout the years. The evolution of telecommunications has led to thousands of devices being connected, including nodes of the Internet of Things (IoT), autonomous vehicles, user devices, sensors, etc. Routing optimization plays an important role in guiding data packets between network nodes to follow the best end-to-end paths (from the source to the destination) according to certain network conditions. Regardless of the type of network, a poor routing strategy may prove harmful to the overall performance of the network. Routing strategies involve the definition of a set of one or more paths over which communication between end devices takes place over a network.

Typically, in conventional multi-hop networks, only one of the desired objectives is optimized, whereas other objectives are assumed to be constraints of the problem. Multi-objective algorithms have been previously explored in the literature, with the discussion focused on comparing complexities and rates of convergence with respect to the number of network nodes. An example is [1], where the authors proposed two approaches: (i) an algorithm based on the non-dominated sorting-based genetic algorithm-II (NSGA-II), and (ii) a multi-objective differential evolution (MODE) algorithm. Many single-objective optimization techniques have also been proposed; even though they are in their majority (based on classical solutions), some remarkable examples of quantum approaches exist.

In [2], the authors proposed the non-dominated quantum iterative optimization (NDQIO) algorithm, which exploits the parallelism property in quantum mechanics for finding the optimum of a multi-objective routing problem in wireless multi-hop networks. Another algorithm, the tree-based quantum algorithm (TQA) in [3], solves the delay-constraint multicast tree problem. The most significant advantage of TQA is that the solving speed is much faster and more noticeable when the network topology scale becomes more considerable.

Furthermore, inspired by the IoT concept, with millions of interconnected wireless devices acquiring data ubiquitously, ref. [4] presented a novel quantum computing-inspired (IoT-QciO) optimization technique for wireless networks. The approach is based on the maximization of data accuracy (DA) in a real-time environment of IoT applications. In another recent paper [5], a quantum particle swarm optimization (PSO) algorithm is proposed, which outperformed existing optimization algorithms in terms of precision and convergence speed for smart IoT parking applications. The same authors applied an enhanced version of the PSO to routing in an industrial IoT (IIoT) scenario [6]. Formulations of routing problems on quantum computers appear in [7], where the authors focused on vehicle routing problems with time window(s) (VRPTW) and investigated and compared the VRPTW from a quantum computing perspective. Additionally, in [8,9], the quantum approximate optimization algorithm (QAOA) was tested with good results for the vehicle routing problem (VHP), a generalization of the traveling salesman problem (TSP). Finally, quantum optimization for 6G wireless communication is addressed through two use cases, MIMO detection and LDPC decoding in [10]. The case studies focus on quantum annealing (QA) technology, but the gate model processors are also described as potential quantum computations for communication networks.

Motivated by 6G connectivity requirements, the possibilities offered by the quantum computation, and the crucial role of routing strategies in wireless communications, we present a multi-objective routing optimization use case using QAOA and quantum systems. We integrated the parameterized lexicographic heuristic method to solve routing problems with quantum computing as a novel approach in the literature, and the results are provided to argue quantum supremacy.

In this context, the contributions of the present paper are as follows:

- We solved single and multi-objective network routing problems in a 6G network scenario, formulating the first QAOA [11] generalization to multi-objective optimization problems applied to vehicular ad hoc networks (VANETs) to the best of the authors' knowledge. As these problems are known to be Non-deterministic polynomial-time hardness (NP-hard) [12], quantum algorithms may help speed up the problem-solving process when the problem's size makes classical algorithms struggle to find optimal solutions in a reasonable time.
- We presented a single-objective routing optimization, providing the steps for the problem solution through quadratic unconstrained binary optimization (QUBO) and the Ising model to create the Hamiltonian problem;
- We discuss the performance of QAOA concerning the number of layers.
- We proposed a multi-objective routing optimization based on the parameterized lexicographic heuristic method and QAOA;
- We conclude quantum supremacy expectations from insights on estimations collected from the literature review.

The rest of the paper is structured as follows: Section 2 exposes firstly a brief overview of Quantum Optimization Algorithms, focused on QAOA. In Section 2.2, an example of a single-objective routing problem executed on IBM quantum experience is presented. After that, Section 2.3 provides a multi-objective routing problem using QAOA and parameterized lexicographic method. Finally, in Section 3, a discussion about supremacy expectations is presented. The main conclusions are drawn in Section 4.

2. Quantum Routing Optimization

In most cases, when network operators or service providers design and control their networks in real-world settings, they first formulate an optimization problem corresponding to the desired communication network with the required parameters and then solve the problem using a computer. These problems are mainly integer optimization problems whose complexities require high computational resources. To solve these problems on classic computers, the GNU linear programming kit (GLPK) package can be used (as was the approach of [13]). This kit is intended to solve linear programming (LP), integer LP, and mixed-integer LP programming. Since the main result of the decision-making logic of a routing process is the recommended path to be followed, an integer variable may be assigned to each connection between each connected node. If the variable takes the value of 1, which means the path is taken. In other cases, the variable will be 0 if that way is not optimal to reach the desired objective. Primary objectives could involve minimizing battery costs for remote nodes, maximizing the throughput for each involved node, or minimizing the number of jumps (which have advantageous impacts on latency). When the number of nodes increases, the LP problem can become so big that classic computers struggle to find the optimal route. This happens because the number of combinations between nodes is so significant that the variable numbers scale rapidly, meaning that high computational resources are needed to land on a solution. This will be true for future 6G communication networks, which are expected to provide global coverage and space–air–ground–sea [14]. Considering the multiple requirements that 6G will need to address simultaneously, quantum computers and quantum algorithms could play even more significant roles in such optimization problems.

2.1. Quantum Optimization Algorithms

Quantum computation takes advantage of the properties of quantum mechanics to tackle optimization problems radically differently. Ideally, due to the superposition of quantum states, quantum computers can process all the data simultaneously to find the solution that optimizes the objective function. In contrast, present-day "near-term" quantum computing or "noisy intermediate-scale quantum" (NISQ) technology implement at most 50–100 qubits, and while they might be able to perform tasks that exceed the capabilities of classical computers, they also exhibit noise-related inaccuracies that complicate the demonstration of the advantages of practical quantum computers and limit the sizes of quantum circuits [15]. One of the goals in the NISQ era is to extract the maximum quantum computational power from current devices while developing techniques that will suit the "long-term" goal of fault-tolerant quantum computations. Consequently, new classes of algorithms have been developed for this kind of system. Most of the current NISQ algorithms are based on a hybrid quantum-classic arrangement, such as the variational quantum eigensolver (VQE) and QAOA [16].

VQE was introduced in 2014 [17] for chemistry applications and quantum mechanics to estimate the ground state energy of a molecule using shallow depth circuits. The ground state of energy is equivalent to finding the minimum eigenvalue and/or eigenvector of a matrix (Hamiltonian), which characterizes the molecule. Apart from being applicable in these fields, it has spread up its functionality to optimization problems; one can also use the VQE for optimizing a cost function by encoding it as a matrix whose ground state (minimum eigenvector) corresponds to the optimal solution of the problem. This idea also lies at the heart of QAOA.

Since these algorithms require a smaller circuit (a few quantum gates), it better preserves the coherent evolution of the system, allowing a higher probability of successful results, also beneficial for the available systems with just a few noisy qubits.

QAOA is a variational quantum algorithm (quantum–classical hybrid algorithm) due to its implementation through quantum circuits that depend on a set of variational parameters (β, γ). It was introduced by Farhi and Goldstone in 2014 [11] to solve the problem of finding out a cut whose size was at least the size of any other cut (MaxCut) on a regular graph. This algorithm is characterized by a lower bound for the ratio between the result obtained by the algorithm and the optimal cost (the "approximation ratio") and depends on an integer $p(layers) \geq 1$. The quality of the approximation improves as p is increased, and the depth of the quantum circuit grows linearly p times the number of constraints. In fact, in [11], QAOA always found a cut that was at least 0.6924 times the size of the optimal cut.

QAOA uses a unitary U(β, γ) characterized by the parameters (β, γ) to prepare a quantum state $|\psi(\beta,\gamma)\rangle$. The goal of the algorithm is to find optimal parameters (β_{opt}, γ_{opt}), such that the quantum state $|\psi(\beta_{opt},\gamma_{opt})\rangle$ encodes the optimal solution to the problem [18].

To summarize, QAOA's principle is to extract (measure) the quantum solution prepared by a quantum state in a variational quantum circuit. Then, a classical optimizer is used to tune the circuit parameters and minimize the measured expectation value. Figure 1 graphically represents this principle of operation.

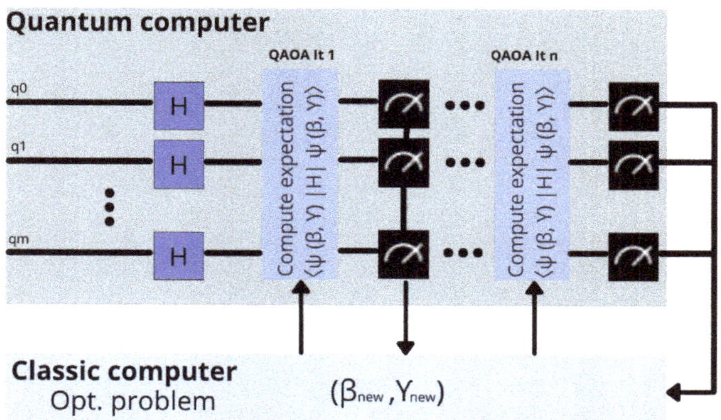

Figure 1. Graphical representation of the operation principle of a QAOA scheme.

2.2. Single-Objective Quantum Routing Optimization

Since communication networks consist of nodes and links; one of the main objectives is to find the minimum cost (in terms of battery consumption) to transmit the traffic from an origin node to a destination node. A network is represented by a graph $G(V,E)$, where V is the set of vertices (nodes), E is the set of links (weights), and the link from node i to node j is expressed as $(i,j) \in E$. Figure 2 shows an example of a network. If node 1 is the source node and node 4 is the destination node, then the problem consists of finding the shortest path from node 1 to node 4 depending on the requirements, constraints, and/or objectives.

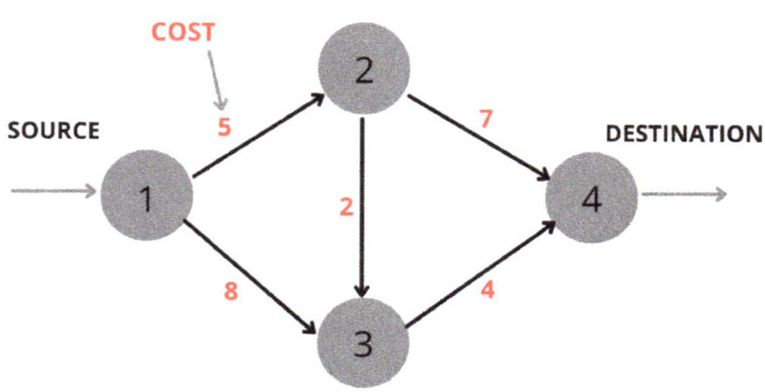

Figure 2. Scheme of an example network with four nodes.

Generally, this kind of problem can be formulated as a cost (objective) function (1), which is minimized or maximized according to constraints (2) and (3) and variables' bound (4) since they involve seeking the best configuration among a set of parameters to achieve the desired objectives. In this example, the cost values are not representative of a real scenario and take values from 1 to 10, implying a higher value and bigger battery cost. This simplification is made because the goal is not to accurately define a cost function but to test QAOA and elaborate supremacy predictions on routing problems. The following equations can be found in [13].

Objective:

$$\min / \max\left(\sum_{(i,j)\in E} C_{ij}X_{ij}\right) \tag{1}$$

Subject to:

$$\sum_{j:(i,j)\in E} X_{ij} - \sum_{j:(i,j)\in E} X_{ji} = 1, if \quad i = p, \tag{2}$$

$$\sum_{j:(i,j)\in E} X_{ij} - \sum_{j:(i,j)\in E} X_{ji} = 0, \forall_i \neq p, q \in V, \tag{3}$$

$$0 \leq X_{ij} \leq 1, \forall_{(i,j)} \in E. \tag{4}$$

For this example, the problem formulation is presented below:

$$\min\ (5X_{12} + 8X_{13} + 2X_{23} + 7X_{24} + 4X_{34}) \tag{5}$$

Subject to:

$$X_{12} + X_{13} = 1, \tag{6}$$

$$X_{12} - X_{23} - X_{24} = 0, \tag{7}$$

$$X_{13} + X_{23} - X_{34} = 0, \tag{8}$$

One common model that is suitable for solving combinatorial optimization problems in quantum computers is the quadratic unconstrained binary optimization, or QUBO for short. QUBO can embrace many models in combinatorial optimization; QUBO models were shown to be equivalent to the Ising model, which plays a crucial role in physics and particle interactions [19].

A formal definition of the QUBO model is given by:

$$\min / \max\ (\mathbf{X}^T Q\mathbf{X} + \mathbf{C}^T\mathbf{X} + c) \tag{9}$$

where $\mathbf{X}$ is a vector of binary decision variables, Q is a square matrix of quadratic coefficients, and $\mathbf{C}$ is a vector of linear coefficients.

Before solving our problem with QAOA, it should be cast in QUBO form. Although our problem includes additional constraints, it can be effectively reformulated as a QUBO model by introducing quadratic penalties (P) into the objective function (10).

$$\begin{aligned} \min\ (5X_{12} + 8X_{13} + 2X_{23} + 7X_{24} + 4X_{34} \\ +P(X_{12} + X_{13} - 1)^2 \\ +P(X_{12} - X_{23} - X_{24})^2 \\ +P(X_{13} + X_{23} - X_{34})^2) \end{aligned} \tag{10}$$

Arbitrarily choosing P to be equal to 27, the Q matrix and $\mathbf{C}$ vector are given by:

$$Q = \begin{bmatrix} 54 & 27 & -27 & -27 & 0 \\ 27 & 54 & 27 & 0 & -27 \\ -27 & 27 & 54 & 27 & -27 \\ -27 & 0 & 27 & 27 & 0 \\ 0 & -27 & -27 & 0 & 27 \end{bmatrix} \quad (11)$$

$$\mathbf{C}^T = \begin{bmatrix} -49 & -46 & 2 & 7 & 4 \end{bmatrix} \quad (12)$$

As mentioned before, the cost function can be mapped to a Hamiltonian in order to find the ground state energy of the system that is equivalent to the optimal solution. QAOA is defined by the problem Hamiltonian (H_P) (13), which contains the cost function, and the mixer Hamiltonian (H_M) [18], defined as the sum of single Pauli X-operators on all qubits (14).

$$H_P|x\rangle = \left(x^T Q x + c^T x\right)|x\rangle = \left(\sum_{i,j=1}^{n} x_i Q_{ij} x_j + \sum_{i=1}^{n} c_i x_i\right)|x\rangle \quad (13)$$

$$H_M = \sum_{i=1}^{n} X_i \quad (14)$$

To define the H_P by Pauli Z-operators, the objective function should be formulated as the Ising spin model: $x_i = \frac{1-Z_i}{2}$

$$\begin{aligned} H_P = {} & 11(IIIIZ_0) - 17.5(IIIZ_1I) - 28(IIZ_2II) - 17(IZ_3III) + 11.5(Z_4IIII) \\ & +13.5(IIIZ_1Z_0 - IIZ_2IZ_0 - IZ_3IIZ_0 + IIZ_2Z_1I - Z_4IIZ_1I + IZ_3Z_2II - Z_4IZ_2II) \end{aligned} \quad (15)$$

The small, single-objective, four-node example (Figure 2) is represented by its corresponding variational quantum circuit based on H_P and H_M ($H_P + H_M$) in Figure 3. The initial prepared state is the equal superposition state through Hadamard (H) gates. The iterations required to reach the optimal results depend on the quantum system used. It was tested on "ibm_perth", a seven-qubit IBM Quantum System, and 59 iterations were necessary using COBYLA as a classical optimizer to find $\beta_{opt} = 0.28517317$, and $\gamma_{opt} = -5.05969577$.

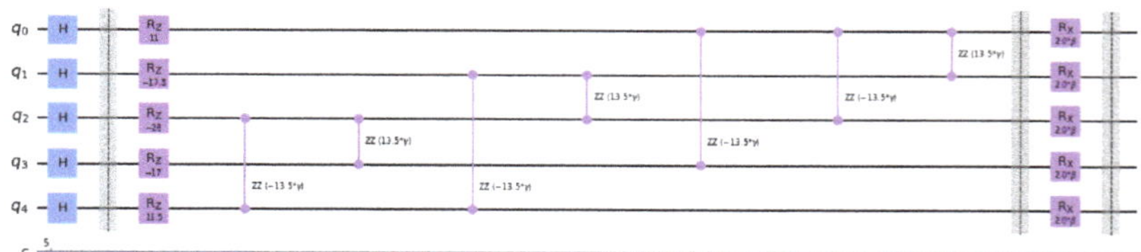

Figure 3. Representation of the single-objective, four-node example QAOA circuit.

It is quite simple to note that the optimum path for minimizing costs would be 1–2–3–4, resulting in 11 ($5 + 2 + 4$). Figure 4 shows the probabilities results according to the possible paths and β and γ values. As expected, the $|10101\rangle$ state has the higher probability that corresponds with the $X_{12}, X_{13}, X_{23}, X_{24}, X_{34}$ where $X_{12} = 1, X_{23} = 1, X_{34} = 1$.

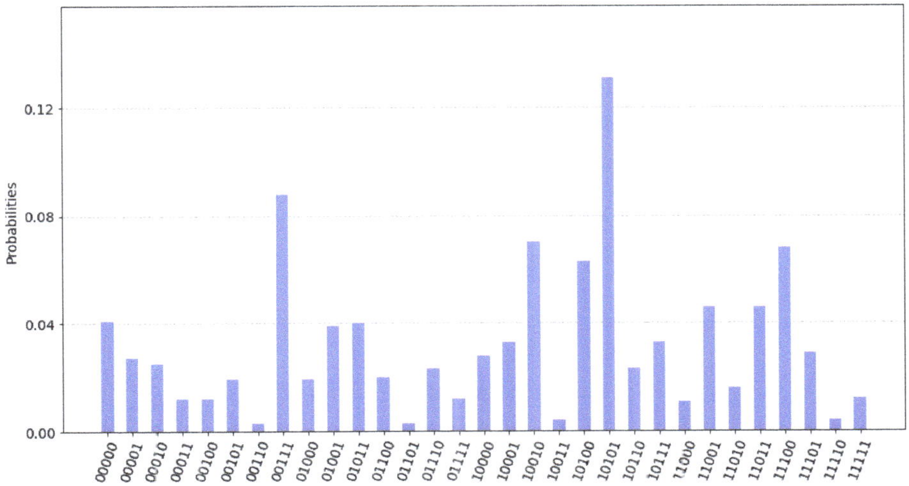

Figure 4. Probability results for the example network.

While this small example may seem simple (this specific case took five qubits) when the number of nodes increases, the problem becomes classically intractable. Note that while a bigger network could not be solved on available quantum computer hardware because of the number of qubits required; this does not mean that larger problems cannot be solved. It depends on the number of available qubits. Furthermore, it must be mentioned that despite the QAOA result being 11, identical to the classical solution, in more complex problems by its nature, QAOA might provide only a good approximate near-optimum solution.

Additionally, to test how the algorithm performs according to the number of QAOA layers, simulations with 500 shots were executed and $p = 1, 2, 3$. The theoretical accuracy provided by QAOA improves with higher values of p as was already anticipated in Section 2.1. Figure 5 illustrates how the probability of obtaining the right solution increases with higher values of p. However, it turns out that for implementation on a real device, this improvement is not remarkable because the depth of the quantum circuit has a negative impact on the noise level.

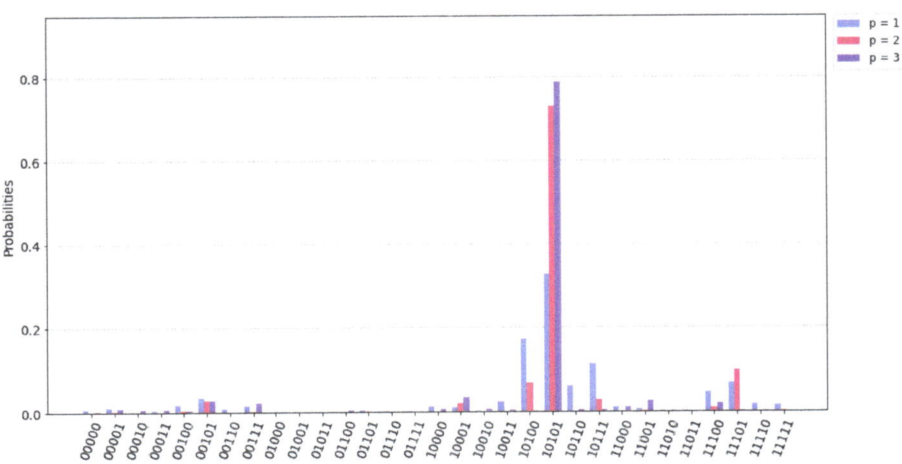

Figure 5. QAOA performance comparison according to p values.

Furthermore, Algorithm 1 below, shows the pseudocode of the algorithm proposed for the routing when following the steps to programming the single target routing problem.

Algorithm 1 Routing Optimization using QAOA.

from pyqubo **import** *Array*, *Constraint*, *Placeholder*

Design the network graph:
$edges = [(1,2), (1,3), (2,3), (2,4), (3,4)]$, $weights = [5,8,2,7,4]$
$x = Array.create(shape = \text{len}(edges))$
for $iteration = 1, 2, \ldots$ **in** range(len($edges$)) **do**
    f_{cost} +=Constraint($weights[i] * x[i]$)
end for
f_{cost} += $p*$Constraint($(x[0] + x[1] - 1)^2$)
f_{cost} += $p*$Constraint($(x[0] - x[2] - x[3])^2$)
f_{cost} += $p*$Constraint($(x[1] + x[2] - x[4])^2$)

Create the problem Hamiltonian and mixer Hamiltonian: $x_i = \frac{1-Z_i}{2}$

$$H_P + H_M \tag{16}$$

Create the QAOA circuit according to the linear and quadratic coefficients of H_P:
$linear\_coefficients(lc) = [11.0, -17.5, -28.0, -17.0, 11.5]$
$quadratic\_coefficients(qc) = \{(0,1) : 13.5, (0,2) : -13.5, (0,3) : -13.5, (1,2) : 13.5, (1,4) : -13.5, (2,3) : 13.5, (2,4) : -13.5\}$
Circuit ($num\_qubits, param, n\_layers, lc, jc$):
circ = QuantumCircuit(num_qubits)

Initial state (H gates):
for qubit **in** range($circ.num\_qubits$) **do**
    circ.h($qubit$)
end for

Problem Hamiltonian:
for qubit **in** range($circ.num\_qubits$)) **do**
    circ.rz($lc[qubit] * param, qubit$)
end for
for key **in** $qc.keys()$) **do**
    circ.rzz($qc[key] * param, key[0], key[1]$)
end for

Mixer Hamiltonian:
for qubit **in** range($circ.num\_qubits$) **do**
    circ.rx($param, qubit$)
end for
circ.measure(range(num_qubits), range(num_qubits))

Compute the expectation values according to the measurement results.

Optimize classically to find β and γ, with scipy.optimize.

Repeat the process until β_{opt} and γ_{opt} optimum are found.

2.3. Multi-Objective Quantum Routing Optimization

When different performance metrics need to be considered in communication networks, the optimization problem becomes multi-objective. In this regard, those metrics could be to reduce the cost to provide a better quality of service, improve the throughput, or minimize the number of hops taking care of the network's latency. If only one parameter is considered in the link, the other parameters will probably not follow the requirements for a determined service. As a result, multi-objective problems are purposed to obtain solutions satisfying the multiple criteria for each scenario.

Solving multi-objective problems does require more computation power than single-objective problems. In addition, multi-objective problems do not have global optima. If one objective is optimized, it is likely that the others will not be. Equilibrium needs to be found to cover all the requirements desired; each equilibrium point is known as a Pareto optimal point.

From heuristics, there are multiple ways to find Pareto's solutions. The parameterized lexicographic method is presented in this paper. This method needs to order objectives by importance. The optimization will have as many stages as objectives formulated. The first stage will solve the first objective without including any other one. The second stage will solve the second objective, including a margin constraint inherited from the first stage. The third stage will include inherited constraints from the first and second and so on. For example, if minimizing the battery cost is the most important objective, the cost result obtained from the first stage will have a deviation from the optimum allowed in the second stage. This deviation is parameterized using slack parameters α. If an objective wants to be minimized, the slack parameter of the inherited constraint, α_n will be higher than 1 (allowing its increase from the optimal) and if maximized lower than 1, (allowing its decrease from the optimal).

This subsection presents a practical case of multi-objective routing optimization using QAOA and the lexicographic method. Note that before solving our problem with QAOA, it should be cast in QUBO form as outlined previously in Section 2.2; the same steps presented in the pseudocode were followed, not included here for brevity. This case included six nodes following the scheme shown in Figure 6. Three parameters were optimized: i) the battery cost, ii) the available throughput for the served nodes, and iii) the number of hops from the origin to the destiny. The battery cost and the number of hops were minimized, and the throughput maximized. Since the number of qubits available was not enough, the problem was simulated on IBM Quantum first. In addition, it was executed on IBM Cloud through Qiskit Runtime [20] available on IBM Quantum systems and IBM Cloud. Qiskit Runtime is an architecture and programming service offered by IBM that allows users to optimize workloads and efficiently execute them on quantum systems. Runtime works based on programming via primitives such as Sampler, Estimator, and in our case, QAOA. The system used was imb_algiers, which has 27 qubits.

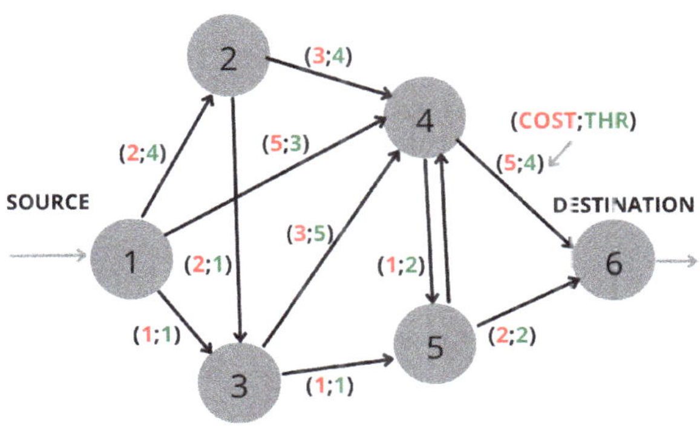

Figure 6. Scheme of an example network with six nodes.

The first objective of the lexicographic approach was to minimize the battery cost, according to (1). All the constraints (2), (3), and (4), shown in Section 2.2 were added to the model. Following the same procedure, the solution gave the path 1–3–5–6, offering services to four nodes, which meant three hops. The total battery cost was 4, and the throughput obtained was 4.

Once the battery cost was optimized, the solution found was added as an inequity constraint with the alpha parameter, giving some clearance to the cost while optimizing the throughput. This parameter in this example was set to $\alpha_c = 2$. Equation (17) was the added constraint to the model in this second step

$$C_{ij}X_{ij} \leq (C_{ij}X^*_{ij})\alpha_c. \tag{17}$$

The next objective was the throughput, calculated as the sum of the throughput values for all of the nodes served. The objective is shown in Equation (18). As was done with cost, a simplification was also made to the calculus of the throughput values for each path, not being the ones used representative of a real scenario. In this case, higher values are better since this objective wants to be maximized.

$$\max \sum_{(i,j)\in E} Thr_{ij}X_{ij}. \tag{18}$$

The results changed, increasing the throughput to 10 and the cost to 7. In addition, the number of hops increased to 4. The path taken was 1–3–4–5–6. This means a sacrifice for both cost and hops to increase the throughput. Since multi-objective optimization does not have a correct result, this result is as valid as the one obtained in the first stage, and the final decision is made by the decision-maker.

As was done with the cost, an alpha parameter was added to the model, giving clearance to the throughput in the minimization of the last objective, the hops. This parameter was fixed to $\alpha_{thr} = 0.4$. The reason for this parameter being that small was due to the tiny size of the problem, which required big clearances to change from one solution to another. This means that Equation (19) needed to be included in the model. Equation (17) was also added to the constraints for the last step of the multi-objective optimization process, the minimization of the number of hops.

$$Thr_{ij}X_{ij} \geq (Thr_{ij}X^*_{ij})\alpha_{Thr}. \tag{19}$$

The last objective, the number of hops, was minimized. The objective is shown in Equation (20).

$$min \sum_{(i,j)\in E} H_{ij}X_{ij}. \tag{20}$$

All constraints in (2), (3), (4), (17), and (19) were added to the model.

The result obtained for the hops was 3, the cost was also reduced to 4, but throughput decreased to 4. This gave the same result as the one obtained in the first stage of the optimization, following the path 1–3–5–6.

The first step of the lexicographic method only considered minimizing the cost. As a result, even though only three hops were performed, throughput was not really good. When maximizing throughput, the cost was slightly increased and so were the hops. Finally, for this particular case, the solution taking into account the three objectives was equivalent to the solution that we would obtain from only considering the cost objective. The result-deriving process was forced to showcase the operation of the method. However, other combinations of alpha parameters or a change in the order of the objectives would change the results. Problems of a bigger size, which unfortunately could neither be run on a QC nor simulated, will be more dependent on changes of alpha, altering the final result upon small variations of alpha. The final results obtained from this method can be summarized in Table 1.

Table 1. Results obtained for the six-node lexicon optimization.

Objectives	Min Cost	Max Thr st Cost	Min Hops st Cost+Thr
Cost	4	7	4
Throughput	4	10	4
Hops	3	4	3

3. QAOA Supremacy Expectations

QAOA is one of the most promising candidates for achieving quantum supremacy in optimization problems, which means outperforming the best-known classical algorithms on a given problem. This algorithm can solve scheduling, data analysis, and machine learning optimization problems. This paper used it to solve a hard integer routing optimization problem with good results. However, a higher number of qubits with relatively low noise is necessary to outperform the classic computation solution.

In the last few years, quantum computers have entered a new phase, where qubit noise was more considered and could benefit from error correction techniques. Quantum volume (QV) has turned into an important parameter apart from the qubit number since quantum computer developers considered that it was also important to have well-calibrated controlled noise qubits rather than ti have a huge number of noisy qubits. QV measures the performance of gate-based quantum computers [21]. Therefore, to achieve supremacy, it is necessary to have a higher number of qubits that are well-calibrated. Another crucial metric to measure quantum device performances were introduced by IBM recently. CLOPS (circuit layer operations per second) corresponds to the number of quantum circuits a quantum processing unit (QPU) can execute per unit of time. In this regard, and according to the available systems (ibm_perth, ibm_lagos, ibm_jakarta with seven qubits; ibm_manila, ibm_bogota, ibm_quito, ibm_belem, and ibm_lima with five qubits), a representation of CLOPS vs time is presented in Figure 7.

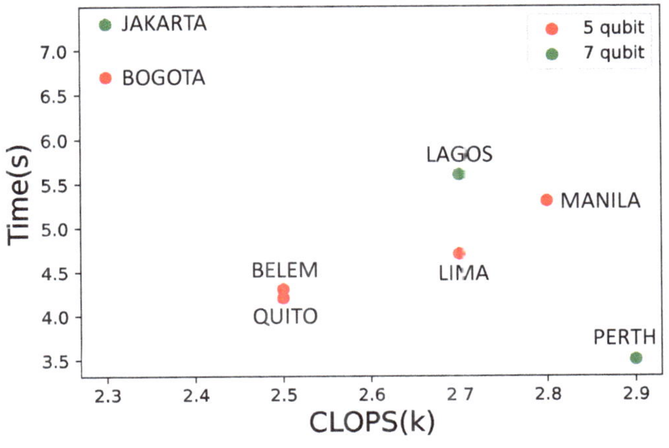

Figure 7. Time for solving the four-node problem depending on QC CLOPS.

The trend is that a system with higher CLOPS solves the problem faster; however, this statement is not always true. This is due to the internal architecture of each quantum computer and how the connections between the qubits are done. A low CLOPS QC can be fast for a determined problem but slow for another one. For each problem, the performance can vary since CLOPS is determined using a general purpose test created by IBM. As a result, for this particular routing problem, the Belem and Quito architectures and qubit connections seem to perform better than the Lima or Manilla ones, even though their CLOPS systems in the IBM tests were higher.

The results correspond to the single objective problem. The same examination could be applied to different problems, depending on the qubits available. The outcomes reveal that CLOPS also play a role, as higher CLOPS systems tend to solve the problem faster on average, even though more qubits can affect the system coherence. With 7 qubits and 2.9K as CLOPS, the time decreases as Figure 6 reveals.

Moreover, simulations performed in [22] conclude that QAOA will achieve a quantum speedup with hundreds of qubits since classic computers have begun to struggle with the size of the problem. [23] also discusses an approach to supremacy for optimization problems making use of QAOA. This algorithm is designed to run purely on a gate model quantum computer, and it is hard to simulate by any simulator when a high number of qubits are involved. Although there exist classical natural algorithms that have better success chances, QAOA has a performance guarantee. As a result, it is argued if QAOA can exhibit any form of "quantum supremacy", with the conclusion being positive due to the theoretical complexity assumptions. The paper postulates QAOA as an excellent candidate to run on near-term quantum computers, not only because of the potential use in optimization for some particular problems but for the very probable supremacy demonstration once the hardware has small hundreds of intermediate-noise qubits. Since the device availability only allows tests on IBM quantum systems with few qubits and good noise levels, theoretical estimations need to be performed.

Figure 8 shows supremacy expectations on node routing problems considering IBM quantum computers, which are the ones used in this research.

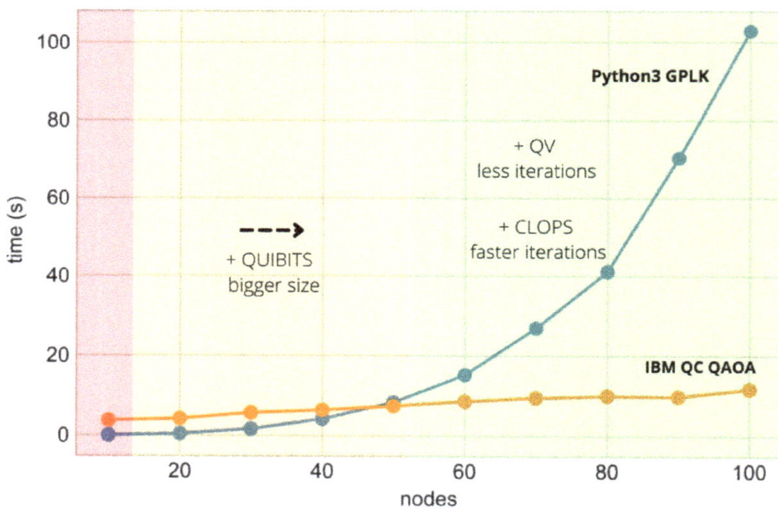

Figure 8. QAOA Supremacy estimations for network routing problems.

The red bar indicates the executions done in this paper on open-for-research IBM QC, where the time taken to find the solution is relatively low but still much higher compared to the classic solution. The yellow space indicates where IBM is working now, testing systems up to 65 qubits (Hummingbird quantum processor) and acceptable levels of noise to allow accurate calculus. The green space indicates the estimations of supremacy, making use of QAOA to solve the routing problem. The plot in Figure 8 providing estimates was calculated based on the complexity analysis performed on [23]. The mathematical and quantum physics principles behind the QAOA algorithm make such low complexities possible for the application to this kind of problem, where the main handicap is the difficulty in simulating it on a classical computer. However, the tests performed so far on IBM hardware proved the small complexity of the problem; to perform the supremacy estimations, the trend was extrapolated to bigger problems using regression techniques. This was compared with a classical implementation of the problem on a Python general purpose linear programming kit (GPLK) to check the computational cost of this implementation compared with the quantum solution. Afterward, supremacy expectations were achieved by comparing costs and positioning the quantum implementation in the IBM real hardware road map. As a result, the research determines that a few hundred qubits (well-calibrated) are enough to reach supremacy for this kind of problem, where classic algorithms really struggle to reach optimal solutions. This is expected to be reached when an Eagle 127 qubits processor achieves stability in a 2–3 year vista. Some factors that may move the exact supremacy point are the CLOPS and the volume achieved in the Eagle 127 qubits processor. Higher CLOPS will decrease the time used to calculate the solution, and a higher quantum volume will increase the performance of the algorithm by obtaining better solutions with fewer iterations.

4. Conclusions

Optimization problems have always been challenging for computers due to the inherent difficulty involved. Quantum computers have already shown their potential in many application fields, with optimization tools being an advanced area. The next generation of computers will be able to use other methods to solve hard optimization problems. Quantum computers have shown their potential for these purposes, making use of new algorithms designed to take profit from quantum properties. QAOA is an example of those algorithms and has shown strong potential since its release. This results in fast, powerful algorithms for quantum computers (that nowadays lack the hardware to reach their true potential). In this case, QAOA has been used to solve some routing problems applicable to the next generation of wireless communication systems. Since 6G will include the massive interconnections between multiple nodes, optimal routing will have a large importance in resource optimization. This paper has shown that it is already feasible to solve routing problems with QAOA. The first example introduced the entire procedure followed by solving the single-target routing problem; the second case involved a higher number of nodes and multi-objectives to test QAOA and provide conclusions. Moreover, as better devices are available, fewer iterations of the algorithm will be needed. Routing problems have additional difficulty since integer programming is harder to solve than linear programming, but QAOA still managed to find optimal routes for the problems proposed. These examples were small demos of what a bigger problem would look like, but the actual quantum computers can only solve limited-size problems. The scalability of the problems solved using QAOA rely on the new quantum hardware available. A higher number of qubits will allow bigger problem resolutions, and soon a quantum computer could reach the size where it outperforms classical computations in terms of solving time. The higher quantum volume will enhance accuracy and, as a result, will decrease the number of iterations of QAOA needed to solve the problem. Higher CLOPS will speed up any process in the quantum computer and, as a result, will enable real-time calculations. The research topic is still open. In this paper, the results were obtained from the lexicographic method to find the Pareto point in the multi-objective problem. Another method, such as goal programming, could also be used in this framework. The number of qubits required for a 6-node example was 50 for this method. It could not be run on a quantum computer, but its implementation will be under consideration for further study.

Author Contributions: H.U. and P.P.-M.; methodology. H.U. and P.P.-M.; software, H.U. and P.P.-M.; validation, H.U., P.P.-M., D.G.-R. and J.F.M.; formal analysis, J.F.M.; investigation, H.U. and P.P.-M.; resources, H.U., P.P.-M., D.G.-R. and J.F.M.; writing—original draft preparation, H.U., P.P.-M., D.G.-R.; writing—review and editing, D.G.-R. and J.F.M.; supervision, J.F.M.; project administration, J.F.M.; funding acquisition, J.F.M. All authors have read and agreed to the published version of the manuscript.

Funding: This research received no external funding.

Institutional Review Board Statement: Not applicable.

Informed Consent Statement: Not applicable.

Data Availability Statement: Not applicable.

Conflicts of Interest: The authors declare no conflict of interest.

References

1. Yetgin, H.; Cheung, K.T.K.; Hanzo, L. Multi-objective routing optimization using evolutionary algorithms. In Proceedings of the 2012 IEEE Wireless Communications and Networking Conference (WCNC), Paris, France, 1–4 April 2012; pp. 3030–3034.
2. Alanis, D.; Botsinis, P.; Babar, Z.; Ng, S.X.; Hanzo, L. Non-Dominated Quantum Iterative Routing Optimization for Wireless Multihop Networks. *IEEE Access* **2015**, *3*, 1704–1728. [CrossRef]
3. Sun, X.; Wang, H.; An, C.; Zheng, Z.; Yao, L. An improved quantum optimization algorithm for multicast routing problem. In Proceedings of the 2011 IEEE 13th International Conference on Communication Technology, Jinan, China, 25–28 September 2011; pp. 182–186.
4. Bhatia, M.; Sood, S.K. Quantum computing-inspired network optimization for IoT applications. *IEEE Internet Things J.* **2020**, *7*, 5590–5598. [CrossRef]
5. Ghorpade, S.N.; Zennaro, M.; Chaudhari, B.S.; Saeed, R.A.; Alhumyani, H.; Abdel-Khalek, S. Enhanced Differential Crossover and Quantum Particle Swarm Optimization for IoT Applications. *IEEE Access* **2021**, *9*, 93831–93846. [CrossRef]
6. Ghorpade, S.N.; Zennaro, M.; Chaudhari, B.S.; Saeed, R.A.; Alhumyani, H.; Abdel-Khalek, S. A Novel Enhanced Quantum PSO for Optimal Network Configuration in Heterogeneous Industrial IoT. *IEEE Access* **2021**, *9*, 134022–134036. [CrossRef]
7. Harwood, S.; Gambella, M.; Trenev, D.; Simonetto, A.; Bernal, D.; Greenberg, D. Formulating and Solving Routing Problems on Quantum Computers. *IEEE Trans. Quantum Eng.* **2021**, *2*, 3100118. [CrossRef]
8. Behera, B.K.; Panigrahi, P.K. Solving vehicle routing problem using quantum approximate optimization algorithm. *arXiv* **2020**, arXiv:2002.01351.
9. Fitzek, D.; Ghandriz, T.; Laine, L.; Granath, M.; Kockum, A.F. Applying quantum approximate optimization to the heterogeneous vehicle routing problem. *arXiv* **2021**, arXiv:2110.06799.
10. Kim, M.; Kasi, S.; Lott, P.A.; Venturelli, D.; Kaewell, J.; Jamieson, K. Heuristic Quantum Optimization for 6G Wireless Communications. *IEEE Netw.* **2021**, 35, 8–15. [CrossRef]
11. Farhi, E.; Goldstone, J.; Gutmann, S. A quantum approximate optimization algorithm. *arXiv* **2014**, arXiv:1411.4028 2014.
12. Ramakrishnan, K.G.; Rodrigues, M.A. Optimal routing in shortest-path data networks. *Bell Labs Tech. J.* **2001**, *6*, 117–138. [CrossRef]
13. Oki, E. *Linear Programming and Algorithms for Communication Networks: A Practical Guide to Network Design, Control, and Management*; CRC Press: Boca Raton, FL, USA, 2012.
14. You, X.; Wang, C.X.; Huang, J.; Gao, X.; Zhang, Z.; Wang, M.; Huang, Y.; Zhang, C.; Jiang, Y.; Wang, J.; et al. Towards 6G wireless communication networks: Vision, enabling technologies, and new paradigm shifts. *Sci. China Inf. Sci.* **2020**, *64*, 110301. [CrossRef]
15. Preskill, J. Quantum Computing in the NISQ era and beyong. *Quantum* **2018**, *2*, 79. [CrossRef]
16. Bharti, K.; Cervera-Lierta, A.; Kyaw, T.H.; Haug, T.; Alperin-Lea, S.; Anand, A.; Degroote, M.; Heimonen, H.; Kottmann, J.S.; Menke, T.; et al. Noisy intermediate-scale quantum (NISQ) algorithms. *APS Phys.* **2022**, *94*, 015004
17. Peruzzo, A.; McClean, J.; Shadbolt, P.; Yung, M.-H.; Zhou, X.-Q.; Love, P.J.; Aspuru-Guzik, A.; O'brien, J.L. A variational eigenvalue solver on a photonic quantum processor. *Nat. Commun.* **2014**, *5*, 4213. [CrossRef]
18. Solving Combinatorial Optimization Problems Using QAOA. Available online: https://qiskit.org/textbook/ch-applications/qaoa.html (accessed on 5 October 2022).
19. Glover, F.; Kochenberger, G.; Hennig, R.; Du, Y. Quantum bridge analytics I: A tutorial on formulating and using QUBO models. *Ann. Oper. Res.* **2022**, *314*, 141–183. [CrossRef]
20. Qiskit Runtime—IBM Quantum. Available online: https://quantum-computing.ibm.com/lab/docs/iql/runtime/ (accessed on 5 October 2022).
21. System Configuration—IBM Quantum. Available online: https://quantum-computing.ibm.com/lab/docs/iql/manage/systems/configuration (accessed on 5 October 2022).
22. Guerreschi, G.G.; Matsuura, A.Y. QAOA for Max-Cut requires hundreds of qubits for quantum speed-up. *Sci. Rep.* **2019**, *9*, 6903. [CrossRef] [PubMed]
23. Farhi, E.; Harrow, A.W. Quantum Supremacy through the Quantum Approximate Optimization Algorithm. *arXiv* **2016**, arXiv:1602.07674.

sensors

MDPI

Article

A Fair Channel Hopping Scheme for LoRa Networks with Multiple Single-Channel Gateways

Alexandre Figueiredo [1], Miguel Luís [1,2,*] and André Zúquete [1,3]

[1] Instituto de Telecomunicações, 3810-193 Aveiro, Portugal; alex.figueiredo@ua.pt (A.F.); andre.zuquete@ua.pt (A.Z.)
[2] ISEL—Instituto Superior de Engenharia Lisboa, Instituto Politécnico de Lisboa, 1959-001 Lisbon, Portugal
[3] Departamento de Eletrónica, Telecomunicações e Informática (DETI), University of Aveiro, 3810-193 Aveiro, Portugal
* Correspondence: nmal@av.it.pt

Abstract: LoRa is one of the most prominent LPWAN technologies due to its suitable characteristics for supporting large-scale IoT networks, as it offers long-range communications at low power consumption. The latter is granted mainly because end-nodes transmit directly to the gateways and no energy is spent in multi-hop transmissions. LoRaWAN gateways can successfully receive simultaneous transmissions on multiple channels. However, such gateways can be costly when compared to simpler single-channel LoRa transceivers, and at the same time they are configured to operate with pure-ALOHA, the well-known and fragile channel access scheme used in LoRaWAN. This work presents a fair, control-based channel hopping-based medium access scheme for LoRa networks with multiple single-channel gateways. Compared with the pure-ALOHA used in LoRaWAN, the protocol proposed here achieves higher goodput and fairness levels because each device can choose its most appropriate channel to transmit at a higher rate and spending less energy. Several simulation results considering different network densities and different numbers of single-channel LoRa gateways show that our proposal is able to achieve a packet delivery ratio (PDR) of around 18% for a network size of 2000 end-nodes and one gateway, and a PDR of almost 50% when four LoRa gateways are considered, compared to 2% and 6%, respectively, achieved by the pure-ALOHA approach.

Keywords: low power wide-area networks; large-scale LoRa networks; single-channel LoRa gateways

Citation: Figueiredo, A.; Luís, M.; Zúquete, A. A Fair Channel Hopping Scheme for LoRa Networks with Multiple Single-Channel Gateways. *Sensors* **2022**, *22*, 5260. https://doi.org/10.3390/s22145260

Academic Editors: Federico Tramarin and Peter Han Joo Chong

Received: 16 May 2022
Accepted: 12 July 2022
Published: 14 July 2022

1. Introduction

The emergence of the Internet of Things (IoT) [1,2], potentiated by its ability to connect every device to the Internet, increased the demand for low power wide-area networks (LPWANs). LPWANs are a category of wireless communication technologies with unique characteristics, such as long-range communications, low power consumption, and low deployment costs, making them suitable for IoT applications. Therefore, to satisfy the wide range of IoT applications, several communication technologies were developed, such as LoRa, Sigfox [3], Ingenu [4], and NB-IoT [5].

Long-range (LoRa) [6] has been one of the most popular and exciting LPWAN technologies thanks to its robustness to noise, which allows long-range transmissions, the non-destructive property of colliding packets, and its power efficiency. LoRa, initially only a physical layer, was extended by adding a medium access control (MAC) layer, LoRaWAN, standardized and open-sourced by the LoRa Alliance. This MAC layer defines the network architecture, along with a channel access scheme and the adaptive data rate (ADR) mechanism. This mechanism aims to optimize the transmissions by dynamically changing the transmission parameters according to the signal quality with the receivers.

LoRa operates on the sub-GHz Industrial, Scientific and Medical (ISM) unlicensed bands, the 433 and 868 MHz bands in Europe. These bands are free to use and have less retention than other bands, such as the 2.4 and 5 GHz bands. However, in most situations, the transmissions are limited to 1% of the duty cycle. LoRa signals are modulated using chirp spread spectrum (CSS) modulation, which provides excellent resistance to noise. LoRa allows the customization of several transmission parameters, such as spreading factor (SF), coding rate (CR), bandwidth (BW), and transmission power (TP), resulting in different and orthogonal LoRa communication channels, with specific communication ranges, robustness, and data rates. For example, a larger SF improves the communication range but also increases the time on air (ToA), which increases the energy consumption and reduces the data rate.

Multiple-channel LoRaWAN GWs can simultaneously receive multiple packets with different spreading factors; however, they can be expensive and operate according to the pure-ALOHA channel access scheme, which presents poor network performance, especially in high-density networks. Additionally, it is also known that when two or more concurrent LoRa transmissions occur, the one with the highest signal quality is the one to be decoded by the GW, which makes the network unfair [7]. For these reasons, in this work we explore the use of single-channel LoRa gateways operating in a channel hopping scheme. We propose a cycle-based medium access strategy for large-scale LoRa networks with multiple single-channel GWs. To increase the network fairness, each GW switches between the different channels, i.e., different combinations of transmission parameters, guaranteeing different ranges and rates of transmission so that every end-device can have its best transmission opportunity. These different channels are assigned to the end devices (ED) according to their quality of signal towards the GWs. A study on the channel allocation time was also performed to analyze its impact in the network fairness. Extensive simulation results considering different network densities, different numbers of LoRa GWs, and different channel allocation times showed that our proposal clearly outperforms the channel access scheme used by the LoRaWAN standard with respect to network goodput, packet delivery ratio, number of collisions, and access fairness.

The remainder of this paper is organized as follows. Section 2 addresses the related work regarding network management and medium access control in LoRa networks. Section 3 presents the channel hopping protocol for single-channel LoRa networks, along with the pure-ALOHA behavior for comparison purposes. Section 4 describes the simulation environment and presents preliminary results about the performance of the proposed protocol, and in Section 5 we relate an extensive study on the impact of the channel allocation time in the network performance, using goodput and fairness as performance metrics. Finally, Section 6 enumerates the main conclusions and points out directions for future work.

2. Related Work

Some experiments have already been executed regarding the performance of single-channel LoRa GWs in real-life environments, such as the study by Hanaffi et al. [8], in which a network with a low-cost single-channel LoRaWAN GW was tested. They analyzed the packet loss and the received signal strength of packets transmitted by sensor devices connected to the referred GW from different floors of a building. Despite the performance analysis, no strategy was presented to select the LoRa channel in use by the LoRa GW.

With the goal of improving LoRaWAN network performance, Cuomo et al. [9] proposed two different algorithms to outperform the ADR strategy. First, they proposed EXPLoRa-SF, which uses, in addition to the distance and received signal strength indicator (RSSI) values, the density of the network, i.e., the number of EDs that are in it, to assign the SFs. Next, they proposed and tested EXPLoRa-AT, a more complex algorithm than the previous one, whose goal is to assign the SFs providing a balanced distribution of the channel load among the network EDs through the equalization of the ToA. According to the tests performed, EXPLoRa-AT outperformed both ADR and EXPLoRa-SF, not only in goodput but also in data extraction rate (DER).

The work in [10] explores the idea of EXPLoRa-AT, but extends it to a multiple GW scenario. This scenario raises a situation that must be taken into account in comparison to the single GW scenario: a data packet can be received simultaneously by more than one GW. To prevent a channel being overused, giving rise to a large number of collisions, this work proposes adaptive mitigation of the air-time pressure in LoRa (AD MAIORA). This method allocates the SFs, distributing the network load by the channels and the GWs. Performance tests have shown that AD MAIORA presents a substantial improvement over the basic ADR approach.

Additionally, considering a multi-GW LoRa network, Liao et al. [11] proposed a dynamic method that selects the serving GW and allocates the most appropriate channel, considering the packet error rate (PER) for each network node. The PER value is calculated for each node using ACK/NAK signaling after every up-link transmission. Thus, the proposed method re-selects the serving GW of a node when there is a GW with better transmission conditions, and re-allocates a new SF value to the ED when the number of recent NAK signals surpasses a certain threshold. The method is based on the RSSI value between each GW and node pair to proceed to the GW re-selection, and to do the SF re-allocation, to minimize the probability of data packet collisions, the nodes are distributed in a balanced way by the several SF values. Performance tests regarding the BER and PER showed improvements of 35% and 29%, respectively, compared to the traditional SF allocation, based only on the RSSI values.

Kim et al. [12] proposed a scheme in which the LoRa gateway changes its operation channel to avoid congestion. Thus, nodes sending data marked as urgent follow the gateway as it changes its channel. However, the other nodes sending non-critical information do not change; instead, they wait until the gateway returns to the original channel to transmit again. According to the performed tests, the proposed scheme avoids congestion and improves transmission efficiency by reducing the accumulated transmission delays caused by channel congestion.

Iglesias-Rivera et al. [13] proposed a time-slotted spreading factor hopping (TSSFH) mechanism to tackle blind spots and performance issues in traditional LoRaWAN solutions. In this mechanism, nodes in a blind spot need to find a relay node connected to the gateway, resulting in a two-hop LoRa-based network that uses all available SFs, taking advantage of the orthogonality of LoRa transmissions. The communication opportunity for blind nodes increases with the listening windows opened by the relay nodes. Some performance tests of the proposed mechanism were conducted using the OMNeT++ simulator, concluding that nodes in a blind spot might reach moderate packet delivery rate values when using TSSFH in very high-density LoRa-based WANs.

TSCH-over-LoRa et al. [14] is a layer that connects the implementation of time-slotted channel hopping in Contiki-NG to a LoRa radio driver. Originally, TSCH is a synchronous MAC protocol in which all node transmissions are scheduled. The scheduling defines the timeslots each node can operate in and the channel to enable orthogonal communications. Furthermore, using a pseudo-random channel hopping sequence, TSCH is resilient to external interference and multi-path fading.

Kim et al. [15] proposed a contention-aware ADR to obtain optimal throughput. Thus, it optimizes the pure ALOHA with the gradient projection method for an ideal distribution of SFs in the network, using LoRaWAN with FHSS (frequency-hopping spread spectrum). Hence, it adjusts the SFs to increase the number of nodes using smaller values, which guarantees higher data rates.

Adelantado et al. [16] used the pseudo-random channel hopping method in LoRaWAN to tackle reliability constraints. This method allocates the transmissions for the different available channels, decreasing the number of collisions in the network. Additionally, to minimize the collision probability even further, new adaptive hopping methods were introduced.

Benkhla et al. [17] developed a mechanism called Enhanced-ADR to enhance the ADR mechanism by taking into account the position and trajectory of the EDs to reconfigure the several communication channels. The allocation model is driven by the network server, which calculates the corresponding RSSI and searches for the best RSSI interval in which it can be located to determine the most suitable channel. According to the performed tests, Enhanced-ADR improves the quality of service of the overall networks, solving the issues of ADR, such as low adaptation speed, low performance, and power consumption.

The work presented here is differentiated from the previous work for several reasons. First, instead of considering multi-channel LoRa GWs, it admits that a LoRa GW is only capable of decoding one LoRa channel at a time. This assumption is important, since it allows the possibility of a simple LoRa transceiver being used as LoRa GW. Then, the LoRa protocol here presented does not follow the channel access behavior of the LoRaWAN standard (the pure-ALOHA scheme), which is known to perform poorly in high-density scenarios. Instead, and following previous works that have shown that control packets can still be used in LoRa networks without harming the energy consumption of the EDs [18] (lower energy spent per packet delivered), we present a fair and efficient channel hopping scheme for multiple LoRa GW networks. GWs switch between different communication channels, as they can only operate one at a time, to allow all EDs in their range to successfully transmit data packets, regardless of the signal strength. Finally, the performance results achieved through simulation consider a more realistic LoRa packet capture effect [7], instead of the outdated 6 dB collision model that is still used in the literature.

3. A Fair MAC Channel Hopping Scheme for LoRa Networks

This work presents a MAC protocol for LoRa networks based on channel hopping, which was tested through simulation. It uses a reservation strategy, where EDs use a control packet for advertising their neighboring EDs about their intent to transmit a data packet. Moreover, each ED uses the most favorable channel available to send their data packets.

Therefore, to address every communication situation and keep a low level of complexity, +three channels were chosen, i.e., three combinations of transmission parameters, each one with a different transmission range and rate. They are:

- **Fast-Rate Channel**: Due to its short communication range capability, this channel is for EDs closer to a GW, and has excellent signal quality. It is the fastest of the three, so the EDs that use it take less time to transmit a packet. Consequently, the duty-cycle restriction time is shorter, increasing the number of transmissions allowed per ED;
- **Slow-Rate or Standard Channel**: This channel is for EDs within reach of a GW but has poor signal quality. The ToA of the data packets is the longest. The GWs use it to send synchronization packets, which will be introduced later in this chapter, as it is the one with the longest range;
- **Mid-Rate Channel**: Used by EDs with intermediate signal strength with a GW, i.e., worse than those using fast-rate but better than those using slow-rate. The ToA of the data packets is longer than in the fast-rate channel due to the more extended range, but shorter than in the slow-rate channel.

An ED only transmits data packets in its ideal channel. This is the one that guarantees the highest data rate and lowest ToA, based on the RSSI to the GWs. Before each transmission, it transmits a ready-to-send (RTS) control packet to advertise to neighboring EDs its intent. An RTS is also transmitted using the ideal channel; therefore, only the EDs in its reach that are using the same channel can decode it.

The RTS packet structure, depicted in Figure 1, is based on the one defined by the SX1272 LoRa module support library and used in [18]. In this structure, 5 bytes are the minimum packet header, and the remaining 4 bytes are for data payload, used directly by the EDs. Of these, 2 bytes are reserved for the address of the GWs that might receive the advertised packet, 1 byte for the type of packet transmitted (control, in this case), and 1 byte for the size of the advertised packet.

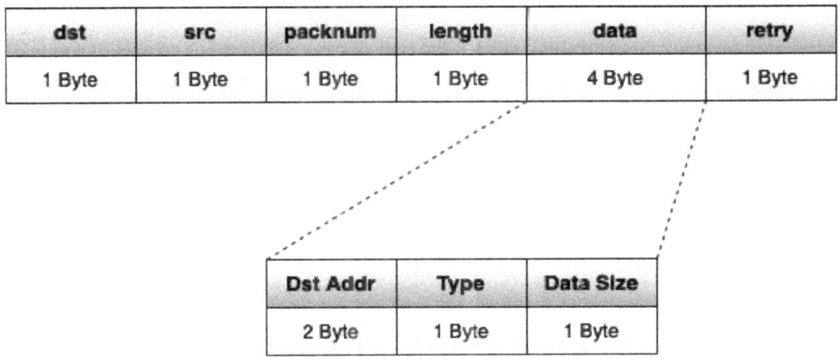

Figure 1. RTS packet structure.

The EDs can discard RTS packets based on the information regarding the targeted GWs. If none of the targeted GWs reach the ED or do not provide it with the ideal transmission rate available, the RTS is discarded. Otherwise, the ED calculates the backoff time to take, being prevented from transmitting during it. The backoff time is calculated using the data packet size described by the RTS, plus an additional time, given by a random amount of backoff slots, to reduce collisions upon backoff periods. The duration of each backoff slot is equal to the ToA of an RTS packet. After each transmission, an ED must calculate the mandatory self-restriction period, as LoRa transmissions are restricted to 1% of duty cycle. This period also includes a random number of backoff slots to desynchronize EDs that might have transmitted simultaneously the same amount of data.

As we are dealing with networks with multiple single-channel GWs, an ED can be located in an overlap zone of the GWs' coverage. In those areas, an ED can have different ideal channels for each GW. However, it will use exclusively one of those channels, namely, the fastest one.

The GWs change their communication parameters over time to give equal chances to all EDs in their reach to transmit. When the operation channel hops from the Standard to a non-Standard, a GW sends a synchronization packet, the change mode (CM) packet, which advertises to the EDs in its reach about the non-Standard channel to which it will change and for how long it will use it. This packet is always transmitted using the Standard channel, as it has the most significant reach. The execution flow of each GW using our channel hopping protocol is shown in Algorithm 1.

Algorithm 1: LoRa channel hopping protocol: GW execution flow.

```
// End_CT: The time in the actual channel comes to an end;
// Using_Standard: The GW is using Standard channel;
// End_BT: The mandatory backoff time comes to an end;
// CM_BT: CM packet backoff time;
// Overheard_DP: Overhears a Data packet;
if End_CT then
    if Using_Standard then
        if End_BT then
            Send a CM packet ;
            Calculate CM_BT ;
            Change to the advertised channel ;
    else
        Change to Standard channel ;
else
    if Overheard_DP then
        Decode Data Packet ;
```

The structure of CM packets is similar to the one used in RTS packets, but with a payload of 6 bytes. Of these, 2 bytes are for the address of the transmitting GW, 1 byte for the packet type identification, 1 byte for identifying the non-Standard channel to which the GW will hop to, and 2 bytes for the period that the GW will remain in that channel.

As in the LoRa networks all the devices have to respect the duty-cycle restriction, the GWs have to change their operation channels accordingly. Hence, they have to wait for ninety-nine times the ToA of a CM packet between transitions to non-Standard channels. During that period, the GWs hop back to the Standard channel without further advertisement, upon the end of the non-Standard time advertised by a CM packet. As soon as the duty-cycle restriction ceases, a GW can change again to a non-Standard channel. GWs do not change to the same non-Standard channel consecutively; they alternate between Mid-Rate and Fast-Rate. Therefore, if a GW hops from the Standard to Mid-Rate channel in a first instance, in the next hop to a non-Standard channel it will chose the Fast-Rate channel, and vice versa. The GW channel hopping cycle is shown in Figure 2.

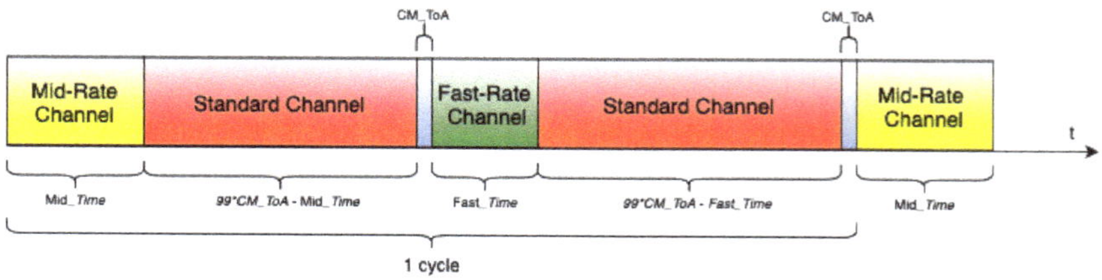

Figure 2. LoRa channel hopping protocol: GW hopping cycle.

Algorithm 2 shows the complete behavior of the EDs with our channel hopping protocol. As we can see, control packets (RTS and CM) are filtered by EDs according to the targeted GWs and the used channel. Thus, EDs only consider RTS packets that advertise a transmission in their ideal channel and to the same GW, or GWs, that they use. With this, as soon as a GW in their reach changes to their ideal channel, they can transmit right away, without having to wait for the conclusion of transmissions with which they do not interfere.

Algorithm 2: LoRa channel hopping protocol: ED idle state execution flow.

```
// Scheduled_CM: CM packet about to be received;
// CM_RecBT: Backoff time until receiving a CM packet;
// Overheard_RTS: Overhears an RTS packet;
// GW_inUse: Advertised destination is GW in reach;
// RTS_BT: RTS packet backoff time;
// Overheard_CM: Overhears a CM packet;
// CM_BT: CM packet backoff time;
// Ideal_Channel_GW: GW, or GWs, using the ED Ideal Channel at the moment;
// Random_BT: Random backoff time;
// DC_Restriction: Duty-cycle restriction left to be respected;
// Data_isReady: Data packets ready to be transmitted
// Mandatory_BT: Mandatory backoff time;
if any Scheduled_CM then
    if using Standard_Channel then
        Calculates CM_RecBT ;
    else
        Change to Standard_Channel ;
        Calculate CM_RecBT ;
    Enter Backoff State ;
else
    if Overheard_RTS then
        if GW_inUse then
            Calculate RTS_BT ;
            Enter Backoff State ;
    if Overheard_CM then
        if advertised Ideal_Channel then
            Change to Ideal_Channel ;
            Calculate CM_BT ;
            Enter Backoff State ;
        else
            if any Ideal_Channel_GW then
                Change to Ideal_Channel ;
                Calculate Random_BT ;
                Enter Backoff State ;
            else
                Enter Backoff State ;
    else
        if any DC_Restriction then
            Enter Idle State ;
        else
            if Data_isReady then
                if using Ideal_Channel then
                    if any Ideal_Channel_GW then
                        Send RTS packet ;
                        Send Data packet ;
                        Calculate Mandatory_BT ;
                        Enter Backoff State ;
                    else
                        Enter Idle State ;
                else
                    Enter Idle State ;
            else
                Enter Idle State ;
```

4. Performance Evaluation

4.1. Setup and Methodology

The results presented here were obtained using MATLAB simulators developed specifically to represent the behavior of each LoRa protocol: the pure-ALOHA, in which all devices use the same transmission parameters; and the channel hopping, where the EDs transmit using their most appropriate transmission channels.

The simulation time was chosen to guarantee that every ED has the opportunity to transmit, regardless of the network size. Therefore, the chosen value was 3×10^8 ms, approximately 83.33 h. Data packets are generated periodically so that every ED has always a packet ready to be transmitted. Regarding packet sizes, data packets have always 100 bytes, RTS packets 9 bytes, and CM packets 11 bytes, whose ToA are detailed in Table 1 (given the channel details in Table 2). The non-destructive property of LoRa was modeled following the work in [7].

Table 1. ToA of each control packet.

	Standard Channel	Mid-Rate Channel	Fast-Rate Channel
RTS Packet	286.55 ms	75.52 ms	12.39 ms
CM Packet	307.05 ms	-	-

Table 2. Radio parameters characterizing the different LoRa channels used in this work.

	BW (kHz)	CR	SF	Sensitivity (dBm)
Standard Channel	125	4/5	10	−129
Mid-Rate Channel	250	4/5	9	−123
Fast-Rate Channel	500	4/5	7	−114

For the pure-ALOHA protocol, as every device uses the same channel, the radio parameters never change, corresponding to the Standard channel ones in Table 2, as they are the ones that enable longer-range communications. The radio parameters used were based on the SX1272 LoRa module operation modes [19], and the selected frequency of operation was 868 MHz.

For each performance test, the number of EDs varied between 100 and 2000. The circumferences centered in each one of the GWs represent the maximum communication reach of each available channel. Moreover, the times allocated for the Fast-Rate and Mid-rate channels, per cycle, were 7.5 and 10 s, respectively.

Figure 3 shows the positioning of the GWs in the different tested scenarios. GW1, represented as a red cross, is the only one considered in all tested scenarios. Its maximum reached using the Standard channel is represented as a red circumference; all the EDs were located in the circle limited by this circumference. Regarding the remaining GWs, 2 through 5, they were considered only for multiple GW scenarios.

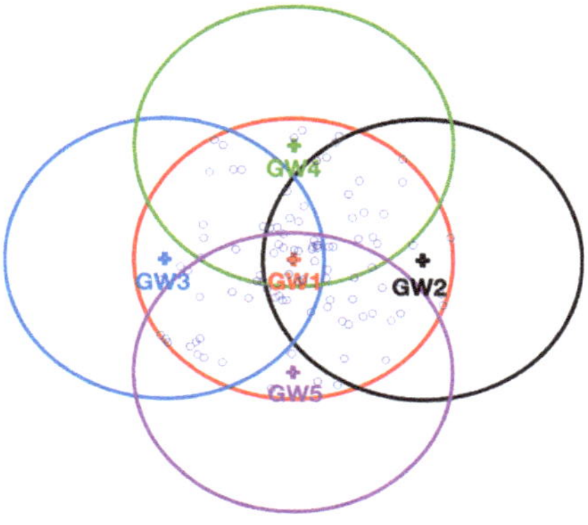

Figure 3. LoRa network with 5 gateways, GW1, GW2, GW3, GW4, and GW5, represented by red, black, blue, green, and purple crosses, respectively, and 100 EDs, represented as blue circles.

Regarding the quality of each ED transmission, we decided to resort to measurements obtained by Oliveira et al. [20]. Thus, the maximum communication and signal strength ranges, for each channel, are presented in Table 3.

Table 3. Maximum communication and RSSI range for each LoRa channel.

	Range (m)	RSSI Range (dBm)
Fast-Rate	1210	[−100, −90]
Mid-Rate	2890	[−110, −101]
Standard	4030	[−125, −111]

4.2. Performance Results

Figure 4 shows the network goodput for both access protocols, for different network sizes and scenarios. With the addition of more GWs, the goodput of pure-ALOHA improves, as a higher number of EDs have a better chance to transmit their packets successfully, namely, those which are furthest from GW1. However, our scheme, when compared with the pure-ALOHA access scheme in a network of 2000 EDs, further improved the goodput by 157.07% with the installation of a second gateway (GW2), 62.01% upon including a third one (GW3), and 76.33% when five GWs were considered.

The packet delivery ratio (PDR), illustrated in Figure 5, also increased with the number of GWs; this increase was more significant for smaller networks. For the same network configuration, our scheme always outperformed the pure-ALOHA one.

Regarding the collision percentage, illustrated in Figure 6, we can see that for pure-ALOHA this value is close to 100% no matter the number of gateways in the network; this is justified by the high number of EDs in the experiments, leading to a network saturation scenario. With our scheme, the percentage of collision increases with the network size. For 2000 EDs, it ranges from 72% with five GWs up to 97% with only one GW. Nevertheless, in spite of the high percentage values observed in both schemes, the network goodput is not equal to zero, thanks to the non-destructive property of LoRa.

The percentage of duplicate packets, i.e., packets received simultaneously by more than a single GW, increases with the number of GWs, for networks with less than approximately 1000 EDs, as seen in Figure 7. From this density up, the percentages are very similar for the networks with different sizes.

As shown in Figure 8, the Jain's network fairness of pure-ALOHA improves with the introduction of new GWs, as there is a better chance for the EDs, namely, those further from GW1, to have a better signal strength to a given GW. As the network increases, the number of collisions also increases, and the EDs with better signal quality have higher chances to deliver their packets. Therefore, the fairness decreases, regardless of the introduction of GWs to the network.

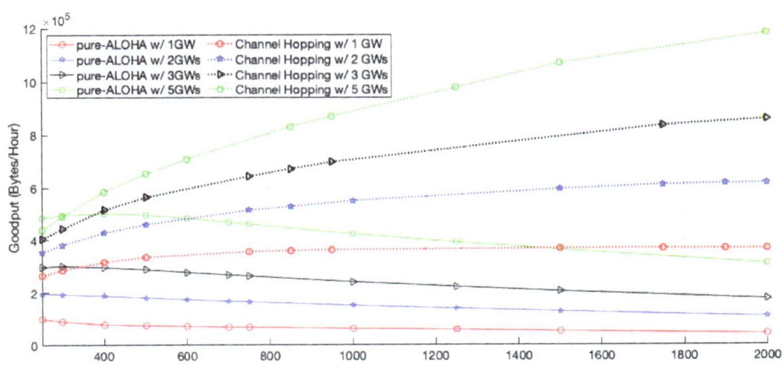

Figure 4. Network goodput.

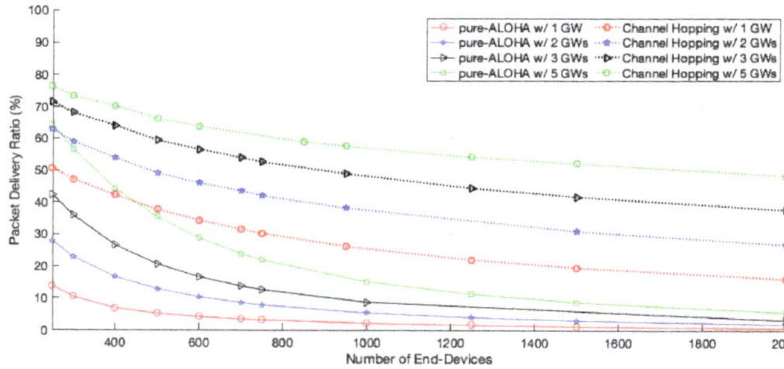

Figure 5. Network packet delivery ratio (PDR).

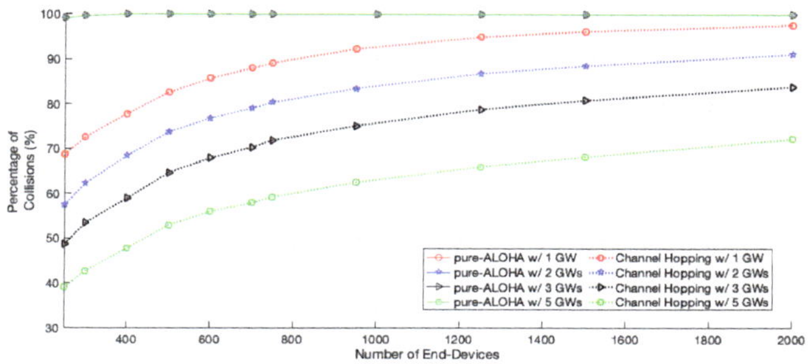

Figure 6. Percentage of collisions.

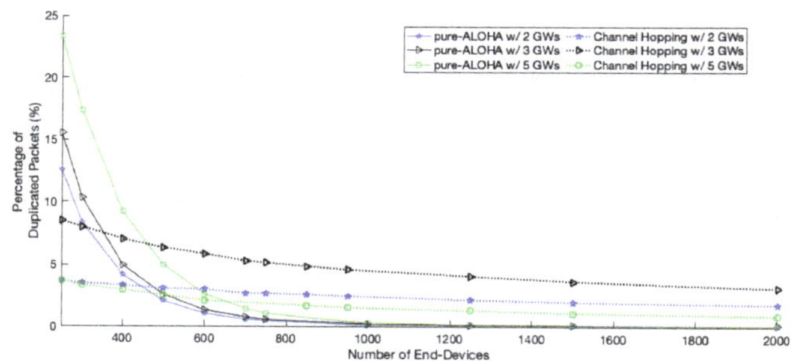

Figure 7. Percentage of duplicated packets among all packets received.

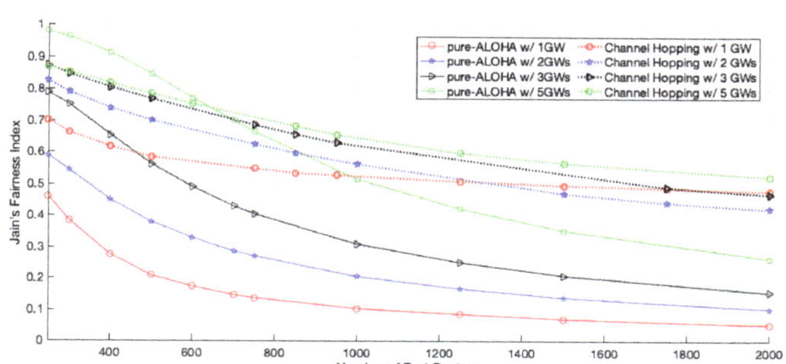

Figure 8. Jain's network fairness.

As for the channel hopping protocol, the network goodput also increases with the number of GWs. This time, the increase is more pronounced than with pure-ALOHA, as the EDs can use faster channels to transmit, and thus have more opportunities to access the medium. Moreover, as the competition to access the medium of each GW is lower, there is a better chance for the transmissions to be successful, which leads to an increase in the PDR, as shown in Figure 5. For a network size of 2000 EDs, the goodput of channel hopping increases by 68.12%, 39.14%, or 37.11%, with two, three, and five GWs, respectively, when adding GWs to the network.

The percentage of duplicated packets per medium access is not as high as that obtained in pure-ALOHA for small networks; however, as the network scales up, the percentage becomes higher than pure-ALOHA's because each transmission is more likely to be delivered, as shown in Figure 5. Moreover, with this protocol, the percentage registered for 5 GWs is the lowest of the three scenarios. The network EDs further from GW1 do not transmit duplicated packets so often, as they have closer GWs and can transmit their packets using a faster channel. This behavior is shown in Figure 9.

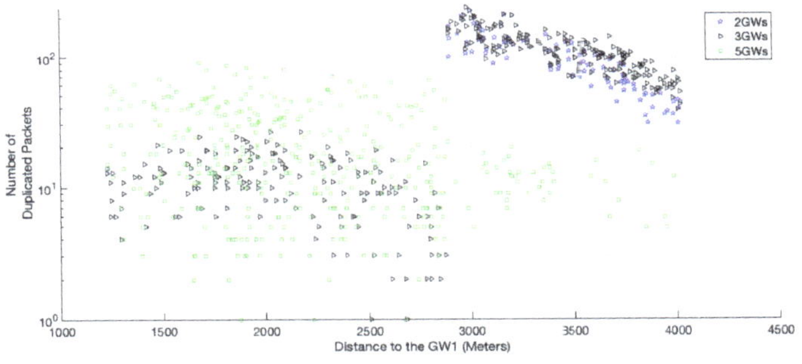

Figure 9. Number of duplicated packets per ED for a network with 2000 EDs using channel hopping protocol.

Regarding the network fairness of the channel hopping protocol, shown in Figure 8, for small networks, the scenario with one GW started to be the unfairer. However, for larger-scale networks, the same one became fairer and surpassed those with two and three GWs from 1313 EDs and 1800 EDs, respectively. This was due to the better transmission conditions that are guaranteed to some EDs with the addition of GWs, to the detriment of others, causing the performance differences to increase in the network. In the scenario with fiver GWs, as the GWs are distributed more equally around GW1, the network is fairer than in the others where the network is unbalanced. The EDs which use the channels with the fastest data rates can transmit packets more often than the ones that use the slowest channel. Thus, the time allocation for the channels, by the GWs, must be according to the transmission rate and density of EDs using each one of them, giving similar chances to each ED to transmit a data packet, regardless of the channel they use.

5. Channel Time Analysis

Up to this point, the performance analysis of both pure-ALOHA and channel hopping protocols has been focused on the number of GWs on the network. In this section, we will focus our attention on the impact of the channel time allocation on the network performance. In the previous section, the time allocated by the GWs, per cycle, to each channel was static. Thus, despite the varying number of EDs and the number of GWs covering the network, the amount of time set to each channel remained unchanged, distributing the time unfairly.

5.1. Channel Time Allocation for a Single GW Network

To obtain the time allocation that grants the best network fairness, we first considered a single GW network with a variable number of EDs. For each network size, several channel time distributions were evaluated in order to find the ones that grant similar chances of channel access opportunities. Table 4 presents the channel time allocations for each network size which yield the best fairness indicator. As expected, the channel time allocated to the Fast-Rate must be lower than for the other channels, regardless the network size, as the data rate is faster, and thus the EDs have more chances to access the medium.

Table 4. Channel times for a single GW scenario, per cycle.

Number of EDs	Mid-Rate Channel (ms)	Fast-Rate Channel (ms)
100	9205	4336
250	9133	2805
500	8774	2211
1000	7957	1820
1500	7568	1663
2000	7483	1628

Figure 10 illustrates the Jain's network fairness according to the channel allocation times presented in Table 4, where we can see that by using a customized time channel allocation time, according to the number of EDs in the network, the protocol is considerably fair when compared to a blindsided version of it, with a fixed allocation time. In particular, for a network of 2000 EDs, the network fairness index increased from 0.48 to 0.76. As explained before, by using a fair channel allocation time we are reducing the channel time allocated to the fastest channel because the data rate is higher, thereby granting the same channel access opportunities to EDs in other zones. Naturally, such a reduction has an impact on the network goodput, as illustrated in Figure 11, because EDs in faster zones are the ones that contribute more to the network goodput.

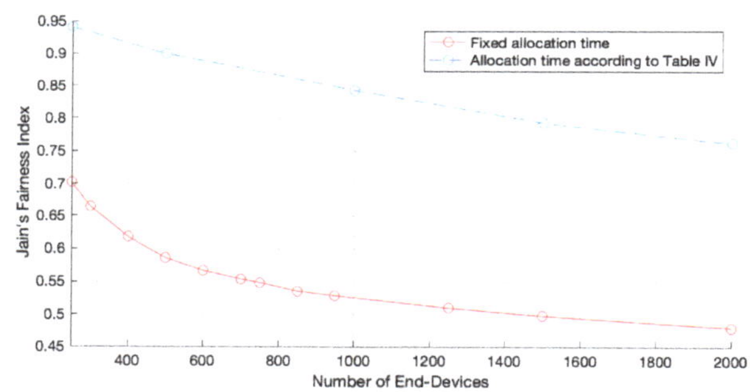

Figure 10. Network fairness for one GW and fair channel allocation times.

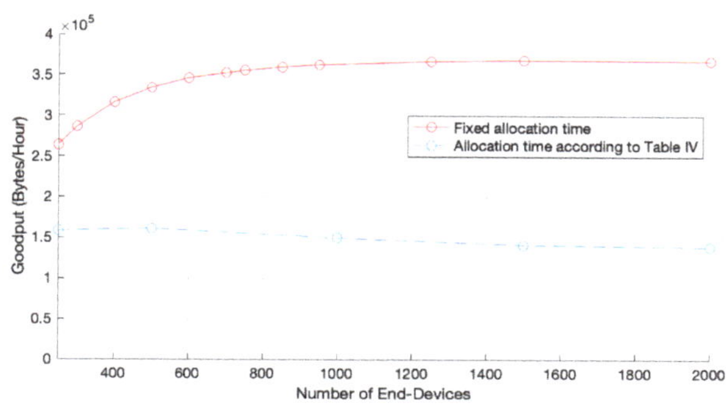

Figure 11. Network goodput for 1 GW and fair channel allocation times.

5.2. Channel Time Variation for Multiple GW Networks

By increasing the number of GWs, the number of EDs optimally using the Standard channel decreases. Therefore, in order to keep the network fair, the time allocated to the Standard channel must be reduced. To analyze the impact of the channel time variation in the scenario of multiple GWs, we decreased the amount of time initially given to the Standard channel and distributed to the non-Standard channels following the proportion of non-Standard times obtained for a single GW (hereafter simply denoted as percentage of decrease, or simply PoD). Therefore, the Standard channel time was decreased from 2.5% to 50%.

Let us start by analyzing a network with 2 GWs. As shown in Figure 12, for large networks, the fairness index increases with the PoD until a maximum value is observed. For example, for a network size of 1500 EDs, this maximum is achieved when the PoD is around 15%, and for a network size of 100 EDs, the fairness peak is registered when the PoD is around 2.5%. When we remove more time from the Standard channel and give it to the non-Standard channels, i.e., when we increase the PoD, the fairness drops due to the lack of channel access opportunities of the EDs using the Standard channel as the optimal one.

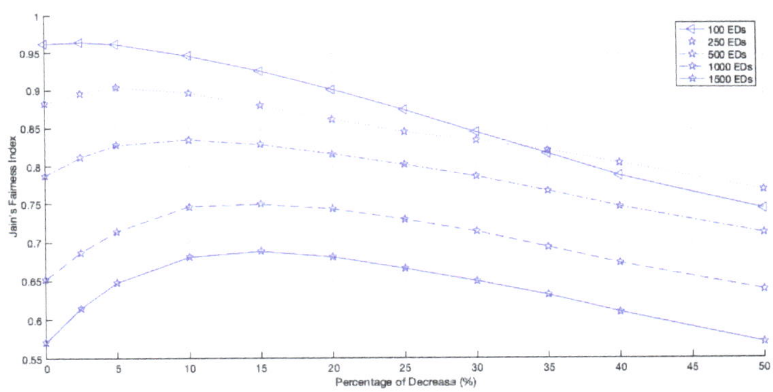

Figure 12. Network fairness for 2 GWs, different network sizes and different PoD.

As the EDs can transmit their packets at a higher data rate, competing for the medium more often, the network goodput improves with an increase in the PoD, as shown in Figure 13. However, as the EDs that use channels with slower data rates have fewer transmission opportunities, network fairness is negatively affected.

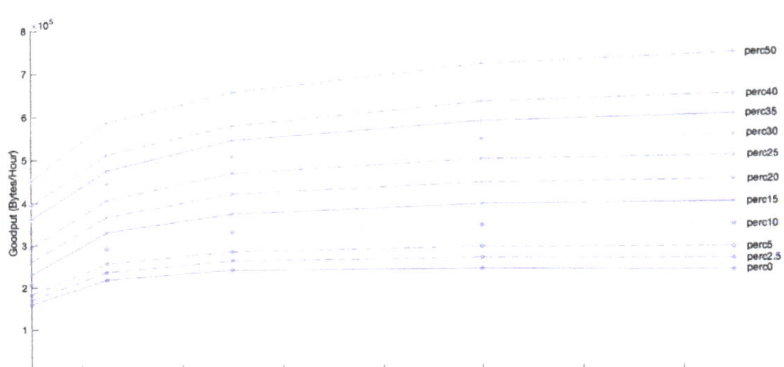

Figure 13. Network goodput for 2 GWs, different network sizes and different PoD.

When we install a third GW, the behavior of the network fairness, as shown in Figure 14, is identical to the prior behavior. The PoD from which the network fairness starts to decrease is about 15% when the network has 1500 EDs, the same value registered for the 2 GWs scenario. However, the peak occurrence with the network growth does not always have increasing behavior. The network with 100 EDs, regardless of initially being the fairest scenario, was surpassed when the decrease surpassed 38%. This new threshold occurred later than in the 2 GWs network (34%).

Similarly to what happened in the scenario with 2 GWs, the network goodput for a 3 GWs scenario, illustrated in Figure 15, increased with the time allocated to the faster channels, namely, the Fast-Rate channel, which allowed the EDs to access the medium more often.

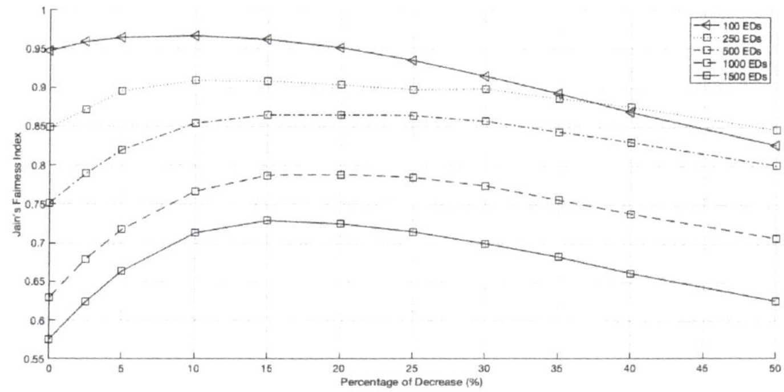

Figure 14. Network fairness for 3 GWs, different network sizes and different PoD.

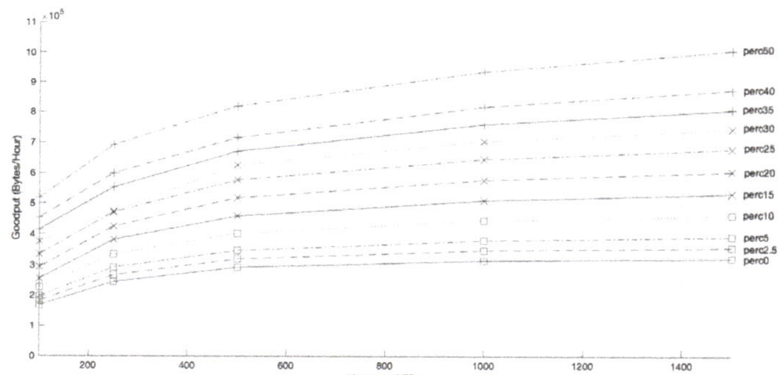

Figure 15. Network goodput for 3 GWs, different network sizes and different PoD.

At last, we present the results regarding the time allocation in a 5-GWs scenario. As shown in Figure 16, we can see an increase in the network fairness until the PoD of the Standard channel time surpasses 25%, for a network with 1500 EDs. This time, the fairness in less populated networks, namely, 100 EDs, increases for a PoD below 25%.

The interception between the curves of 100 EDs and 250 EDs happened for a higher PoD, approximately 44%, as shown in Figure 16. Furthermore, PoD above 40% had a more negligible influence on the fairness of networks with 1500 EDs. Like the networks with three GWs, the fairness peak does not always increase with the network growth; it starts to decrease above about 500 EDs.

Regarding the network goodput, the behavior is very similar to that registered for scenarios with two and three GWs: increasing with the time allocated to faster channels, as shown in Figure 17.

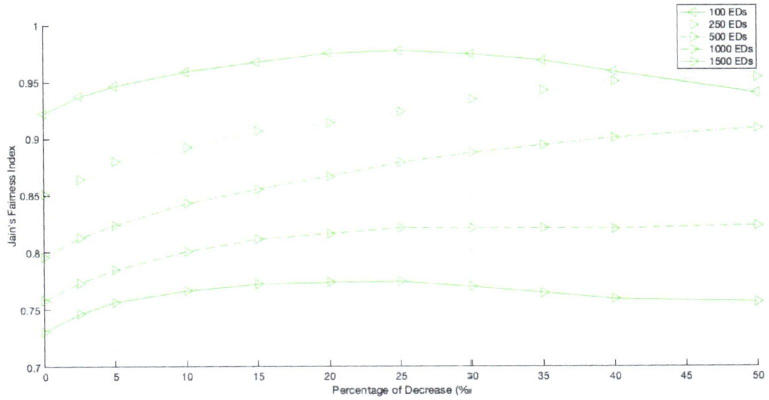

Figure 16. Network fairness for 5 GWs, different network sizes and different PoD.

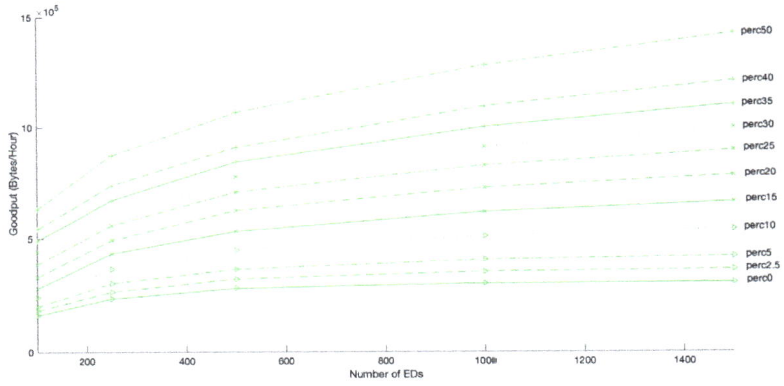

Figure 17. Network goodput for 5 GWs, different network sizes and different PoD.

To easily compare the network fairness, Figure 18 presents the fairness behavior for each tested scenario and the different PoD of the Standard channel time. Due to the fairer distribution of GWs by the network, the scenario with five GWs was only outperformed for percentages below 15% and in networks with under 500 EDs.

The results also show an overlap between the three tested scenarios. First, the scenario with two GWs achieved better fairness than the others, followed by the network with three GWs. and last, the one with five GWs. This behavior is supported by Figure 19, in which the curve representing the 0.95 fairness index moves towards higher PoD as the number of GWs increases. As seen, for a network with 100 EDs, the higher fairness achieved with the original time allocation occurred with a scenario with two GWs. However, as the number of EDs increased, better fairness was guaranteed by the scenario with three GWs when the PoD was around 5%. Then, above a PoD of about 13.4%, as the 0.95 fairness lines of three and five GWs intersect, the fairness of the latter scenario became the best.

When the scenario with five GWs had 1500 EDs, the fairness peak was at 0.7745, as the PoD was equal to 25%. For the same PoD and network size, the fairness values verified for the other scenarios were 0.6662 and 0.7153, for two GWs and three GWs, respectively. This is a clear difference, which is as marked as the time allocated to faster rate channels, per GW cycle.

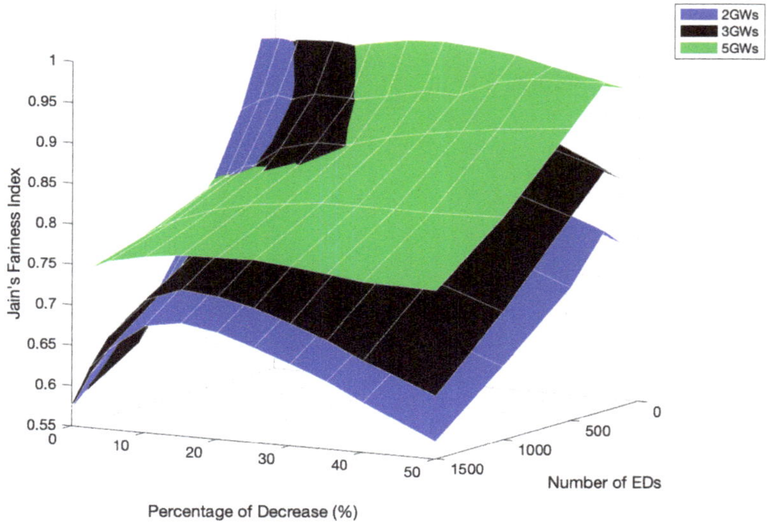

Figure 18. Network fairness comparison for different numbers of GWs, network sizes, and PoD.

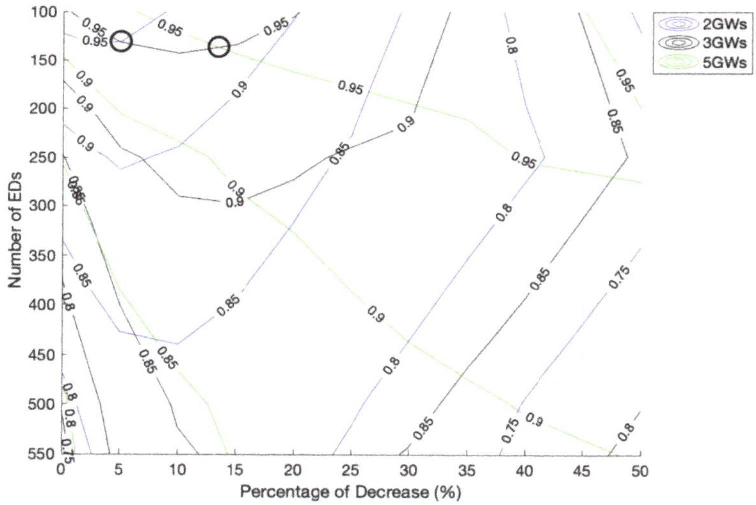

Figure 19. Detailed analysis of the fairness behavior regarding Figure 18.

As shown in Figure 20, by comparing the goodput for all the tested scenarios, an increase was noticeable with the number of GWs, which became more accentuated with increases in the PoD and the network size, due to the decrease in the channel access competition of each GW. Along with this, the number of EDs in the ideal channel increases in the Fast-Rate channel, which results in an increase in the number of channel accesses for the same period of time. The transmissions are faster than in slower channels, leading to smaller restriction periods.

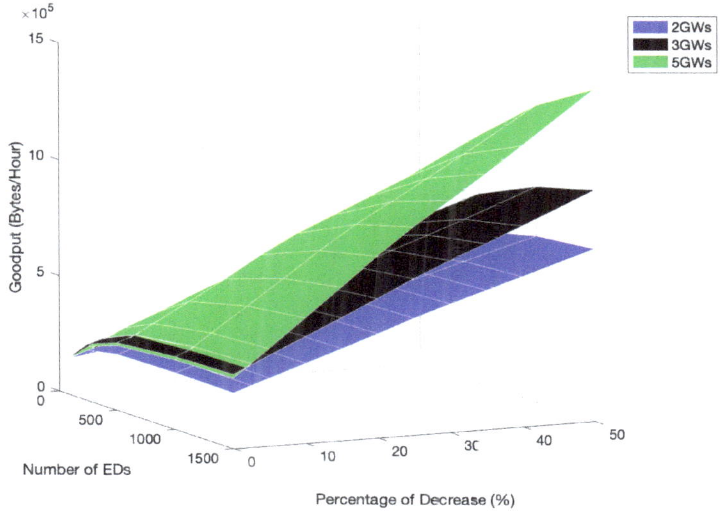

Figure 20. Network goodput comparison for different numbers of GWs, network sizes, and PoD.

Figure 21 illustrates the percentage of collision for every tested scenario. The network with five GWs had the lowest number of channel access collisions when considering every medium access performed by the EDs. As referred to before, the main reason for this behavior is the reduction in the number of EDs sharing the channel and the lower probability of two or more EDs choosing the same time slot to transmit. Another reason is the decreasing possibility of the hidden-terminal problem. As the number of GWs increases, the EDs transmitting to a given GW, with a specific channel, are more within reach of the other EDs that also transmit to the same GW. Thus, the control packets are more likely to be received by the generality of the EDs, decreasing the number of situations where two EDs transmit at the same time because none of them receive the control packet sent by the other.

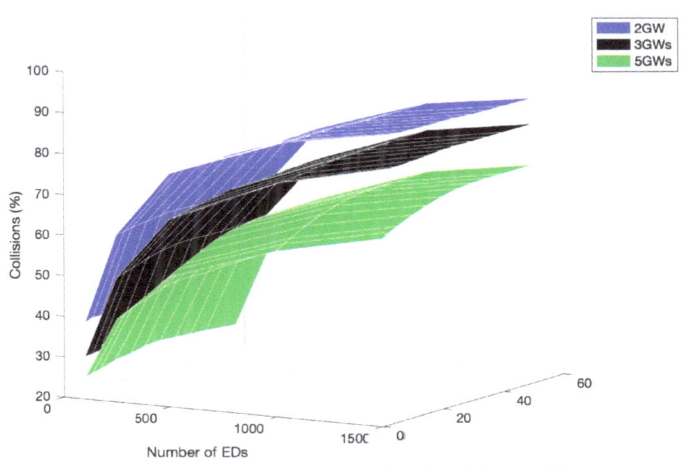

Figure 21. Percentage of collisions comparison for different numbers of GWs, network sizes, and PoD.

6. Conclusions

This study presented a new scheme for large-scale LoRa networks with multiple, low-cost single-channel GWs. The GWs switch between different operation parameters, so-called channels, to serve the EDs in their communication range, regardless of their signal strength. When comparing the developed scheme with pure-ALOHA, where every ED transmits using the same channel, the LoRa network improved regarding goodput and fairness in all tested scenarios for an arbitrary channel time allocation. The fairness for a scenario with five GWs and 2000 EDs improved from 0.2644 to 0.5245. The primary goal was to use low-cost gateways and keep the network fair, giving the same medium access opportunity to all EDs. Therefore, this paper also presented a study regarding the time allocation, per GW cycle, for the different channels, first only addressing single GW scenarios and then scenarios with multiple GWs. We concluded that the time allocation to the channel that can transmit faster must be lower to prevent an increase in the disparity of performance (goodput) between EDs using different channels. Otherwise, the network becomes unfair, giving more transmission opportunities to some EDs to the detriment of others.

In the future, to increase the performance levels of the developed protocol, we aim to add new features, such as a dynamic maximum number of backoff slots according to the number of EDs and the channels in use. We will also target the introduction of mobility in the scenario to simulate real-life scenarios more accurately and add adaptive channel periods calculated based on the number of EDs using each channel at the moment. At last, a comparison between simulated and real-life tests is planned to verify the applicability of the developed protocol.

Author Contributions: Conceptualization, M.L.; formal analysis, A.F., M.L. and A.Z.; investigation, A.F. and M.L.; methodology, A.F. and A.Z.; supervision, M.L. and A.Z.; validation: M.L.; visualization, A.F.; writing—original draft preparation, A.F.; writing—review and editing, M.L. and A.Z. All authors have read and agreed to the published version of the manuscript.

Funding: This work was funded by FCT/MEC through national funds and when applicable co-funded by the FEDER—PT2020 partnership agreement under the project UIDB/50008/2020-UIDP/50008/2020.

Institutional Review Board Statement: Not applicable.

Informed Consent Statement: Not applicable.

Conflicts of Interest: The authors declare no conflict of interest.

References

1. Barillaro, S.; Rhee, S.; Escudero, G.; Kacker, R.; Badger, L.; Kuhn, D.R. Low-Power Wide Area Networks (LPWAN) for Communications of Mobile Sensor Data. In Proceedings of the 2nd ACM/EIGSCC Symposium on Smart Cities and Communities (SCC '19), Portland, OR, USA, 10–12 September 2019; Association for Computing Machinery: New York, NY, USA, 2019; pp. 1–8.
2. Zourmand, A.; Kun Hing, A.L.; Hung, C.W.; AbdulRehman, M. Internet of Things (IoT) using LoRa technology. In Proceedings of the 2019 IEEE International Conference on Automatic Control and Intelligent Systems (I2CACIS), Selangor, Malaysia, 29 June 2019; pp. 324–330. [CrossRef]
3. Lavric, A.; Petrariu, A.I.; Popa, V. SigFox Communication Protocol: The New Era of IoT? In Proceedings of the 2019 International Conference on Sensing and Instrumentation in IoT Era (ISSI), Lisbon, Portugal, 29–30 August 2019; pp. 1–4. [CrossRef]
4. Ingenu. How RPMA Works? The Making of RPMA. Available online: https://www.ingenu.com/technology/rpma/how-rpma-works/ (accessed on 13 July 2022).
5. Ayoub, W.; Samhat, A.E.; Nouvel, F.; Mroue, M.; Prévotet, J. Internet of Mobile Things: Overview of LoRaWAN, DASH7, and NB-IoT in LPWANs Standards and Supported Mobility. *IEEE Commun. Surv. Tutorials* **2019**, *21*, 1561–1581. [CrossRef]
6. Devalal, S.; Karthikeyan, A. LoRa Technology—An Overview. In Proceedings of the 2018 Second International Conference on Electronics, Communication and Aerospace Technology (ICECA), Coimbatore, India, 29–31 March 2018; pp. 284–290. [CrossRef]
7. Fernandes, R.; Oliveira, R.; Luís, M.; Sargento, S. On the Real Capacity of LoRa Networks: The Impact of Non-Destructive Communications. *IEEE Commun. Lett.* **2019**, *23*, 2437–2441. [CrossRef]
8. Hanaffi, H.; Mohamad, R.; Suliman, S.I.; Kassim, M.; Anas, N.M.; Bakar, A.Z.A. Single-Channel LoRaWAN Gateway for Remote Indoor Monitoring System: An Experimental. In Proceedings of the 2020 8th International Electrical Engineering Congress (iEECON), Chiang Mai, Thailand, 4–6 March 2020; pp. 1–4. [CrossRef]

9. Cuomo, F.; Campo, M.; Caponi, A.; Bianchi, G.; Rossini, G.; Pisani, P. EXPLoRa: Extending the performance of LoRa by suitable spreading factor allocations. In Proceedings of the 2017 IEEE 13th International Conference on Wireless and Mobile Computing, Networking and Communications (WiMob), Rome, Italy, 9–11 October 2017; pp. 1–8. [CrossRef]
10. Cuomo, F.; Campo, M.; Bassetti, E.; Cartella, L.; Sole, F.; Bianchi, G. Adaptive mitigation of the Air-Time pressure in LoRa multi-gateway architectures. In Proceedings of the European Wireless 2018 24th European Wireless Conference, Aarhus, Denmark, 2–4 May 2019; pp. 1–6.
11. Liao, W.; Zhao, O.; Ishizu, K.; Kojima, F. Adaptive Parameter Adjustment for Uplink Transmission for Multi-gateway LoRa Systems. In Proceedings of the 2019 22nd International Symposium on Wireless Personal Multimedia Communications (WPMC), Lisbon, Portugal, 24–27 November 2019; pp. 1–5. [CrossRef]
12. Kim, D.Y.; Kim, S. Gateway Channel Hopping to Improve Transmission Efficiency in Long-range IoT Networks. *KSII Trans. Internet Inf. Syst.* **2019**, *13*, 1599–1640. [CrossRef]
13. Iglesias-Rivera, A.; Van Glabbeek, R.; Guerra, E.O.; Braeken, A.; Steenhaut, K.; Cruz-Enriquez, H. Time-Slotted Spreading Factor Hopping for Mitigating Blind Spots in LoRa-Based Networks. *Sensors* **2022**, *22*, 2253. [CrossRef] [PubMed]
14. Haubro, M.; Orfanidis, C.; Oikonomou, G.; Fafoutis, X. TSCH-over-LoRA: Long Range and Reliable IPv6 Multi-hop Networks for the Internet of Things. *Internet Technol. Lett.* **2020**, *3*, e165. [CrossRef]
15. Kim, S.; Yoo, Y. Contention-Aware Adaptive Data Rate for Throughput Optimization in LoRaWAN. *Sensors* **2018**, *18*, 1716. [CrossRef] [PubMed]
16. Adelantado, F.; Vilajosana, X.; Tuset-Peiro, P.; Martinez, B.; Melia-Segui, J.; Watteyne, T. Understanding the Limits of LoRaWAN. *IEEE Commun. Mag.* **2017**, *55*, 34–40. [CrossRef]
17. Benkahla, N.; Tounsi, H.; Song, Y.; Frikha, M. Enhanced ADR for LoRaWAN networks with mobility. In Proceedings of the 2019 15th International Wireless Communications Mobile Computing Conference (IWCMC), Tangier, Morocco, 24–28 June 2019; pp. 1–6. [CrossRef]
18. Fernandes, R.; Oliveira, R.; Luís, M.; Sargento, S. Exploring the Use of Control Packets in LoRa Medium Access: A Scalability Analysis. In Proceedings of the 21st International Symposium on a World of Wireless, Mobile and Multimedia Networks (IEEE WoWMoM 2020), Cork, Ireland, 31 August–3 September 2020. [CrossRef]
19. Waspmote-LoRa-868MHz_915MHz-SX1272 Networking Guide. Available online: https://usermanual.wiki/Document/waspmotelora868mhz915mhzsx1272networkingguide.548782060/view/ (accessed on 2 June 2022).
20. Oliveira, R.; Guardalben, L.; Sargento, S. Long range communications in urban and rural environments. In Proceedings of the IEEE Symposium on Computers and Communications (ISCC 2017), Heraklion, Greece, 3–6 July 2017. [CrossRef]

sensors

MDPI

Article

Energy-Aware Dynamic DU Selection and NF Relocation in O-RAN Using Actor–Critic Learning

Shahram Mollahasani [1], Turgay Pamuklu [1], Rodney Wilson [2] and Melike Erol-Kantarci [1,*]

[1] School of Electrical Engineering and Computer Science, University of Ottawa, Ottawa, ON K1N 6N5, Canada; sh.mollahasani@gmail.com (S.M.); turgay.pamuklu@uottawa.ca (T.P.)
[2] Ciena, Ottawa, ON K2K 0L1, Canada; rwilson@ciena.com
* Correspondence: melike.erolkantarci@uottawa.ca

Abstract: Open radio access network (O-RAN) is one of the promising candidates for fulfilling flexible and cost-effective goals by considering openness and intelligence in its architecture. In the O-RAN architecture, a central unit (O-CU) and a distributed unit (O-DU) are virtualized and executed on processing pools of general-purpose processors that can be placed at different locations. Therefore, it is challenging to choose a proper location for executing network functions (NFs) over these entities by considering propagation delay and computational capacity. In this paper, we propose a Soft Actor–Critic Energy-Aware Dynamic DU Selection algorithm (SA2C-EADDUS) by integrating two nested actor–critic agents in the O-RAN architecture. In addition, we formulate an optimization model that minimizes delay and energy consumption. Then, we solve that problem with an MILP solver and use that solution as a lower bound comparison for our SA2C-EADDUS algorithm. Moreover, we compare that algorithm with recent works, including RL- and DRL-based resource allocation algorithms and a heuristic method. We show that by collaborating A2C agents in different layers and by dynamic relocation of NFs, based on service requirements, our schemes improve the energy efficiency by 50% with respect to other schemes. Moreover, we reduce the mean delay by a significant amount with our novel SA2C-EADDUS approach.

Keywords: actor–critic learning; energy-efficiency; O-RAN; RAN optimization

Citation: Mollahasani, S.; Pamuklu, T.; Wilson, R.; Erol-Kantarci, M. Energy-Aware Dynamic DU Selection and NF Relocation in O-RAN Using Actor–Critic Learning. *Sensors* **2022**, 22, 5029. https://doi.org/10.3390/s22135029

Academic Editor: Peter Han Joo Chong

Received: 22 May 2022
Accepted: 26 June 2022
Published: 3 July 2022

1. Introduction

The next generation wireless networks include a fundamental transformation which is the next generation radio access networks (NG-RAN) [1]. The NG-RAN protocol stack can be split into eight different disaggregated options which are combined within three network units: radio unit (RU), distributed unit (DU), and centralized unit (CU). Furthermore, unlike the traditional RAN, NG-RAN functions can be virtualized on top of general-purpose hardware. In Open RAN (O-RAN), the concept of virtualized RAN (vRAN) and the disaggregation of network units reaches its full interoperability by using open interfaces among multiple vendors [2,3]. However, the placement of the virtual network functions can be challenging due to multiple constraints, such as routing path, RAN protocol splits, bandwidth limitations, maximum latency tolerance requirements, heterogeneous computational resources, and so on. The main objective of this work is to develop an RL-based network function placement scheme in a way that the energy consumption is minimized while the expected quality of service (QoS) for that network function is satisfied. In this work, resource block (RB) scheduling is considered as a network function (NF). The proposed scheme dynamically relocates this NF among DUs based on their location and processing power by targeting minimum energy consumption and satisfying the QoS requirements of users.

The idea of disaggregation in O-RAN is allowing RAN functions to be placed at different computing devices in a distributed manner [4]. Therefore, it is vital to identify how this disaggregationcan be executed and what metrics need to be satisfied to run these disaggregated components correctly. By considering the concept of function splits in O-RAN, the requirements for network functions such as minimum bit rate and the maximum latency for communication among O-RAN components (RU, DU, and CU) need to be satisfied. This work is developed based on the O-RAN architecture and split option 7.2 where the functional split is between RU and DU [1,5].

Since the amount of energy consumption in DUs is proportional to the time they are active, i.e., processing tasks, mobile network operators (MNO) are seeking an intelligent assignment model where RUs use minimum required resources at DUs, while QoS requirements such as packet delivery ratio and latency are satisfied [6]. Note that a part of DU energy consumption is fixed due to the cooling system, idle power, etc. Therefore, reducing the active time of DUs can decrease the overall energy consumption in the network. This can be achieved by optimizing load distribution among DUs with respect to their processing power and their location and setting redundant DUs to sleep mode. The main focus of this paper is to highlight the effect of deploying an AI-enabled network function relocation in a dynamic environment on energy consumption and network performance. To this end, we examined our model by considering resource block scheduling as a network function and applied our AI-based framework on it. We evaluate this model by generating User Datagram Protocol (UDP) and Transmission Control Protocol (TCP) packets (video and ITS) with different delay budgets and QoS requirements.

In this paper, we propose an RL-based resource block allocation in 5G and DU selection in O-RAN architecture. In the proposed model, we employed energy awareness as a key performance indicator and provide a multi-agent actor–critic (A2C) method as the primary novelty of this study. In this work, we employed two agents; one is responsible for allocating RB to UEs by considering the type and priority of packets, while the other agent is integrated for reducing the energy consumption by considering processing power, capacity, and propagation delay among DUs in the network. Furthermore, performance evaluation includes a mixed-integer linear programming (MILP) solver comparison to determine the gap between this novel approach and the lower bound of that multi-objective minimization problem. Thus, we demonstrate the feasibility of this advanced machine learning implementation in a disaggregated RAN scheme.

In our prior work [7], we developed an A2C-based algorithm for invoking NFs at the CU or DU by considering traffic requirements. This work extends [7] by considering the propagation delay between O-RAN components and the energy consumption, and in addition, we consider DUs with different processing capacities. The proposed model is examined for a different number of UEs, and it is compared with a heuristic model developed in our previous works and two other recent works. Our results show that the proposed model can reduce energy consumption by 50% with respect to models where network functions are executed locally under the given settings. Apart from energy conservation, the proposed model can reduce the overall mean delay for an intelligent transport system (ITS) and video traffic down to 7 ms and 12 ms, respectively. In contrast, the packet delivery ratio for ITS and video traffics will be increased up to 80% and 30%, respectively.

The rest of this paper is organized as follows: In Section 2, we discuss the recent works in this area. In Section 3, we formulate the problem and propose an MILP solution. The A2C-based algorithm is comprehensively explained in Section 4. In Section 5, the proposed scheme is compared with four baseline algorithms, and in Section 6, we conclude the paper.

2. Related Work

Energy-efficiency algorithms for traditional RANs have been comprehensively presented in [8]. Most of the works covered in [8] are based on on-off techniques for BSs. Some of these models rely on traffic prediction to estimate low traffic intervals, while others use cell zooming techniques to expand the BSs' coverage with respect to their neighbors [9–11]. These studies consider RAN equipment as a monolithic equipment.

Energy efficiency is also considered in centralized RAN (C-RAN), where just the radio unit of BSs is disaggregated at remote radio heads (RRHs), and the rest is implemented at the baseband unit (BBU). For instance, in [12], BBU placement across physical machines in a cloud site is formulated as a bin-packing problem, and the authors tackle this problem by proposing a heuristic algorithm. Additionally, in [13,14], the authors improve the work in [12] by proposing a BBUs virtualization technique with a linear computational complexity order that reduces the power consumption in the network. Furthermore, in [15], the authors show how traffic forecasting at each BS can be used in dynamically switching (on/off) RRHs. Moreover, ref. [16] shows an end-to-end energy-efficient model by activating and placing virtual network functions (VNFs) on physical machines and distributing the traffic among them.

The aforementioned C-RAN scenarios mainly reflect the fixed functional split, and they require high fronthaul bitrate and consequently incur high deployment cost. To this end, the impact of the flexible function split is examined in several recent studies [17]. For instance, in [18], savings in power and computational resources with respect to different function splits are analytically modeled. In [19], the authors aim to optimize the energy efficiency and the midhaul bandwidth in C-RAN.

Different from these works, in this paper we evaluate the energy consumption for dynamic NF migration among edge clouds, i.e., not only CU allocation as in C-RAN but also DU allocation in O-RAN is considered. Finally, refs. [20,21] show that the migration of NFs has a non-negligible impact on energy consumption, which has not been addressed in previous works. Since we aim to address the placement of resource allocation function using machine learning, it is important to give a brief overview on the existing studies on RL-based resource allocation [22]. For instance, an RL-based resource block allocation technique is employed in a vehicle-to-vehicle network in [23].

In [24], an RL-based algorithm is proposed for optimizing the energy consumption and cost in a disaggregated RAN. Moreover, in [25], the authors develop an RL-based user-beam association and resource allocation scheme using transfer reinforcement learning. In [26], a deep RL-based resource block allocation is introduced in which RBs are allocated in a way that the mean delay is reduced. Another Deep RL approach is proposed in [27] to solve a two-tier resource allocation problem in a standalone base station. In our previous work [28], we developed an RL-based NF to schedule URLLC and eMBB packets with respect to the delay budget and channel conditions of UEs in the network. We also developed an RL-based algorithm for invoking NFs at the CU or DU by considering traffic requirements [7]. However, unlike [7], in this paper we develop a comprehensive scheme that considers the propagation delay between O-RAN components and the energy consumption, and in addition we consider DUs with different processing capacities. Furthermore, in [29], we introduced an optimization-based solution for the DU selection problem under delay constraints. This paper extends [29] by modeling the problem as a multi-objective minimization problem for jointly addressing energy-efficiency and delay.

Unlike previous works, in this paper we develop an actor–critic based DU selection scheme for a RAN with disaggregated layers (such as O-RAN) to dynamically relocate network functions among available DUs (edge servers) by considering their processing power and propagation delay in a way that the overall energy consumption in the network is reduced, while packet delivery ratio and delay budget of user traffic are satisfied.

3. System Model

Figure 1 shows the overall architecture that is structured as a time-interval-based model. In each time interval ($t \in \mathcal{T}$), I numbers of user equipment (UEs, $i \in \mathcal{I}$) request demands which are defined as a tuple $\langle i, t \rangle$. These demands may belong to one of two ($K = 2$) different types of traffic ($k \in \mathcal{K}$), such as video and ITS. These traffic types have different demand sizes ($U_{\langle i,t \rangle k}$) and delay budgets ($\Delta_{\langle i,t \rangle k}$). On the infrastructure side, we consider L low processing power DUs (DU^{LP}, $l \in \mathcal{L}$) that service these UEs. These DUs can house network protocols from the lowest level to the packet data convergence protocol [30]. Moreover, each one has a dedicated RU to perform lower layer RF functions.

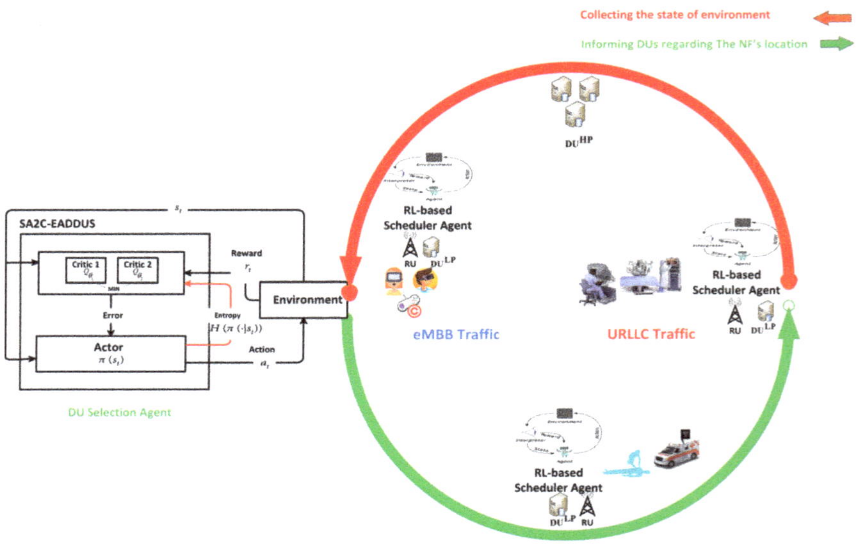

Figure 1. The overall architecture of the proposed model. Agents are located at DUs and interact with the DU selection algorithm by sending some feedback during each time interval (the red arrow). Then the DU selection scheme informs agents (the green arrow) regarding the location of NF during the next time interval based on the types of packets (URLLC or eMBB), available processing capacity, scheduling delay, and propagation delay among DUs to minimize the overall energy consumption in the network.

Moreover, in our architecture, DUs have an option to migrate NFs to a DU with higher processing power (DU^{HP}). This migration option has two essential benefits in our system. First, we can switch off the digital units in the local DUs and save energy. Second, aggregation of NFs from multiple DUs in a common DU^{HP} allows the resource allocation function to observe multiple RB maps in the same DU and mitigate inter-cell interference. This can lead to a lower scheduling delay and reduce the number of retransmissions in the network. However, it should be noted that DU^{HP} has a limited processing capacity (ξ), and due to its location, migrating NFs to it may face higher propagation delay (D_l^P) with respect to other DUs, which can negatively affect packets with a lower latency delay budget. The details of the notations are given in Notations .

3.1. Energy Consumption and Delay Models

Equation (1) calculates the energy consumption in DU^{LP} l in time interval t'. That value equals zero when NFs are migrated to DU^{HP} in this time interval ($b_{lt'} = 1$) (Note that a DU may need to consume energy for other reasons in that time interval. However, these energy consumptions do not change with the decision variables; thus, we do not include them in the energy consumption model. Meanwhile, they can be easily added as constant energy consumption.). The equation provides that case by the rightmost multiplicand ($1 - b_{lt'}$). Processing NFs has two energy consumption terms; the first one is the fixed energy consumption (E_l^F), which does not change by the activity of the processing unit in this DU. The second one is the dynamic energy consumption which is considered when the processing unit is active for user traffic demand. Thus, it depends on the processing unit energy consumption per RB (E_l^D) and the DU utilization that is calculated by the number of allocated RBs r in the time interval t'. Here $a_{\langle i,t\rangle rt'}$ equals one if the traffic demand $\langle i,t\rangle$

is processed in RB r in the time interval t'. Lastly, Equation (2) calculates the total energy consumption in DUs.

$$E_{lt'} = \left(E_l^F + E_l^D * \sum_{\substack{\langle i,t \rangle \in \langle \mathcal{I},\mathcal{T} \rangle \\ r \in \mathcal{R}}} a_{\langle i,t \rangle rt'} \right) * (1 - b_{lt'}) \quad (1)$$

$$E^{TOT} = \sum_{t' \in \mathcal{T}} \left(E^{HP} + \sum_{l \in \mathcal{L}} E_{lt'} \right) \quad (2)$$

The delay of each traffic demand ($\langle i,t \rangle$) is the summation of two terms in which the first one is the scheduling delay ($\delta^S_{\langle i,t \rangle}$), and the second one is the propagation delay ($\delta^P_{\langle i,t \rangle}$). Equation (3) calculates the scheduling delay in which $y_{\langle i,t \rangle t'}$ is a binary decision variable equaling '1', if the traffic demand $\langle i,t \rangle$ scheduled/assigned to process in an RB in time interval t'. However, for a large traffic demand, RBs in a single time interval may not be enough to finish that demand; thus, multiple time intervals might be assigned to demand $\langle i,t \rangle$. For that reason, we find the most recent assigned time interval ($\max_{t' \in \mathcal{T}}(y_{\langle i,t \rangle t'} * t')$) for that demand ($\langle i,t \rangle$). Then, we subtract the demand time interval (t) from that value to find the number of time intervals waited for that demand ($\langle i,t \rangle$). Finally, we multiply that value with the length of transmission time interval (TTI) to find the scheduling delay.

$$\delta^S_{\langle i,t \rangle} = \left(\max_{t' \in \mathcal{T}}(y_{\langle i,t \rangle t'} * t') - t \right) * TTI \quad (3)$$

$$\delta^P_{\langle i,t \rangle} = \frac{D^P_{M(i)}}{\sum_{t' \in \mathcal{T}} y_{\langle i,t \rangle t'}} \sum_{t' \in \mathcal{T}} y_{\langle i,t \rangle t'} * b_{M(i)t'} \quad (4)$$

$$\delta^M = \frac{\sum_{\substack{k \in \mathcal{K} \\ \langle i,t \rangle \in \langle \mathcal{I},\mathcal{T} \rangle}} (\delta^S_{\langle i,t \rangle} + \delta^P_{\langle i,t \rangle}) * u_{\langle i,t \rangle k}}{\sum_{\substack{k \in \mathcal{K} \\ \langle i,t \rangle \in \langle \mathcal{I},\mathcal{T} \rangle}} u_{\langle i,t \rangle k}} \quad (5)$$

Equation (4) defines the propagation delay for demand $\langle i,t \rangle$. As explained, a user demand may be scheduled for more than one time interval ($y_{\langle i,t \rangle t'} > 1$). Some of these time intervals may belong to the times that NFs processed in DU^{LP} ($b_{M(i)t'} = 0$) ($M(i)$ is a given value that maps the user i with its $DU^{LP}, (l = M(i))$.). In those time intervals, propagation delay equals zero due to negligible distance between UEs and DU^{LP}. On the other hand, in some intervals, NFs may be processed in DU^{HP} ($b_{M(i)t'} = 1$). To calculate the number of time intervals in the latter case, we first multiply the scheduled time intervals with the NF migration decision variable ($y_{\langle i,t \rangle t'} * b_{M(i)t'}$). That value will equal one if and only if NFs are migrated to DU^{HP} in time interval t'. Second, we divide that value by the total time intervals needed to process that demand ($\sum_{t' \in \mathcal{T}} y_{\langle i,t \rangle t'}$). That value gives us the percentage of time we process that demand in DU^{HP}. Finally, we multiply that value with the propagation delay between DU_l^{LP} and DU^{HP} ($D^P_{M(i)}$) to find the total propagation delay to finish that user demand.

Equation (5) shows the mean delay in the network. Here, the dividend is the total delay, and the divisor is the number of total demand in the network. In addition, $u_{\langle i,t \rangle k}$ is the traffic demand indicator, which equals one if there is a demand ($\langle i,t \rangle$) in type k; otherwise, it equals zero.

3.2. System Constraints

Equation (6) guarantees that each UE gets its service demand from the system. Equation (7) prevents the allocation of the same RB (rt') to multiple user traffic demands

($\langle i,t \rangle$). Equation (8) correlates the $y_{\langle i,t \rangle t'}$ decision variable with $a_{\langle i,t \rangle rt'}$, indicating that at least one resource block $r \in \mathcal{R}$ in time interval t' is allocated to the user demand $\langle i,t \rangle$. Thus, we can simplify our delay calculation by using $y_{\langle i,t \rangle t'}$ instead of a more complex decision variable $a_{\langle i,t \rangle rt'}$. Equation (9) generates the user traffic indicator $u_{\langle i,t \rangle k}$ from demand size $U_{\langle i,t \rangle k}$. A UE may demand only one type of traffic in a specific time interval which Equation (10) ensures. In addition, the total number of DU_l^{LP} that can migrate their NFs to DU^{HP} is limited by ξ in Equation (11). Lastly, the delay of each demand is limited by $\Delta_{\langle i,t \rangle k}$ according to their traffic type k in Equation (12).

$$\sum_{t'=t}^{\mathcal{T}} \sum_{r \in \mathcal{R}} S_{irt'} * a_{\langle i,t \rangle rt'} \geq \sum_{k \in \mathcal{K}} (U_{\langle i,t \rangle k} * u_{\langle i,t \rangle k}), \qquad \forall \langle i,t \rangle \in \langle \mathcal{I}, \mathcal{T} \rangle \quad (6)$$

$$\sum_{\langle i,t \rangle \in \langle \mathcal{I}, \mathcal{T} \rangle} a_{\langle i,t \rangle rt'} \leq 1, \qquad \forall l \in \mathcal{L}, \forall r \in \mathcal{R}, \forall t' \in \mathcal{T} \quad (7)$$

$$\mathcal{M} * y_{\langle i,t \rangle t'} - \sum_{r \in \mathcal{R}} a_{\langle i,t \rangle rt'} \geq 0, \qquad \forall \langle i,t \rangle \in \langle \mathcal{I}, \mathcal{T} \rangle, \forall t' \in \mathcal{T} \quad (8)$$

$$\mathcal{M} * u_{\langle i,t \rangle k} - U_{\langle i,t \rangle k} \geq 0, \qquad \forall \langle i,t \rangle \in \langle \mathcal{I}, \mathcal{T} \rangle, \forall k \in \mathcal{K} \quad (9)$$

$$\sum_{k \in \mathcal{K}} u_{\langle i,t \rangle k} \leq 1, \qquad \forall \langle i,t \rangle \in \langle \mathcal{I}, \mathcal{T} \rangle \quad (10)$$

$$\sum_{l \in \mathcal{L}} b_{lt} \leq \xi, \qquad \forall t \in \mathcal{T} \quad (11)$$

$$\delta^S_{\langle i,t \rangle} + \delta^P_{\langle i,t \rangle} \leq \sum_{k \in \mathcal{K}} u_{\langle i,t \rangle k} * \Delta_{\langle i,t \rangle k}, \qquad \forall \langle i,t \rangle \in \langle \mathcal{I}, \mathcal{T} \rangle \quad (12)$$

3.3. Problem Definition

We have a multi-objective minimization (P1) problem in which we aim for a balance between the energy consumption in DUs and the mean delay with the objective function (Equation (13)). W is a weighting factor between these key performance indicators (KPIs) in this equation, and Ω is a scaling factor to normalize the ranges.

$$\text{(P1) Minimize: } W * E^{TOT} + \Omega * (1 - W) * \delta^M \quad (13)$$

Subject to: Equations (6)–(12)

Let us consider a special case of the problem (P1) in which the propagation delay, D_l^P, between any DU^{LP} and DU^{HP} is remarkably huge and makes the network choose only DU^{LP} for NFs. Thus, P1 reduces to an RB allocation problem, which is proved as an NP-hard problem by Yu et al. [31]. Therefore, we propose an actor–critic solution for this problem which is explained in the next section.

4. Actor–Critic Solution

4.1. RL Model

In this work, we employ two nested A2C agents which are developed based on O-RAN architecture. While one agent is responsible for scheduling resource blocks during each time interval, the other is designed to dynamically choose a proper DU for executing the scheduler agent by considering energy consumption, processing power, scheduling, and propagation delay with respect to each DU's traffic load and location. Processing resource block allocation decisions of multiple RUs on a single DU (DU^{HP}) can expand the observation level of decision agents, which can lead to applying more accurate actions. More precisely, when an A2C-based scheduler agent can access other RUs' resource block map, subcarriers can be allocated to edge UEs in a way that the inter-cell interference among RUs are minimized as shown in [28]. The goal of the proposed actor–critic agent is improving the performance of the A2C-based scheduler and reducing the overall energy consumption by dynamically selecting DUs by considering their processing power and propagation delay in the network.

Since, in the O-RAN architecture, intelligence is integrated at multiple layers of O-RAN, it is natural to split the intelligence into different layers to take advantage of higher processing power or to be able to apply real-time actions for delay-sensitive applications. Accordingly, each O-RAN component should abide by a specific delay bound based on the tolerable delay within its control loop. To this end, in this work, we deploy the DU selection agent at CU to access a higher perspective with respect to DUs, accessing higher processing power to optimize the more complex problem and its higher tolerable delay bound (10 ms to 1 s) which provides adequate time for processing the model. Additionally, the RL-based scheduler agents are located at DUs to schedule UEs close to real-time time intervals (less than 10 ms) [32]. The tolerable delay bound at each layer closely depends on the midhaul/fronthaul bandwidth and the range. In this work, this delay is assumed as the AI feedback's round trip time (RTT), and as shown in Figure 2, it includes processing, propagation, and switching delay over the XHaul. As it is discussed in [33], the maximum tolerable delay for an XHaul with a 3 Gbps bandwidth and 400 km range is at most 10 ms; therefore, in the system model, we considered the propagation delay in this range (2 ms to 5 ms) and left some room for switching delay to make it close to the real-life condition. It should be noted that the main focus of this work is on propagation delay, and we assumed the switching delay and processing delay constant by considering the constant bitrate for all DUs and defining the processing power of DU^{HP} proportional to DU^{LP} (the processing power of DU^{HP} is considered two to four times higher than DU^{LP} which makes it capable of processing up to four DUs' load at the same time).

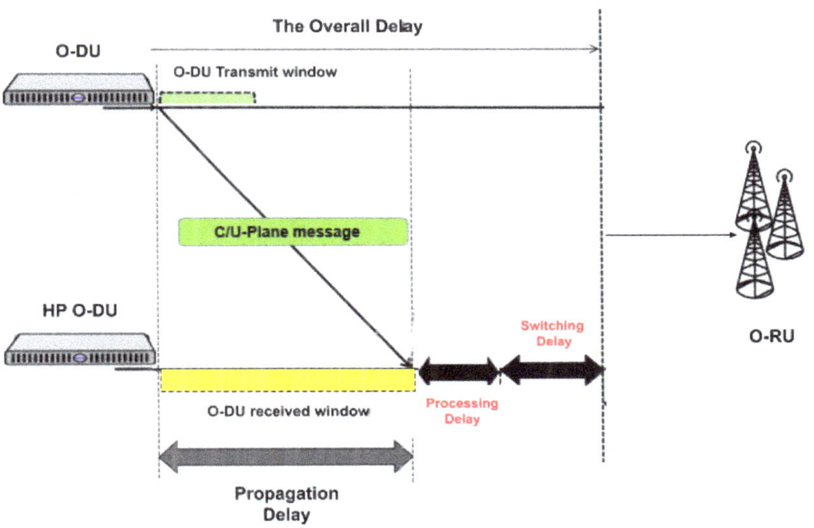

Figure 2. The overall delay in the network.

4.2. Soft Actor–Critic Energy Aware Dynamic DU Selection/NF Placement Scheme (SA2C-EADDUS)

In this work, a soft actor–critic approach is employed since it provides hierarchical control during the learning procedure. In actor–critic models, to reduce the gradient, the corresponding value function needs to be learned to update and assist the system policy. In the proposed scheme, based on the amount of energy consumed at each DU, the DU's processing power, priority of traffic, and its delay tolerance level, the DU for executing the RL-based scheduling agent will be selected. This selection is also performed by an A2C agent. To this end, a neural network (NN) with three layers of neurons is employed. Furthermore, every TTI neurons' weights are updated by interactions occurring between the actor and the critic. In the exploitation stage, the NN works as a non-linear function, and it will be tuned by updating the weights. During each state s, the main goal of SA2C agents is to maximize their received reward r, by applying more accurate actions, a, which can be obtained by action–value (Q(s,a)) and state–value (V(s)) functions. The action–value function is used to estimate the effect of applying actions during states. Consequently, estimation of the outcome is obtained by the state–value function. In this work, for improving the convergence and increasing the stability of the model (by reducing the overestimation bias), the soft actor–critic model (SA2C) is used. The overall architecture of SA2C is depicted in Figure 1. In the SA2C model, instead of evaluating action or state value functions, we just need to estimate $V(s)$. Moreover, the error parameter in SA2C is a metric for examining the effect of performed action by considering the expected value $V(s)$, which can be demonstrated by $A(s_t, a_t) = Q(s_t, a_t) - V(s_t)$. In addition, the SA2C model is a synchronous model (unlike asynchronous actor–critic models) which makes it more consistent and suitable for disaggregated implementations. In the proposed scheme, each of the actors and critics contain independent neural networks which are demonstrated as follows:

- The critic's neural network is used to estimate the corresponding value function for aligning the actor's actions. In the proposed scheme we used two critics to minimize the overestimation bias.
- The actor's neural network is used to estimate proper action (choosing the best DU for executing NF) during each time interval.

The states of the proposed DU selection scheme are extracted from the environment through a tuple, $S_t = \{DU_{type}, QCI, CQI, HOL_{delay}\}$, by which DUs' location and their amount of available processing power can be identified by the DU's type, traffic type and its priorities with a QoS class identifier (QCI) channel quality indicator (CQI) for examining signal strength, and the amount of time packets stay in the scheduler queue or head-of-the-line delay (HOL_{delay}), respectively. Meanwhile there are two independent actions which are generated by agents during each time interval. The action space of the A2C-based scheduler is the location of the assigned resource block in the RB map. Additionally, the action space of the DU selection agent is the DU type which should be capable to handle NFs (in our case, we only consider the placement of the A2C-based scheduler[] however, our scheme can be extended to other NFs in the 5G stack [26,28]). The DU selection agent needs to collaborate with the A2C-based RB allocation agent in a way that the overall energy consumption in the network is reduced, while the expected QoS metrics (delay and packet delivery ratio) of the A2C-based RB allocation agent can be met.

We consider two separate reward functions for our agents since their objectives and their action spaces are different. In this work, the reward function for the A2C-based scheduler is defined as follows as in [28]:

$$Reward_{scheduler} = \beta R_1 + \gamma R_2 + \Phi R_3, \tag{14}$$

$$R_1 = max\left(sgn\left(cqi_k - \frac{\sum_{i=0}^{I} cqi_j}{K}\right), 0\right) \tag{15}$$

$$R_3 = sinc(\pi \left\lfloor \frac{Packet_{Delay}}{Traffic_{Budget}} \right\rfloor) \tag{16}$$

Here, UE_i feedback is considered as cqi_i, R_2 is an extra reward of 1 which is given for URLLC traffic to prioritize, the traffic delay budget and packet HOL delay are presented as $Traffic_{Budget}$ and $Packet_{Delay}$, respectively. β, γ, and Φ are weighting/scaling factors. The scheduling agent consists of an actor and a critic. The actor is located at the BSs and is responsible for allocating resource blocks to UEs, and the critic is integrated into the DUs to inspect actors and improve their decisions. In this model, the reward function is defined based on the CQI feedback, the amount of time the packet stays in the scheduling queue, and the UEs' satisfaction. To this end, the reward function is defined in a way that actors can receive the reward when the received SINR becomes higher (higher CQI), scheduling delay is reduced, and UEs' satisfaction ratio is increased (mean delay should be less than the delay budget). Therefore, when the RB map filled by an actor causes an inter-cell interference, the received SINR will be reduced, and the mean delay will be increased, leading to the reduction of the reward and punishment for the agent, which can improve the learning process in agents.

We define the reward for the DU selection agent as follows:

$$Reward_{DUselector} = \tau(R_1{}' + R_2{}') - \omega E^{TOT}, \tag{17}$$

$$R_1{}' = sinc(\pi \left\lfloor \frac{\delta^S_{\langle i,t\rangle} + \delta^P_{\langle i,t\rangle}}{\Delta_{\langle i,t\rangle k}} \right\rfloor) \tag{18}$$

$$R_2{}' = \frac{n_{URLLC}}{n_{tot}} sinc(\pi \left\lfloor \frac{\alpha \times (\delta^S_{\langle i,t\rangle} + \delta^P_{\langle i,t\rangle})}{\Delta_{\langle i,t\rangle k}}) \right\rfloor) \tag{19}$$

Here, n_{URLLC} shows the number of UEs which generate URLLC traffic. This can be obtained through the QCI value which is assigned by the EPS bearer, and it shows packet priorities, types and their delay budget [34]. In this reward function, $sinc(\pi\lfloor\rfloor)$ output is a binary value which is used to produce discrete actions in each time interval (0 or 1). As it was explained previously, RUs select their local DUs to perform higher layer processing. Our system model includes one DU with higher processing power than the others. For energy efficiency, one would consider offloading NFs of local DUs to the high-processing power DU and switching the local DUs off. This may also allow expanding the inter-cell interference observation capability of agents. However, it is important to note that this approach would have an adverse impact on delay since local DUs can provide access with less propagation delay than the distant high-power DU site. This is in addition to scheduling delay. Moreover, DU^{HP} has a limited processing capacity, and we cannot transfer all of DUs' load to it. Therefore, in our reward function, we want to make sure during each time interval that we choose a proper DU (priority of its packet and propagation delay with respect to DU^{HP}) in a way that the overall delay which each packet experiences (scheduling delay (δ^S_{it}) + propagation delay (δ^P_{it})) always remains below the predefined delay budget (Δ_{itk}) (Equation (18)). Moreover, since URLLC UEs are delay sensitive, we need to make sure that the mean delay of URLLC traffic is kept as low as possible with respect to other UEs (Equation (19)). To this end, by increasing the α the overall delay ($\delta^S_{it} + \delta^P_{it}$) can be proportionally reduced with respect to the predefined delay budget. In addition, we want to ensure the total energy consumption (E^{TOT}) is reduced by a third term which is calculated by Equation (2). τ and ω are scalar weights which are defined based on the priority for enhancing the packet delivery ratio ($R_1{}'$), reducing the overall delay ($R_2{}'$), and reducing energy consumption in the network (E^{TOT}). As a final remark, in this algorithm, unlike our previous work [7], we improve our reward function by considering the propagation delay between O-RAN components, fixed energy consumption, and dynamic energy consumption with respect to DUs processing capacity.

5. Performance Evaluation

The SA2C-EADDUS scheme is implemented in ns3-gym, which is a framework where OpenAI Gym (a tool for using machine learning libraries) is integrated into ns3 [35,36]. The neural network of the proposed scheme is developed by Pytorch. In simulations, we assume the number of UEs is between 40 to 80, the UEs are randomly distributed, and they are associated to closets DU among 4 DU^{LP}s. The URLLC UE ratio is 10%, and we employ numerology zero with 12 subcarriers, which has 14 symbols per subframe and 15 KHz subcarrier spacing. Furthermore, scheduling decisions are applied every TTI.

The SA2C-EADDUS performance is evaluated by considering different processing capacities of DUs and two different traffic types. We assume local DUs processing capacity cannot handle more than one RU's functions. Additionally, we assume there is a DU with higher processing capacity (DU^{HP}) which is located far from the other RUs. DU^{LP} can reduce its energy consumption by offloading its tasks to a DU with higher processing power. However, based on its location with respect to DU^{HP}, the propagation delay would vary.

We examine two processing capacity levels for DU^{HP} to evaluate the effect of increasing the observation level of agents over network performance. We also consider two traffic types (live stream video as eMBB traffic and intelligent transport system (ITS) as URLLC traffic) with different delay budgets. It should be noted that when the overall delay of a packet (scheduling and propagation delay) is higher than the predefined delay budget, we assume the packet is dropped. In Table 1, QoS metrics of each traffic type is depicted.

Table 1. Traffic properties [34].

QCI	Resource Type	Priority	Packet Delay Budget	Service Example
2	GBR	40	150 ms	Live stream Video
84	GBR	24	30 ms	ITS

We use the proposed MILP solution as a benchmark for SA2C-EADDUS. Furthermore, we compare SA2C-EADDUS with three baselines. The first baseline was proposed in our prior work [28]. The second baseline is based on a deep reinforcement learning (DRL) scheduler as in [26], and the last baseline is a heuristic method. We integrated these approaches to our energy-aware dynamic DU selection scheme to evaluate their performance with respect to the proposed scheme. In the following, we briefly present our baselines.

5.1. Delay and Priority Aware Actor–Critic RB Allocation Algorithm (A2C-RBA)

In our previous work [28], we implemented an actor–critic learning-based resource block scheduler where RBs are allocated to UEs by considering the delay budget and the priority of the user traffic. The following algorithm is developed for reconfigurable wireless networks where actions can be autonomously applied over the network. The A2C-RBA can adapt itself to the dynamicity of the environment to increase the utility of the available resources. However, the A2C-RBA algorithm is solely running at DU^{LP}, and it cannot take advantage of DU^{HP}.

5.2. Deep-Reinforcement Learning-Based Energy-Aware Dynamic DU Selection Scheme (DRL-EADDUS)

In [26], the main objective of DRL agents is allocating RBs in a way that the mean delay of packets be minimized. The authors in this work considered UEs with different packet request patterns, and they developed a deep-reinforcement learning algorithm to schedule RBs during each time interval. This work is integrated with the proposed framework to reduce the energy alongside the RB allocation in the network. We integrate this work into our framework to have a fair comparison with the proposed scheme.

5.3. Heuristic DU Selection Method (Heuristic-DUS)

In the heuristic method, decisions are made based on the propagation delay between local DUs and DU^{HP} with respect to a predefined threshold value and type of packets. In this algorithm, traffic in local DUs will be transferred to DU^{HP} when the propagation delay is below the predefined threshold and packets are not URLLC. Therefore, to reduce the packet drop ratio, if a packet's scheduling delay is higher than the threshold or URLLC traffic is scheduled for the next TTI, RBs will be assigned locally. Additionally, when the packets are scheduled at DU^{HP}, RBs will be assigned by considering other DUs RB maps to reduce the interference in the network. It should be noted that assigning a proper threshold value can be challenging due to the high dynamicity of the network parameters. The proposed heuristic algorithm is illustrated in Algorithm 1.

Algorithm 1: Heuristic RB Allocation Algorithm.

```
nDU ← Number of DUs;
nRB ← Number of RBs;
PD_i ← Propagation delay between DU_i^LP and DU^HP;
while nRB >0 do
    for i = 0; i < nDU; i++ do
        if PD_i < Threshold && Traffic! = URLLC then
            Transfer the load to DU^HP
            Examine other DUs RB maps
            Select RBs which are not interfere with other cells
            Add the predefined propagation delay to the packets
        else
            Apply RB allocation locally
        end
    end
end
```

5.4. Simulation Results

In Table 2. the corresponding simulation and NNs parameters are illustrated in detail.

Table 2. Simulation parameters.

Parameters	Value
Number of neurons	1024 × 512 layers (Actor + Critic)
DU selec./scheduler algorithms	SA2C-EADDUS, DRL-EADDUS, A2C-RBA
Number of BSs	4
Number of UEs	40, 60, 80
Maximum Traffic load per UE (Downlink)	512 kbps
Traffic types	Video, ITS
Propagation delay	2–5 ms
Reward's weights	$\tau = 0.5$ and $\omega = 0.5$
Number of RBs	12, 100
Discount factor	0.9
Actor learning-rate	0.01
Critic learning-rate	0.05

5.4.1. Convergence

Before discussing the obtained results, in Figure 3a we present the convergence of the reward function for SA2C-EADDUS. In this figure, the number of UEs is equal to 70, of which 10% is assigned with URLLC traffic load. Additionally, we employed the epsilon-greedy policy, forcing actors to assign RBs with the highest weight or randomly choosing actions during the exploration phase. As it is shown, the algorithm will converge after almost 100 episodes where each episode contains 500 iterations.

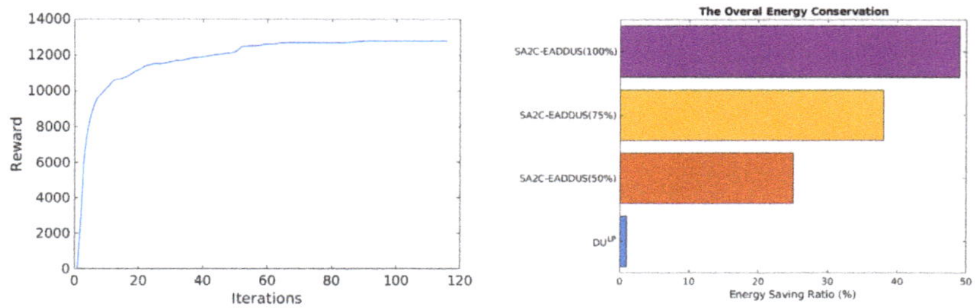

(a) (b)

Figure 3. The convergence of the reward function and the overall energy conservation in the network when the SA2C-EADDUS with different processing capacity level is employed: (**a**) the convergence of the reward function; (**b**) impact of changing ξ in SA2C-EADDUS.

5.4.2. Energy Efficiency

As mentioned previously, in this work, the energy consumption is reduced by migrating a NF dynamically to a DU with a higher processing power by considering its available capacity, priority of packets, and propagation delay during each time interval. We assume that when a DU transfer its load to DU^{HP}, it can be deactivated, and its overall power consumption (dynamic + fix) will be zero during that time interval. We set the amount of fixed energy consumption at DU^{LP} and DU^{HP} as 5 KWh and 10 KWh, respectively [24]. We examine our model by considering three processing capacity levels for DU^{HP}. In the first one, we can transfer the processing load of all DUs (DU^{LP}) to DU^{HP} (100%DRL-EADDUS), while in the second and third ones, DU^{HP} has a limited processing power and the load of only 75% (75%-DRL-EADDUS) or 50% (50%-DRL-EADDUS) of DU^{LP} can be transferred to DU^{HP}. As we can see in Figure 3b, based on the available processing power in DU^{HP}, by dynamically transferring local DUs' load, the SA2C-EADDUS scheme can increase energy conservation up to 50%. Therefore, by dynamically transferring the load among DUs, we can reduce the energy consumption dramatically in the network.

As shown previously, the formulated problem is NP-hard. To this end, we first compare our approach with an optimization model in a smaller scale network, then in the following, we increase the size of the network and evaluate the proposed scheme in comparison to our baselines.

5.4.3. Comparison with the MILP Solution

In this subsection, we compare the performance of our SA2C-EADDUS scheme with the MILP solution presented in Section 3.1. Due to the NP-hard property of our problem, we use only 12 RBs in this comparison. Therefore, we can obtain the optimum solutions with MILP solver in a reasonable run time.

Figure 4a shows the change of energy consumption and mean delay with the weighting factor of the energy consumption. As seen, $W = 0.7$ and $W = 0.6$ are the breaking points for 8 UEs and 16 UEs, respectively, in this multi-objective minimization problem. We observe that choosing a lower weight is crucial to prevent a drastically higher mean delay. In Figure 4b, we detail the impact of the increasing number of UEs when weight W_1 is set to =0.5. The mean delay remains in a reasonable range with the higher number of UEs, while the energy consumption increases due to the decentralization of the NFs. One of the main reasons for this decentralization comes from balancing the increase of scheduling delays due to increasing competition for RBs. Thus, NFs prefer to stay in local DUs to reduce the propagation delay, which also reduces the overall mean delay.

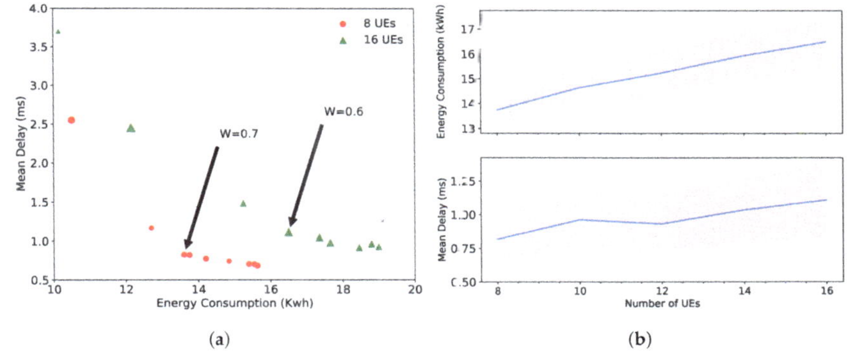

Figure 4. Energy consumption and mean delay performance with MILP solver: (**a**) the tradeoff between these two KPIs; (**b**) impact of increasing number of UEs.

Figure 5a compares the mean delay results of the SA2C-EADDUS scheme and the MILP solution. The SA2C-EADDUS scheme provides a reasonable mean delay for the UEs lower than 14. However, due to the high contention, a higher number of UEs causes package drops, and then they cause the increase of mean delay. On the other hand, the MILP solver is not affected by this due to the ideal conditions and predefined given data. Figure 5b compares the energy consumption of two solutions.The SA2C-EADDUS scheme performs very close to the MILP solution.

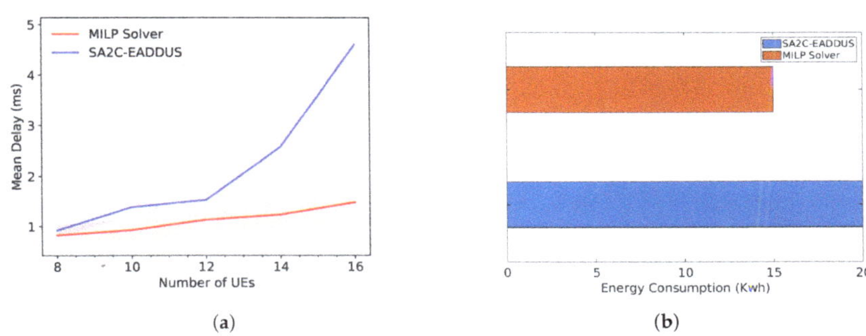

Figure 5. Lower bound comparisons of the proposed method SA2C-EADDUS: (**a**) mean delay comparison; (**b**) energy consumption comparison.

5.4.4. Delay

Hereafter, we examine the performance of the SA2C-EADDUS scheme with respect to three baselines and use a larger network where the number of UEs are varied between 40 to 80. As shown in Table 1, we evaluated the proposed model by generating User Datagram Protocol (UDP) and Transmission Control Protocol (TCP) packets (video and ITS) with different delay budgets (150 ms and 30 ms, respectively) and QoS requirements. In Figure 6a,b we presented the mean delay of video packets and ITS packets independently to illustrate how the proposed model managed to keep the mean delay of each traffic type below its delay budget threshold. The proposed SA2C-EADDUS scheme is compared with an A2C-based scheduler (A2C-RBA) [28], a DRL-based scheduler with different processing capacity levels [26], and a heuristic scheme (Heuristic-DUS). Here, to examine the effect of the processing capacity of DU^{HP} over the network performance, we assume that DU^{HP} has two processing capacity levels. In the first one, we can transfer the processing load of all DUs (DU^{LP}) to DU^{HP} (100%DRL-EADDUS), while in the second one, the load of only 50% of DU^{LP} can be transferred to DU^{HP} (50%-DRL-EADDUS). The energy consumption corresponding to the delay results in Figure 6 is depicted in Figure 3b. In Figure 3b, we presented the overall energy conversation when employing the SA2C-EADDUS with different capacity levels. The maximum energy will be consumed (20 KWh) when UEs are scheduled locally, and we can conserve energy consumption by up to 50% when the processing power increases.

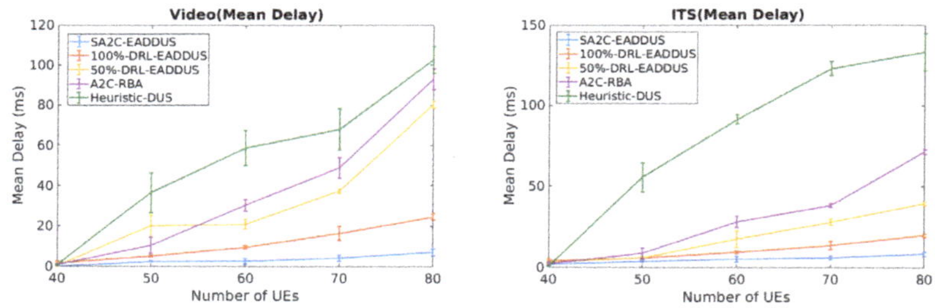

(a) (b)

Figure 6. The overall mean delay by considering different processing capacity levels and algorithms: (**a**) the video packets' mean delay; (**b**) the ITS packets' mean delay.

As observed in Figure 6, the proposed algorithm reduces the mean delay in both ITS and video traffic with respect to baselines. Additionally, by increasing the processing power, the observation level of agents will be increased, and actions become more accurate; therefore, the 100%DRL-EADDUS agent performs better by reducing the mean delay in the network in comparison with the 50%-DRL-EADDUS agent. Finally, the SA2C-EADDUS can reduce the mean delay dramatically with respect to the A2C-RBA and the Heuristic-DUS algorithm, since the SA2C-EADDUS can allocate RBs in a way that the inter-cell interference in the network is reduced. Additionally, the SA2C-EADDUS approach performs better with respect to the DRL-EADDUS algorithm because, unlike DRL-EADDUS, the RB allocation agent in the SA2C-EADDUS scheme, in addition to packet delay, also considers the mean CQI level of UEs and the priority of URLLC packets in its algorithm.

5.4.5. Packet Delivery Ratio

As explained previously, the proposed scheme reduces the inter-cell interference in the network by expanding agents' observation level and applying actions with respect to other RUs' RB maps when the NFs are transferred to DU^{HP}. Therefore, as shown in Figure 7, the SA2C-EADDUS scheme can improve the packet delivery ratio significantly with respect to the A2C-RBA and Heuristic-DUS algorithms when the number of UEs are high. Similarly, by increasing the processing level of DU^{HP}, actions become more accurate, and the number of packets which are successfully delivered will increase.

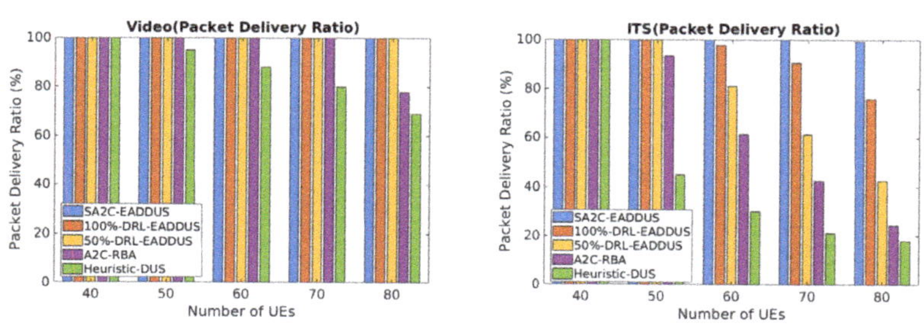

(a) (b)

Figure 7. The overall packet delivery ratio by considering different processing capacity levels and algorithms: (**a**) the video packets' delivery ratio; (**b**) the ITS packets' delivery ratio.

6. Conclusions

As future mobile networks become more complex, the need for intelligence and participation of more players is emerging, eventually leading to the need for openness. As these goals are defining the initiatives such as O-RAN and several others, there is a dire need to explore intelligence capabilities. In this paper, we evaluated the significance of expanding the observation level in O-RAN architecture for NFs. To this end, we consider resource allocation function as an example NF and propose a two nested A2C-based algorithms, which contain two A2C techniques that are working together. The first layer dynamically transfers NFs to a DU with higher processing power to hit the balance between saving energy and improving actions' accuracy. Meanwhile, the second layer contains an A2C-based scheduler algorithm, which allocates RB by considering user traffic type and their delay budget. The simulation results show that the proposed scheme can significantly increase energy efficiency and reduce the overall mean delay and packet drop ratio with respect to the case where the NF is solely executed at local DUs with limited processing power. In future works, we will employ the delay and energy consumption in the fronthaul links and the switching networks.

Author Contributions: Conceptualization, S.M. and T.P.; methodology, S.M. and T.P.; software, S.M. and T.P.; validation, S.M. and T.P.; formal analysis, S.M. and T.P.; investigation, S.M. and T.P.; resources, S.M., T.P. and R.W.; writing—original draft preparation, S.M. and T.P.; writing—review and editing, S.M. and T.P.; visualization, S.M. and T.P.; supervision, M.E.-K.; project administration, M.E.-K.; funding acquisition, Ontario Centers of Excellence (OCE) 5G ENCQOR program and Ciena. All authors have read and agreed to the published version of the manuscript.

Funding: This work is supported by Ontario Centers of Excellence (OCE) 5G ENCQOR program and Ciena.

Conflicts of Interest: The authors declare no conflict of interest.

Notations

The following notations are used in this manuscript:

Sets	Explanation
$r \in \mathcal{R}$	set of RBs in one time interval
$t \in \mathcal{T}$	set of time intervals
$i \in \mathcal{I}$	set of UEs
$\langle i,t \rangle \in \langle \mathcal{I}, \mathcal{T} \rangle$	tuple of demands
$l \in \mathcal{L}$	set of DU^{LP}
$k \in \mathcal{K}$	type of traffic
Variables	Explanation
$a_{\langle i,t \rangle rt'}$	indicates rt' is assigned for $\langle i,t \rangle$
b_{lt}	indicates NFs of DU^{LP} l migrates to DU^{HP} in t
$y_{\langle i,t \rangle t'}$	indicates any r in t' is assigned for $\langle i,t \rangle$
$u_{\langle i,t \rangle k}$	user traffic indicator
Given Data	Explanation
D^P_{lt}	propagation delay
E^F_l	fixed energy consumption at l
E^D_l	dynamic energy consumption coefficient at l
E^{HP}	total energy consumption at DU^{HP}
$S_{irt'}$	max service rate
$U_{\langle i,t \rangle k}$	user traffic demand size
$\Delta_{\langle i,t \rangle k}$	delay budget
ξ	max. number of DU^{LP} that can migrate their NFs
$M(i)$	user & DU^{LP} map
TTI	length of transmission time interval
W	energy consumption weight in objective function
Ω	scaling factor between en. cons. and mean delay

References

1. Klinkowski, M. Latency-Aware DU/CU Placement in Convergent Packet-Based 5G Fronthaul Transport Networks. *Appl. Sci.* **2020**, *10*, 7429. [CrossRef]
2. Semov, P.; Koleva, P.; Tonchev, K.; Poulkov, V.; Cooklev, T. Evolution of mobile networks and C-RAN on the road beyond 5G. In Proceedings of the 2020 43rd International Conference on Telecommunications and Signal Processing (TSP), Milan, Italy, 7–9 July 2020; IEEE: Piscataway, NJ, USA, 2020; pp. 392–398.
3. Dryjański, M.; Kułacz, Ł.; Kliks, A. Toward Modular and Flexible Open RAN Implementations in 6G Networks: Traffic Steering Use Case and O-RAN xApps. *Sensors* **2021**, *21*, 8173. [CrossRef] [PubMed]
4. Yi, B.; Wang, X.; Li, K.; Huang, M. A comprehensive survey of network function virtualization. *Comput. Netw.* **2018**, *133*, 212–262. [CrossRef]
5. Gilson, M.; Mackenzie, R.; Sutton, A.; Huang, J. *NGMN Overview on 5G RAN Functional Decomposition*; NGMN Alliance: Frankfurt am Main, Germany, 2018.
6. Pamuklu, T.; Ersoy, C. GROVE: A Cost-Efficient Green Radio Over Ethernet Architecture for Next Generation Radio Access Networks. *IEEE Trans. Green Commun. Netw.* **2021**, *5*, 84–93. [CrossRef]
7. Mollahasani, S.; Erol-Kantarci, M.; Wilson, R. Dynamic CU-DU Selection for Resource Allocation in O-RAN Using actor–critic Learning. In Proceedings of the IEEE Global Communications Conference (GLOBECOM), Madrid, Spain, 7–11 December 2021.
8. Wu, J.; Zhang, Y.; Zukerman, M.; Yung, E.K.N. Energy-efficient base-stations sleep-mode techniques in green cellular networks: A survey. *IEEE Commun. Surv. Tutor.* **2015**, *17*, 803–826. [CrossRef]
9. Oh, E.; Son, K.; Krishnamachari, B. Dynamic base station switching-on/off strategies for green cellular networks. *IEEE Trans. Wirel. Commun.* **2013**, *12*, 2126–2136. [CrossRef]
10. Niu, Z. TANGO: Traffic-aware network planning and green operation. *IEEE Wirel. Commun.* **2011**, *18*, 25–29. [CrossRef]
11. Mollahasani, S.; Onur, E. Density-aware, energy-and spectrum-efficient small cell scheduling. *IEEE Access* **2019**, *7*, 65852–65869. [CrossRef]
12. Qian, M.; Hardjawana, W.; Shi, J.; Vucetic, B. Baseband processing units virtualization for cloud radio access networks. *IEEE Wirel. Commun. Lett.* **2015**, *4*, 189–192. [CrossRef]
13. Wang, X.; Thota, S.; Tornatore, M.; Chung, H.S.; Lee, H.H.; Park, S.; Mukherjee, B. Energy-efficient virtual base station formation in optical-access-enabled cloud-RAN. *IEEE J. Sel. Areas Commun.* **2016**, *34*, 1130–1139. [CrossRef]
14. Sahu, B.J.; Dash, S.; Saxena, N.; Roy, A. Energy-efficient BBU allocation for green C-RAN. *IEEE Commun. Lett.* **2017**, *21*, 1637–1640. [CrossRef]
15. Saxena, N.; Roy, A.; Kim, H. Traffic-aware cloud RAN: A key for green 5G networks. *IEEE J. Sel. Areas Commun.* **2016**, *34*, 1010–1021. [CrossRef]
16. Malandrino, F.; Chiasserini, C.F.; Casetti, C.; Landi, G.; Capitani, M. An Optimization-Enhanced MANO for Energy-Efficient 5G Networks. *IEEE/ACM Trans. Netw.* **2019**, *27*, 1756–1769. [CrossRef]
17. Larsen, L.M.; Checko, A.; Christiansen, H.L. A survey of the functional splits proposed for 5G mobile crosshaul networks. *IEEE Commun. Surv. Tutor.* **2018**, *21*, 146–172. [CrossRef]
18. Shehata, M.; Elbanna, A.; Musumeci, F.; Tornatore, M. Multiplexing gain and processing savings of 5G radio-access-network functional splits. *IEEE Trans. Green Commun. Netw.* **2018**, *2*, 982–991. [CrossRef]
19. Alabbasi, A.; Wang, X.; Cavdar, C. Optimal processing allocation to minimize energy and bandwidth consumption in hybrid CRAN. *IEEE Trans. Green Commun. Netw.* **2018**, *2*, 545–555. [CrossRef]
20. Akoush, S.; Sohan, R.; Rice, A.; Moore, A.W.; Hopper, A. Predicting the performance of virtual machine migration. In Proceedings of the 2010 IEEE International Symposium on Modeling, Analysis and Simulation of Computer and Telecommunication Systems, Miami Beach, FL, USA, 17–19 August 2010; IEEE: Piscataway, NJ, USA, 2010; pp. 37–46.
21. Zhan, Z.H.; Liu, X.F.; Gong, Y.J.; Zhang, J.; Chung, H.S.H.; Li, Y. Cloud computing resource scheduling and a survey of its evolutionary approaches. *ACM Comput. Surv. (CSUR)* **2015**, *47*, 1–33. [CrossRef]
22. Elsayed, M.; Erol-Kantarci, M. AI-enabled future wireless networks: Challenges, opportunities, and open issues. *IEEE Veh. Technol. Mag.* **2019**, *14*, 70–77. [CrossRef]
23. Şahin, T.; Khalili, R.; Boban, M.; Wolisz, A. Reinforcement learning scheduler for vehicle-to-vehicle communications outside coverage. In Proceedings of the 2018 IEEE Vehicular Networking Conference (VNC), Taipei, Taiwan, 5–7 December 2018; IEEE: Piscataway, NJ, USA, 2018; pp. 1–8.
24. Pamuklu, T.; Erol-Kantarci, M.; Ersoy, C. Reinforcement Learning Based Dynamic Function Splitting in Disaggregated Green Open RANs. In Proceedings of the IEEE International Conference on Communications, Montreal, QC, Canada, 14–23 June 2021.
25. Elsayed, M.; Erol-Kantarci, M.; Yanikomeroglu, H. Transfer Reinforcement Learning for 5G-NR mm-Wave Networks. *IEEE Trans. Wirel. Commun.* **2020**, *20*, 2838–2849. [CrossRef]
26. Zhang, T.; Shen, S.; Mao, S.; Chang, G.K. Delay-aware Cellular Traffic Scheduling with Deep Reinforcement Learning. In Proceedings of the GLOBECOM 2020—2020 IEEE Global Communications Conference, Taipei, Taiwan, 7–11 December 2020; pp. 1–6.
27. Chen, G.; Zhang, X.; Shen, F.; Zeng, Q. Two Tier Slicing Resource Allocation Algorithm Based on Deep Reinforcement Learning and Joint Bidding in Wireless Access Networks. *Sensors* **2022**, *22*, 3495. [CrossRef]

28. Mollahasani, S.; Erol-Kantarci, M.; Hirab, M.; Dehghan, H.; Wilson, R. actor–critic Learning Based QoS-Aware Scheduler for Reconfigurable Wireless Networks. *IEEE Trans. Netw. Sci. Eng.* **2021**, *9*, 45–54. [CrossRef]
29. Pamuklu, T.; Mollahasani, S.; Erol-Kantarci, M. Energy-Efficient and Delay-Guaranteed Joint Resource Allocation and DU Selection in O-RAN. In Proceedings of the 5G World Forum (5GWF), Montreal, QC, Canada, 13–15 October 2021.
30. O-RAN Alliance. *O-RAN-WG1-O-RAN Architecture Description—v04.00.00*; Technical Specification; O-RAN Alliance: Alfter, Germany, 2021.
31. Yu, Y.J.; Pang, A.C.; Hsiu, P.C.; Fang, Y. Energy-efficient downlink resource allocation for mobile devices in wireless systems. In Proceedings of the 2013 IEEE Global Communications Conference (GLOBECOM), Atlanta, GA, USA, 9–13 December 2013; IEEE: Piscataway, NJ, USA, 2013; pp. 4692–4698.
32. Bonati, L.; D'Oro, S.; Polese, M.; Basagni, S.; Melodia, T. Intelligence and Learning in O-RAN for Data-Driven NextG Cellular Networks. *IEEE Commun. Mag.* **2021**, *59*, 21–27. [CrossRef]
33. ITU. ITU-T Recommendation G Suppl. 66. In *5G Wireless Fronthaul Requirements in a Passive Optical Network Context*; Technical Report; International Telecommunications Union: Geneva, Switzerland, 2018.
34. 3GPP. *Table 6.1.7-A: Standardized QCI Characteristics from 3GPP TS 23.203 V16 1.0*; Technical Report; 3GPP: Sophia Antipolis, France, 2020.
35. Gawłowicz, P.; Zubow, A. NS-3 meets openai gym: The playground for machine learning in networking research. In Proceedings of the 22nd International ACM Conference on Modeling, Analysis and Simulation of Wireless and Mobile Systems, Miami Beach, FL, USA, 25–29 November 2019; pp. 113–120.
36. Brockman, G.; Cheung, V.; Pettersson, L.; Schneider, J.; Schulman, J.; Tang, J.; Zaremba, W. Openai gym. *arXiv* **2016**, arXiv:1606.01540.

sensors

MDPI

Article

Electronic Beam-Scanning Antenna Based on a Reconfigurable Phase-Modulated Metasurface

Zakaria Zouhdi [1,2], Badreddine Ratni [1] and Shah Nawaz Burokur [1,*]

1 LEME, UPL, Université Paris Nanterre, F92410 Ville d'Avray, France; zakaria.zouhdi@naval-group.com (Z.Z.); beratni@parisnanterre.fr (B.R.)
2 Naval Group, Naval Research, 83190 Ollioules, France
* Correspondence: sburokur@parisnanterre.fr

Abstract: Metasurfaces (MSs) have enabled the emergence of new ideas and solutions in the design of antennas and for the control of electromagnetic waves. In this work, we propose to design a directional high-gain reconfigurable planar antenna based on a phase-modulated metasurface. Reconfigurability is achieved by integrating varactor diodes into the elementary meta-atoms composing the metasurface. As a proof of concept, a metasurface prototype that operates around 5 GHz is designed and fabricated to be tested in an antenna configuration. The metasurface is flexibly controlled by different bias voltages applied to the varactor diodes, thus allowing the user to control its phase characteristics. By assigning judiciously calculated phase profiles to the metasurface illuminated by a feeding primary source, different scenarios of far-field patterns can be considered. Different phase profiles are tested, allowing us to, firstly, achieve a highly directive boresight radiation and, secondly, to steer the main radiated beam towards an off-normal direction. The whole design process is verified by numerical simulations and is validated experimentally by far-field antenna measurements. The proposed metasurface enables the design of directive flat antennas with beam-scanning characteristics without complex feeding systems and power-consuming phase shifters, and thus provides potential interests for next generation antenna hardware.

Keywords: reconfigurable antenna; metasurface; varactor diode; beam-scanning; high gain

Citation: Zouhdi, Z.; Ratni, B. Burokur, S.N. Electronic Beam-Scanning Antenna Based on a Reconfigurable Phase-Modulated Metasurface. *Sensors* **2022**, *22*, 4990. https://doi.org/10.3390/s22134990

Academic Editor: Yue Li

Received: 21 April 2022
Accepted: 30 June 2022
Published: 1 July 2022

1. Introduction

Connectivity technologies continue to evolve year after year to meet growing global demand for increasing data rates and number of connected devices. Indeed, today, not only are humans connected to the global network but objects, sensors, industrial machines, drones and even satellites are all connected to the same network, which is pushing further the development of new technologies that allow all these needs to be met. As such, the development of a new generation of reconfigurable antennas enabling real-time monitoring of users is necessary.

The emergence of artificially structured materials, called metamaterials, has revolutionized the domain of electromagnetism. These engineered structures were first used in their three-dimensional (3D) configuration and demonstrated several phenomena and concepts [1–5]. Over the years, the community has become more and more interested in the development of two-dimensional (2D) versions of metamaterials, commonly known as metasurfaces. Such planar versions reduce the volume and complexity of realization and losses in comparison to the bulky 3D counterparts [6]. Thanks to these structures, new properties, such as anomalous reflection and refraction, have been proposed and demonstrated [7–9]. Metasurfaces also facilitated the generation of complex wavefronts, such as Airy beams [10,11], vortex beams carrying orbital angular momentum (OAM) [12,13], Bessel beams [14–16] and image holograms [17–21]. In the field of antennas, metasurfaces have aided in developing different topologies of structures. For instance, the dimensions of antennas can be drastically reduced, particularly when they are used as partially reflective surfaces (PRS) in reflex-type Fabry–Perot (FP) cavity antennas to achieve high directivity from a single radiating element [22,23]. Flat lenses [24,25], leaky-wave antennas [26] and polarizers [27,28] have also been designed based on the use of metasurfaces. In order to reach real-time control of the electromagnetic properties, as required in reconfigurable antennas, lumped electronic components are loaded in meta-atoms composing the metasurfaces [29]. Several types of components have been considered for reconfigurability mechanism, such as liquid crystals [30,31], PIN diodes [32,33], varactor diodes [15,34], MEMS [35] and graphene [36]. The electronic elements are used to tune the resonant properties of the meta-atom, thereby helping to dynamically tailor electromagnetic wavefronts. In the field of antennas, one of the applications that necessitates reconfigurable metasurfaces is beam scanning, in which the main beam is electronically steered through tailoring the phase profile implemented in the metasurface [37–39].

In this work, we propose a dynamic metasurface loaded with varactor diodes for an operation around 5 GHz frequency. The metasurface controlled by a direct current (DC) bias voltage is exploited as a reconfigurable planar reflector with a parabolic phase profile in order to design a highly directive antenna. Such antenna allows reconfigurability in frequency, thus making it possible to cover a non-negligible operating frequency band. It further allows control of the direction of the radiated main beam in order to achieve beam steering. A prototype of the reconfigurable metasurface is fabricated and experimentally tested in a reflect-array antenna configuration in a microwave anechoic chamber in order to validate its performances and functionality. Such a dynamic metasurface-based antenna platform is an interesting alternative to transmit highly directive but also continuously steerable beams at large angles, thus showing great potential in wireless communications and smart antennas where real-time control is desired.

2. Design Principle

The main goal of this work is to design a flat metasurface reflector (or meta-reflector) antenna. The meta-reflector is intended to have a parabolic phase profile and to be illuminated by a radiating element used as primary feed, as schematically illustrated in Figure 1. In order to emulate the response of a parabolic reflector, the reflection caused by the metasurface must be identical to that of a parabola. For this, we start by calculating the necessary phase profile that allows to mimic a parabolic reflector. The parabolic phase profile $\varphi(x, y)$ for a given operating wavelength λ and focal distance F is given as [40]:

$$\varphi(x,y) = \frac{2\pi}{\lambda}\left(\frac{(x-x_0)^2+(y-y_0)^2}{4F}\right)+\varphi_0 \tag{1}$$

where (x_0, y_0) is the focal point position and φ_0 is the reflection phase at $(x = 0, y = 0)$.

Equation (1), therefore, describes a parabolic phase profile, as illustrated in Figure 1. Considering the reconfigurability mechanism of the metasurface, the phase profile can be tuned according to the wavelength, focal distance and focal point position. Consequently, the meta-reflector antenna would cover a non-negligible frequency band of operation, as well as steer the reflected beam. For smart systems, it is preferable that the antenna should be as compact as possible with a small focal distance. In the case of parabolic antennas, in order to maximize the gain, it is further desirable to maximize the illumination on the reflector by the primary feed source.

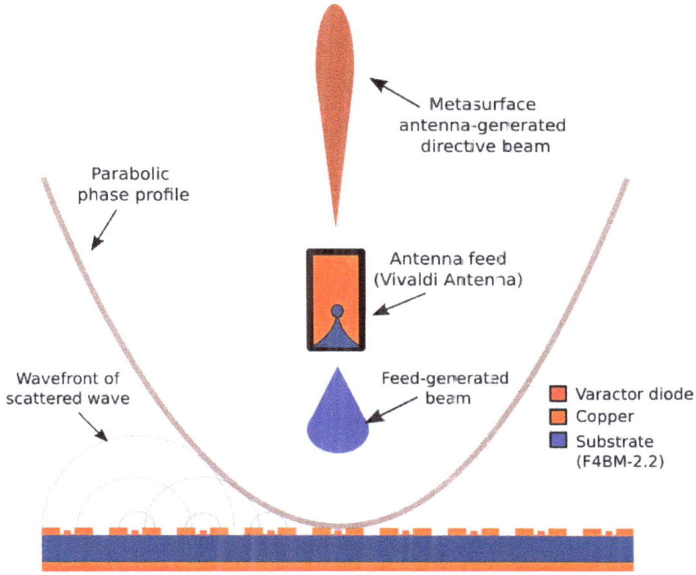

Figure 1. Schematic illustration of the operating principle of the flat meta-reflector antenna. The meta-reflector presents a parabolic phase profile and is illuminated by a radiating element used as primary feed. The resulting far-field radiation pattern of such an antenna configuration is a directive beam.

3. Design of the Metasurface Reflector

For the design of the metasurface that will serve as a reconfigurable reflector, we start by designing the elementary meta-atom. The main requirement is that the meta-atom presents an inductive-capacitive (*LC*) resonance that allows achieving a quasi-360° phase shift together with a high reflectivity. The considered unit cell structure has a periodicity p = 13 mm and is composed of two metal layers printed on the faces of a F4BM dielectric substrate with a relative permittivity ε_r = 2.2 and thickness t = 3 mm. The top layer is composed of two parallel copper strips of width w_p = 3 mm, which are separated by a gap g = 5 mm, while the bottom layer is composed of a continuous ground plane, as shown in the schematic view presented in Figure 2a. The separation gap g between the parallel strips of the top layer allows for a capacitive response when the electric field is oriented perpendicular to the strips. Additionally, the continuous ground plane contributes to produce an inductive response and to fully reflect the incident electromagnetic wave. The combination of the capacitive and inductive layers therefore achieves the desired *LC*-resonant circuit.

In order to electronically control the response of such a *LC*-resonant meta-atom, a varactor diode that serves to modify the capacitance is loaded in the elementary meta-atom. A DC bias voltage is applied to the varactor diode through the two parallel copper strips of the capacitive layer. A MAVR-000120-1411 varactor diode model [41] is considered due to its low losses and high tuning factor. In order to simulate the behavior of the elementary controllable meta-atom, the varactor diode is modelled as a *RLC* series circuit, where R = 3.5 Ω represents the ohmic losses, L = 0.9 nH is the inductance due to the packaging and C is the overall capacitance of the structure. The meta-atom structure is simulated using the finite element method (FEM) of Maxwell's equations from the commercially available high-frequency structure simulator (HFSS) code by Ansys [42]. Periodic boundary conditions and Floquet ports are utilized in the simulation setup in order to consider an infinite array of meta-atoms, as presented in Figure 2b. The influence of the variable capacitance on the meta-atom is illustrated in Figure 2c, where the frequency response for different capacitance values, are presented. The simulation results show that the varactor diode allows a shift of the resonance frequency from 4.2 GHz to 5.5 GHz, when the capacitance varies from 0.6 pF to 0.2 pF. The *LC* resonance feature can also be clearly observed with a reflection phase varying from +180° to −180° and passing through 0° at the resonance frequency.

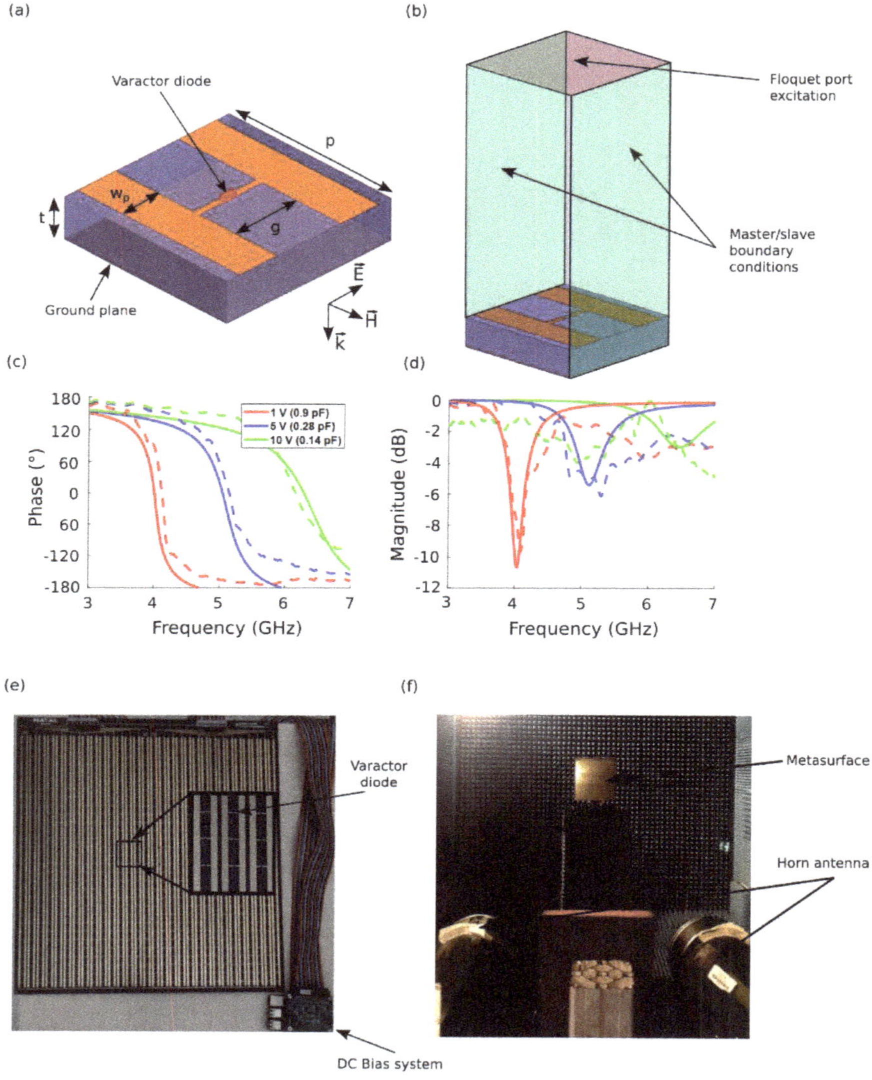

Figure 2. Design of the reconfigurable metasurface reflector. (**a**) Schematic view of the elementary meta-atom incorporating a voltage-controlled varactor diode and (**b**) schematic view of the simulation environment. The substrate used has a relative permittivity $\varepsilon_r = 2.2$ and a thickness $t = 3$ mm. The geometrical dimensions are: $p = 13$ mm, $w_p = 3$ mm and $g = 5$ mm. Simulated and measured reflection responses for different applied stimuli signals (capacitance values in simulations and bias voltages in experiments): (**c**) phase and (**d**) magnitude. Photograph of the fabricated metasurface (**e**) and the experimental characterization setup (**f**).

A prototype of the reconfigurable metasurface is realized using a conventional printed circuit board (PCB) manufacturing process. A photograph of the fabricated prototype is shown in Figure 2e, together with the experimental characterization setup (Figure 2f). The metasurface is composed of 30 columns, each containing 30 resonant unit cells (30 × 30 cells), and has lateral dimensions 390 mm × 390 mm ($6.5\lambda_0 \times 6.5\lambda_0$ at 5 GHz). In this voltage bias configuration, where a similar voltage will be applied to all the meta-atoms in a column, only a one-dimensional phase profile can be tailored. As such, we are restricted to cylindrical parabolic phase profile instead of a full parabolic phase profile. However, such cylindrical parabolic phase distribution is considered enough to meet the requirements of a proof-of-concept prototype.

The metasurface is characterized in a microwave anechoic chamber, where two broadband FLANN DP240 horn antennas [43] operating in the 2–18 GHz frequency band are connected to the vector network analyzer (VNA) and positioned side by side in front of the metasurface in order to measure its reflection coefficient, as shown in Figure 2f. The reflection coefficient is obtained by measuring the amplitude of the transmitted wave between the two horn antennas after being reflected by the metasurface. The reflection measurements performed on the metasurface are referenced with respect to that on a metal plate with similar lateral dimensions. The simulation and measurement results are presented in Figure 2c, where the correspondence between bias voltage and capacitance value is shown. It is important to note that in such a characterization procedure, all bias voltages are set at the same value throughout the metasurface. As illustrated for a different set of values, 1 V (C = 0.6 pF), 4 V (C = 0.28 pF) and 8 V (C = 0.2 pF), a good qualitative agreement is obtained, thus validating the concept of the reconfigurability mechanism implemented in the metasurface. A decrease in the capacitance value of the varactor diode induced by an increase in bias voltage leads to a resonance shift toward higher frequencies, which enables a shift in the phase response of the metasurface. This phase shift is very important since it will allow the tailoring of the phase distribution required to mimic the profile of a parabola. A high phase shift $\Delta\varphi$ (above 280°) is achieved within the 4.4–5 GHz due to the intrinsic design of the meta-atom. With such phase-tuning capability, it is therefore possible to generate a cylindrical parabolic phase profile from the metasurface.

4. Primary Feed Design

A primary feed is required to illuminate the metasurface reflector. This radiating element positioned at a certain distance in front of the metasurface will launch electromagnetic waves, which will be reflected by the meta-reflector. It is therefore necessary that the feeding source does not mask the reflected waves. Here, we therefore propose to use a Vivaldi radiating element as primary source. The choice of such a type of source is made for two reasons; the first one is due to its end-fire radiation characteristics, i.e., the maximum radiation lies in the plane of the radiating element, which considerably reduces the masking effect mentioned above, while the second reason is that such antennas present broadband features and, therefore, allow covering a wide frequency band of operation. A schematic view of the proposed Vivaldi radiating element printed on a F4BM dielectric substrate with relative permittivity ε_r = 2.2 and thickness h = 1 mm is shown in Figure 3a. The antenna is optimized in numerical simulations for an operation around 5 GHz. The geometry of the source is based on the coplanar Vivaldi antennas with the Vivaldi design on the upper layer of the substrate composed of an exponentially tapered slot [44,45]. The feed of the Vivaldi antenna is composed of a microstrip line transition to a radial slot of diameter D_1 with a circular stub of diameter D_2. The optimized geometrical dimensions of the radiating element on the top layer are D_1 = 17 mm, L = 83 mm and W = 72 mm for the top layer. The bottom layer is composed of the transmission-line excitation system, which is composed of a 50 Ω line with length l_1 = 6 mm and width w_1 = 3 mm. In order to match the input impedance of the excitation line, a quarter-wavelength transformer of length l_2 = 11 mm and width w_2 = 1.8 mm is added. The last part of the line, with l_3 = 7 mm, l_4 = 21 mm and w_3 = 0.9 mm, is connected to an offset feed of diameter D_2 = 9.2 mm.

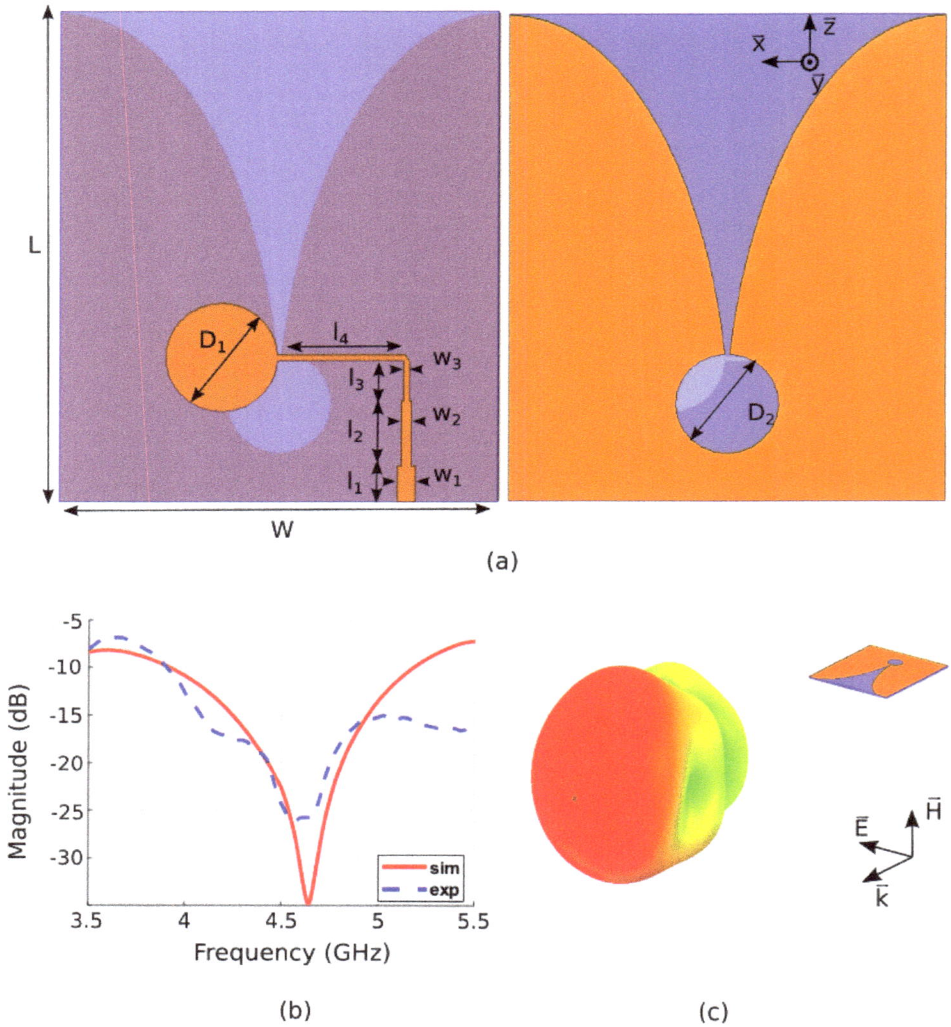

Figure 3. Feeding source used to illuminate the meta-reflector. (**a**) Schematic view of the Vivaldi antenna structure (top and bottom layers). The geometrical dimensions are: L = 83 mm, W = 72 mm, l_1 = 6 mm, l_2 = 11 mm, l_3 = 7 mm, l_4 = 21 mm, w_1 = 3 mm, w_2 = 1.8 mm, w_3 = 0.9 mm, D_1 = 17 mm and D_2 = 9.2 mm. (**b**) Simulated (blue trace) and measured (red trace) reflection responses (S_{11} coefficients) of the proposed primary source. (**c**) Schematic view of the Vivaldi antenna along with its 3D radiation pattern.

It should also be noted that in the reflector antenna scenario, the Vivaldi radiating element will be in close proximity to the metasurface. Therefore, it has to be designed and optimized by taking into account the coupling with the metasurface. The antenna is fabricated using PCB technique and measured in an anechoic chamber. The experimental reflection coefficient is compared to the simulated one in Figure 3b. A good agreement is observed between the two results and a good impedance matching ($S_{11} < -10$ dB) is observed over a wide frequency, ranging from 3.9 GHz to 5.2 GHz in simulation, while the measured result shows a good impedance matching from 3.9 GHz to above 5.5 GHz. This result is excellent in our case since the desired frequency band of operation of our meta-reflector antenna is from 4.4 GHz to 5 GHz. It will be shown in the next section that when placed in front of the metasurface, the reflection coefficient of the reflector antenna system remains quasi-similar to the feeding Vivaldi element alone. The simulated 3D radiation pattern of the Vivaldi radiating element is shown in Figure 3c, where a unidirectional beam can be observed.

In Figure 4, the 2D far-field radiation patterns are shown at 4.4, 4.7 and 5 GHz, where the blue solid line and red dotted line are the simulated and measured co-polarized gain (*E*-plane patterns), respectively. The simulation shows good agreement with the experiment as the maximum gain is measured to be 7 dBi at 4.4 GHz, 9.8 dBi at 4.7 GHz and 5.7 dBi at 5 GHz. The half-power beamwidth is measured to be approximately 60° at the three tested frequencies. The cross-polarized gains (*H*-plane patterns), represented by the cyan and green traces, show relatively low level.

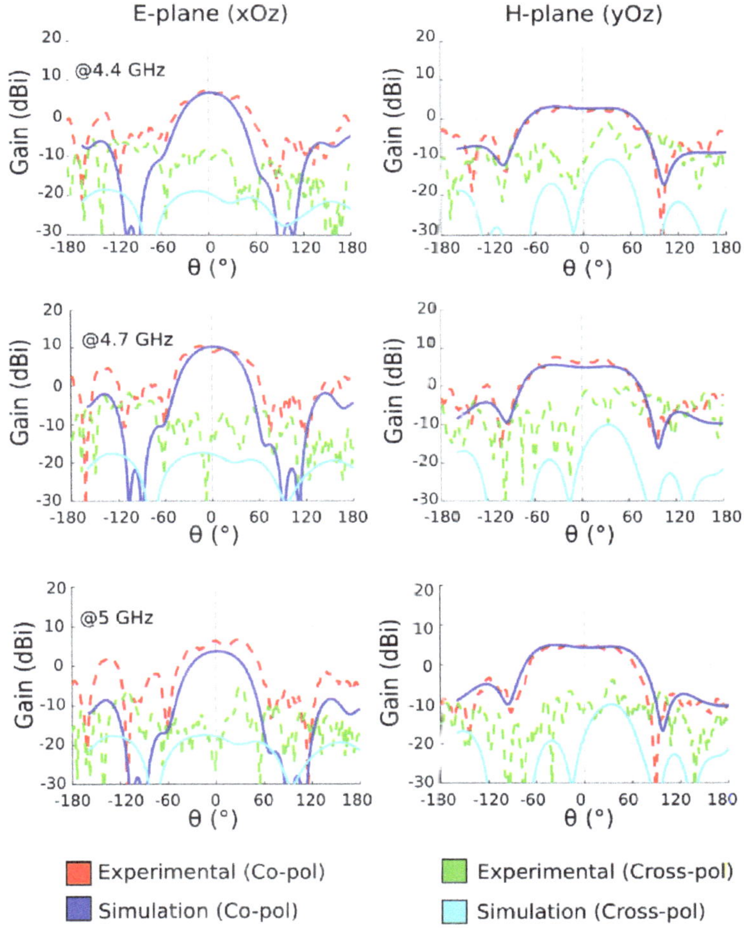

Figure 4. Simulated and measured *E*-plane and *H*-plane radiation patterns of the Vivaldi radiating element at 4.4 GHz, 4.7 GHz and 5 GHz.

The difference in side-lobes between simulated and experimental results in the *E*-plane is caused by the coaxial cable feeding the Vivaldi antenna. The end-fire configuration of such a Vivaldi antenna seems to be the reason for such a phenomenon, which then has some influence on the side lobes of the metasurface antenna.

5. Design of the Flat Meta-Reflector Antenna

Once the primary source and the metasurface have been designed and experimentally characterized separately, the meta-reflector antenna is assembled, as illustrated in Figure 5a. For an experimental proof-of-concept prototype of the reconfigurable meta-reflector, we limit the parabolic phase profile to only a one-dimensional (1D) plane, as restricted by the design of the metasurface, which can be controlled in a single plane. Therefore, the phase profile is applied only along the x-axis. The full parabolic phase distribution of Equation (1) is then simplified to a cylindrical parabolic one as:

$$\varphi(x) = \frac{2\pi}{\lambda}\frac{(x - x_0)^2}{4F} + \varphi_0 \tag{2}$$

In our case, as the metasurface has lateral dimensions of 390 mm × 390 mm, and with the HPBW (half-power beamwidth) of the antenna being 60°, a focal distance should be considered such that the metasurface is correctly illuminated by the main beam. The antenna system is simulated using Ansys HFSS and after several optimization simulations, the focal distance F is set to 120 mm. Imposing 120 mm as a focal distance is mainly motivated by a trade-off between two opposing factors. A higher F/D ratio (D being the metasurface dimension) would lead to an improvement in gain, whereas a lower one would reduce the antenna's profile but with a reduced gain. It is important to note that while a large focal distance would yield an increase in aperture efficiency, the structure is being designed for a low-profile perspective in the present study.

The reflection coefficient of the meta-reflector antenna system is measured and compared to the simulation results, as presented in Figure 5b. A good agreement between measurements and simulations is obtained and a good impedance matching is observed in our frequency band of interest between 4.4 GHz and 5 GHz, where the reflection coefficient shows amplitude values lower than −10 dB.

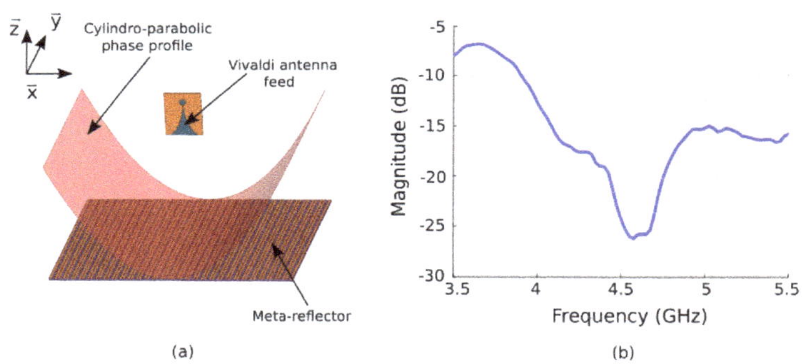

Figure 5. (**a**) Schematic view the reconfigurable flat meta-reflector antenna. (**b**) Measured reflection coefficient.

In order to validate the proposed flat meta-reflector antenna, measurements of the radiation patterns are performed in an anechoic chamber. A schematic view of the measurement setup is presented in Figure 6a and a photograph of the antenna in the test environment is shown in Figure 6b. The antenna under test is placed on a turntable and is connected to one port (port 1) of the VNA, while the (2–18 GHz) FLANN broadband horn antenna connected to the other port (port 2) of the VNA is used as a receiving antenna at a distance of 6 m. The antenna system under test on the turntable is rotated between from −180° to +180° in order to measure the transmitted power between the two antennas and therefore the antenna far-field radiation patterns. Several antenna configurations are tested. The cylindrical parabolic phase profiles allowing a boresight radiation to be obtained at 4.4 GHz, 4.7 GHz and 5 GHz are calculated from Equation (2) by fixing $\varphi_0 = 0°$ and are plotted in Figure 7a. These phase profiles are applied to the metasurface by judiciously applying the correct bias voltage corresponding to the required capacitance value to each column of meta-atoms. For large antennas, typical full-wave solvers require significantly large simulation time and memory constraints. As a way to deal with this limitation, the metasurface is approximated by its ideal case, i.e., a metallic cylindrical parabolic reflector antenna. In order to verify the accurateness of this approximation, the phase-modulated metasurface is fully simulated at 4.7 GHz and compared to the subsequent metallic cylindrical parabolic case, as well as the experimental measurements. The results, as shown in Figure 7b, display a clear correlation between the cylindrical parabolic reflector antenna, the fully simulated gradient metasurface and experimental results, thus validating the approximation, which will be used henceforward.

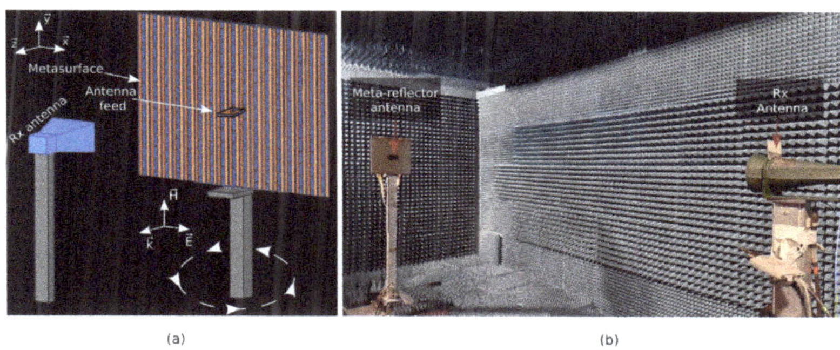

Figure 6. Antenna radiation patterns measurement. (**a**) Schematic illustration of the experimental measurement setup. The meta-reflector is illuminated by the Vivaldi radiating element and a receiving horn antenna allows measurement of the antenna radiation patterns. (**b**) Photograph of the far-field measurement setup in an anechoic chamber.

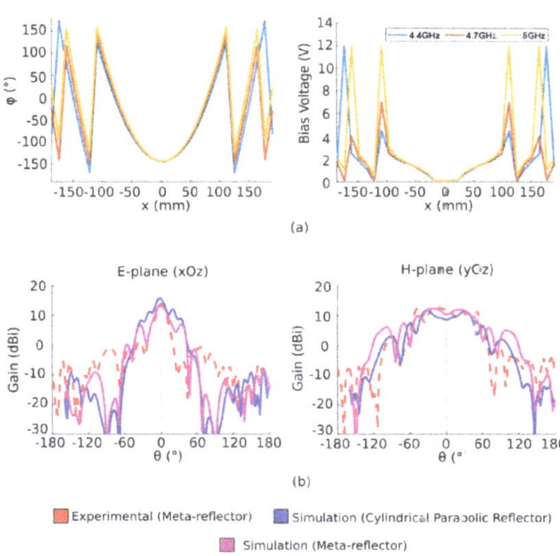

Figure 7. (**a**) Calculated phase profiles and applied bias voltages at 4.4 GHz, 4.7 GHz and 5 GHz. (**b**) Radiation patterns of the fully simulated gradient metasurface structure at 4.7 GHz, the metallic cylindrical reflector antenna simulation and the experimental measurements of the gradient meta-reflector.

The measured radiation patterns are presented in Figure 8 and show a highly directive beam in the *E*-plane (*xOz* plane), where the phase profile is applied. The maximum gain reaches 16 dBi at the central frequency of 4.7 GHz. The half-power beamwidth of the antenna is found to be around 32°. In the *H*-plane (*yOz* plane), the patterns are similar to those of the Vivaldi feeding source since no parabolic phase profile is applied in this plane. A difference of minimum 12 dB can be observed between the co- and cross-polarized gains, indicating a sufficient decoupling. The presented results allow the validation of the proposed flat parabolic reflector antenna concept by comparing the experimental measurements with a simulated metallic cylindrical parabolic reflector antenna.

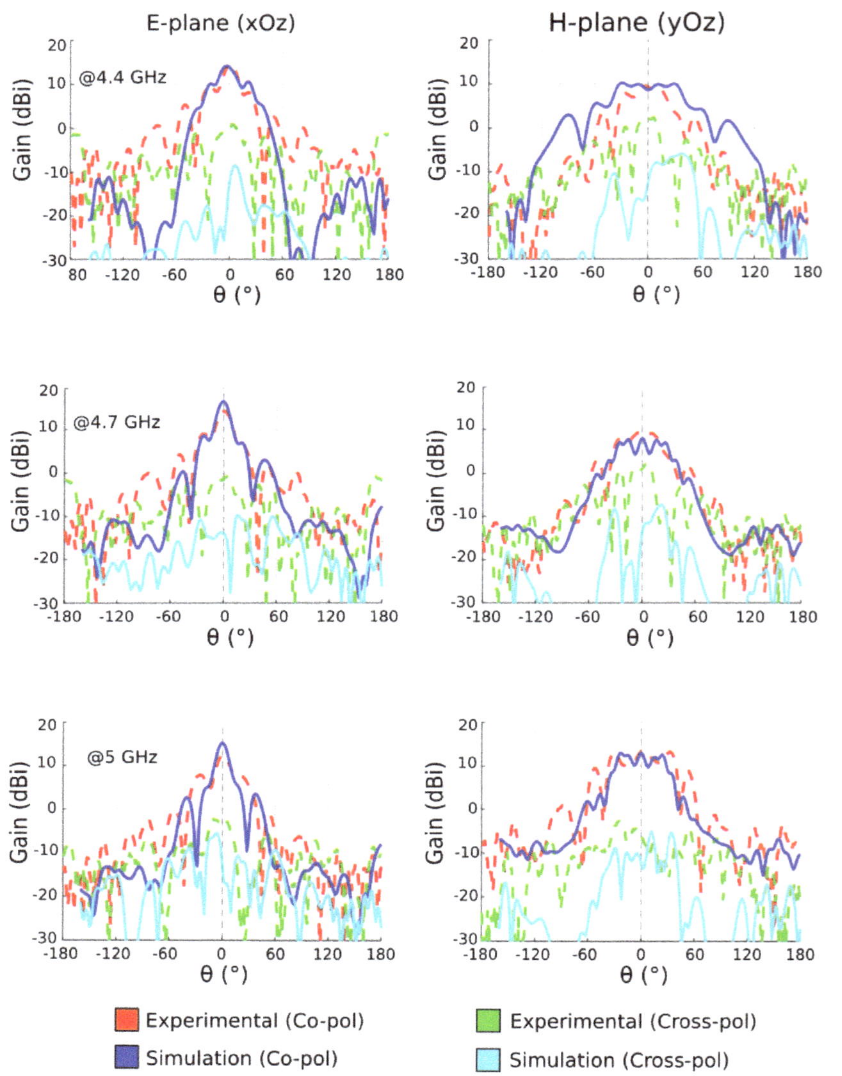

Figure 8. Simulated and measured *E*-plane radiation patterns of the meta-reflector antenna at 4.4 GHz, 4.7 GHz and 5 GHz.

Moreover, the gain at boresight is measured and is presented in Figure 9. At 4.7 GHz, the gain reaches 16 dBi. The aperture efficiency η can be calculated as [44]:

$$\eta = \frac{A_e}{A} = \frac{G\lambda^2}{4\pi A} \quad (3)$$

with A_e being the effective aperture, A the physical aperture (390 mm × 390 mm) and G the experimentally measured gain at wavelength λ. The calculated aperture efficiency at 4.4 GHz, 4.7 GHz and 5 GHz are, respectively, 6%, 7% and 8%. Comparatively, the aperture efficiency of a parabolic reflector is typically around 50% to 70%. The reason for this big difference is explained by different factors. Firstly, the proposed metasurface antenna presents gain values ranging from 14 dBi to 16 dBi. These low gains are mostly due to the fact that the meta-reflector has been engineered to produce a cylindrical parabolic phase profile (i.e., a parabolic phase profile in only one plane) instead of a full parabolic phase profile, where the gain would be above 20 dBi. Another important issue is that we use electronic components, namely varactor diodes, in the meta-reflector. These varactor diodes have a certain parasitic resistance that will partly absorb electromagnetic waves. Thus, the gain is lower with the reconfigurable meta-reflector. The second factor is justified by our choice of focal distance F = 120 mm so as to achieve a low-profile antenna. Consequently, the best trade-off that allows the gain to be maximized while reducing the profile of the antenna as much as possible is to place the illuminating source at a focal distance of 120 mm from the reflector.

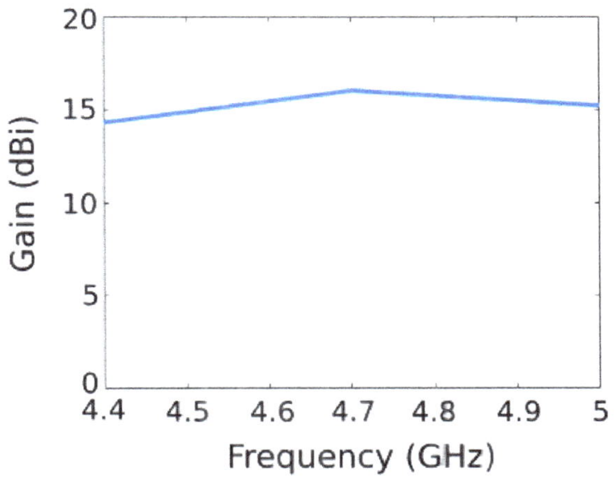

Figure 9. Experimental gain at boresight versus frequency of the meta-reflector antenna.

The illumination efficiency η_i and spillover efficiency η_s have also been calculated using [46]:

$$\eta_i = \frac{\left(\frac{1+\cos^{q+1}\theta}{q+1} + \frac{1+\cos^{q}\theta}{q}\right)^2}{2\tan^2\theta\left(\frac{1-\cos^{2q+1}\theta}{2q+1}\right)} \quad (4)$$

$$\eta_s = 1 - \cos^{2q+1}\theta \quad (5)$$

where q is the exponent of a $\cos^{q}\theta$ (with q = 3) radiation pattern that approximates the experimentally measured pattern of the Vivaldi source and θ is half of the subtend angle from the feed to the reflectarray aperture for a 120 mm focal distance (θ = 57° in our case).

At 4.7 GHz, it is found that η_i = 45% and η_s = 98%. This is consistent with the fact that the focal distance is at a sub-optimal configuration and is, therefore, unable to fully illuminate the structure (low illumination efficiency). Meanwhile such a small focal distance leads to a high spillover efficiency.

The possibility of controlling the direction of the main radiated beam is also proposed. In the conventional case, it is possible to perform beam steering by physically shifting the feeding source away from the focal point or by turning the parabolic reflector toward the desired direction. However, given that the metasurface allows an electronic control of the phase profile, we are able to introduce an offset x_0 in order to virtually move the parabola with respect to the source and, therefore, achieve beam steering. In order to validate the beam-steering capabilities, the phase profile is shifted progressively. Three different beam-steering angles (30°, 45° and 60°) are tested at 4.7 GHz. The corresponding phase and bias voltage profiles that need to be implemented on the metasurface are shown in Figure 10a. Simulation and experimental results presented in Figure 10b show high beam-steering capabilities up to 60° with side-lobe levels (SLL) lower than 10 dB. The high beam-steering range of the proposed antenna is an improvement over the capabilities of previous gradient metasurfaces presented in literature, where beam steering was achieved up to 50° [8,37]. The parabolic phase profile also enables a significant increase in gain compared to classic beam-steering gradient metasurfaces. Though the radiated beam is steered, it can be clearly observed that a high maximum gain is maintained. However, due to the 1-D property of the phase profile, the half-power beamwidth remains large in the *H*-plane.

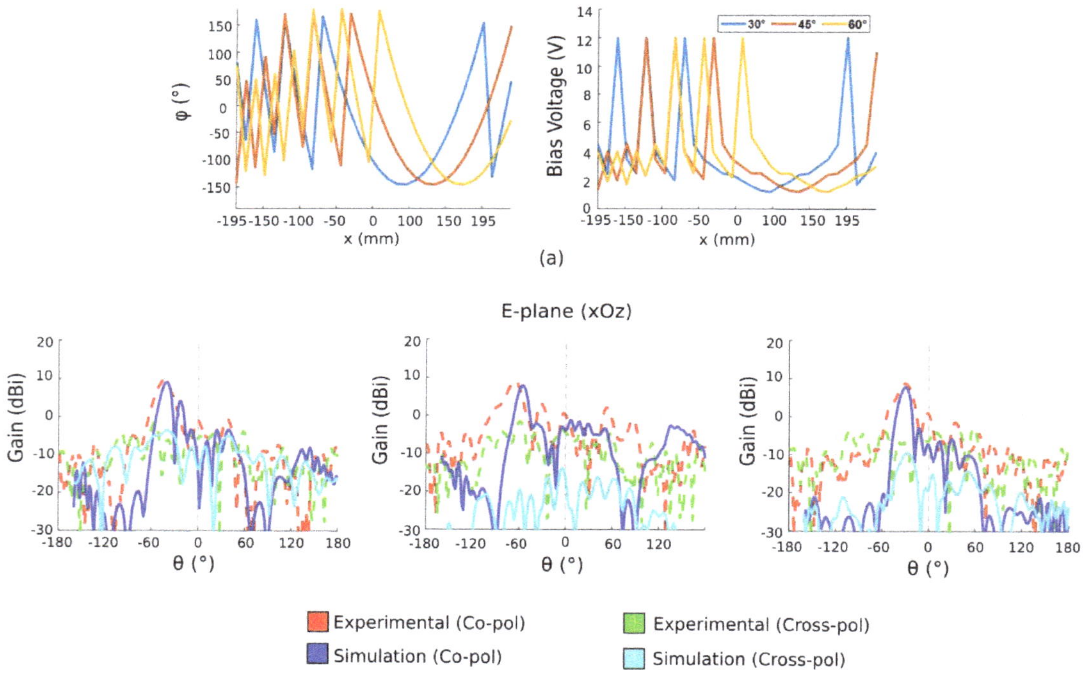

Figure 10. (**a**) Calculated phase profiles and applied bias voltage allowing beam steering at 30°, 45°, 60°. (**b**) Measured radiation patterns along the *E*-plane showing beam steering at 30°, 45°, 60°.

In order to improve the half-power beam width in the *H*-plane and acquire higher gain suitable for 5G applications, a 1 × 4 Vivaldi array of identical antenna elements is proposed to be aligned in the *H*-plane of the antenna platform, as shown in Figure 11a,b.

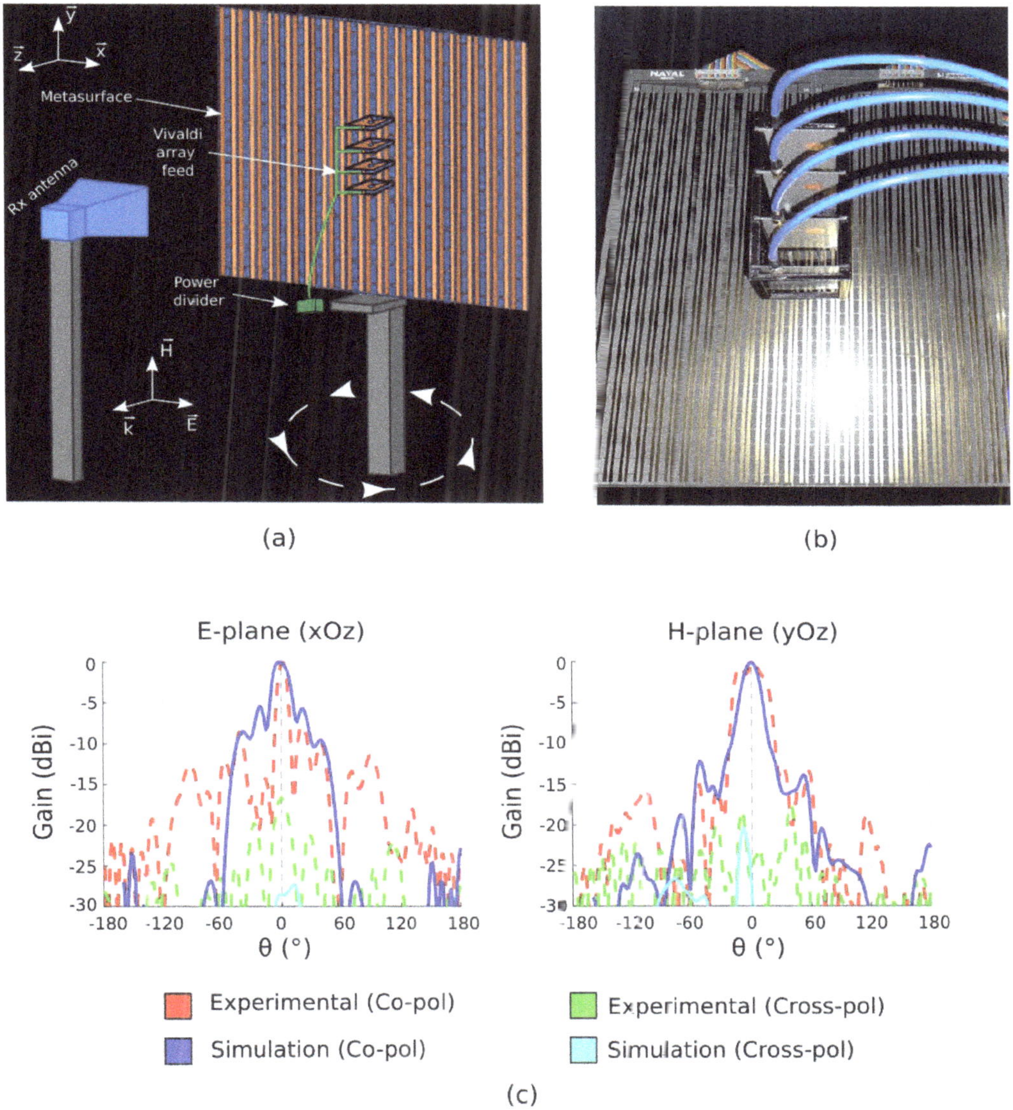

Figure 11. (**a**) Schematic illustration of the experimental measurement setup. (**b**) Photograph of the metasurface with its Vivaldi array feed. (**c**) Simulated and measured far-field radiation patterns at 4.7 GHz in the *E*- and *H*-planes for the meta-reflector fed by the Vivaldi antenna array.

Experimental results presented in Figure 11c show a high directivity in both *E*- and *H*-planes in the case of boresight radiation configuration. The measured maximum half-power beam width is around 20° in both planes, showing a 65% decrease of the HPBW in the *H*-plane as well as a 38% decrease in the *E*-plane. Consequently, this solution allows the design of a highly directive antenna where only a single beam-scanning plane is desired. Furthermore, the antenna can possibly show beam-steering capabilities in both planes by using the reconfigurable meta-reflector in the *E*-plane and phase shifters in the *H*-plane or designing a metasurface where control of the phase can be achieved in the two planes.

6. Conclusions

In summary, a reconfigurable meta-reflector including varactor diodes is proposed. The use of varactor diodes allows continuous control of the phase response of the meta-reflector. The proposed reflector structure is associated with a Vivaldi antenna used as a primary source in order to realize a highly directive reconfigurable antenna. The primary source is designed specifically to take into account the metasurface/feed coupling. A prototype of the meta-reflector antenna has been fabricated and characterized. Far-field measurements have been performed in an anechoic chamber, where several different desired configurations were tested with judiciously calculated phase profiles. Firstly, boresight radiation at different frequencies (4.4 GHz, 4.7 GHz and 5 GHz) were tested, and the experimental results show a highly directive beam with a gain of around 16 dBi. Then, beam-steering functionality was also verified in one radiation plane through the introduction of a lateral shift in the parabolic phase profile. Different phase profiles for different beam-steering capabilities were tested and the results show that remarkably high beam steering can be achieved, with up to 60° in steering while maintaining side-lobe levels under 10 dB. The obtained results for these different tested configurations allowed the validation of the proposed concept. Finally, due to the use of a Vivaldi radiating element as feed to illuminate the reflector, the radiated beam in the *H*-plane shows a large half-power beamwidth. As such, a solution of using a 1×4 Vivaldi array of identical antenna elements to reach high directivity also in *H*-plane is proposed.

The resulting performances, particularly the directive nature of the antenna's radiated beam, pave the way for several applications in the field of 5G, satellite and naval communications. The high beam-steering capabilities allow a hemispheric coverage using a reduced number of antennas. Furthermore, owing to its relatively low profile and simple design, the proposed antenna is a good candidate for integration on planar surfaces, such as walls and ship topside hulls.

Future works will be dedicated to electronic beam-scanning in both *E*- and *H*-planes. Two different solutions can be considered. The first one consists of using a phase shifter on the 1×4 array of feeding elements. The second solution consists of developing a meta-reflector platform where the meta-atoms can be controlled individually in such a way that a full parabolic phase profile can be achieved, as well as beam-scanning in both radiation planes.

Author Contributions: Conceptualization, Z.Z. and B.R.; methodology, B.R. and S.N.B.; software, Z.Z.; validation, Z.Z. and B.R.; formal analysis, Z.Z., B.R. and S.N.B.; resources, S.N.B.; writing—original draft preparation, Z.Z. and B.R.; writing—review and editing, Z.Z., B.R. and S.N.B.; supervision, B.R. and S.N.B.; project administration, S.N.B. All authors have read and agreed to the published version of the manuscript.

Funding: This research work was supported by Naval Group and the French Ministry of Higher Education, Research and Innovation, in the framework of an ANRT Cifre Agreement No. 2020/0163.

Institutional Review Board Statement: Not applicable.

Informed Consent Statement: Not applicable.

Data Availability Statement: Not applicable.

Acknowledgments: Authors would also like to acknowledge Marie-Hélène Billot for her assistance in the project administration and funding and also for fruitful discussions.

Conflicts of Interest: The authors declare no conflict of interest.

References

1. Pendry, J.B.; Schurig, D.; Smith, D.R. Controlling Electromagnetic Fields. *Science* **2006**, *312*, 1780–1782. [CrossRef] [PubMed]
2. Ma, H.F.; Cui, T.J. Three-Dimensional Broadband and Broad-Angle Transformation-Optics Lens. *Nat. Commun.* **2010**, *1*, 124. [CrossRef]
3. Tichit, P.-H.; Burokur, S.N.; Qiu, C.-W.; de Lustrac, A. Experimental Verification of Isotropic Radiation from a Coherent Dipole Source via Electric-Field-Driven *LC* Resonator Metamaterials. *Phys. Rev. Lett.* **2013**, *111*, 133901. [CrossRef] [PubMed]

4. Yi, J.; Burokur, S.N.; Piau, G.-P.; de Lustrac, A. 3D Printed Broadband Transformation Optics Based All-Dielectric Microwave lenses. *J. Opt.* **2016**, *18*, 044010. [CrossRef]
5. Tian, H.W.; Jiang, W.; Li, X.; Chen, Z.P.; Cui, T.J. An Ultrawideband and High-Gain Antenna Based on 3-D Impedance-Matching Metamaterial Lens. *IEEE Trans. Antennas Propag.* **2021**, *69*, 3084–3093. [CrossRef]
6. Holloway, C.; Kuester, E.; Gordon, J.; O'Hara, J.; Booth, J.; Smith, D. An Overview of the Theory and Applications of Metasurfaces: The Two-Dimensional Equivalents of Metamaterials. *IEEE Antennas Propag. Mag.* **2012**, *54*, 10–35. [CrossRef]
7. Yu, N.; Genevet, P.; Kats, M.A.; Aieta, F.; Tetienne, J.-P.; Capasso, F.; Gaburro, Z. Light Propagation with Phase Discontinuities: Generalized Laws of Reflection and Refraction. *Science* **2011**, *334*, 333–337. [CrossRef] [PubMed]
8. Ratni, B.; de Lustrac, A.; Piau, G.-P.; Burokur, S.N. Active Metasurface for Reconfigurable Reflectors. *Appl. Phys. A* **2018**, *124*, 104. [CrossRef]
9. Ratni, B.; de Lustrac, A.; Piau, G.-P.; Burokur, S.N. Reconfigurable Meta-Mirror for Wavefronts Control: Applications to Microwave Antennas. *Opt. Express* **2018**, *26*, 2613–2624. [CrossRef]
10. Hao, W.; Deng, M.; Chen, S.; Chen, L. High-Efficiency Generation of Airy Beams with Huygens' Metasurface. *Phys. Rev. Appl.* **2019**, *11*, 054012. [CrossRef]
11. Feng, R.; Ratni, B.; Yi, J.; Zhang, K.; Ding, X.; Zhang, H.; de Lustrac, A.; Burokur, S.N. Versatile Airy-Beam Generation Using a 1-Bit Coding Programmable Reflective Metasurface. *Phys. Rev. Appl.* **2020**, *14*, 014081. [CrossRef]
12. Zhang, K.; Wang, Y.; Yuan, Y.; Burokur, S.N. A Review of Orbital Angular Momentum Vortex Beams Generation: From Traditional Methods to Metasurfaces. *Appl. Sci.* **2020**, *10*, 1015. [CrossRef]
13. Zhang, K.; Yuan, Y.; Ding, X.; Li, H.; Ratni, B.; Wu, Q.; Liu, J.; Burokur, S.N.; Tan, J. Polarization-engineered Noninterleaved Metasurface for Integer and Fractional Orbital Angular Momentum Multiplexing. *Laser Photon. Rev.* **2021**, *15*, 2000351. [CrossRef]
14. Feng, R.; Ratni, B.; Yi, J.; Jiang, Z.; Zhang, H.; Lustrac, A.; Burokur, S.N. Flexible Manipulation of Bessel-like Beams with a Reconfigurable Metasurface. *Adv. Opt. Mater.* **2020**, *8*, 2001084. [CrossRef]
15. Feng, R.; Ratni, B.; Yi, J.; Zhang, H.; de Lustrac, A.; Burokur, S.N. Versatile Metasurface Platform for Electromagnetic Wave Tailoring. *Photonics Res.* **2021**, *9*, 1650–1659. [CrossRef]
16. Wang, Z.; Dong, S.; Luo, W.; Jia, M.; Liang, Z.; He, Q.; Sun, S.; Zhou, L. High-Efficiency Generation of Bessel Beams with Transmissive Metasurfaces. *Appl. Phys. Lett.* **2018**, *112*, 191901. [CrossRef]
17. Wang, Z.; Ding, X.; Zhang, K.; Ratni, B.; Burokur, S.N.; Gu, X.; Wu, Q. Huygens Metasurface Holograms with the Modulation of Focal Energy Distribution. *Adv. Opt. Mater.* **2018**, *6*, 1800121. [CrossRef]
18. Ratni, B.; Wang, Z.; Zhang, K.; Ding, X.; de Lustrac, A.; Piau, G.-P.; Burokur, S.N. Dynamically Controlling Spatial Energy Distribution with a Holographic Metamirror for Adaptive Focusing. *Phys. Rev. Appl.* **2020**, *13*, 034006. [CrossRef]
19. Guan, C.; Liu, J.; Ding, X.; Wang, Z.; Zhang, K.; Li, H.; Jin, M.; Burokur, S.N.; Wu, Q. Dual-Polarized Multiplexed Meta-Holograms Utilizing Coding Metasurface. *Nanophotonics* **2020**, *9*, 3605–3613. [CrossRef]
20. Ding, X.; Wang, Z.; Hu, G.; Liu, J.; Zhang, K.; Li, H.; Ratni, B.; Burokur, S.N.; Wu, Q.; Tan, J.; et al. Metasurface Holographic Image Projection Based on Mathematical Properties of Fourier Transform. *PhotoniX* **2020**, *1*, 16. [CrossRef]
21. Shang, G.; Wang, Z.; Li, H.; Zhang, K.; Wu, Q.; Burokur, S.N.; Ding, X. Metasurface Holography in the Microwave Regime. *Photonics* **2021**, *8*, 135. [CrossRef]
22. Ghasemi, A.; Burokur, S.N.; Dhouibi, A.; de Lustrac, A. High Beam Steering in Fabry–Pérot Leaky-Wave Antennas. *IEEE Antennas Wirel. Propag. Lett.* **2013**, *12*, 261–264. [CrossRef]
23. Ratni, B.; Merzouk, W.A.; de Lustrac, A.; Villers, S.; Piau, G.-P.; Burokur, S.N. Design of Phase-Modulated Metasurfaces for Beam Steering in Fabry–Perot Cavity Antennas. *IEEE Antennas Wirel. Propag. Lett.* **2017**, *16*, 1401–1404. [CrossRef]
24. Li, H.; Wang, G.; Xu, H.-X.; Cai, T.; Liang, J. X-Band Phase-Gradient Metasurface for High-Gain Lens Antenna Application. *IEEE Trans. Antennas Propag.* **2015**, *63*, 5144–5149. [CrossRef]
25. Dhouibi, A.; Burokur, S.N.; de Lustrac, A.; Priou, A. Low-Profile Substrate-Integrated Lens Antenna Using Metamaterials. *IEEE Antennas Wirel. Propag. Lett.* **2013**, *12*, 43–46. [CrossRef]
26. Minatti, G.; Caminita, F.; Martini, E.; Sabbadini, M.; Maci, S. Synthesis of Modulated-Metasurface Antennas with Amplitude, Phase, and Polarization Control. *IEEE Trans. Antennas Propag.* **2016**, *64*, 3907–3919. [CrossRef]
27. Ratni, B.; de Lustrac, A.; Piau, G.-P.; Burokur, S.N. Electronic Control of Linear-to-Circular Polarization Conversion Using a Reconfigurable Metasurface. *Appl. Phys. Lett.* **2017**, *111*, 214101. [CrossRef]
28. Wu, Z.; Ra'di, Y.; Grbic, A. Tunable Metasurfaces: A Polarization Rotator Design. *Phys. Rev. X* **2019**, *9*, 011036. [CrossRef]
29. He, Q.; Sun, S.; Zhou, L. Tunable/Reconfigurable Metasurfaces: Physics and Applications. *Research* **2019**, *2019*, 1849272. [CrossRef]
30. Lininger, A.; Zhu, A.Y.; Park, J.-S.; Palermo, G.; Chatterjee, S.; Boyd, J.; Capasso, F.; Strangi, G. Optical Properties of Metasurfaces Infiltrated with Liquid Crystals. *Proc. Natl. Acad. Sci. USA* **2020**, *117*, 20390–20396. [CrossRef]
31. Wang, Q.; Zhang, X.G.; Tian, H.W.; Jiang, W.X.; Bao, D.; Jiang, H.L.; Luo, Z.J.; Wu, L.T.; Cui, T.J. Millimeter-Wave Digital Coding Metasurfaces Based on Nematic Liquid Crystals. *Adv. Theory Simul.* **2019**, *2*, 1900141. [CrossRef]
32. Cui, T.J.; Qi, M.Q.; Wan, X.; Zhao, J.; Cheng, Q. Coding Metamaterials, Digital Metamaterials and Programming Metamaterials. *Light Sci. Appl.* **2014**, *3*, e218. [CrossRef]
33. de Lustrac, A.; Ratni, B.; Piau, G.-P.; Duval, Y.; Burokur, S.N. Tri-State Metasurface-Based Electromagnetic Screen with Switchable Reflection, Transmission, and Absorption Functionalities. *ACS Appl. Electron. Mater.* **2021**, *3*, 1184–1190. [CrossRef]

34. Popov, V.; Ratni, B.; Burokur, S.N.; Boust, F. Non-local Reconfigurable Sparse Metasurface: Efficient Near-field and Far-field Wavefront Manipulations. *Adv. Opt. Mater.* **2021**, *9*, 2170014. [CrossRef]
35. Dirdal, C.A.; Thrane, P.C.V.; Dullo, F.T.; Gjessing, J.; Summanwar, A.; Tschudi, J. MEMS-Tunable Dielectric Metasurface Lens Using Thin-Film PZT for Large Displacements at Low Voltages. *Opt. Lett.* **2021**, *47*, 1049–1052. [CrossRef] [PubMed]
36. Cheng, J.; Fan, F.; Chang, S. Recent Progress on Graphene-Functionalized Metasurfaces for Tunable Phase and Polarization Control. *Nanomaterials* **2019**, *9*, 398. [CrossRef]
37. Rotshild, D.; Abramovich, A. Ultra-wideband reconfigurable X-band and Ku-band metasurface beam-steerable reflector for satellite communications. *Electronics* **2021**, *10*, 2165. [CrossRef]
38. Tian, H.W.; Zhang, X.G.; Jiang, W.X.; Li, X.; Liu, Y.K.; Qiu, C.-W.; Cui, T.J. Programmable Controlling of Multiple Spatial Harmonics via a Nonlinearly-Phased Grating Metasurface. *Adv. Funct. Mater.* **2022**, 2203120. [CrossRef]
39. Boyarsky, M.; Sleasman, T.; Imani, M.F.; Gollub, J.N.; Smith, D.R. Electronically Steered Metasurface Antenna. *Sci. Rep.* **2021**, *11*, 4693. [CrossRef]
40. Yao, W.; Yang, H.; Huang, X.; Tian, Y.; Guo, L. An X-Band Parabolic Antenna Based on Gradient Metasurface. *AIP Adv.* **2016**, *6*, 075013. [CrossRef]
41. MACOM MAVR-000120-1411 Varactor Diode. Available online: https://www.macom.com/products/product-detail/MAVR-000120-14110P (accessed on 29 June 2022).
42. Ansys HFSS (High Frequency Structure Simulator). Available online: http://www.ansys.com/products/electronics/ansys-hfss (accessed on 29 June 2022).
43. FLANN DP240 Horn Antenna. Available online: https://flann.com/products/antennas/dual-polarised-horn-series-dp240/ (accessed on 29 June 2022).
44. Balanis, C.A. *Antenna Theory: Analysis and Design*, 3rd ed.; John Wiley & Sons: Hoboken, NJ, USA, 2005.
45. Schantz, H.G. *The Art and Science of Ultra-Wideband Antennas*, 1st ed.; Artech House: Norwood, MA, USA, 2005.
46. Huang, J.; Encinar, J.A. *Reflectarray Antennas*; John Wiley & Sons: Hoboken, NJ, USA, 2007.

sensors

MDPI

Article

Geodesic Path Model for Indoor Propagation Loss Prediction of Narrowband Channels

Abdil Kaya [1,*], Brecht De Beelde [2], Wout Joseph [2], Maarten Weyn [1] and Rafael Berkvens [1]

[1] IDLab-IMEC, Faculty of Applied Engineering, University of Antwerp, Sint-Pietersvliet 7, 2000 Antwerp, Belgium; maarten.weyn@uantwerpen.be (M.W.); rafael.berkvens@uantwerpen.be (R.B.)
[2] WAVES-IMEC, Department of Information Technology, Ghent University, Technologiepark-Zwijnaarde 126, 9052 Ghent, Belgium; brecht.debeelde@ugent.be (B.D.B.); wout.joseph@ugent.be (W.J.)
* Correspondence: abdil.kaya@uantwerpen.be

Abstract: Indoor path loss models characterize the attenuation of signals between a transmitting and receiving antenna for a certain frequency and type of environment. Their use ranges from network coverage planning to joint communication and sensing applications such as localization and crowd counting. The need for this proposed geodesic path model comes forth from attempts at path loss-based localization on ships, for which the traditional models do not yield satisfactory path loss predictions. In this work, we present a novel pathfinding-based path loss model, requiring only a simple binary floor map and transmitter locations as input. The approximated propagation path is determined using geodesics, which are constrained shortest distances within path-connected spaces. However, finding geodesic paths from one distinct path-connected space to another is done through a systematic process of choosing space connector points and concatenating parts of the geodesic path. We developed an accompanying tool and present its algorithm which automatically extracts model parameters such as the number of wall crossings on the direct path as well as on the geodesic path, path distance, and direction changes on the corners along the propagation path. Moreover, we validate our model against path loss measurements conducted in two distinct indoor environments using DASH-7 sensor networks operating at 868 MHz. The results are then compared to traditional floor-map-based models. Mean absolute errors as low as 4.79 dB and a standard deviation of the model error of 3.63 dB is achieved in a ship environment, almost half the values of the next best traditional model. Improvements in an office environment are more modest with a mean absolute error of 6.16 dB and a standard deviation of 4.55 dB.

Keywords: radio channel; path loss; signal strength; receivers; transmitters; wireless communication; computational modeling; path planning; electromagnetic propagation; loss measurement; propagation loss

Citation: Kaya, A.; De Beelde, B.; Joseph, W.; Weyn, M.; Berkvens, R. Geodesic Path Model for Indoor Propagation Loss Prediction of Narrowband Channels. *Sensors* **2022**, 22, 4903. https://doi.org/10.3390/s22134903

Academic Editor: Peter Han Joo Chong

Received: 27 May 2022
Accepted: 27 June 2022
Published: 29 June 2022

1. Introduction

Radio channel models characterize radio propagation for a certain frequency and type of environment and are valuable for the design of wireless communication systems. Path loss (PL) models characterize signal attenuation between a transmitting (TX) and receiving (RX) antenna and allow signal strength prediction and coverage calculations during the network planning phase [1], but can also be used for sensing [2] and localization [3] applications.

1.1. Related Work

In the past decade, a lot of research has focused on creating path loss models for indoor environments.

A recent and comprehensive survey on specifically indoor propagation models channel models is provided in [4]. The distance losses of the empirical or site-specific models discussed in this survey paper either use the direct path [5,6]. Geometric information about the environment is used in [6,7], but not without the use of explicit knowledge on the type of walls.

In [8], propagation measurements are presented in a corridor and office environment for frequencies ranging from 8 GHz to 11 GHz. In [9], models are created based on measurements up to 22 GHz in a corridor environment, for different angles of arrival and different antenna heights. Parameters are added to the close-in free space reference distance model and the floating-intercept model to better serve the ability to tune parameters. At such high frequencies, wall penetrations are not often taken into consideration. In [10], the linearity of attenuation due to obstructions or lossy media is discussed, based on measurements at 858 MHz and 1.935 GHz. The authors implicitly account for wall losses, under the assumption that the average wall loss in a particular environment is known, but instead of explicitly incorporating the number of walls crossing between transmitting (TX) and receiving (RX) antennas, an additional parameter is used, defined based on general information about the environment. An indoor path loss model for wireless local area networks accounting for wall attenuation is presented in [11]. The authors derive their model from the Average Wall Model. Instead of using the average wall attenuation within the environment, however, their model considers the different types of walls on the direct path between TX and RX and then averages the attenuation based on the wall crossed. This approach requires the wall types on a floor map and does not estimate non-direct propagation paths.

In [12], indoor path loss measurements in a residential living room at 60 GHz are presented. Radio propagation in office environments is well studied [8,13–20]. On the other hand, only limited literature is available for path loss models on board of naval vessels [21–23]. Previously, geodesic paths were used for ear-to-ear propagation in the 2.4 GHz frequency band by the authors of [24]. The distances considered in this work are at least two orders of magnitude smaller and are intended to consider geodesics without interruptions in path-connected space.

Aforementioned models and methods make no distinction in path-connectedness of the TX and RX antennas. Models and algorithms that are based on path finding are presented in [25–27]. The methodologies herein are based on finding the dominant path between a transmitter and a receiver. This dominant path is defined as the least attenuation accumulating path from transmitter to receiver. In order to do this, every possible path must be considered. The paths are considered consecutive connections between concave corners and center points of walls on a floor map. The connector points are used in a tree to search for a single path.

In this research, we present a novel path loss model for predicting signal strength based on a 2-dimensional binary floor map (representing walls and empty space, but not their types). The PL modeling approach is validated using experimental PL measurements in the cabin environment of a freight ship and in an office environment.

1.2. Background

In this section, we provide an overview of the different PL modeling approaches that are typically used and in terms of the required input, i.e., a floor map, best compared to our proposed model. We provide an implementation of the floor-map-based models under the constraint that floor maps do not make a distinction between wall types. It is often the case in real environments that a binary floor map is readily available, but wall types are not.

1.2.1. One-Slope PL Model

The one-slope floating-intercept (FI) model from (1) describes a linear relation between the logarithmic distance and PL, with PL_0 the PL in dB at reference distance d_0 = 1 m, n the PL exponent and d_d the Euclidean distance in meter between the TX and RX antennas.

$$\mathrm{PL}_{\mathrm{FI}}(d_{\mathrm{d}}) = \mathrm{PL}_0 + \mathrm{n} \cdot 10\log_{10}\left(\frac{d_d}{\mathrm{d}_0}\right) \quad (1)$$

The model parameters PL_0 and n depend on the frequency and environment, and are fitted based on measurement data.

1.2.2. Floor-map-based Path Loss Prediction

Several approaches exist to predict PL based on the floor map of the environment, taking into account attenuation due to objects obstructing the Line-of-Sight path, as well as floor and wall penetration.

The average wall model (AWM) from (2), also known as the Keenan-Motley model is a simplified model that only adds floor and wall attenuation to the free space PL model without distinguishing between different wall types, and without accounting for non-linearity of floor losses. In (2), $\overline{\mathbf{L}_\mathrm{w}}$ represents the average value for attenuation due to a wall crossing, and k_w is the number of walls crossed on the direct line between TX and RX.

$$\mathrm{PL_{AWM}} = \mathrm{PL_0} + \mathrm{n}\;10\log_{10}\frac{d_d}{\mathrm{d_0}} + \overline{\mathbf{L}_\mathrm{w}}\;k_w \tag{2}$$

The COST 231 multi-wall model from (3) uses a free space PL model to which losses due to wall and floor intersections are added [28]. In this equation, d_d is the Euclidean distance between the antennas, f is the frequency in Hz, c = 3×10^8 m/s is the speed of light, $\mathbf{L}_\mathrm{c}$ is a regression parameter representing a constant loss, typically close to zero, k_{wi} is the number of walls of type i that are penetrated, $\mathbf{L}_\mathrm{wi}$ is the loss coefficient for wall type i, and $\mathbf{I}$ is the number of wall types, k_f is the number of penetrated floors with floor attenuation $\mathbf{L}_\mathrm{f}$, and b is an empirical parameter to adapt to the non-linearity of floor losses.

$$\mathrm{PL_{COST}} = 20\log_{10}\left(4\pi d_\mathrm{d}\frac{\mathrm{f}}{\mathrm{c}}\right) + \mathbf{L}_\mathrm{c} + \sum_{i=1}^{\mathbf{I}} k_\mathrm{wi}\mathbf{L}_\mathrm{wi} + k_f^{\left[\frac{k_\mathrm{f}+2}{k_f+1}-\mathrm{b}\right]}\mathbf{L}_\mathrm{f} \tag{3}$$

The recommendation P.1238-11 from the International Telecommunication Union (ITU) presents both a site-general and ray-tracing-based site-specific model [29]. We will consider the site-general model from (4), because it best compares to the model we propose in this paper in terms of the input requirements. While this site-general model does not consider wall crossings, it implies the wall losses by making a distinction between coefficients for Line-of-Sight (LOS) and non-Line-of-Sight (NLOS) scenarios. This means that the model also requires a floor map. In this equation, d_d is the Euclidean distance between the antennas, f is the operating frequency in GHz, α is the path loss exponent, β is an offset value, similar to PL_0 in the one-slope PL model, γ is a coefficient related to the increasing transmission loss with frequency f.

$$\mathrm{PL_{ITU}} = 10\;\alpha\log_{10}(d_\mathrm{d}) + \beta + 10\;\gamma\log_{10}(\mathrm{f}) \tag{4}$$

1.2.3. Ray Tracing Based Models

In ray tracing algorithms, rays are launched for different azimuth and elevation angles, and interactions with the environment are determined. The drawback is the high computational complexity, as well as the requirement of having an accurate description of the environment [30,31].

1.2.4. Pathfinding Based Models

The indoor dominant path prediction (IDP) model provides ray tracing accuracy at a significantly lower complexity. IDP searches for only a single path, i.e., the dominant path, which contributes to the most received power at the receiver antenna RX. The IDP model justifies the use of a single path between TX and RX because more than 95% of the contributed power is contained in 2–3 rays [25,26]. To find the dominant path, without the computational intensity of ray tracing, IDP initially limits the number of 'passages' that a radio wave can supposedly go through when TX and RX are not in LOS. It does so by relying on a graph that connects distinct rooms by connector points in the center of shared walls. We refer to rooms as path-connected spaces in the remainder of this paper. Non-convex path-connected spaces are connected similarly by placing connector points in the concave corners. In doing so, the model considers an exhaustive list of all possible paths from TX to RX. A pre-defined heuristic then decides on the least loss-inducing path. Parameters along this path are then used to determine wall losses, diffraction losses, reflection loss, waveguiding, and distance losses. While the original IDP model used a neural network to find the path losses, the authors of [27] provide a thoroughly documented version of IDP with an intuitive model equation instead of a neural network. Even though IDP is conceptually similar to our proposed model, it requires detailed information such as the type of walls on the floor map to populate crucial reflection, diffraction, and waveguiding parameters. For this reason, we will not include IDP in the comparison and limit the comparison to floor-map-based models which can be used without wall type information.

1.3. Contributions

In this paper, we propose a path loss model and parameter estimation algorithm for path loss predictions. The proposed model only requires the input of a binary floor map, without the need for detailed wall type information. As opposed to the models in the previous sections, the proposed model is based on the combination of the direct path and the shortest geodesic path [32] between a transmitter and receiver. Instead of the euclidean distance, the distance along the points of the geodesic path is used. While the Euclidean distance of the direct link is not used as the distance metric, the number of wall crossings on the direct link is still used as a parameter. We find that in the proposed model, the impact of walls crossed on the direct link has a logarithmic relationship with the path loss, while the walls crossed on the geodesic paths (thus between distinct path-connected spaces) show a linear relationship with path loss. Using only the geodesic path, without performing ray-tracing, interaction losses are represented by the direction changes along the path. The model is implemented in a tool that automatically extracts the parameters of the model to estimate path loss based on a 2-dimensional floor map and predefined transmitter locations. The implementation is validated against path loss measurements using DASH-7 transceivers operating at 868 MHz in two different environments, i.e., the metallic environment of a ship and an office environment. This validation is then compared to three conventional floor-map-based path loss models. We evaluate the most commonly used floor-map-based models by tuning the coefficients to our measurement environments in our PL model tool. Given that the models compared range from site-general to site-specific and from zero to three tunable coefficients, the goal of the evaluation comparison is not to gauge a head-to-head performance difference, but rather to provide a set of broad, yet commonly used model performances in conjunction with our PL model tool. The compared results are however intended to contribute to an informed trade-off decision when selecting a PL model.

2. Methods

2.1. Path Loss Model

The proposed model in this paper is based on the approximated propagation paths of the radio signal. Many floor-map-based path loss prediction models either consider the direct path (AWM, COST 231, ITU-R P.1238) or choose a single most likely propagation path of the signal, discarding the direct path altogether (IDP). In the latter case, all the attenuation is implied to be a result of the approximated propagation path and the losses incurred along it. From our measurements, we find that even when considering an approximated propagation path, it is still useful to consider wall losses on the direct path with regard to the prediction of path losses. In our model, we make a distinction between walls crossed on the direct path and walls crossed on the propagation path. The former has a logarithmic relation to the path loss and the latter a linear one. We find and report on the importance of making a distinction between propagation paths crossing walls that divide path-disconnected spaces and those that cross walls that merely reside in the same path-connected space. We propose the geodesic path loss model (GPM) from (5). In this model, we use geodesics or shortest distances constrained by walls to determine the shortest paths within path-connected spaces. In the remainder of the paper, we specify the approximated propagation path as the geodesic paths. Finding geodesic paths can only be done in path-connected spaces, as such, we apply a systematic methodology to cross distinct path-connected spaces and concatenate multiple geodesic paths into one resulting propagation path. The extraction of GPM parameters is detailed in Algorithm 1. The coefficients and parameters of the model are as follows. PL_0 is the reference path loss in dB at distance d_0, n the path loss exponent, $d_{\mathcal{P}}$ the path distance in meter of the (concatenated) geodesic path. $\mathbf{L}_{\mathrm{wd}}$ and $\mathbf{L}_{\mathrm{wp}}$ are the coefficients for the average wall losses on the direct path and geodesic path respectively. k_{wd} and k_{wp} are the number of walls crossed on the direct path and geodesic path respectively. $\mathbf{L}_{\alpha}$ is the coefficient for the interaction loss expressed as the sum of direction changes α_i along the path.

$$\mathrm{PL}_{\mathrm{GPM}} = \underbrace{\mathrm{PL}_0}_{\text{Loss at 1 m}} + \begin{cases} 10\,\mathrm{n}\log_{10}\left(\frac{d_{\mathcal{P}}}{\mathrm{d}_0}\right) & \text{if LOS,} \\ \underbrace{10\,\mathrm{n}\log_{10}\left(\frac{d_{\mathcal{P}}}{\mathrm{d}_0}\right)}_{\text{Distance loss}} \underbrace{+\mathbf{L}_{\mathrm{wd}}10\log_{10}\left(k_{\mathrm{wd}}-k_{\mathrm{wp}}\right)+\mathbf{L}_{\mathrm{wp}}k_{\mathrm{wp}}}_{\text{Wall losses}} \underbrace{+\mathbf{L}_{\alpha}\sum_i sin^2\left(\frac{\alpha_i}{2}\right)}_{\text{Interaction losses}} & \text{otherwise.} \end{cases} \tag{5}$$

The subtraction $k_{\mathrm{wd}} - k_{\mathrm{wp}}$ in the logarithmic term serves to avoid the double counting of path-disconnected wall crossings k_{wp}. If $k_{\mathrm{wd}} = k_{\mathrm{wp}}$, then the logarithmic term is removed.

A visual representation of the parameters described is shown in Figure 1.

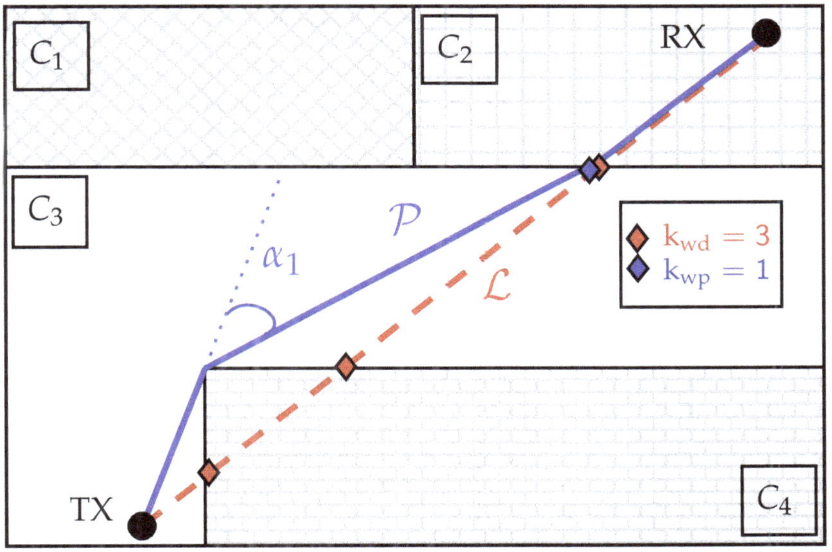

Figure 1. The blue line is the concatenated geodesic path $\mathcal{P}$ from TX to RX. Along it, the blue and red diamonds represent the points at which walls are crossed along the geodesic path and direct path $\mathcal{L}$ respectively. The number of walls crossings are enumerated by k_{wd} and k_{wp}. The direction changes along the path $\mathcal{P}$ are indicated by $\alpha_i, i \in \{0, \ldots,$ corner count$\}$. The distinct path-connected spaces are indicated by $C_j, j \in \{1, \ldots, J\}$.

Algorithm 1: Tool to determine geodesic paths and other PL model parameters

Data: Floor map $F \subset \mathbb{Z}^2$, as a set of couples (2-tuple coordinates) $p_i \leftarrow (x_i, y_i)$. f is a binary mapping of F into $\{0,1\}$ such that $f(p)$ is either 0 (open space) or 1 (walls). The set of wall coordinates $W \leftarrow \{p \in F | f(p) = 1\}$, antenna locations p_{TX} and $p_{\mathrm{RX}} \in (F - W)$.

Result: Shortest (concatenated) geodesic path $\mathcal{P}$, path distance $d_{\mathcal{P}}$, direct distance d_{d}, the number of walls on the direct link k_{wd}, the number of walls on the geodesic path k_{wp}, an ordered list of direction changes across the geodesic path $\forall_i \alpha_i$ (rad).

Functions:

$[\mathcal{D}_{p,C}] \longleftarrow GeoTransform(p, C)$: Returns the geodesic transform $\mathcal{D}_{p,C}$ with seed point p in connected space (mask) C, using a quasi-euclidean 8-connected kernel, for which each element of $\mathcal{D}_{p,C}$ represents a coordinate in C and its distance to p.

$[\mathcal{P}_{p,q,C}, d_{\mathcal{P}}] \longleftarrow GeoPath(p, q, C)$: Returns the geodesic shortest path from p to q in C by considering the thinned, minimal distance coordinates between p and q, resulting from the sum of the two geodesic transforms. $\mathcal{P}_{p,q,C} \longleftarrow \mathcal{D}\{\mathcal{D} = \min(\mathcal{D}_{p,C} + \mathcal{D}_{q,C})\}$. The shortest path distance is then equal to the value of any coordinate element in $\mathcal{P}$. $d_{\mathcal{P}} \longleftarrow \mathcal{P}\{1\}$

$\mathcal{P} \longleftarrow DouglasPeucker(\mathcal{P}, \varepsilon)$: Returns the coordinate couples sequence of path $\mathcal{P}$, by recursively decimating points which do not deviate more than tolerance ε from the current line segment under evaluation. This tolerance value is set to the max radius of the first Fresnel zone.

begin
- Let $\forall_j C_j \subseteq (F - W)$ be the ordered set of unique subsets, each containing the coordinates of path-connected spaces in F, labeled as $j \in \{1, 2, 3, ..., J\}$. For any given point $p \in (F - W)$, its corresponding connected space label is denoted as j_p and its connected space subset as C_{j_p}.
- Let $L_{\mathrm{TR}} \subseteq F$ denote the set points on the direct link line $\mathcal{L} \longleftarrow |p_{\mathrm{TX}}, p_{\mathrm{RX}}|$.
- $d_{\mathrm{d}} \longleftarrow ||p_{\mathrm{TX}}, p_{\mathrm{RX}}||$.
- **if** $L_{TR} \cap W = \emptyset$ **then**
  - $|p_{\mathrm{TX}}, p_{\mathrm{RX}}|$ in LOS.
  - $d_{\mathcal{P}} = d_{\mathrm{d}}$
- **else**
  - $k_{wd} \longleftarrow |label(L_{\mathrm{TR}} \cap W)|$. Assigning set cardinality *after* labeling connected components avoids overestimating the number of walls crossed due to stretches of connected wall components.
  - **if** $j_{p_{TX}} = j_{p_{RX}}$ **then**
    - $\mathcal{L}$ is not a LOS link, but both transceivers are in the same connected space, thus the link is considered as a path-connected NLOS link (NLOS_{PC}).
    - p_{TX} and p_{RX} are not in LOS, but they are objects in the same path-connected space C_j such that $C_{p_{\mathrm{TX}}} = C_{p_{\mathrm{RX}}}$.
    - $[\mathcal{P}, d_{\mathcal{P}}] \longleftarrow GeoPath(p_{\mathrm{TX}}, p_{\mathrm{RX}}, C_j)$.
  - **else**
    - $\mathcal{L}$ is not a LOS link and both transceivers are in a different distinct connected space, thus the link is considered as a path-disconnected NLOS link (NLOS_{PD}).
    - The shortest geodesic paths through the least number of neighboring spaces $N \subseteq J$ are searched. In case of multiple possible sets of neighboring spaces, the subset of C_N is chosen according to $min(|centroid(C_N), \mathcal{L}|)$. Within these C_N, the connector points $p_c \in N$ are then used to find intermediate geodesic paths $\mathcal{P}_N$, which are then concatenated to result in the final $\mathcal{P}$ and $d_{\mathcal{P}}$.
    - $\mathcal{P} \longleftarrow concat(\mathcal{P}_N)$
    - $d_{\mathcal{P}} \longleftarrow \sum_{n=1}^{|N|} d_n$
    - $k_w p \longleftarrow |\mathcal{P}_N| - 1$
  - $\mathcal{P} \longleftarrow DouglasPeucker(\mathcal{P}, \varepsilon)$
  - $\alpha_i \longleftarrow diff(atan2(\nabla\mathcal{P}_{i,y}, \nabla\mathcal{P}_{i,x})), \forall i \in \{1, ..., |\nabla\mathcal{P}|\}$

2.2. Model Implementation

Algorithm 1 presents the algorithm for our tool, used to determine PL parameters.

2.3. Experiment Sites and Setup

The proposed GPM is a site-specific model, just like the AWM and the COST 231 model. For these three models, the coefficients and terms need to be tuned or chosen on a per-environment basis. The ITU-R model is the only site-general model in the comparison and requires no tuning of coefficients. The coefficients are chosen based on the type of environment the path loss prediction is performed for.

The path loss measurements from two measurement campaigns are independently used to validate the different PL models. The first measurement environment, shown in Figure 2a, is the superstructure main cabin floor of a freight ship. The walls of this environment are all made of either steel for the superstructure's load-bearing walls or aluminum for doors and thin compartmentalization walls. A second measurement environment, shown in Figure 2b, is a regular office floor in a 10-story building. The building materials of this environment range vastly from windowed walls to reinforced concrete. Typical materials such as timber and plasterboard for compartment walls, different types of metal alloys for the elevator shafts (near transceiver 17), and glass can be found in the environment. However, we explicitly do not take this information into account for any of the implemented models, as the assumption is that wall-type information is not available. Open doors are indicated on these two-floor maps by an actual opening, while closed doors are indicated as wall lines. Furthermore, we have not taken the thickness of the walls on the floor map into account. Any wall crossing is counted as such, a single wall crossing, with no distinction in wall types, even if the floor maps suggest differences in wall thickness.

Figure 2. Floor map of the measurement environments and wireless sensor network deployment. The ship environment is show in (**a**) and the office environment in (**b**).

To acquire the training and validation data from these environments, we held multiple measurement campaigns. The transceiver locations are shown in Figure 2. The specific link distinction between training and validation links are shown in Figure A2 of Appendix A and the number of links per environment are provided in Table 1. The deployed wireless sensor network (WSN) shown in Figure 2 is a highly connected DASH-7 WSN, operating at 868 MHz. The communication model, hardware, and network setup from the communication perspective are detailed in a previous work in [33]. The deployed transceivers on tripods, as well as the environment, are shown in Figure 3.

2.4. Model Parameter Algorithm Output

A small and random set of the resulting paths and parameters are shown in Figure 4. This shows the geodesic paths through path-connected spaces ($NLOS_{PC}$ links), path-disconnected spaces ($NLOS_{PD}$ links) as well as the corners and direction changes expressed in degrees (for illustration). Corners are determined using the Douglas-Peucker algorithm [34].

This algorithm requires a tolerance value, which we set at the maximum radius of the first Fresnel zone. In Appendix A, Figure A1, we provide a floor map with all measurement links drawn and represented as direct links. The algorithmically found geodesic paths are overlaid on this map.

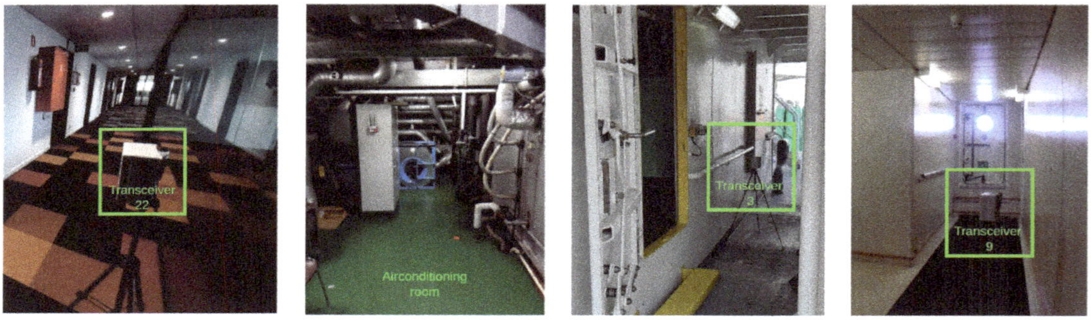

(**a**) (**b**) (**c**) (**d**)

Figure 3. This figure shows the measurement environments. The office environment is a typical looking office, with a mix of unknown wall types of all sorts in (**a**), of which the location corresponds to that of transceiver 22 in Figure 2b. The ship environment is highly metallic one. The air conditioning room in (**b**) corresponds to the location of the transceivers 6 and 12 in Figure 2a. The location of transceivers 3 and 9 in (**c**,**d**) can also be referred to in Figure 2a.

Table 1. Collected number of PL samples per link per measurements environment. A split is made between distinct links used for training and evaluation.

	Training Measurements		Validation Measurements	
	Link Count	**Samples per Link**	**Link Count**	**Samples per Link**
Ship	45	463	90	248
Office	63	121	210	89

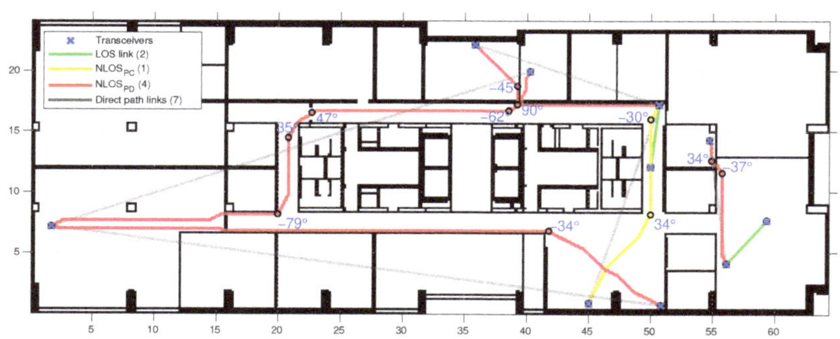

Figure 4. A visualization of the parameters produced by the algorithm, showing direction changes along the geodesic paths expressed in degrees, wall crossings by the direct paths and wall crossings by the geodesic paths. A randomly selected subset (approximately 3% of all links) is generated for visualization.

2.5. Model Coefficients and Terms

Usually the path loss exponent n is either fitted to LOS measurements at a range of distances or set to the free space path loss (FSPL) exponent of 2. We prefer the latter, but because certain environments have a lower PL exponent than 2, such as highly metallic industrial environments or in this case a ship, both our own measurements as well as findings in literature indicate a PL exponent lower than 2 [1,35–37]. As such, in our case, we will use a PL exponent of $n = 2$ for the office environment and $n = 1.15$ for the ship environment. The reference path loss PL_0 is set as 31.2 dB, which is the FSPL at an operating frequency of 868 MHz and 1 m distance from the TX antenna. The frequency dependent reference path loss term can be translated to other sub-6 GHz frequency bands. The frequency dependency thus stems from (6), but also from the corner selection process in which the maximum radius of the first Fresnel zone is used as the tolerance value in the iterative-end-point-fit algorithm.

$$PL_0 = 20\log_{10}(d_0) + 20\log_{10}(f) + 20\log_{10}\left(\frac{4\pi}{c}\right) \quad (6)$$

All other model-specific parameters, unless prescribed by the model itself, were attained using multivariate nonlinear least squares. The GPM uses three coefficients to be fitted and both the AWM and the COST 231 models only one. The ITU-R model is not a site-specific model, but instead a site-general model with predetermined coefficients and terms based on the type of environment, making it the only model that doesn't need any coefficient fitting.

The number of unique measurement links between TX and RX is 117 for the ship environment and 273 for the office environment, with a total number of respectively 39,012 and 17,046 path loss measurements across those links. A third is used for coefficient tuning, presented in Table 2, while the remainder is used for model validation and error analysis. Table 2 does not include the coefficients for ITU-R P.1238, because those are pre-determined based on the type of environment and whether or not the current link between TX and RX is in LOS [29].

Table 2. GPM, AWM and COST231 model coefficients for 868 MHz in a ship and office room environment, fitted using a multivariate non-linear least squares algorithm. The bold values indicate the coefficient estimates, accompanied by the lower limits (LL) and upper limits (UL) for the 95% confidence intervals (CI).

		GPM			AWM	COST231
		L_{wd}	L_{wp}	L_{α}	L_{wd}	L_{wd}
Ship	Estimate	0.5588	17.79	9.6895	4.8137	2.9484
	95% CI [LL]	0.5357	17.5993	9.5737	3.9177	2.071
	95% CI [UL]	0.5819	17.9806	9.8053	5.7096	3.8258
Office	Estimate	2.2929	3.6716	4.5151	3.09	3.09
	95% CI [LL]	2.2628	3.2456	4.3487	2.7307	2.7307
	95% CI [UL]	2.323	4.0977	4.6814	3.4472	3.4472

3. Results

After obtaining the various model coefficients, tuned against a training sample set of PL measurements, we evaluate the resulting predictions per model, per environment. First, we present the statistical error analysis. Paired with the PL prediction on every point of the maps, we can interpret and discuss the model behavior in conjunction with the error results with respect to the measured validation links.

3.1. Prediction Errors

Figure 5 shows different perspectives on how well the predictions from the different models fit. In Figure 5a,c, the PL is shown as a function of the direct distances.

For the model prediction comparisons for the ship environment in Figure 5a, we can see that the GPM predictions follow the measurements fairly well irrespective of the distance between TX and RX antenna, whereas the traditional floor-map-based models tend to have larger prediction errors for larger distances. The model prediction comparisons of the office environment in Figure 5c on the other hand show that predictions across all models tend to follow the measurements in a similar manner.

In Figure 5b,d, a scatter plot is shown. The x-axis corresponds to the measured PL on a communication link, with its respective PL prediction on the y-axis. This is done for two PL models, the GPM and AWM. The first bisector line represents a perfect prediction relation between the measured PL (dB) on the x-axis and the predicted PL (dB) on the y-axis.

In the office environment, AWM and COST231 are essentially the same model, given that, in this environment, the AWM uses FSPL PL_0 and PL exponent $n = 2$. The ITU-R P.1238 model (site-general version), which uses predetermined coefficients, makes predictions with very large errors of up to 41.72 dB as reported in Table 3.

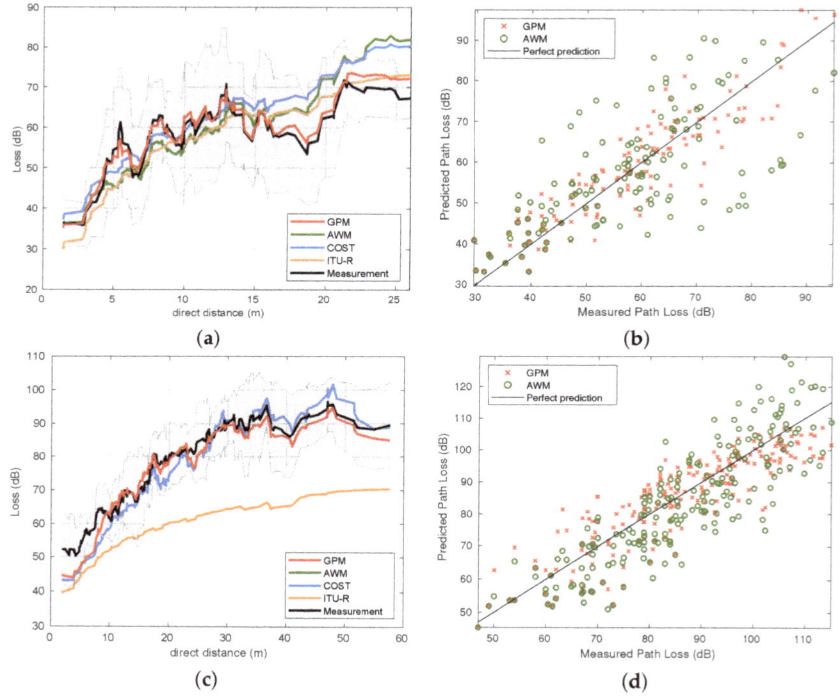

Figure 5. PL is outlined with respect to the direct path distance between TX and RX for all the considered models in (**a**) for the ship environment and in (**c**) for the office environment. A comparison between measured and predicted PL is made visible more explicitly in (**b**) for the ship environment and in (**d**) for the office environment.

Table 3. Statistical error analysis summary for the GPM, AWM, COST 231 and ITU-R P.1238.

	Model	MAE (dB)	$\sigma_{\lvert\epsilon\rvert}$ (dB)	$\lvert\epsilon_{max}\rvert$ (dB)	RMSE (dB)	R^2
Ship	GPM	4.79	3.63	16.38	6.0	0.83
	AWM	8.88	7.80	29.37	11.79	0.43
	COST 231	9.19	7.17	27.07	11.64	0.41
	ITU-R P.1238	8.79	7.83	33.65	11.75	0.38
Office	GPM	6.1637	4.55	20.783	7.66	0.77
	AWM	7.93	5.61	28.19	9.69	0.74
	COST 231	7.93	5.60	27.33	9.7	0.74
	ITU-R P.1238	18.73	9.92	41.72	21.19	0.67

Figure 6a,c show the empirical cumulative distribution functions (CDFs) for the absolute errors of the path loss predictions of all considered models. Similarly, Figure 6b,d show the CDFs, only now with the signed Errors, ($PL_{PREDICTION} - PL_{MEASUREMENT}$), such that the negative side of the x-axis represent the underestimations from the models and the positive side of the x-axis represents overestimations from the models.

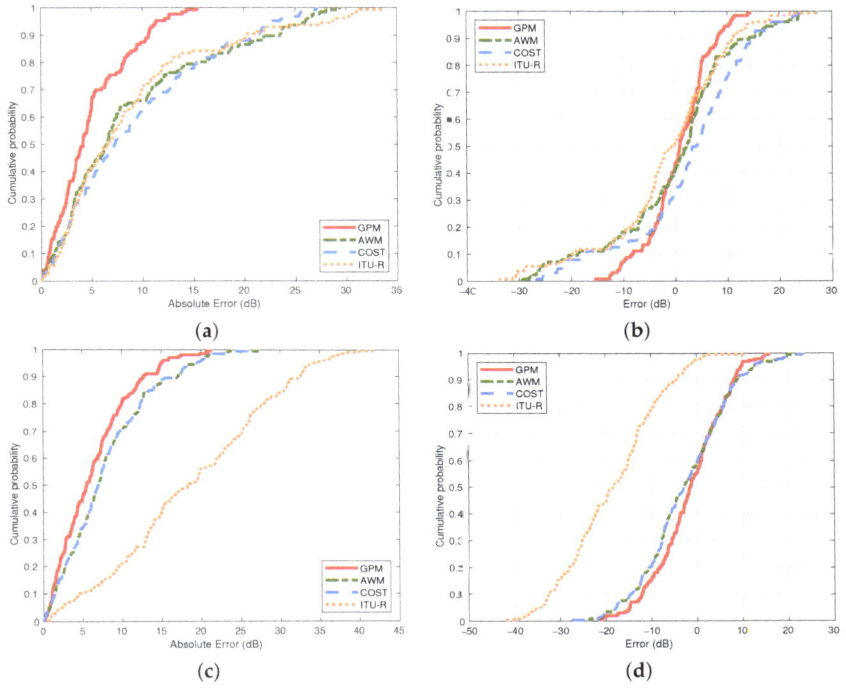

Figure 6. CDF plots of the PL estimation errors. The CDF of the absolute errors is shown in (**a**) for the ship environment and in (**c**) for the office environment.The CDF of the signed errors is shown in (**b**) for the ship environment and in (**d**) for the office environment.

Lastly, a summary of the statistical error analysis is provided in Table 3. A large disparity between GPM and other floor-map-based models is apparent in the results from the ship environment. The mean absolute error (MAE) is 4.79 dB for GPM, whereas that of the next best prediction model for the ship environment (ITU-R P.1238) is almost double at 8.79 dB. The standard deviation of the error is only 3.63 dB for the GPM. Standard deviations between 3 and 6 dB are excellent according to [38], and referred to in the works of [27]. Moreover, the coefficient of determination (R^2) of the GPM is much better than that of the others. While the GPM outperforms other models in the office environment, the observed results are more comparable to each other. One exception is the ITU-R model, which under-performs in comparison to the site-general models in this specific office environment. While it is interesting to assess the performance of the ITU-R model, it remains a site-general model in which the coefficients are not tuned to the specific environment, unlike the other models. The GPM, AWM, and COST 231 models are trained, and their respective coefficients are tuned from the same training set. Since these three are site-specific models, the tuning is done separately for the ship and office environment, but equally within each environment.

3.2. PL Prediction Results

From the environment parameters and model coefficients, we predict the PL at every location on the floor map from a given TX antenna for both the ship and office environment. The presented prediction is limited to the GPM and AWM. The apparent differences between these two models are instructive and relay the shortcomings of traditional floor-map-based PL models, which we discuss in Section 4. Figure 7 provides an overview of the simulations.

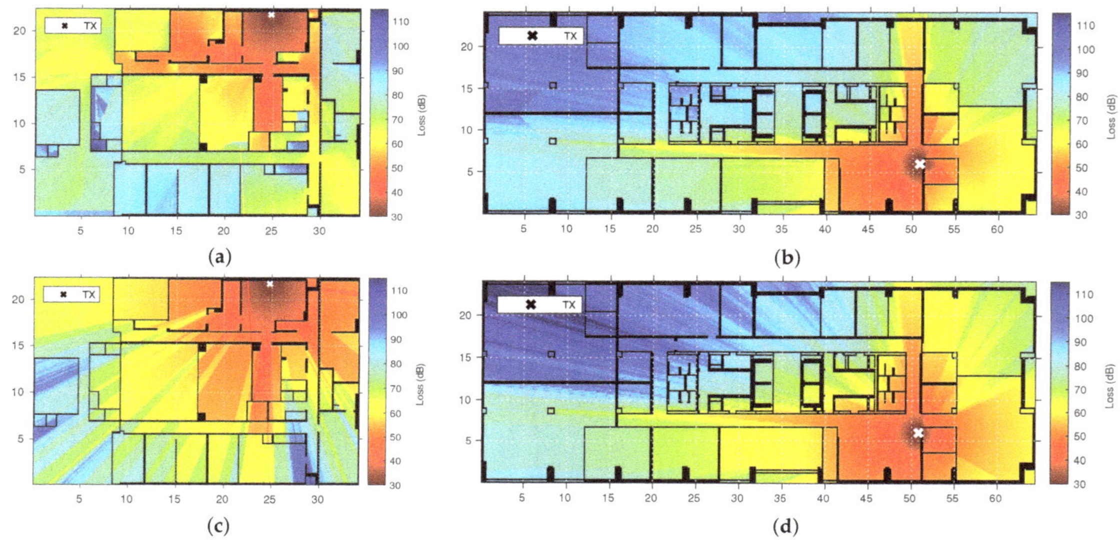

Figure 7. Predicted GPM PL is presented in (**a**) for the ship environment and in (**b**) for the office environment. For the purpose of comparison, the same is provided for the next best performing traditional model, AWM, in (**c**) for the ship environment and (**d**) for the office environment. We can see larger differences between GPM and AWM in the ship environment than in the office environment. The AWM considers walls and distance as the only path loss causes, while the GPM takes diffraction and differences in path-connectedness of wall crossings into account as well.

4. Discussion

Originating from the need for a more representative PL model for the localization of transceivers in an indoor ship environment, we set out to find a pathfinding-based PL model. The goal of this model is to alleviate the systematic errors induced by map-based models that ignore propagation paths beyond the direct path, without requiring additional information such as wall types or predetermined connector points between rooms and around concave corners. The only input required for the model is a simple binary floor map, without any additional details about wall types. Such detailed information is seldom readily available, even when computer-aided design files are present. The model algorithm to find the parameters is part of the model. The algorithm is based on finding shortest paths using geodesics with a quasi-Euclidean kernel, the crossing between path-disconnected spaces using heuristics, and lastly, the finding of angles on corners, which are first determined by path point reduction using the Douglas–Peucker algorithm.

4.1. Experimental Validation

In Table 2, the GPM coefficients are presented along with the 95% confidence interval bounds for both the ship and office environment. Unsurprisingly, we see that for the ship environment, the impact of wall crossings between path-disconnected spaces has a very large impact, as indicated by $\mathbf{L}_{wp|GPM}$. Conversely, the impact of wall crossings across the direct path has a significantly lower impact, as indicated by $\mathbf{L}_{wd}$. However, the accumulated impact thereof is still sizable, because the number of walls crossed on the direct path is approximately ten times the number of walls crossed on the geodesic path, both for the ship and office environments. From that aspect, we can see that for the average link with average parameter values, the impact of walls crossed on the direct path in the office environment have a significantly larger impact on PL than the walls crossed on the geodesic path.

4.2. Errors

The traditional floor-map-based models have difficulty predicting PL in the ship environment as the distance between TX and RX antennas grows larger. This is again due to improper valuation of the wall crossings. Surprising however is the comparable performance of the ITU-R P.1238 model to the other traditional multi-wall models in this environment. It does not consider walls, but rather only whether or not links are in LOS or NLOS and uses model coefficients prescribed by the model itself. For the ship environment, it is clear that a pathfinding model is a must. The GPM prediction errors show an MAE of 4.79 dB, with a high coefficient of determination (R^2) of 0.83, whereas the traditional floor-map-based models perform comparably among each other with an average MAE of 8.95 dB and an R^2 value of 0.41.

The large under- and over-estimations in the ship environment from models such as AWM and COST231 stem directly from not being able to find a path. Especially in an environment with such a low PL exponent and at the same time, metal walls that attenuate significantly if TX and RX are not in the same path-connected space and a wall that divides path-disconnected spaces must be crossed. The significance of path-connectedness becomes apparent in the differences in wall loss coefficients from the GPM. This is however not something the AWM, COST231, or ITU-R P.1238 (site-general version) models can account for. For these traditional floor-map-based models, it's near impossible to deal with the more prominent interaction losses as opposed to the very small distance losses. We would like to note that the availability of wall types would not solve this problem for the traditional floor-map-based models, because the wall types in the ship environment are all similar to each other. The important consideration of path-connectedness upon wall crossings is a necessary one, given an environment characterized by a low PL exponent.

However, in the office environment, we see that the average wall models perform significantly better than in the ship environment, although not better than the GPM. The GPM predictions show an MAE of 6.16 dB and an R^2 value of 0.77. AWM and COST231 predictions have an MAE of 7.93 dB and an R^2 value of 0.74. The ITU-R P.1238 (site-general version) model has the largest prediction errors and the model would need a PL prediction offset value far from zero. We would like to note however that the ITU-R P.1238 (site-general version) model does not consider walls, but rather applies different terms and coefficients based on environment type and whether or not the TX and RX antenna are in LOS.

5. Conclusions

In this research, we presented a novel path loss model and tool for predicting path loss. The model only requires the input of a simple binary 2-dimensional floor map, without the need for additional wall type information. The model is based on the use of shortest constrained distance, or geodesics, to find paths within path-connected spaces. Crossing walls to path-disconnected spaces is done systematically based on a heuristic. We presented an algorithm that automatically determines the parameters used in the model. Furthermore, the model was validated using path loss data from a considerably large measurement campaign in two vastly different environments, one being the main cabin floor in the superstructure of a ship and the other in an ordinary office environment. The path loss prediction results of the proposed Geodesic Path Model are mean absolute errors as low as 4.79 dB and a standard deviation of the error of 3.63 dB in the ship environment and a mean absolute error of 6.16 dB and standard deviation of 4.55 dB in the office environment. The most significant improvements are observed in the ship environment, an environment characterized by its low path loss exponent (1.15, as opposed to 2 for free space). We also find the importance of making a distinction between the crossing of walls that divide path-disconnected spaces as opposed to the crossing of walls that reside in the same path-connected space.

Future Work

Validating our Geodesic Path Model with in-field measurements, not only allows us to have a better input for our active device-based localization efforts, but it opens the gate to other sensing objectives such as our device-free crowd or people sensing efforts. If determined propagation paths are reliable to a certain degree, then time-variable PL changes along those paths can facilitate the detection of people counts and presence in complex environments, whereas radio-frequency based device-free people counting traditionally requires a mostly unobstructed line-of-sight between TX and RX antennas.

We are interested in the reliability of the model in higher frequencies such as the millimeter-wave bands. At higher frequencies, the reflected signals become more diffuse and diffraction losses increase as well. In order to compensate for the incurred losses, directional antennas are used. This specific working of antenna directionality will need to be implemented in the path loss parameter determination tool.

Author Contributions: Conceptualization, A.K.; methodology, A.K.; software, A.K.; validation, A.K.; investigation, A.K. and B.D.B.; resources, M.W., R.B., B.D.B. and A.K.; data curation, A.K.; writing—original draft preparation, A.K.; writing—review and editing, B.D.B., W.J., R.B. and A.K.; visualization, A.K.; supervision, R.B. and M.W.; project administration, R.B., W.J. and M.W.; funding acquisition, R.B., W.J., M.W. and A.K. All authors have read and agreed to the published version of the manuscript.

Funding: Abdil Kaya is funded by Research Foundation-Flanders (FWO) under grant number 1S99720N. Part of this research was funded by the ICON project Internet of Shipping (IoS). IoS is realized in collaboration with imec, with project support from VLAIO. Project partners are imec, Exmar, Ovinto, and Aloxy.

Institutional Review Board Statement: Not applicable.

Informed Consent Statement: Not applicable.

Data Availability Statement: Not applicable.

Acknowledgments: This work was executed within the imec ICON project Internet of Shipping (IoS). The authors would like to thank Ivan Renette from Exmar Shipmanagement for facilitating the measurement campaign of the ship and Marwan Yusuf from the research group imec-WAVES for their help during the measurement campaign on the ship.

Conflicts of Interest: The authors declare no conflict of interest.

Abbreviations

The following abbreviations are used in this manuscript:

AWM	Average Wall Model
CDF	Cumulative Distribution Function
CI	Confidence Interval
eCDF	Empirical Cumulative Distribution Function
FI	Floating-intercept
FSPL	Free-Space Path Loss
GPM	Geodesic Path Model
IDP	Indoor Dominant Path Model

LL	Lower Limit
LOS	Line-of-Sight
MAE	Mean Absolute Error
MDPI	Multidisciplinary Digital Publishing Institute
NLOS	Non-Line-of-Sight
PL	Path loss
R^2	Coefficient of Determination
RMSE	Root Mean Square Error
RX	Receiver antenna
TX	Transmitter antenna
UL	Upper Limit
WSN	Wireless Sensor Network

Appendix A

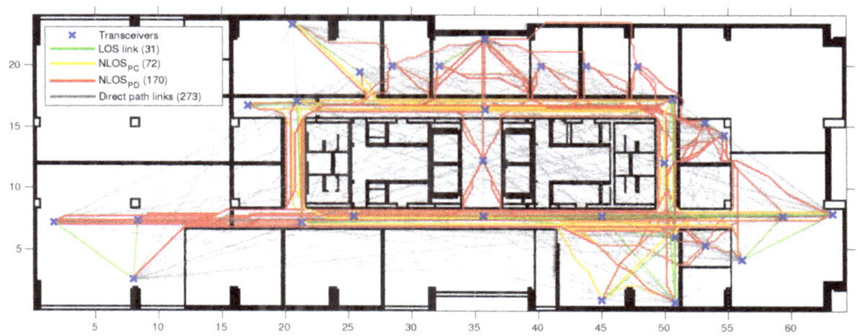

Figure A1. The resulting estimated propagation paths are presented in this figure. The resulting (concatenated) geodesics indicated as paths in green, yellow or red are a result of applying the algorithm in Algorithm 1 to the direct paths indicated as the gray lines.

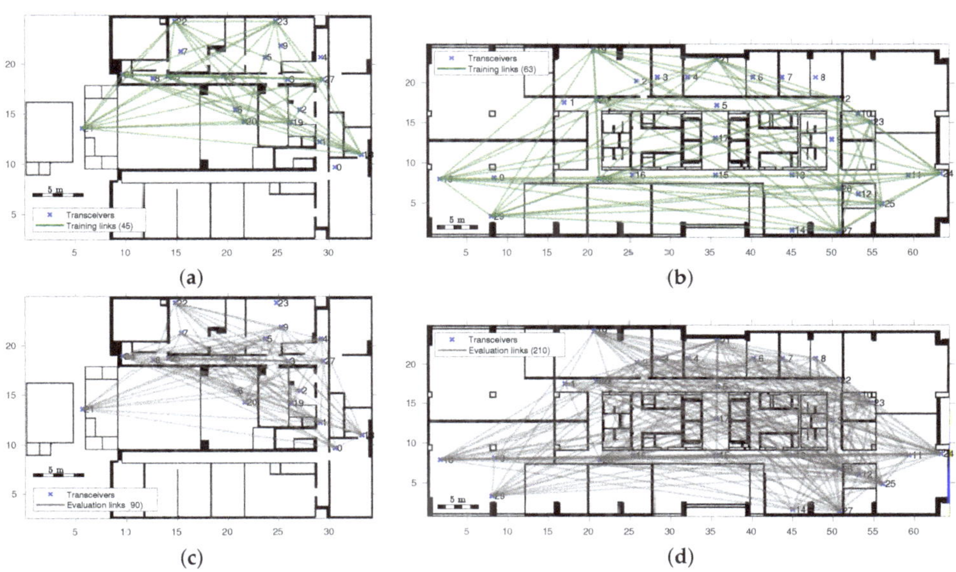

Figure A2. The training links for the ship environment and office environment is shown in (**a**,**b**) respectively. Floor map of the measurement environments and wireless sensor network deployment. The evaluation links for the ship environment and office environment is shown in (**c**,**d**) respectively.

References

1. De Beelde, B.; Plets, D.; Joseph, W. Wireless Sensor Networks for Enabling Smart Production Lines in Industry 4.0. *Appl. Sci.* **2021**, *11*, 11248. [CrossRef]
2. Yusuf, M.; Tanghe, E.; De Beelde, B.; Laly, P.; Ridolfi, M.; De Poorter, E.; Martens, L.; Gaillot, D.P.; Lienard, M.; Joseph, W. Human Sensing in Reverberant Environments: RF-Based Occupancy and Fall Detection in Ships. *IEEE Trans. Veh. Technol.* **2021**, *70*, 4512–4522. [CrossRef]
3. Zhang, J.; Han, G.; Sun, N.; Shu, L. Path-Loss-Based Fingerprint Localization Approach for Location-Based Services in Indoor Environments. *IEEE Access* **2017**, *5*, 13756–13769. [CrossRef]
4. Diago-Mosquera, M.E.; Aragón-Zavala, A.; Castañón, G. Bringing It Indoors: A Review of Narrowband Radio Propagation Modeling for Enclosed Spaces. *IEEE Access* **2020**, *8*, 103875–103899. [CrossRef]
5. Fraiha, S.G.C.; Rodrigues, J.C.; Barbosa, R.N.S.; Gomes, H.S.; Cavalcante, G.P.S. An empirical model for propagation-loss prediction in indoor mobile communications using the Padé approximant. *Microw. Opt. Technol. Lett.* **2006**, *48*, 255–261. [CrossRef]
6. Yu, Y.; Liu, Y.; Lu, W.J.; Zhu, H.B. Measurement and empirical modelling of root mean square delay spread in indoor femtocells scenarios. *IET Commun.* **2017**, *11*, 2125–2131. [CrossRef]
7. Austin, A.C.M.; Neve, M.J.; Rowe, G.B. Modeling Propagation in Multifloor Buildings Using the FDTD Method. *IEEE Trans. Antennas Propag.* **2011**, *59*, 4239–4246. [CrossRef]
8. Batalha, I.D.S.; Lopes, A.V.R.; Araújo, J.P.L.; Castro, B.L.S.; Barros, F.J.B.; Cavalcante, G.P.D.S.; Pelaes, E.G. Indoor Corridor and Office Propagation Measurements and Channel Models at 8, 9, 10 and 11 GHz. *IEEE Access* **2019**, *7*, 55005–55021. [CrossRef]
9. Elmezughi, M.K.; Afullo, T.J. An Efficient Approach of Improving Path Loss Models for Future Mobile Networks in Enclosed Indoor Environments. *IEEE Access* **2021**, *9*, 110332–110345. [CrossRef]
10. Degli-Esposti, V.; Falciasecca, G.; Fuschini, F.; Vitucci, E.M. A Meaningful Indoor Path-Loss Formula. *IEEE Antennas Wirel. Propag. Lett.* **2013**, *12*, 872–875. [CrossRef]
11. Obeidat, H.A.; Asif, R.; Ali, N.T.; Dama, Y.A.; Obeidat, O.A.; Jones, S.M.R.; Shuaieb, W.S.; Al-Sadoon, M.A.; Hameed, K.W.; Alabdullah, A.A.; et al. An indoor path loss prediction model using wall correction factors for wireless local area network and 5G indoor networks. *Radio Sci.* **2018**, *53*, 544–564. [CrossRef]
12. De Beelde, B.; Almarcha, A.; Plets, D.; Joseph, W. V-Band Channel Modeling, Throughput Measurements, and Coverage Prediction for Indoor Residential Environments. *Electronics* **2022**, *11*, 659. [CrossRef]
13. De Beelde, B.; Tanghe, E.; Desset, C.; Bourdoux, A.; Plets, D.; Joseph, W. Office Room Channel Modeling and Object Attenuation at Sub-THz Frequencies. *Electronics* **2021**, *10*, 1725. [CrossRef]
14. Ju, S.; Xing, Y.; Kanhere, O.; Rappaport, T.S. Millimeter Wave and Sub-Terahertz Spatial Statistical Channel Model for an Indoor Office Building. *IEEE J. Sel. Areas Commun.* **2021**, *39*, 1561–1575. [CrossRef]
15. Kim, M.D.; Liang, J.; Lee, J.; Park, J.; Park, B. Path loss measurements and modeling for indoor office scenario at 28 and 38 GHz. In Proceedings of the 2016 International Symposium on Antennas and Propagation (ISAP), Okinawa, Japan, 24–28 October 2016; pp. 64–65.
16. Wu, X.; Wang, C.X.; Sun, J.; Huang, J.; Feng, R.; Yang, Y.; Ge, X. 60-GHz Millimeter-Wave Channel Measurements and Modeling for Indoor Office Environments. *IEEE Trans. Antennas Propag.* **2017**, *65*, 1912–1924. [CrossRef]
17. Maccartney, G.R.; Rappaport, T.S.; Sun, S.; Deng, S. Indoor Office Wideband Millimeter-Wave Propagation Measurements and Channel Models at 28 and 73 GHz for Ultra-Dense 5G Wireless Networks. *IEEE Access* **2015**, *3*, 2388–2424. [CrossRef]
18. Choi, J.; Kang, N.G.; Sung, Y.S.; Kang, J.S.; Kim, S.C. Frequency-Dependent UWB Channel Characteristics in Office Environments. *IEEE Trans. Veh. Technol.* **2009**, *58*, 3102–3111. [CrossRef]
19. Tsukada, H.; Kumakura, K.; Tang, S.; Kim, M. Millimeter-Wave Channel Model Parameters for Various Office Environments. *IEEE Access* **2022**, *10*, 60387–60396. [CrossRef]
20. Wyne, S.; Singh, A.P.; Tufvesson, F.; Molisch, A.F. A statistical model for indoor office wireless sensor channels. *IEEE Trans. Wirel. Commun.* **2009**, *8*, 4154–4164. [CrossRef]
21. De Beelde, B.; Tanghe, E.; Yusuf, M.; Plets, D.; Joseph, W. Radio Channel Modeling in a Ship Hull: Path Loss at 868 MHz and 2.4, 5.25, and 60 GHz. *IEEE Antennas Wirel. Propag. Lett.* **2021**, *20*, 597–601. [CrossRef]
22. Kdouh, H.; Brousseau, C.; Zaharia, G.; Grunfelder, G.; Zein, G.E. Measurements and path loss models for shipboard environments at 2.4 GHz. In Proceedings of the 2011 41st European Microwave Conference, Manchester, UK, 10–13 October 2011; pp. 408–411. [CrossRef]
23. Mariscotti, A.; Sassi, M.; Qualizza, A.; Lenardon, M. On the propagation of wireless signals on board ships. In Proceedings of the 2010 IEEE Instrumentation & Measurement Technology Conference Proceedings, Austin, TX, USA, 3–6 May 2010; pp. 1418–1423. [CrossRef]
24. Kammersgaard, N.P.B.; Kvist, S.H.; Thaysen, J.; Jakobsen, K.B. Ear-to-Ear Propagation Model Based on Geometrical Theory of Diffraction. *IEEE Trans. Antennas Propag.* **2019**, *67*, 1153–1160. [CrossRef]
25. Wolfle, G.; Landstorfer, F. Dominant paths for the field strength prediction. In Proceedings of the VTC'98. 48th IEEE Vehicular Technology Conference. Pathway to Global Wireless Revolution (Cat. No.98CH36151), Ottawa, ON, Canada, 21–21 May 1998; Volume 1, pp. 552–556. [CrossRef]

26. Wolfle, G. A recursive model for the field strength prediction with neural networks. In Proceedings of the Tenth International Conference on Antennas and Propagation (Conf. Publ. No. 436), Edinburgh, UK, 14–17 April 1997. [CrossRef]
27. Plets, D.; Joseph, W.; Vanhecke, K.; Tanghe, E.; Martens, L. Coverage prediction and optimization algorithms for indoor environments. *EURASIP J. Wirel. Commun. Netw.* **2012**, *2012*, 123. [CrossRef]
28. Commission, E.; for the Information Society, D.G.; Media. *COST Action 231: Digital Mobile Radio towards Future Generation Systems: Final Report*; Publications Office: Luxembourg 1999.
29. ITU-R. *Recommendation ITU-R P.1238-11—Propagation Data and Prediction Methods for the Planning of Indoor Radiocommunication Systems and Radio Local Area Networks in the Frequency Range 300 MHz to 450 GHz*; Technical Report; International Telecommunication Union: Geneva, Switzerland, 2021.
30. Hassan-Ali, M.; Pahlavan, K. A new statistical model for site-specific indoor radio propagation prediction based on geometric optics and geometric probability. *IEEE Trans. Wirel. Commun.* **2002**, *1*, 112–124. [CrossRef]
31. Aguado Agelet, F.; Formella, A.; Hernando Rabanos, J.; Isasi de Vicente, F.; Perez Fontan, F. Efficient ray-tracing acceleration techniques for radio propagation modeling. *IEEE Trans. Veh. Technol.* **2000**, *49*, 2089–2104. [CrossRef]
32. Soille, P. *Morphological Image Analysis : Principles and Applications*; Springer: Berlin/Heidelberg, Germany, 2004; pp. 1–463. [CrossRef]
33. Kaya, A.; Denis, S.; Bellekens, B.; Weyn, M.; Berkvens, R. Large-Scale Dataset for Radio Frequency-Based Device-Free Crowd Estimation. *Data* **2020**, *5*, 52. [CrossRef]
34. Douglas, D.H.; Peucker, T.K., Algorithms for the Reduction of the Number of Points Required to Represent a Digitized Line or its Caricature. In *Classics in Cartography*; John Wiley & Sons, Ltd.: Hoboken, NJ, USA 2011; Chapter 2, pp. 15–28. [CrossRef]
35. Cheffena, M. Propagation Channel Characteristics of Industrial Wireless Sensor Networks [Wireless Corner]. *IEEE Antennas Propag. Mag.* **2016**, *58*, 66–73. [CrossRef]
36. De Beelde, B.; Tanghe, E ; Yusuf, M.; Plets, D.; De Poorter, E.; Joseph, W. 60 GHz Path Loss Modelling Inside Ships. In Proceedings of the 2020 14th European Conference on Antennas and Propagation (EuCAP), Copenhagen, Denmark, 15–20 March 2020. [CrossRef]
37. Estes, D.R.; Welch, T.B.; Sarkady, A.A.; Whitesel, H. Shipboard radio frequency propagation measurements for wireless networks. In Proceedings of the 2001 MILCOM Proceedings Communications for Network-Centric Operations: Creating the Information Force (Cat. No.01CH37277), McLean, VA, USA, 28–31 October 2001; Volume 1, pp. 247–251. [CrossRef]
38. Wagen, J.F. Indoor service coverage predictions: How good is good enough? In Proceedings of the Fourth European Conference on Antennas and Propagation, Barcelona, Spain, 12–16 April 2010; pp. 1–5.

sensors

MDPI

Article

A Secure Blockchain-Based Authentication and Key Agreement Scheme for 3GPP 5G Networks

Man Chun Chow [1] and Maode Ma [2,*]

[1] School of Electrical and Electronic Engineering, Nanyang Technological University, Singapore 639798, Singapore; manchun001@e.ntu.edu.sg
[2] College of Engineering, Qatar University, Doha P.O. Box 2713, Qatar
* Correspondence: mamaode@qu.edu.qa

Abstract: The futuristic fifth-generation cellular network (5G) not only supports high-speed internet, but must also connect a multitude of devices simultaneously without compromising network security. To ensure the security of the network, the Third Generation Partnership Project (3GPP) has standardized the 5G Authentication and Key Agreement (AKA) protocol for mutually authenticating user equipment (UE), base stations, and the core network. However, it has been found that 5G-AKA is vulnerable to many attacks, including linkability attacks, denial-of-service (DoS) attacks, and distributed denial-of-service (DDoS) attacks. To address these security issues and improve the robustness of the 5G network, in this paper, we introduce the Secure Blockchain-based Authentication and Key Agreement for 5G Networks (5GSBA). Using blockchain as a distributed database, our 5GSBA decentralizes authentication functions from a centralized server to all base stations. It can prevent single-point-of-failure and increase the difficulty of DDoS attacks. Moreover, to ensure the data in the blockchain cannot be used for device impersonation, our scheme employs the one-time secret hash function as the device secret key. Furthermore, our 5GSBA can protect device anonymity by mandating the encryption of device identities with Subscription Concealed Identifiers (SUCI). Linkability attacks are also prevented by deprecating the sequence number with Elliptic Curve Diffie–Hellman (ECDH). We use Burrows–Abadi–Needham (BAN) logic and the Scyther tool to formally verify our protocol. The security analysis shows that 5GSBA is superior to 5G-AKA in terms of perfect forward secrecy, device anonymity, and mutual Authentication and Key Agreement (AKA). Additionally, it effectively deters linkability attacks, replay attacks, and most importantly, DoS and DDoS attacks. Finally, the performance evaluation shows that 5GSBA is efficient for both UEs and base stations with reasonably low computational costs and energy consumption.

Keywords: 5G; 5G-AKA; authentication; blockchain; BAN logic; Scyther

Citation: Chow, M.C.; Ma, M. A Secure Blockchain-Based Authentication and Key Agreement Scheme for 3GPP 5G Networks. *Sensors* **2022**, *22*, 4525. https://doi.org/10.3390/s22124525

Academic Editor: Luis Velasco

Received: 23 May 2022
Accepted: 12 June 2022
Published: 15 June 2022

1. Introduction

In recent years, the exponential growth of mobile subscribers and smart devices has fostered the rapid development of the fifth-generation cellular network (5G). Unlike the conventional 4G networks that only support limited numbers and types of devices, the 5G network is designed to connect as many devices as possible within one network. All devices such as mobile phones, autonomous vehicles, and Internet of Things (IoT) can now connect to the same 5G network with optimal speed and latency. To cater to these stringent requirements, the 5G network is constructed by many tiny femtocells to serve many users [1]. In this way, limited spectrum resources can be reused effectively to provide services to more devices simultaneously. Additionally, having more base stations installed, 5G wireless networks can alleviate traffic congestion in wireless channels. Hence, the futuristic 5G networks improve wireless connections with faster speed, lower latency, and greater capacity.

Although the 5G network is said to provide numerous benefits, there are also many new security challenges. In view of these potential security issues, the Third Generation

Partnership Project (3GPP) has standardized a new Authentication and Key Agreement (AKA) protocol known as 5G-AKA in TS 33.501 [2]. 5G-AKA can mutually authenticate base stations, 5G core networks, and user equipment (UE). It has resolved some pre-existing security issues found in the 4G Long-Term Evolution (LTE) networks. For example, by encrypting the permanent identity of the UE using the Subscription Concealed Identifier (SUCI), 5G-AKA can prevent International Mobile Subscriber Identification (IMSI), catching attacks and rogue base station attacks [3]. However, as some of the authentication methods in 5G-AKA are inherited from the 4G EPS-AKA [4], security issues in 4G networks remain unsolved in 5G-AKA. For example, 5G-AKA suffers from linkability attacks, in which malicious users can track a specific device by using synchronization error messages [5]. Additionally, it lacks perfect session key forward secrecy that guarantees data confidentiality, even if the long-term key is stolen in the future [3]. Furthermore, 5G-AKA is a centralized protocol that relies heavily on two functional entities, namely the Authentication Server Function (AUSF), and the Authentication credential Repository and Processing Function (ARPF) located inside the Unified Data Management (UDM) server. As there will be a tremendous number of devices connecting to the same 5G core network, 5G-AKA would be highly vulnerable to Denial of Service (DoS) attacks and Distributed Denial of Service (DDoS) attacks which aim to paralyze functional entities in the core network. Consequently, the existing 5G-AKA protocol is subject to many security threats that affect the robustness and reliability of the 5G network.

On the other hand, blockchain is a new technology for decentralized applications. Initially proposed by the creator of Bitcoin cryptocurrency, blockchain is a practical way to construct and manage a trustworthy decentralized ledger database across the network. By storing transactions into data blocks and linking them together using cryptographic hash functions, blockchain ensures all blocks in the chain reach the consensus effectively. Additionally, blockchain ensures the data in the database becomes computationally infeasible to mutate. In recent years, researchers have envisioned that blockchain can be used in a distributed way to solve many challenging problems in the 5G network [6]. As the number of 5G network infrastructures and mobile devices is growing exponentially, the benefits of blockchain would become more prominent in the future.

In this paper, we propose a Secure Blockchain-based Authentication and Key Agreement scheme for the 3GPP 5G network (5GSBA). Our 5GSBA protocol offers these benefits: first, it provides a distributed way to store all subscribers' information safely. By employing the one-time hash secret, authentication-related entries are the hashed digests that work similarly as public keys. Hence, even if the database is disclosed in the future, adversaries cannot use it to impersonate any UEs. Moreover, 5GSBA prevents not only typical network attacks such as eavesdropping, man-in-the-middle attacks, replay attacks, and IMSI-catching attacks, but also prevents DoS and DDoS attacks effectively and provides perfect session key forward/backward secrecy. The main contributions of this paper are as follows:

1. We design a novel Authentication and Key Agreement protocol for the 3GPP 5G network. 5GSBA works based on the improvement of the existing system architecture of the 5G core network. It can be easily adopted to the 3GPP access scenario, in which all UEs are connected to the home network via nearby gNBs;
2. Our proposed 5GSBA protocol is secure and efficient. Using blockchains and other state-of-the-art cryptographic functions, 5GSBA can guarantee device unlinkability, mutual authentication, and data confidentiality with low computational and energy costs. Most importantly, not only can all typical network attacks be prevented, but DoS and DDoS attacks can be deterred;
3. The security of the protocol is verified with BAN logic and the formal verification tool Scyther. The performance evaluation and simulations also demonstrate its resistivity to DoS and DDoS attacks.

The rest of the paper is organized as follows. Section 2 reviews the existing works on 5G authentication and some blockchain-based 5G applications. Section 3 introduces the system and security model of 5GSBA. Section 4 discusses the motivations for and the details of 5GSBA. Section 5 presents the work of security evaluation and Section 6 presents the work of performance evaluation with some simulation results under different attacks. Finally, a conclusion is drawn in Section 7.

2. Related Work

Recently, various solutions have been proposed to improve security in the Authentication and Key Agreement (AKA) process in the 5G network. In this section, we first discuss the security vulnerabilities in the existing 5G-AKA protocol. Then, we briefly review the major research work related to our work, including blockchain-related 5G authentication schemes and AKA schemes against DoS attacks.

2.1. Security Vulnerabilities in 5G-AKA

5G-AKA is the standardized Authentication and Key Agreement protocol in the latest 3GPP 5G security architecture TS 33.501 [2]. Evolved from the architecture of EPS-AKA in the LTE security architecture, 5G-AKA aims to ensure the authenticity between UE, the serving network, and the home network. However, some security vulnerabilities in the 5G-AKA have recently been disclosed, making it less secure than has been claimed. For example, Ref. [3,5] found that 5G-AKA suffers from linkability attacks, by which adversaries can use synchronization error messages (MAC_FAIL and SYNC_FAIL) to detect if the UE is currently located in a certain area. Additionally, 3GPP TR 33.846 [7] found that 5G-AKA fails to prevent denial-of-service (DoS) attacks because the 5GC has no way to justify if the SUCI is a replayed message. The 5G-AKA is a centralized protocol that heavily relies on the authentication functional entities of AUSF/UDM, so it could be vulnerable to Distributed DoS (DDoS) attacks and single-point-of-failure issues in the AUSF/UDM. Moreover, [4] found that 5G-AKA fails to provide perfect forward secrecy and post-compromise secrecy due to the use of the long-term symmetric keys and sequence numbers. In fact, according to the 3GPP TS 33.501 [2], the device anonymity protection of UE in 5G-AKA is also vulnerable. For example, network operators can opt out of the encryption in the Subscription Concealed Identifier (SUCI) that encrypts the Subscription Permanent Identifier (SUPI) of UE. It is also known as a "null-scheme". Thus, the UE will send the cleartext of its SUPI through wireless channels, which could be dangerous for IMSI-catching attacks. In some emergent situations, UE also sends its SUPI directly to initiate authentication procedures. To conclude, 5G-AKA is vulnerable to many network attacks, including but not limited to linkability attacks, DoS attacks, and IMSI-catching attacks. The lack of perfect forward secrecy also makes 5G-AKA vulnerable to session data recovery if the long-term key (LTK) is compromised at any time.

2.2. Blockchain in 5G Authentication

Linking data blocks into a chain, blockchain technology is essentially a secure decentralized database solution that guarantees data immutability and practical consensus across multiple network nodes. There are three different types of blockchain platforms [8]: permission-less, permissioned, and consortium blockchains. Among all three types of blockchains, it is envisioned that private and consortium blockchains are the most suitable distributed solutions to solve the security challenges in 5G because of their high efficiency [6].

In recent years, some proposals combining blockchains with 5G authentication have surfaced. For example, Yang et al. [9] introduced the idea of a blockchain-based anonymous access (BAA) scheme that allows equipment manufacturers, network operators, and users to access the blockchain-based database and perform mutual authentication. However, there is no formally proved protocol presented in the proposal. Haddad et al. proposed a blockchain-based 5G authentication protocol based on a public blockchain in [10,11]. They suggested that all 5G access points (APs) can use the UE public keys listed in the blockchain to perform mutual authentication between the AP and the UE. However, this misses a mechanism for UE to retrieve the public keys of the surrounding APs. Xu et al. [12] proposed the use of redactable blockchains to store all subscriber's information. The redactable blockchain provides key deletion and revocation functions. It is beneficial for network operators to protect the privacy of their users. However, the proposal lacks an authentication protocol for UEs and core networks to secure user data using the keys in the blockchain. Jia et al. [13] proposed a decentralized authentication scheme for 5G IoT devices. This protocol suggests that authentication entities in all domains can upload their device registration records to the same alliance blockchain. However, the protocol uses an identity-based cryptosystem. It introduces high computational overhead to mobile devices and edge servers, making them energy inefficient and prone to request flooding. Liu et al. [14] proposed an efficient authentication protocol based on 5G extensible authentication protocol (5G EAP-AKA') and a private blockchain. However, the security functionality of the proposed scheme has not been formally analyzed. Moreover, the EAP framework could also introduce more signaling overhead than the existing 5G-AKA scheme.

Some recent solutions have been designed to accelerate handover authentication in 5G wireless networks by using blockchains [15–19]. While all of them are providing a fast way to share the secret keys among base stations, most of them did not discuss how to prevent DoS and DDoS attacks during the UE authentication phase in a fast and efficient way. As a result, this shows that most of the existing blockchain authentication works are incomplete, and almost all of them could not alleviate the threats of DoS and DDoS attacks effectively during the UE authentication. In other words, designing a blockchain-based authentication protocol that provides adequate attack prevention, is energy-efficient, and computationally fast at the same time is a challenging research work. Overall, Table 1 summarizes all recent blockchain-based 5G authentication schemes and their challenges.

Table 1. Recent works on blockchain-based 5G authentication protocols.

	Type	Highlights	Security Features					
			FV	DA	PFS	LA	DoS	DDoS
Our Work	5G Initial Authentication	Decentralized authentications with low overhead	✓	✓	✓	✓	✓	✓
[9]		No single trust authority					✓	✓
[10,11]		Decentralized authentication to nearby gNBs		✓	✓		✓	✓
[12]		Removal of obsolete data in blockchains		✓	✓		✓	✓
[13]		Inter-domain authentication					✓	✓
[14]		Improving 5G EAP-AKA′ protocol security				✓	✓	✓
[19]		Formally verified protocol using chameleon signature	✓	✓	✓		✓	✓
[15]	5G Handover Authentication	Efficient handover authentication	✓	✓	✓	✓	✓	
[16]		Optimized for frequent handover				✓		
[17]		Lightweight handover authentication				✓		
[18]		Traceability for base stations to record malicious devices	✓	✓	✓	✓		

FV = formally verified protocol; DA = device anonymity; PFS = perfect forward secrecy; LA = lightweight authentication; DoS = DoS attack prevention; DDoS = DDoS attack prevention.

2.3. AKA Schemes against DoS and DDoS Attacks

On the other hand, many solutions have been designed to alleviate DoS and DDoS attacks during the UE authentication. Some recent 5G AKA schemes aimed at preventing DoS and DDoS are presented in Table 2. For example, by allowing multiple devices to choose a group leader as an agent, many group-based authentication protocols such as [20,21] aim to relieve computational burden and signaling overheads while authenticating a mass of devices. Although they can alleviate DoS attacks by including secret keys and short delays in the authentication requests, adversaries can bypass the security mechanism by launching the attacks individually. Leu et al. [22] have proposed to construct an AUSF pool and add a mediator to monitor all AUSFs. Although this provides disruption-free 5G-AKA authentication, it cannot prevent DoS attacks effectively because it is uses 5G-AKA, which is the protocol vulnerable to DoS attacks. Additionally, the bandwidth in the AUSF pool is still finite and expensive, and thus needs more investment. Yan et al. [23] have proposed a lightweight and secure handover authentication scheme based on a prediction of the potential target gNB from all neighbor gNBs in the 5G wireless network. The proposal facilitates fast 5G handover authentication by using time-to-live (TTL) attributes and encrypting the next hop chaining counter (NCC) with the Chinese remainder theorem. There are also some other schemes, including [24–28], proposed recently to fix other security issues in EPS-AKA and 5G-AKA. However, all of them suffer from single-point-of-failure due to the centralized protocol designs. Thus, all the existing DoS attack prevention schemes cannot provide adequate DoS and DDoS attack prevention.

Table 2. Recent works of AKA schemes with DoS and DDoS prevention.

	Method	Advantages	Drawbacks
[20]	Group Authentication	Prevention of DoS using timestamp	Weak DDoS prevention for individual authentication
[21]		Efficient group-based authentication	
[22]	Computation Pool	Fault-tolerant 5G-AKA authentication	5G-AKA is inherently vulnerable to DoS attacks
[23]	One-to-One Authentication	Lightweight and formally verified handover authentication	Vulnerable to DDoS due to centralized design
[24]		Formally verified protocol	
[25]		Lightweight symmetric key-based protocol	
[26]		Formally verified protocol	
[27]		Backward compatibility with 5G-AKA	
[28]		DDoS prevention using zero-knowledge proof	Centralized design, lack of formally verified protocol

3. System and Security Model

3.1. System Model

Our system model follows the 3GPP 5G system architecture listed in TS 23.501 [29] and the security architecture listed in TS 33.501 [2]. We are adding some new features to the existing functional entities. Figure 1 shows the 3GPP 5G core network (5GC) consisting of many functional entities.

In the existing 3GPP 5GC system model, the Next Generation Node B (gNB) is the base station that directly communicates with the UE. All gNBs in the 5G network are connected to the nearby Access and Mobility Function (AMF) servers. During conventional 5G-AKA authentication, when UE is under the coverage of 3GPP access (i.e., under the signal coverage of gNBs), it should send an authentication request to gNBs. Then, the gNB forwards it to the AMF, and the AMF forwards it to the Authentication Server Function (AUSF) server. After that, AUSF fetches the device secret keys from the Unified Data Management (UDM) server to continue the subsequent authentication steps. Normally, one AUSF serves many AMFs across the 5G core network, and each AMF serves many gNBs nearby.

In our proposal, we follow the same system model with a decentralization of the authentication entities as follows. All gNBs in the 5G core network become the members of a private blockchain that stores subscribers' information. AMF and AUSF should no longer need to forward authentication requests, but they can optionally be one of the members in the blockchain. UDM is the protected repository that stores private keys of the blockchain and other important secrets. For each authentication request, UE uses the one-time hash secret stored in the Universal Subscriber Identity Module (USIM, or commonly known as a SIM card) to initiate an authentication request. Then, the gNB hashes the received secret and compares it with the entries in the blockchain. Since only legitimate UE would have the original secret, the gNB can immediately grant or deny the UE's authentication request without forwarding the request to AMF, AUSF, or UDM.

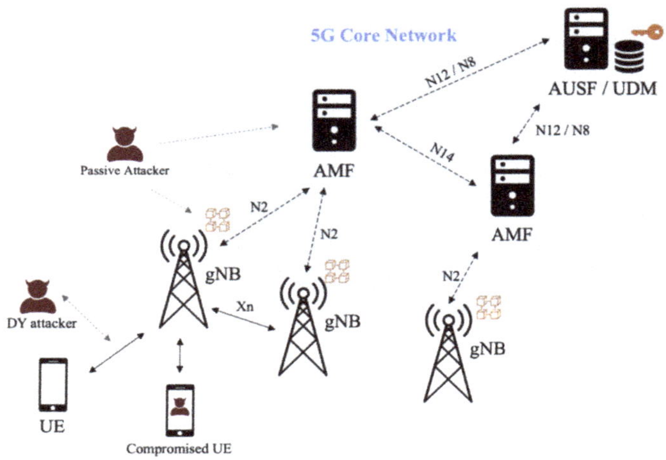

Figure 1. System and security model.

Considering that 5G devices can connect to the 5GC by many different approaches such as non-3GPP access, 3GPP access in the home network, and 3GPP access in a visiting network, designing a universal authentication protocol could be very complicated. In this work, we focus on the most common scenarios to simplify the designed authentication protocol. It is assumed that all UE uses 3GPP access to connect to the gNBs in the home network. These gNBs are connected to the 5GC. Additionally, the connections between gNBs and 5GC are secured by wired connections protected by IPSec tunnels. Thus, if a gNB can mutually authenticate with UE using the private key of the home network and information stored in the blockchain, it can be regarded that the UE has joined a legitimate 5GC network.

3.2. Security Model

Our security model is also displayed in Figure 1. It is assumed that the 5G wireless channels follow the Dolev-Yao model [30], which assumes that there could be some neighboring active and passive attackers. Passive attackers eavesdrop and interpret the messages sent from both UEs and gNBs. Then, they can analyze the intercepted data to figure out the messages sent from both parties. Active attackers not only eavesdrop on the wireless channels, but also modify the intercepted messages, replay them, or even fabricate new messages to impersonate legal UEs or gNBs to disrupt the network. All these attackers are labeled as "Dolev-Yao (DY) Attackers" in Figure 1. Moreover, although it is assumed that the communication channels within 5GC are trustworthy, some accidents or misconfiguration could happen to the functional entities. For example, information in the data repositories of the 5GC functional entities could be leaked in some rare cases. In this scenario, these passive attackers would try to use the exposed permanent keys to decrypt previous communication data. Additionally, it is possible that there is compromised botnet UE in the wireless network. UE could launch DoS and DDoS attacks at any scheduled time, attempting to flood or paralyze the authentication-related functional entities in the 5GC.

Due to all the assumptions above, a desired 5G authentication protocol should provide security functionalities including device anonymity, mutual authentication, secure data transmission, and session key perfect forward secrecy. Moreover, it should be able to prevent active attacks such as impersonation, linkability attacks [5], replay attacks, man-in-the-middle (MITM) attacks, and DoS attacks. For passive attacks, there could be eavesdropping and location tracking attacks. These attacks must also be deterred.

4. The Proposed 5GSBA Scheme

4.1. Motivation

The existing 5G authentication protocol, 5G-AKA, is a vulnerable protocol that creates a vast burden to the AUSF. Since there will be more connected devices in the future, attackers are likely to launch DoS and DDoS attacks to flood the AUSF and other 5G authentication entities. However, as there could also be some essential utilities using 5G, the 5G network has to be stable at all times. Therefore, to prevent DoS and DDoS attacks from paralyzing the 5G network, there is an urgent and critical need to design an authentication protocol that ensures no DoS and DDoS attacks can be successful. This protocol should work in a decentralized manner, such that it will not suffer from a single point of failure due to request flooding. Given all these constraints, we believe that an authentication protocol combined with a blockchain would be an ideal solution. Although a blockchain could introduce more overhead during database synchronizations, it helps reduce the opportunity of system overloading by decentralizing authentication tasks. In this paper, we propose a 5GSBA scheme that uses a blockchain to decentralize the subscription repository from UDM to all gNBs, such that authentication tasks can be decentralized to the gNBs. By doing so, we can prevent DoS and DDoS attacks from impacting the quality of service (QoS) of the entire 5G network.

4.2. Details of the 5GSBA

This section presents the Secure Blockchain-based Authentication and Key Agreement Protocol (5GSBA) in detail, which is a two-step protocol designed to mutually authenticate between UEs and gNBs in the 5G network. To satisfy all aforementioned security requirements, the 5GSBA combines the one-time hash function, Elliptic Curve Diffie–Hellman (ECDH), the Elliptic Curve Integrated Encryption Scheme (ECIES), and the keyed-hash message authentication code (HMAC) at different steps of the protocol. The 5GSBA uses a private blockchain network across all gNBs as a distributed subscriber data repository. Therefore, all gNBs connected to the 5GC may approve authentication requests from UE autonomously without taxing the AUSF and UDM. The 5GSBA has four phases: the system initialization phase, the USIM registration phase, the gNB broadcast phase, and the mutual authentication phase. All notations used in this paper are listed in Table 3, and Figure 2 shows a sequence diagram explaining different phases in 5GSBA.

Table 3. Notations.

Notation	Description
SUPI	Subscription Permanent Identifier of the UE
ID_{gNB}	Permanent Identifier of gNB
P	Generator of the Elliptic Curve
Y/Y_2	One-Time Hash Secret
H (msg)	Cryptographic Hash Function
HMAC (msg, K)	Keyed-hash Message Authentication Code
σ	Generated HMAC code
PK_{core}	Public Key of 5G Core
SK_{core}	Private Key of 5G Core

Table 3. *Cont.*

Notation	Description
TS	Timestamp
K_{hmac}	Symmetric Key for HMAC Generation
E_{PKCore} (msg)	Encrypt Message with the Public Key of 5GC

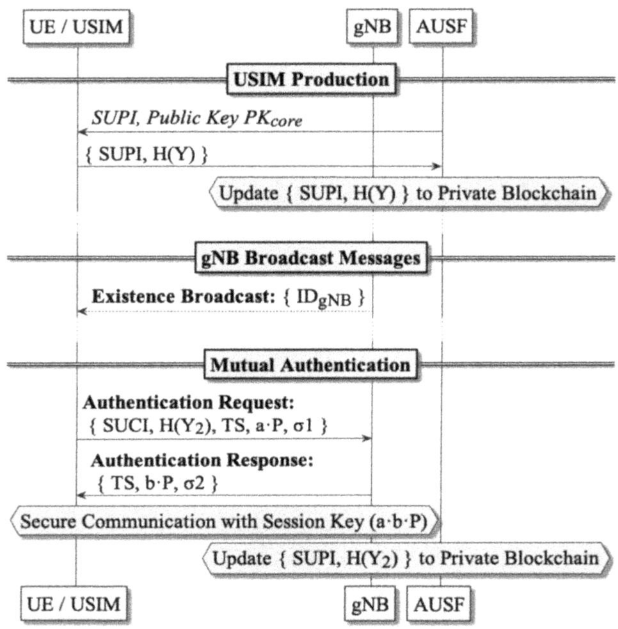

Figure 2. Sequence Diagram of 5GSBA.

Phase 1—System Initialization: Let p be the modulus, $E(\mathbb{F}_p)$ be the elliptic curve over a finite field $\mathbb{F}_p$, P be the generator point on $E(\mathbb{F}_p)$ with an order n, and $\mathbb{G}$ be the generated subgroup which multiplies the generator point P. Additionally, we let the cryptographic hash function be $H \subseteq \mathbb{Z}_n^*$. Having the assumptions above, gNBs and AUSF run the following procedures:

1. AUSF generates a new ECIES private key SK_{core} representing the 5GC by choosing a random input k. Then, the ECIES public key $PK_{core} = k \cdot P$ is stored at UDM.
2. When there is a new gNB joining the 5GC, they should mutually authenticate with any existing approach such as IPSec tunnels. After that, AUSF installs the ECIES private key SK_{core} from UDM into the secure enclave of the newly joined gNB.
3. Finally, the authenticated gNB downloads the latest private blockchain from the 5GC. gNB may also index the transactions in the blockchain locally for faster access.
4. Whenever the blockchain has any updates, the gNB will download and index the new blocks accordingly.

Regarding the private blockchain administered by AUSF, Figure 3 illustrates the structure of a block in the whole blockchain. Every block should contain the following elements:

- **Block Header:** It contains the block version for future maintenance and upgrades.
- **Previous Block Hash:** It is the hash of the previous block. It guarantees the immutability of the blockchain.

- **Timestamp:** It is the block creation time for tracking purposes. All transaction timestamps within the block should never be larger than this timestamp.
- **Transactions:** Each transaction is a subscription record for one device. To reduce the storage overhead of the blockchain, each block contains multiple transaction records. This size should be adjustable according to the preference of network operators. By default, we follow the block size of bitcoin as 1 MB.

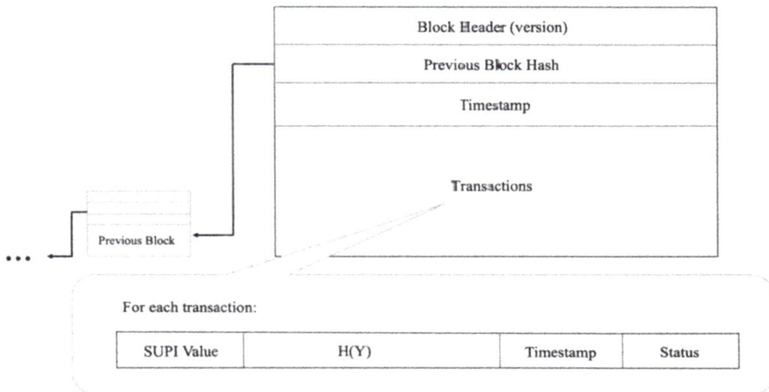

Figure 3. Structure of a block and a transaction.

Since it is a private blockchain, any efficient algorithm can be used, such as Practical Byzantine Fault Tolerance (PBFT) [31], to reach a consensus for all gNBs. The details about blockchain consensus implementation are omitted in this paper.

Phase 2—USIM Production: Network operators should install a one-time hash secret Y and the ECIES public key of the 5GC PK_{core} to the USIM during USIM production. Additionally, the one-time secret hash digest $H(Y)$ should be posted to the private blockchain. In this way, when the UE sends the collision of the hash function (i.e., the secret Y) to the gNB, it can prove to the gNB that it is the legitimate UE. These are the detailed procedures:

1. The operator generates a one-time hash secret Y and the digest of the one-time hash secret $H(Y)$;
2. USIM stores its permanent identity (i.e., SUPI), elliptic curve parameters, one-time hash secret Y, and the ECIES public key of 5GC PK_{core} into its non-volatile storage;
3. AUSF creates a new blockchain transaction including the SUPI, $H(Y)$, timestamp, and a status code. The status code is the activation status of the SUPI. For example, "activated" can be 1, "suspended" can be 2, "revoked" can be 3, and so on. The format of one transaction in the blockchain is also shown in Figure 3.
4. If we need to revoke the access of a specific USIM, AUSF can post a new transaction with a "revoked" activation status code and a timestamp to the blockchain. Therefore, when gNBs retrieve the latest transactions from the blockchain, they will follow the last record to deny access from that USIM.

Phase 3—gNB Broadcast: after initialization, gNBs broadcast their identities ID_{gNB} through the air. Since 5GSBA makes SUCI mandatory during UE authentication, the identity request procedures in 3GPP TS 33.501 [2] are no longer needed. Hence, UE can freely choose when to start authentication and when to prepare SUCI without having to respond to possibly forged identity requests from gNBs.

Phase 4—Mutual Authentication: whenever UE is powered up to start authentication with the 5G network, the following two-step protocol will be executed:

1. **UE → gNB:** UE sends an authentication request to the gNB with these steps:
   a. Generate a new random HMAC key K_{hmac}, random ECDH public key $a \cdot P$, and timestamp TS;
   b. Generate SUCI by encrypting {SUPI, Y, K_{hmac}} with PK_{core};
   c. Generate the next one-time hash secret Y_2, and calculate its hash $H(Y_2)$;
   d. Update the Y in the local storage as Y_2, as it will become the Y for the next authentication;
   e. Calculate σ1 = HMAC ($\{ID_{gNB}, SUPI, Y, H(Y_2), TS, a \cdot P\}$ K_{hmac});
   f. Send the authentication request = $\{SUCI, H(Y_2), TS, a \cdot P, \sigma 1\}$ to gNB.
2. **gNB** checks the incoming authentication request with these steps:
   a. Check the validity of the timestamp TS, and then decrypt the SUCI into {SUPI, Y, K_{hmac}} using the private key SK_{core};
   b. Verify the HMAC of the message σ1 = HMAC ($\{ID_{gNB}, SUPI, Y, H(Y_2), TS, a \cdot P\}$ K_{hmac});
   c. Fetch the latest transaction of the SUPI from the private blockchain locally;
   d. Compare the hash of the received Y with the $H(Y)$ value stored in the blockchain. If there is a collision (i.e., two values are equal), send an authentication response. Otherwise, gNB should stop the protocol;
   e. Create a new blockchain transaction containing the value of $H(Y_2)$, and upload the block containing this transaction when the gNB is idle.
3. **gNB → UE:** gNB issues an authentication response to the UE with these steps:
   a. Generate a new random ECDH public key $b \cdot P$;
   b. Calculate σ2 = HMAC ({SUPI, TS, $b \cdot P$}, K_{hmac}), where TS is the received timestamp;
   c. Send the authentication response = {TS, $b \cdot P$, σ}.
4. **UE** checks the incoming authentication response with these steps:
   a. Calculate σ′ = HMAC ({SUPI, TS, $b \cdot P$}, K_{hmac}). If σ equals to σ′, the UE continues to calculate the common ECDH session key using formula $a \cdot b \cdot P$;
   b. Similarly, gNB calculates the common ECDH session key using formula $b \cdot a \cdot P$. Since $a \cdot b \cdot P = b \cdot a \cdot P$, a common session key is derived. Both parties are now mutually authenticated.

5. Security Evaluation

In this section, we firstly justify the logical correctness of the 5GSBA using Burrows–Abadi–Needham (BAN) logic. Then, we provide formal verification on the security of the 5GSBA using the Scyther formal verification tool. Moreover, we present an extensive qualitative security analysis based on the discussion in Section 3.2 to show that the 5GSBA is secure to fight against various malicious attacks.

5.1. Burrows–Abadi–Needham (BAN) Logic

Burrows–Abadi–Needham (BAN) logic is a set of logic rules to verify the logical correctness of an authentication protocol [32]. Assuming that the cryptographic functions in the protocol are perfect, BAN logic can systematically find out all incorrect designs in an authentication protocol. To apply BAN logic to our 5GSBA protocol, we formalize our protocol into the idealized form. Then, we use BAN logic symbols and rules [32] such as the message meaning rule, belief rule, nonce verification rule, jurisdiction rule, etc., to validate if our 5GSBA protocol fulfills the targeted security goals.

5.1.1. Formalized 5GSBA Protocol

In our idealized protocol, U refers to UE and C refers to one of the gNBs in the 5G cellular network (CN). All cleartext and identities in the protocol are omitted as they can be easily forged. For the notations, SUCI can be regarded as a message encrypted by the public key of 5GC (i.e., the PK_{core}). The timestamp token is represented by TS, and the one-time hash token is represented by Y. In addition, all message content protected by the HMAC can be viewed as a message encrypted by the HMAC key (i.e., K_{hmac}). Therefore, the idealized 5GSBA protocol is shown below:

Message 1: $U \rightarrow C : C \triangleleft \left\{ U \overset{K_{hmac}}{\leftrightarrow} C,\ Y_Y \right\}_{PK_{core}}, \{\ TS,\ a{\cdot}G\ \}_{K_{hmac}}$

Message 2: $C \rightarrow U : U \triangleleft \{\ TS,\ b{\cdot}P\ \}_{K_{hmac}}$

5.1.2. Logical Assumptions

We made the following assumptions according to the nature of the protocol. First, the CN believes that the UE should control the HMAC key issued by themselves:

$$C \mid\equiv U \mid\Rightarrow U \overset{K_{hmac}}{\leftrightarrow} C \tag{1}$$

Second, since both the CN and the UE check the timestamp in the protocol, they should believe that the timestamps are fresh:

$$C|\equiv\#(TS) \tag{2}$$

$$U|\equiv\#(TS) \tag{3}$$

Third, the CN and the UE should also believe that their locally generated keys are trustworthy to themselves:

$$U \mid\equiv U \overset{K_{hmac}}{\leftrightarrow} C \tag{4}$$

$$U \mid\equiv a \tag{5}$$

$$U \mid\equiv a{\cdot}P \tag{6}$$

$$C \mid\equiv b \tag{7}$$

$$C \mid\equiv b{\cdot}P \tag{8}$$

Fourth, the UE should believe that the key generated by the CN is controlled and trusted by himself. Similarly, the CN should also believe that the keys generated by UE are controlled by himself:

$$U \mid\equiv C \mid\Rightarrow b{\cdot}P \tag{9}$$

$$C \mid\equiv U \mid\Rightarrow a{\cdot}P \tag{10}$$

$$U \mid\equiv C \mid\equiv b \tag{11}$$

$$C \mid\equiv U \mid\equiv a \tag{12}$$

Fifth, the CN should believe the secret of the one-time hash sent from the UE by validating it with the records in the blockchain. Additionally, since it is only valid once, it can be viewed as a fresh nonce:

$$C \mid\equiv U \overset{Y}{\rightleftharpoons} C \tag{13}$$

$$C \mid\equiv \#(Y) \tag{14}$$

Finally, since ECDH is used, it can be assumed that for the UE (U), the session key $U \overset{K_{UC}}{\leftrightarrow} C = a{\cdot}b{\cdot}P$ can be calculated with the received $b{\cdot}P$ and the locally generated a. Similarly, for the CN (C), the session key $U \overset{K_{UC}}{\leftrightarrow} C = b{\cdot}a{\cdot}P$ can be calculated with the received $a{\cdot}P$ and the locally generated b.

5.1.3. Protocol Goal

The goal of the 5GSBA is to achieve mutual authentication between two sides (UE and CN). Hence, we need to create a mutually trusted common session key after the execution of the protocol. We can express the goal with these four equations:

$$U \mid\equiv U \overset{K_{UC}}{\leftrightarrow} C \tag{15}$$

$$C \mid\equiv U \overset{K_{UC}}{\leftrightarrow} C \tag{16}$$

$$U \mid\equiv C \mid\equiv U \overset{K_{UC}}{\leftrightarrow} C \tag{17}$$

$$C \mid\equiv U \mid\equiv U \overset{K_{UC}}{\leftrightarrow} C \tag{18}$$

5.1.4. Protocol Verification

The detailed verification steps are listed below. Using the rule with Message 1, we have Equation (19):

$$\frac{C \mid\equiv \overset{PK_{core}}{\mapsto} U,\ C \triangleleft \left\{ \langle U \overset{K_{hmac}}{\leftrightarrow} C,\ Y \rangle_Y \right\}_{PK_{core}}}{C \triangleleft \langle U \overset{K_{hmac}}{\leftrightarrow} C, Y \rangle_Y} \tag{19}$$

Using the message meaning rule with Equations (13) and (19), we have Equation (20):

$$\frac{C \mid\equiv U \overset{Y}{\rightleftharpoons} C,\ C \triangleleft \langle U \overset{K_{hmac}}{\leftrightarrow} C,\ Y \rangle_Y}{C \mid\equiv U \mid\sim \left(U \overset{K_{hmac}}{\leftrightarrow} C,\ Y \right)} \tag{20}$$

Using the freshness rule with Equations (14) and (19), we have Equation (21):

$$\frac{C \mid\equiv \#(Y)}{C \mid\equiv \#\left(U \overset{K_{hmac}}{\leftrightarrow} C,\ Y \right)} \tag{21}$$

Using the nonce verification rule with Equations (20) and (21), we have Equation (22):

$$\frac{C \mid\equiv \#\left(U \overset{K_{hmac}}{\leftrightarrow} C,\ Y \right),\ C \mid\equiv U \Big| \sim \#(K_{hmac}, Y)}{C \mid\equiv U \mid\equiv \left(U \overset{K_{hmac}}{\leftrightarrow} C,\ Y \right)} \tag{22}$$

Using the belief rule with Equation (22), we have Equation (23):

$$\frac{C \mid\equiv U \mid\equiv \left(U \overset{K_{hmac}}{\leftrightarrow} C,\ Y \right)}{C \mid\equiv U \mid\equiv U \overset{K_{hmac}}{\leftrightarrow} C} = C \mid\equiv U \mid\equiv U \overset{K_{hmac}}{\leftrightarrow} C \tag{23}$$

Using the jurisdiction rule with Equations (1) and (23), we have Equation (24):

$$\frac{C \mid\equiv U \mid\Rightarrow U \overset{K_{hmac}}{\leftrightarrow} C,\ \ C \mid\equiv U \mid\equiv U \overset{K_{hmac}}{\leftrightarrow} C}{C \mid\equiv U \overset{K_{hmac}}{\leftrightarrow} C} = C \mid\equiv U \overset{K_{hmac}}{\leftrightarrow} C \tag{24}$$

As a result, the CN believes the received HMAC key, so the CN continues to process Message 1. Using the message meaning rule with Equation (24) and Message 1, we have Equation (25):

$$\frac{C \mid\equiv U \overset{K_{hmac}}{\leftrightarrow} C,\ C \triangleleft \{\, TS,\ a{\cdot}P \,\}_{K_{hmac}}}{C \mid\equiv U \mid \sim\ (TS,\ a{\cdot}P)} = C \mid\equiv U \mid \sim (TS,\ a{\cdot}P) \tag{25}$$

Using the freshness rule with Equations (2) and (25), we have Equation (26):

$$\frac{C \mid\equiv \#(TS)}{C \mid\equiv \#(TS,\ a{\cdot}P)} = C \mid\equiv \#(TS,\ a{\cdot}P) \tag{26}$$

Using the nonce verification rule with Equations (25) and (26), we have Equation (27):

$$\frac{C \mid\equiv \#(TS,\ a{\cdot}P),\quad C \mid\equiv U \mid\sim (TS,\ a{\cdot}P)}{C \mid\equiv U \mid\equiv (TS,\ a{\cdot}P)} \tag{27}$$

Using the belief rule with Equation (27), we have Equation (28):

$$\frac{C \mid\equiv U \mid\equiv (TS,\ a{\cdot}P)}{C \mid\equiv U \mid\equiv a{\cdot}P} = C \mid\equiv U \mid\equiv a{\cdot}P \tag{28}$$

Using the jurisdiction rule with Equations (10) and (28), we have Equation (29):

$$\frac{C \mid\equiv U \mid\Rightarrow a{\cdot}P,\ C \mid\equiv U \mid\equiv a{\cdot}P}{C \mid\equiv a{\cdot}P} = C \mid\equiv a{\cdot}P \tag{29}$$

As a result, the CN believes the received ECDH public key from UE, and the protocol continues with Message 2. Using the message meaning rule with Equation (4) and Message 2, we have Equation (30):

$$\frac{U \mid\equiv U \overset{K_{hmac}}{\leftrightarrow},\ P \triangleleft \{TS,\ b{\cdot}P\}_{K_{hmac}}}{U \mid\equiv C \mid \sim\ (TS,\ b{\cdot}P)} = U \mid\equiv C \mid \sim\ (TS,\ b{\cdot}P) \tag{30}$$

Using the freshness rule with Equations (3) and (30), we have Equation (31):

$$\frac{U \mid\equiv \#(TS)}{U \mid\equiv \#(TS,\ b{\cdot}P)} = U \mid\equiv \#(TS,\ b{\cdot}P) \tag{31}$$

Using the nonce verification rule with Equations (30) and (31), we have Equation (32):

$$\frac{U \mid\equiv \#(TS,\ b{\cdot}P),\ U \mid\equiv C \mid \sim\ (TS,\ b{\cdot}P)}{U \mid\equiv C \mid\equiv (TS,\ b{\cdot}P)} \tag{32}$$

Using the belief rule with Equation (32), we have Equation (33):

$$\frac{U \mid\equiv C \mid\equiv (TS,\ b{\cdot}P)}{U \mid\equiv C \mid\equiv b{\cdot}P} = U\ \equiv C \mid\equiv b{\cdot}P \tag{33}$$

Using jurisdiction rule with Equations (9) and (33), we have Equation (34):

$$\frac{U \mid\equiv C \mid\Rightarrow b{\cdot}P,\ U \mid\equiv C \mid\equiv b{\cdot}P}{U \mid\equiv b{\cdot}P} = U \mid\equiv b{\cdot}P \tag{34}$$

As a result, the UE also believes the received ECDH public key from the gNB.

For the UE, since we know $U \mid\equiv a$ in Equation (5) and $U \mid\equiv b{\cdot}P$ in Equation (34), the common key $U \overset{K_{UC}}{\leftrightarrow} C$ can be derived in Equation (35):

$$U \mid\equiv a{\cdot}(b{\cdot}P) = U \mid\equiv a{\cdot}b{\cdot}P = U \mid\equiv U \overset{K_{UC}}{\leftrightarrow} C \tag{35}$$

For the CN, since we know $C \mid\equiv b$ in Equation (7) and $C \mid\equiv a{\cdot}P$ in Equation (29), the common key $U \overset{K_{UC}}{\leftrightarrow} C$ can be derived in Equation (36):

$$U \mid\equiv a{\cdot}(b{\cdot}P) = U \mid\equiv a{\cdot}b{\cdot}P = U \mid\equiv U \overset{K_{UC}}{\leftrightarrow} C \tag{36}$$

Moreover, to finish the protocol, the CN has to believe Message 1 to continue the protocol and send Message 2. Hence, we can say that if the UE has received Message 2, the UE can be sure that the CN has believed Message 1, and therefore $U \mid\equiv C \mid\equiv a{\cdot}P$. Combining this with Equation (11), we can conclude that:

$$U \mid\equiv C \mid\equiv b{\cdot}(a{\cdot}P) = U \mid\equiv C \mid\equiv a{\cdot}b{\cdot}P = U \mid\equiv C \mid\equiv U \overset{K_{UC}}{\leftrightarrow} C \tag{37}$$

Similarly, the UE has to believe Message 2 to start the subsequent data transmission. Hence, we can say that if the CN receives the subsequent data correctly, CN can be sure that the UE has believed Message 2, and therefore $C \mid\equiv U \mid\equiv b{\cdot}P$. Combining this with Equation (12), we can conclude that:

$$C \mid\equiv U \mid\equiv a{\cdot}(b{\cdot}P) = C \mid\equiv U \mid\equiv a{\cdot}b{\cdot}P = C \mid\equiv U \mid\equiv U \overset{K_{UC}}{\leftrightarrow} C \tag{38}$$

Consequently, all the security goals are satisfied. Hence, the security of 5GSBA is logically verified.

5.2. *Scyther Tool*

The Scyther tool [33] is an automated formal verification tool for analyzing authentication protocols. Under the perfect cryptography assumption and the Dolev–Yao adversary model [30], Scyther searches for all potential security vulnerabilities of a protocol efficiently. Perfect cryptography assumption refers that the cryptographic functions used in the protocol are assumed to be secure. Adversaries should know nothing about the encrypted content unless they hold the decryption key. The Dolev–Yao adversary model, as mentioned in Section 3.2, assumes that there are neighboring attackers in the network. In this section, we model the 5GSBA with the Security Protocol Description Language (SPDL) and let Scyther find all security issues automatically.

The Scyther tool has six different security claims: **Aliveness** or **Alive** ensure the protocol instances complete their steps with any active responders. That is, all replies should be active replies from living partners, not replayed messages. **Niagree** ensures a protocol instance receives the expected variables without consideration of a one-to-one relationship (i.e., non-injective agreement). **Nisynch** ensures the protocol can complete a run as expected without a one-to-one relationship (i.e., non-injective synchronization). **Weakagree** ensures all protocol instances communicate with their same set of initiators or responders (i.e., injective). **Reachable** is the checkpoint claim indicating that the protocol can reach the specific line, which can be used for code debugging. **Secrecy** or **SKR** check if the specified variables or session keys remain secret to adversaries throughout the execution of the protocol. By combining all these security claims, Scyther ensures the injective agreement of the protocol and detects most of the network protocol attacks including message fabrication, message replay, and MITM attacks.

Our formal verification result with Scyther is shown in Figure 4. Specifically, we created a SPDL model that has two roles: the gNB and the UE. First, the 5GSBA can achieve a mutual key agreement using ECDH. To emulate ECDH key exchange between two parties, we define functions g1 and g2, and then set alpha to be g1(a) and beta to be g1(b). We firstly use the claims of Secret a and Secret b to check the secrecy of the ECDH private keys. Then, we emulate the derivation of the ECDH common key with g2(beta, a) (i.e., a · b · G) and g2(alpha, b) (i.e., b · a · G). We also verify the secrecy of this ECDH common key with *SKR* claims. Second, the 5GSBA uses HMAC to ensure the integrity of all messages. To make sure adversaries cannot obtain the HMAC key K_{1mac} (i.e., only gNBs can receive the key from the UE), we check it with Secret Khmac claims for both the gNB and the UE. Third, the 5GSBA can guarantee device anonymity by encrypting the SUPI into SUCI. To ensure the adversaries have no way to retrieve the SUPI of the UE through the air, we check the secrecy of the SUPI using the Secret MSIN claim. In this scenario, Mobile Subscriber Identification Number (MSIN) is equivalent to SUPI, because SUPI should contain mobile country code (MCC), mobile network code (MNC), and MSIN. While MCC and MNC are not the unique identifiers of a device, we only mandate MSIN to be secret. Fourth, our protocol assumes that the UE needs to use its one-time hash secret Y to prove its identity to gNB. To ensure that the adversaries cannot obtain the secret Y with any means, we checked its secrecy using Secret Y claims. Additionally, the secrecy of Y2 was checked to ensure that the UE could not reveal the next one-time hash secret by any means. Fifth, to further ensure the correctness of our SPDL model, a *Reachable* claim was put at the end of every role. This made sure that every line of our code had been executed. Finally, by testing all security claims including *Aliveness*, *Niagree*, *Nisynch*, and *Weakagree* in both parties, this showed that the 5GSBA guarantees injective agreement with active initiator and responders. In other words, we conclude that no network attacks were found in the 5GSBA.

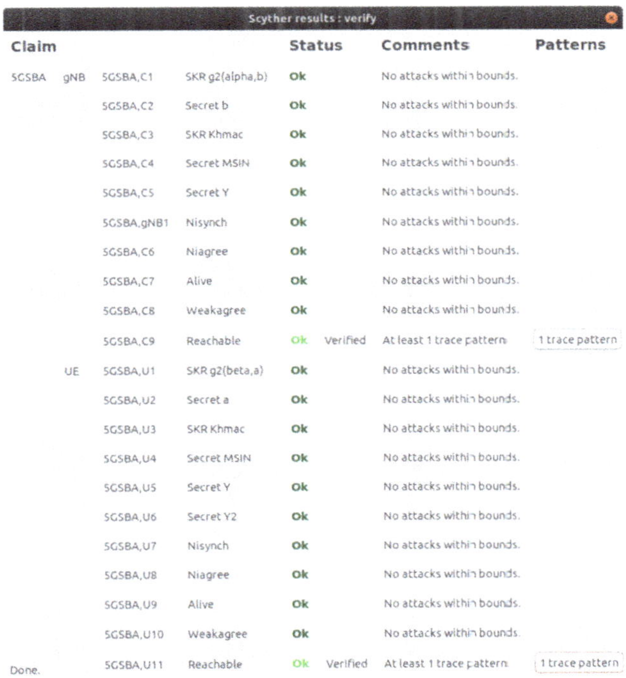
Scyther results : verify

Claim				Status		Comments	Patterns
5GSBA	gNB	5GSBA,C1	SKR g2(alpha,b)	Ok		No attacks within bounds.	
		5GSBA,C2	Secret b	Ok		No attacks within bounds.	
		5GSBA,C3	SKR Khmac	Ok		No attacks within bounds.	
		5GSBA,C4	Secret MSIN	Ok		No attacks within bounds.	
		5GSBA,C5	Secret Y	Ok		No attacks within bounds.	
		5GSBA,gNB1	Nisynch	Ok		No attacks within bounds.	
		5GSBA,C6	Niagree	Ok		No attacks within bounds.	
		5GSBA,C7	Alive	Ok		No attacks within bounds.	
		5GSBA,C8	Weakagree	Ok		No attacks within bounds.	
		5GSBA,C9	Reachable	Ok	Verified	At least 1 trace pattern.	1 trace pattern
	UE	5GSBA,U1	SKR g2(beta,a)	Ok		No attacks within bounds.	
		5GSBA,U2	Secret a	Ok		No attacks within bounds.	
		5GSBA,U3	SKR Khmac	Ok		No attacks within bounds.	
		5GSBA,U4	Secret MSIN	Ok		No attacks within bounds.	
		5GSBA,U5	Secret Y	Ok		No attacks within bounds.	
		5GSBA,U6	Secret Y2	Ok		No attacks within bounds.	
		5GSBA,U7	Nisynch	Ok		No attacks within bounds.	
		5GSBA,U8	Niagree	Ok		No attacks within bounds.	
		5GSBA,U9	Alive	Ok		No attacks within bounds.	
		5GSBA,U10	Weakagree	Ok		No attacks within bounds.	
Done.		5GSBA,U11	Reachable	Ok	Verified	At least 1 trace pattern.	1 trace pattern

Figure 4. Formal verification with Scyther.

5.3. Security Analysis

We qualitatively justify that our 5GSBA provides the following security features.

Mutual Authentication and Key Agreement: By the 5GSBA, UE uses a one-way hash secret Y stored in the USIM to prove that it is the legitimate party. A gNB verifies it by comparing it to the one-way hash value $H(Y)$ stored in the private blockchain. Since finding the original secret of a hash is a computationally infeasible problem, the UE can effectively prove to the gNB that its authentication request message is legitimate. Besides, the UE encrypts the one-way hash secret Y and a randomly generated HMAC key K_{hmac} using the public key PK_{core}. Since only legitimate gNBs possessing the private key SK_{core} can read the one-way hash secret and the HMAC key, by issuing a correct HMAC code $\sigma 2$, the gNB effectively prove to the UE that the authentication response is also legitimate. Thus, both gNB and UE can be mutually authenticated to derive a common key using ECDH parameters in the authentication messages.

Secure Data Transmission: The 5GSBA assumes that all subsequent user data will be encrypted with the session key generated in the authentication procedure. For the session key generation, a UE sends an ECDH public key $a \cdot P$ to a gNB, and the gNB replies to another ECDH public key $b \cdot P$ to the UE. Hence, the actual session key $a \cdot b \cdot P$ is never transmitted anywhere. In fact, the ECDH relies on the Computational Diffie–Hellman (CDH) problem, which means it is difficult to find the a from $a \cdot P$. Even if an adversary captures all the authentication messages, it is computationally infeasible for him to recover the session key by finding either a or b to calculate the $a \cdot b \cdot P$. Consequently, the 5GSBA can ensure only legitimate parties (i.e., UE and gNB) can read the user data.

Session Key Perfect Forward/Backward Secrecy: The 5GSBA uses the public key PK_{core} of the 5GC, the randomly generated HMAC key K_{hmac}, and the one-way hash secret Y for authentication purposes only. The session key generation relies on the randomly generated ECDH parameters. Hence, when the permanent keys are stolen in the future, attackers still could not derive the session key to recover the content of a specific session. Additionally, since the newly generated ECDH parameters for each session are irrelevant to the previous or future sessions, compromising the current session key will only affect the current session. The secrecy of the previous or future sessions will remain unaffected.

Device Anonymity: Since the SUCI is mandatory by the 5GSBA, the SUPIs of requesting UE devices are always concealed with ECIES. Hence, eavesdroppers cannot use authentication request messages to identify or trace any devices.

Protocol Attack Resistance: The 5GSBA outperforms most proposals, including the standardized 5G-AKA with the stronger resistibility to many attacks. For example, thanks to the aforementioned security properties, the 5GSBA prevents common attacks such as eavesdropping, location tracking, and man-in-the-middle (MITM) attacks. Moreover, it is highlighted that some critical attacks can be prevented including DoS/DDoS attacks, linkability attacks, UE impersonation attacks, rogue base station attacks, and replay attacks:

- **DoS Semantic Attack Prevention:** With the 5GSBA, semantic attacks exploiting the weaknesses of the protocol are impracticable, because the authentication request contains a timestamp and a one-time hash secret Y. Specifically, when the gNB receives an authentication request, it first checks if the timestamp is fresh. Then, it decrypts the SUCI and compares the Y with the records stored in the private blockchain. If the calculated hash value does not match, it will reject the session immediately. In this way, adversaries cannot hoard multiple sessions in the gNB by replaying the same authentication messages (i.e., the original SUCI in 5G-AKA). Additionally, since the authentication request involves only the computationally inexpensive ECIES and hash functions, adversaries cannot exhaust the computational resources of gNBs easily;
- **DDoS Flooding Attack Prevention:** By decentralizing the authentication tasks from the AUSF/UDM to all gNBs, the 5GSBA lessens the effects of flooding attacks that paralyze the network with authentic requests. In the 5G era, the inter-site distance (ISD) of gNBs is getting smaller, and the number of gNBs deployed keeps growing. Hence, the total computational power of gNBs is growing steadily, making it increasingly difficult to flood or even paralyze the entire 5G network. Furthermore, since the 5GSBA shifts the authentication tasks from AUSF/UDM to gNBs, it can also prevent the single-point-of-failure. One gNB failure or one AUSF failure will not affect the entire 5G network. Thus, DDoS attacks in the 5G authentication can be prevented, and the quality of service (QoS) across the 5G network can be maintained. The performance

analysis will show that the 5GSBA can serve much more incoming authentication requests than the existing centralized schemes;

- **Linkability Attack Prevention:** Unlike conventional symmetric key-based protocols such as 5G-AKA, the 5GSBA generates session keys using ECDH instead of the sequence number. Hence, the 5GSBA does not have the MAC failure or synchronization failure commonly found in symmetric key-based AKA protocols. Adversaries can no longer use these error messages as a loophole to trace a specific device;
- **UE Impersonation Attack Prevention:** By the conventional 5G-AKA protocol, users have to trust the network operator implicitly. Since the network operator owns a copy of users' symmetric keys and sequence numbers, insider attackers in the network can impersonate the UE by abusing these keys. By the 5GSBA, since the one-time hash secret is only stored at the USIM, there is no way for network operators to impersonate the UE using the data stored in the private blockchain. Hence, if the network operators cannot provide the one-time hash secret used in the authentication, users can simply deny all malicious behavior for that session;
- **Rogue Base Station Attack Prevention:** By the 5GSBA, since only the legitimate gNBs can decrypt the SUCI, rogue base stations cannot produce genuine authentication responses by generating the correct HMAC code σ. Thus, it can prevent UE from establishing connections with rogue base stations;
- **Replay Attack Prevention:** The authentication requests and responses by the 5GSBA are all tagged with a timestamp TS, and the HMAC key K_{hmac} should also work once only. Therefore, by checking the timestamp in both UEs and gNBs, replayed messages can be easily identified and discarded;
- **Battery Depletion Attack Prevention:** With the 5G-AKA, UE must respond to identity requests from the serving network by generating a fresh SUCI. If some rogue base stations frequently send the identity requests, the batteries of UE devices could deplete faster. In the 5GSBA, only UE can take the initiative to generate a fresh SUCI for authentication. Hence, the serving network cannot force the UE to create a fresh SUCI, and it prevents the battery depletion attacks effectively.

6. Performance Evaluation

Most UE has limited computational resources and battery life. Additionally, the network resources in the 5G network are always finite. Hence, a good authentication scheme should have low computational overhead, communication overhead, and energy consumption. In this section, we analyze the performance of the 5GSBA from these three aspects. Additionally, we evaluate its effectiveness against DoS/DDoS attacks using two simulation experiments. For the comparison, 5G-AKA [4] is selected because it is the 3GPP standardized protocol. SE-AKA [20] is chosen because it supports perfect forward secrecy with the similar security functions as the 5GSBA. Another blockchain-based 5G authentication protocol, BB-AKA 5G [10] is also included to show that the 5GSBA can achieve a better performance than the existing blockchain-based scheme.

6.1. Computational Overheads

6.1.1. Theoretical and Experimental Delays

A smaller computational delay means the protocol can run faster, so it is always preferred. In this subsection, computation overheads are evaluated through simulation experiments. All simulations are conducted on a computer with Intel® Core™ i5-3210M CPU @ 2.5 GHz CPU and 16 GB of RAMs. The Charm Crypto v0.5 with Python 3.7.5 is used to create and run the simulation. Charm Crypto [34] is a Python wrapper of Pairing-Based Cryptography (PBC) libraries. It allows the easy prototyping of cryptographic functions based on elliptic curve cryptosystems. Among all the selected protocols, each of them uses various cryptographic functions, and each of them has a unique key size requirement. To holistically and equally evaluate their performance at the same security level, we conformed to the recommendation from NIST [35] to use 256-bit equivalent key strength throughout the simulations. Hence, for all elliptic curve cryptography (ECC)-based operations, secp256k1 was chosen as the default elliptic curve. For HMAC operations, HMACSHA256 was selected. We ran each cryptographic function 2000 times to measure its average time. Moreover, to evaluate the data access delays incurred in blockchains, we followed the specifications in Section 4.2 to create a blockchain prototype. This blockchain had many 1-megabyte-sized blocks. Every block was stored as one file, and all transactions in the blockchain were indexed using Python Dictionary. To compare the performance difference between blockchain and traditional database, we also built another centralized database with MariaDB v10.4.14. Finally, Table 4 shows the experimental time of execution for the different functions. Then, all this information is combined with the theoretical computational times in Table 5 to find the experimental computational overheads shown in Figure 5.

Table 4. Experimental time for different functions.

Notation	Description	Time (ms)
$T_{ECDSA.sign}$	ECDSA Sign	0.7286
$T_{ECDSA.ver}$	ECDSA Verify	1.3442
$T_{ECIES.enc}$	ECIES Encryption	2.0572
$T_{ECIES.dec}$	ECIES Decryption	0.7851
T_{HMAC}	KDF/HMAC Calculation	0.0495
$T_{HMAC.ver}$	HMAC Verification	0.0281
$T_{ECDH.gen}$	ECDH Key Generation (1 Exp)	0.6945
$T_{ECDH.CK}$	ECDH Common Key	0.7099
T_{hash}	SHA256 Calculation Time (Hash Time)	0.0206
$T_{sym.enc}$	Symmetric Encryption	0.0925
T_{xor}	XOR Operation	0.0084
$T_{BC.read}$	Blockchain Transaction Read	0.2914
$T_{BC.write}$	Blockchain Transaction Write	0.0434
$T_{DB.read}$	Centralized Database Read	0.4956

Table 5. Theoretical computational overheads.

Protocol	Entity	Authentication Computational Overhead	Execution Time (ms)
5GSBA	UE	$T_{ECIES.enc} + T_{ECDH.gen} + T_{hash} + T_{HMAC} + T_{HMAC.ver}$	2.8499
	CN	$T_{ECIES.dec} + T_{ECDH.gen} + T_{hash} + T_{HMAC} + T_{HMAC.ver} + T_{BC.read} + T_{BC.write}$	1.9126
	Both	$T_{ECDH.CK}$	0.7099
5G-AKA	UE	$T_{ECIES.enc} + 2T_{sym.enc} + T_{xor} + 2T_{HMAC}$	2.3496
	CN	$T_{ECIES.dec} + 2T_{sym.enc} + T_{xor} + 2T_{HMAC} + 2T_{hash} + T_{DB.read}$	1.6143
SE-AKA	UE	$T_{ECIES.enc} + 4T_{HMAC} + T_{HMAC.ver} + T_{ECDH.gen} + T_{ECDH.CK}$	3.6877
	CN	$T_{ECIES.dec} + 2T_{HMAC.ver} + 3T_{HMAC} + T_{ECDH.gen} + T_{ECDH.CK} + T_{DB.read}$	2.8898
BB-AKA 5G	UE	$T_{ECDSA.sign} + 2T_{ECDSA.ver} + T_{ECDH.gen}$	4.1115
	CN	$3T_{ECDSA.sign} + 2T_{ECDSA.ver} + T_{ECDH.gen} + T_{BC.read} + T_{BC.write}$	5.9035
	Both	$T_{ECDH.CK} + T_{hash}$	0.7305

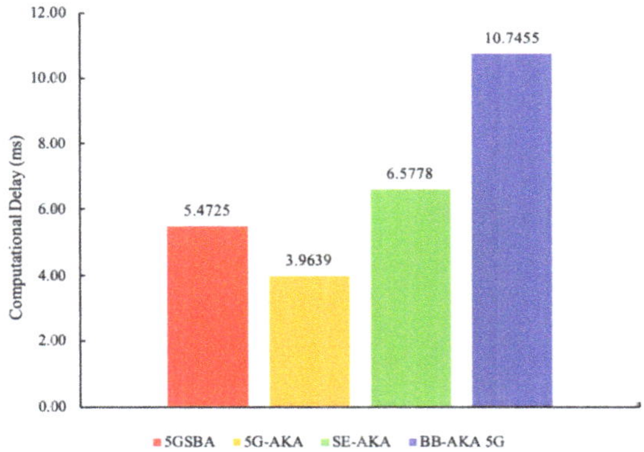

Figure 5. Experimental total computational delays.

It is noteworthy that, for all blockchain-based schemes, we omitted the blockchain indexing time in the computational overhead calculation. Blockchain indexing is a process for gNBs to categorize recently downloaded blocks. It can improve the blockchain random access speed to a constant time. Since indexing can be performed in the background parallelly at any time, it does not impose a significant negative effect on authentication. Hence, only the average blockchain transaction read time $T_{BC.read}$ and write time $T_{BC.write}$ were included in the calculation.

For the experimental computational delays shown in Figure 5, the 5GSBA was slower than the standardized 5G-AKA by 5.1377 − 3.4683 = 1.6694 milliseconds. This is because the 5GSBA employs asymmetric ECDH to replace the less secure symmetric key-based key derivation function (KDF) in the 5G-AKA [4]. By introducing this tiny computational overhead, the 5GSBA can ensure session key perfect forward secrecy (PFS), backward secrecy, and linkability attack prevention. On the other hand, the computational time of the 5GSBA and the SE-AKA [20] are comparable because both schemes employ ECDH, but the 5GSBA offers an enhanced security function in terms of DoS attack prevention and DDoS alleviation. The BB-AKA 5G [10] consumes much more time than the 5GSBA because the BB-AKA 5G needs two ECDSA signatures and verifications at the gNB and the AMF, while in contrast, the 5GSBA only requires one lightweight one-way hash secret verification in the gNB. Thus, the 5GSBA saves (10.4107 − 5.1377)/10.4107 = 50.6% of computational overhead compared to the BB-AKA 5G. Consequently, although the 5GSBA protocol has unavoidably introduced tiny computational overhead, it is still the most efficient protocol in terms of balancing between security and performance.

6.1.2. Average Delays under Unknown Attacks

Although the 5GSBA could resist several malicious attacks, as shown in the security analysis, it is possible that new unknown attacks in the 5G network could interrupt the authentication process. To evaluate the performance under unknown attacks, it was assumed that, for each step of the protocol, the network faced either known or unknown attacks. Specifically, there will be a probability that the protocol will encounter an unknown attack. If the incoming attack is known, the protocol should continue smoothly until completion. However, if the incoming attack is unknown, the protocol would be unavoidably interrupted. In this case, it must restart from the first step until it completes the last step. Based on the assumptions above, we created a simulation model on MATLAB 2020b. Using 1 million threads to run the protocol, we found the average execution time to complete the protocol under different unknown attacks, as shown in Figure 6. The results showed that although the 5GSBA had slightly higher delays (from 1.67 to 4.17 milliseconds) than those of the 5G-AKA [4] in all scenarios, its performance was still much better than another blockchain-based protocol of the BB-AKA 5G [10]. Considering that the 5GSBA ensured the best security among all protocols, we can still conclude that it can achieve a reasonable balance between high security and reasonable delays.

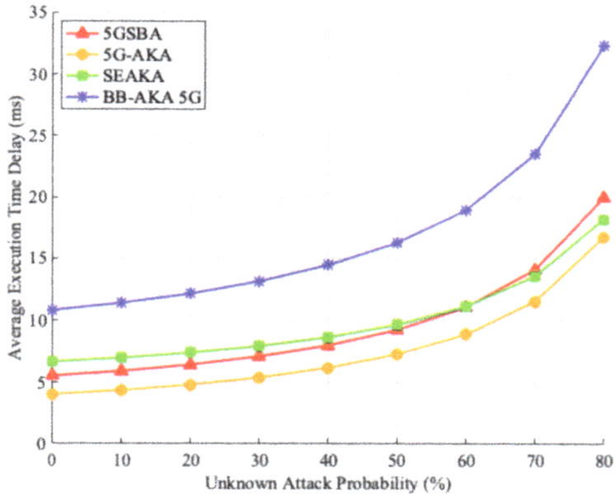

Figure 6. Average delays under unknown attacks.

6.2. Communication Overhead

As bandwidth resources in the 5G network are invaluable, smaller communication overhead is always preferred. Thus, we further evaluated the communication overhead in terms of bandwidth consumption and transmission overhead.

For the bandwidth consumption based on five different security levels suggested by NIST [35], the total message sizes of the related protocols are derived in Figure 7. It shows that our 5GSBA outperformed all other schemes by achieving the lowest bandwidth consumption. This is because the 5GSBA authenticates the incoming UE locally using the two-step protocol, which can save redundant forwarding messages.

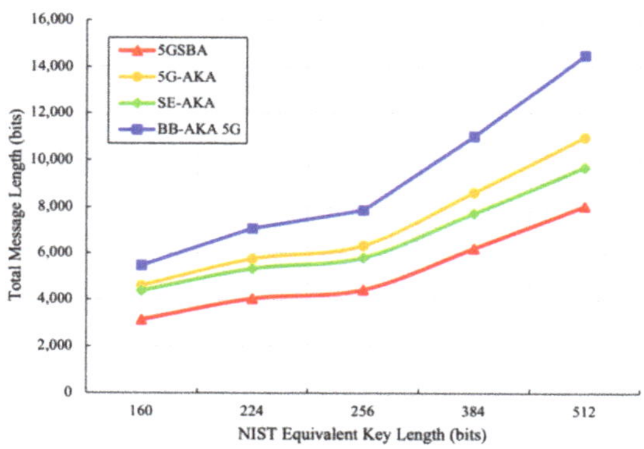

Figure 7. Total bandwidth consumption.

For communication delays there are two components: propagation delays and transmission delays. Propagation delay calculations consist of both wired and wireless connections. For the wired connections, we referred to the link between the gNB and the nearest 5GC functional entities. It is assumed that optical fiber is deployed for all wired connections, and the 5GC functional entities are geographically distant to the gNB. To simulate a real-life situation, we assumed the distance between the gNB and the nearest 5GC functional entity to be 1 km, and the distance between two functional entities within the 5GC to be less than 0.5 km. For the wireless connection, according to the dense urban 5G service requirement in TS 22.261 [36], we assumed that the inter-site distance (ISD) between two gNBs was 200 m, and the requesting UE was located to the edge of the coverage of the gNB. Therefore, the distance between the gNB and UE was about 100 m. The transmission delay calculation also consisted of wired and wireless connections. For the wired connections, due to the link of optical fiber, it was assumed that the uplink and downlink speed were both 1 Gbit/s. For the wireless connections, similarly, by following TS 22.261 [36], it was assumed that the uplink speed for UE was 50 Mbit/s and the downlink speed for UE was 300 Mbit/s. For the blockchain-based schemes, it was assumed that blockchain nodes could synchronize new blocks parallelly during the idle time of gNBs. To better reflect the extra communication overhead introduced by blockchains, we included the average delay of downloading one old transaction and uploading one new transaction into the calculation. Finally, by combining all the assumptions above, the sums of propagation and transmission delays are derived in Figure 8.

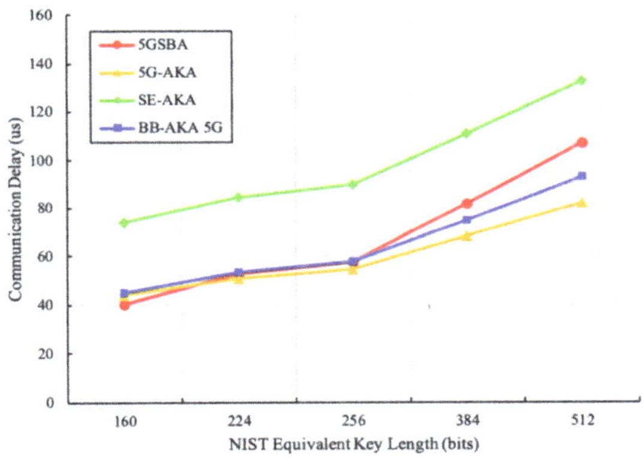

Figure 8. Total communication delays.

Figure 8 shows that the 5GSBA achieves similar total communication delays from 160-bit to 256-bit key length. Although the total communication delay of the 5GSBA was marginally higher than that of the 5G-AKA [4] from 384 bit to 512 bit, the 5GSBA still consumed much less time than that of the BB-AKA 5G [10]. This is because the 5GSBA saved some propagation delays by consuming the lengthy authentication requests locally in gNBs. Considering that the 256-bit ECC key was accepted in the NIST recommendation [35], the communication delays of the 5GSBA were very close to that of the 5G-AKA. Overall, the 5GSBA achieved the right balance between strong security and relatively low communication overhead.

6.3. Energy Consumption for UE

Since mobile devices have limited battery life, a security protocol with the strongest security and the least energy consumption is always preferable. To evaluate the energy consumption for UE two factors should be considered: data transmission energy and the energy for the execution of the cryptographic functions. For the data transmission energy, the data transfer power model in [37] is adopted to estimate the energy consumption. Thus, the energy cost for uplink transmission of the UE is in Equation (39) and the energy cost for the downlink is in Equation (40), where α_u= 438.39 mW/Mbps, α_d = 51.97 mW/Mbps, β = 1288.04 mW, t_{udr} is the uplink throughput, t_{ddr} is the downlink throughput, t_{ul} is the transmission time spent on the uplink, and t_{dl} is the transmission time spent on the downlink. We further assume that the uplink and downlink throughput follow the dense urban 5G service requirement in TS 22.261 [36], so that t_{udr} = 50 Mbps and t_{ddr} = 300 Mbps.

$$E_{ul} = (\alpha_u t_{udr} - \beta) \cdot t_{ul} \tag{39}$$

$$E_{dl} = (\alpha_d t_{ddr} + \beta) \cdot t_{dl} \tag{40}$$

On the other hand, for the energy consumption for the execution of cryptographic functions, the measure of energy cost approximation in [38] is taken. In [38], all experiments were conducted using a battery-powered Compaq iPAQ H3670 PDA. It was equipped with an Intel SA-1110 StrongARM processor clocked at 206 MHz, 64 MB of RAM, and 16 MB of FlashROM. Moreover, the energy cost of ECDH public key generation $E_{ECDH.gen}$ was 276.7 mJ, the ECDH common key derivation $E_{ECDH.ck}$ was 163.5 mJ, the ECDSA signature generation $E_{ECDSA.sign}$ was 134.2 mJ, and the ECDSA signature validation $E_{ECDSA.ver}$ was 196.23 mJ. For the energy cost of symmetric key-based operations, the symmetric key-based encryption $E_{sym.}$ was $9.92 + 2.29l$ uJ, where l was the number of bytes of the cleartext. The HMAC operation E_{HMAC} was 1.16 mJ, and hash operation E_{hash} was 0.76 mJ. For the ECIES operations, since it was a hybrid encryption combining both CDH problems and a key encapsulation mechanism (KEM), the cost of the ECIES encryption was estimated as $E_{ECDH.gen} + E_{ECDH.ck} + E_{sym} = 440.20992 + 0.00229l$ mJ.

Finally, by combining all parameters above with the unknown attack model stated in Section 6.1.2, the average UE energy consumption for different protocols is simulated in Figure 9. It shows that all the schemes providing perfect forward and backward secrecy, including the 5GSBA [4], SE-AKA [20], and BB-AKA 5G [10], consume a similar amount of energy in a situation without any unknown attacks. When the unknown attack probability increases, the 5GSBA unavoidably uses more energy because of the increased SUCI generation. However, the 5GSBA still consumes less energy than the BB-AKA 5G. On the other hand, although the 5G-AKA uses the least energy, its power-saving symmetric key-based cryptosystem makes it vulnerable to many other known attacks.

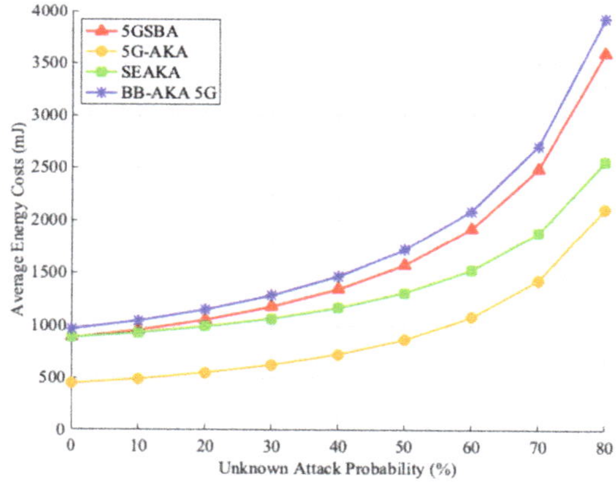

Figure 9. Energy consumption for UE.

6.4. *Resistivity to DoS and DDoS Attacks*

To show that our 5GSBA can fight against DoS and DDoS attacks, we further design two simulations to compare all related protocols. All simulation results show that the 5GSBA can achieve a high level of security functionality with reasonable delays.

6.4.1. Average Delays under DDoS Flooding Attacks

One of the outstanding features of the 5GSBA is the alleviation of the flooding of DDoS attacks. The 5GSBA, since all authentication tasks are distributed at various gNBs, avoids congestion and processing at the single authentication entity AUSF. Hence, to illustrate the performance of the protocols under the flooding of many authentication requests, we can model all centralized protocols, such as 5G-AKA [4] and SE-AKA [20], as M/M/1 queuing models, while the blockchain-based protocol of the 5GSBA can be modeled as a M/M/c queuing model (c is the number of gNBs). For simplicity, it is assumed that the arrival and location of UE devices follow Poisson distribution, such that all devices within 1 km^2 are located uniformly and can initiate authentication requests randomly. For the centralized protocols, since the AUSF in the 5GC should be a computationally powerful server, it is assumed that they run on the server 80% faster than the performance listed in Table 2. For blockchain-based protocols, as the computational resources for each gNB are limited, it is assumed that its computational time is the same as that in Table 2. Furthermore, one AUSF only serves the gNBs within the 1km^2 area. and the ISD of every gNB is 200 m according to the TS 22.261 5G dense urban service requirement [36]. Consequently, each server needs to process the requests from $1\ km^2/\pi 0.1^2 = 31.8 \approx 32$ gNBs (i.e., c = 32).

Finally, the average time to complete the execution of the protocol is shown in Figure 10. It shows that although the decentralized 5GSBA takes slightly longer time than the other centralized schemes when the network traffic is low, as the number of requests increases, the 5GSBA outperforms the others by maintaining an almost constant delay. In other words, the 5GSBA can serve more authentication requests with the same amount of delay than the centralized schemes. In future, as the number of gNBs keeps increasing, it is expected that the 5GSBA will be able to perform even better by accommodating more users. Therefore, this shows that the 5GSBA is more robust in terms of serving more users, and thus is more effective in lessening the effects of flooding attacks. Moreover, since the 5GSBA can accommodate more requests, this also gives network operators more time to detect DDoS attacks and deploy countermeasures. For example, network operators can use the time to analyze network traffic and block potential DDoS attackers without impacting overall service quality. By doing so, attackers would find it more challenging to perform a successful DDoS attack in the 5G network.

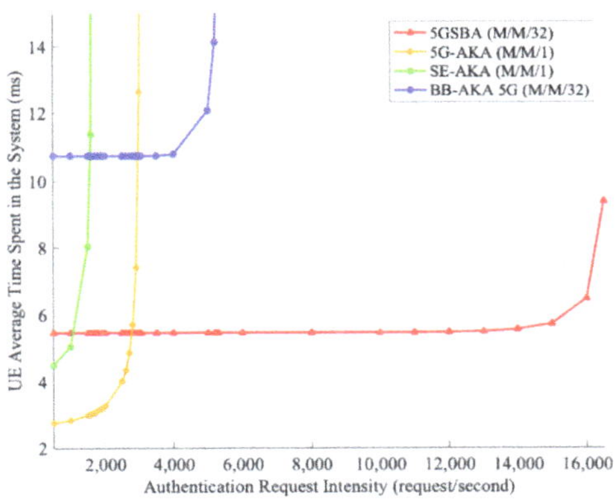

Figure 10. Average delay under many attacks.

6.4.2. Average Successful Authentications under DoS Attacks

The 5GSBA can alleviate the effect of DoS attacks by verifying the timestamp and the HMAC code in the authentication request. To prove that the 5GSBA can provide the best DoS attack resistivity among all related schemes, a simulation model was built in MATLAB 2020b with the following assumptions. There are 1000 mobile devices sending authentication requests to a server at every one millisecond, which can hold—at most—1000 sessions simultaneously without any performance degradation. Then, for each authentication request, there could be a percentage chance that it is a fabricated DoS request. In case the request is legitimate, the server will complete the protocol operation normally following the protocol specifications. However, if the request is a DoS attack, either of the following results could happen. If the server can identify it as a DoS request, it will terminate the session immediately to accept another new request. If the server cannot identify it, the session will hold until it reaches the protocol expiry time at 3 s, which is a common default value for RADIUS server timeout. This simulation is run for 60 s. The number of successful authentications of different schemes is recorded in Figure 11.

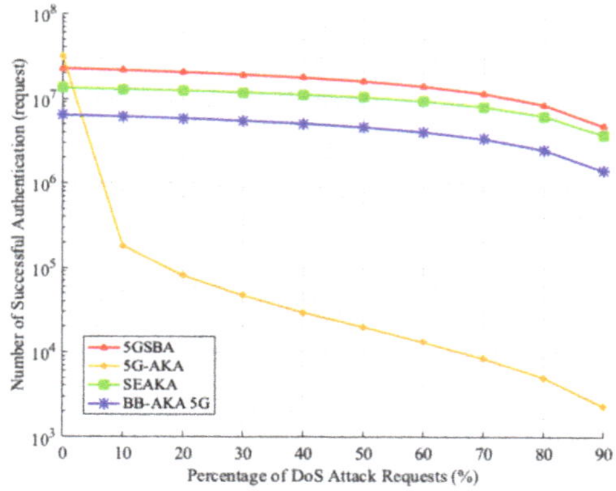

Figure 11. Successful authentications under DoS attacks.

Figure 11 shows that when there is no DoS attack in the network, the 5G-AKA performs the best among all schemes with the highest number of successful authentications. However, when the percentage of DoS attacks increases, the performance of the 5G-AKA [4] drops drastically, while the 5GSBA maintains the highest number of successful authentications. This is because the 5G-AKA has no way to identify if the authentication requests are fabricated, so the server wastes some sessions for holding until their expiry. On the other hand, the SE-AKA [20] can also provide DoS attack prevention, similar to our 5GSBA. However, the 5GSBA runs faster, so it has a higher number of successful authentications.

6.5. Discussion of the Results

The performance evaluation shows that when there are DoS or DDoS attacks in the 5G networks, the 5GSBA is the better protocol with a higher performance in terms of a stable authentication delay and a high authentication success rate. The only limitation of the 5GSBA would be the slightly added overhead in computational overhead, communication overhead, and energy consumption compared with the 3GPP 5G-AKA [4]. However, given that there will be an exponential growth of 5G mobile devices in the future, the risk of DoS and DDoS attacks will become more prominent. Thus, we believe that these imperceptible overheads are justifiable to safeguard future 5G networks.

7. Conclusions

In 5G networks, DoS and DDoS attacks have become a critical issue due to the increasing number of mobile devices. To ensure the robustness and security of the 5G network, we have proposed a Secure Blockchain-based 5G Authentication and Key Agreement (5GSBA) protocol in this paper. The 5GSBA protocol holds many security features that the existing 3GPP 5G-AKA scheme fails to achieve, including perfect forward secrecy, device anonymity, and most importantly, the resistivity to DoS and DDoS attacks. By the decentralization of authentication functions with blockchain technology, the 5GSBA delivers the best quality of service (QoS) among all schemes in relation to DoS and DDoS attacks. Our performance evaluation also demonstrates that the overhead added by 5GSBA is imperceptible compared to other existing solutions. Therefore, we believe that the 5GSBA protocol is ideal for balancing strong security functionality and high performance.

Author Contributions: Conceptualization, M.C.C. and M.M.; data curation, M.C.C.; formal analysis, M.C.C.; investigation, M.C.C.; methodology, M.C.C.; software, M.C.C.; visualization, M.C.C.; writing—original draft, M.C.C.; funding acquisition, M.M.; project administration, M.M.; project administration,

M.M.; resources, M.M.; supervision, M.M.; validation, M.M.; writing—review and editing, M.M. All authors have read and agreed to the published version of the manuscript.

Funding: This research received no external funding.

Informed Consent Statement: Not applicable.

Conflicts of Interest: The authors declare no conflict of interest.

References

1. Agiwal, M.; Roy, A.; Saxena, N. Next generation 5G wireless networks: A comprehensive survey. *IEEE Commun. Surv. Tutor.* **2016**, *18*, 1617–1655. [CrossRef]
2. 3GPP. Security Architecture and Procedures for 5G System (Release 16.3.0); TS 33.501. 2018. Available online: https://www.3gpp.org/ftp/Specs/archive/33_series/33.501/33501-g30.zip (accessed on 11 June 2022).
3. Cao, J.; Ma, M.; Li, H.; Ma, R.; Sun, Y.; Yu, P.; Xiong, L. A survey on security aspects for 3GPP 5G networks. *IEEE Commun. Surv. Tutor.* **2020**, *22*, 170–195. [CrossRef]
4. Basin, D.; Radomirovic, S.; Dreier, J.; Sasse, R.; Hirschi, L.; Stettler, V. A formal analysis of 5g authentication. In Proceedings of the 2018 ACM SIGSAC Conference on Computer and Communications Security, New York, NY, USA, 15 October 2018; pp. 1383–1396. [CrossRef]
5. Liu, F.; Peng, J.; Zuo, M. Toward a Secure Access to 5G Network. In Proceedings of the 17th IEEE International Conference on Trust, Security and Privacy in Computing and Communications and 12th IEEE International Conference on Big Data Science and Engineering, Trustcom/BigDataSE 2018, New York, NY, USA, 1–3 August 2018; pp. 1121–1128. [CrossRef]
6. Tahir, M.; Habaebi, M.H.; Dabbagh, M.; Mughees, A.; Ahad, A.; Ahmed, K.I. A Review on Application of Blockchain in 5G and beyond Networks: Taxonomy, Field-Trials, Challenges and Opportunities. *IEEE Access* **2020**, *8*, 115876–115904. [CrossRef]
7. 3GPP. 3rd Generation Partnership Project; Technical Specification Group Services and System Aspects; Study on authentication enhancements in 5G System; (Release 17); 3GPP TR 33.846 V0.7.0. 2020. Available online: https://www.3gpp.org/ftp/Specs/archive/33_series/33.846/33846-070.zip (accessed on 11 June 2022).
8. Chaer, A.; Salah, K.; Lima, C.; Ray, P.P.; Sheltami, T. Blockchain for 5G: Opportunities and challenges. In Proceedings of the 2019 IEEE Globecom Workshops, GC Wkshps, Waikoloa, HI, USA, 9–13 December 2019. [CrossRef]
9. Yang, H.; Zheng, H.; Zhang, J.; Wu, Y.; Lee, Y.; Ji, Y. Blockchain-based trusted authentication in cloud radio over fiber network for 5G. In Proceedings of the 16th International Conference on Optical Communications and Networks (ICOCN), Wuzhen, China, 7–10 August 2017; pp. 1–3.
10. Haddad, Z.; Fouda, M.M.; Mahmoud, M.; Abdallah, M. Blockchain-based Authentication for 5G Networks. In Proceedings of the 2020 IEEE International Conference on Informatics, IoT, and Enabling Technologies, ICIoT 2020, Doha, Qatar, 2–5 February 2020; pp. 189–194. [CrossRef]
11. Haddad, Z.; Baza, M.; Mahmoud, M.M.E.A.; Alasmary, W.; Alsolami, F. Secure and Efficient AKA Scheme and Uniform Handover Protocol for 5G Network Using Blockchain. *IEEE Open J. Commun. Soc.* **2021**, *2*, 2616–2627. [CrossRef]
12. Xu, J.; Xue, K.; Tian, H.; Hong, J.; Wei, D.S.L.; Hong, P. An Identity Management and Authentication Scheme Based on Redactable Blockchain for Mobile Networks. *IEEE Trans. Veh. Technol.* **2020**, *69*, 6688–6698. [CrossRef]
13. Jia, X.; Hu, N.; Yin, S.; Zhao, Y.; Zhang, C.; Cheng, X. A2Chain: A Blockchain-Based Decentralized Authentication Scheme for 5G-Enabled IoT. *Mob. Inf. Syst.* **2020**, *2020*, 8889192. [CrossRef]
14. Liu, J.; Huang, C.T. Efficient and Trustworthy Authentication in 5G Networks Based on Blockchain. In Proceedings of the International Conference on Computer Communications and Networks, ICCCN, Las Vegas, NV, USA, 19–22 July 2021; pp. 1–6. [CrossRef]
15. Zhang, Y.; Deng, R.; Bertino, E.; Zheng, D. Robust and Universal Seamless Handover Authentication in 5G HetNets. *IEEE Trans. Dependable Secur. Comput.* **2019**, *18*, 858–874. [CrossRef]
16. Chen, Z.; Chen, S.; Xu, H.; Hu, B. A security authentication scheme of 5G ultra-dense network based on block chain. *IEEE Access* **2018**, *6*, 55372–55379. [CrossRef]
17. Sharma, V.; You, I.; Palmieri, F.; Jayakody, D.N.K.; Li, J. Secure and Energy-Efficient Handover in Fog Networks Using Blockchain-Based DMM. *IEEE Commun. Mag.* **2018**, *56*, 22–31. [CrossRef]
18. Yu, F.; Ma, M.; Li, X. A Blockchain-Assisted Seamless Handover Authentication for V2I Communication in 5G Wireless Networks. In Proceedings of the IEEE International Conference on Communications, Montreal, QC, Canada, 14–23 June 2021; pp. 1–5. [CrossRef]
19. Chow, M.C.; Ma, M. A Blockchain-Enabled 5G Authentication Scheme Against DoS Attacks. In Proceedings of the 2020 International Conference on Electronics, Communications and Information Technology, Coimbatore, India, 5–7 November 2020; IOP Publishing: Bristol, UK, 2020; pp. 3–8.
20. Lai, C.; Li, H.; Lu, R.; Shen, X. SE-AKA: A secure and efficient group authentication and key agreement protocol for LTE networks. *Comput. Netw.* **2013**, *57*, 3492–3510. [CrossRef]
21. Cao, J.; Ma, M.; Li, H. LPPA: Lightweight privacy-preservation access authentication scheme for massive devices in fifth Generation (5G) cellular networks. *Int. J. Commun. Syst.* **2019**, *32*, e3860. [CrossRef]

22. Leu, F.Y.; Tsai, K.L.; Susanto, H.; Gu, C.Y.; You, I. A Fault Tolerant Mechanism for UE Authentication in 5G Networks. *Mob. Netw. Appl.* **2020**, *26*, 1650–1667. [CrossRef]
23. Yan, X.; Ma, M. A lightweight and secure handover authentication scheme for 5G network using neighbour base stations. *J. Netw. Comput. Appl.* **2021**, *193*, 103204. [CrossRef]
24. Braeken, A.; Liyanage, M.; Kumar, P.; Murphy, J. Novel 5G Authentication Protocol to Improve the Resistance against Active Attacks and Malicious Serving Networks. *IEEE Access* **2019**, *7*, 64040–64052. [CrossRef]
25. Gharsallah, I.; Smaoui, S.; Zarai, F. A secure efficient and lightweight authentication protocol for 5G cellular networks: SEL-AKA. In Proceedings of the 15th International Wireless Communications and Mobile Computing Conference, IWCMC 2019, Tangier, Morocco, 24–28 June 2019; pp. 1311–1316. [CrossRef]
26. Hu, X.; Liu, C.; Liu, S.; Cheng, X. A security enhanced 5g authentication scheme for insecure channel. *IEICE Trans. Inf. Syst.* **2020**, *103*, 711–713. [CrossRef]
27. Xiao, Y.; Wu, Y. 5G-IPAKA: An Improved Primary Authentication and Key Agreement Protocol for 5G Networks. *Information* **2022**, *13*, 125. [CrossRef]
28. Ramezan, G.; Abdelnasser, A.; Liu, B.; Jiang, W.; Yang, F. EAP-ZKP: A Zero-Knowledge Proof based Authentication Protocol to Prevent DDoS Attacks at the Edge in beyond 5G. In Proceedings of the 2021 IEEE 4th 5G World Forum, 5GWF 2021, Montreal, QC, Canada, 13–15 October 2021; pp. 259–264. [CrossRef]
29. 3GPP. System Architecture for the 5G System (5GS) (Release 15.7.0); TS 23.501. p. 353. 2019. Available online: https://www.etsi.org/deliver/etsi_ts/123500_123599/123501/15.07.00_60/ts_123501v150700p.pdf (accessed on 11 June 2022).
30. Dolev, D.; Yao, A.C. On the Security of Public Key Protocols. *IEEE Trans. Inf. Theory* **1983**, *29*, 198–208. [CrossRef]
31. Castro, M. Practical Byzantine Fault Tolerance. In Proceedings of the Third Symposium on Operating Systems Design OSDI '99, New Orleans, LA, USA, 22–25 February 2001; pp. 1–172. Available online: http://pmg.csail.mit.edu/papers/osdi99.pdf (accessed on 11 June 2022).
32. Burrows, M.; Abadi, M.; Needham, R. A logic of Authentication. *ACM Trans. Comput. Syst.* **1990**, *8*, 18–36. [CrossRef]
33. Cremers, C. The Scyther Tool. CISPA Helmholtz Center for Information Security. 2020. Available online: https://people.cispa.io/cas.cremers/scyther/ (accessed on 18 March 2020).
34. Akinyele, J.A.; Garman, C.; Miers, I.; Pagano, M.W.; Rushanan, M.; Green, M.; Rubin, A.D. Charm: A framework for rapidly prototyping cryptosystems. *J. Cryptogr. Eng.* **2013**, *3*, 111–128. [CrossRef]
35. Barker, E. *Recommendation for Key Management—Part 1: General*; NIST Special Publication 800-57; National Institute of Standards and Technology, Technology Administration: Gaithersburg, MD, USA, 2016; pp. 1–142. [CrossRef]
36. 3GPP. Technical Specification Group Services and System Aspects; Service requirements for the 5G system; Stage 1 (Release 17); TS 22.261. 2019. Available online: https://www.3gpp.org/ftp/Specs/archive/22_series/22.261/22261-ha0.zip (accessed on 11 June 2022).
37. Huang, J.; Qian, F.; Gerber, A.; Mao, Z.M.; Sen, S.; Spatscheck, O. A close examination of performance and power characteristics of 4G LTE networks. In Proceedings of the 10th International Conference on Mobile Systems, Applications, and Services, Low Wood Bay Lake District, UK, 25 June 2012; pp. 225–238. [CrossRef]
38. Potlapally, N.R.; Ravi, S.; Raghunathan, A.; Jha, N.K. A study of the energy consumption characteristics of cryptographic algorithms and security protocols. *IEEE Trans. Mob. Comput.* **2006**, *5*, 128–143. [CrossRef]

sensors

MDPI

Article

An Optimization Model for Appraising Intrusion-Detection Systems for Network Security Communications: Applications, Challenges, and Solutions

Mohamed Abdel-Basset [1], Abduallah Gamal [1], Karam M. Sallam [2,*], Ibrahim Elgendi [2], Kumudu Munasinghe [2] and Abbas Jamalipour [3]

1 Faculty of Computers and Informatics, Zagazig University, Zagazig 44519, Egypt; mohamedbasset@ieee.org (M.A.-B.); abduallahgamal@fci.zu.edu.eg (A.G.)
2 School of IT and Systems, University of Canberra, Canberra ACT 2601, Australia; ibrahim.elgendi@canberra.edu.au (I.E.); kumudu.munasinghe@canberra.edu.au (K.M.)
3 School of Electrical and Information Engineering, The University of Sydney, Sydney, NSW 2006, Australia; abbas.jamalipour@sydney.edu.au
* Correspondence: karam.sallam@canberra.edu.au

Abstract: Cyber-attacks are getting increasingly complex, and as a result, the functional concerns of intrusion-detection systems (IDSs) are becoming increasingly difficult to resolve. The credibility of security services, such as privacy preservation, authenticity, and accessibility, may be jeopardized if breaches are not detected. Different organizations currently utilize a variety of tactics, strategies, and technology to protect the systems' credibility in order to combat these dangers. Safeguarding approaches include establishing rules and procedures, developing user awareness, deploying firewall and verification systems, regulating system access, and forming computer-issue management groups. The effectiveness of intrusion-detection systems is not sufficiently recognized. IDS is used in businesses to examine possibly harmful tendencies occurring in technological environments. Determining an effective IDS is a complex task for organizations that require consideration of many key criteria and their sub-aspects. To deal with these multiple and interrelated criteria and their sub-aspects, a multi-criteria decision-making (MCMD) approach was applied. These criteria and their sub-aspects can also include some ambiguity and uncertainty, and thus they were treated using q-rung orthopair fuzzy sets (q-ROFS) and q-rung orthopair fuzzy numbers (q-ROFNs). Additionally, the problem of combining expert and specialist opinions was dealt with using the q-rung orthopair fuzzy weighted geometric (q-ROFWG). Initially, the entropy method was applied to assess the priorities of the key criteria and their sub-aspects. Then, the combined compromised solution (CoCoSo) method was applied to evaluate six IDSs according to their effectiveness and reliability. Afterward, comparative and sensitivity analyses were performed to confirm the stability, reliability, and performance of the proposed approach. The findings indicate that most of the IDSs appear to be systems with high potential. According to the results, Suricata is the best IDS that relies on multi-threading performance.

Keywords: cyber-attacks; intrusion-detection system; MCDM; q-rung orthopair fuzzy sets; q-ROFWG

Citation: Abdel-Basset, M.; Gamal, A.; Sallam, K.M.; Elgendi, I.; Munasinghe, K.; Jamalipour, A. An Optimization Model for Appraising Intrusion-Detection Systems for Network Security Communications: Applications, Challenges, and Solutions. *Sensors* **2022**, *22*, 4123. https://doi.org/10.3390/s22114123

Academic Editor: Peter Han Joo Chong

Received: 13 April 2022
Accepted: 25 May 2022
Published: 29 May 2022

Publisher's Note: MDPI stays neutral with regard to jurisdictional claims in published maps and institutional affiliations.

1. Introduction

The continuous development of computer systems has led to the increasing dependence of companies, organizations, and people on computer networks in performing their functions and offering their services in modern ways [1]. However, at the same time, it has become vulnerable to penetration by attackers with the aim of making illegal gains by exploiting some security vulnerabilities, which led to an increase in interest in issues of protection and security of these systems. Today there are many methods used within this field, and there are many intrusion-detection systems (IDSs) available [2]. IDSs are a necessity for the stability of an organization's normal system performance. In this regard, traditional intrusion-detection techniques are highly unrewarding and ineffective due to the multiplicity of attack methods and their different forms. Among the traditional methods used previously are obfuscation, transformation, and polymorphism techniques, which lead to malware resistance [3]. Despite the prominent role it plays, it still has some shortcomings. Therefore, there was a need to continue conducting research on intrusion-detection systems in order to reach an optimal structure that achieves a high protection rate.

The internet changed the concept of computing as we know it. The possibilities and opportunities available became unlimited, and with it, the risks and opportunities for breakthroughs increased. Computer security primarily focuses on protecting a specific source or valuable data and information within a single computer device. Security is defined as the reaction taken to security threats resulting from a harmful act by some people. The value of the data can be violated in three ways: privacy, integrity, and availability of information [4]. Computer protection is generally referred to by the term CIA, which is represented by the following three concepts:

- Confidentiality: Preventing unauthorized persons from disclosing or accessing information; i.e., accessing information only by authorized persons.
- Integrity: Maintaining information integrity by preventing unauthorized modification.
- Availability: It is the ability of a computer to work and provide the resources and services expected of it to legitimate people upon request.

Network security includes all actions or activities taken by organizations and companies in order to protect resources and ensure the integrity and continuity of operations across networks [5]. Security policies also define the permissions available to users in the way they use network components and resources. In order to build an effective network-protection strategy, all potential security threats must be identified, and then the most effective set of tools to combat them must be selected [6]. Preventing all exploits for vulnerabilities in networks and systems is not possible [7]. Network protection is achieved through the use of a set of components at several levels, with the aim of protecting organizations from internal attacks and external attacks as much as possible. A firewall is a component that achieves the most basic level of protection but is not sufficient on its own. Designing and implementing a completely secure system is very difficult in practice, but it is possible to detect intrusions and take appropriate measures to protect against them. This is what the IDS basically does, as it is used as an alert system, within the security and protection system, that gives an alert when it detects an attempt by someone to penetrate the computer system or network [8]. As a result, IDSs are important in a network security solution. The primary goal of IDSs is to detect an intrusion while it is occurring rather than after it has ended, and then alert the person responsible for the problem by sending an email or setting off an alert. It must be able to take any action to minimize harm to the system due to the hack. The second goal is to collect data from the system, record all important events, and determine the source of the attack, and these data are used for legal purposes as evidence or proof against the attacker.

IDSs must be placed in strategic places so that they are able to see network traffic in order to analyze it and thus achieve the maximum benefit from it [9]. In this regard, several ways to classify IDSs are described using different analysis and control methods. The most common way to classify intrusion-detection systems is to group them according to the location of the information source. Basic information sources are network packets captured from a network backbone or local network segments, operating systems, and critical files. Intrusion-detection systems can be classified into host-based detection-intrusion systems, network-based intrusion-detection systems, and hybrid systems. On the user's computer, a host-based intrusion-detection system (HIDS) is installed. HIDSs operate on information collected from within a single computer system [10]. HIDSs employ monitoring sensors, also called clients, on each host to be monitored. In general, the most common forms of information sources for HIDSs are operating system audit logs, system logs, and critical system files [10]. The customer checks these sources for unauthorized changes or patterns of suspicious activity. This allows HIDSs to reliably analyze activities, accurately identifying which users and which processes are participating in a specific attack on the operating system. The most common form of IDSs is network-based intrusion-detection systems (NIDSs) [11]; also, most companies and organizations are often supported by NIDSs along with firewalls. These systems detect attacks by capturing and analyzing network packets by listening to network segments or switches [11]. In this case, the system is placed on an entire network segment and not on a single device within the network, or it is placed to monitor a gateway on the switch. Thus, it can monitor all mobile packets between groups of computers connected to the network, by matching one or more packets with the database of signatures of attacks, or by analyzing the traffic to detect anomalies. NIDSs can be taken advantage of by placing it outside of firewalls, thus alerting the responsible person to incoming packets that might circumvent the firewall. Both HIDSs and NIDSs have strengths and benefits that complement each other [12]. Figure 1 introduces the general architecture of an HIDS and NIDS. The next generation of IDSs must combine the two technologies in order to improve the network's resistance to attacks and abuse. In addition, they should enhance security policy and provide greater flexibility in application and deployment options. A hybrid IDS is a mixture of an HIDS and NIDS. It provides a combination of the strengths of the two methods. Their modus operandi varies from product to product, making it difficult to define and determine hybrid intrusion systems in a more accurate manner.

Afterward, detection methods are the basis of intrusion-detection techniques, which are the engine in detecting the malicious activities of the information source. Detection methods analyze the information they monitor and trigger alerts if malicious traffic is detected. Accordingly, IDSs can be categorized, according to the detection methods used, into anomaly-based intrusion-detection systems (AIDSs) [13] and signature-based intrusion-detection systems (SIDSs) [14]. In this regard, AIDSs operate on the assumption that malicious events are different from normal actions, and thus the differences are sought to detect the attack. These systems constitute profiles of historical data collected during a period of normal operation. It then collects event data to determine when the monitored activity deviates from normal behavior and triggers an alarm. SIDSs also are called signature-based detection because alerts are generated based on the signatures of specific attacks. This type of signature attack involves specific traffic or activity based on known hacking activity. It is also called misuse-based detection. The basic premise is the model that has been written to describe bad behavior, after which the system compares the sequence of information with this model to decide what is normal and what is malicious. These systems are accurate and emit fewer false alarms, but they do not detect a hack unless there is a predetermined model for it.

Nowadays, Internet networks are vulnerable to a wide range of threats and attacks, such as impersonation, privilege breaches, data loss, altered and fraudulent data units, and denial of connections [15]. Therefore, IDSs have an essential task to protect the normal system performance of an organization. It is, therefore, necessary to add new security requirements and additional networking measures to the network security requirements [16]. IDSs must constantly change and adapt to all these new threats and assault technologies. Therefore, the process of determining the best IDS for network protection, threat warning, and cyber-attacks is a very difficult task in light of the various criteria on which an IDS is developed. Thus, IDSs should not be chosen to quickly secure the network without a thorough understanding of the technology, solutions, and potential consequences.

In this study, a set of criteria were adopted to evaluate IDSs according to previous studies and expert opinions. The criteria that have been adopted were divided into four basic criteria—protected system, audit source location, alerts, and types—and each main criterion includes several sub-aspects. The set of sub-aspects are as follows: HIDS, NIDS, hybrids, host log files, network packets, application log files, IDS sensor application, network, host, open-source, closed source, and freeware. In order to solve such complex problems related to the evaluation of IDSs, multi-criteria decision-making (MCDM) has been proven to be one of the best tools for the effective evaluation of IDS [16]. MCDM is popular in complex problems because it enables the decision-maker to take care of all the available criteria and take an appropriate decision as per the priority [17]. Since the ideal choice is governed by multiple criteria, a good decision-maker, in certain situations, may look for criteria of high impact on which to focus.

Consequently, due to the assessment of IDSs under multiple criteria and a pluralistic viewpoint, the assessment process is tainted by ambiguity and uncertainty, which is difficult to deal with in real numbers. Hence, the q-rung orthopair fuzzy sets (q-ROFSs) theory has been applied to deal with such complex problems [18]. The q-ROFS proved to be effective in solving ambiguous and uncertain problems as it came as a generalization of the intuitionistic fuzzy sets (IFSs) [19] and Pythagorean fuzzy sets (PFSs) theories [20].

Finally, to deal with the problem of evaluating the effectiveness of IDSs, a hybrid approach consisting of two multi-criteria decision-making methods, the entropy method [21] and the combined compromised solution (CoCoSo) method [22], was adopted. The proposed hybrid approach is presented under the q-rung orthopair fuzzy environment and by utilizing the q-rung orthopair fuzzy numbers (q-ROFNs) numbers. Firstly, the entropy method was adopted to evaluate the main and sub-aspects and to determine the final weights. Secondly, the CoCoSo method was applied to evaluate the available alternatives and determine the best alternative.

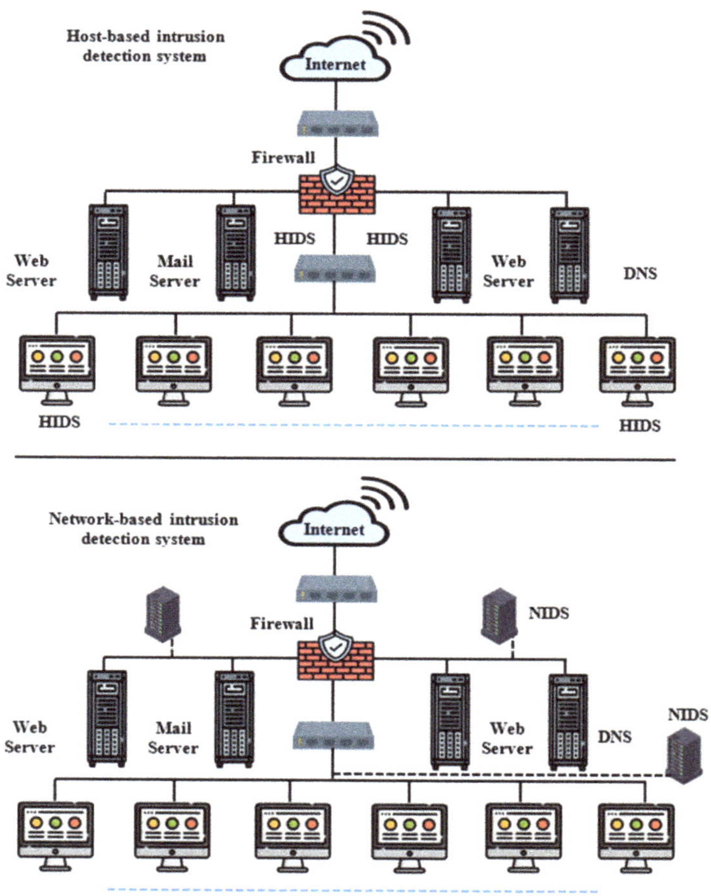

Figure 1. Comparison of the HIDS and NIDS structure [23].

The main contributions of this study can be listed as follows: (i) a novel hybrid MCDM approach is proposed for the evaluation of the IDS; (ii) this hybrid MCDM approach is named q-ROF entropy-CoCoSo, which give assessments of subjective and impartial expert insights; (iii) the q-ROFSs method is conducted to handle the uncertainty in experts' evaluations; (iv) a q-ROF entropy is utilized to compute the criteria weights; and (v) a q-ROF CoCoSo is suggested to evaluate the selected IDSs.

The article is organized as follows: Section 2 highlights the IDS and insights into MCDM; Section 3 introduces some preliminaries of q-ROFS and the proposed research approach; Section 4 illustrates the application of the MCDM approach in evaluating the IDS and discussion; and Section 5 contains the conclusions.

2. Background Information

This section provides basic knowledge about IDSs to enable a deeper understanding of this topic; also, some basic information related to MCDM is presented. Afterward, some studies related to q-ROF theory are introduced. Alyami et al. presented a study based on a hybrid MCDM approach consisting of the fuzzy analytical hierarchy process (AHP) and the fuzzy technique for order performance by similarity to ideal solution (TOPSIS) to evaluate the effectiveness of the IDSs [16]. In their study, they used four main criteria and thirteen sub-aspects to evaluate five IDSs. Their results indicate that Suricata is the most effective IDS. Their results also indicate that most of the IDSs that were evaluated in the study are effective and close in their results. Abushark et al. developed a study to evaluate the optimization of machine learning-based IDSs using a hybrid MCDM approach that comprises the AHP and TOPSIS in a fuzzy environment [24]. Their findings aim to identify attributes related to cyber security, allowing the design of more effective and efficient IDSs. Al-Harbi et al. presented a study for an optimal evaluation of machine learning-based IDSs using a hybrid MCDM model that includes AHP and TOPSIS under hesitant fuzzy conditions [8]. Their findings aim to identify features related to cyber security, allowing the design of more effective and IDSs. Almotiri presented an evaluation system to detect malicious traffic based on system performance [25]. They adopted an MCDM approach consisting of AHP–TOPSIS methods to rank the impact of alternatives according to their overall performance. Their study aims to be a reference for practitioners working in the field of evaluating and selecting the most effective traffic detection approach.

Afterward, some studies related to MCDM approaches and their applications in different fields were presented. Sharma and Kaul presented a study to deal with network performance delays and disruptions due to cluster-based communications that place a significant burden on the cluster head (CH) [26]. They used an MCDM approach including two AHP–TOPSIS methods to reduce the overburden on a single CH through a multi-CH scheme. Ogundoyin and Kamil presented a study to address security and privacy issues where fog servers can be used to process private and respond to time-sensitive information [27]. They applied the fuzzy AHP MCDM approach to determine and prioritize confidence parameters in fog computing. Their results indicate that quality of service is the best priority parameter that a service requester can use to evaluate the trusted standard of a service provider. Kumar et al. introduced a study to evaluate the impact of various malware analysis methods on the perspective of web applications [28]. They applied an MCDM approach that comprises AHP and TOPSIS methods under a fuzzy environment. Their results indicate that reverse engineering is the most effective method for analyzing complex malware.

There are many theories dealing with uncertainty, including the q-ROFS theory. Thus, we present some related studies as follows. Duane et al. introduced a study to deal with risks in information sharing and software piracy, which poses a threat to any system [29]. They applied the q-rung orthopair double hierarchy linguistic term set (q-RODHLTS) in the MCDM process. To prove the validity of their results, they applied the proposed approach to many information security systems. Panetikul et al. presented a study to analyze computer security threat analysis and control under a q-ROF environment [30].

They applied an approach based on the combination of the Heronian mean (HM) operator with complex q-ROFS is to initiate the complex q-rung orthopair fuzzy HM (Cq-ROFHM) operator. To prove the reliability and efficiency of the techniques used, some illustrative examples are introduced. Cheng et al. presented a study for evaluating sustainable enterprise risk management in manufacturing small and medium-sized enterprises [31]. They adopted a new extended Vlse Kriterijumska Optimizacija Kompromisno Resenje (VIKOR) approach using q-ROFSs. Peng et al. introduced a study for presenting a new score function of q-ROFN for solving the failure issues when comparing two q-ROFNs [32].

Finally, given the importance of IDSs and the importance of implementing them in a manner appropriate to their specific situation, choosing the most effective and appropriate one is a great challenge. Hence, there is an urgent and great need to evaluate IDSs. In this regard, a set of main and sub-criteria affecting the selection of the most effective IDS is identified; also, a set of alternatives are identified to be evaluated according to these criteria, using an MCDM approach and under a q-ROF environment.

3. Proposed Research Approach

3.1. Prelimianries

In this section, we list some concepts, procedures, and fundamental definitions related to IFSs, PFSs, and q-ROFSs.

3.1.1. Intuitionistic Fuzzy Sets

Atanassov developed IFSs as an extension of fuzzy set theory in 1986. IFSs are distinguished by the grade of membership and the grade of non-membership when their total is 1 or less than 1. It is explained as stated in Definition 1 [19].

Definition 1. *Let be* X *a fixed set. An IFS* $\tilde{I}$ *in* X *is an entity having the form given by*

$$\tilde{I} = \{(x,\ \mu_{\tilde{I}}(x),\ \upsilon_{\tilde{I}}(x)) \mid x \in X\} \tag{1}$$

where the function $\mu_{\tilde{I}}$: $X \rightarrow [0, 1]$ *describes the grade of membership of an element to the sets* $\tilde{I}$ *and* $\upsilon_{\tilde{I}}$: $X \rightarrow [0, 1]$ *describes the grade of non-membership of an element to the sets* $\tilde{I}$, *with the condition that*

$$0 \leq \mu_{\tilde{I}}(x) + \upsilon_{\tilde{I}}(x) \leq 1, \text{ for } \forall\ x \in X \tag{2}$$

The grade of hesitancy is computed as follows:

$$\pi_{\tilde{I}}(x) = 1 - \mu_{\tilde{I}}(x) - \upsilon_{\tilde{I}}(x) \tag{3}$$

Definition 2. *Let* $\tilde{A} = (\mu_{\tilde{A}}, \upsilon_{\tilde{A}})$ *and* $\tilde{B} = (\mu_{\tilde{B}}, \upsilon_{\tilde{B}})$ *be two intuitionistic fuzzy numbers (IFNs), then the addition and multiplication operations on these two IFNs as follows:*

$$\tilde{A} \oplus B = (\mu_{\tilde{A}} + \widetilde{\mu_{\tilde{B}} - \mu_{\tilde{A}}\mu_{\tilde{B}}},\ \upsilon_{\tilde{A}}\upsilon_{\tilde{B}}) \tag{4}$$

$$\tilde{A} \otimes \tilde{B} = (\mu_{\tilde{A}}\mu_{\tilde{B}},\ \upsilon_{\tilde{A}} + \upsilon_{\tilde{B}} - \upsilon_{\tilde{A}}\upsilon_{\tilde{B}}) \tag{5}$$

3.1.2. Pythagorean Fuzzy Sets

Yager developed PFSs as an extension of the IFSs [20]. They are distinguished by two membership grades termed as membership and non-membership. The total membership and non-membership grades in PFSs are unlike in IFSs. In PFSs, the membership and non-membership grade may be more than 1, but the total of their squares has to be at most 1. It is explained as stated in Definition 3.

Definition 3. *Let be X a fixed set. A PFS $\widetilde{P}$ in X is an entity having the form given by*

$$\widetilde{P} = \{(x, \mu_{\widetilde{P}}(x), \upsilon_{\widetilde{P}}(x)) \mid x \in X\} \quad (6)$$

where the function $\mu_{\widetilde{P}}$: X → [0, 1] describes the grade of membership of an element x ∈ X to the sets $\widetilde{P}$ and $\upsilon_{\widetilde{P}}$: X → [0, 1] describes the grade of non-membership of an element x ∈ X to the sets $\widetilde{P}$, with the condition that

$$0 \leq (\mu_{\widetilde{P}}(x))^2 + (\upsilon_{\widetilde{P}}(x))^2 \leq 1, \text{ for } \forall\, x \in X \quad (7)$$

The grade of uncertainty is computed as follows:

$$\pi_{\widetilde{P}}(x) = \sqrt{1 - \mu_{\widetilde{P}}(x)^2 - \upsilon_{\widetilde{P}}(x)^2} \quad (8)$$

Definition 4. *Let $\widetilde{P}_1 = (\mu_{\widetilde{P}_1}, \upsilon_{\widetilde{P}_1})$ and $\widetilde{P}_2 = (\mu_{\widetilde{P}_2}, \upsilon_{\widetilde{P}_2})$ be two Pythagorean fuzzy numbers (PFNs), then the addition and multiplication operations on these two PFNs as follows:*

$$\widetilde{P}_1 \oplus \widetilde{P}_2 = \left(\sqrt{\mu_{\widetilde{P}_1}{}^2 + \mu_{\widetilde{P}_2}{}^2 - \mu_{\widetilde{P}_1}{}^2\mu_{\widetilde{P}_2}{}^2}, \upsilon_{\widetilde{P}_1}\upsilon_{\widetilde{P}_2}\right) \quad (9)$$

$$\widetilde{P}_1 \otimes \widetilde{P}_2 = \left(\mu_{\widetilde{P}_1}\mu_{\widetilde{P}_2}, \sqrt{\upsilon_{\widetilde{P}_1}{}^2 + \upsilon_{\widetilde{P}_2}{}^2 - \upsilon_{\widetilde{P}_1}{}^2\upsilon_{\widetilde{P}_2}{}^2}\right) \quad (10)$$

3.1.3. Q-Rung Orthopair Fuzzy Sets

Yager presented q-ROFSs in 2018 with the grade of membership and non-membership. In q-ROFSs, the total of the *q*th power of the membership and non-membership grades should be at most equal to 1 [18]. In Figure 2, it is readily noted that q-ROFSs have a reasonable membership degree extent greater than that of the IFSs and PFSs. q-ROFSs are explained as stated in Definition 5.

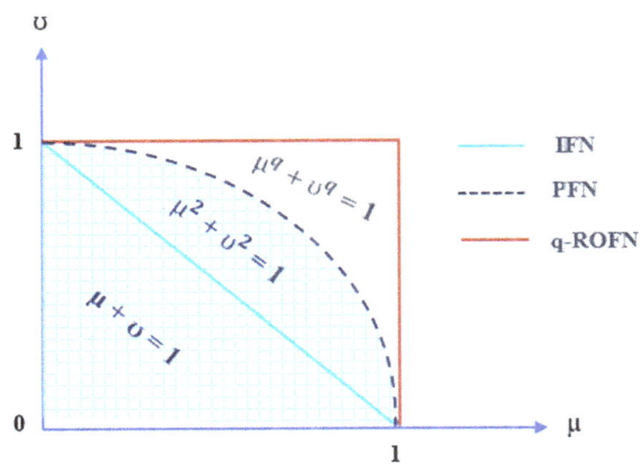

Figure 2. Comparison of the geometric area of various fuzzy membership degrees: IFNs, PFNs, and q-ROFNs.

Definition 5. *A q-ROFS $\breve{Q}$ in a finite universe of discourse X = $x_1, x_2, \ldots, x_n$ is defined by Yager as follows [18]:*

$$\breve{Q} = \{(x, \mu_{\breve{Q}}(x), \upsilon_{\breve{Q}}(x)) \mid x \in X\} \quad (11)$$

where the function $\mu_{\widetilde{\mathbb{Q}}}$: $X \rightarrow [0, 1]$ *defines the grade of membership of an element* $x \in X$ *to the sets* $\widetilde{\mathbb{Q}}$ *and* $\upsilon_{\widetilde{\mathbb{Q}}}$: $X \rightarrow [0, 1]$ *defines the grade of non-membership of an element* $x \in X$ *to the sets* $\widetilde{\mathbb{Q}}$, *with the condition that*

$$0 \leq \mu_{\widetilde{\mathbb{Q}}}(x)^q + \upsilon_{\widetilde{\mathbb{Q}}}(x)^q \leq 1, \text{ for } \forall\, x \in X \tag{12}$$

The grade of uncertainty is computed as follows

$$\pi_{\widetilde{\mathbb{Q}}}(x) = \sqrt[q]{1 - \mu_{\widetilde{\mathbb{Q}}}(x)^q - \upsilon_{\widetilde{\mathbb{Q}}}(x)^q} \tag{13}$$

Definition 6. *Let* $\breve{\mathbb{Q}} = (\mu_{\widetilde{\mathbb{Q}}}, \upsilon_{\widetilde{\mathbb{Q}}})$, $\breve{\mathbb{Q}}_1 = (\mu_{\widetilde{\mathbb{Q}}_1}, \upsilon_{\widetilde{\mathbb{Q}}_1})$, $\breve{\mathbb{Q}}_2 = (\mu_{\widetilde{\mathbb{Q}}_2}, \upsilon_{\widetilde{\mathbb{Q}}_2})$, *be three q-ROFNs, then their procedures can be well-defined as follows [18]:*

$$\breve{\mathbb{Q}}_1 \cap \breve{\mathbb{Q}}_2 = \left(\min\{\mu_{\widetilde{\mathbb{Q}}_1}, \mu_{\widetilde{\mathbb{Q}}_2}\}, \max\{\upsilon_{\widetilde{\mathbb{Q}}_1}, \upsilon_{\widetilde{\mathbb{Q}}_2}\}\right) \tag{14}$$

$$\breve{\mathbb{Q}}_1 \cup \breve{\mathbb{Q}}_2 = \left(\max\{\mu_{\widetilde{\mathbb{Q}}_1}, \mu_{\widetilde{\mathbb{Q}}_2}\}, \min\{\upsilon_{\widetilde{\mathbb{Q}}_1}, \upsilon_{\widetilde{\mathbb{Q}}_2}\}\right) \tag{15}$$

$$\breve{\mathbb{Q}}_1 \oplus \breve{\mathbb{Q}}_2 = \left(\left(\mu_{\widetilde{\mathbb{Q}}_1}{}^q + \mu_{\widetilde{\mathbb{Q}}_2}{}^q - \mu_{\widetilde{\mathbb{Q}}_1}{}^q \mu_{\widetilde{\mathbb{Q}}_2}{}^q\right)^{\frac{1}{q}}, \upsilon_{\widetilde{\mathbb{Q}}_1}\upsilon_{\widetilde{\mathbb{Q}}_2}\right) \tag{16}$$

$$\breve{\mathbb{Q}}_1 \otimes \breve{\mathbb{Q}}_2 = \left(\mu_{\widetilde{\mathbb{Q}}_1}\mu_{\widetilde{\mathbb{Q}}_2}, \left(\upsilon_{\widetilde{\mathbb{Q}}_1}{}^q + \upsilon_{\widetilde{\mathbb{Q}}_2}{}^q - \upsilon_{\widetilde{\mathbb{Q}}_1}{}^q \upsilon_{\widetilde{\mathbb{Q}}_2}{}^q\right)^{\frac{1}{q}}\right) \tag{17}$$

$$\lambda\breve{\mathbb{Q}} = \left((1 - (1 - \mu_{\breve{\mathbb{Q}}}{}^q)^\lambda)^{\frac{1}{q}}, \upsilon_{\breve{\mathbb{Q}}}{}^\lambda\right), \lambda > 0 \tag{18}$$

$$\breve{\mathbb{Q}}^\lambda = \left(\mu_{\breve{\mathbb{Q}}}{}^\lambda, (1 - (1 - \upsilon_{\breve{\mathbb{Q}}}{}^q)^\lambda)^{\frac{1}{q}}\right), \lambda > 0 \tag{19}$$

Definition 7. *Let* $\breve{\mathbb{Q}} = (\mu_{\widetilde{\mathbb{Q}}}, \upsilon_{\widetilde{\mathbb{Q}}})$ *be a q-ROFN; the score function S(*$\breve{\mathbb{Q}}$*) of* $\breve{\mathbb{Q}}$ *can be expressed as in [33], and the accuracy function A(*$\breve{\mathbb{Q}}$*) of* $\breve{\mathbb{Q}}$ *can be well defined, as in [34], shown by Equations (20) and (21), respectively.*

$$S(\breve{\mathbb{Q}}) = \frac{1}{2}(1 + \mu_{\breve{\mathbb{Q}}}{}^q - \upsilon_{\breve{\mathbb{Q}}}{}^q) \tag{20}$$

$$A(\breve{\mathbb{Q}}) = \mu_{\breve{\mathbb{Q}}}{}^q + \mathbb{Q}_{\breve{\mathbb{Q}}}{}^q \tag{21}$$

Definition 8. *Let* $\breve{\mathbb{Q}}_i = (\mu_{\widetilde{\mathbb{Q}}_i}, \upsilon_{\widetilde{\mathbb{Q}}_i})$ *(i = 1, 2, ... n) be set of q-ROFNs and* $W = (w_1, w_2, \ldots, w_n)^T$ *be weight vector of* $\breve{\mathbb{Q}}_i$ *with* $\sum_{i=1}^{n} W_i = 1$ *and* $W_i \in [0, 1]$. *Q-rung orthopair fuzzy weighted average (q-ROFWA) and q-rung orthopair fuzzy weighted geometric (q-ROFWG) operators can be expressed as in [34], shown by Equations (22) and (23), respectively.*

$$\text{q-ROFWA}\left(\breve{\mathbb{Q}}_1, \breve{\mathbb{Q}}_2, \ldots, \breve{\mathbb{Q}}_n\right) = \left(\left(1 - \prod_{i=1}^{n}\left(1 - \mu_{\widetilde{\mathbb{Q}}_i}{}^q\right)^{W_i}\right)^{\frac{1}{q}}, \prod_{i=1}^{n} \upsilon_{\widetilde{\mathbb{Q}}_i}{}^{W_i}\right) \tag{22}$$

$$\text{q-ROFWA}\left(\breve{\mathbb{Q}}_1, \breve{\mathbb{Q}}_2, \ldots, \breve{\mathbb{Q}}_n\right) = \left(\prod_{i=1}^{n} \mu_{\widetilde{\mathbb{Q}}_i}{}^{W_i}, \left(1 - \prod_{i=1}^{n}\left(1 - \upsilon_{\widetilde{\mathbb{Q}}_i}{}^q\right)^{W_i}\right)^{\frac{1}{q}}\right) \tag{23}$$

Definition 9. *Darko and Liang developed an operator named the weighted q-rung orthopair fuzzy Hamacher average (Wq-ROFHA) [35] as in Equations (24) and (25). Let* $\breve{\mathbb{Q}}_i = (\mu_{\widetilde{\mathbb{Q}}_i}, \upsilon_{\widetilde{\mathbb{Q}}_i})$ *(i = 1, 2,*

... n) be set of q-ROFNs and $W = (w_1, w_2, \ldots, w_n)^T$ *be a weight vector of* $\breve{\mathfrak{Q}}_i$ *with* $\sum_{i=1}^{n} W_i = 1$ *and* $W_i \in [0, 1]$.

$$\text{Wq-ROFHA}(\breve{\mathfrak{Q}}_1, \breve{\mathfrak{Q}}_2, \ldots, \breve{\mathfrak{Q}}_n) = w_1(\breve{\mathfrak{Q}}_1) \oplus w_2(\breve{\mathfrak{Q}}_2) \oplus \ldots \oplus w_n(\breve{\mathfrak{Q}}_n) = \underset{i=1}{\overset{n}{\oplus}} w_i(\breve{\mathfrak{Q}}_i) \tag{24}$$

$$\text{Wq-ROFHA}(\breve{\mathfrak{Q}}_1, \breve{\mathfrak{Q}}_2, \ldots, \breve{\mathfrak{Q}}_n) = \left(\sqrt[q]{\frac{\prod_{i=1}^{n}\left(1+(\gamma-1)\left(\mu_{\breve{\mathfrak{Q}}_i}\right)^q\right)^{W_i} - \prod_{i=1}^{n}\left(1-\left(\mu_{\breve{\mathfrak{Q}}_i}\right)^q\right)^{W_i}}{\prod_{i=1}^{n}\left(1+(\gamma-1)\left(\mu_{\breve{\mathfrak{Q}}_i}\right)^q\right)^{W_i} + (\gamma-1)\prod_{i=1}^{n}\left(1-\left(\mu_{\breve{\mathfrak{Q}}_i}\right)^q\right)^{W_i}}} \frac{\sqrt[q]{\gamma}\prod_{i=1}^{n}\left(\upsilon_{\tilde{\mathfrak{Q}}_i}\right)^{W_i}}{\sqrt[q]{\prod_{i=1}^{n}\left(1+(\gamma-1)\left(1-\left(\upsilon_{\tilde{\mathfrak{Q}}_i}\right)^q\right)\right)^{W_i} + (\gamma-1)\prod_{i=1}^{n}\left(\upsilon_{\tilde{\mathfrak{Q}}_i}\right)^{qW_i}}} \right) \tag{25}$$

where $\gamma > 0, q \geq 0$.

Definition 10. *Darko and Liang presented an operator named the weighted q-rung orthopair fuzzy Hamacher geometric mean (Wq-ROFHGM) [35], as in Equations (26) and (27). Let* $\breve{\mathfrak{Q}}_i = (\mu_{\tilde{\mathfrak{Q}}_i}, \upsilon_{\tilde{\mathfrak{Q}}_i})$ *(i = 1, 2, ... n) be set of q-ROFNs and* $W = (w_1, w_2, \ldots, w_n)^T$ *be the weight vector of* $\breve{\mathfrak{Q}}_i$ *with* $\sum_{i=1}^{n} W_i = 1$ *and* $W_i \in [0, 1]$.

$$\text{Wq-ROFHGM}(\breve{\mathfrak{Q}}_1, \breve{\mathfrak{Q}}_2, \ldots, \breve{\mathfrak{Q}}_n) = w_1(\breve{\mathfrak{Q}}_1) \oplus w_2(\breve{\mathfrak{Q}}_2) \oplus \ldots \oplus w_n(\breve{\mathfrak{Q}}_n) = \underset{i=1}{\overset{n}{\oplus}} w_i(\breve{\mathfrak{Q}}_i) \tag{26}$$

$$\text{Wq-ROFHGM}(\breve{\mathfrak{Q}}_1, \breve{\mathfrak{Q}}_2, \ldots, \breve{\mathfrak{Q}}_n) = \left(\frac{\sqrt[q]{\gamma}\prod_{i=1}^{n}\left(\mu_{\tilde{\mathfrak{Q}}_i}\right)^{W_i}}{\sqrt[q]{\prod_{i=1}^{n}\left(1+(\gamma-1)\left(1-\left(\mu_{\tilde{\mathfrak{Q}}_i}\right)^q\right)\right)^{W_i} + (\gamma-1)\prod_{i=1}^{n}\left(\mu_{\tilde{\mathfrak{Q}}_i}\right)^{qW_i}}} \sqrt[q]{\frac{\prod_{i=1}^{n}\left(1+(\gamma-1)\left(\upsilon_{\breve{\mathfrak{Q}}_i}\right)^q\right)^{W_i} - \prod_{i=1}^{n}\left(1-\left(\upsilon_{\breve{\mathfrak{Q}}_i}\right)^q\right)^{W_i}}{\prod_{i=1}^{n}\left(1+(\gamma-1)\left(\upsilon_{\breve{\mathfrak{Q}}_i}\right)^q\right)^{W_i} + (\gamma-1)\prod_{i=1}^{n}\left(1-\left(\upsilon_{\breve{\mathfrak{Q}}_i}\right)^q\right)^{W_i}}} \right) \tag{27}$$

where $\gamma > 0, q \geq 0$.

3.2. Suggested Approach

In this part, a sequential multi-step approach is presented to evaluate several intrusion-detection systems by combining two MCDM methods, namely, Entropy-CoCoSo. The proposed approach was performed under the q-rung orthopair fuzzy environment and by using q-ROFNs. The proposed approach was divided into three main parts. The data aggregation part includes identifying experts, selecting the criteria used, and determining the IDSs alternatives available. Then, the criteria evaluation part assesses the selected criteria using the q-ROF Entropy method. After that, the alternatives evaluation part assesses the available IDSs using the q-ROF CoCoSo method. The steps of the proposed approach are shown in Figure 3. Consequently, the steps of the suggested approach used are presented in detail as follows:

Step 1. The problem was studied and its main and sub-aspects were identified. The basic criteria for selecting the participating experts also were established. The criteria for selecting experts were determined as follows: the participants should have sufficient experience in the field of cyber security and the field of information security in general; also, their experience in the field of information security should not be less than 10 years. In addition, the participants should have practical experience in the field of information security technology and in the academic field. Next, the number of experts (EX_s) participating in the study was considered. After that, the participating experts were divided into several groups and the appropriate weight for each group was determined according to the measure of experience. Finally, the most appropriate means of communication with the participating experts were determined.

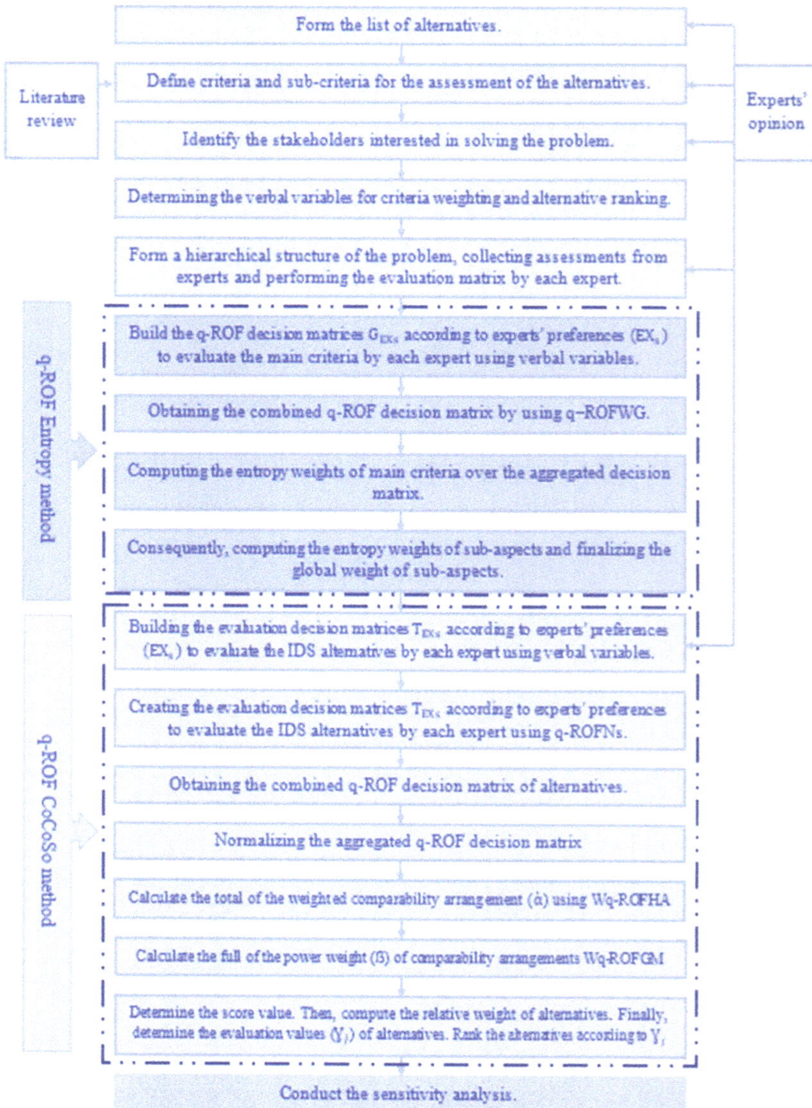

Figure 3. Decision framework for IDS evaluation.

Step 2. The main criteria and their sub-aspects used in the study were determined based on an analysis of the relevant literature, as well as insight from the participating experts. $C_j = \{C_1, C_2, \ldots, C_n\}$, with $j = 1, 2 \ldots n$. Let $W = (w_1, w_2, \ldots, w_n)$ be the vector set utilized for determining the criteria weights, where $w_j > 0$ and $\sum_{j=1}^{n} w_j = 1$.

Step 3. After studying the problem and its details and identifying the most important criteria, the available alternatives were determined to be used in the study. After that, experts' opinions were taken on the selected alternatives and a final list of alternatives to be used in the evaluation process was prepared. The set $A_i = \{A_1, A_2, \ldots, A_m\}$, having i = 1, 2,..,m alternatives, was evaluated by n decision criteria of set $C_j = \{C_1, C_2, \ldots, C_n\}$, with j = 1, 2, ... , n.

Step 4. After defining the main criteria and their sub-aspects and adopting a set of final alternatives, all these aspects were organized in the form of a hierarchical structure. This hierarchy shows the main objective of the problem, the criteria used, and the alternatives determined.

Step 5.Verbal variants and their equivalent q-ROFNs were identified. These variables were used in the evaluation process to assist the participating experts. These variants were divided into two parts in the same table. The first part refers to the variables that are used in evaluating the main criteria and their sub-aspects. The second part refers to the variables that are used in evaluating the available alternatives, as shown in Table 1.

Table 1. Verbal variables and their corresponding q-ROFNs for the weighting criteria and ranking alternatives.

Verbal Variables for Criteria	Abbreviations for Criteria	Verbal Variables for Alternatives	Abbreviations for Alternatives	q-ROFNs	
				μ	υ
Extremely poor	ELP	Extremely low	EXO	0.11	0.99
Very poor	VPO	Very low	VLO	0.22	0.88
Poor	PCO	Low	LLO	0.33	0.77
Medium poor	MDP	Medium low	MEL	0.44	0.66
Fair	FAR	Medium	MED	0.55	0.55
Medium good	MDG	Medium high	MEH	0.66	0.44
Good	GCO	High	HGH	0.77	0.33
Very good	VGO	Very high	VEH	0.88	0.22
Extremely good	EXG	Extremely high	EXH	0.99	0.11

Step 6. Build the q-ROF decision matrices, G_{EXs}, according to experts' preferences (EX_s) to evaluate the criteria by each expert using verbal variables in Table 1, and then by using q-rung orthopair fuzzy scale q-ROFNs, as shown in Table 2.

Table 2. The evaluation matrix for criteria based on q-ROFN with respect to experts.

Criteria	Experts			
	Ex_1	Ex_2	Ex_3	Ex_4
C_1	$[\mu_{1Ex_1}, \upsilon_{1Ex_1}]$	$[\mu_{1Ex_2}, \upsilon_{1Ex_2}]$	$[\mu_{1Ex_3}, \upsilon_{1Ex_3}]$	$[\mu_{1Ex_4}, \upsilon_{1Ex_4}]$
C_2	$[\mu_{2Ex_1}, \upsilon_{2Ex_1}]$	$[\mu_{2Ex_2}, \upsilon_{2Ex_2}]$	$[\mu_{2Ex_3}, \upsilon_{2Ex_3}]$	$[\mu_{2Ex_4}, \upsilon_{2Ex_4}]$
$\vdots$	$\vdots$	$\vdots$	$\vdots$	$\vdots$
C_n	$[\mu_{nEx_1}, \upsilon_{nEx_1}]$	$[\mu_{nEx_2}, \upsilon_{nEx_2}]$	$[\mu_{nEx_3}, \upsilon_{nEx_3}]$	$[\mu_{nEx_4}, \upsilon_{nEx_4}]$

Step 7. Aggregate the evaluations on criteria weights. The individual expert evaluations are collected by using q−ROFWG given in Equation (23). Here $(w_j)_{1\times n}$, presents the q-ROF weight of the jth criterion.

Step 8. The steps of the entropy method based on q-ROFSs are applied to evaluate and weight the main criteria and their sub-aspects [36]. Compute the entropy values of each q-ROFN of the aggregated experts' evaluations by applying Equations (28) and (29).

$$KE_{q,ij}(x)= \frac{1}{\sqrt{2}} \sqrt{((\mu(x))^q)^2 + ((\upsilon(x))^q)^2 + ((\mu(x))^q + (\upsilon(x))^q)^2} \tag{28}$$

$$EN_{q,ij}(x)= 1 - KE_{q,ij}(x) = 1- \frac{1}{\sqrt{2}} \sqrt{((\mu(x))^q)^2 + ((\upsilon(x))^q)^2 + ((\mu(x))^q + (\upsilon(x))^q)^2} \tag{29}$$

Step 9. The main criteria weights are calculated based on the entropy values using Equation (30).

$$w_j = \frac{1 - \mathcal{E}_j}{\sum_{j=1}^{n} (1 - \mathcal{E}_j)};\ j = 1,\ 2, \ldots\ n \tag{30}$$

where $\mathcal{E}_j = \frac{\sum_{i=1}^{m} EN_{q,ij}}{\sum_{i=1}^{m} \sum_{j=1}^{n} \cdot EN_{q,ij}}$ refers to the q-ROF entropy value.

Step 10. In the same way, the weights of the sub-aspects of the main criteria are calculated as in Steps 6–9.

Step 11. A q-ROF evaluation decision matrix (T_{EX}) is generated by each expert (EX) individually among the selected sub-aspects and alternatives to determine the best intrusion-detection system through the use of verbal variables, as shown in Table 1, and then by using the q-ROFNs in Table 1, as shown in Table 3. Here, $\check{T}_{EX} = \left(\check{t}_{ijEX}\right)_{n \times m}$, in which $\check{t}_{ijEX} = \left[\mu_{ijEX}, \upsilon_{ijEX}\right]$ is created by applying the verbal variables in Table 1. Consequently, $\check{t}_{ijEX}$ refers the performance of intrusion-detection systems (alternatives) A_i according to criteria C_j of the EX^{th} expert.

Table 3. Decision evaluation matrix for alternatives in terms of criteria based on q-ROFN.

Criteria	Alternatives (Intrusion-Detection Systems)			
	A_1	A_2	…	A_m
C_1	$[\mu_{11Ex}, \upsilon_{11Ex}]$	$[\mu_{12Ex}, \upsilon_{12Ex}]$	…	$[\mu_{1mEx}, \upsilon_{1mEx}]$
C_2	$[\mu_{21Ex}, \upsilon_{21Ex}]$	$[\mu_{22Ex}, \upsilon_{22Ex}]$	…	$[\mu_{2mEx}, \upsilon_{2mEx}]$
⋮	⋮	⋮	⋱	⋮
C_n	$[\mu_{n1Ex}, \upsilon_{n1Ex}]$	$[\mu_{n2Ex}, \upsilon_{n2Ex}]$	…	$[\mu_{nmEx}, \upsilon_{nmEx}]$

Step 12. After the q-ROF evaluation decision matrix (T_{EX}) is generated by each expert (EX) between the sub-aspects and the available alternatives by all experts, the q-ROF evaluation decision matrices (T_{EXs}) were aggregated into one matrix by utilizing q−ROFWG, as presented in Equation (23). A combined q-ROF evaluation decision matrix ($\check{T}$) was created as in Table 4. Accordingly, $\check{T} = \left(\check{t}_{ij}\right)_{n \times m}$ in which $\check{t}_{ij} = \left[\mu_{ij}, \upsilon_{ij}\right]$ is utilized to refer to the combined q-ROFN of the ith substitute with regard to the jth criteria.

Table 4. Combined evaluation matrix for alternatives in terms of the criteria based on q-ROFN.

Criteria	Alternatives (Intrusion-Detection Systems)			
	A_1	A_2	…	A_m
C_1	$[\mu_{11}, \upsilon_{11}]$	$[\mu_{12}, \upsilon_{12}]$	…	$[\mu_{1m}, \upsilon_{1m}]$
C_2	$[\mu_{21}, \upsilon_{21}]$	$[\mu_{22}, \upsilon_{22}]$	…	$[\mu_{2m}, \upsilon_{2m}]$
⋮	⋮	⋮	⋱	⋮
C_n	$[\mu_{n1}, \upsilon_{n1}]$	$[\mu_{n2}, \upsilon_{n2}]$	…	$[\mu_{nm}, \upsilon_{nm}]$

Step 13. Compute the normalized aggregated q-ROF evaluation decision matrix ($\check{H}$) by applying Equation (31).

$$\check{H} = \left(\check{h}_{ij}\right)_{m \times n} = \left[\check{\mu}_{ij}, \check{\upsilon}_{ij}\right] = \begin{cases} \left(\check{\mu}_{ij}, \check{\upsilon}_{ij}\right) if\ i\ \in Ƀ \\ \left(\check{\upsilon}_{ij}, \check{\mu}_{ij}\right) if\ i\ \in \textcent \end{cases} \tag{31}$$

where Ƀ refers to the set of benefit criteria and ₡ refers to the set of cost criteria.

Step 14. Calculate the total of the weighted comparability arrangement ($\acute{\alpha}$) for all substitutes by applying the Wq-ROFHA operator as presented in Equations (24) and (25).

Step 15. Calculate the full of the power weight (ẞ) of comparability arrangements for all substitutes by applying the Wq-ROFHGM operator as exhibited in Equations (26) and (27).

Step 16. Determine the score values of the substitutes by applying the values of values of the Wq-ROFHA and Wq-ROFHGM operators for each substitute by applying Equation (20).

Step 17. Calculate the proportional weight of the substitutes with the assistance of Equations (32)–(34).

$$Ɣ_{ja} = \frac{\acute{\alpha}_j + ẞ_j}{\sum_{j=1}^{n}(\acute{\alpha}_j + ẞ_j)} \tag{32}$$

$$Ɣ_{jb} = \frac{\acute{\alpha}_j}{\min(\acute{\alpha}_j)} + \frac{ẞ_j}{\min(ẞ_j)} \tag{33}$$

$$Ɣ_{jc} = \frac{\Psi\acute{\alpha}_j + (1-\Psi)ẞ_j}{\Psi\max(\acute{\alpha}_j) + (1-\Psi)\max(ẞ_j)}; 0 \le \Psi \le 1 \tag{34}$$

where $Ɣ_{ja}$ refers to the arithmetic mean of sums of the weighted sum method (WSM) and weighted product model (WPM) scores. Then, $Ɣ_{jb}$ indicates the sum of proportional scores of WSM and WPM. $Ɣ_{jc}$ also refers to the stable adjustment of the WSM and WPM models scores.

Step 18. Determine the evaluation values ($Ɣ_j$) of the substitutes by applying Equation (35). Then, rank the available intrusion-detection systems according to the most possible value of the evaluation values ($Ɣ_j$).

$$Ɣ_j = \sqrt[3]{Ɣ_{ja}Ɣ_{jb}Ɣ_{jc}} + \frac{Ɣ_{ja} + Ɣ_{jb} + Ɣ_{jc}}{3} \tag{35}$$

4. Empirical Results and Interpretation

In this section, the steps of the proposed multistep hybrid MCDM approach consisting of Entropy-CoCoSo methods are applied to evaluate the efficiency and reliability of some IDSs. The proposed approach was applied under the q-rung orthopair fuzzy environment and by using q-ROFNs. In this regard, the process of evaluating intrusion-detection systems and selecting the most effective and reliable one is necessary and inevitable in light of the recent cyber-attacks and intrusion methods. In this regard, the intrusion-detection systems are revealed in the next sub-section.

4.1. Description of Intrusion-Detection Systems

In this subsection, a brief description of the intrusion-detection systems are considered; also, Figure 4 demonstrates the general structure of the network and IDS.

- **Suricata (IDS_1):** Suricata was developed by the Open Information Security Foundation in 2010. Suricata is the main alternative to Snort because the design of Suricata is very close to that of Snort [37]. Suricata has an advantage over Snort, which is that it collects data at the application layer. Suricata consists of so-called threads, thread units, and queues. Suricata is a multi-threaded program, so there will be multiple threads running at the same time [37]. Thread units are divided according to functions; for example, one unit is used to analyze data packets, and the other unit is used to discover data packets. Each data packet can be processed by several different threads, and the queue is used to transfer the data packet from one thread to another. At the same time, a thread can contain several thread units, but only one unit runs at a given time.

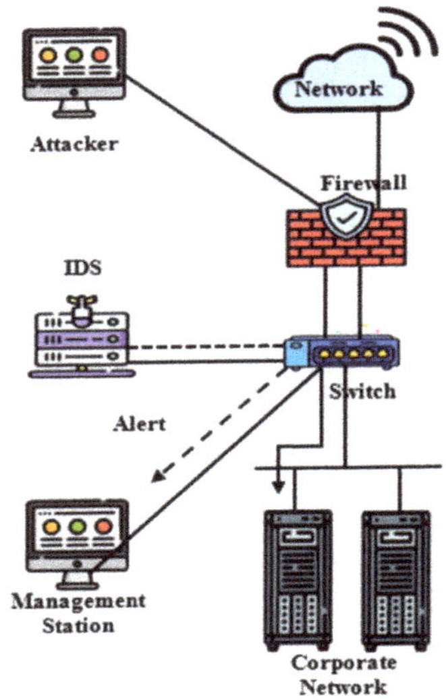

Figure 4. The general structure of the network and IDS.

- **Zeek (IDS_2):** Zeek (previously known as Bro until 2019) is a network intrusion-detection system that is compatible with Linux, Unix, and Mac OS [38]. Zeek uses network-based intrusion-detection methods by tracking the network and searching for malicious activities. The Zeek intrusion detection function is realized in two stages: traffic logging, and analysis. As with Suricata, Zeek has a significant advantage over Snort in that its analysis runs at the application layer, resulting in a broader analysis of network protocol activity.
- **Security onion (IDS_3):** Security Onion is a Linux-based IDS that is a mixture of several IDS that are both HIDS and NIDS solutions [16]. Although Security Onion is classified as NIDS, it includes HIDS functionality as well. It monitors log and configuration files for suspicious activity and checks those files for any unexpected changes. One of the downsides to Security Onion's comprehensive network monitoring system is its complexity. Thus, the Security Onion analysis engine is where things get complicated because there are so many different tools with different operating procedures that most of them may end up being overlooked.
- **Snort (IDS_4):** Snort is a Linux-based lightweight cross-platform network intrusion-detection system that can be used to monitor TCP/IP networks [39]. Snort is easy to deploy and can be configured to monitor network traffic for intrusion attempts, log intrusion behavior, and perform specific actions when intrusion attempts are detected. It is one of the most widely deployed IDS tools and can also be used as an intrusion-prevention system. Snort can be traced back to 1998, and there are still no signs of disappearing. There are some active communities that offer good help and support. The high level of personalization that Snort provides makes it a good choice for many different organizations.
- **Wazuh (IDS_5):** Wazuh is an IDS used to detect security and monitor compliance with security rules. Wazuh is an open-source intrusion-detection system project. It was developed as a fork part of OSSEC HIDS and was later integrated with Elastic Stack and Opens CAP. It relies on a cross-platform approach that redirects system data such as log messages, file tables, and detected anomalies to a central manager, where it is further analyzed and processed, resulting in security alerts. It monitors the file system and identifies changes in content, permissions, ownership, and file properties that

need to be monitored. It monitors configuration files to ensure that they comply with security policies, standards, or hardening guides.

- **OSSEC (IDS_6):** OSSEC is an open source IDS developed by Daniel B. Cid, who had sold the system to Trend Micro in 2008 [39]. Its detection methods are based on checking log files, making it a host-based IDS. OSSEC works on Unix, Linux, Mac OS, and Windows. There is no front end for this tool, but you can connect with Kibana or Graylog. OSSEC disclosure rules are called 'Policies'. You can write your own policies or get packages of them for free from the user community. It is also possible to define actions that should be performed automatically when specific warnings appear.

4.2. Application of the Proposed Approach

In this sub-section, the steps for evaluating the selected intrusion-detection systems through the Entropy-CoCoSo approach are presented as follows:

Step 1. Initially, a set of standards was identified to select the experts involved with the researchers in the study to evaluate the IDSs. The standards were as follows: the number of years of experience should not be less than 10 years in the field of cyber security and the field of information security in general; also, the scientific degree of the participating experts must not be lower than M.Sc. Accordingly, 60 experts were selected to participate in the IDSs evaluation process. After that, the participating experts were divided into four groups. Each group included a certain number of experts. The first and fourth groups included 12 experts. Whereas, the second and third groups included 18 experts. Accordingly, the appropriate weight was assigned to each group according to the years of experience and the number of experts. So, the first, second, third, and fourth groups had weights of 0.20, 0.30, 0.30, and 0.20, respectively. In addition, a leader was assigned to each group to express the final opinion in the evaluation process. Finally, the experts were contacted online.

Step 2. Based on the literature analysis and expert review, a set of main and sub-criteria were defined to evaluate the effectiveness and reliability of the IDSs. Initially, four main criteria were defined, which are as follows: protected system, PSC_1; audit source location, ASC_2; targets, TGC_3; and types, TPC_4. The main criteria also contained several sub-criteria, as follows: HIDS (HIC_{1_1}), NIDS (NIC_{1_2}), hybrids (HYC_{1_3}), host log files (HLC_{2_1}), network packets (NPC_{2_2}), application log files (ALC_{2_3}), IDS sensors alerts (ISC_{2_4}), applications (APC_{3_1}), network (NEC_{3_2}), host (HOC_{3_3}), open source (OSC_{4_1}), closed source (CSC_{4_2}), and freeware (CSC_{4_2}).

Step 3. A definitive list of available IDSs for use was prepared, as follows: Suricata (IDS_1), Zeek (IDS_2), Security onion (IDS_3), Snort (IDS_4), Wazuh (IDS_5), and OSSEC (IDS_6).

Step 4. A final hierarchical form of the problem was prepared, defining the main objective of the study, which was to evaluate the effectiveness of several IDSs, as shown in Figure 5; this in addition to regulating the relationship between the basic criteria and their sub-aspects, with the IDSs used as alternatives.

Step 5. A set of verbal variants and their equivalent q-ROFNs were prepared by reviewing the previous literature and expert opinions. Verbal variants were divided into two parts. The first part of the verbal variants is presented in Table 1, to assess the main criteria and their sub-aspects. The second part of the verbal variants is presented in Table 1, to evaluate the alternatives used.

Step 6. The decision matrix was built with the help of Table 2 by the four experts to assess the main criteria using the verbal variables as shown in Table 5.

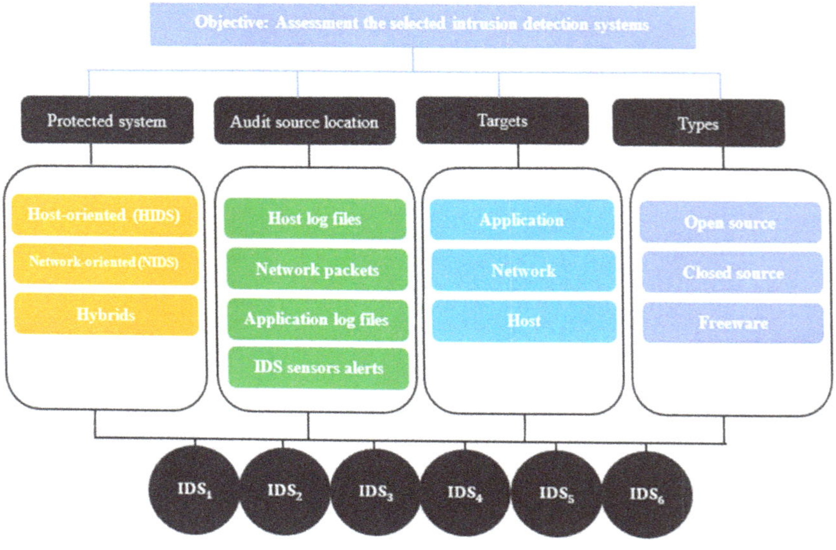

Figure 5. The hierarchy structure of the problem.

Table 5. Verbal evaluations of the main criteria by each expert and the aggregated main criteria weights.

Main Criteria	Ex_1	Ex_2	Ex_3	Ex_4	Aggregated Results	$KE_{q,ij}(x)$	$EN_{q,ij}(x)$	ε_j	w_j
PSC_1	GOO	VGO	EXG	VGO	[0.887, 0.253]	0.549576	0.450424	0.166067	0.278
ASC_2	FAR	MDG	MDG	GOO	[0.656, 0.461]	0.133121	0.866879	0.319610	0.226
TGC_3	GOO	VGO	VGO	VGO	[0.856, 0.260]	0.460183	0.539817	0.199025	0.266
TPC_4	MDG	POO	VGO	MDP	[0.538, 0.651]	0.144820	0.855180	0.315296	0.230

Step 7. Individual expert evaluations of the main criteria were compiled using a q-ROFWG operator in Equation (23), and using the weights assigned to the four experts, namely, 0.2, 0.3, 0.3, and 0.2, respectively, as exhibited in Table 5. The parameter was determined in a discretionary manner to reflect the position of the experts in terms of optimism and pessimism. In this case, q = 5 was introduced for a stronger illustration of the uncertainty.

Step 8. The entropy method was applied to calculate the entropy values for each q-ROFN from the aggregated experts' evaluations by applying Equations (28) and (29), as shown in Table 5.

Step 9. The weights of the main criteria were calculated based on the entropy values using Equation (30), as presented in Table 5.

Step 10. Likewise, the weights of the sub-aspects of the main criteria were calculated, as presented in Tables 6–9. Accordingly, the global weights of the sub-aspects were calculated, as in Table 10.

Table 6. Verbal evaluations of the protected system's criteria and aggregated main criteria weights.

Sub-Criteria	Ex_1	Ex_2	Ex_3	Ex_4	Aggregated Results	$KE_{q,ij}(x)$	$EN_{q,ij}(x)$	ε_j	w_j
HIC_{1_1}	MDP	FAR	VPO	EXG	[0.449, 0.748]	0.243794	0.756206	0.360331	0.319
NIC_{1_2}	ELP	VGO	MDP	MDG	[0.445, 0.862]	0.484883	0.515117	0.245452	0.377
HYC_{1_3}	VGO	GOO	GOO	POO	[0.667, 0.576]	0.172681	0.827319	0.394216	0.304

Table 7. Verbal evaluations of the audit source location's criteria and aggregated main criteria weights.

Sub-Criteria	Ex_1	Ex_2	Ex_3	Ex_4	Aggregated Results	$KE_{q,ij}(x)$	$EN_{q,ij}(x)$	ε_j	w_j
HLC_{2_1}	VPO	EXG	MDP	FAR	[0.510, 0.922]	0.684180	0.315820	0.151557	0.282
NPC_{2_2}	ELP	VGO	MDP	MDG	[0.445, 0.862]	0.484883	0.515117	0.247197	0.251
ALC_{2_3}	VGO	MDG	MDG	POO	[0.609, 0.588]	0.133588	0.866412	0.415779	0.196
ISC_{2_4}	VGO	MDP	ELP	MDG	[0.361, 0.906]	0.613525	0.386475	0.185464	0.271

Table 8. Verbal evaluations of the targets' criteria and aggregated main criteria weights.

Sub-Criteria	Ex_1	Ex_2	Ex_3	Ex_4	Aggregated Results	$KE_{q,ij}(x)$	$EN_{q,ij}(x)$	ε_j	w_j
APC_{3_1}	FAR	MDG	MDG	GOO	[0.656, 0.461]	0.133121	0.866879	0.431377	0.284
NEC_{3_2}	VGO	MDP	ELP	MDG	[0.361, 0.906]	0.613525	0.386475	0.192318	0.404
HOC_{3_3}	MDP	FAR	VPO	EXG	[0.449, 0.748]	0.243794	0.756206	0.376304	0.312

Table 9. Verbal evaluations of the types' criteria and aggregated main criteria weights.

Sub-Criteria	Ex_1	Ex_2	Ex_3	Ex_4	Aggregated Results	$KE_{q,ij}(x)$	$EN_{q,ij}(x)$	ε_j	w_j
OSC_{4_1}	MDP	FAR	VPO	GOO	[0.427, 0.748]	0.241568	0.758432	0.356283	0.322
CSC_{4_2}	MDG	POO	VGO	MDP	[0.538, 0.651]	0.144820	0.855180	0.401732	0.299
FRC_{4_3}	ELP	VGO	MDP	MDG	[0.445, 0.862]	0.484883	0.515117	0.241983	0.379

Table 10. The global weights of the sub criteria for evaluating intrusion-detection systems.

Main Criteria		**PSC_1** (0.278)			**ASC_2** (0.226)		
Sub-criteria	HIC_{1_1}	NIC_{1_2}	HYC_{1_3}	HLC_{2_1}	NPC_{2_2}	ALC_{2_3}	ISC_{2_4}
Local weights	0.319	0.377	0.304	0.282	0.251	0.196	0.271
Global weights	0.089	0.105	0.085	0.064	0.056	0.045	0.061
Main Criteria		**TGC_3 (0.266)**			**TPC_4 (0.230)**		
Sub-criteria	APC_{3_1}	NEC_{3_2}	HOC_{3_3}	OSC_{4_1}	CSC_{4_2}	FRC_{4_3}	
Local weights	0.284	0.404	0.312	0.322	0.299	0.379	
Global weights	0.075	0.107	0.083	0.074	0.069	0.087	

Step 11. An evaluation decision matrix was established to evaluate the IDSs according to the sub-aspects by the four experts and with the assistance of Table 3, as presented in Table 11.

Step 12. Individual expert evaluations of the alternatives between the sub-aspects and the available alternatives were compiled using the q-ROFWG operator in Equation (23), and using the weights assigned to the four experts, namely, 0.2, 0.3, 0.3, and 0.2, respectively, as exhibited in Table 12. The parameter also was determined in a discretionary manner to reflect the position of the experts in terms of optimism and pessimism. In this case, q = 5 and γ =1 were introduced for a stronger illustration of the uncertainty.

Table 11. Evaluations of the IDSs in terms of criteria.

IDS	Ex$_s$	HIC$_{1_1}$	NIC$_{1_2}$	HYC$_{1_3}$	HLC$_{2_1}$	NPC$_{2_2}$	ALC$_{2_3}$	ISC$_{2_4}$	APC$_{3_1}$	NEC$_{3_2}$	HOC$_{3_3}$	OSC$_{4_1}$	CSC$_{4_2}$	FRC$_{4_3}$
IDS$_1$	Ex$_1$	VEH	VEH	VEH	VEH	MEH	MEH	HGH	HGH	HGH	VLO	VLO	VLO	HGH
	Ex$_2$	VEH	VEH	VEH	HGH	VEH	HGH	HGH	VEH	VEH	MEL	HGH	HGH	MED
	Ex$_3$	HGH	HGH	HGH	HGH	HGH	HGH	HGH	VEH	VEH	MED	MEL	MEL	MED
	Ex$_4$	HGH	HGH	HGH	VEH	EXH	HGH	HGH	EXH	VEH	HGH	VLO	HGH	MED
	Ex$_s$	HIC$_{1_1}$	NIC$_{1_2}$	HYC$_{1_3}$	HLC$_{2_1}$	NPC$_{2_2}$	ALC$_{2_3}$	ISC$_{2_4}$	APC$_{3_1}$	NEC$_{3_2}$	HOC$_{3_3}$	OSC$_{4_1}$	CSC$_{4_2}$	FRC$_{4_3}$
IDS$_2$	Ex$_1$	HGH	LLO	EXO	VEH	LLO	MEL	HGH	VEH	VEH	VLO	VLO	VLO	VEH
	Ex$_2$	VEH	LLO	EXO	HGH	HGH	MEH	VEH	VEH	VEH	MEL	HGH	HGH	VLO
	Ex$_3$	HGH	LLO	EXO	HGH	MEL	MED	HGH	VEH	VEH	MED	MEL	MEL	MED
	Ex$_4$	HGH	LLO	EXO	EXH	VEH	MED	HGH	EXH	VEH	HGH	VLO	HGH	LLO
	Ex$_s$	HIC$_{1_1}$	NIC$_{1_2}$	HYC$_{1_3}$	HLC$_{2_1}$	NPC$_{2_2}$	ALC$_{2_3}$	ISC$_{2_4}$	APC$_{3_1}$	NEC$_{3_2}$	HOC$_{3_3}$	OSC$_{4_1}$	CSC$_{4_2}$	FRC$_{4_3}$
IDS$_3$	Ex$_1$	HGH	VEH	VEH	VEH	HGH	MED	HGH	VEH	MEL	VLO	VLO	VLO	EXO
	Ex$_2$	VEH	VEH	VEH	HGH	EXH	VEH	VEH	VEH	MEL	MEL	HGH	HGH	EXO
	Ex$_3$	HGH	HGH	HGH	HGH	VEH	MEH	HGH	VEH	MEL	MED	MEL	MEL	EXO
	Ex$_4$	HGH	HGH	HGH	VEH	EXH	HGH	HGH	VEH	MEL	HGH	VLO	HGH	EXO
	Ex$_s$	HIC$_{1_1}$	NIC$_{1_2}$	HYC$_{1_3}$	HLC$_{2_1}$	NPC$_{2_2}$	ALC$_{2_3}$	ISC$_{2_4}$	APC$_{3_1}$	NEC$_{3_2}$	HOC$_{3_3}$	OSC$_{4_1}$	CSC$_{4_2}$	FRC$_{4_3}$
IDS$_4$	Ex$_1$	VEH	VEH	VEH	VEH	LLO	MEL	HGH	HGH	MED	VLO	VLO	VLO	MEL
	Ex$_2$	VEH	VEH	VEH	HGH	HGH	MEH	VEH	HGH	MED	MEL	HGH	HGH	MEL
	Ex$_3$	HGH	HGH	HGH	HGH	MEL	MED	HGH	HGH	MED	MED	MEL	MEL	MEL
	Ex$_4$	HGH	HGH	HGH	EXH	VEH	MED	HGH	HGH	MED	HGH	VLO	HGH	MEL
	Ex$_s$	HIC$_{1_1}$	NIC$_{1_2}$	HYC$_{1_3}$	HLC$_{2_1}$	NPC$_{2_2}$	ALC$_{2_3}$	ISC$_{2_4}$	APC$_{3_1}$	NEC$_{3_2}$	HOC$_{3_3}$	OSC$_{4_1}$	CSC$_{4_2}$	FRC$_{4_3}$
IDS$_5$	Ex$_1$	HGH	VEH	VLO	VEH	HGH	MED	HGH	VLO	VEH	VLO	VLO	VLO	HGH
	Ex$_2$	VEH	VEH	HGH	HGH	EXH	VEH	VEH	VLO	VEH	MEL	HGH	HGH	MED
	Ex$_3$	HGH	HGH	MEL	HGH	VEH	MEH	HGH	VLO	VEH	MED	MEL	MEL	MED
	Ex$_4$	HGH	HGH	VLO	VEH	EXH	HGH	HGH	VLO	VEH	HGH	VLO	HGH	MED
	Ex$_s$	HIC$_{1_1}$	NIC$_{1_2}$	HYC$_{1_3}$	HLC$_{2_1}$	NPC$_{2_2}$	ALC$_{2_3}$	ISC$_{2_4}$	APC$_{3_1}$	NEC$_{3_2}$	HOC$_{3_3}$	OSC$_{4_1}$	CSC$_{4_2}$	FRC$_{4_3}$
IDS$_6$	Ex$_1$	HGH	VEH	VEH	VEH	HGH	MED	HGH	VEH	MEH	VLO	VLO	VLO	VLO
	Ex$_2$	VEH	VEH	VEH	HGH	EXH	VEH	VEH	VEH	MEH	MEL	HGH	HGH	VLO
	Ex$_3$	HGH	HGH	HGH	HGH	VEH	MEH	HGH	VEH	MEH	MED	MEL	MEL	VLO
	Ex$_4$	HGH	HGH	HGH	VEH	EXH	HGH	HGH	VEH	MEH	HGH	VLO	HGH	VLO

Table 12. The aggregated evaluations matrix of the IDSs.

IDS	HIC$_{1_1}$	NIC$_{1_2}$	HYC$_{1_3}$	HLC$_{2_1}$	NPC$_{2_2}$	ALC$_{2_3}$	ISC$_{2_4}$
IDS$_1$	[0.823, 0.295]	[0.823, 0.295]	[0.823, 0.295]	[0.823, 0.295]	[0.817, 0.342]	[0.747, 0.365]	[0.770, 0.330]
IDS$_2$	[0.801, 0.318]	[0.330, 0.770]	[0.110, 0.990]	[0.832, 0.301]	[0.564, 0.630]	[0.555, 0.562]	[0.801, 0.311]
IDS$_3$	[0.801, 0.318]	[0.823, 0.295]	[0.823, 0.295]	[0.832, 0.301]	[0.909, 0.248]	[0.715, 0.438]	[0.801, 0.311]
IDS$_4$	[0.823, 0.295]	[0.823, 0.295]	[0.823, 0.295]	[0.832, 0.301]	[0.817, 0.342]	[0.555, 0.562]	[0.801, 0.311]
IDS$_5$	[0.801, 0.318]	[0.823, 0.295]	[0.394, 0.780]	[0.812, 0.303]	[0.909, 0.284]	[0.715, 0.438]	[0.801, 0.318]
IDS$_6$	[0.801, 0.318]	[0.823, 0.295]	[0.823, 0.295]	[0.812, 0.303]	[0.909, 0.284]	[0.715, 0.438]	[0.801, 0.318]

IDS	APC$_{3_1}$	NEC$_{3_2}$	HOC$_{3_3}$	OSC$_{4_1}$	CSC$_{4_2}$	FRC$_{4_3}$
IDS$_1$	[0.877, 0.259]	[0.857, 0.260]	[0.458, 0.713]	[0.394, 0.780]	[0.507, 0.704]	[0.588, 0.528]
IDS$_2$	[0.901, 0.211]	[0.880, 0.220]	[0.458, 0.715]	[0.441, 0.661]	[0.507, 0.706]	[0.414, 0.765]
IDS$_3$	[0.880, 0.220]	[0.440, 0.660]	[0.458, 0.714]	[0.394, 0.780]	[0.507, 0.705]	[0.110, 0.990]
IDS$_4$	[0.770, 0.330]	[0.550, 0.550]	[0.458, 0.714]	[0.394, 0.780]	[0.507, 0.705]	[0.440, 0.432]
IDS$_5$	[0.220, 0.880]	[0.880, 0.220]	[0.458, 0.714]	[0.394, 0.780]	[0.507, 0.705]	[0.588, 0.528]
IDS$_6$	[0.880, 0.220]	[0.660, 0.440]	[0.458, 0.714]	[0.394, 0.780]	[0.507, 0.705]	[0.220, 0.880]

Step 13. The normalized aggregated q-ROF evaluation decision matrix was calculated by applying Equation (31), as presented in Table 13.

Table 13. The normalized evaluation matrix of the IDSs.

IDS	HIC_{1_1}	NIC_{1_2}	HYC_{1_3}	HLC_{2_1}	NPC_{2_2}	ALC_{2_3}	ISC_{2_4}
IDS_1	[0.823, 0.295]	[0.823, 0.295]	[0.823, 0.295]	[0.823, 0.295]	[0.817, 0.342]	[0.747, 0.365]	[0.770, 0.330]
IDS_2	[0.801, 0.318]	[0.330, 0.770]	[0.110, 0.990]	[0.832, 0.301]	[0.564, 0.630]	[0.555, 0.562]	[0.801, 0.311]
IDS_3	[0.801, 0.318]	[0.823, 0.295]	[0.823, 0.295]	[0.832, 0.301]	[0.909, 0.248]	[0.715, 0.438]	[0.801, 0.311]
IDS_4	[0.823, 0.295]	[0.823, 0.295]	[0.823, 0.295]	[0.832, 0.301]	[0.817, 0.342]	[0.555, 0.562]	[0.801, 0.311]
IDS_5	[0.801, 0.318]	[0.823, 0.295]	[0.394, 0.780]	[0.812, 0.303]	[0.909, 0.284]	[0.715, 0.438]	[0.801, 0.318]
IDS_6	[0.801, 0.318]	[0.823, 0.295]	[0.823, 0.295]	[0.812, 0.303]	[0.909, 0.284]	[0.715, 0.438]	[0.801, 0.318]

IDS	APC_{3_1}	NEC_{3_2}	HOC_{3_3}	OSC_{4_1}	CSC_{4_2}	FRC_{4_3}
IDS_1	[0.877, 0.259]	[0.857, 0.260]	[0.458, 0.713]	[0.394, 0.780]	[0.507, 0.704]	[0.588, 0.528]
IDS_2	[0.901, 0.211]	[0.880, 0.220]	[0.458, 0.715]	[0.441, 0.661]	[0.507, 0.706]	[0.414, 0.765]
IDS_3	[0.880, 0.220]	[0.440, 0.660]	[0.458, 0.714]	[0.394, 0.780]	[0.507, 0.705]	[0.110, 0.990]
IDS_4	[0.770, 0.330]	[0.550, 0.550]	[0.458, 0.714]	[0.394, 0.780]	[0.507, 0.705]	[0.440, 0.432]
IDS_5	[0.220, 0.880]	[0.880, 0.220]	[0.458, 0.714]	[0.394, 0.780]	[0.507, 0.705]	[0.588, 0.528]
IDS_6	[0.880, 0.220]	[0.660, 0.440]	[0.458, 0.714]	[0.394, 0.780]	[0.507, 0.705]	[0.220, 0.880]

Step 14. Calculate the total of the weighted comparability arrangement for all alternatives by applying the Wq-ROFHA operator, as presented in Equations (24) and (25), and as exhibited in Table 14.

Table 14. The $\acute{\alpha}_j$ and $ß_j$ values of the IDSs.

IDSs	Wq-ROFHA Operator		Wq-ROFHGM Operator	
	$\acute{\alpha}$	$ß$	$\acute{\alpha}$	$ß$
IDS_1	0.784	0.382	0.139	0.567
IDS_2	0.741	0.489	0.102	0.789
IDS_3	0.767	0.438	0.113	0.777
IDS_4	0.741	0.421	0.127	0.578
IDS_5	0.760	0.483	0.119	0.671
IDS_6	0.771	0.419	0.125	0.651

Step 15. Calculate the full power weight of the comparability arrangements for all alternatives by applying the Wq-ROFHGM operator, as exhibited in Equations (26) and (27), and as exhibited in Table 14.

Step 16. The score values of the all alternatives are computed by applying the values of the Wq-ROFHA and Wq-ROFHGM operators for each alternative by applying Equation (20), as presented in Table 15.

Table 15. The score values of the IDSs for $\acute{\alpha}_j$ and $ß_j$.

IDS	$\acute{\alpha}_j$	$ß_j$
IDS_1	0.701	0.286
IDS_2	0.626	0.156
IDS_3	0.665	0.168
IDS_4	0.660	0.275
IDS_5	0.639	0.224
IDS_6	0.676	0.237

Step 17. The proportional weight of the alternatives is calculated with the assistance of Equations (32)–(34), as shown in Table 16.

Table 16. The proportional importance and the final ranking of the IDSs.

IDS	Ɣ_{ja}	Ɣ_{jb}	Ɣ_{jc}	Ɣ_{j}	Rank
IDS_1	0.186	2.952	1.000	2.398	1
IDS_2	0.147	2.000	0.792	1.595	6
IDS_3	0.157	2.136	0.843	1.701	5
IDS_4	0.176	2.814	0.947	2.089	2
IDS_5	0.162	2.455	0.874	1.867	4
IDS_6	0.172	2.597	0.925	1.976	3

Step 18. The evaluation values (Ɣ_j) of the alternatives are identified by applying Equation (35). Then, the six intrusion-detection systems are rated according to the most possible value of the evaluation values (Ɣ_j), as presented in Table 16 and in Figure 6.

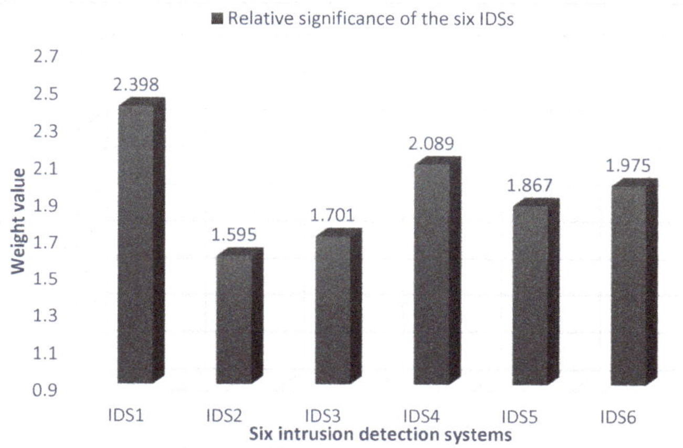

Figure 6. Ranking of the IDSs using the CoCoSo method.

4.3. Results Interpretation

In this subsection, some interpretations are introduced of the results obtained from applying the proposed approach, Entropy-CoCoSo, under a q-rung orthopair fuzzy environment. The results obtained are divided into two parts. The first part relates to the results of the main criteria weights and their sub-aspects. The second part relates to the results of the intrusion-detection systems evaluation used in the study. Initially, the four main criteria were evaluated by the participating experts. The results obtained indicate that the PSC_1 criterion has the highest weight, with a weight of 0.278, followed by the TGC_3 criterion with a weight of 0.266. Whereas, the ASC_2 criterion has the lowest weight, 0.226, and occupies the last rank in the ranking of the main criteria. Accordingly, the sub-criteria related to each main criterion were evaluated. Thus, the sub-criteria related to the PSC_1 criterion were evaluated as follows: the NIC_{1_2} criterion has the top weight with a weight of 0.377, followed by the HIC_{1_1} criterion with a weight of 0.319; the HYC_{1_3} criterion has the lowest weight, 0.304. The sub-criteria related to the ASC_2 criterion were calculated as follows: the HLC_{2_1} criterion has the maximum weight, with a weight of 0.282, followed by the ISC_{2_4} criterion, with a weight of 0.271; the ALC_{2_3} criterion has the minimum weight, 0.196. In addition, the sub-criteria related to the TGC_3 criterion were estimated as follows: the NEC_{3_2} criterion has the highest weight, with a weight of 0.404, followed by the HOC_{3_3} criterion with a weight of 0.312; the APC_{3_1} criterion has the lowest weight, 0.284. Afterward, the sub-criteria related to the TPC_4 criterion were assessed as follows:

the FRC_{4_3} criterion has the largest weight, with a weight of 0.379, followed by the OSC_{4_1} criterion with a weight of 0.322; the CSC_{4_2} criterion has the smallest weight, 0.299.

In the end, the results of the intrusion-detection systems used in the evaluation process were revealed as follows: Suricata (IDS_1) has the top rank with a weight of 2.398 followed by Snort (IDS_4) with a weight of 2.089. In turn, Zeek (IDS_2) has the lowest rank with a weight of 1.595.

4.4. Comparative Analysis

In this sub-section, a comparative analysis is demonstrated to test and verify the effectiveness of the developed approach, q-ROF Entropy-CoCoSo. Consequently, the assessment results were compared with Alyami et al.'s [16] fuzzy AHP–TOPSIS approach. In this regard, the same weights of the main criteria and sub-aspects obtained by applying the proposed approach were used, as shown in Table 10. Accordingly, the results of ranking the alternatives used in the study using the two approaches are presented in Table 17 and in Figure 7. The results of the comparison show that IDS_1 is the best alternative according to the results of the two approaches. IDS_2 is the least alternative in the order. According to the results, it can be seen that there are some changes in the order of some alternatives, such as IDS_3, IDS_4, IDS_5, and IDS_6. The presence of some differences in the order of the alternatives can be explained by the difference in the mathematical basis for each approach. Finally, the results of the comparative analysis and the reliability of the proposed approach can be verified by the experts.

Table 17. Comparative analysis with other approach for ranking the IDSs.

Approaches	IDS_1	IDS_2	IDS_3	IDS_4	IDS_5	IDS_6
q-ROF Entropy-CoCoSo	2.398	1.595	1.701	2.089	1.867	1.976
Ranking	1	6	5	2	4	3
Fuzzy AHP-TOPSIS	0.927	0.287	0.582	0.675	0.497	0.723
Ranking	1	6	4	3	5	2

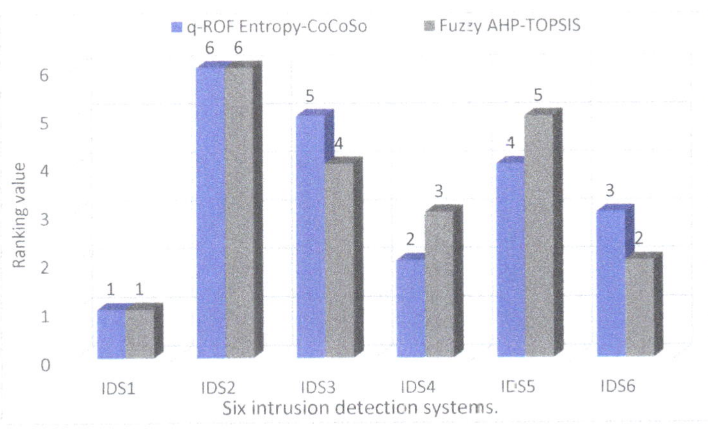

Figure 7. Final ranking of the six IDSs using various approaches.

4.5. Sensitivity Analysis

We have conducted a sensitivity analysis from the three perspectives of changes in parameter q, parameter γ, and parameter Ψ. Sensitivity analysis was conducted on the results obtained to confirm their reliability and stability and to examine the change that occurred to them as a result of the change in some inputs and parameters. In decision-making approaches, some parameters are defined subjectively based on the perception of the problem by decision-makers and the extent of the risks in the environment. Consequently, these parameters change according to the circumstances in which the decision-making system is being modeled. In our proposed Entropy-CoCoSo q-ROF approach, three parameters—q, γ, and Ψ—are defined, which are determined based on the personal preferences of the experts. Accordingly, several changes were made to these parameters to show their decisive influence on the final IDS's ranking results. These changes were divided into four scenarios. The first scenario refers to the change in the values of parameter q. The second scenario indicates the change in the values of parameter γ. The third scenario refers to the change in the values of parameters q and γ. Lastly, the fourth scenario refers to the change in the values of parameter Ψ.

The first scenario is the effect of a change in parameter q on the evaluation of IDSs. Accordingly, the value of the parameter q was changed several times, from q = 2 to q = 20, to show its impact on the evaluation of IDSs, as presented in Figure 8. Although the value of the q parameter has been changed several times, the order of the IDSs has not changed at all. IDS_1 remains the best alternative throughout the sensitivity analysis and parameter value change q, followed by IDS_4. By contrast, IDS_2 remains the lowest in order despite the change in the value of the parameter q. The changes that can be observed based on the change in the value of the parameter q in the order of the IDSs, show there is a large convergence between the values of the assessment of IDS_4 and IDS_6 at the value of q = 2. Significant convergence occurs between the IDS_4 and IDS_1 assessment values at q = 8; otherwise, the order of the IDSs remains the same despite the presence of some increases in the weights of the IDSs.

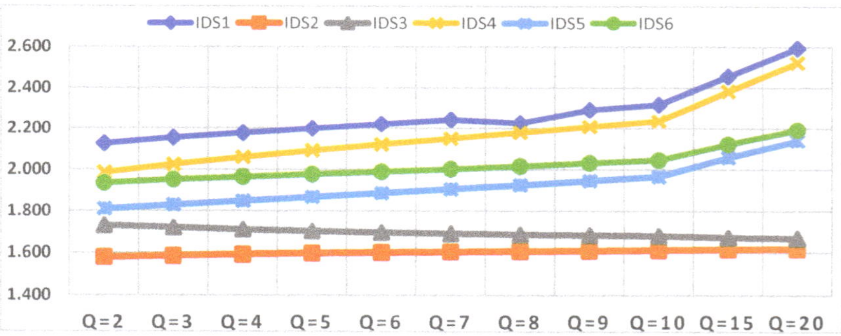

Figure 8. Closeness coefficient values of IDSs in terms of different values of q.

The second scenario is the effect of a change in parameter γ on the evaluation of IDSs. Accordingly, the value of the parameter γ was changed several times, from γ = 0.1 to γ = 1.0, to show its effect on the evaluation of IDSs, as shown in Figure 9. Although the value of the parameter γ was changed several times, the order of the IDSs changed only when the value of parameter γ = 1.0. IDS_1 remains the best alternative throughout the sensitivity analysis and parameter value change γ = 0.1 to γ = 0.9 followed by IDS_4, except when parameter value γ = 1.0, then IDS_6 becomes the second rank in the analysis process. In contrast, IDS_2 remains lowest in order throughout the change of the value of the parameter γ = 0.0 to γ = 0.9, except when the value of γ = 1.0 is changed, then IDS_2 becomes the fifth rank, penultimate. The changes that can be observed based on the change in the value of the parameter q in the order of IDSs, is that when the value of the parameter γ = 1.0, the order of the IDSs changes so that IDS_1 is in the first order, while IDS_6 is in the second order, and IDS_4 is in the third order. On the contrary, the rank of some IDSs was changed, such as the rank of IDS_2 and the IDS_5, which became the fifth and sixth, respectively.

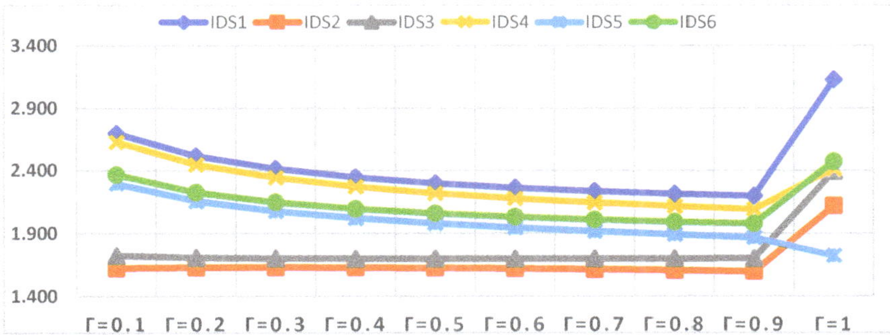

Figure 9. Closeness coefficient values of IDSs in terms of different values of γ.

The third scenario is the effect of a change in parameter q and γ on the evaluation of IDSs. Accordingly, the values of q and γ were changed several times, from q = 2 to q = 15, and $\gamma = 0.1$ to $\gamma = 1$ to show their combined effect on the assessment of the IDSs, as shown in Figure 10. Although the values of the parameters q and γ changed many times, the order of the IDSs has not changed at all. IDS_1 remains the best alternative during sensitivity analysis and changing the values of q and γ parameters, followed by IDS_4. In contrast, IDS_2 remains the lowest in terms of rank despite the change in the values of the two parameters q and γ. Changes that can be observed based on the change in the value of parameters q and γ for the order of IDSs, is that there is a great convergence between the values of the evaluation process weights for the IDSs used in the study. The convergence between the weights of the IDSs is difficult to see in Figure 10, and this is one of the shortcomings of Figure 10.

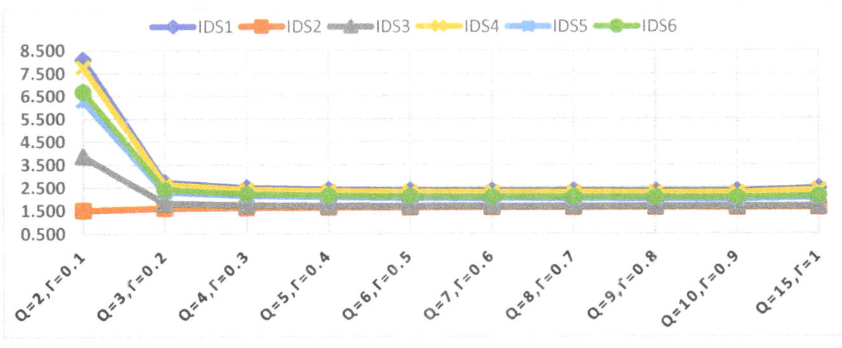

Figure 10. Closeness coefficient values of IDSs in terms of the different values of q and γ.

The fourth scenario is the effect of a change in parameter Ψ on the evaluation of IDSs. Accordingly, the value of parameter Ψ was changed several times from $\Psi = 0.1$ to $\Psi = 1.0$, to show its effect on the evaluation of the IDSs, as shown in Figure 11. Although the value of parameter Ψ was changed several times, the order of the IDSs did not change at all. IDS_1 remains the best alternative throughout the sensitivity analysis and parameter value change $\Psi = 0.1$ to $\Psi = 0.9$, followed by IDS_4. On the contrary, IDS_2 remains in the lowest order by changing the parameter value $\Psi = 0.1$ to $\Psi = 1.0$.

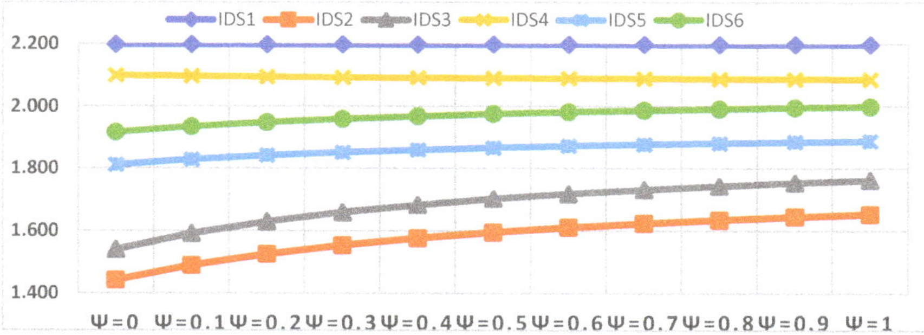

Figure 11. Closeness coefficient values of the IDSs in terms of the different values of Ψ.

5. Conclusions

Given the spread of computer networks and the dependence of public and private institutions on their efficiency and quality of work, any disruption or sabotage of them may lead to great losses. Information systems and networks are constantly subject to cyber-attacks. Thus, firewalls and antivirus are not enough to fend off all these attacks, as they are only able to protect the "front entrance" of computer systems and networks. IDSs can help protect your corporation from malicious activities. There are different types of IDSs to protect networks, as intrusion attacks are becoming more and more common on a global scale. In addition, hackers using new technologies are trying to penetrate systems. An IDS is a tool that identifies these attacks and will take an immediate step to get the system back to normal, as the IDS can also detect network traffic and send an alarm if a breach is found.

In this regard, this study discusses the most effective and used IDSs. This study was conducted with the participation of many experts under the q-ROF environment to deal with the uncertainty that may occur as a result of different circumstances and differences in evaluation frameworks. Six intrusion-detection systems, namely, Suricata (IDS_1), Zeek (IDS_2), Security onion (IDS_3), Snort (IDS_4), Wazuh (IDS_5), and OSSEC (IDS_6), were evaluated according to four key criteria and thirteen sub-aspects. The main criteria were protected system, audit source location, targets, and types. The sub-aspects, on the basis of which the effectiveness of the intrusion-detection systems was evaluated, were HIDS, NIDS, hybrids, host log files, network packets, application log files, IDS sensors alerts, applications, network, host, open-source, closed source, and freeware. A hybrid MCDM approach, including q-ROF entropy-CoCoSo techniques, was proposed, where entropy was applied to evaluate the main criteria and their sub-aspects. The CoCoSo method is applied to rate six IDSs according to their effectiveness. Afterward, comparative and sensitivity analyses were performed to confirm the stability, reliability, and performance of the proposed approach. The findings indicate that most of the IDSs appear to be systems with high potential. According to the results, Suricata is the best IDS that relies on multi-threading performance. Although the results here confirm that the proposed method is applicable and effective, it has some limitations. The key limitation of the approach is the difficult mathematical algorithm for the computation of Hamacher functions.

Author Contributions: Conceptualization, M.A.-B., A.G., K.M.S., I.E., K.M. and A.J.; methodology, M.A.-B., A.G., K.M.S., I.E., K.M. and A.J.; software, M.A.-B., A.G.; validation, M.A.-B., A.G., K.M.S., I.E., K.M. and A.J.; formal analysis, M.A.-B., A.G.; investigation, M.A.-B., A.G., K.M.S., I.E., K.M. and A.J.; resources, M.A.-B., A.G.; writing—original draft preparation, M.A.-B., A.G., K.M.S.; writing—review and editing, M.A.-B., A.G., K.M.S., I.E., K.M. and A.J.; visualization, M.A.-B., A.G., K.M.S., I.E., K.M.; supervision, M.A.-B., K.M. and A.J.; project administration, M.A.-B. funding acquisition K.M. and A.J. All authors have read and agreed to the published version of the manuscript.

Funding: This research received no external funding.

Institutional Review Board Statement: The study did not involve humans or animals.

Informed Consent Statement: The study did not involve humans.

Conflicts of Interest: The authors declare no conflict of interest.

References

1. Ahmad, T.; Zhang, D. Using the internet of things in smart energy systems and networks. *Sustain. Cities Soc.* **2021**, *68*, 102783. [CrossRef]
2. Jaafar, G.A.; Abdullah, S.M.; Ismail, S. Review of Recent Detection Methods for HTTP DDoS Attack. *J. Comput. Netw. Commun.* **2019**, *2019*, 1283472. [CrossRef]
3. Harter, G.T.; Rowe, N.C. *Testing Detection of K-Ary Code Obfuscated by Metamorphic and Polymorphic Techniques BT—National Cyber Summit (NCS) Research Track 2021*; Choo, K.-K.R., Morris, T., Peterson, G., Imsand, E., Eds.; Springer International Publishing: Cham, Switzerland, 2022; pp. 110–123.
4. Malhotra, P.; Singh, Y.; Anand, P.; Bangotra, D.K.; Singh, P.K.; Hong, W.-C. Internet of Things: Evolution, Concerns and Security Challenges. *Sensors* **2021**, *21*, 1809. [CrossRef] [PubMed]
5. Mullet, V.; Sondi, P.; Ramat, E. A Review of Cybersecurity Guidelines for Manufacturing Factories in Industry 4.0. *IEEE Access* **2021**, *9*, 23235–23263. [CrossRef]
6. Wu, Z.; Shen, S.; Zhou, H.; Li, H.; Lu, C.; Zou, D. An effective approach for the protection of user commodity viewing privacy in e-commerce website. *Knowl.-Based Syst.* **2021**, *220*, 106952. [CrossRef]
7. Quincozes, S.E.; Albuquerque, C.; Passos, D.; Mossé, D. A survey on intrusion detection and prevention systems in digital substations. *Comput. Netw.* **2021**, *184*, 107679. [CrossRef]
8. Alharbi, A.; Seh, A.H.; Alosaimi, W.; Alyami, H.; Agrawal, A.; Kumar, R.; Khan, R.A. Analyzing the Impact of Cyber Security Related Attributes for Intrusion Detection Systems. *Sustainability* **2021**, *13*, 12337. [CrossRef]
9. Carta, S.; Podda, A.S.; Recupero, D.R.; Saia, R. A Local Feature Engineering Strategy to Improve Network Anomaly Detection. *Futur. Internet* **2020**, *12*, 177. [CrossRef]
10. Lu, Y.; Teng, S. Application of Sequence Embedding in Host-based Intrusion Detection System. In Proceedings of the 2021 IEEE 24th International Conference on Computer Supported Cooperative Work in Design (CSCWD), Dalian, China, 5–7 May 2021; pp. 434–439.
11. Oliveira, N.; Praça, I.; Maia, E.; Sousa, O. Intelligent Cyber Attack Detection and Classification for Network-Based Intrusion Detection Systems. *Appl. Sci.* **2021**, *11*, 1674. [CrossRef]
12. Bhati, N.S.; Khari, M. *A Survey on Hybrid Intrusion Detection Techniques BT—Research in Intelligent and Computing in Engineering*; Kumar, R., Quang, N.H., Kumar Solanki, V., Cardona, M., Pattnaik, P.K., Eds.; Springer: Singapore, 2021; pp. 815–825.
13. Zachos, G.; Essop, I.; Mantas, G.; Porfyrakis, K.; Ribeiro, J.C.; Rodriguez, J. An Anomaly-Based Intrusion Detection System for Internet of Medical Things Networks. *Electronics* **2021**, *10*, 2562. [CrossRef]
14. Díaz-Verdejo, J.; Muñoz-Calle, J.; Estepa Alonso, A.; Estepa Alonso, R.; Madinabeitia, G. On the Detection Capabilities of Signature-Based Intrusion Detection Systems in the Context of Web Attacks. *Appl. Sci.* **2022**, *12*, 852. [CrossRef]
15. Sikora, M.; Fujdiak, R.; Kuchar, K.; Holasova, E.; Misurec, J. Generator of Slow Denial-of-Service Cyber Attacks. *Sensors* **2021**, *21*, 5473. [CrossRef] [PubMed]
16. Alyami, H.; Ansari, M.T.; Alharbi, A.; Alosaimi, W.; Alshammari, M.; Pandey, D.; Agrawal, A.; Kumar, R.; Khan, R.A. Effectiveness Evaluation of Different IDSs Using Integrated Fuzzy MCDM Model. *Electronics* **2022**, *11*, 859. [CrossRef]
17. Abdel-Basset, M.; Gamal, A.; ELkomy, O.M. Hybrid Multi-Criteria Decision Making approach for the evaluation of sustainable photovoltaic farms locations. *J. Clean. Prod.* **2021**, *328*, 129526. [CrossRef]
18. Yager, R.R. Generalized Orthopair Fuzzy Sets. *IEEE Trans. Fuzzy Syst.* **2017**, *25*, 1222–1230. [CrossRef]
19. Atanassov, K. Intuitionistic Fuzzy Sets. *Fuzzy Sets Syst.* **1999**, *20*, 110–115.
20. Yager, R.R. Pythagorean fuzzy subsets. In Proceedings of the 2013 Joint IFSA World Congress and NAFIPS Annual Meeting (IFSA/NAFIPS), Edmonton, AB, Canada, 24–28 June 2013; pp. 57–61.
21. Mishra, A.R.; Rani, P. A q-rung orthopair fuzzy ARAS method based on entropy and discrimination measures: An application of sustainable recycling partner selection. *J. Ambient Intell. Humaniz. Comput.* **2021**, 1–22. [CrossRef]
22. Shang, C.; Saeidi, P.; Goh, C.F. Evaluation of circular supply chains barriers in the era of Industry 4.0 transition using an extended decision-making approach. *J. Enterp. Inf. Manag.* 2022, *in press*. [CrossRef]
23. Meng, W.; Tischhauser, E.W.; Wang, Q.; Wang, Y.; Han, J. When Intrusion Detection Meets Blockchain Technology: A Review. *IEEE Access* **2018**, *6*, 10179–10188. [CrossRef]
24. Abushark, Y.B.; Irshad Khan, A.; Alsolami, F.; Almalawi, A.; Mottahir Alam, M.; Agrawal, A.; Kumar, R.; Ahmad Khan, R. Cyber Security Analysis and Evaluation for Intrusion Detection Systems. *Comput. Mater. Contin.* **2022**, *72*, 1765–1783. [CrossRef]
25. Almotiri, S.H. Integrated Fuzzy Based Computational Mechanism for the Selection of Effective Malicious Traffic Detection Approach. *IEEE Access* **2021**, *9*, 10751–10764. [CrossRef]
26. Sharma, S.; Kaul, A. Hybrid fuzzy multi-criteria decision making based multi cluster head dolphin swarm optimized IDS for VANET. *Veh. Commun.* **2018**, *12*, 23–38. [CrossRef]

27. Ogundoyin, S.O.; Kamil, I.A. A Fuzzy-AHP based prioritization of trust criteria in fog computing services. *Appl. Soft Comput.* **2020**, *97*, 106789. [CrossRef]
28. Kumar, R.; Alenezi, M.; Ansari, M.; Gupta, B.; Agrawal, A.; Khan, R. Evaluating the Impact of Malware Analysis Techniques for Securing Web Applications through a Decision-Making Framework under Fuzzy Environment. *Int. J. Intell. Eng. Syst.* **2020**, *13*, 94–109. [CrossRef]
29. Duan, W.-Q.; Gulistan, M.; Abbasi, F.H.; Khurshid, A.; Al-Shamiri, M.M. q-Rung double hierarchy linguistic term set fuzzy AHP; applications in the security system threats features of social media platforms. *Int. J. Intell. Syst.* **2021**, 1–34. [CrossRef]
30. Panityakul, T.; Mahmood, T.; Ali, Z.; Aslam, M. Analyzing and controlling computer security threats based on complex q-rung orthopair fuzzy heronian mean operators. *J. Intell. Fuzzy Syst.* **2021**, *41*, 6949–6981. [CrossRef]
31. Cheng, S.; Jianfu, S.; Alrasheedi, M.; Saeidi, P.; Mishra, A.R.; Rani, P. A New Extended VIKOR Approach Using q-Rung Orthopair Fuzzy Sets for Sustainable Enterprise Risk Management Assessment in Manufacturing Small and Medium-Sized Enterprises. *Int. J. Fuzzy Syst.* **2021**, *23*, 1347–1369. [CrossRef]
32. Peng, X.; Dai, J.; Garg, H. Exponential operation and aggregation operator for q-rung orthopair fuzzy set and their decision-making method with a new score function. *Int. J. Intell. Syst.* **2018**, *33*, 2255–2282. [CrossRef]
33. Wei, G.; Gao, H.; Wei, Y. Some q-rung orthopair fuzzy Heronian mean operators in multiple attribute decision making. *Int. J. Intell. Syst.* **2018**, *33*, 1426–1458. [CrossRef]
34. Liu, P.; Wang, P. Some q-Rung Orthopair Fuzzy Aggregation Operators and their Applications to Multiple-Attribute Decision Making. *Int. J. Intell. Syst.* **2018**, *33*, 259–280. [CrossRef]
35. Darko, A.P.; Liang, D. Some q-rung orthopair fuzzy Hamacher aggregation operators and their application to multiple attribute group decision making with modified EDAS method. *Eng. Appl. Artif. Intell.* **2020**, *87*, 103259. [CrossRef]
36. Liu, Z.; Liu, P.; Liang, X. Multiple attribute decision-making method for dealing with heterogeneous relationship among attributes and unknown attribute weight information under q-rung orthopair fuzzy environment. *Int. J. Intell. Syst.* **2018**, *33*, 1900–1928. [CrossRef]
37. Radoglou-Grammatikis, P.I.; Sarigiannidis, P.G. Securing the Smart Grid: A Comprehensive Compilation of Intrusion Detection and Prevention Systems. *IEEE Access* **2019**, *7*, 46595–46620. [CrossRef]
38. Martinez, C.V.; Sollfrank, M.; Vogel-Heuser, B. A Multi-Agent Approach for Hybrid Intrusion Detection in Industrial Networks: Design and Implementation. In Proceedings of the 2019 IEEE 17th International Conference on Industrial Informatics (INDIN), Helsinki-Espoo, Finland, 22–25 July 2019; Volume 1, pp. 351–357.
39. Badotra, S.; Panda, S.N. SNORT based early DDoS detection system using Opendaylight and open networking operating system in software defined networking. *Clust. Comput.* **2021**, *24*, 501–513. [CrossRef]

sensors

MDPI

Article

A Perspective on Passive Human Sensing with Bluetooth

Giancarlo Iannizzotto [1,*], Miryam Milici [1], Andrea Nucita [1] and Lucia Lo Bello [2]

[1] Department of Cognitive Sciences, Psychology, Education and Cultural Studies (COSPECS), University of Messina, 98122 Messina, Italy; miryam.milici@studenti.unime.it (M.M.); anucita@unime.it (A.N.)
[2] Department of Electrical, Electronic and Computer Engineering (DIEEI), University of Catania, 95124 Catania, Italy; lucia.lobello@unict.it
* Correspondence: ianni@unime.it

Abstract: Passive human sensing approaches based on the analysis of the radio signals emitted by the most common wireless communication technologies have been steadily gaining momentum during the last decade. In this context, the Bluetooth technology, despite its widespread adoption in mobile and IoT applications, so far has not received all the attention it deserves. However, the introduction of the Bluetooth direction finding feature and the application of Artificial Intelligence techniques to the processing and analysis of the wireless signal for passive human sensing pave the way for novel Bluetooth-based passive human sensing applications, which will leverage Bluetooth Low Energy features, such as low power consumption, noise resilience, wide diffusion, and relatively low deployment cost. This paper provides a reasoned analysis of the data preprocessing and classification techniques proposed in the literature on Bluetooth-based remote passive human sensing, which is supported by a comparison of the reported accuracy results. Building on such results, the paper also identifies and discusses the multiple factors and operating conditions that explain the different accuracy values achieved by the considered techniques, and it draws the main research directions for the near future.

Keywords: Bluetooth; wireless passive human sensing; wireless sensor networks

Citation: Iannizzotto, G.; Milici, M.; Nucita, A.; Lo Bello, L. A Perspective on Passive Human Sensing with Bluetooth. *Sensors* **2022**, *22*, 3523. https://doi.org/10.3390/s22093523

Academic Editor: Antonio Lázaro

Received: 4 April 2022
Accepted: 1 May 2022
Published: 5 May 2022

1. Introduction

In general terms, radio-based passive human sensing refers to the ability to remotely discern the presence and, possibly, the activities, of human beings, without the need for them to bring or wear any device and for the sensing device to emit a probing signal [1]. In fact, the sensing device relies on external sources for the generation of the probing signal by exploiting the radio signal emitted by the most common wireless communication devices (e.g., WiFi and Bluetooth). More specifically, the radio signal is received, processed and the required information is extracted from the deformation and degradation that affect the signal when the presence of a human body obstructs its trajectory (see Figure 1a).

Comparing with conventional, computer-vision based approaches, such as those described in [2–5], passive remote sensing offers several important advantages. First, there is easier user acceptance, as wireless networks are not able to record images, and therefore, they are not perceived as privacy-invasive by the perspective users. Second, they have wider applicability, as wireless networks do not suffer from Line-of-Sight (LoS) obstruction and bad illumination conditions and may also support through-the-wall sensing. LoS sensing is shown in Figure 1a), whereas non-LoS sensing and through-the-wall visibility are illustrated in Figure 1b).

Finally, deployment efforts and costs may be reduced, as passive sensing may take advantage of the existing wireless communication infrastructures and sensor networks [6], such as the ones used for public utility and safety [7].

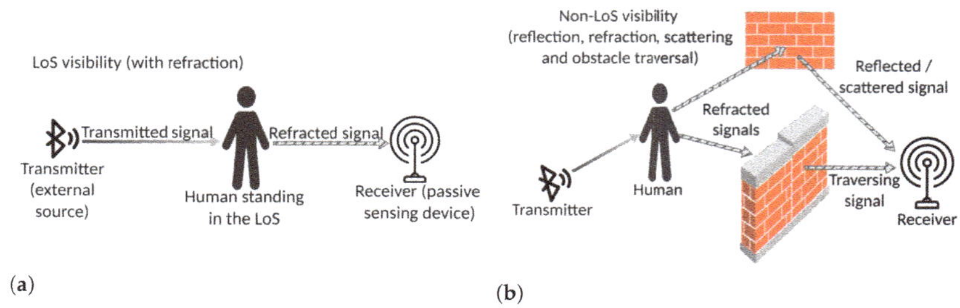

Figure 1. Line of Sight (LoS) visibility (**a**). Non-Line of Sight and through-the-wall (**b**).

In the literature, passive (also called deviceless) remote sensing of human beings was first developed using ad hoc RF transmitters and receivers. Next, WiFi off-the-shelf technology, already successfully applied to a variety of other applications, such as patient monitoring [8–10] and factory automation [11], was adopted for this goal. The reason for relying on WiFi mainly resides in the high transmission rate, which allows high-frequency sampling, and in the availability of very detailed Channel State Information (CSI), which allows a stable and reliable extraction of the presence, position and activity information of the human subject. Nevertheless, the Bluetooth technology offers some interesting advantages over WiFi, which suggest the investigation of its application to the passive remote sensing area. In fact, recent studies demonstrated that Bluetooth represents a valid alternative for several reasons:

- Bluetooth is integrated in most portable devices (such as tablet, smartphone, PDA, etc.), and it is often used for personal health monitoring [12–14].
- It is energy-efficient (its power consumption is lower than WiFi), in particular after the introduction of the Bluetooth Low Energy specifications in Bluetooth 4.0.
- Its deployment in business, industrial and home environments is simple and flexible, as Bluetooth devices are small, minimally invasive and less expensive than other solutions [15].

Despite the aforementioned advantages, the number of studies that have investigated the application of Bluetooth to passive human sensing so far is significantly lower than the number of works based on WiFi. This consideration motivates this paper, which presents an overview of the most relevant works dealing with Bluetooth-based passive human sensing. The paper provides a thorough analysis of the data preprocessing and classification techniques proposed in the literature, which is followed by a comparison of the reported accuracy results. Building on the presented results, the paper identifies and discusses the multiple factors and operating conditions that explain the different accuracy results achieved by the considered techniques. On the basis of this discussion, the paper draws the main research directions for the near future.

To the best of our knowledge, there are no other reviews, in the literature, dealing with Bluetooth-based wireless passive human sensing at the time of writing this work.

The paper is structured as follows. Section 2 discusses related work on human body passive sensing through wireless communication technologies, with a special focus on Bluetooth.

Section 3 addresses the Bluetooth protocol features that are relevant to the considered application, describing how the human body affects the received Bluetooth signal.

Section 4 discusses data collection, analysis and preprocessing techniques, while Section 5 analyzes the classification methods adopted in the relevant literature.

Section 6 proposes a performance comparison of the results presented in the considered literature, while Section 7 discusses the results obtained by each described technique, deriving insights on the factors that affect their accuracy. Finally, Section 8 provides conclusive remarks and hints for future research.

2. Related Work

Most of the literature dealing with Bluetooth indoor signal propagation and the effect of the human body focuses on active (or device-based) indoor localization [16–27], where the human to be localized carries, or wears, a Bluetooth device that can track, or can be tracked by, other Bluetooth devices in the environment. This approach belongs to the active sensing class, where either the sensing devices are equipped with both a transmitter and a receiver (see Figure 2 on the left), or the sensed human carries an active device (see Figure 2 on the right).

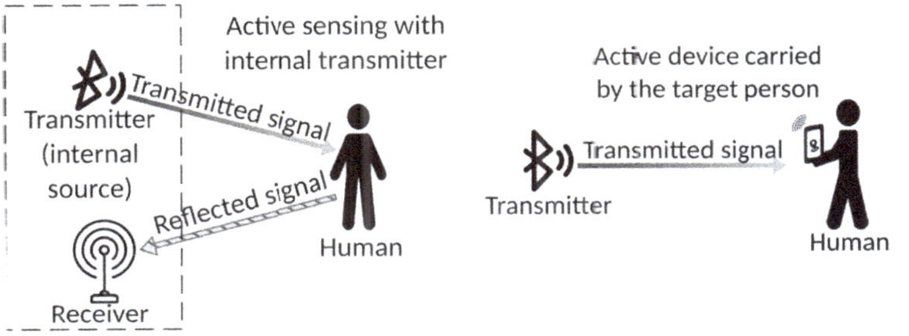

Figure 2. Two classes of active sensing. On the **left**, the sensing device is equipped with both internal probing signal transmitter and receiver. On the **right**, the sensed human brings an active device that receives the probing signal and produces the sensing result.

Device-based sensing approaches can as well be used to detect and track human beings and their activities [28,29]. However, requesting the subjects to carry a traceable device all the time is not always feasible or desirable. Passive, device-free sensing systems, instead, take advantage of the human body's property of disturbing the radio signals when partially or totally obstructing their trajectory, thus causing signal fluctuations. In principle, by analyzing such fluctuations, it could be possible to separate the influence of the human body from both the original signal and other causes of noise, thus allowing human detection, positioning and activity recognition. However, modeling the "baseline" noise and signal fluctuations that can be expected in the absence of the obstructing human body, thus allowing a reliable detection and tracking, is not a trivial process [30]. The Bluetooth standard stack only allows to measure the Received Signal Strength Information (RSSI), which is a scalar indication of the intensity of the received signal at a given time and is subject to fluctuations and fading during normal operation. The WiFi standard, instead, allows the measurement of the much more informative Channel State Information that, for the most recent devices, is a 3D matrix of complex values representing the amplitude attenuation and phase shift of multi-path WiFi channels [31]. Moreover, the CSI provides more stable information than the RSSI [32,33]. Consequently, the literature regarding the use of WiFi for passive human detection and activity tracking is much wider than the literature on Bluetooth for similar applications.

Human presence can influence a wireless signal in its spatial, temporal and frequency domains [34]. In particular, human motion can affect the frequency and temporal domains, whereas the location of the human body in the environment can affect the signal's phase. The RSSI alone does not provide phase information and therefore, some other source of information is needed when the CSI is not directly available. Recently, Bluetooth v5.1 [35] introduced the possibility to measure the signal's Angle of Arrival and Angle of Departure, which can be used to improve both the positioning and activity tracking accuracy. However, to the best of the authors' knowledge, such an approach has not been reported in the literature so far.

Multiple RSSI samples can be used to extract more information, either in time or in space. In time, for example, by measuring through a fixed, static receiver a time sequence of RSSI samples emitted by a fixed, static transmitter, it is possible to track human body motion and recognize specific activities (see Figure 3a).

In space, instead, by deploying several transmitters and receivers, a very accurate positioning of a human body can be obtained by exploiting geometrical considerations [36] (see Figure 3b).

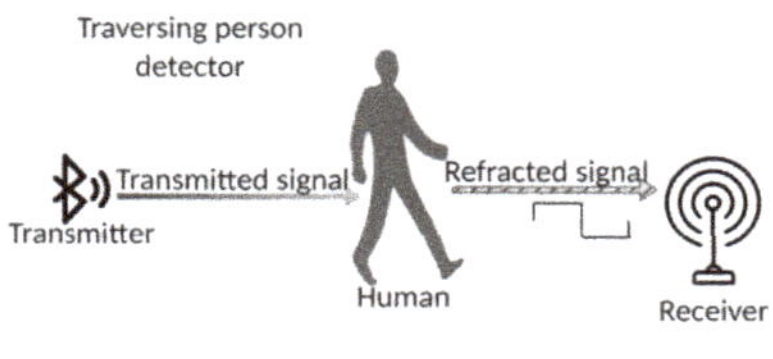

(a)

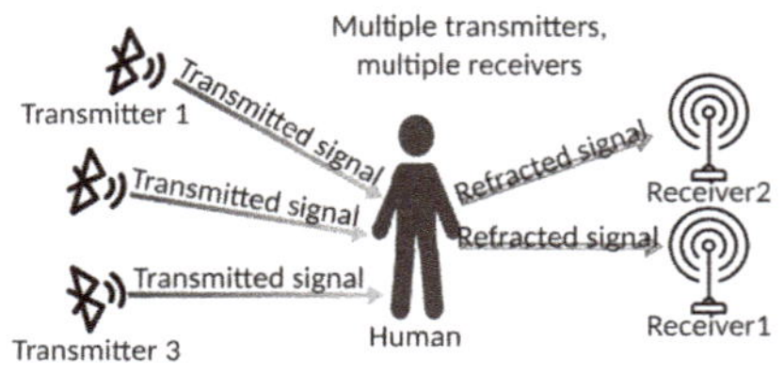

(b)

Figure 3. Moving humans counter and multi-transmitter, multi-receiver configurations. (**a**) Moving humans counter. The RSSI undergoes an abrupt change when a human body traverses the signal trajectory. (**b**) Multiple transmitters, multiple receivers.

A special case is based on the radio tomographic approach, where a very high number of transmitter/receiver couples over a spacial grid (see Figure 4) allows to use imaging techniques to obtain an extremely detailed and accurate positioning of the subject [37].

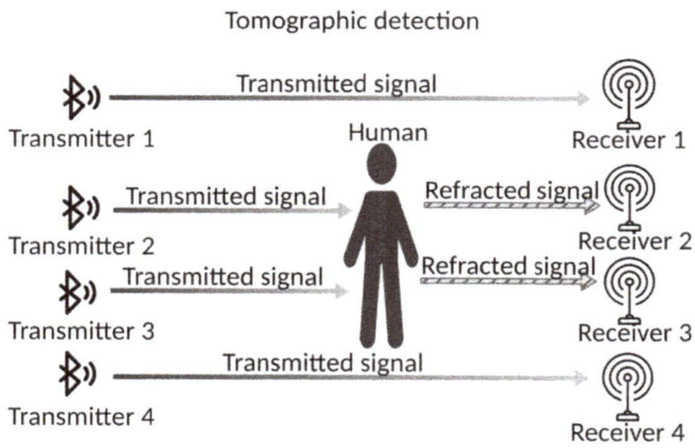

Figure 4. Tomographic sensing.

In principle, any wireless transmission technology can support some form of remote human sensing approach. In particular, an interesting comparison is reported in [6], where, curiously, Bluetooth is not mentioned. Table 1 summarizes the main pros and cons of exploiting some of the most popular communication technologies for wireless remote human sensing, such as the ones listed below.

Radio Frequency Identification (RFID) is a technology for contactless, short-range, two-way communications that is mainly used for tag detection and identification as well as for low datarate short message exchange. The main advantages of RFID technology are the reasonable resilience to radio frequency noise and the low cost of RFID tags. Conversely, the RFID reader is quite expensive [6]. The main disadvantage of using RFID for passive human sensing is the very short sensing range.

Frequency Modulated Continuous Wave (FMCW) radars are instead active sensing devices in which the sensor contains both a transmitter and a receiver [38,39]. The main advantages of applying FMCW to human sensing are the high sensitivity and distance accuracy [6]. However, FMCW-based devices are specifically aimed at sensing, and therefore, a human sensing approach based on this technology cannot leverage existing communication infrastructures, but a separate deployment is needed. Moreover, FMCW-based human sensing is not in the scope of this study, as the focus here is on passive human sensing.

Bluetooth and its low-energy version (BLE) are among the most widely adopted wireless communication technologies. Their main advantages are low power consumption, which allows the installation of battery-operated nodes, and the relatively low deployment costs, even in case a specific infrastructure is required. Moreover, such technologies have the ability to support both regular communications and passive human sensing at the same time, as regular advertisement signals are exploited for sensing (see, for example, [40,41]). The main disadvantages of Bluetooth and BLE are that they currently do not provide CSI and an effective way to cope with the abrupt RSSI changes due to Frequency Hopping (see Section 3.2).

WiFi is the wireless communication technology that is most widely adopted for passive human sensing. Early works in the literature exploited the WiFi RSSI values, but the most accurate and effective applications are based on the ability of some recent devices to provide Channel State Information (CSI) multidimensional samples, thus supporting a more detailed extraction of human activity information [31]. The main advantages of using WiFi for human sensing are the wide diffusion of the WiFi communication technology and the reliable and fine-grained information provided by the CSI, which allows sensing low-amplitude activities (e.g., hand gestures) and vital signs (e.g., breathing). However, WiFi devices cannot be used for communications and passive sensing at the same time [6], and therefore, a separate infrastructure is needed for sensing or the same infrastructure must be multiplexed in time, thus reducing the time available for both services. Moreover, WiFi power consumption is sensibly higher than Bluetooth power consumption (this is particularly true for Bluetooth Low-Energy), and the deployment cost of an ad hoc infrastructure is also higher. Finally, WiFi is relatively less robust to changes in the environment and noise than other signals [6].

Visible Light Communication (VLC) technology is also being exploited for human sensing [42]. VLC technology uses low-cost, high-efficiency photodiodes (LED), is resilient to RF noise, and it can leverage the existing lighting infrastructure. However, such a technology requires a complex light sensing infrastructure that is hardly justifiable by its application to human sensing, as a large number of photodiodes need to be installed on the sensorized environment floor [6].

LoRa [43] is a radio frequency transmission technique based on a spread spectrum modulation, which enables long-range transmissions with low power consumption [44]. Both properties are interesting for human sensing. In the work in [45], LoRa is used for active remote human sensing in a radar-like fashion, by equipping the sensing device with both a transmitter and a receiver modified and adapted to the specific application. However, LoRa requires a dedicated infrastructure for sensing.

The *LTE* (long-term evolution) mobile communication system [46], which seamlessly covers almost all areas, both indoor and outdoor, can act as a diffused external illuminator for wireless human sensing [47].

The 6G communication networks also promise localization and human activity detection among their most relevant services [48]. The two main disadvantages of using LTE and 6G for remote human sensing are the long distance between the transmitter, the receiver and the human, which require complex filtering techniques to remove the noise and separate the signals [6], and the significant risk of privacy violation due to both the non-locality of the probing signal and the potential for receiving the sensing signal from a long distance, which would allow for mass monitoring.

Table 1. Pros and cons of the most relevant applications of wireless communication technologies to wireless remote human sensing, as reported in the literature.

Wireless Technology	Advantages	Disadvantages
RFID	Resilience to RF noise	Very short range
FMCW	High sensitivity; High distance resolution	Active approach; Ad hoc infrastructure
Bluetooth, BLE	Widespread use for communications; Low power consumption; Low deployment cost; Simultaneous communication and sensing	No support for CSI; Abrupt changes in RSSI due to FHSS
WiFi	Widely adopted for communications; High spatial resolution; Reliability	Not all devices provide CSI; Non-simultaneous communications and sensing; Relatively high deployment cost; Relatively high power consumption
VLC	Resilience to RF noise; Relatively cheap sensing devices	Complex ad hoc sensing infrastructure
LoRa	Long sensing range; Low power consumption	Active approach
LTE, 6G	Wide availability of the illuminating signal; Stability and reliability	Complex noise filtering and signal separation techniques; Severe privacy concerns

One of the earliest and simplest applications of Bluetooth to deviceless human sensing is to support remote elderly care, by allowing the remote caregivers to monitor that their patients regularly take their prescriptions [40,41]. The described system consists of three principal components:

- A Bluetooth beacon positioned under a medicine calendar (i.e., a tool used to norm the assumption of the correct daily dose of drugs).
- A computer tablet positioned at fixed, short distance from the beacon.
- A smartphone for the remote caregiver.

In this application, the Bluetooth beacon periodically transmits a message with a unique ID, while the tablet measures the received signal RSSI. When a human approaches the medicine calendar, thus disturbing the signal sent by the beacon, the RSSI suddenly drops, and the dedicated application running on the tablet records the event and sends a notification to the caregiver's smartphone. The detection accuracy obtained by the system in a real environment is higher than 90%.

More recently, by measuring and analyzing the fluctuations in the received signal RSSI, a passive device-free system based on a network of Bluetooth Low Energy (BLE) beacons was able to detect the presence [49] and estimate the number [50] of humans in a lecture room. With 24 beacons positioned under the seats in the room and four receivers positioned on the ceiling laterally to the seats, the detection accuracy was higher than 95%, while the accuracy of the number of people estimated was about 82%, with an average room occupancy of about 30 attendees.

In [51], a system to detect the presence of humans in a waiting queue is presented. Usually, waiting queues are controlled by barrier poles with retractable belts, which can be moved if needed, and the system described in [51] is able to detect the transit of people between two barrier poles so as to allow to estimate the queue waiting time. To this aim, the transmitting and receiving devices are positioned on the barrier poles, facing from the two opposite sides of the queue path, and two detection methods are studied:

- Analysis of RSSI variance;
- Analysis of RSSI average.

The results reported in [51] show that the RSSI variance is better suited for detecting the transit of people between the two poles, i.e., the motion of a body traversing the ideal line connecting the two poles (see Figure 3a), while the RSSI average is more suitable for detecting people standing between the two poles. The accuracy of this system in detecting walking people is about 98% using variance and 96% using average. However, the second algorithm is also able to detect motionless people standing in the monitored area, whereas the first algorithm is only suitable for detecting moving people.

3. Human Body Influence on the Bluetooth Signal

In order to better understand the advantages and limitations of using the Bluetooth technology for passive human detection and tracking, this section recapitulates some relevant Bluetooth features.

3.1. Bluetooth Recapitulation

Bluetooth is a short-range wireless protocol that supports connections within a range from a few meters to a few dozen meters (depending on the antenna gain, the environment and the specific PHY adopted) [52].

The Bluetooth Classic protocol, also referred to as Bluetooth Basic Rate/Enhanced Data Rate (BR/EDR), works in the globally unlicensed Industrial, Scientific and Medical (ISM) 2.4 GHz short-range radio frequency band, which is divided into 79 channels with 1 MHz spacing. The Bluetooth technology is designed to work well even in very noisy environments, copying with fading and interference. To this end, Bluetooth uses the Frequency Hopping Spread Spectrum (FHSS) [53] technique, that forces two connected devices (master and slave) to frequently change the dedicated communication channel, so the devices hop from one channel to another according to a pseudo-random sequence. The channel sequence is maintained by the master device through a map in which the channels are marked as 'in use' if they work properly and 'unused' otherwise. This map is updated after a channel is found working well for a given time interval and is shared with the secondary devices so as to have the same information at both ends of the communication.

Two types of Bluetooth connections are available:

1. Asynchronous Connection-Less: a uni-directional communication, in which a slave device (also called advertiser or broadcaster) periodically sends packets, while a master device (also called hub or scanner) continuously scans the channels while waiting for packets.
2. Synchronous Connection-Oriented: a bi-directional communication, in which a connection between a master and a slave is established over a dedicated channel.

The most relevant advantages of the Bluetooth technology are the resilience to noise and interference, thanks to the FHSS technology, the low cost of the devices, and the low power consumption compared to other technologies. In particular, the introduction of Bluetooth Low Energy (BLE) has further reduced the power consumption, thus improving the lifetime of battery-powered devices. For this reason, BLE is also widespread in Internet of Things (IoT) applications.

3.1.1. Bluetooth Low Energy

Bluetooth Low Energy divides the ISM band into 40 channels with 2 MHz spacing. Three channels are used as advertising channels (or primary channels) for broadcasting, while the remaining 37 data channels (or secondary channels) are used for data exchange after a connection event. Mobile devices can be designed to support both Bluetooth Classic and BLE (dual-mode device) or to only support BLE (single-mode device). Single-mode devices are generally used in applications that require low power consumption as a major constraint. Moreover, the BLE stack is thinner compared to the Bluetooth Classic one, to reduce the firmware footprint and protocol management complexity for the applications that run directly on sensors [54].

The first version of BLE protocol was introduced with Bluetooth v4.0, and then, it was updated up to Bluetooth v5.3.

Comparing with the previous versions, in Bluetooth v5.0, major improvements were introduced. First, the coverage range was extended from about 50 m to more than 200 m for outdoor environments, whereas in indoor environments, the range changed from about 10 m to about 40 m. Furthermore, the Bluetooth Core Specification v5.0 [55] introduced a new way to perform advertising, called Extended Advertising, which allows the 37 channels previously reserved for data communication to be also used as secondary advertising channels. Traditional advertising transmits the same payload on the three primary channels, whereas the Extended Advertising transmits payload data only once on a secondary channel. This way, the total amount of transmitted data is lower, and therefore, the duty cycle is reduced. Another benefit of the Extended Advertising is that using a secondary channel to transmit the payload, 255-byte-long packets can be broadcast. In the previous versions of the protocol, instead, only 31-byte-long packets could be broadcast. With this version of BLE, it is also possible to chain packets together and transmit each chained packet on a different channel.

Bluetooth v5.0 also introduced the Periodic Advertising, which allows the receivers to synchronize their scanning for packets with the schedule of the transmitter device. In the previous versions, the advertising process included a degree of randomness in the timing of the advertising packet transmission to avoid repeated packet collisions. However, this implies that scanners could lose some packets, i.e., those transmitted out of their round of scan. In the Periodic Advertising, scanning is performed within the transmission window of the transmitter, thus avoiding such packet losses, and therefore in a more power-efficient way.

In addition, Bluetooth v5.0 reduced the minimum allowed Advertising Interval from 100 ms to 20 ms, thus allowing a rapid recognition of the advertising beacons.

Bluetooth v5.1 [35] introduced a new feature that allows Bluetooth devices to determine the direction of a Bluetooth incoming transmission.By equipping either the receiver or the transmitter with an array of antennas, the receiver can determine either the Angle of Arrival (AoA) or the Angle of Departure (AoD), respectively [56].

In both methods, a special signal, called a direction-finding signal, is transmitted by the transmitting device and used by the receiving device to calculate the direction of the received signal, which in ideal conditions corresponds to the direction along which the transmitting device lays. In the AoA method, the receiving device (that is connected to an array of antennas) receives different copies of the same signal from different consecutive antennas in the array. The received signals are phase-shifted due to the different distances of the receiving antennas to the single transmitting antenna. The Angle of Arrival θ is computed from the phase difference according to Equation (1), where λ is the wavelength, ψ is the phase difference and d is the distance between two consecutive antennas in the array [56].

$$\theta = arccos(\frac{\psi\lambda}{2\pi d}) \tag{1}$$

In the AoD method, instead, the transmitting device is equipped with an antenna array which emits a signal from each of the antennas. The receiving device, which is equipped with a single antenna, given that λ is the wavelength and d is the distance between two consecutive antennas in the transmitting array, determines the phase difference ψ from two received signals and computes the direction θ according to Equation (2).

$$\theta = arcsin(\frac{\psi\lambda}{2\pi d}) \tag{2}$$

In Bluetooth 5.1, the direction-finding signals are generated by both defining a new Link Layer Protocol Data Unit (PDU) for direction finding between two connected devices, and a way to use the existing advertising PDUs for connectionless direction finding. In both cases, a special field, known as the Constant Tone Extension (CTE), is added to the end of the PDUs.

Bluetooth v5.1 also introduced the Randomized Advertising Channel Indexing, which allows the devices in advertising state to randomize the selection of advertising channels so that they are not selected in strict order (37, 38, 39) as in the previous Bluetooth versions but in a random order. This feature improves packet collision avoidance.

In Bluetooth v5.2 [57], the LE Power Control feature was introduced. This new feature allows the transmitting devices to dynamically change their transmission power level and inform the receiving device that this has happened. This process is useful to keep high the signal quality and low the error rates, respectively. Moreover, the process also improves coexistence in general, thus benefiting all the protocols working in the 2.4 GHz frequency band.

Bluetooth v5.3 [58] introduced some improvements in the Periodic Advertising, i.e., the host of a receiving device can inform the controller that a packet has been found to contain data that has already been received in an earlier packet. The controller can, therefore, disregard the current periodic advertising packet and immediately switch to another channel.

3.2. *Influence on Bluetooth Signal*

Numerous studies proved that radio signals in the 2.4 GHz frequency band are easily influenced by the human body. An experiment explicitly aimed at demonstrating the effects of a human body occluding the Line-of-Sight between a transmitter and a receiver (see Figure 1a) was reported in 2014 [59], where the received signal power was measured with and without obstruction, and the measured attenuation was approximately 10 dB. However, signal fluctuations were exploited to detect the presence of people in an environment already in 2006 [60].

Received signal fluctuations can be caused by changes in the environment, in the state of the transmitter or the receiver, or by design, e.g., because the protocol requires actions that produce such fluctuations. In order to exploit the signal fluctuations for passive human sensing, it is mandatory to exclude the unwanted causes or, at least, to find a way to separate their effects from the effects of the targeted human body. Compared to WiFi, Bluetooth appears to be better suited for this purpose, because it is less subject to electromagnetic noise thanks to the FHSS technique. However, the FHSS itself by design generates fluctuations in the received signal due to the communication channel switch. In fact, measurements at the receiver side of the signal emitted by a Bluetooth beacon in advertising mode showed that different channels have different noise and attenuation characteristics [20]. To reduce the influence of the channel hopping on the signal RSSI, each Bluetooth channel should be modeled and processed separately [16]. In the reported experiments, four BLE beacons were used. In particular, three beacons were modified to allow a separate RSSI measurement for each advertising channel, while the other one was not modified. After measuring the RSSI values of all beacons, the RSSI variance was calculated for each beacon. The modified beacons allowed the calculation of separate variances for each channel, while the non-modified beacon only allowed the calculation of the variance of the hybrid signal resulting from the channel hopping. As the Bluetooth protocol does not allow the determination of the time instant at which the hopping takes place, and it does not report the current advertising channel, it was not possible to filter out the RSSI fluctuations produced by the hopping. Consequently, the hopping contributed to the overall RSSI variance. As it was expected, the RSSI variance calculated over the signal transmitted by the unmodified beacon was significantly larger than the per-channel RSSI variances obtained by the modified beacons.

The described study demonstrated that the FHSS causes a significant increase of the RSSI fluctuations, which would heavily impact the reliability of a passive human detection approach based on a set of unmodified Bluetooth devices. For this reason, most of the research works in this area rely on some way to manage each advertising channel separately.

4. Data Preprocessing

The demonstration that the wireless signal generally suffers from the noise due to the surrounding environment is introduced in [61]. Moreover, endogenous fluctuations, i.e., the ones coming from inside the device, always affect the received signal. The aim of the preprocessing stage is ideally the elimination, and more in general the attenuation, of noise and endogenous fluctuations, so that the only fluctuations present in the signal after this step are those generated by the human presence between the connected devices.

In [62], a Kalman filter was used for signal noise reduction. The Kalman filter uses linear models, i.e., linear transformations both in the transitions from the current state to the next state and in the transformation from state to measurement, and it also assumes that the noise associated to both the measurements and state of the system is Gaussian. Unfortunately, the hypotheses of linearity and Gaussian noise are not always satisfied in wireless passive human sensing, as shown in [61], where the Kalman filter and the Moving Average filter were compared.

In [62], the Kalman filter performed well with the available data, whereas in [61], such a filter performed poorly, flattening the signal to the point where relevant peaks corresponding to the presence of a human body were deleted. The Moving Average filter, instead, is simpler and less accurate than the Kalman filter under linearity and Gaussian noise hypotheses, but it performed better on RSSI data [61].

An advanced version of the Moving Average filter, i.e., the Exponential Weighted Moving Average filter, was adopted in [22]. Unlike the Moving Average filter, this filter assigns different weights to the values of the considered time series according to an approximated exponential law. The filtered RSSI value is computed according to Equation (3), where $\alpha \in [0,1]$ is the smoothing factor. In particular, the value of α in [22] is 0.05.

$$f_{i,j}[n] = \alpha RSSI_{i,j}[n] + (1-\alpha) f_{i,j}[n-1] \tag{3}$$

A similar principle is used in [51], where data preprocessing consists of computing the weighted average according to Equation (4), where $\alpha = 0.9$ produces a larger weight for the previous values than for the current value, thus reducing the effect of noisy outliers.

$$Mean_{current} = \alpha Mean_{old} + (1-\alpha) RSSI_{current} \tag{4}$$

In [63], a passive detection system is described, in which the signal is preprocessed using an α-trimmed mean filter. The α-trimmed filter is generally used in very high noise conditions. In this case, calculating the mean of the signal as the average of the samples is not recommended, because an outlier might significantly alter it. Instead, given a sliding window with q RSSI values, this filter sorts the RSSI values and then removes α extremes. After this process, the average of the remaining values is computed according to Equation (5), where $0 \le \alpha \le 0.5$.

$$f(q;a) = \frac{1}{q - 2\lceil \alpha q \rceil} \sum_{i=\lceil \alpha q \rceil + 1}^{q - \lceil \alpha q \rceil} RSS_i \tag{5}$$

The described process has an offline phase, which produces an estimate of the initial parameters of the system, and an online phase, which produces an estimate of the detected human bodies location, according to the received RSSI samples. Since the environment may significantly change between the offline estimation phase and the online phase, the Analysis of Variance (ANOVA) [64] is adopted to check if the estimate of the initial parameters is still valid.

In [16], a localization system is discussed that uses BLE beacons with a modified firmware able to select a specific channel (either 37, 38, or 39) to send the advertising packets through. This way, one of the causes of increased RSSI variance is removed, but a new problem is introduced. In fact, some packets may get lost in some channels during a scan round due to the fact that the transmitter might send its packet out of the receiver scan interval. At each scan round, the receiver stores the RSSI values from each channel in a vector of RSSI values that represents the RSSI sample for that round. If a packet is lost, the vector has one component missing, and this leads to errors in the analysis steps. As a consequence, some method to fill the missing component of the vector is needed. First, the missing sample component must be detected, and then, an adequate value must be selected to replace the missing value. The chosen solution is that the missing sample component is replaced by the median value of the last WL_1 samples from the same channel.

Other preprocessing methods based on statistic techniques are available in the literature. The works in [49,50] describe a passive system that is able to detect the presence, and count the number, of human beings in a lecture room. To train and test this system, three datasets were used:

- DS1, containing raw RSSI values.
- DS2 and DS3, containing preprocessed data.

The first dataset was produced by collecting five measurement sessions for each lecture, thus obtaining a total of $80 \times 5 = 400$ sessions. The installation was composed of 24 transmitting beacons and four receivers. Provided that each beacon sent one frame per second, and that each session lasted 300 s, for each session, a total of $24 \times 4 \times 300 = 28{,}800$ samples were collected. To reduce the dataset size, a summarized dataset DS2 was built by calculating, for each measurement session, four features, i.e., mean, variance, trimmed-mean, and trimmed-variance, which were calculated according to Equation (5). Next, the position of the beacons was taken into account by introducing a weighting factor depending on the receiver–transmitter distance. The weighting factor was a normalization matrix W, which is used to apply a Min/Max normalization to the data. The set of normalized data was the third dataset DS3. Experimental results show that with the described preprocessing approach, the performance improved from 70% to 98%.

Recently, Deep Neural Networks are being applied also to passive human sensing. The main advantage of deep learning techniques is that they do not need, in general, to manually build specific features for the classification process. Instead, such features are calculated by a neural network that is adequately trained over large amounts of data [65], thus consistently reducing the amount of preprocessing. In [66], for example, raw RSSI values are simply averaged over one-minute intervals and then fed to a deep neural network that is trained to output an estimate of the initial and final position of a human moving in the monitored area.

5. Classification Methods

The lack of protocol support to gather the Channel State Information (CSI) sensibly reduces the amount of information that can be obtained by analyzing the received signal. Most likely for this reason, to the best of the authors' knowledge, as reported in Section 6, the current literature regarding human body sensing through Bluetooth mainly focuses on indoor passive people detection, counting, and approximate motion tracking, whereas complex activities recognition and vital signs sensing have not yet been investigated.

For the considered applications, statistic techniques, Machine Learning techniques, and Artificial Neural Networks are adopted in the literature, as described in the following subsections.

5.1. Statistical Methods

The approaches defined as "statistic" use methods based on the analysis of either the mean or the variance of the RSSI values. In [51], two algorithms based on these techniques are presented. The first algorithm, based on the analysis of the variance of a stream (sequence) of RSSI values, is more suitable to detect the motion of a human body, for example, when entering an area of interest. In this case, the RSSI values are detected in subsequent time instants, and therefore, their sequence represents the evolution of the Bluetooth signal perturbation when a human body moves in the area between the receiver and the transmitter devices.

In general, the RSSI variance is calculated over a sliding window always containing the last n samples according to Equation (6), where x_i represents the current RSSI value and μ is the average of all the RSSI values within the sliding window.

$$Var(x) = \frac{1}{n}\sum_{i=1}^{n}(x_i - \mu)^2 \tag{6}$$

The size n of the sliding window must be chosen according to the specific application. If n is too large, too old RSSI values are considered in the calculation, and this could lead to delayed detection. If n is too small, the smoothing effect of previous samples on the calculation is not sufficient to remove the measurement noise, thus leading to false positives or false negatives in the classification. In [51], the optimal value of n is 10. It was determined considering the time it takes a human to cross the monitored area and the number of messages received for second. Moreover, the mean μ in Equation (6) is replaced with a weighted average computed according to Equation (4).

In the second algorithm reported in [51], the weighted average calculated according to Equation (4) is subtracted from each new RSSI sample and, if the resulting value exceeds a given threshold T_m, then a detection event is triggered. The weighted average is not updated when a detection event is triggered, because the average is intended to represent the "background" condition, where no human body is present in the monitored area.

The threshold T_m is critical and requires a complex calculation involving a preliminary calibration, which is repeated every time something changes in the system deployment, and an initial stabilization time interval that is needed every time the system is turned on or reset. Moreover, the formulation of T_m includes some constants that were experimentally determined on the basis of the available data and setup and that might potentially undermine the generality of the proposed approach.

5.2. Machine Learning Methods

Machine Learning techniques include both classification and regression algorithms, so they can be used both for presence detection and counting people. For example, K-Means and Gradient Boosting were adopted in [67], whereas Artificial Neural Networks, regression models and Decision Trees (Random Forest) were adopted in [68], and Random Forest were applied in [69]. Passive, BLE-based presence detection systems in the literature exploit Logistic Regression, k-Nearest Neighbor (KNN), and Support Vector Machine (SVM) with linear, polynomial and Radial Basis Function kernel. In [49], a presence detection system was developed, and the algorithms mentioned above were compared on three different datasets. Such datasets, here called DS1, DS2, and DS3, were built according to the process formerly described in Section 4, i.e.,

- DS1, containing the raw RSSI data.
- DS2, containing data resulting from the first preprocessing step.
- DS3, containing data resulting from the second preprocessing step.

The best accuracy was obtained with the (SVM) algorithm [70] using a Radial Basis Function (RBF) kernel applied to the third dataset. An RBF kernel [71] is a function used when the boundaries of the classes are hypothesized to be curve-shaped and nonlinear. An RBF kernel involves the choice of two hyperparameters, i.e., the penalty parameter and the kernel width. In [49], such parameters are auto-optimized in the SVM implementation. In 98.97% of cases, this algorithm recognized the presence of a subject between two nodes. The KNN algorithm, instead, was more efficient with the first and second datasets and obtained higher accuracy than the other algorithms.

In general, all the analyzed algorithms performed similarly when applied to the same dataset, apart from a few exceptions. In particular, for the dataset DS1, the highest accuracy obtained was 70.15% with the KNN algorithm, while the Logistic Regression algorithm obtained 69.81%. If we look only at the accuracy that the two algorithms achieve, the KNN performed better than the Logistic Regression, but the KNN is a parametric algorithm in which the execution time increases with an increasing amount of data. The Logistic Regression algorithm, instead, is a linear algorithm and is faster than the KNN. The SVM algorithm performed slightly worse than the other algorithms on DS1, especially with the linear kernel (62.67%). The reason for this could lay in the distribution of the data. The linear SVM algorithm assumes that a linear boundary can be found between two classes, whereas the kernel trick supports different, nonlinear models of the data. However, in some cases, even the nonlinear kernels struggle in approximating the strongly nonlinear distribution of real data.

For the dataset DS2, containing four features for every transmitter–receiver couple, the KNN resulted as the best performing algorithm with 96.90% accuracy, whereas the SVM with an RBF kernel produced 96.46% accuracy.

The situation changed for dataset DS3, which was normalized to keep in account the distance between the transmitter and the receiver. The SVM with the RBF kernel performed better than the other algorithms (98.97%).

An analysis of the accuracy obtained by applying the described algorithms to all the datasets evidences that the preprocessing stage is always useful to increase the classification accuracy. In fact, as it is shown in Figure 5, the classification accuracy obtained with DS1 is 70%, with DS2 approaches 97%, and with DS3 approaches 99%.

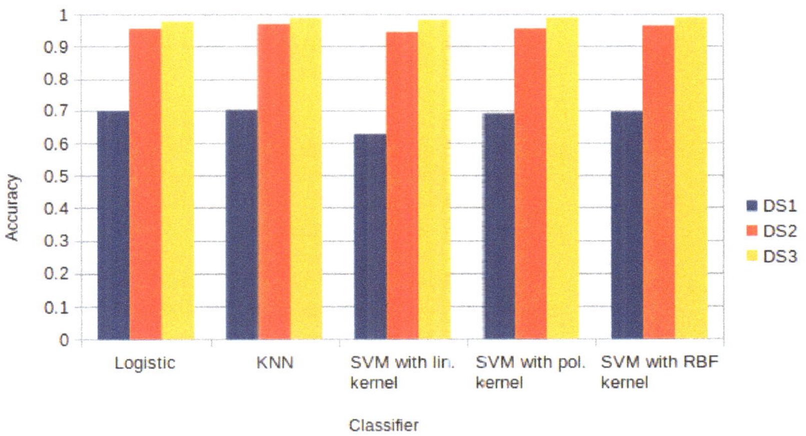

Figure 5. Comparison of the classification models used in [49] with the datasets DS1 (without any processing), DS2 (after aggregation and preprocessing), and DS3 (after normalization with weight factors).

The regression algorithms used to develop the counting system, instead, are the KNN Regression, the Least Squares Regression (LSR), the Polynomial Regression, and the Support Vector Regression. These algorithms were compared in [50], in which a counting system was developed and the same preprocessing approach of the previous paper was adopted, obtaining similar performance improvements. In fact, for the dataset DS1, the lowest Root Mean Square Error (RMSE) among all the considered approaches was 18.78, for the second dataset DS2, it was 6.40, and for DS3, it was 5.42. Notably, for DS2 and DS3, the LSR algorithm performed considerably worse than the others, as shown in Figure 6.

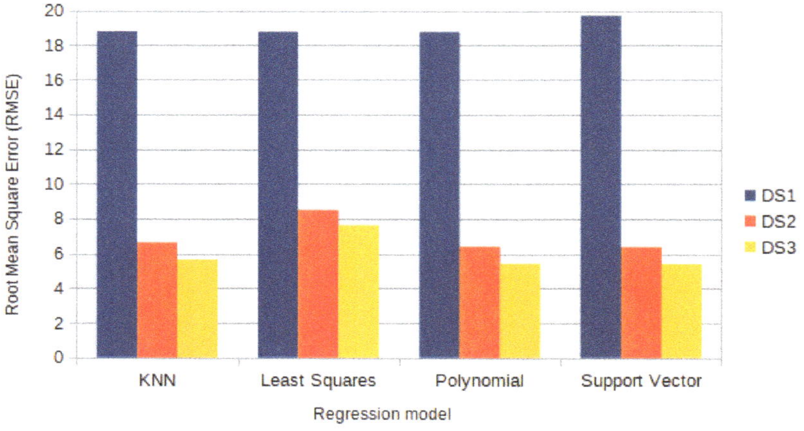

Figure 6. Comparison of the regression models used in [50] with datasets DS1 (without any processing), DS2 (after aggregation and preprocessing), and DS3 (after normalization with weight factors). The error measure is the Root Mean Square Error (RMSE).

As a consequence, the selection of the best algorithm requires further considerations regarding the resource requirements or other specific characteristics of the intended application. In particular, the computational complexity of the considered approach, and therefore its execution time and memory requirements, can sensibly influence the choice of the approach, depending on the hardware that will be used for implementing the system.

5.3. *Artificial Neural Networks*

Artificial Neural Networks are still little used in Bluetooth-based passive counting and detection, whereas the development of a Bluetooth-based localization system exploiting Artificial Neural Networks is described in [66].

In [18], a system for obstruction detection between two nodes, aimed at RSSI signal correction, was developed. Two Artificial Neural Networks were compared to detect obstructions, i.e., a Multi-Layer Perceptron (MLP) network, with two hidden layers of 20 neurons, and a Radial Basis Function (RBF) network, with only one layer of 20 neurons. The comparison demonstrated that both the neural networks detected the obstructions between nodes with high accuracy, with the MLP performing better (91%) than the RBF (89%) according to an estimate based on the data reported in the paper. The detection rate of the proposed approaches in discriminating clear Line-of-Sight, i.e., when no human body was present in the LoS between transmitter and receiver, was also measured. The MLP obtained a 94% detection rate, while the RBF obtained a 92% detection rate.

6. Overall Performance Comparison

Table 2 provides a comparison of the different approaches addressed in this paper. Presence detection refers to methods able to reveal the presence of a human body either still or in motion, whereas motion detection refers to methods that only reveal moving human bodies. People counting, instead, refers to methods able not only to reveal the presence of a human body but also to count the number of human bodies present in the monitored area. Finally, clean Line-of-Sight (LoS) detection refers to approaches able to detect the presence of a human body obstructing the line of sight between the transmitter and the receiver (see Figure 1a).

The approaches discussed in this section were trained and tested on different datasets and with different sensors set up. As the codebase and the datasets of the cited studies are not publicly available, the comparison is based on the data reported in the cited works.

Table 2. Performance comparison of the considered approaches.

Application	Preprocessing	Classification/ Regression Techniques	Accuracy
Presence detection [40]	-	Statistical techniques	90%
Presence detection [51]	Weighted Moving Avg.	RSSI-mean	96%
Motion detection [51]	Weighted Moving Avg.	RSSI-variance	98%
Presence detection [49]	α-trimmed filter and normalization	Logistic Regression, KNN, SVN (linear, polynomial and RBF kernel)	$\approx$ 98%
People counting [50]	α-trimmed filter and normalization	Least Squares Regression, NN Regression, Support Vector Regression, Polynomial Regression	±5.42 miscalculation [1]
Presence detection [18]	-	Neural Networks: MLP and RBF	(MLP) 91%, (RBF) 89%
Clean LoS detection [18]	-	Neural Networks: MLP and RBF	(MLP) 94%, (RBF) 92% clean LoS detection rate [2]

[1] This value is not an accuracy value, but the Root Mean Square Error. In [50], such an error metrics was chosen because it represents the misestimation in counting the number of people in the room. [2] These percentages are detection rates and not accuracy values. Accuracy is the sum of the true positives and true negatives over the whole set of samples, whereas here, the true positives over the whole set are considered.

The comparison shows that the accuracy of all the techniques is above 87%, and in some cases, it approaches 100%. Nevertheless, some significant differences need to be taken into account. In [40], a simple statistical approach was used for human detection, i.e., when the value of the average RSSI was above a threshold, a detection event was triggered. With this approach, the detection rate was slightly above 90%. Conversely, in [51], a more complex statistical approach was used, and two algorithms were implemented based on the RSSI-variance and RSSI-mean. The first algorithm is more suitable to detect a human crossing the LOS between two nodes, and it obtains an accuracy of 98%. The second algorithm, instead, is best suited for detecting a human that stands between two devices, and it obtains an accuracy of 96%.

In [49], different Machine Learning algorithms for human detection were compared. The best result was obtained with the SVM algorithm using an RBF kernel (98.97%) on a set of preprocessed data. However, the other algorithms perform similarly on the same type of preprocessed data ($\approx$98%). On raw RSSI data, all the tested algorithms performed worse, with accuracy values between $\approx$70% and 62%. As stated in Section 5, this result supports the observation that appropriate preprocessing is generally needed to improve the accuracy of the classification and regression algorithms and reduce the size of the final dataset, thus shortening the execution time of more complex algorithms, such as the k-Nearest Neighbor one. In [50], a very similar approach was applied to human counting, obtaining an RMSE of 5.42.

In [18], Artificial Neural Networks were used to detect a human body obstructing the space between a transmitter and a receiver, although the system was indeed intended for the correction of the RSSI signal in case of obstructions. The proposed networks, an MLP and an RBF, reached 91% and 89% accuracy, respectively, thus underperforming compared to the statistical approaches. However, the proposed Artificial Neural Networks were only tested on raw data. In principle, they could benefit from data preprocessing, thus further improving their performance. Consequently, more experiments are needed to better understand which approach has the higher potential over the others.

7. Discussion

In the previous sections, some applications of the Bluetooth technology (in particular, BLE) to passive human detection, counting and tracking in indoor environment were described and analyzed. Moreover, a comparative overview of their performance was provided.

As the signal RSSI depends on the position and distance between two Bluetooth devices and on the geometric and physical characteristics of the environment, the development of a "generic" passive human sensing system able to cope with all the possible variations in the environment and spatial deployment of the transmitting and receiving devices would require a generalization ability that is hardly obtainable.

Current approaches, instead, rely on some training or initialization process that builds a model of the received signal RSSI that is able to explain all the possible fluctuations of the RSSI values, thus allowing their classification. Such a model can be either explicit, e.g., defined in terms of statistical parameters, or implicit, e.g., obtained by training an ANN.

Building a data-driven model, however, requires that a suitable dataset is gathered from the specific environment where the system will be deployed; i.e., a long data collection process is needed to initialize the system. Furthermore, if Neural Networks are exploited, the data collection process becomes longer and longer, because large datasets are needed to train Neural Networks. The lack of universal datasets to train and test the developed systems makes the comparison of existing techniques difficult.

Another issue often highlighted in the literature is the impossibility of independently extracting the RSSI signal values from each advertising channel of the BLE beacons. The BLE beacons need to be modified at the hardware or firmware level in order to transmit on a certain preset channel and to allow the researcher to discriminate the variation in the signal due to the presence of a human body from other fading effects.

According to Table 2, the RSSI-variance and RSSI-mean approaches used in [51] might be considered as the best performers for presence detection. The reported accuracy results are even better than those reached by more recent Artificial Neural Networks approaches. However, in this study, the BLE transmitters were modified to transmit only on one advertising channel, so that the channel information can be transmitted to the receiver into the payload of the advertising message. Such a modification allows to separately analyze each advertising channel, thus substantially reducing the variance in the RSSI signal and dramatically improving the detection stability and reliability. However, the proposed modification, although being frequently adopted in the relevant literature, is not allowed by the Bluetooth standard, and therefore, it represents a substantial violation of the protocol. In other words, the proposed approach would not work with off-the-shelf beacons.

The Machine Learning techniques used in [49] obtained about 98% accuracy. In particular, as already stated in Section 5, 98% accuracy was reached by using the SVM algorithm with an RBF kernel on a reduced and normalized dataset. In this case, unlike the previous approach, only standard devices were used, and only four receivers were used with 24 transmitters, instead of requiring the same number of receivers and transmitters. As a consequence, the approach proposed in [49] can be applied to a much wider range of cases and environments while performing close to best in class.

8. Conclusions

This paper provided a reasoned overview of the Bluetooth-based approaches for passive remote sensing of the human body reported in the literature. The work illustrated and commented on the pros and cons of each approach. Moreover, the experimental results and performance of the considered approaches were described, compared and discussed.

Although the inherent limitations imposed by the Bluetooth protocol may affect its applicability to passive human sensing, a number of recent research works proposed specific preprocessing and non-parametric classification and regression techniques, and they demonstrated the effectiveness and accuracy of the proposed approaches. In general, the adoption of BLE-based remote passive wireless human sensing approaches should be preferred whenever the power consumption of the sensing devices, the deployment cost, and the simultaneous communication and sensing capability are very relevant. However, some interesting aspects still need to be investigated. In particular, the FHSS mechanism significantly affects the reliability of RSSI-based human sensing, and therefore, several works in the literature address modifications of the transmitting device original firmware in order to either bypass the FHSS mechanism or include the channel information in the transmitted frame. Such modifications, however, violate the Bluetooth protocol and should be avoided, also because bypassing FHSS reduces the robustness to noise of the proposed approach. A possible alternative approach can be to estimate, at the receiver side, the transmitter channel mapping and, therefore, to foresee the next hop. Effective techniques are available to obtain such an estimate, such as the ones in [72–74].

Furthermore, the recent introduction in the Bluetooth standard of the direction-finding technology has a potential for supporting the development of novel approaches able to extract more information from the BLE signal, thus allowing for more accurate and detailed human sensing applications, similarly to what happened with WiFi. In particular, when determining the Direction of Arrival (DoA) of the received signal (see Section 3.1.1), the receiver can compute a suitable estimate of the phase shift of the reflected, scattered and refracted signal components, and leverage such information to produce a more accurate sensing, thus allowing for complex activity recognition, vital signs detection, and multi-person tracking.

Author Contributions: Conceptualization, methodology, investigation, formal analysis, writing—original draft preparation, writing—review and editing, G.I. and L.L.B.; investigation, M.M.; investigation, data curation, A.N. All authors have read and agreed to the published version of the manuscript.

Funding: This research received no external funding.

Data Availability Statement: Not applicable.

Conflicts of Interest: The authors declare no conflict of interest.

References

1. Youssef, M.; Mah, M.; Agrawala, A. Challenges: Device-Free Passive Localization for Wireless Environments. In Proceedings of the 13th Annual ACM International Conference on Mobile Computing and Networking, Montreal, QC, Canada, 9–14 September 2007; Association for Computing Machinery: New York, NY, USA, 2007; pp. 222–229. [CrossRef]
2. Aggarwal, J.; Ryoo, M. Human Activity Analysis: A Review. *ACM Comput. Surv.* **2011**, *43*, 1–43. [CrossRef]
3. Iannizzotto, G.; Lo Bello, L.; Patti, G. Personal Protection Equipment detection system for embedded devices based on DNN and Fuzzy Logic. *Expert Syst. Appl.* **2021**, *184*, 115447. [CrossRef]
4. Beddiar, D.R.; Nini, B.; Sabokrou, M.; Hadid, A. Vision-based human activity recognition: A survey. *Multimed. Tools Appl.* **2020**, *79*, 30509–30555. [CrossRef]
5. Dondi, P.; Porta, M.; Donvito, A.; Volpe, G. A gaze-based interactive system to explore artwork imagery. *J. Multimodal User Interfaces* **2022**, *16*, 55–67. [CrossRef]
6. Liu, J.; Teng, G.; Hong, F. Human Activity Sensing with Wireless Signals: A Survey. *Sensors* **2020**, *20*, 1210. [CrossRef] [PubMed]
7. Iannizzotto, G.; La Rosa, F.; Lo Bello, L. A wireless sensor network for distributed autonomous traffc monitoring. In Proceedings of the 3rd International Conference on Human System Interaction, Rzeszow, Poland, 13–15 May 2010; pp. 612–619. [CrossRef]
8. Bayrakdar, M.E. Priority based health data monitoring with IEEE 802.11 af technology in wireless medical sensor networks. *Med. Biol. Eng. Comput.* **2019**, *57*, 2757–2769. [CrossRef]
9. Mishra, A.; Kumari, A.; Sajit, P.; Pandey, P. Remote web based ECG Monitoring using MQTT Protocol for IoT in Healthcare. *Development* **2018**, *5*, 1096–1109.
10. Leonardi, L.; Lo Bello, L.; Patti, G.; Ragusa, O. A Network Architecture and Routing Protocol for the MEDIcal WARNing System. *J. Sens. Actuator Netw.* **2021**, *10*, 44. [CrossRef]
11. Fedullo, T.; Tramarin, F.; Vitturi, S. The Impact of Rate Adaptation Algorithms on Wi-Fi-Based Factory Automation Systems. *Sensors* **2020**, *20*, 5195. [CrossRef]
12. Depari, A.; De Dominicis, C.M.; Flammini, A.; Rinaldi, S.; Vezzoli, A. Integration of Bluetooth HandsFree Sensors into a Wireless Body Area Network Based on Smartphone. In *Sensors*; Baldini, F., D'Amico, A., Di Natale, C., Siciliano, P., Seeber, R., De Stefano, L., Bizzarri, R., Andò, B., Eds.; Springer: New York, NY, USA, 2014; pp. 547–551.
13. Depari, A.; Flammini, A.; Rinaldi, S.; Vezzoli, A. Multi-sensor system with Bluetooth connectivity for non-invasive measurements of human body physical parameters. *Sens. Actuators A Phys.* **2013**, *202*, 147–154. [CrossRef]

14. De Dominicis, C.M.; Mazzotti, D.; Piccinelli, M.; Rinaldi, S.; Vezzoli, A.; Depari, A. Evaluation of Bluetooth Hands-Free profile for sensors applications in smartphone platforms. In Proceedings of the 2012 IEEE Sensors Applications Symposium, Brescia, Italy, 7–9 February 2012; pp. 1–6. [CrossRef]
15. Collotta, M.; Pau, G. An Innovative Approach for Forecasting of Energy Requirements to Improve a Smart Home Management System Based on BLE. *IEEE Trans. Green Commun. Netw.* **2017**, *1*, 112–120. [CrossRef]
16. Huang, B.; Liu, J.; Sun, W.; Yang, F. A Robust Indoor Positioning Method based on Bluetooth Low Energy with Separate Channel Information. *Sensors* **2019**, *19*, 3487. [CrossRef] [PubMed]
17. Madhavapeddy, A.; Tse, A. A Study of Bluetooth Propagation Using Accurate Indoor Location Mapping. In Proceedings of the 7th International Conference on Ubiquitous Computing, Tokyo, Japan, 11–14 September 2005; Springer: Berlin/Heidelberg, Germany, 2005; pp. 105–122. [CrossRef]
18. Naghdi, S.; O'Keefe, K. Detecting and Correcting for Human Obstacles in BLE Trilateration Using Artificial Intelligence. *Sensors* **2020**, *20*, 1350. [CrossRef] [PubMed]
19. Faragher, R.; Harle, R. Location Fingerprinting With Bluetooth Low Energy Beacons. *IEEE J. Sel. Areas Commun.* **2015**, *33*, 2418–2428. [CrossRef]
20. Zhuang, Y.; Yang, J.; Li, Y.; Qi, L.; El-Sheimy, N. Smartphone-Based Indoor Localization with Bluetooth Low Energy Beacons. *Sensors* **2016**, *16*, 596. [CrossRef]
21. Kaltiokallio, O.; Bocca, M.; Patwari, N. Follow @grandma: Long-term device-free localization for residential monitoring. In Proceedings of the 37th Annual IEEE Conference on Local Computer Networks—Workshops, Clearwater, FL, USA, 22–25 October 2012; pp. 991–998. [CrossRef]
22. Kaltiokallio, O.; Bocca, M. Real-Time Intrusion Detection and Tracking in Indoor Environment through Distributed RSSI Processing. In Proceedings of the 2011 IEEE 17th International Conference on Embedded and Real-Time Computing Systems and Applications, Toyama, Japan, 28–31 August 2011; Volume 1, pp. 61–70. [CrossRef]
23. Surian, D.; Kim, V.; Menon, R.; Dunn, A.G.; Sintchenko, V.; Coiera, E. Tracking a moving user in indoor environments using Bluetooth low energy beacons. *J. Biomed. Inform.* **2019**, *98*, 103288. [CrossRef]
24. Demrozi, F.; Bragoi, V.; Tramarin, F.; Pravadelli, G. An indoor localization system to detect areas causing the freezing of gait in Parkinsonians. In Proceedings of the 2019 Design, Automation Test in Europe Conference Exhibition (DATE), Florence, Italy, 25–29 March 2019; pp. 952–955. [CrossRef]
25. Filippoupolitis, A.; Oliff, W.; Loukas, G. Occupancy Detection for Building Emergency Management Using BLE Beacons. In *Computer and Information Sciences*; Czachórski, T., Gelenbe, E., Grochla, K., Lent, R., Eds.; Springer International Publishing: Cham, Switzerland, 2016; pp. 233–240.
26. Tekler, Z.D.; Low, R.; Blessing, L. An alternative approach to monitor occupancy using bluetooth low energy technology in an office environment. *J. Phys. Conf. Ser.* **2019**, *1343*, 012116. [CrossRef]
27. Daniş, F.S.; Cemgil, A.T. Model-Based Localization and Tracking Using Bluetooth Low-Energy Beacons. *Sensors* **2017**, *17*, 2484. [CrossRef]
28. Baronti, P.; Barsocchi, P.; Chessa, S.; Mavilia, F.; Palumbo, F. Indoor Bluetooth Low Energy Dataset for Localization, Tracking, Occupancy, and Social Interaction. *Sensors* **2018**, *18*, 4462. [CrossRef]
29. Čakić, S.; Šandi, S.; Nedić, D.; Krčo, S.; Popović, T. Human Activity Detection Using Deep Learning and Bracelet with Bluetooth Transmitter. In Proceedings of the 2021 29th Telecommunications Forum (TELFOR), Belgrade, Serbia, 23–24 November 2021; pp. 1–4. [CrossRef]
30. Hussain, S.; Peters, R.; Silver, D. Using received signal strength variation for surveillance in residential areas. *Proc. Spie -Int. Soc. Opt. Eng.* **2008**, *6973*, 213–218. [CrossRef]
31. Ma, Y.; Zhou, G.; Wang, S. WiFi Sensing with Channel State Information: A Survey. *ACM Comput. Surv.* **2019**, *52*. [CrossRef]
32. Chen, H.; Zhang, Y.; Li, W.; Tao, X.; Zhang, P. ConFi: Convolutional Neural Networks Based Indoor Wi-Fi Localization Using Channel State Information. *IEEE Access* **2017**, *5*, 18066–18074. [CrossRef]
33. Chen, Z.; Zhang, L.; Jiang, C.; Cao, Z.; Cui, W. WiFi CSI Based Passive Human Activity Recognition Using Attention Based BLSTM. *IEEE Trans. Mob. Comput.* **2019**, *18*, 2714–2724. [CrossRef]
34. Liu, Y.; Wang, T.; Jiang, Y.; Chen, B. Harvesting Ambient RF for Presence Detection Through Deep Learning. *IEEE Trans. Neural Netw. Learn. Syst.* **2020**, *33*, 1–13. [CrossRef] [PubMed]
35. Woolley, M. Bluetooth Core Specification v5.1 Feature Overview. Available online: https://www.bluetooth.com/bluetooth-resources/bluetooth-core-specification-v5-1-feature-overview/ (accessed on 1 April 2022).
36. Zhou, B.; Sun, C.; Ahn, D.; Kim, Y. A Novel Passive Tracking Scheme Exploiting Geometric and Intercept Theorems. *Sensors* **2018**, *18*, 895. [CrossRef]
37. Wilson, J.; Patwari, N. Radio Tomographic Imaging with Wireless Networks. *IEEE Trans. Mob. Comput.* **2010**, *9*, 621–632. [CrossRef]
38. Muñoz-Ferreras, J.M.; Gómez-García, R.; Li, C. Chapter 5—Human-aware localization using linear-frequency-modulated continuous-wave radars. In *Principles and Applications of RF/Microwave in Healthcare and Biosensing*; Li, C., Tofighi, M.R., Schreurs, D., Horng, T.S.J., Eds.; Academic Press: Cambridge, MA, USA, 2017; pp. 191–242. [CrossRef]
39. Adib, F.; Hsu, C.Y.; Mao, H.; Katabi, D.; Durand, F. Capturing the Human Figure through a Wall. *ACM Trans. Graph.* **2015**, *34*, 1–3. [CrossRef]

40. Sugino, K.; Katayama, S.; Niwa, Y.; Shiramatsu, S.; Ozono, T.; Shintani, T. A Bluetooth-Based Device-Free Motion Detector for a Remote Elder Care Support System. In Proceedings of the 2015 IIAI 4th International Congress on Advanced Applied Informatics, Okayama, Japan, 12–16 July 2015; pp. 91–96. [CrossRef]
41. Sugino, K.; Niwa, Y.; Shiramatsu, S.; Ozono, T.; Shintani, T. Developing a human motion detector using bluetooth beacons and its applications. *Inf. Eng. Express* **2015**, *1*, 95–105. [CrossRef]
42. Li, T.; An, C.; Tian, Z.; Campbell, A.T.; Zhou, X. Human Sensing Using Visible Light Communication. In Proceedings of the 21st Annual International Conference on Mobile Computing and Networking, Paris, France, 11 September 2015; Association for Computing Machinery: New York, NY, USA, 2015; pp. 331–344. [CrossRef]
43. Augustin, A.; Yi, J.; Clausen, T.; Townsley, W.M. A Study of LoRa: Long Range Low Power Networks for the Internet of Things. *Sensors* **2016**, *16*, 1466. [CrossRef]
44. Leonardi, L.; Lo Bello, L.; Battaglia, F.; Patti, G. Comparative Assessment of the LoRaWAN Medium Access Control Protocols for IoT: Does Listen before Talk Perform Better than ALOHA? *Electronics* **2020**, *9*, 553. [CrossRef]
45. Chen, L.; Xiong, J.; Chen, X.; Lee, S.I.; Chen, K.; Han, D.; Fang, D.; Tang, Z.; Wang, Z. WideSee: Towards Wide-Area Contactless Wireless Sensing. In Proceedings of the 17th Conference on Embedded Networked Sensor Systems, New York, NY, USA, 10–13 November 2019; Association for Computing Machinery: New York, NY, USA, 2019; pp. 258–270. [CrossRef]
46. Remy, J.G.; Letamendia, C. LTE Standards and Architecture. In *LTE Standards*; John Wiley Sons, Ltd.: Hoboken, NJ, USA, 2014; Chapter 1, pp. 1–112. [CrossRef]
47. Xu, S.; Tian, Y. Device-Free Motion Detection via On-the-Air LTE Signals. *IEEE Commun. Lett.* **2018**, *22*, 1934–1937. [CrossRef]
48. Wang, Q.; Kakkavas, A.; Gong, X.; Stirling-Gallacher, R.A. Towards Integrated Sensing and Communications for 6G. *arXiv* **2022**, arXiv:2201.04498.
49. Münch, M.; Huffstadt, K.; Schleif, F. Towards a device-free passive presence detection system with Bluetooth Low Energy beacons. In Proceedings of the 27th European Symposium on Artificial Neural Networks, ESANN 2019, Bruges, Belgium, 24–26 April 2019.
50. Münch, M.; Schleif, F.M. Device-Free Passive Human Counting with Bluetooth Low Energy Beacons. In *Advances in Computational Intelligence, Proceedings of the 15th International Work-Conference on Artificial Neural Networks, IWANN 2019, Gran Canaria, Spain, 12–14 June 2019*; Rojas, I., Joya, G., Catala, A., Eds.; Springer International Publishing: Cham, Switzerland, 2019; pp. 799–810. [CrossRef]
51. Brockmann, F.; Figura, R.; Handte, M.; Marrón, P.J. RSSI Based Passive Detection of Persons for Waiting Lines Using Bluetooth Low Energy. In Proceedings of the 2018 International Conference on Embedded Wireless Systems and Networks, Madrid, Spain, 14–16 February 2018; pp. 102–113.
52. Bluetooth SIG, Inc. Understanding Bluetooth Range. Available online: https://www.bluetooth.com/learn-about-bluetooth/key-attributes/range/ (accessed on 1 April 2022).
53. Woolley, M. Understanding Reliability in Bluetooth Technology. Available online: https://www.bluetooth.com/wp-content/uploads/2020/10/EN-Understanding_Reliability.pdf (accessed on 1 April 2022).
54. Zhang, M.; Xia, W.; Shen, L. Bluetooth Low Energy based motion sensing system. In Proceedings of the 2014 Sixth International Conference on Wireless Communications and Signal Processing (WCSP), Hefei, China, 23–25 October 2014; pp. 1–5. [CrossRef]
55. Woolley, M. Bluetooth Core Specification Version 5.0 Feature Enhancements. Available online: https://www.bluetooth.com/bluetooth-resources/bluetooth-5-go-faster-go-further/ (accessed on 1 April 2022).
56. Woolley, M. Bluetooth Direction Finding: A Technical Overview. Available online: https://www.bluetooth.com/bluetooth-resources/bluetooth-direction-finding/ (accessed on 1 April 2022).
57. Woolley, M. Bluetooth Core Specification Version 5.2 Feature Overview. Available online: https://www.bluetooth.com/wp-content/uploads/2020/01/Bluetooth_5.2_Feature_Overview.pdf (accessed on 1 April 2022).
58. Woolley, M. Bluetooth Core Specification Version 5.3 Feature Enhancements. Available online: https://www.bluetooth.com/bluetooth-resources/bluetooth-core-specification-version-5-3-feature-enhancements/ (accessed on 1 April 2022).
59. Faragher, R.; Harle, R.K. An Analysis of the Accuracy of Bluetooth Low Energy for Indoor Positioning Applications. In Proceedings of the 27th International Technical Meeting of the Satellite Division of The Institute of Navigation (ION GNSS+ 2014), Tampa, FL, USA, 8–12 September 2014; pp. 201–210.
60. Woyach, K.; Puccinelli, D.; Haenggi, M. Sensorless Sensing in Wireless Networks: Implementation and Measurements. In Proceedings of the International Symposium on Modeling and Optimization in Mobile, Ad Hoc and Wireless Networks, Boston, MA, USA, 26 February–2 March 2006; pp. 1–8. [CrossRef]
61. Habaebi, M.; Rosli, R. RSSI-based Human Presence Detection System for Energy Saving Automation. *Indones. J. Electr. Eng. Inform.* **2017**, *5*, 339–350. [CrossRef]
62. Bulten, W.; Rossum, A.C.V.; Haselager, W.F.G. Human SLAM, Indoor Localisation of Devices and Users. In Proceedings of the 2016 IEEE First International Conference on Internet-of-Things Design and Implementation (IoTDI), Berlin, Germany, 4–8 April 2016; pp. 211–222. [CrossRef]
63. Sabek, I.; Youssef, M.; Vasilakos, A.V. ACE: An Accurate and Efficient Multi-Entity Device-Free WLAN Localization System. *IEEE Trans. Mob. Comput.* **2015**, *14*, 261–273. [CrossRef]
64. Kaufmann, J.; Schering, A. Analysis of Variance ANOVA. In *Wiley StatsRef: Statistics Reference Online*; John Wiley & Sons, Ltd.: Hoboken, NJ, USA, 2014. [CrossRef]

65. Alom, M.Z.; Taha, T.M.; Yakopcic, C.; Westberg, S.; Sidike, P.; Nasrin, M.S.; Hasan, M.; Van Essen, B.C.; Awwal, A.A.S.; Asari, V.K. A State-of-the-Art Survey on Deep Learning Theory and Architectures. *Electronics* **2019**, *8*, 292. [CrossRef]
66. Abdull Sukor, A.S.; Kamarudin, L.M.; Zakaria, A.; Abdul Rahim, N.; Sudin, S.; Nishizaki, H. RSSI-Based for Device-Free Localization Using Deep Learning Technique. *Smart Cities* **2020**, *3*, 444–455. [CrossRef]
67. Tekler, Z.D.; Low, R.; Gunay, B.; Andersen, R.K.; Blessing, L. A scalable Bluetooth Low Energy approach to identify occupancy patterns and profiles in office spaces. *Build. Environ.* **2020**, *171*, 106681. [CrossRef]
68. Beato Gutiérrez, M.E.; Sánchez, M.M.; Berjón Gallinas, R.; Fermoso García, A.M. Capacity Control in Indoor Spaces Using Machine Learning Techniques Together with BLE Technology. *J. Sens. Actuator Netw.* **2021**, *10*, 35. [CrossRef]
69. Rahaman, M.S.; Pare, H.; Liono, J.; Salim, F.D.; Ren, Y.; Chan, J.; Kudo, S.; Rawling, T.; Sinickas, A. OccuSpace: Towards a Robust Occupancy Prediction System for Activity Based Workplace. In Proceedings of the 2019 IEEE International Conference on Pervasive Computing and Communications Workshops (PerCom Workshops), Kyoto, Japan, 11–15 March 2019; pp. 415–418. [CrossRef]
70. Cortes, C.; Vapnik, V. Support-vector networks. *Mach. Learn.* **1995**, *20*, 273–297. [CrossRef]
71. Keerthi, S.S.; Lin, C.J. Asymptotic Behaviors of Support Vector Machines with Gaussian Kernel. *Neural Comput.* **2003**, *15*, 1667–1689. [CrossRef]
72. Albazrqaoe, W.; Huang, J.; Xing, G. A Practical Bluetooth Traffic Sniffing System: Design, Implementation, and Countermeasure. *IEEE/ACM Trans. Netw.* **2019**, *27*, 71–84. [CrossRef]
73. Lee, J.; Park, C.; Roh, H. Revisiting Adaptive Frequency Hopping Map Prediction in Bluetooth with Machine Learning Classifiers. *Energies* **2021**, *14*, 928. [CrossRef]
74. Mototolea, D.; Youssef, R.; Radoi, E.; Nicolaescu, I. Non-Cooperative Low-Complexity Detection Approach for FHSS-GFSK Drone Control Signals. *IEEE Open J. Commun. Soc.* **2020**, *1*, 401–412. [CrossRef]

sensors

MDPI

Article

Multilink Operation in IEEE 802.11be Wireless LANs: Backoff Overflow Problem and Solutions

Wisnu Murti and Ji-Hoon Yun *

Department of Electrical and Information Engineering, Seoul National University of Science and Technology, Seoul 01811, Korea; wisnumurti@seoultech.ac.kr
* Correspondence: jhyun@seoultech.ac.kr

Abstract: The next-generation wireless LAN standard named IEEE 802.11be supports a multilink operation to cost-efficiently boost throughput performance, for which an efficient multilink channel scheme is essential. The synchronous channel access scheme with an enhancement allowing multilink transmission before backoff completion greatly enhances the performance of multilink devices with no simultaneous transmit and receive capability, for which, however, backoff count compensation is necessary for coexistence with legacy and other multilink devices. In this paper, we identify the backoff count overflow problem of the enhanced synchronous channel access scheme with backoff compensation, which becomes aggravated once triggered due to repeated compensations. Then, we propose four solutions to mitigate this problem: limiting consecutive free-riding transmissions, limiting a compensated backoff value, using the contention window value of a main link, and balancing transmissions between links. Through comparative evaluation and analyses for dense single-spot and indoor random deployment scenarios, we demonstrate in terms of throughput and latency that the proposed solutions successfully mitigate the problem while preserving the coexistence performance.

Keywords: IEEE 802.11be; multilink operation; channel access; backoff; coexistence

Citation: Murti, W.; Yun, J.-H. Multilink Operation in IEEE 802.11be Wireless LANs: Backoff Overflow Problem and Solutions. *Sensors* **2022**, *22*, 3501. https://doi.org/10.3390/s22093501

Academic Editor: Peter Han Joo Chong

Received: 29 March 2022
Accepted: 1 May 2022
Published: 4 May 2022

1. Introduction

A multilink operation (MLO) is a salient feature of the upcoming next-generation wireless LAN (WLAN) standard, i.e., IEEE 802.11be (also called extremely high throughput (EHT)) [1,2], that enables the simultaneous utilization of multiple links using individual frequency channels for both transmission and reception. The benefit of MLO is the simultaneous utilization of multiple frequency bands at a lower hardware cost than the approach of using a single multiband radio. Designing an efficient channel access operation for the presence of radio frequency (RF) power leakage between links to best exploit the benefit of MLO is still a challenge. In the IEEE 802.11be working group (WG), it was agreed that the multiple links involved in MLO may or may not have RF power leakage between them. A multilink device (MLD), which has links with power leakage between them, is unable to simultaneously transmit and receive (STR) across different links, as the power leakage from one link will make the other links unable to sense each's channel medium. This type of MLD is called a non-STR MLD. On the other side, an MLD with no RF power leakage between its links will be able to simultaneously transmit and receive signals and thus is called an STR MLD.

The channel access scheme of IEEE 802.11be is classified into an asynchronous operation (Async) and a synchronous operation (Sync). In Async, each of the links belonging to an MLD independently performs a channel access process (backoff procedure, BO). Therefore, different links finish their BOs at different times; thus, their transmissions will be asynchronous between them, which is ideal for STR MLDs. In Sync, all links also perform individual BOs. When a link finishes its BO earlier than other links, it waits until the other links also finish their BOs. If the waiting link sees its channel busy, the link must rerun its BO. Because only simultaneous transmission initiation is allowed on all links, the RF power leakage between links does not affect the operation of Sync. However, Sync does not utilize multiple channels better than Async, since it requires all links to have zero backoff counts for transmission.

There has been a notable discussion on the enhancements of Sync for better multilink exploitation and a hybrid of the Sync and Async schemes for faster transmission (FT), called Sync-FT [3]. In this work, the coexistence issue of Sync-FT was discussed since legacy single-link devices (SLDs) may coexist with EHT MLDs in practical WLANs, and multiple solutions were proposed. This is because, with Sync-FT, a link of the MLD can transmit before its BO completion, which is called free-riding. This behavior prioritizes channel access for the MLD, resulting in long and frequent wait times for the legacy device. It was shown in the work that compensating the backoff count for each free-riding transmission achieved the best coexistence performance while having a marginal throughput decrease in MLDs. However, there may arise a backoff count overflow problem where the accumulated backoff count of an MLD resulting from repeated compensations becomes too large; thus, the MLD is not able to obtain a transmission opportunity by its own BO completion once an overflow occurs.

This paper aims to identify the backoff count overflow problem of the enhanced multilink channel access scheme (Sync-FT) with backoff compensation in an IEEE 802.11be WLAN and propose solutions applicable to both STR and non-STR MLDs. First, we describe the Sync scheme of an MLO and its design variants including Sync-FT. Next, we demonstrate the backoff count overflow problem of backoff compensation for Sync-FT and identify that the problem is aggravated once triggered due to continuing free-riding transmissions and compensations. Subsequently, we propose four solutions to mitigate this problem: limiting consecutive free-riding transmissions, limiting a compensated backoff value, using the contention window (CW) value of a main link, and balancing transmissions between links. Through comparative evaluation and analyses for dense single-spot and 3GPP indoor random deployment scenarios, we demonstrate in terms of throughput and latency that the proposed solutions successfully mitigate the backoff count overflow problem. We also investigate the coexistence performance of the solutions and demonstrate that the proposed solutions preserve coexistence performance with other MLDs and legacy SLDs while solving the backoff count overflow problem.

In summary, the main contributions of our work are listed as follows:

- Identification of the backoff count overflow problem of the Sync-FT channel access scheme of EHT with backoff count compensation;
- Design of potential solutions to mitigate the problem with minimal modification of standards;
- Comprehensive simulation work to show the performance gain of the proposed solutions with coexistence of multiple networks with different channel access mechanisms and legacy devices.

The rest of the paper is organized as follows. Section 2 explains the multilink Sync operation and Sync-FT modification, and its coexistence issue and a backoff compensation solution are presented in Section 3. Section 4 describes the backoff count overflow problem. Section 5 details the proposed solutions, and Section 6 evaluates them via simulation compared with other channel access schemes in coexistence scenarios. Finally, Section 7 concludes the paper.

2. Synchronous Multilink Channel Access

Sync is one of the multilink channel access mechanisms for EHT that has been considered to be used for transmissions of non-STR MLD. This operation uses both links in the backoff process, but a transmission is triggered only if the backoff counters of both links are zero. Once a link finishes its BO, it waits until the other link finishes BO. If another device occupies the channel while an MLD is in the waiting mode, the backoff count of the waiting link should be rechosen, and its BO is initiated with the new count. This means that a multilink transmission aligns the transmission start time in each link; thus, RF power leakage does not affect channel access and transmission. However, the requirement of the availability of both links at the same time leads to severe performance degradation especially when the channels are crowded. The inability to perform independent transmission in individual links may cause an MLD to suffer from waiting. Therefore, the use of Sync may not be able to exploit the potential of MLO for throughput enhancement.

In what follows, we describe two design variants of Sync.

2.1. Synchronous Operation with a Primary Link

The Sync with a primary link scheme (Sync-PL) is similar to the conventional wideband operation of IEEE 802.11 [4,5] In this scheme, an MLD runs a single BO in a primary link (PL) only and does not run in other links. When the primary link is about to reach a zero backoff count, the MLD performs a short clear channel assessment (CCA) for a point coordination function interframe space (PIFS) on each of the other links to check each's availability. The links that are clear to send (idle) in a given short CCA period will then be able to send simultaneously with the primary link. The links in which the MLD transmits frames are the primary link and the other idle link(s). The MLD starts transmission in these links at the same time and ends transmission at the same time as well to avoid RF power leakage between the links. If all other links are found to be busy, the MLD transmits in the primary link only. Since only the primary link performs BO, a single link transmission is available only in the primary link. Sync-PL is expected to have better throughput performance than the basic Sync since at least a single link transmission is enabled. One concern is that a transmission is triggered only by the backoff completion of the primary link; thus, transmission opportunities highly depend on the load condition of the corresponding channel.

2.2. Synchronous Operation for Faster Transmission

Another enhancement of Sync is called Sync for faster transmission (Sync-FT) and has proven to be better in throughput performance than Sync. Sync-FT is a hybrid of Sync-PL and Async. As in Async, it lets the links of an MLD run individual backoff processes. When a link finishes its BO earlier than the others, the MLD performs a short CCA in each of the other links and starts a multilink transmission in a set of links sensed idle as in Sync-PL while freezing the backoff counts of the links. Once the links finish transmission, they resume individual backoff processes. The links that completed backoff at the time of transmission choose a new backoff count and initiate a new backoff process, while the other links resume BO with their remaining counts. We call the transmission of a link with BO completion as a main transmission and the transmission of a link before BO completion as a free-riding transmission. This operation enhances the multilink utilization of non-STR MLDs by avoiding the failure of link utilization due to RF power leakage. The operation is applicable to both STR and non-STR MLDs. Sync-FT enables both single-link and multilink transmissions in all links.

3. Coexistence Issue of Sync-FT and Backoff Count Compensation Solution

In this section, we describe the coexistence issue of Sync-FT and the backoff compensation solution.

3.1. Coexistence Issue

There may coexist legacy SLDs and EHT MLDs with different channel access schemes in networks. Therefore, the main coexistence scenarios are given below:

- MLD vs. legacy SLD;
- MLD vs. MLD with different channel access schemes.

Suppose that a legacy device starts a backoff process with the same backoff count as an MLD with Sync-FT. However, while the legacy station obtains a single transmission opportunity (TXOP), the MLD may obtain more TXOPs. This is because Sync-FT enables the MLD to transmit in a link before the backoff completion of the link. Such behavior prioritizes the channel access of the MLD and gives more TXOPs to it, resulting in long and frequent waiting times of the legacy device. When two MLDs, one with Sync-PL (MLD1) and the other with Sync-FT (MLD2) coexist, MLD2 is likely to obtain more TXOPs than MLD1 due to the aggressiveness of Sync-FT against Sync-PL.

In the basic design of Sync-FT, a link performing transmission by another link's BO completion, which we call a free-riding link, resumes its ongoing backoff process after finishing transmission, with the remaining count value. This behavior makes the links under Sync-FT transmit more frequently than others that have to complete BO before transmission (e.g., legacy SLDs). In other words, this behavior makes the free-riding link go through a relatively deflated backoff count on average.

3.2. *Sync-FT with Backoff Compensation*

One of the proposed designs to tackle the coexistence issue of Sync-FT is having backoff count compensation after free-riding of a link [3]. This is a straightforward modification where the backoff count of a free-riding link is compensated for by an appropriate amount such that it finally goes through the backoff time given by its backoff counts on average. Before the additional backoff count value is added to the current count by compensation, a new backoff count should first be chosen. After a free-riding transmission finishes, the link stores the current backoff count value and rechooses a new backoff count. Then, it performs a new backoff process with an initial count as the sum of the stored backoff count and the rechosen one. This solution is simple in implementation, since it reuses existing functions of the backoff mechanism and can resolve the coexistence issue caused by Sync-FT's aggressive behavior.

4. Backoff Count Overflow Problem

The backoff compensation solution described in the previous section leads to another problem called a backoff count overflow problem. This problem is that a link has a very large value of its backoff count, thus preventing it from finishing its countdown and transmission. In general WLANs, a greater backoff value is usually caused by consecutive transmission failures, causing a CW value to rise and in turn increasing a random backoff value chosen within the CW. This backoff count overflow problem, however, happens in Sync-FT with backoff compensation implementation even without consecutive transmission failures.

The problem is illustrated in Figure 1. Upon completion of each free-riding transmission, the link compensates for its backoff count value by the remaining amount of the previous value. There is a possibility that one link may become stuck in consecutive free-riding transmissions and keep compensating its backoff count value, thus making the count increase indefinitely. This phenomenon happens especially when the channels of the links have heterogeneous load conditions; hence, some links in low load conditions keep decreasing their backoff values and make the other links keep free-riding with no chance of transmitting via their own BO completion. This is the case when links tend to have different backoff stages (i.e., different CW values) and thus heterogeneous backoff counts. As a free-riding transmission makes a link rechoose a backoff count from its current CW value (without a change in its backoff stage), a link with a higher backoff stage will statistically choose a larger backoff count value and, thus, have a higher chance of free-riding again. In other words, once a link performs a free-riding transmission, it will retain half of the current backoff count value on average (assuming that the BO completions of links are independent from each other and, thus, randomized between them) and add a new rechosen value, which is highly likely to make the resulting backoff count larger than the other links; If this happens consecutively, the link will always be a free-riding link.

Figure 2 shows the time evolution of the backoff count values of the two links of an MLD. In the figure, Link 1 and 2 have similar backoff count ranges in the beginning, but

Link 2 suddenly keeps increasing its backoff count, while Link 2 still has a similar backoff count range. This is due to consecutive free-riding transmissions of Link 2. As mentioned above, a free-riding transmission of a link inflates the link's backoff count when backoff count compensation is applied, which in turn increases the possibility of the link's recurrent free-riding transmissions. This happens even when links are in the same load condition, but it is aggravated when they are in different load conditions. In such a condition, the backoff stages of links may be different from each other because of different occurrence rates of collisions.

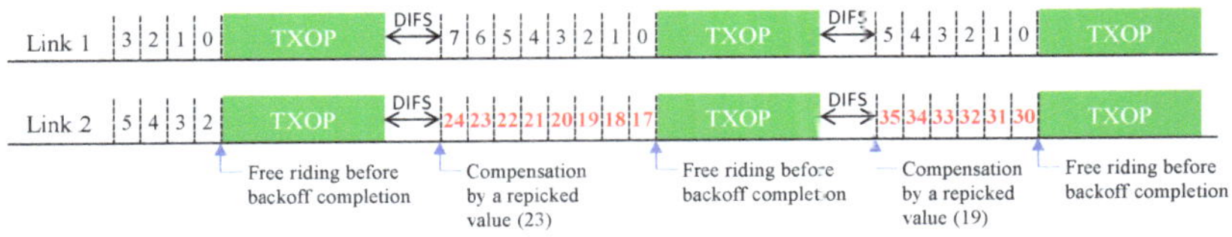

Figure 1. Illustration of the backoff count overflow problem (backoff count values after compensation are colored red).

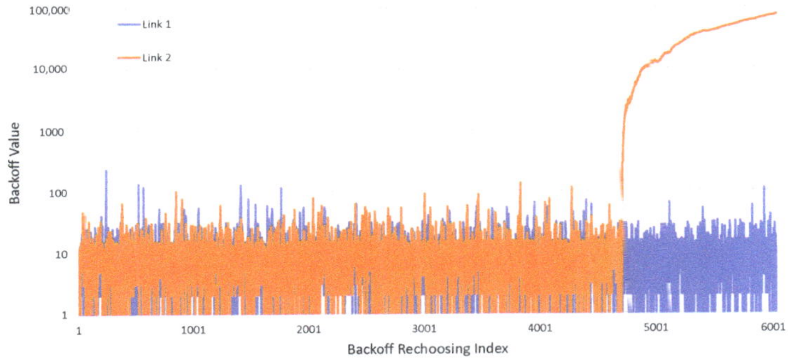

Figure 2. Backoff count evolution example for two links of an MLD.

5. Solutions to Backoff Count Overflow Problem

In this section, we describe the proposed solutions with additional design variants. Figures 3–5 (Proposals 1, 2, and 3, respectively) illustrate the changed operation of Link 2 against the Sync-FT operation with compensation illustrated in Figure 1 (the operation of Link 1 remains the same as Figure 1, since only Link 2 performs free-riding) while Figure 6 (Proposal 4) illustrates another example operation of Sync-FT (with backoff compensation) and the changed operation side by side. We assume in the figures that Link 1 is in the first backoff stage (CW = 16) and Link 2 is in the second backoff stage (CW = 32).

5.1. *Proposal 1: Limiting Consecutive Free-Riding Transmissions*

The main cause of an indefinite increase of a backoff count value, resulting in a backoff overflow, is consecutive free-riding transmissions of a link. This solution is made to alleviate the problem by limiting the number of allowed consecutive free-riding transmissions of each link. As illustrated in Figure 3, if a link performs a free-riding transmission, it starts its consecutive counter with value one. The next free-riding transmission right after the first one will result in an increment of the consecutive counter by one. The counter will keep increasing through consecutive free-riding transmissions of the link. Once the counter reaches its limit, the link will block its next free-riding transmission and reset the counter to zero. The counter will also be reset if there is a transmission caused by the link's own BO completion. The blocked free-riding transmission will keep its remaining backoff count value and continue to decrease the value after a main transmission finishes without a compensation. Setting the limit to one is the same as disabling any consecutive free-riding transmission, i.e., one link can free-ride only once, then must finish its own backoff process to transmit.

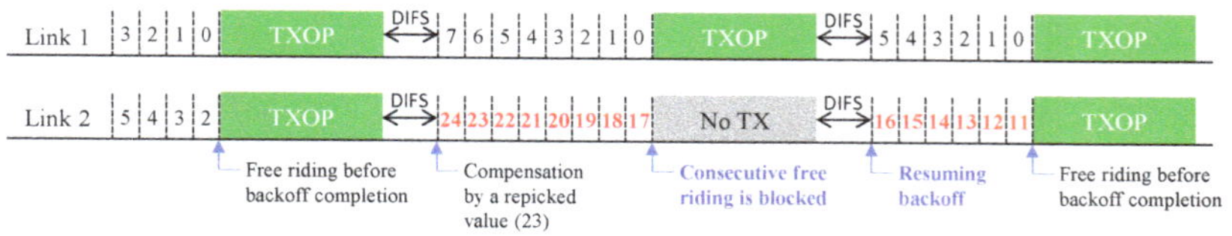

Figure 3. Proposal 1: limiting consecutive free-riding transmissions (backoff count values after compensation are colored red).

5.2. Proposal 2: Limiting a Compensated Backoff Value

A backoff overflow is caused by adding compensation to what is already a large backoff count value. In order to prevent that from happening, we designed the second solution to limit the compensated backoff count value itself. There are two ways of limiting the compensated backoff count value:

- (Option 1) limiting the total backoff count: The compensated backoff count, which is the sum of a rechosen number and a compensation value, is limited by a certain value (e.g., current CW value or that multiplied by a certain factor). It is illustrated in Figure 4.
- (Option 2) limiting a compensation value: Instead of limiting the total backoff value, only the compensation value to be added to the new rechosen number is limited. This is to prevent adding an overlarge value to a large remaining backoff count.

In both solutions, we can ensure the new backoff value remains at a smaller value and increase the chance of a free-riding link to complete its backoff process the next time.

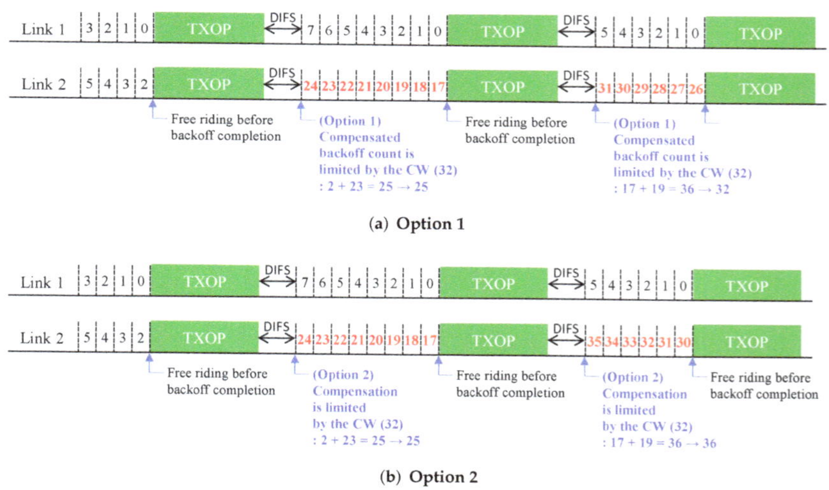

Figure 4. Proposal 2: limiting a compensated backoff value.

5.3. Proposal 3: Using the CW Value of a Main Link

As explained previously, the backoff overflow problem is aggravated if the links of an MLD have different backoff stages. The case is especially true when a free-riding link has a higher backoff stage than the link of the main transmission, because this can cause the randomly chosen count value of the free-riding link to become higher. In the solution illustrated in Figure 5, a free-riding link will use the same CW value as that of the link performing a main transmission, without considering its own backoff stage or current CW value. By using the same CW value as the main link, the free-riding link will have a higher probability to finish its own backoff process at a similar time as the main link. The same CW value with the main link will only apply to the new compensated rechosen value and will not affect transmissions by other means.

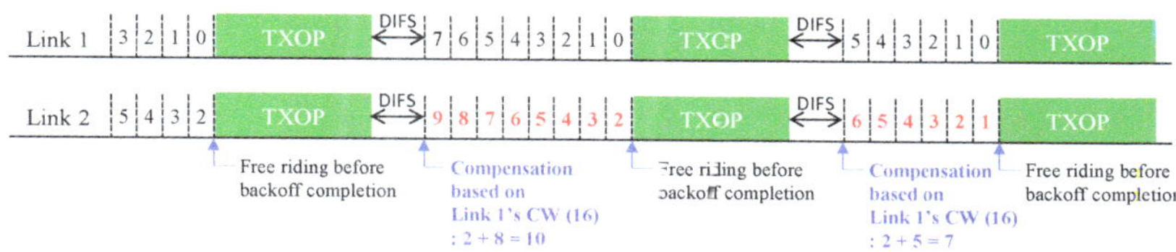

Figure 5. Proposal 3: using the CW value of a main link (backoff count values after compensation are colored red).

5.4. Proposal 4: Balancing Transmissions between Links

The main idea of this solution is to balance the number of TXOPs between links so that equal opportunities are given between them. The solution introduces a FR_COUNT value. This is a variable that increases when a link performs a free-riding transmission and decreases when it skips a transmission caused by its own BO completion, but it can never be lower than zero. Skipped transmissions will be excluded from compensation, thus mitigating the backoff overflow problem. The solution has several optional features to be implemented according to what to skip when the FR_COUNT of a link becomes higher than a limit as illustrated in Figure 6:

Figure 6. Proposal 4: balancing transmissions between links (backoff count values after compensation are colored red).

- Basic: A single link transmission of the link is skipped when its backoff count becomes zero.
- Option 1: A free-riding transmission is skipped when another link finishes BO.
- Option 2: Both main and free-riding transmissions are skipped together when the link finishes BO.
- Option 3: Only a main transmission is skipped while a free-riding transmission of another link is still allowed when the link finishes BO.

6. Performance Evaluation

We evaluated and compared the performance of MLO with Sync-FT with the proposed solutions. Both throughput and latency performance were observed in the evaluation. The coexistence of MLDs with legacy SLDs was also considered. In addition, backoff count values were observed as a way to determine the effectiveness of the proposed solutions to the backoff overflow problem.

6.1. Environmental Setup

Each MLD had two links (Link 1 and Link 2) working in individual channels. RF power leakage between the links of an MLD was considered, so all MLDs were non-STR MLDs. Each EHT basic service set (BSS) was composed of an access point (AP) and connected device(s) (a single device in a single-spot scenario and multiple devices in a 3GPP indoor scenario). APs and MLDs were equipped with MLO capability, while legacy devices were not. All devices used the modulation and coding scheme (MCS) 7 in 80 MHz (i.e., data bit rate of 680.6 Mbps) with the aggregation MAC service data unit (AMPDU) of 64 MPDUs, where each MPDU was 1500 bytes long. We considered full-buffer traffic conditions. Proposal 1 used one as a consecutive free-riding transmission limit, Proposal 2 used Option 1, and Proposal 4 used Option 1 with the FR_COUNT limit set to five. The operation frequency band of the network was 5 GHz, and the channel bandwidth was 20 MHz. Other system parameters followed those for the simulation scenario [6] and evaluation methodology [7] used for IEEE 802.11ax as listed in Table 1. All simulation results were obtained by averaging over five runs; in each run, 50 s were simulated.

Table 1. Simulation parameters.

Parameter	Value
Number of links per MLD	2
Channel bandwidth of a link	80 MHz
Number of STAs per BSS	1
MCS	7 (680.6 Mbps)
Traffic generation	Full buffer
Max aggregation size	64 MPDUs
MPDU size	1500 bytes
Slot length	9 us
SIFS	16 us
DIFS	34 us
CWmin	16
CWmax	1024
Center frequency	5 GHz
STA transmit power	18 dBm
AP transmit power	21 dBm
STA antenna gain	−4 dBi
AP antenna gain	+2 dBi
Noise figure	7 dB

We considered two deployment scenarios:

- *Single-spot deployment*: We placed all BSSs in one spot with no distance between them. Overlapping BSSs (OBSSs) were added in each of the links at the same time, i.e., one OBSS means one legacy BSS in either channel of Link 1 and Link 2.
- *3GPP indoor deployment* [8]: We followed Figure 7 to deploy the BSSs in the field. Each BSS consisted of 20 connected devices working in the respective link mentioned in the figure. The devices were randomly placed within their serving AP's coverage area.

Each deployment scenario had two coexistence scenarios:

- *MLD vs. MLD*: We considered EHT BSSs, each with its own multilink channel access scheme. One half of EHT BSSs was set using Sync-PL, and the rest used various channel access schemes, including the proposed solutions. The goal of this scenario was to examine the performance of the proposed solutions compared to the Sync-PL channel access scheme.
- *MLD vs. Legacy SLD*: We considered two legacy BSSs in each of the channels per 1 EHT BSS. The EHT BSS used various channel access schemes, including the proposed designs. The goal was to examine the performance of legacy BSSs, which is affected by the EHT BSSs.

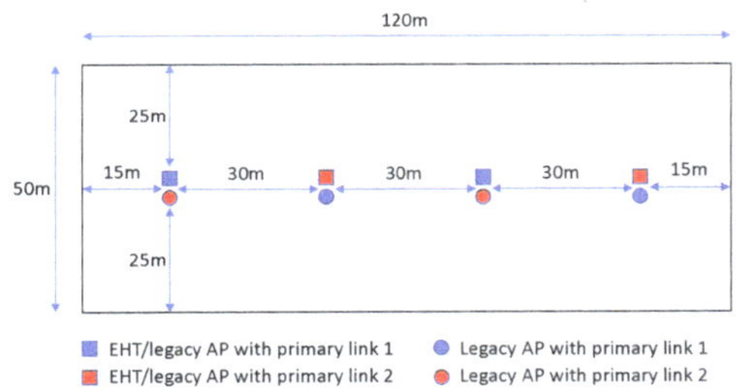

Figure 7. 3GPP indoor deployment scenario.

6.2. *Single-Spot Deployment*

6.2.1. MLD vs. MLD

Figure 8 shows the performance results with one MLD BSS with Sync-PL and the other with various channel access schemes. Sync-PL was chosen as the benchmark, because it suits a non-STR MLD better than the basic Sync operation. In the figure, along with the proposed solutions, we also considered Sync-FT (with no backoff compensation), Sync-FT with a rechoose option (Sync-FT-Repick), and with backoff compensation (Sync-FT-Repick+Comp). From the throughput and latency results, we see that Sync-FT and its modified rechoose design (MLD1) had a significantly higher performance than Sync-PL, while the other schemes (except Proposal 4) showed only slightly better performance than Sync-PL. This is because the earlier designs are highly aggressive against Sync-PL. On the contrary, the proposed solutions (except Proposal 4) coexisted better with Sync-PL. One point to note is that Proposal 4 had a worse performance than Sync-PL, which was caused by the excessive penalties given to the free-riding link. Figure 8c shows that our solutions alleviated the backoff overflow problem. The problem was severe, with Sync-FT-Repick+Comp showing the largest value of the average backoff count. Sync-FT and Sync-FT-Repick did not show the problem since they did not use backoff compensation and, thus, showed aggressive behavior. The proposed solutions achieved both harmonized coexistence and alleviation of backoff overflow.

6.2.2. MLD vs. Legacy SLD

The results of this scenario are presented in Figure 9. The case of legacy BSSs only (with no MLD BSS) is shown first labeled as "Legacy Only" in the graphs as a comparative measure to examine whether an EHT BSS has a positive or negative effect on the legacy performance. Compared to Sync-FT and Sync-FT-Repick, Sync-FT-Repick+Comp had a lower MLD performance in throughput and latency. This can be considered as due to the conservative behavior of Sync-FT-Repick+Comp; however, it is due to the backoff overflow problem as shown in Figure 9c. All the proposed solutions succeeded in suppressing the Sync-FT's channel access aggressiveness, while still maintaining good performance. Their average backoff count values were well limited compared to Sync-FT-Repick+Comp. The performance enhancement by mitigating the problem was especially noticeable in terms of latency; Figure 9b shows that Proposal 2 had 21.5% lower latency than Sync-FT-Repick+Comp. This implies that the proposed solutions were effective in solving the backoff overflow problem, while supporting the coexistence with legacy SLDs. On the other hand, it is also shown that the performance of legacy BSSs was also higher when half of them were replaced with an EHT BSS. This is due to the higher effectiveness of MLO in utilizing the channel medium and, thus, more room given to remaining legacy BSSs for channel access.

6.3. 3GPP Indoor Deployment

The indoor path loss model of the IEEE 802.11ax simulation scenario was used as given below:

$$PL(d) = 40.05 + 20\log_{10}(f_c/2.4) + 35\log_{10}(d/5) \quad (1)$$

where d is a transmitter–receiver distance in meters, and f_c is the center frequency in GHz. Then, the receive power (P_r) is obtained as

$$P_r = P_t + A_t - PL(d) + A_r \quad (2)$$

where P_t is the transmit power, and A_t and A_r are the antenna gains of the transmitter and receiver, respectively. The noise figure is 7 dB, and the noise floor is −94 dBm. We considered the transmission bit rates of IEEE 802.11ac (6.5 to 78 Mbps). The signal–to–noise ratio (SNR) vs. packet error rate (PER) curves of the IEEE 802.11ax's evaluation methodology [7] were used for packet error generation. For link adaptation, the highest bit rate with a PER lower than 10% was selected [9,10]. The control frames were transmitted at the lowest bit rate.

6.3.1. MLD vs. MLD

With the same reasoning as the single-spot deployment, Sync-PL was also used as a comparative benchmark in this scenario. The evaluation results are given in Figure 10. From both throughput and latency results, we can see that Sync-FT-Repick+Comp had the worst performance compared to Sync-PL because of the backoff overflow problem. All of the proposed solutions showed increased performance that was similar to Sync-FT and Sync-FT-Repick. This means that they alleviated the backoff overflow problem. Another proof that the proposed solutions effectively alleviated the problem is shown in Figure 10c. The problem happened with Sync-FT-Repick+Comp, showing a significantly large value of the average backoff count. The problem was considerably alleviated by the proposed solutions, as they showed small average backoff count values. Among the solutions, Proposal 2 (limiting a compensated backoff value) showed a relatively high backoff count value, meaning it was less effective than the other solutions.

6.3.2. MLD vs. Legacy SLD

The evaluation results of this scenario are presented in Figure 11. In the figure, Sync-FT-Repick+Comp showed lower performance than Sync-FT and Sync-FT-Repick, especially in the latency performance, which was possibly caused by backoff overflow. Figure 11c shows that Sync-FT-Repick+Comp suffered the backoff overflow problem. All proposed solutions mitigated the problem without ruining coexistence with the legacy BSSs. Proposals 1 and 4 showed slightly worse MLD performance, which was more noticeable in the latency performance. It is shown in Figure 11c that the average backoff count values of all proposed solutions were as small as Sync-FT and Sync-FT-Repick, thus proving that they were effective in solving the backoff overflow problem. As observed in the results of the single-spot deployment scenario, the performance enhancement in this deployment scenario was also noticeable in terms of latency; Figure 11b shows that Proposal 2 had 13.5% lower latency than Sync-FT-Repick+Comp.

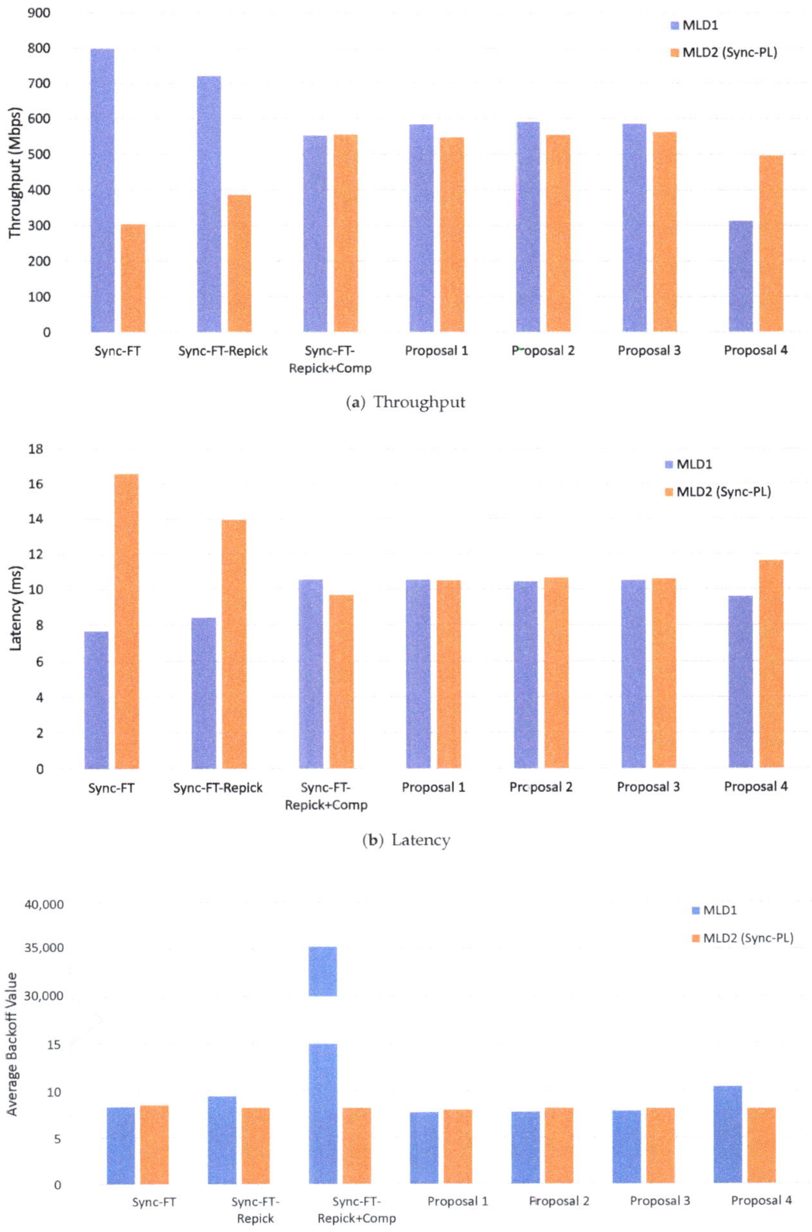

(a) Throughput

(b) Latency

(c) Average backoff count

Figure 8. MLD vs. MLD performance in the single-spot deployment scenario.

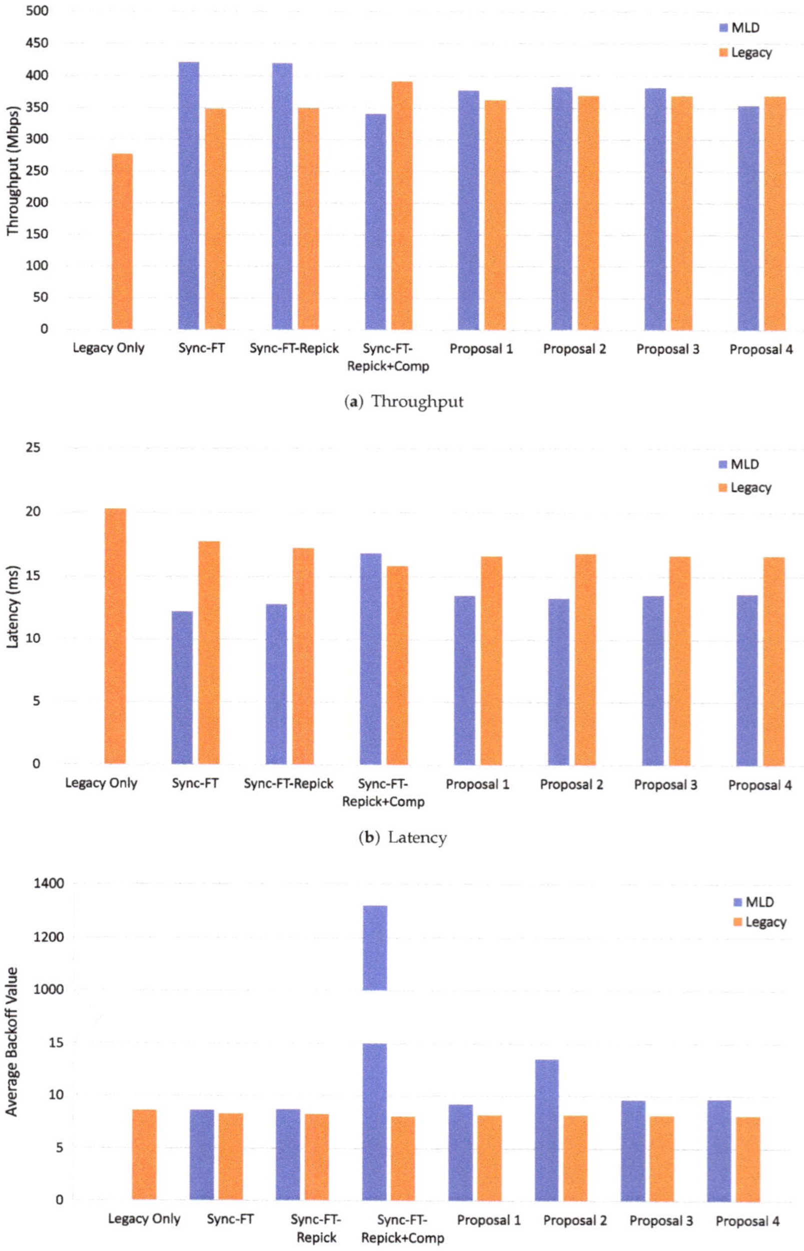

(**a**) Throughput

(**b**) Latency

(**c**) Average backoff count

Figure 9. MLD vs. legacy SLD performance in the single-spot deployment scenario.

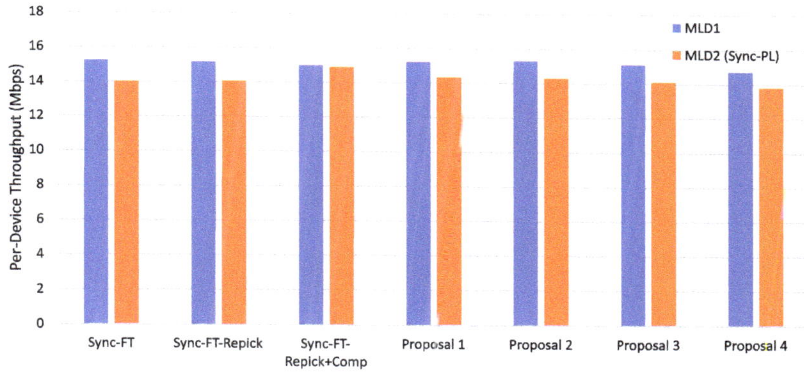

(a) Throughput

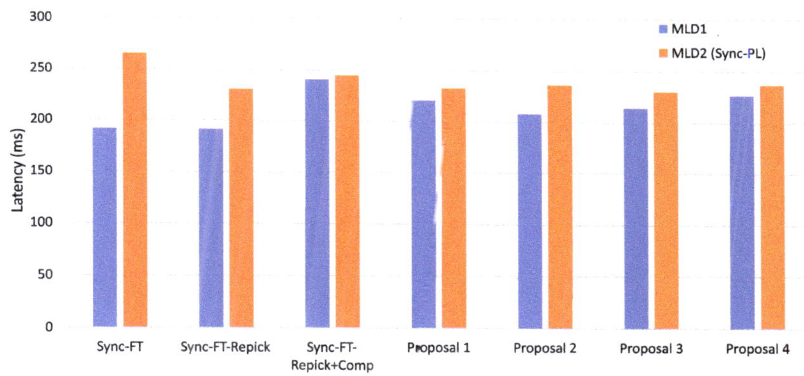

(b) Latency

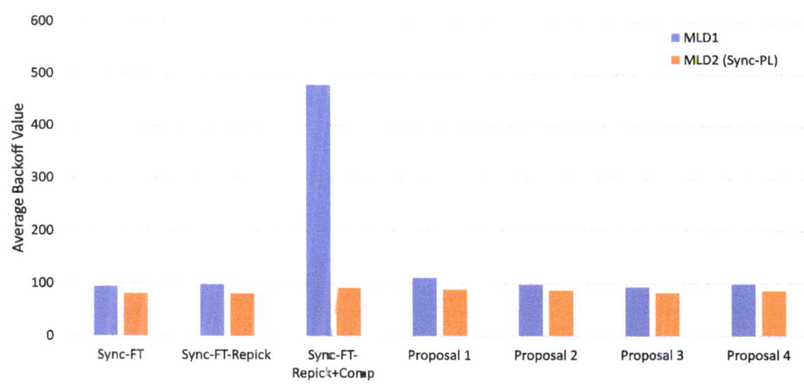

(c) Average backoff count

Figure 10. MLD vs. MLD performance in the 3GPP indoor deployment scenario.

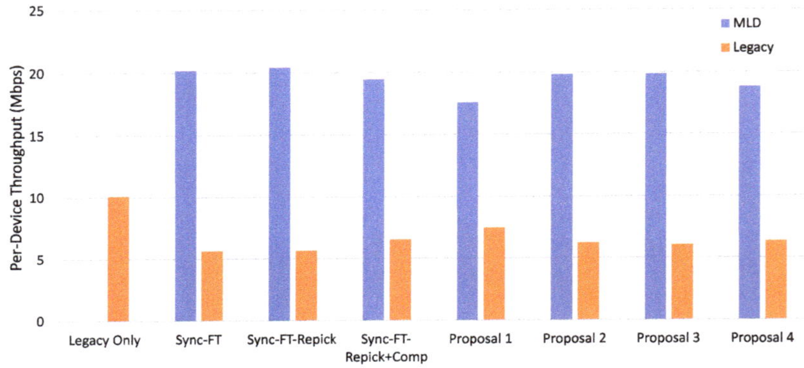

(**a**) Throughput

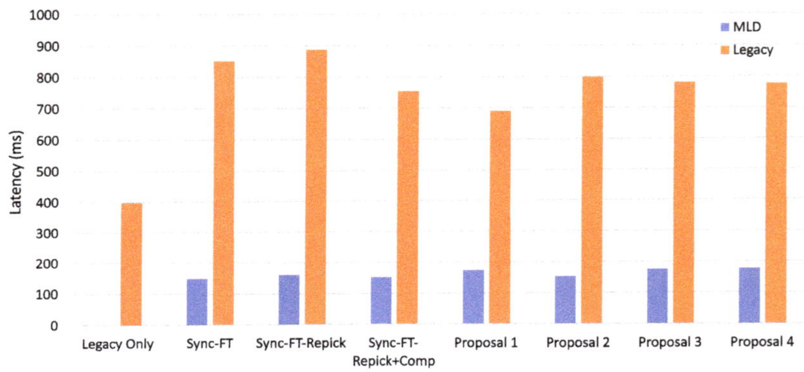

(**b**) Latency

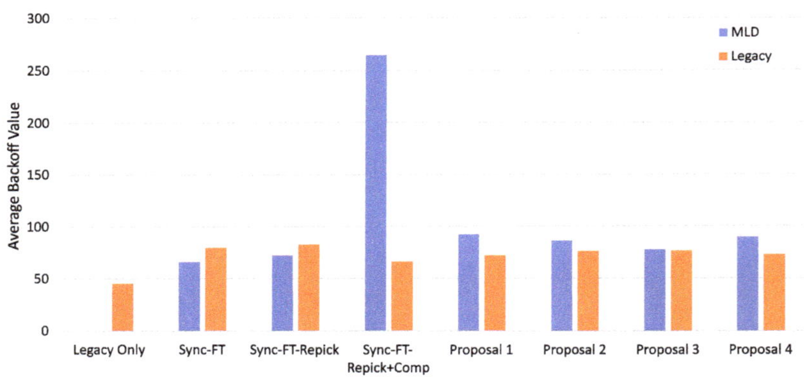

(**c**) Average backoff count

Figure 11. MLD vs. legacy SLD performance in the 3GPP indoor deployment scenario.

7. Conclusions

Sync-FT is one of the best options for multilink channel access in IEEE 802.11be WLANs, especially for non-STR MLDs. Due to the raised coexistence issue of Sync-FT resulting from its aggressive channel access behavior, the backoff count compensation solution can be applied to it. In this paper, we identified the backoff count overflow problem of Sync-FT with backoff count compensation, which happens due to continuing free-riding transmissions and backoff count compensations, thus causing an unlimited increase in the backoff count value. We proposed four solutions to alleviate the problem and showed through comprehensive evaluation that all proposed solutions were effective in mitigating the problem while still preserving coexistence with other MLDs and legacy SLDs.

Author Contributions: This work was realized through the collaboration of all authors. W.M. contributed to the main results and simulation. J.-H.Y. organized the work, provided the funding, supervised the research, and reviewed the draft of the paper. All authors have read and agreed to the published version of the manuscript.

Funding: This work was supported by the Research Program funded by the SeoulTech (Seoul National University of Science and Technology).

Data Availability Statement: Not applicable.

Conflicts of Interest: The authors declare no conflict of interest.

Abbreviations

The following abbreviations are used in this manuscript:

AP	Access Point
ASync	Asynchronous channel access operation
BO	BackOff procedure
BSS	Basic Service Set (BSS)
CCA	Clear Channel Assessment
CW	Contention Window
CWmax	Contention Window maximum
CWmin	Contention Window minimum
DIFS	Distributed coordination function InterFrame Space
EHT	Extremely High Throughput
FT	Faster Transmission
MAC	Medium Access Control
MCS	Modulation and Coding Scheme
MLO	MultiLink Operation
MLD	MultiLink Device
OBSS	Overlapping Basic Service Set
PIFS	Point coordination function InterFrame Space
RF	Radio Frequency
STR	Simultaneous Transmission and Reception
SLD	Single-Link Device
SLO	Single-Link Operation
Sync	Synchronous channel access operation
WLAN	Wireless Local Area Network

References

1. *IEEE 802.11-19/1262r23*; Specification Framework for TGbe. IEEE: Hoboken, NJ, USA, 2021.
2. *IEEE P802.11be Draft 1.0*; Wireless LAN Medium Access Control (MAC) and Physical Layer (PHY) Specifications Amendment 8: Enhancements for Extremely High Throughput (EHT). IEEE: Hoboken, NJ, USA, 2021.
3. Murti, W.; Yun, J.H. Multi-Link Operation with Enhanced Synchronous Channel Access in IEEE 802.11be Wireless LANs: Coexistence Issue and Solutions. *Sensors* **2021**, *21*, 7974. [CrossRef] [PubMed]
4. *IEEE Std. 802.11*; Wireless LAN Medium Access Control (MAC) and Physical Layer (PHY) Specifications. IEEE: Hoboken, NJ, USA, 2020.
5. Kim, S.; Yun, J.H. Wider-Bandwidth Operation of IEEE 802.11 for Extremely High Throughput: Challenges and Solutions for Flexible Puncturing. *IEEE Access* **2020**, *8*, 213840–213853. [CrossRef]
6. *IEEE 802.11-14/0980r16*; TGax Simulation Scenarios. IEEE: Hoboken, NJ, USA, 2015.
7. *IEEE 802.11-14 0571r12*; 11ax Evaluation Methodology. IEEE: Hoboken, NJ, USA, 2016.

8. 3GPP. *Feasibility Study on Licensed-Assisted Access to Unlicensed Spectrum*; Technical Report 36.889; 3GPP: Sophia Antipolis, France, 2015.
9. *IEEE 802.11 TGax Contribution 802.11-14/0620r0*; Link Adaptation for PHY SLS Calibration. LG Electronics: Tokyo, Japan, 2014.
10. *IEEE 802.11 TGax Contribution 802.11-15/1284r0*; Simulation results for spatial reuse in 11ax. LG Electronics: Tokyo, Japan, 2015.

sensors

Article

Popularity-Aware Closeness Based Caching in NDN Edge Networks

Marica Amadeo [1,2], Claudia Campolo [1,2,*], Giuseppe Ruggeri [1,2] and Antonella Molinaro [1,2,3]

1 DIIES Department, University Mediterranea of Reggio Calabria 89100 Reggio Calabria, Italy; marica.amadeo@unirc.it (M.A.); giuseppe.ruggeri@unirc.it (G.R.); antonella.molinaro@unirc.it (A.M.)
2 National Inter-University Consortium for Telecommunications (CNIT), 43124 Parma, Italy
3 Laboratoire des Signaux et Systémes (L2S), CentraleSupélec, Université Paris-Saclay, 91190 Gif-sur-Yvette, France
* Correspondence: claudia.campolo@unirc.it

Abstract: By enabling name-based routing and ubiquitous in-network caching, Named Data Networking (NDN) is a promising network architecture for sixth generation (6G) edge network infrastructures. However, the performance of content retrieval largely depends on the selected caching strategy, which is implemented in a distributed fashion by each NDN node. Previous research showed the effectiveness of caching decisions based on content popularity and network topology information. This paper presents a new distributed caching strategy for NDN edge networks based on a metric called popularity-aware closeness (PaC), which measures the proximity of the potential cacher to the majority of requesters of a certain content. After identifying the most popular contents, the strategy caches them in the available edge nodes that guarantee the higher PaC. Achieved simulation results show that the proposed strategy outperforms other benchmark schemes, in terms of reduced content retrieval delay and exchanged data traffic.

Keywords: Named Data Networking; information centric networking; caching; edge networks; 6G

Citation: Amadeo, M.; Campolo, C.; Ruggeri, G.; Molinaro, A. Popularity-Aware Closeness Based Caching in NDN Edge Networks. *Sensors* **2022**, *22*, 3460. https://doi.org/10.3390/s22093460

Academic Editor: Peter Han Joo Chong

Received: 29 March 2022
Accepted: 28 April 2022
Published: 2 May 2022

1. Introduction

A multitude of innovative applications, ranging from holographic telepresence to extended reality (XR), are expected to be delivered on top of sixth-generation (6G) networks, which would highly challenge the existing Internet infrastructure. Disruptive solutions are needed to cope with the demands of such future bandwidth-hungry and low-latency applications. By supporting name-based routing and ubiquitous in-network caching, the Named Data Networking (NDN) [1] paradigm is identified as a key enabler of 6G network architectures aimed at improving content distribution.

NDN is an information centric networking (ICN) architecture that promotes a communication model directly based on topology-independent content names, instead of internet protocol (IP) addresses. Content retrieval is based on the exchange of two named packets: the Interest, transmitted by the end-clients to retrieve the content, and the Data, transmitted by any node owning a copy of the content. Each Data packet is uniquely named and secured and, therefore, it can be cached by any NDN node in the path between the requester and the original source.

More specifically, NDN nodes are provided with a Content Store (CS) to cache incoming Data packets. The default caching strategy in NDN is Cache Everything Everywhere (CEE) coupled with Least Recently Used (LRU) replacement, i.e., each node caches each incoming Data packet and, if the CS is full, the LRU policy is applied to remove an existing item and make room for the new one. Although this strategy can speed up the data retrieval, several studies have showed that better performance can be obtained if selective decision strategies are implemented that improve the cache diversity [2].

One of the most important caching decision metrics is the content popularity. Typically, Internet contents show a skewed popularity distribution [3]: only a few contents are highly requested and deserve to be cached. Another crucial caching decision metric is the centrality of the nodes. The work in [4] focuses on the betweenness centrality metric and demonstrates that caching at the most central nodes can increase the cache hit probability and decrease the cache eviction rate, as these nodes are traversed by the majority of content requests. However, from the perspective of content retrieval, the node centrality should be re-defined by taking into account the capability of the node of satisfying the content requests, in addition to its topological feature [5,6].

So far, existing work [5–7] defined a popularity-aware betweenness centrality metric to enable coordinated caching decisions between distributed cachers. Basically, these strategies weight the betweenness centrality of a node with the popularity of the contents that can be transmitted through it. By introducing a relevant signalling among the nodes, collaborative decisions are deployed, where the most popular contents are cached in the most central nodes. However, in edge topologies, typically characterized by hierarchical topologies [6,8], the nodes with the highest betweenness centrality are also far away from the end-clients, i.e., the content consumers. Therefore, they cannot guarantee a low retrieval delay and a small network traffic. Instead, it would be crucial to cache the content as close as possible to the consumers to meet the proximity requirements of upcoming 6G applications.

In this paper, we define a new caching scheme based on a popularity-aware closeness (*PaC*) metric, which allows to cache the most popular contents in the edge nodes according to their proximity to the majority of requesters. Thanks to lightweight signalling piggybacked in Interest and Data packets, the potential cachers are ranked according to the *PaC* metric, and the best available cacher per path is selected. Performance evaluation shows that the conceived strategy is able to limit the retrieval delay while maintaining low exchanged traffic and the signalling overhead, compared to related literature based on the betweenness centrality metric [6].

The remainder of this paper is organized as follows. Section 2 provides background material on caching strategies in NDN. The proposal is discussed in Section 3 and evaluated in Section 4. Final remarks are reported in Section 5.

2. Background and Motivations

2.1. NDN in a Nutshell

In NDN, caching operations are embedded in the forwarding process of the Interest and Data packets, which is shown is Figure 1.

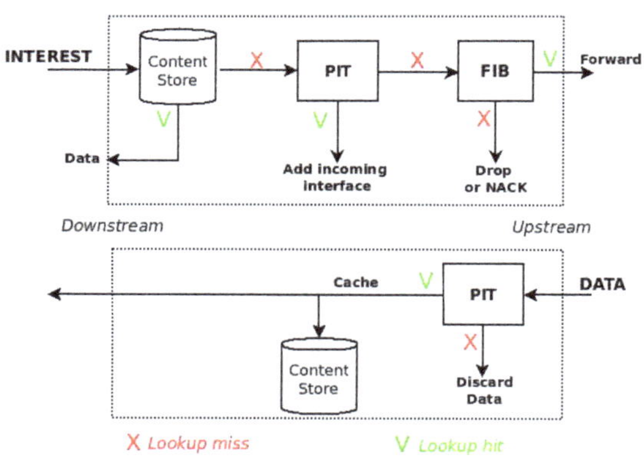

Figure 1. Forwarding process in NDN.

At the reception of an Interest packet, the node looks in the CS and, in case of a name matching, it immediately sends the Data back. Vice versa, the node checks if an equal request is pending in the Pending Interest Table (PIT) and, in case of a positive outcome, the incoming interface of the Interest is added to the PIT entry and the packet is discarded, thus reducing the traffic in the network. If the PIT matching fails, the Interest is forwarded according to named-based forwarding rules towards the original source.

When the Data packet arrives, the node checks for a name matching in the PIT and, in case of a positive outcome, the packet is cached and then forwarded towards the consumer(s). Vice versa, if the PIT matching fails, the Data is considered unsolicited and it is discarded.

The vanilla NDN caching strategy, CEE, typically leads to poor performance due to the lack of cache diversity [2]. A conventional scheme that limits the cache redundancy without introducing much complexity is Fixed Probability-based caching, where the Data packets are cached according to a fixed probability, usually set to 0.5 [9].

2.2. Caching Strategies in the Literature

To improve the content retrieval performance, several NDN caching strategies have been proposed in the last few years [2]. Among them, we can identify two major categories, namely popularity-based schemes and centrality-based schemes, which leverage, respectively, content popularity information and network topology information to select which contents to cache and where.

In popularity-based schemes [10–14], NDN nodes cache the most popular contents according to the locally measured rate of received Interests, which is stored in a Popularity Table. Typically, a threshold based mechanism is considered: contents that are requested a number of times higher than the threshold are cached.

In centrality-based schemes [4], instead, the caching decision depends on the *importance* of the nodes, expressed in terms of topological centrality. For instance, the pioneering proposal called *Betw* [4] leverages the betweenness centrality of the nodes as decision metric and replaces contents with the least recently used (LRU) policy. According to the graph theory, given a set V of network nodes, the betweenness centrality (C_B) of a node $v_i \in V$ is defined as:

$$C_B(v_i) = \sum_{v_k \neq v_j \neq v_i \in V} \frac{\sigma_{v_k,v_j}(v_i)}{\sigma_{v_k,v_j}}, \tag{1}$$

with $\sigma_{v_k,v_j}(v_i)$ and σ_{v_k,v_j} being, respectively, the number of shortest content delivery paths from the two endpoints v_k and v_j that pass through v_i and the total number of shortest content delivery paths between the same endpoints. *Betw* strategy is built on the following intuition: if a node v_i is traversed by many content delivery paths, then it is more likely to get a cache hit. Therefore, contents are cached in the nodes with the higher betweenness centrality.

In [5], however, the authors observe that the topological centrality metric alone does not reflect the importance of a node from the content delivery perspective. Therefore, the authors define a new popularity-aware centrality metric, which aims at placing the most popular contents at high central nodes, and the remaining contents with decreasing popularity at nodes with decreasing centrality score. With the same target, the Betweenness and Edge Popularity caching (BEP) strategy [6] leverages a coordinated signalling mechanism piggybacked into Interest and Data packets. In BEP, the edge nodes (i.e., the leaf nodes directly connected to the consumers) track the number of received requests and periodically compute the content popularity with an exponential weighted moving average (EWMA) formula. The information is maintained in a Popularity Table, where contents are also ranked in terms of popularity. When a content request arrives, the edge node includes the correspondent popularity ranking in the Interest and forwards it towards the origin source. In addition, the Interest carries an array of betweenness centrality values, which is filled by all on-path routers. When the origin source receives the Interest, it compares the popularity ranking against the available betweeneess values and identifies the cacher by matching the two metrics, i.e., if the content has the highest popularity, it will be cached in the node with the highest centrality, and so on.

Despite their differences, all the noted approaches consider the betweenness centrality in the caching decision. However, this metric does not guarantee the minimum retrieval delay for the consumers. Indeed, especially when considering edge domains, typically characterized by hierarchical topologies [8], the nodes with the highest betweenness centrality are usually far away from the leaf nodes, where the consumers are attached. Instead, a peripheral edge node, e.g., a base station (BS), has typically low betweenness centrality, but it can be very close to the consumers and cover a key role as cacher.

Given a content x, the closest node to the requesters of x would be able to deliver the content with the lowest delay. Therefore, to take advantage of the limited cache capacity at the edge, it is necessary to select the most popular contents and cache them in the closest nodes along the delivery paths. In addition to reducing the content retrieval latency, caching contents close to the consumers allows for the reduction of intra-domain traffic. Indeed, contents may traverse a lower number of hops and free bandwidth resources over edge links.

Table 1 compares the main features of the related caching strategies available in the literature and our proposal.

Table 1. Comparison of caching strategies based on popularity and/or topology metrics.

Work	Popularity	Topology	Domain	Decision Strategy
[10]	✓	-	Edge/Core	Caching popular contents based on a fixed popularity threshold
[11]	✓	-	Edge	Caching popular and fresh contents based on a flexible popularity threshold
[12]	✓	-	Edge/Core	Caching only popular long-lasting contents (in the core network); caching popular short-lasting contents only once per each delivery path (in the edge network)
[13]	✓	-	Edge/Core	Caching popular contents based on a flexible popularity threshold
[14]	✓	-	Edge	Caching popular contents based on a strict hierarchical coordination between the nodes
[4]	-	✓	Edge/Core	Caching contents in the most central nodes based on the betweenness centrality metric
[5]	✓	✓	Edge	Caching based on a popularity-weighted content-based centrality
[6]	✓	✓	Edge/Core	Caching based on popularity and betweenness centrality metric
Our work	✓	✓	Edge	Caching based on a popularity aware consumer proximity metric

3. PaC-Based Caching

3.1. Main Pillars and Assumptions

To capture the aforementioned needs and overstep the limitations of existing solutions, we propose a different topological metric that accounts for *the proximity of potential cachers to the consumers and is weighted by the content popularity*. We leverage the resulting metric, namely *PaC*, in a new strategy aimed at caching the most popular contents as close as possible to the majority of consumers in order to limit the data retrieval delay and the exchanged data traffic.

As shown in Figure 2, the reference scenario of our study is an edge domain, e.g., the backhaul network of a mobile network operator, a campus network, composed of a set V of NDN nodes [8]. A subset of nodes denoted as $I \subset V$ act as ingress nodes that consumers are connected to. A few other nodes, instead, act as egress nodes towards the content sources, i.e., remote servers hosting the contents. A catalogue X of cacheable contents is considered.

For ease of reference, the key notations used in the paper are summarized in Table 2.

Table 2. Summary of the main notations.

Symbol	Description
V	set of NDN edge nodes
I	set of NDN ingress nodes, with $I \subset V$
X	catalogue of contents
v_j	generic ingress node
v_i	generic edge node
x_n	generic content
$R_{v_i}(x_n)$	average request rate for content x_n at node v_i
Θ_P	popularity threshold
$\overline{R_{v_j}}$	average content request rate at ingress node v_j
T	time interval for updating caching decision parameters
M_{v_j}	number of distinct contents received at the ingress node v_j
$\hat{I}_i(x_n)$	set of ingress nodes forwarding Interests for content x_n to node v_i
$PaC(v_i, x_n)$	popularity-aware closeness metric for content x_n at node v_i
$h_{(v_i,v_j)}$	hop distance between nodes v_i and v_j

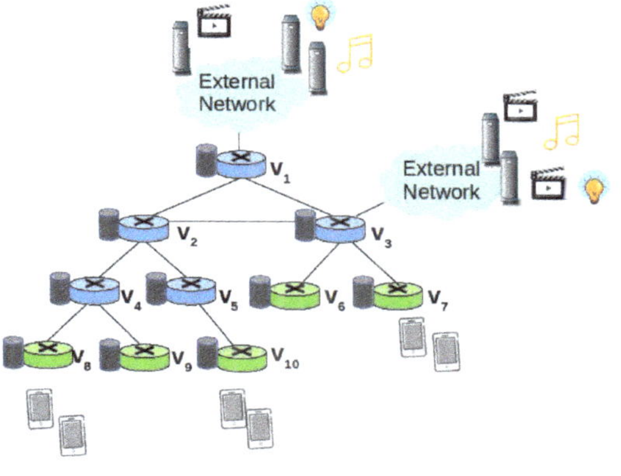

Figure 2. Reference scenario.

All the nodes have caching capabilities and implement the traditional NDN forwarding fabric with the *best route* strategy, i.e., Interests are forwarded along the shortest path between ingress and egress nodes. A routing protocol, e.g., named-data link state routing protocol (NLSR) [15], is enabled for intra-domain dissemination of both connectivity and name prefix information.

To enable the PaC-based caching (PaCC), the following main modifications are foreseen to the legacy NDN routines, data structures, and packet fields:

- Content popularity is tracked at the edge nodes in terms of locally perceived content request rate. Values are maintained in a *Popularity Table* and properly advertised in the new RATE field of the Interest packet to account for the actual content request number over each edge link.
- Each node tracks the distance, in terms of hop count, from the on-path ingress nodes through a newly added field HOPCOUNT. This information, combined with the content request rate, is used to compute the PaC metric that is then advertised in the new PAC field of the Interest packet by the forwarding nodes.
- The highest PaC metric, discovered during the Interest forwarding, is carried by the returning Data packet in a new PAC field, and it is used to select the cacher.

Figure 3 shows the new fields introduced in the Interest and Data packets, respectively, that enable the PaC-based caching thanks to an implicit signalling mechanism, which will be clarified subsequently.

NAME	SELECTORS	RATE	PAC	HOP COUNT	NONCE

(**a**) Interest

NAME	METAINFO	PAC	CONTENT	SIGNATURE

(**b**) Data

Figure 3. Overhauled NDN packets (fields with text in orange are those added by the PaC-based caching).

3.2. Tracking Content Popularity

Similarly to other schemes [6,10,11,13], the edge nodes track in the Popularity Table the received requests in order to identify the most popular contents.

Basically, the ingress nodes (green devices in Figure 2) counts the Interests received from consumers they are connected to, in order to determine the average request rate of each content. At the ingress node $v_j \in I$, the average request rate for content x_n is denoted as $R_{v_j}(x_n)$.

The node periodically updates the average request rate, with a time interval T set to one minute, similarly to [13], to properly infer potential changes in the request patterns. Therefore, we assume each entry in the Popularity Table includes three fields: content name, average request rate, and current counter of requests.

In our design, the ingress nodes are also in charge of identifying the most popular contents that should be cached along a delivery path. More specifically, a content is considered popular by the ingress node v_j if it is requested a higher number of times than a popularity threshold Θ_P. This latter is computed as the EWMA of the average request rate per each requested content:

$$\Theta_P = (1-\alpha)\Theta_{P_{Old}} + \alpha\overline{R_{v_j}}, \tag{2}$$

with

$$\overline{R_{v_j}} = \frac{\sum_{n=1}^{M_{v_j}} R_{v_j}(x_n)}{M_{v_j}}, \tag{3}$$

with M_{v_j} being the number of distinct contents requested during the last time interval T, as perceived by node v_j, and $\alpha \in (0,1)$ set to 0.125 to avoid large fluctuations in the computation and give relevance to the historical values.

Because the Interest aggregation in the PIT hides the actual number of consumers requesting the same contents, a specific signalling mechanism is deployed to let intermediate nodes effectively track the request rate. More specifically, each time an ingress node v_j forwards an Interest for content x_n to the next on-path node v_i, it includes the average request rate information in the RATE field. Of course, the same edge node v_i can be in multiple shortest delivery paths towards the same content. For instance, node v_4 in Figure 2 is traversed by two shortest paths from the ingress nodes v_8 and v_9. Instead, node v_2 is traversed by three shortest delivery paths from the ingress nodes v_8, v_9, and v_{10}.

We denote as $\hat{I}_i(x_n) \subset I$ the set of ingress nodes forwarding Interests for content x_n to v_i. In other words, v_i belongs to the shortest paths connecting the ingress nodes belonging to $\hat{I}_i(x_n)$, which receive the requests from the consumers, and the egress node towards the origin source. Since NDN implements only on-path caching, v_i is a candidate cacher for a content x_n for which requests are received by ingress nodes in $\hat{I}_i(x_n)$. For instance, a content cached at v_2, in Figure 2, may serve the requests coming from the three ingress nodes, v_8, v_9, v_{10}.

The intermediate node v_i (e.g., v_4 in Figure 2) collects the $R_{v_j}(x_n)$ values from the Interests received through the incoming interfaces and calculates the local average request rate as:

$$R_{v_i}(x_n) = \sum_{j \in \hat{I}_i(x_n)} R_{v_j}(x_n). \tag{4}$$

When re-transmitting the Interest packet, v_i, in its turn, overwrites the RATE field with the new cumulative value, thus the next-hop node will be aware of the average number of requests that can be satisfied with a Data packet over that incoming interface. The next-hop node (e.g., v_2 in Figure 2) also calculates the cumulative request rate, and so on.

3.3. Popularity-Aware Closeness Metric

The PaC metric of a potential cacher v_i for content x_n, $PaC(v_i, x_n)$, considers the distance, in terms of hop count, between v_i and the ingress nodes connected to the consumers. Since, in NDN, v_i receives Interests for x_n from an ingress node v_j only if it is in the forwarding path between v_j and the origin source, $PaC(v_i, x_n)$ takes into account only the set of ingress nodes $\hat{I}_i(x_n)$.

In parallel, the metric considers the number of consumers that could be satisfied by the potential cacher. The higher is the number of requests for x_n that cross v_i, the higher should be the PaC metric.

In mathematical terms, the PaC metric for a node $v_i \in V$ and a content x_n can be expressed as:

$$PaC(v_i, x_n) = \frac{R_{v_i}(x_n)|\hat{I}_i(x_n)|}{\sum_{j \in \hat{I}_i(x_n)}(h_{(v_i,v_j)} + 1)} \tag{5}$$

where $|\hat{I}_i(x_n)|$ is the cardinality of the set of ingress nodes in $\hat{I}_i(x_n)$, used to normalize the metric, and $h_{(v_i,v_j)}$ is the hop distance between v_i and the ingress node v_j.

It can be observed that $PaC(v_i, x_n)$ increases with the request rate of x_n and it is equal to zero if v_i does not receive any request for x_n. At the same time, the metric decreases if v_i is far from the consumers.

3.4. Caching Algorithm

When an Interest for content x_n arrives at an ingress node v_j, the latter updates the corresponding entry in the Popularity Table, checks if the content is popular, and then accesses the CS to find a matching Data packet to send back immediately. If the CS lookup fails, then v_j checks the PIT. If this lookup also fails, then v_j acts differently depending whether the content is popular, i.e., its average request rate is higher than the popularity threshold, or not. If it is not popular, then there is no need to cache it along the path. Therefore, v_j increases by one the HOPCOUNT field of the Interest, fills the RATE field, in order to allow the next hop updating the Popularity Table, but leaves the PAC field to the default zero value, to indicate that the content is not popular. The request will be forwarded towards the origin server according to the standard NDN forwarding fabric. Of course, it may happen that an edge on-path node belonging to multiple delivery paths has cached the content (because it is considered popular by another ingress node) and, therefore, the request can be still satisfied at the edge.

The returning Data packet, being not popular, should not be cached. However, to make the best of the available storage space, unpopular Data packets can be cached in case the CS is not full, e.g., during the network bootstrap phases.

Vice versa, if x_n is popular, then v_j updates all the new fields of the Interests, i.e., HOPCOUNT, RATE, and PAC, and transmits the packet to the next-hop, according to the FIB entry.

The subsequent node v_i receiving the Interest performs a slightly different processing. First, it accesses the information from the Interest header fields and checks if the Popularity

Table needs to be updated. In NDN, Interests cannot loop, therefore a newly received Interest carries consistent information in the header fields. The latter can be equal to the previously recorded one, if the request pattern has not changed. The node then performs a lookup in the CS for a match. If the Data packet is found, then v_i computes its PaC metric and compares it with the value in the PAC field. If its value is greater, then it overwrites the field and sends the packet back. Vice versa, if its PaC value is smaller, it simply sends the Data packet without altering it. The receiving node with the highest PaC will also cache the Data, thus moving the copy closer to the consumers.

In case the CS and the consequent PIT matching fail, v_i computes its PaC and, if its value is greater than the current one, it updates the PAC, HOPCOUNT, and RATE fields. Conversely, if the PaC is lower than the current one, it only updates the HOPCOUNT and RATE fields. It then re-transmits the request according to the FIB and waits for the Data packet.

When finally receiving the Interest, the origin server, or an intermediate cacher, copies the PaC value from Interest to the corresponding field in the Data packet header and transmits the packet. The first node with the corresponding PaC will cache the Data. If the CS if full, an existing item is replaced according to the LRU policy.

The flowcharts summarizing the Interest processing and Data processing for the PaC-based caching are depicted in Figure 4 and Figure 5, respectively.

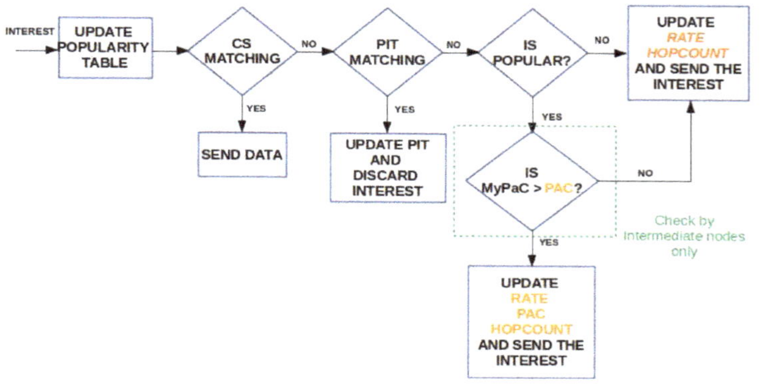

Figure 4. Interest processing in the presence of PaC-based caching.

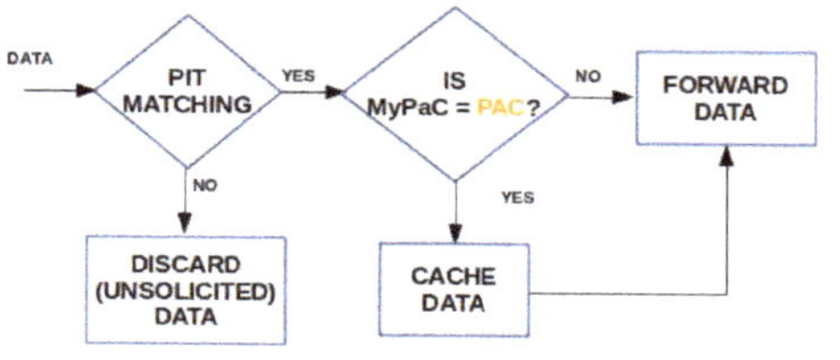

Figure 5. Data processing in the presence of PaC-based caching.

3.5. A Toy Example

To better understand the PaC metric, we consider the toy example in Figure 6 that includes eight edge nodes in a hierarchical topology. The nodes $v_1, v_2, \ldots, v_5$ are ingress nodes tracking the average content request rate. For the sake of simplicity, we assume that three distinct popular contents, namely x_1, x_2, x_3, should be cached at the edge and their request rate is stable. The available storage space at each node allows for caching only one content.

It can be observed that the average request rate for x_1, the most requested content at the edge, is 22 at v_1 and 20 at v_2, v_3, whereas it is 6 at v_5. When computing the PaC metric for the nodes traversed by the Interests for x_1, according to Equation (5), it results in $PaC(v_1, x_1) = 22$, $PaC(v_2, x_1) = PaC(v_3, x_1) = 20$, $PaC(v_5, x_1) = 6$, $PaC(v_6, x_1) = \frac{62 \cdot 3}{6} = 31$, $PaC(v_7, x_1) = \frac{6 \cdot 1}{2} = 3$, and $PaC(v_8, x_1) = \frac{68 \cdot 4}{12} = 22.66$. The node with the highest PaC metric for x_1 is v_6, which is indeed the node closer to the majority of consumers that are attached to the ingress nodes $v_1 - v_3$. If instead we consider the path $\{v_5 \rightarrow v_7 \rightarrow v_8\}$, the node with the highest PaC is v_8. However, caching x_1 at v_6 would imply that the node at the upper layer, v_8, will not receive further Interests for x_1 and therefore, its local request rate will decrease. As a consequence, in a subsequent time window, x_1 will be cached at v_5.

Vice versa, an approach based on the betweenness centrality, like BEP, would be considered as best cacher for x_1 node v_8, which has the highest centrality, but it is also the farthest away from the consumers. Therefore, caching x_1 at v_8 would highly increase the retrieval delay and the intra-domain traffic.

Content x_2 is requested only at v_4 and v_5 and it results in $PaC(v_4, x_2) = 15$, $PaC(v_5, x_2) = 3$, $PaC(v_7, x_2) = \frac{18 \cdot 2}{4} = 9$, $PaC(v_8, x_2) = \frac{18 \cdot 2}{6} = 6$. The node with the highest PaC metric is v_4 which, again, is the one closer to the majority of consumers. Therefore, the majority of requests will be served with minimum delay and with minimum intra-domain traffic. To serve requests coming from path $\{v_5 \rightarrow v_7 \rightarrow v_8\}$, instead, v_7 would be selected as cacher.

By following the same procedure, it can be found that the higher PaC for content x_3 is obtained at v_8.

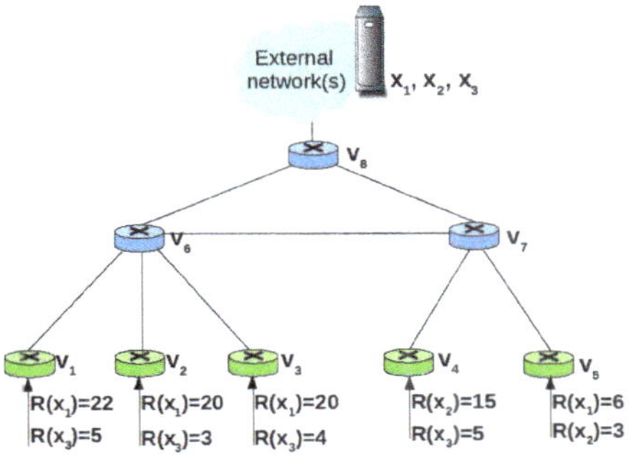

Figure 6. Toy example: an edge network with eight nodes.

4. Performance Evaluation

4.1. Simulation Settings

The proposed caching strategy has been implemented in ndnSIM, the official simulator of the NDN research community [16]. As representative edge domain, we consider a tree topology randomly generated with the Georgia Tech Internetwork Topology Models (GT-ITM) [17]. The topology includes 20 intermediate nodes and 8 leaves acting as ingress nodes. The root node connects the edge domain with a remote server acting as content producer. Because the performance assessment is focused on the edge domain, the external network is simply simulated as a link, with latency of 30 ms [18], between the root node and the server. The latency of edge links is instead uniformly distributed in the range [2–5] ms.

We consider a catalog of 15,000 contents, each one consisting of 1000 Data packets 1 Kbyte-long. A variable number of consumers attached to the leaf nodes of the topology, request contents according to the Zipf's law [3], with skewness parameter α set to 1. In these settings, we consider two distinct simulation scenarios.

- In the first scenario, we assume that the total caching capacity of the edge domain, uniformly distributed among the nodes, is varying from 0.25% to 2% of the overall catalog size, similarly to the values reported in [19]. The number of consumers is set to 60.

- In the second scenario, we assume that the cache capacity is fixed to 0.5% of the overall catalog size, whereas the number of consumers range from 20 to 70.

The main simulation settings are summarized in Table 3.

Table 3. Main simulation settings.

Parameter	Value
Content catalog size	15,000 contents
Content size	1000 Data packets
Data packet size	1000 bytes
Content Popularity	Zipf-distributed with $\alpha = 1$
Scenario	GT-ITM [17]
Edge link latency	Uniformly distributed in [2, 5] ms
Number of consumers	20–70
Number of edge nodes	28
Caching capacity	From 0.25% to 2% of the catalogue size

We compare the proposed model (labeled as *PaCC* in the plots) against the following benchmark solutions:

- Cache everything everywhere (labeled as *CEE* in the plots). It is the vanilla NDN caching strategy where all the incoming Data packets are cached.
- Fixed probability-based caching (labeled as *Prob* in the plots). It caches incoming Data according to a fixed probability set to 0.5 [9].
- Betweenness and edge popularity caching (labeled as *BEP* in the plots). It implements a caching scheme based on the popularity-aware betweenness centrality metric, as in [6].

For all the noted schemes, the replacement policy is LRU.

The following performance metrics are considered:

- Retrieval delay: it is computed as the average time taken by a consumer to retrieve a content.
- Number of hops: it is computed as the average number of hops traveled by the Interest packets for retrieving the corresponding Data packets.
- Exchanged NDN packets: it is the total number of Interest and Data packets transmitted by all the nodes, i.e., consumers, providers, and edge nodes, during the simulation, to retrieve the contents.

The first two metrics capture the effectiveness of the compared schemes in caching content copies in proximity to the consumers. The last metric provides insights about the efficiency of the schemes, in terms of traffic exchanged in the network. Results are averaged over 10 runs and reported with 95% confidence intervals.

4.2. Results

4.2.1. Impact of the Cache Size

Figure 7 shows the performance metrics when varying the cache size of the edge domain from 0.25% to 2% of the content catalogue size. As expected, it can be observed that, as the cache size increases, performance improves for all the considered schemes. Indeed, the higher the storage space at the edge, the higher is the number of contents that can be stored, with consequent advantages in terms of reduced retrieval delay, number of hops, and exchanged traffic.

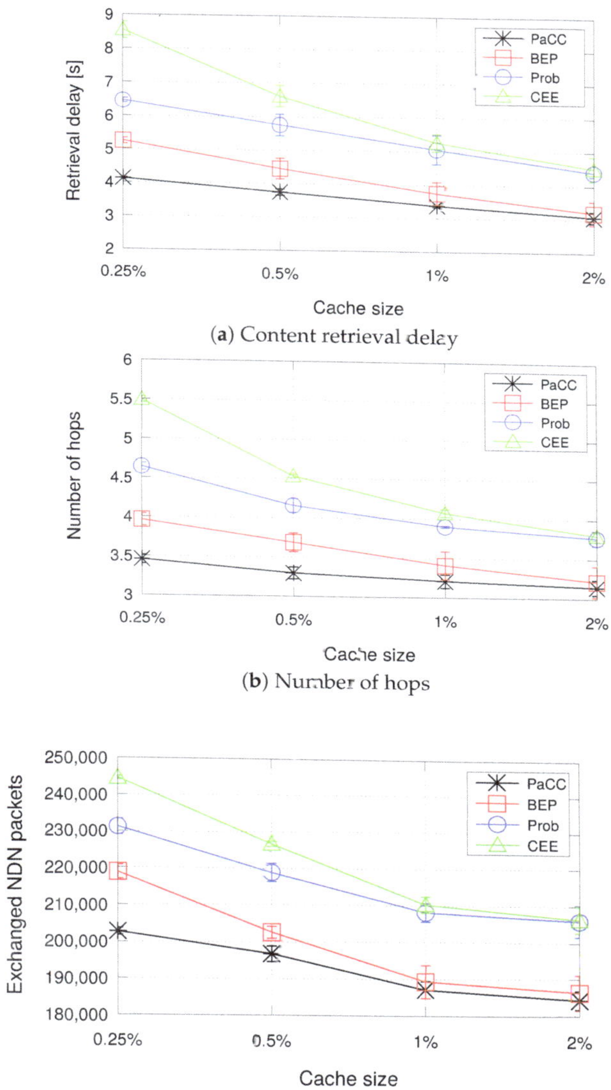

Figure 7. Metrics when varying the cache size of the edge domain from 0.25% to 2% of the content catalogue size (number of consumers equal to 60).

Being oblivious of content popularity and topology information, the simplest CEE solution shows the worst performance in terms of content retrieval delay, due to higher number of hops traversed to reach the requested content. It generates the highest load of exchanged NDN packets. The Prob scheme slightly outperforms CEE, because it does not cache indiscriminately contents but it tries to better distribute them in the CS of edge nodes.

As the cache size of the edge domain increases, differences among the two decrease, because there is higher chance to find storage space for caching contents within nodes.

PaCC outperforms all the other schemes in terms of all the considered metrics. However, the gap with BEP reduces as the edge domain is more capable to cache contents.

To conclude the analysis, Figure 8 shows the percentage of Interest packets that reach the origin server because it has not been found a CS matching at the edge. Reasonably, the traffic to the cloud reduces when the storage space at the edge increases, with PaCC outperforming all the benchmark schemes.

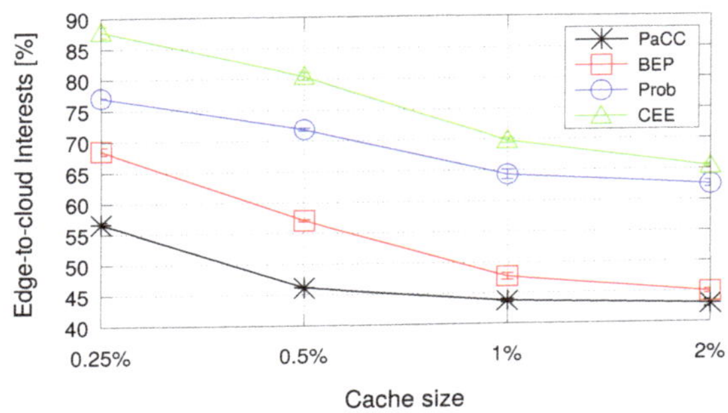

Figure 8. Percentage of Interest packets that are forwarded to the origin server.

4.2.2. Impact of the Number of Consumers

The second set of results, reported in Figure 9, measures the metrics of interest when varying the number of consumers.

Under such settings, the proposed PaCC solution achieves the best performance in terms of content retrieval delay, number of hops, and exchanged NDN packets.

It can be observed that, as the number of consumers increases, all schemes experience a shorter delay and a lower number of hops. Such a behaviour has to be ascribed to the fact that under the simulated settings, due to the Zipf distribution, a higher number of requests, from different consumers, concentrate on the same few contents, the most popular ones. Hence, such contents are more likely to be cached in the edge domain, instead of being retrieved from the original producer. The number of exchanged packets, instead, reasonably increases with the number of consumers, to account for the increasing number of Interests issued by more consumers and number of Data packets forwarded to each single consumer.

4.2.3. Overhead Analysis

To better show the pros and cons of the conceived solution, we summarize in Table 4 the main differences between the compared schemes in terms of incurred overhead, i.e., additional signalling bytes piggybacked in the exchanged NDN packets. A high signalling overhead introduced by a caching strategy could deteriorate the content retrieval performance, e.g., by increasing the content retrieval delay due to the transmission of larger packets that occupy the channel for a longer time and may generate a higher channel load and resulting congestion. It is worth observing that the blind CEE and Prob schemes incur no additional overhead. Similarly to PaCC, BEP foresees one additional field in the Data packet to convey the maximum betweenness to be compared with that of the nodes along the backforwarding path. It is the PaC metric in our proposal. Three additional fields per Interest are foreseen by PaCC, for a total of 10 bytes (less than 1% overhead per packet). For BEP, the overhead per Interest is a function of the number of nodes traversed along the path ($|\Pi|$) by the packet carrying the BETWEENNESSARRAY field. For a path made of 5 nodes the overhead gets equal to 24 bytes. Under the majority of settings ($|\Pi| > 1$), PaCC is also more efficient than BEP in terms of exchanged bytes in the network per packet. Therefore, the signalling overhead introduced by PaCC is extremely low, which contributes to limit the retrieval delay metric, as showed in Figures 7 and 9.

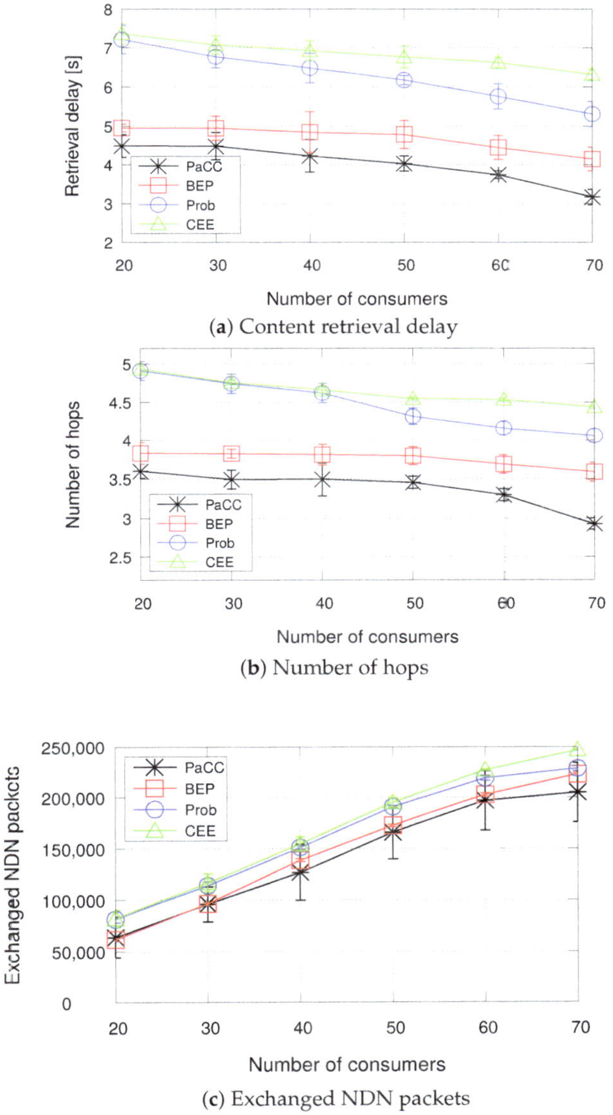

(**a**) Content retrieval delay

(**b**) Number of hops

(**c**) Exchanged NDN packets

Figure 9. Metrics when varying the number of consumers (cache size equal to 0.5% of the catalogue size).

Table 4. Additional fields per packet: PaCC vs. benchmark schemes.

Strategy	Interest (Size in Bytes)	Data (Size in Bytes)
CEE	-	-
Prob	-	-
BEP	POPULARITYRANKING (4) BETWEENNESSARRAY ($4\times|\Pi|$)	BETWEENNESS (4)
PaCC	RATE (4) PAC (4) HOPCOUNT (2)	PAC (4)

5. Conclusions

In this work we have proposed a novel caching strategy for edge domains, called PaCC, which accounts for content popularity and proximity to the consumers. Achieved results, collected under a variety of settings, confirm that the devised solution allows more judicious content caching decisions that are particularly crucial when the storage capabilities of the edge domain are small. As a consequence, the content retrieval delay experienced by PaCC is reduced compared to the considered benchmark schemes.

As a further benefit, the proximity of contents to consumers achieved by PaCC allows for reduction in the overall amount of exchanged data traffic at the expense of a negligible additional overhead per NDN packet. Such a finding is relevant because future networks will be overwhelmed by a myriad of (huge) contents exchanged by massively deployed devices and requested by increasingly demanding users.

The performance of the conceived solution can be further improved by addressing the following aspects: (i) optimizing the method for tracking the content popularity, (ii) optimizing the computation of the popularity-aware closeness metric. In PaCC, a threshold-based mechanism is implemented to track the popularity of contents based on the number of received content requests. However, more accurate mechanisms could be introduced in our design, for instance based on artificial intelligence (AI) content popularity prediction algorithms [20]. In parallel, more accurate metrics could be considered to estimate the proximity of the cachers to the consumers. In our design, we leverage the hop count metric, which, however, cannot reflect the presence of congested links or congested nodes. The proximity information could be improved by taking into account other additional metrics like the round-trip-time over the network links and/or the load on the nodes.

Author Contributions: Conceptualization, M.A., C.C., and G.R.; methodology, M.A. and C.C.; software, M.A.; validation, M.A.; investigation, M.A., C.C., and G.R.; writing—original draft preparation, M.A. and C.C.; writing—review and editing, M.A. and C.C.; supervision, A.M. and G.R. All authors have read and agreed to the published version of the manuscript.

Funding: This research received no external funding.

Institutional Review Board Statement: Not applicable.

Informed Consent Statement: Not applicable.

Data Availability Statement: Not applicable.

Conflicts of Interest: The authors declare no conflict of interest.

References

1. Zhang, L.; Afanasyev, A.; Burke, J.; Jacobson, V.; Claffy, K.; Crowley, P.; Papadopoulos, C.; Wang, L.; Zhang, B. Named data networking. *ACM SIGCOMM Comput. Commun. Rev.* **2014**, *44*, 66–73. [CrossRef]
2. Zhang, M.; Luo, H.; Zhang, H. A Survey of Caching Mechanisms in Information-Centric Networking. *IEEE Commun. Surv. Tutorials* **2015**, *17*, 1473–1499. [CrossRef]

3. Breslau, L.; Cao, P.; Fan, L.; Philips, G.; Shenker, S. Web caching and Zipf-like distributions: Evidence and implications. In Proceedings of the IEEE INFOCOM '99. Conference on Computer Communications, Eighteenth Annual Joint Conference of the IEEE Computer and Communications Societies, The Future is Now (Cat. No.99CH36320), New York, NY, USA, 21–25 March 1999; pp. 126–134.
4. Chai, W.K.; He, D.; Psaras, I.; Pavlou, G. Cache "less for more" in Information-Centric Networks. In Proceedings of the International Conference on Research in Networking, Prague, Czech Republic, 21–25 May 2012, Springer: Berlin/Heidelberg, Germany, 2012; pp. 27–40.
5. Khan, J.A.; Westphal, C.; Ghamri-Doudane, Y. A popularity-aware centrality metric for content placement in information centric networks. In Proceedings of the 2018 International Conference on Computing, Networking and Communications (ICNC), Maui, HI, USA, 5–8 March 2018; pp. 554–560.
6. Zheng, Q.; Kan, Y.; Chen, J.; Wang, S.; Tian, H. A Cache Replication Strategy Based on Betweenness and Edge Popularity in Named Data Networking. In Proceedings of the ICC 2019—2019 IEEE International Conference on Communications (ICC), Shanghai, China, 20–24 May 2019; pp. 1–7.
7. Khan, J.A.; Westphal, C.; Garcia-Luna-Aceves, J.; Ghamri-Doudane, Y. Nice: Network-oriented information-centric centrality for efficiency in cache management. In Proceedings of the 5th ACM Conference on Information-Centric Networking, Boston, MA, USA, 21–23 September 2018; pp. 31–42.
8. Perino, D.; Gallo, M.; Boislaigue, R.; Linguaglossa, L.; Varvello, M.; Carofiglio, G.; Muscariello, L.; Ben Houidi, Z. A high speed information-centric network in a mobile backhaul setting. In Proceedings of the 1st ACM Conference on Information-Centric Networking, Paris, France, 24–26 September 2014; pp. 199–200.
9. Tarnoi, S.; Suksomboon, K.; Kumwilaisak, W.; Ji, Y. Performance of probabilistic caching and cache replacement policies for content-centric networks. In Proceedings of the 39th Annual IEEE Conference on Local Computer Networks, Edmonton, AB, Canada, 8–11 September 2014; pp. 99–106.
10. Bernardini, C.; Silverston, T.; Olivier, F. MPC: Popularity-based Caching Strategy for Content Centric Networks. In Proceedings of the 2013 IEEE International Conference on Communications (ICC), Budapest, Hungary, 9–13 June 2013; pp. 3619–3623.
11. Amadeo, M.; Ruggeri, G.; Campolo, C.; Molinaro, A.; Mangiullo, G. Caching Popular and Fresh IoT Contents at the Edge via Named Data Networking. In Proceedings of the IEEE INFOCOM 2020—IEEE Conference on Computer Communications Workshops (INFOCOM WKSHPS), Toronto, ON, Canada, 6–9 July 2020; pp. 610–615.
12. Amadeo, M.; Campolo, C.; Ruggeri, G.; Molinaro, A. Beyond Edge Caching: Freshness and Popularity Aware IoT Data Caching via NDN at Internet-Scale. *IEEE Trans. Green Commun. Netw.* **2021**, *6*, 352–364. [CrossRef]
13. Ong, M.D.; Chen, M.; Taleb, T.; Wang, X.; Leung, V. FGPC: Fine-Grained Popularity-based Caching Design for Content Centric Networking. In Proceedings of the 17th ACM International Conference on Modeling, Analysis and Simulation of Wireless and Mobile Systems, Montreal, QC, Canada, 21–26 September 2014; pp. 295–302.
14. Li, J.; Wu, H.; Liu, B.; Lu, J.; Wang, Y.; Wang, X.; Zhang, Y.; Dong, L. Popularity-driven coordinated caching in named data networking. In Proceedings of the 2012 ACM/IEEE Symposium on Architectures for Networking and Communications Systems (ANCS), Austin, TX, USA, 29–30 October 2012; pp. 15–26.
15. Wang, L.; Lehman, V.; Hoque, A.M.; Zhang, B.; Yu, Y.; Zhang, L. A secure link state routing protocol for NDN. *IEEE Access* **2018**, *6*, 10470–10482. [CrossRef]
16. Mastorakis, S.; Afanasyev, A.; Moiseenko, I.; Zhang, L. ndnSIM 2: An Updated NDN Simulator for NS-3. *NDN, Technical Report NDN-0028*, Revision 2. Available online: https://named-data.net/publications/techreports/ndn-0028-1-ndnsim-v2/ (accessed on 20 March 2022).
17. GT-ITM: Georgia Tech Internetwork Topology Models. Available online: https://www.cc.gatech.edu/projects/gtitm/ (accessed on 20 March 2022).
18. Al Azad, M.W.; Shannigrahi, S.; Stergiou, N.; Ortega, F.R.; Mastorakis, S. CLEDGE: A Hybrid Cloud-Edge Computing Framework over Information Centric Networking. In Proceedings of the 2021 IEEE 46th Conference on Local Computer Networks (LCN), Edmonton, AB, Canada, 4–7 October 2021; pp. 589–596.
19. Rossi, D.; Rossini, G. Caching performance of content centric networks under multi-path routing (and more). *Relatório Técnico Telecom ParisTech* **2011**, *2011*, 1–6.
20. Skaperas, S.; Mamatas, L.; Chorti, A. Real-time algorithms for the detection of changes in the variance of video content popularity. *IEEE Access* **2020**, *8*, 30445–30457. [CrossRef]

sensors

MDPI

Article

New Results for the Error Rate Performance of LoRa Systems over Fading Channels

Kostas Peppas [1,*], Spyridon K. Chronopoulos [2,3,*], Dimitrios Loukatos [4] and Konstantinos Arvanitis [4]

1 Department of Informatics and Telecommunications, University of Peloponnese, 22100 Tripoli, Greece
2 Department of Speech and Language Therapy, University of Ioannina, 45110 Ioannina, Greece
3 Electronics-Telecommunications and Applications Laboratory, Department of Physics, University of Ioannina, 45110 Ioannina, Greece
4 Department of Natural Resources Management and Agricultural Engineering, Agricultural University of Athens, 75 Iera Odos Street, Botanikos, 11855 Athens, Greece; dlouka@aua.gr (D.L.); karvan@aua.gr (K.A.)
* Correspondence: peppas@uop.gr (K.P.); spychro@uoi.gr (S.K.C.)

Abstract: Long Range (LoRa) systems have recently attracted significant attention within the research community as well as for commercial use due to their ability to transmit data over long distances at a relatively low energy cost. In this study, new results for the bit error rate performance of Long Range (LoRa) systems operating in the presence of Rayleigh, Rice, Nakagami-m, Hoyt, η-μ and generalized fading channels are presented. Specifically, we propose novel exact single integral expressions as well as simple, accurate expressions that yield tight results in the entire signal-to-noise ratio (SNR) region. The validity of our newly derived formulas is substantiated by comparing numerically evaluated results with equivalent ones, obtained using Monte-Carlo simulations and exact analytical expressions.

Keywords: bit error rate; fading channels; Internet of things; LoRa; performance evaluation

Citation: Peppas, K.; Chronopoulos, S.K.; Loukatos, D.; Arvanitis, K. New Results for the Error Rate Performance of LoRa Systems over Fading Channels. *Sensors* **2022**, *22*, 3350. https://doi.org/10.3390/s22093350

Academic Editor: Peter Han Joo Chong

Received: 27 March 2022
Accepted: 25 April 2022
Published: 27 April 2022

Publisher's Note: MDPI stays neutral with regard to jurisdictional claims in published maps and institutional affiliations.

1. Introduction

In recent years, the exponential growth in the number of inexpensive, Internet-connected devices has given birth to the Internet of things (IoT) and its numerous applications, including autonomous farming, wearable health monitoring, smart homes and cities. Nevertheless, the increasing number of connected devices in conjunction with memory, bandwidth and energy availability constraints has revealed the limits of traditional connectivity technologies, namely ZigBee, Bluetooth and WiFi, in terms of energy consumption, scalability and throughput [1].

In order to fulfill the communication requirements of the IoT, the so-called Low-Power Wide Area Networks (LPWAN) have recently attracted significant attention within the research community as well as for commercial use, due to their ability to complement traditional cellular and short-range wireless technologies in an efficient manner [2–5].

Among all available LPWAN protocols, the so-called LoRa (Long Range) technology [6] has emerged as a promising candidate for smart sensing technology for civil (e.g., environment and health monitoring, smart metering, precision agriculture) and industrial applications, in urban and rural environments, due to its long-range and low-power capabilities. LoRa modulation is a 3GPP standard based on the chirp spread-spectrum (CSS) technology [7–20] and uses the industrial, scientific and medical (ISM) frequency bands at 433 MHz, 868 MHz or 915 MHz with data rates of up to 50 kbps.

Although the LoRa technology is well documented in [6], there are still relatively few studies on its theoretical performance. A summary of related works is presented in Table 1.

Table 1. Related works on theoretical performance of LoRa systems in the presence of fading and noise.

Authors	Title	Source	Findings
Vangelista, L.	Frequency shift chirp modulation: The LoRa modulation	[7]	Introduced the LoRa modulation system and provided initial results on its performance over AWGN channels by means of a single integral.
Elshabrawy, T.; Robert, J.	Closed-form approximation of LoRa modulation BER performance	[21]	Provided simple closed-form expressions of LoRa systems in the presence of AWGN and Rayleigh fading.
Dias, C.F.; Lima, E.R.D.; Fraidenraich, G.	Bit error rate closed-form expressions for LoRa systems under Nakagami and Rice fading channels.	[22]	Provided an exact closed-form expression for the BER of LoRa systems under Rayleigh fading as well as analytical expressions for the BER under Nakagami-*m* and Rice fading in terms of a finite sum.
Courjault, J.; Vrigenau, B.; Berder, O.; Bhatnagar, M.	A Computable Form for LoRa Performance Estimation: Application to Ricean and Nakagami Fading.	[23]	Authors elaborate on the properties of the generalized Marcum Q-function to provide accurate expressions for the BER of LoRa systems in the presence of Rice and Nakagami-*m* fading.
Hoeller, A.; et al.	Analysis and Performance Optimization of LoRa Networks With Time and Antenna Diversity	[11]	Authors addressed the performance of LoRa systems operating in the presence of Rayleigh fading, enhanced with antenna and time diversity techniques. The optimization of the performance of such systems has further been addressed.
Ma, H.; Cai, G.; Fang, Y.; Chen, P.; Han, G.	Design and Performance Analysis of a New STBC-MIMO LoRa System	[24]	Authors have proposed a new STBC MIMO LoRa system architecture. Its theoretical performance was analyzed in the presence of Rayleigh fading. A closed-form approximate BER expression of the proposed system under perfect and imperfect channel state information (CSI) was proposed.
Xu, W.; Cai, G.; Chen,	Performance analysis of a two-hop relaying LoRa system	[25]	Authors studied a two-hop opportunistic amplify-and-forward relaying LoRa system employing a best relay-selection protocol and operating over Nakagami-*m* fading.

Specifically, the mathematical representation of the LoRa modulation/demodulation process and its performance in terms of the symbol and bit error rates for additive white Gaussian noise (AWGN) and frequency selective fading channels were addressed in [7]. In [21], a moment matching method was employed to obtain accurate closed-form approximations for AWGN and Rayleigh fading channels. Multi-antenna LoRa systems were addressed in [11,24]. The performance of relay-based LoRa networks was addressed in [25]. A first attempt to provide exact BER expressions for Rayleigh, Rice and Nakagami-*m* fading is available in [22]. Nevertheless, as was pointed out in [21,23], the proposed methodology for channels other than Rayleigh may suffer from numerical stability issues, due to the computation of large values of binomial coefficients. To this end, ref [23] leveraged the properties of the Marcum Q-function to provide accurate approximations for the BER of LoRa systems over Rice and Nakagami-*m* fading channels.

However, several results obtained using this method, i.e., for Nakagami-*m* fading, require the computation of hypergeometric functions with two arguments [23], which in turn are not available as built-in functions in standard mathematical software packages such as Matlab or Mathematica [26]. Moreover, for the numerical evaluation of the underlying mathematical expressions, the computation of an approximation threshold parameter is required. Nevertheless, the exact computation of this threshold is rather complicated and therefore, a heuristic method for its computation was proposed by the authors in the same work. The above facts motivate simpler, yet accurate expressions for the evaluation of the BER of LoRa systems in the presence of noise and fading. On the other hand, analytical results for the error performance of LoRa systems in the presence of fading channels other than Rayleigh, Rice and Nakagami-m, are—to the best of our knowledge—not available in the open technical literature. Indeed, as it was pointed out in [27], the above mentioned classical fading models do not always fit well measured data, especially at the tail portion. This motivates research on performance evaluation over generalized fading models that include the classical ones as special cases.

Motivated by the above facts, in this study we present new analytical expressions for the average bit error rate evaluation of LoRa systems in the presence of fading. More specifically, the novel research contributions of this work can be summarized as follows.

1. Under the assumption of Nakagami-m and Rice fading channels, we present approximate analytical expressions for the SER performance of LoRa systems. These expressions yield accurate results in the entire signal to noise ratio (SNR) region that are practically indistinguishable from the exact solution;
2. For the special case of Nakagami-m fading, using a moment matching method, a simple yet tight approximation to the SER is obtained in closed form;
3. For all fading scenarios, exact analytical SER expressions in terms of a single integral are presented;
4. A novel, accurate analytical expression for the SER of LoRa systems operating in the presence of Hoyt fading is presented. To this end, a new integral involving exponentials, modified Bessel functions and the Marcum-Q function, whose second argument is a linear function of the integration variable, is evaluated;
5. An exact single integral expression for the SER of LoRa systems operating over η-μ fading channel is presented, assuming a propagation environment consisting of a finite number of multi-path clusters;
6. An exact single integral expression for the SER of LoRa systems operating over generalized fading channels is presented, by approximating the PDF of the SNR with a mixture gamma distribution. As a test case, SER results of LoRa systems operating in the presence of κ-μ fading channels are presented.

In order to validate the correctness of the proposed mathematical analysis, all analytical results are substantiated by means of Monte-Carlo simulations. Note that the proposed analytical framework provides accurate results in the entire SNR range, thus circumventing the need for evaluating system performance via time consuming Monte-Carlo simulations (It is a common practice to use Monte-Carlo simulations in order to verify the correctness of analytical results. Please note that although Monte-Carlo simulations may also be used to obtain performance evaluation results, they suffer from two significant disadvantages, as compared to analytical results. First, one has to specify the system model using software defined components, i.e., to simulate channel, noise, modulation, demodulation and detection. Although this process provides further insights on the system structure, it is computationally very intensive, time consuming and requires large amounts of memory to achieve a given accuracy. Specifically, as a rule of thumb, in order to obtain exact BER results of the order of 10^{-6}, random samples of two orders of magnitude larger, namely 10^8, are required. Such large vectors are difficult to be handled by software tools such as Matlab. On the other hand, analytical results in the form of equations yield accurate results within a large range of system level parameters). The remainder of this work is structured as follows. Section 2 presents an overview of the LoRa modulation and its BER performance. Section 3 presents the main results of this work. Numerical results are presented in Section 4 whereas Section 5 concludes the work. *Notations:* A list of mathematical notations used in this work is available in Table A1.

2. Overview of the LoRa Modulation

In this section, an overview of the LoRa modulation and the corresponding bit error probability are presented. LoRa systems employ the shift chirp modulation scheme, also known as spread spectrum modulation. The number of samples within the duration of a symbol, T_s, is determined by the spreading factor (SF). It holds that $T_s = 2^{\mathrm{SF}}/B$, where B is the signal bandwidth. In typical applications, $\mathrm{SF} \in \{6, 7, \ldots, 12\}$. Note that the coverage of LoRa is determined by SF. Specifically, increasing SF results in wider coverage but also in a reduction in the data rate.

The modulation encoder maps a group of SF bits to a symbol, s_k, $k \in \{0, 1, \ldots L\}$ where $L = 2^{\mathrm{SF}} - 1$. The transmitted waveform can be expressed as [21]

$$s_k(nT) = h\sqrt{E_s}\omega_k(nT_s) \tag{1}$$

$$= h\sqrt{\frac{E_s}{N}} \exp\left\{\jmath 2\pi \frac{n}{N}[(k+n) \mod N]\right\} \tag{2}$$

where $N = 2^{\mathrm{SF}}$, $T_s = 1/B$ is the sampling period, $n \in \{0, 1, \ldots L\}$ is the sample index at time nT_s, E_s is the signal energy, h is the fading channel coefficient and $\omega_k(nT_s)$ are orthonormal basis functions.

Figure 1 depicts the main functional blocks of a LoRa non-coherent demodulator. Specifically:

- The input signal is sampled at a period of $T_s = 1/B$;
- The resulting signal is then multiplied with a down chirp signal;
- A Fast Fourier Transform (FFT) is performed at the output of the previous block to retrieve the symbol value;
- The information signal is estimated using maximum likelihood detection.

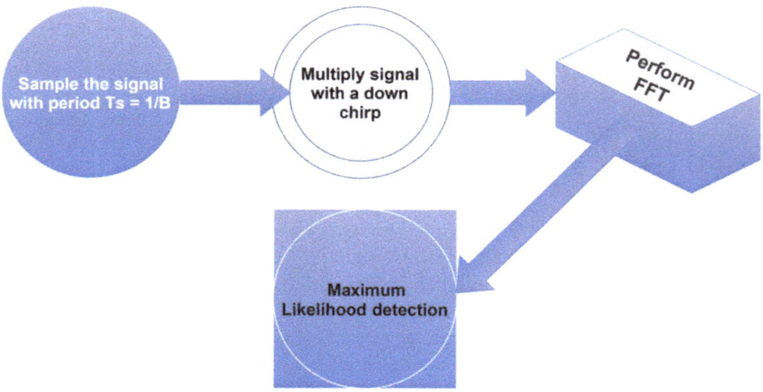

Figure 1. A simplified overview of the LoRa non-coherent demodulator.

Using the orthogonal properties of $s_k(nT)$, the correlator output at the demodulator is given as [21]

$$\sum_{n=0}^{L} r_k(nT)\omega_i^*(nT) = \begin{cases} h\sqrt{E_s} + \phi_i & \text{if } k = i \\ \phi_i & \text{if } k \neq i \end{cases} \tag{3}$$

where $r_k(\cdot)$ is the received signal and ϕ_i is the complex Gaussian noise. The decision rule for the detected index symbol can be expressed as [21]

$$\hat{k} = \{i | \arg_i \max\left(|\delta_{k,i}h\sqrt{E_s} + \phi_i|\right)\}. \tag{4}$$

The conditional symbol error probability given the squared channel coefficient h is given as [21–23]

$$P(e|h) = \Pr\{\rho^2 > |h\sqrt{E_s} + \phi_i|^2\} \tag{5}$$

where $\rho^2 = \max\{|\phi_i|\}|_{i\neq k}$ is the maximum of L independent and identically distributed (i.i.d) exponential random variables with CDF given by

$$F_{\rho^2}(x) = [1 - \exp(-x/2)]^L. \tag{6}$$

Moreover, the RV $R \triangleq |h\sqrt{E_s} + \phi_i|^2$ conditioned to h^2 follows a non-central chi-square distribution with PDF given by [21–23]

$$f_{R|h^2}(x) = \frac{1}{2}\exp\left(-\frac{x + 2Nh^2\gamma}{2}\right) I_0\left(\sqrt{2Nh^2\gamma x}\right), \tag{7}$$

where $\gamma = 1/\mathbb{E}\langle|\phi_i|^2\rangle$ is the SNR. The CDF of R conditioned to h^2 can be expressed in terms of the Marcum Q-function as

$$F_{R|h^2}(x) = 1 - Q_1\left(\sqrt{2Nh^2\gamma}, \sqrt{x}\right). \tag{8}$$

Finally, using (5)–(7), the average symbol error probability is given in terms of the following two-fold integral [21–23]

$$P_s = \frac{1}{2}\int_0^\infty \int_0^\infty \left\{1 - [1 - \exp(-x/2)]^L\right\} \times \exp\left(-\frac{x + 2Ny\gamma}{2}\right) I_0\left(\sqrt{2N\gamma xy}\right) f_{h^2}(y) \mathrm{d}x\mathrm{d}y. \tag{9}$$

The resulting bit error probability can be expressed as [21,23]

$$P_b = \frac{2^{\mathrm{SF}-1}}{2^{\mathrm{SF}} - 1} P_s. \tag{10}$$

3. Main Results

In this section, exact analytical expressions for P_s in terms of a single integral as well as accurate approximations will be obtained for Nakagami-m, Ricean and Hoyt fading channels.

3.1. Symbol Error Probability for Nakagami-m Fading Channels

Under Nakagami-m fading, the RV h^2 follows a gamma distribution with PDF given as [28]

$$f_{h^2}(y) = \frac{m^m}{\Gamma(m)} y^{m-1} \exp(-my) \tag{11}$$

where $m > 0$ is the fading parameter. For $m = 1$, i.e., for Rayleigh fading, (11) reduces to the exponential distribution. An exact analytical expression for P_s is given in the following proposition.

Proposition 1. *The exact symbol error probability of LoRa systems under Nakagami-m fading in terms of a single integral is given as*

$$P_s^{\text{Nak}} = \frac{1}{2}\left(\frac{m}{N\gamma + m}\right)^m \int_0^\infty e^{-x/2}[1 - (1 - e^{-x/2})^L] \times \mathrm{L}_{-m}\left[\frac{N\gamma x}{2(N\gamma + m)}\right]\mathrm{d}x \tag{12}$$

Proof. By substituting (11) into (9) and changing the order of integration, a valid operation according to the Fubini theorem because the resulting integrals are convergent, one obtains

$$P_s = \frac{m^m}{2\Gamma(m)} \int_0^\infty e^{-x/2}[1 - (1 - e^{-x/2})^L] \times \left[\int_0^\infty y^{m-1} e^{-(N\gamma+m)y} I_0\left(\sqrt{2Nxy}\right)\mathrm{d}y\right]\mathrm{d}x. \tag{13}$$

By employing [29] (Equation (3.15.1/2)) and [30] (Equation (8.972/1)), (12) is readily obtained, thus completing the proof. □

Note that (12) yields the exact value of P_s for arbitrary values of m. In addition, it converges rapidly due to its exponentially decaying kernel and can be evaluated numerically in an efficient manner using built-in routines available in popular mathematical software packages such as Matlab or Mathematica. In what follows, we derive accurate approximations for P_s, assuming both arbitrary and integer values of the fading parameter m. The following result holds.

Proposition 2. *For arbitrary values of m, an accurate approximation for the P_s of LoRa systems in the presence of Nakgami-m fading is given as*

$$P_s^{\text{Nak}} \approx \left(\frac{m}{N\gamma + m}\right)^m \exp(-\tilde{x}_N/2) \times \sum_{n=1}^{\infty} \frac{\tilde{x}_N^n}{2^n \Gamma(n+1)} {}_1F_1\left[m; n+1; \frac{N\gamma \tilde{x}_N}{2(N\gamma + m)}\right] \tag{14}$$

where as for integer values of m

$$P_s^{\text{Nak}} \approx 1 - \frac{N\gamma}{N\gamma + m} \exp\left[-\frac{m\tilde{x}_N}{2(N\gamma + m)}\right] \times \sum_{n=0}^{m-1} \epsilon_n \left(\frac{m}{N\gamma + m}\right)^n \mathrm{L}_n\left[-\frac{N\gamma \tilde{x}_N}{2(N\gamma + m)}\right] \tag{15}$$

where

$$\epsilon_n = \begin{cases} 1 & \textit{if } n < m-1 \\ 1 + \frac{m}{N\gamma} & \textit{if } n = m-1 \end{cases} \tag{16}$$

and

$$\tilde{x}_N = 2\sum_{n=1}^{N-1} n^{-1}. \tag{17}$$

Proof. Our starting point to the proof is (5) via which a generic expression for P_s can be obtained. Observe that for large values of SF, i.e., for SF ≥ 6, the RV ρ^2 can be replaced with its mean with sufficient accuracy. Note that this observation has also been reported in [21]. Consequently, using (5) and (8), P_s can be approximated as

$$P_s \approx 1 - \mathbb{E}_{h^2}\left\langle Q_1\left(\sqrt{2N\gamma h^2}, \sqrt{\tilde{x}_N}\right)\right\rangle, \tag{18}$$

where $\tilde{x}_N$ is the expectation of the RV ρ^2, which, by employing the memoryless property of the exponential distribution [21], can be deduced as (17). Using (11), the expectation in (18) can be further written as

$$P_s \approx 1 - \frac{m^m}{\Gamma(m)} \int_0^\infty y^{m-1} e^{-my} Q_1\left(\sqrt{2N\gamma y}, \sqrt{\tilde{x}_N}\right) \mathrm{d}y \tag{19}$$

Using [31] (Equation (10)) and [31] (Equation (11)) (Note that [31] (Equation (11)) has a typo, i.e., N should be replaced with $\Gamma(N)$), (14) and (15) can be deduced for real and integer values of m, respectively, thus completing the proof. □

Next, using a moment matching method, a simpler closed-form expression for P_s will be derived, which holds for arbitrary values of m. Specifically, we propose approximating the statistics of the RV R with those of a gamma distribution with scale parameter a and shape parameter b, using a moment matching method. The following result holds.

Proposition 3. *A closed-form approximation for the P_s of LoRa systems under Nakagami-m fading can be obtained as*

$$P_s^{\mathrm{Nak}} \approx 1 - \frac{\Gamma(a, b\tilde{x}_N)}{\Gamma(a)}. \tag{20}$$

where $\tilde{x}_N$ is given by (17),

$$a = \tilde{\mu}_1^2/(\tilde{\mu}_2 - \tilde{\mu}_1^2),\ b = \tilde{\mu}_1/(\tilde{\mu}_2 - \tilde{\mu}_1^2), \tag{21a}$$

$$\tilde{\mu}_1 = 2(1 + N\gamma), \tag{21b}$$

$$\tilde{\mu}_2 = 8(1 + 2N\gamma) + 4\gamma^2(1 + m)N^2/m. \tag{21c}$$

Proof. Observe that R^2 follows a squared gamma-shadowed Rice distribution and thus, using [32] (Equation (5)), its n-moment is readily obtained as

$$\tilde{\mu}_n = 2^n \left(\frac{m}{N\gamma + m}\right)^m \Gamma(n+1) \times {}_2F_1\left(m, n+1; 1; \frac{N\gamma}{N\gamma + m}\right). \tag{22}$$

Using [33] (Equation (7.3.1/129)), $\tilde{\mu}_1$ and $\tilde{\mu}_2$ can be further simplified as (21). Finally, P_s can be deduced as the CDF of a gamma distribution with parameters a and b that can be obtained in closed form using a moment matching method [34] as (20) and (21), thus completing the proof. □

Using Proposition 2, a closed-form approximation for the P_s under Rayleigh fading will be obtained. Specifically, the following result holds.

Corollary. *Under Rayleigh fading, a closed-form approximation for P_s can be deduced as*

$$P_s^{\mathrm{Ray}} \approx 1 - \exp\left[-\frac{\tilde{x}_N}{2(1 + N\gamma)}\right] \tag{23}$$

Proof. The proof can be readily obtained by setting $m = 1$ to (15). □

3.2. Symbol Error Probability for Rice Fading Channels

Under Rice fading, the RV h^2 follows a non-central chi-square distribution with PDF given as [28]

$$f_{h^2}(y) = \frac{1+K}{\exp(K)}\exp[-(1+K)y]I_0\left[2\sqrt{K(1+K)y}\right], \tag{24}$$

where K is the Rice factor. For $K = 0$, (24) reduces to the exponential distribution, i.e., Rayleigh fading.

An exact analytical expression for P_s is given in the following proposition.

Proposition 4. *The exact symbol error probability of LoRa systems under Rice fading in terms of a single integral is given as*

$$\begin{aligned} P_s^{\text{Rice}} = & \frac{(1+K)\exp(-K)}{2(1+K+N\gamma)}\int_0^{\infty} e^{-x/2}[1-(1-e^{-x/2})^L] \\ & \times e^{\frac{2K+2K^2+\gamma N x}{2+2K+2\gamma N}} I_0\left[\frac{\sqrt{2NK(1+K)\gamma x}}{1+K+N\gamma}\right]\mathrm{d}x \end{aligned} \tag{25}$$

Proof. The proof can be concluded by following a similar line of arguments as in the proof of Proposition 1. Specifically, by substituting (24) into (9) and changing the order of integration, P_s can be expressed as

$$\begin{aligned} P_s = & \frac{1+K}{2\exp(K)}\int_0^{\infty} e^{-x/2}[1-(1-e^{-x/2})^L] \\ & \times \left[\int_0^{\infty} e^{-(N\gamma+K+1)y} I_0\left[2\sqrt{K(1+K)y}\right]\right. \\ & \left.\times\, I_0\left(\sqrt{2Nxy}\right)\mathrm{d}y\right]\mathrm{d}x. \end{aligned} \tag{26}$$

The inner integral, i.e., with respect to y, can be evaluated in closed form by employing [29] (Equation (3.15.17/1)), yielding (25), thus completing the proof. □

Again, (25) converges rapidly due to its exponentially decaying kernel and can be evaluated numerically in an efficient manner. In what follows, an accurate approximation for P_s will be derived. The following result holds.

Proposition 5. *Under Rice fading, an accurate closed-form approximation for P_s can be deduced as*

$$P_s^{\text{Rice}} \approx 1 - Q_1\left[\sqrt{\frac{2N\gamma K}{1+K+N\gamma}}, \sqrt{\frac{\tilde{x}_N(1+K)}{1+K+N\gamma}}\right] \tag{27}$$

Proof. Using (18) and (24), P_s can be approximated as

$$\begin{aligned} P_s \approx & 1 - \frac{1+K}{\exp(K)}\int_0^{\infty} e^{-(1+K)y} I_0\left[2\sqrt{K(1+K)y}\right] \\ & \times Q_1\left(\sqrt{2N\gamma y}, \sqrt{\tilde{x}_N}\right)\mathrm{d}y \end{aligned} \tag{28}$$

By employing [35] (Equation (15)), the resulting integral can be evaluated in closed-form yielding (27), thus completing the proof. □

3.3. Symbol Error Probability for Hoyt Channels

Under Hoyt (Nakagami-q) fading, the RV h^2 follows a non-central chi-square distribution with PDF given as [28]

$$f_{h^2}(y) = \sqrt{\frac{1}{2} + \frac{1}{4\eta} + \frac{\eta}{4}}\exp\left[-\left(\frac{1}{2} + \frac{1}{4\eta} + \frac{\eta}{4}\right)y\right] \times I_0\left[\left(\frac{1}{4\eta} - \frac{\eta}{4}\right)y\right], \tag{29}$$

where $\eta = q^2$, with $0 < q \leq 1$ being a parameter related to the fade intensity. For $q = 1$, (29) reduces to the exponential distribution (Rayleigh fading).

An exact expression for P_s can be deduced using the following proposition.

Proposition 6. *Under Hoyt fading, P_s can be expressed in terms of a single integral as*

$$P_s^{\text{Hoyt}} = \frac{\eta+1}{2\sqrt{A}}\int_0^\infty e^{-x/2}[1-(1-e^{-x/2})^L] \times e^{\frac{N\gamma x[(1+\eta)^2+4N\eta\gamma]}{2A}} I_0\left[\frac{N(\eta^2-1)\gamma x}{2A}\right]\mathrm{d}x \tag{30}$$

where

$$A = 1 + 2N\gamma + \eta^2(1+2N\gamma) + 2\eta(1+2N\gamma+2N^2\gamma^2) \tag{31}$$

Proof. The proof can be concluded by following a similar line of arguments as in the proof of Proposition 1. Specifically, by substituting (29) into (9) and changing the order of integration, P_s can be expressed as

$$P_s = \frac{1}{2}\sqrt{\frac{1}{2} + \frac{1}{4\eta} + \frac{\eta}{4}}\int_0^\infty e^{-x/2}[1-(1-e^{-x/2})^L] \times \left\{\int_0^\infty e^{-\left(N\gamma+\frac{1}{2}+\frac{1}{4\eta}+\frac{\eta}{4}\right)y} I_0\left[\left(\frac{1}{4\eta}-\frac{\eta}{4}\right)y\right] \times I_0\left(\sqrt{2Nxy}\right)\mathrm{d}y\right\}\mathrm{d}x. \tag{32}$$

The inner integral, i.e., with respect to y, can be evaluated in closed form by employing [29] (Equation (3.15.17/15)), yielding (30), thus completing the proof. □

An accurate approximation for P_s can be obtained using the following proposition. The following result holds.

Proposition 7. *Under Hoyt fading, an accurate approximation for P_s can be deduced as*

$$P_s^{\text{Hoyt}} \approx 1 - \frac{1}{\gamma}\sqrt{\frac{1}{4\eta} + \frac{\eta}{4} + \frac{1}{2}} \times \mathcal{I}\left[\sqrt{2N}, \sqrt{\tilde{x}_N}, \frac{1}{\gamma}\left(\frac{1}{4\eta} - \frac{\eta}{4}\right), \frac{1}{\gamma}\left(\frac{1}{4\eta} + \frac{\eta}{4} + \frac{1}{2}\right)\right] \tag{33}$$

where

$$\mathcal{I}(a,b,c,p) = \frac{a^2}{2p+a^2}\exp\left[-\frac{b^2p}{2p+a^2}\right] \times \sum_{k=0}^{\infty}\sum_{n=0}^{2k}\frac{\Gamma(2k+1)c^{2k}}{p^{2k+1}(k!)^2 4^k} \times \zeta_{n,k}\left(\frac{2p}{2p+a^2}\right)^n \mathrm{L}_n\left[-\frac{b^2a^2}{4p+2a^2}\right] \quad (34)$$

and

$$\zeta_{n,k} = \begin{cases} 1 & \text{if } n < 2k \\ 1+\frac{2p}{a^2} & \text{if } n = 2k \end{cases} \quad (35)$$

Proof. Using (18) and (29), P_s can be approximated as

$$P_s \approx 1 - \sqrt{\frac{1}{2}+\frac{1}{4\eta}+\frac{\eta}{4}}\int_0^{\infty} e^{-\left(\frac{1}{2}+\frac{1}{4\eta}+\frac{\eta}{4}\right)y} \times I_0\left[\left(\frac{1}{4\eta}-\frac{\eta}{4}\right)y\right] Q_1\left(\sqrt{2N\gamma y}, \sqrt{\tilde{x}_N}\right)\mathrm{d}y \quad (36)$$

which can be written as (33) with

$$\mathcal{I}(a,b,c,p) = \int_0^{\infty}\exp(-px)I_0(cx)Q_1(a\sqrt{x},b)\mathrm{d}x \quad (37)$$

To the best of our knowledge, however, this integral is not available in related works such as [31,35,36]. Nevertheless, as shown in the Appendix A, $\mathcal{I}(a,b,c,p)$ can be evaluated as (34), thus completing the proof. □

3.4. *Symbol Error Probability for Physical η-μ Fading Channels*

Under η-μ fading, the PDF of h^2 is given by [27]

$$f_{h^2}(y) = \frac{2\sqrt{\pi}\mu^{\mu+0.5}\theta^{\mu}y^{\mu-0.5}}{\Gamma(\mu)H^{\mu-0.5}}\exp(-2\mu\theta y)I_{\mu-0.5}(2\mu_{\ell}H\gamma) \quad (38)$$

where μ is related to the fading severity. The η-μ fading is quite general as it can accurately model small-scale variations of the fading signal under non line-of-sight (NLOS) conditions and includes as special cases both the Nakagami-m and the Hoyt fading models. The PDF of h^2 may be expressed in two formats, namely Format 1, where $\theta = (2+\eta^{-1}+\eta)/4$ and $H = (\eta^{-1}-\eta)/4$ with $0 < \eta < \infty$ and Format 2, where $\theta = 1/(1-\eta^2)$ and $H = \eta/(1-\eta^2)$ with $-1 < \eta < 1$. As pointed out in [27], Format 1 can be converted into Format 2 by employing a bilinear transformation. Thus and without loss of generality, Format 1 will be assumed next. Moreover, the special case of integer μ, termed *physical* η-μ model, assumes a finite number of multipath clusters and has been adopted in several works, e.g., see [27,37–39]

Using [30] (Equation (8.467)), the modified Bessel function $I_{\pm(n+1/2)}(z)$, with $n > 0$ being an integer, can be expressed as a finite sum, namely

$$I_{\pm(n+1/2)}(z) = \frac{1}{\sqrt{\pi}}\sum_{k=0}^{n}\frac{(n+k)!}{n!(n-k)!}\left[\frac{(-1)^k e^z \mp (-1)^n e^{-z}}{(2z)^{k+0.5}}\right]. \quad (39)$$

Substituting (39) into (38) and employing [30] (Equation (8.353/7)), $f_{h^2}(y)$ can be expressed as

$$f_{h^2}(y) = \frac{\mu\theta}{H\Gamma(\mu)} \sum_{k=0}^{\mu-1} a_k y^{\mu-k-1}\left[(-1)^k e^{-\mathcal{A}y} + (-1)^\mu e^{-\mathcal{B}y}\right] \tag{40}$$

where

$$a_k = \frac{(-1)^k(\mu+k-1)!(4\mu H)^{-k}}{k!(\mu-k-1)!} \tag{41a}$$

$$\mathcal{A} = 2\mu(\theta - H),\ \mathcal{B} = 2u(\theta + H). \tag{41b}$$

An exact expression for P_s can be obtained using the following proposition.

$$P_s^{\eta-\mu} = \frac{0.5}{\Gamma(\mu)}\left(\frac{\mu\theta}{H\gamma}\right)^\mu \left\{\sum_{k=0}^{\mu-1} a_k\gamma^k(-1)^k\Gamma(\mu-k)2^{\mu-k}\left[(-1)^k\left(\lambda+\frac{2\mathcal{A}}{\gamma}\right)^{k-\mu}\right.\right.$$
$$\times \int_0^\infty e^{-x/2}[1-(1-e^{-x/2})^L]\mathrm{L}_{k-\mu}\left[\frac{\lambda\gamma x}{2(\lambda\gamma+2\mathcal{A})}\right]\mathrm{d}x$$
$$\left.\left. + (-1)^\mu\left(\lambda+\frac{2\mathcal{B}}{\gamma}\right)^{k-\mu}\int_0^\infty e^{-x/2}[1-(1-e^{-x/2})^L]\mathrm{L}_{k-\mu}\left[\frac{\lambda\gamma x}{2(\lambda\gamma+2\mathcal{B})}\right]\mathrm{d}x\right]\right\}. \tag{42}$$

Proposition 8. *The exact symbol error probability of LoRa systems operating under* physical η-μ *fading channels can be expressed in terms of a single integral as* (42).

Proof. The proof can be readily deduced by following the same steps as in the proof of Proposition 1. □

3.5. Symbol Error Probability for Generalized Fading Channels Using a Mixture Gamma Distribution

In what follows, we present analytical results for the SER of LoRa systems assuming generalized fading channels for which the PDF of the SNR can be expressed as a mixture gamma distribution. As was shown in [40], the proposed approach is valid for a plethora of fading distributions, including the κ-μ, the η-μ and composite fading/shadowing channels such as the generalized-K and the Suzuki ones. The PDF of h^2 can be expressed as [40] (Equation (1))

$$f_{h^2}(y) = \sum_{i=1}^{\mathrm{N_{terms}}} a_i y^{\beta_i-1} e^{-\zeta_i y}, \tag{43}$$

where $\mathrm{N_{terms}}$ is the number of terms required for a given accuracy, and a_i, β_i and ζ_i are the parameters of the ith Gamma component. The parameter $\mathrm{N_{terms}}$ can be selected so that the first k moments of the original and the approximate distributions are matched or the Kullback–Leibler distance of the original and the approximate distributions is minimized [40]. For such channels, an exact expression for P_s can be obtained using the following proposition.

Proposition 9. *The exact symbol error probability of LoRa systems operating under generalized fading channels can be expressed in terms of a single integral as*

$$P_s^{\mathrm{gen}} = \frac{1}{2}\sum_{i=1}^{\mathrm{N_{terms}}} a_i(N\gamma+\zeta_i)^{-\beta_i}\Gamma(\beta_i)\int_0^\infty e^{-x/2}$$
$$\times [1-(1-e^{-x/2})^L]\mathrm{L}_{-\beta_i}\left[\frac{N\gamma x}{2(N\gamma+\zeta_i)}\right]\mathrm{d}x \tag{44}$$

Proof. The proof can be readily deduced by following the same steps as in the proof of Proposition 1. □

4. Numerical Results

In this section, numerical results are presented to validate the proposed error rate analysis. Analytical results are compared with equivalent ones obtained using Monte-Carlo simulations. A number of random samples equal to 10^5 is used to ensure statistical convergence. The simulation methodology is described in Algorithm 1. Unless otherwise specified, values of SF of 7 and 12 have been assumed.

Algorithm 1 Monte-Carlo simulation methodology.

Require: Number of samples ≥ 0
  $E_s \leftarrow$ SNR
  errors $\leftarrow 0$
  Number of samples $\leftarrow 10^5$
  while Number of samples $\neq 0$ **do**
    generate random channel coefficient h for a given fading distribution
    generate noise coefficient ϕ_i from a normal distribution
    generate ρ^2 as the maximum of exponential random variables

   if $\rho^2 > |h\sqrt{E_s} + \phi_i|^2$ **then**
     errors $\leftarrow$ errors $+ 1$
   end if
   Number of samples $\leftarrow$ Number of samples $- 1$
  end while

Figures 2 and 3 depict the BER of LoRa modulation in the presence of Nakagami-m fading as a function of γ for $m \in \{1, 1.5, 2, 3, 3.55\}$, and SF of 7 and 12, respectively. In both figures, approximate BER results for $m > 1$ were obtained using the approximation presented in Proposition 2 as well as the moment matching method in Proposition 3. The exact BER values were obtained using the two-fold integral in (9), the single integral expression in Proposition 1 and Monte-Carlo simulation based on (5), using 10^5 random samples. For $m = 1$, i.e., Rayleigh fading, approximate results were obtained using (23). As it can be observed, the approximate formulas obtained using Proposition 2 match well the exact results in the entire SNR region. In addition, the moment matching method yields accurate results for low values of m; nevertheless, deviations from the exact results are observed for $m > 2$ and high SNR values. Finally, for Rayleigh fading, (23) yields very accurate results that are practically indistinguishable from the exact ones.

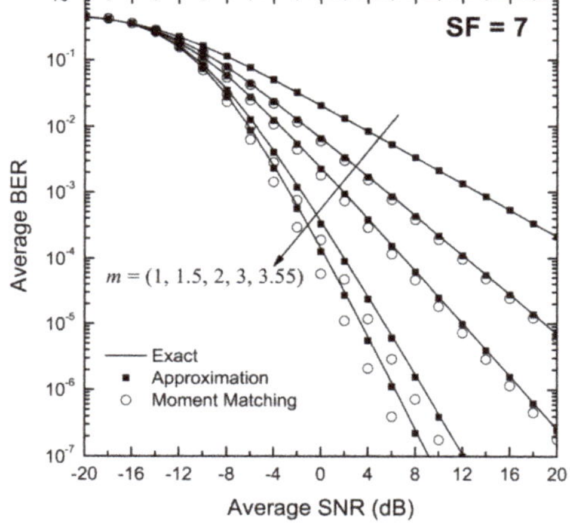

Figure 2. BER of LoRa systems operating in the presence of Nakagami-m fading as a function of the SNR, γ, for SF = 7 and various values of m.

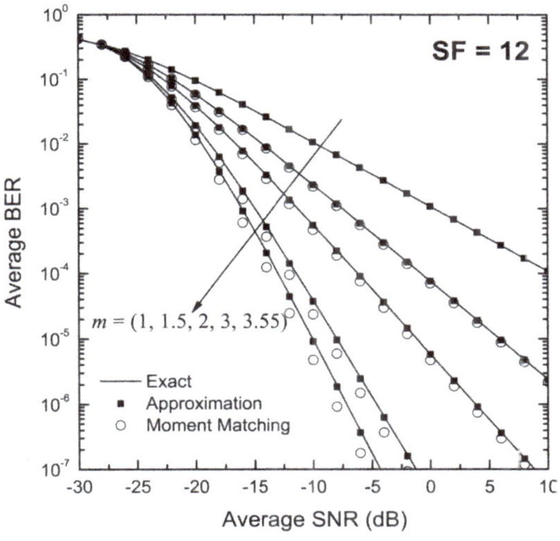

Figure 3. BER of LoRa systems operating in the presence of Nakagami-m fading as a function of the SNR, γ, for SF = 12 and various values of m.

Figures 4 and 5 depict the BER of LoRa modulation in the presence of Rice fading as a function of γ for $K \in \{1, 5, 10\}$, and SF of 7 and 12, respectively. The exact BER values were obtained using both the two-fold integral in (9) as well as the single integral expression in Proposition 4. In both figures, approximate BER results have also been obtained using the approximation presented in Proposition 5. As it can be observed, the approximate formulas obtained using Proposition 2 match well the exact results in the entire SNR region for all values of K.

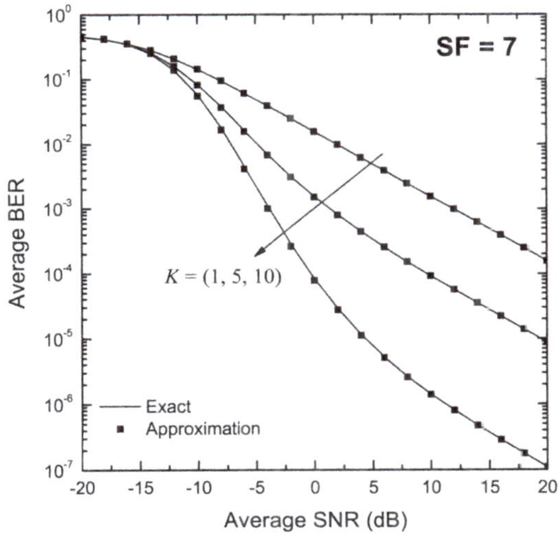

Figure 4. BER of LoRa systems operating in the presence of Rice fading as a function of the SNR, γ, for SF = 7 and various values of K.

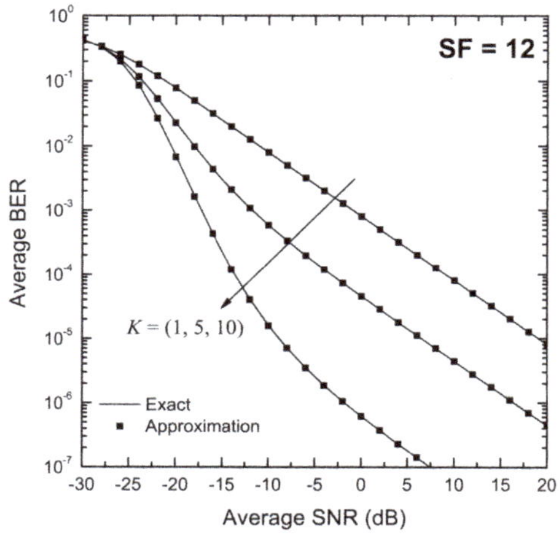

Figure 5. BER of LoRa systems operating in the presence of Rice fading as a function of the SNR, γ, for SF = 12 and various values of K.

Next, we estimate the BER of LoRa systems in an agricultural environment, described in detail in [41]. In that work, an experimental test bed exploiting smartphone components was utilized in a measurement campaign, performed under realistic, in terms of agriculture, conditions. Specifically, part of the measurement campaign focused on measuring the Received Signal Strength Indicator (RSSI) for various distances between the LoRa radio modules participating in the experiments and for various transmit power level settings. Both LoRa radios had their transmit power adjusted to 10 dBm, the SF was set to either 7 or 11 and the bandwidth (BW) was 125 kHz or 250 kHz. It has also further been assumed that small scale fading is modeled by the Rice distribution with $K = 2.63$ dB, a typical value encountered in rural environments [42]. The noise power equals -85 dBm. Figure 6 depicts the estimated BER of the considered propagation scenario as a function of the link distance. Again, an excellent match of exact and approximate results was observed for all test cases under consideration.

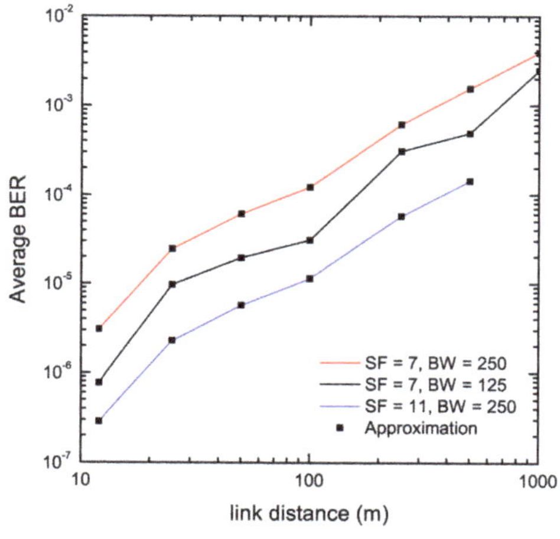

Figure 6. BER estimation of LoRa systems operating in the presence of Rice fading in an agricultural environment using a measurement campaign.

Figures 7 and 8 depict the SER of LoRa modulation in the presence of Hoyt fading as a function of γ for $q \in \{0.1, 0.5, 0.9\}$, and SF of 7 and 12, respectively. Again, the exact BER values were obtained using both the two-fold integral in (9) and the single integral expression in Proposition 6. The approximate BER results were obtained using Proposition 7. In order to derive the approximate BER results, the corresponding infinite series were truncated to $N = 220$ and $N = 350$ terms for $q = 0.1$ and SF of 7 and 12, respectively whereas for $q = 0.5$ and $q = 0.9$, only 20 terms were sufficient to provide a good match with the analytical results for both considered values of SF. Again, the approximate formulas obtained using Proposition 7 match well with the exact results in the entire SNR region for all values of q.

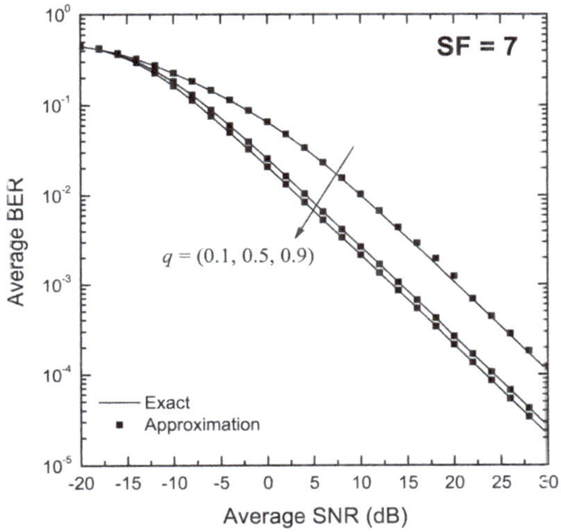

Figure 7. BER of LoRa systems operating in the presence of Hoyt fading as a function of the SNR, γ, for SF = 7 and various values of q.

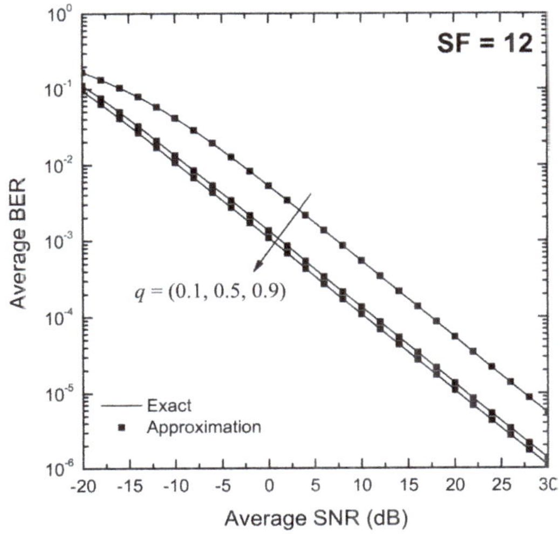

Figure 8. BER of LoRa systems operating in the presence of Hoyt fading as a function of the SNR, γ, for SF = 12 and various values of q.

Figure 9 depicts the SER of LoRa modulation in the presence of η-μ fading as a function of γ for SF $\in \{7, 9, 10\}$. The fading parameters are assumed to be $\mu = 2.065$ and $\eta = 0.00847518$, obtained through a measurement campaign in an indoor environment, as reported in [27]. In order to apply the analytical results obtained in Proposition 8, a *physical* η-μ model with $\mu = 2$ was assumed. Exact results were obtained using Monte-Carlo simulation. As can be observed, analytical results closely approximate exact ones, for all considered values of SF, especially for low and medium values of γ, thus demonstrating the usefulness of the proposed analysis.

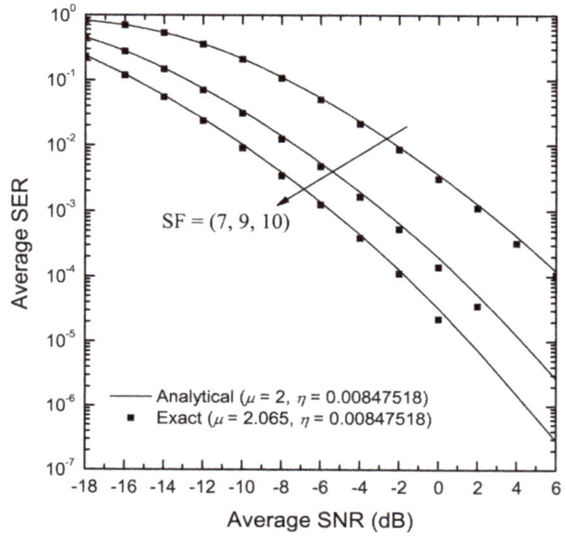

Figure 9. SER of LoRa systems operating in the presence of η-μ fading as a function of the SNR, γ, in an indoor environment, as reported in [27] and various values of SF.

In the following, we consider LoRa systems operating in the presence of κ-μ fading as a function of γ. Note that the κ-μ distribution is a two-parameter fading model that well describes wireless propagation in the presence of a line-of-sight (LoS) component [27]. The PDF of h^2 is given by [27]

$$
\begin{aligned}
f_{h^2}(y) = {} & \frac{\mu(1+\kappa)^{0.5(\mu+1)} y^{0.5(\mu_{\chi,\kappa}-1)}}{\kappa^{0.5(\mu-1)}\exp(\mu\kappa)} \\
& \times \exp[-\mu(1+\kappa)y] I_{\mu-1}\left[2\mu\sqrt{\kappa(1+\kappa)y}\right]
\end{aligned}
\tag{45}
$$

where κ and μ account for the intensity of the LoS component and the Nakagami-m component, respectively. Note that the κ-μ distribution includes both the Nakagami-m $(\kappa = 0)$ and the Rice $(m = 1)$ distributions as special cases.

Using [40] (Equation (19)), the parameters of its mixture gamma approximation can be expressed as

$$
a_i = \psi(\theta_i, \beta_i, \zeta_i),\ \beta_i = \mu + i - 1,\ \zeta_i = \mu(1+\kappa) \tag{46a}
$$

$$
\theta_i = \frac{\mu(1+\kappa)^{0.5(\mu+1)}}{\kappa^{0.5(\mu-1)}\exp(\mu\kappa)} \frac{\mu^{2i+\mu-3}[\kappa(1+\kappa)]^{0.5(2i+\mu-3)}}{(i-1)!\Gamma(\mu+i-1)} \tag{46b}
$$

$$
\psi(\theta_i, \beta_i, \zeta_i) = \frac{\theta_i}{\sum_{j=1}^{\mathrm{N_{terms}}} \theta_j \Gamma(\beta_j)\zeta_j^{-\beta_j}}. \tag{46c}
$$

Figure 10 depicts the SER of LoRa systems over κ-μ fading, assuming $\mu = 2.1$, $\kappa = 10$ and SF $\in \{7, 9, 10\}$. The κ-μ distribution was approximated with a mixture gamma distribution using 37 terms. As it is evident, the results obtained using the mixture gamma approximation match very well with the exact ones, obtained using (9) and (45), for all considered values of SF.

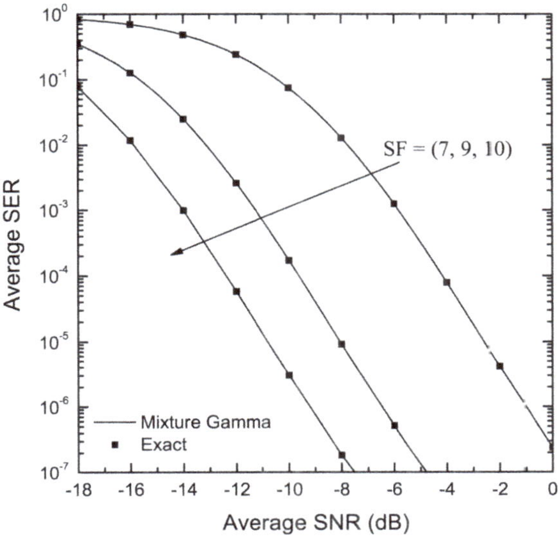

Figure 10. SER of LoRa systems operating in the presence of κ-μ fading as a function of the SNR, γ, for $\kappa = 10$, $\mu = 2.1$ and various values of SF.

Finally, it is worth pointing out that our newly derived formulae for Rice and Nakagami-m fading were tested against the ones proposed in [23] and a close match was reported. Nevertheless, as also mentioned in the introduction section, the proposed analytical framework still provides accurate results with much lower complexity than those reported in [23].

5. Conclusions

In this work, we elaborated the LoRa system model to include an extensive performance analysis in the presence of various types of fading channels, using exact single integral expressions as well as accurate approximations. The results presented herein are valid for most of the well-known fading models available in the open technical literature. Moreover, they are computationally efficient and thus they may serve as a useful tool for system engineers for performance evaluation purposes.

Author Contributions: Conceptualization, K.P., S.K.C. and D.L.; methodology, K.P., S.K.C. and D.L.; software, K.P.; validation, K.P., S.K.C. and D.L.; formal analysis, K.P., S.K.C. and D.L.; investigation, K.P., S.K.C. and D.L.; resources, K.P.; writing—original draft preparation, K.P.; writing—review and editing, S.K.C. and D.L.; visualization, K.P.; supervision, K.A.; project administration, K.A. All authors have read and agreed to the published version of the manuscript.

Funding: This research received no external funding.

Data Availability Statement: Data can be available upon mail request.

Conflicts of Interest: The authors declare no conflict of interest.

Appendix A. Evaluation of $\mathcal{I}(a,b,c,p)$

In order to obtain an analytical expression for $\mathcal{I}(a,b,c,p)$, we first employ an infinite series representation for the modified Bessel function, namely [30] (Equation (8.447/1))

$$I_0(cx) = \sum_{n=0}^{\infty} \frac{c^{2k}x^{2k}}{2^{2k}(k!)^2} \tag{A1}$$

Substituting (A1) into (37) and exchanging the series and integral operators—a valid operation due to the uniform convergence of the resulting integrals—$\mathcal{I}(a,b,c,p)$ can be written as

$$\mathcal{I}(a,b,c,p) = \sum_{n=0}^{\infty} \frac{c^{2k}}{2^{2k}(k!)^2} \int_0^{\infty} x^{2k} \exp(-px) \\ Q_1(a\sqrt{x}, b)\mathrm{d}x. \tag{A2}$$

Because $2k$ is always an integer, (A2) can be evaluated using [31] (Equation (11)), yielding (A2), thus completing the proof.

Table A1. Mathematical Notations.

$\jmath = \sqrt{-1}$	imaginary unit
z^*	conjugate of the complex number z
$\Pr\{\cdot\}$	probability operator
$\mathbb{E}_X\langle\cdot\rangle$	expectation of the random variable (RV)
$f_X(\cdot)$	probability density function of the RV X
$F_X(\cdot)$	cumulative distribution function of the RV X
$\delta_{i,k}$	Kronecker delta function: $\delta_{i,k} = 1$ for $i = k$ and 0 otherwise
$I_a(\cdot)$	modified Bessel function of the first kind and order a [30] (Equation (8.431))
$\Gamma(\cdot)$	Gamma function [30] (Equation (8.310/1))
$\Gamma(\cdot,\cdot)$	incomplete Gamma function [30] (Equation (8.350/2))
${}_pF_q(\cdot)$	generalized hypergeometric function [30] (Equation (9.14/1))
$Q_m(\cdot)$	generalized Marcum-Q function [35]: $Q_m(a,b) = a^{1-m}\int_b^{\infty} x^m \exp\left[-(x^2+a^2)/2\right] I_{m-1}(ax)$, $m \geq 1$
$\mathrm{L}_\nu(\cdot)$	The generalized Laguerre function of order ν [30] (Equation (8.972/1)): $\mathrm{L}_\nu(z) = {}_1F_1(-\nu;1;z)$

References

1. Centenaro, M.; Vangelista, L.; Zanella, A.; Zorzi, M. Long-range communications in unlicensed bands: The rising stars in the IoT and smart city scenarios. *IEEE Wirel. Commun.* **2016**, *23*, 60–67. [CrossRef]
2. Yang, G.; Liang, H. A Smart Wireless Paging Sensor Network for Elderly Care Application Using LoRaWAN. *IEEE Sens. J.* **2018**, *18*, 9441–9448. [CrossRef]
3. Marais, J.M.; Malekian, R.; Abu-Mahfouz, A.M. Evaluating the LoRaWAN Protocol Using a Permanent Outdoor Testbed. *IEEE Sens. J.* **2019**, *19*, 4726–4733. [CrossRef]
4. Raza, U.; Kulkarni, P.; Sooriyabandara, M. Low power wide area networks: An overview. *IEEE Commun. Surv. Tut.* **2017**, *19*, 855–873. [CrossRef]
5. Loukatos, D.; Arvanitis, K.G. Multi-Modal Sensor Nodes in Experimental Scalable Agricultural IoT Application Scenarios. In *IoT-Based Intelligent Modelling for Environmental and Ecological Engineering: IoT Next Generation EcoAgro Systems*; Springer International Publishing: Cham, Switzerland, 2021; pp. 101–128. [CrossRef]
6. Seller, O.; Sornin, N. Low Power Long Range Transmitter. U.S. Patent 14,170,170, 7 August 2014.
7. Vangelista, L. Frequency shift chirp modulation: The LoRa modulation. *IEEE Signal Process. Lett.* **2017**, *24*, 1818–1821. [CrossRef]
8. Chiani, M.; Elzanaty, A. On the LoRa modulation for IoT: Waveform properties and spectral analysis. *IEEE Internet Things J.* **2019**, *6*, 8463–8470. [CrossRef]
9. Elshabrawy, T.; Robert, J. Interleaved chirp spreading LoRa-based modulation. *IEEE Internet Things J.* **2019**, *6*, 3855–3863. [CrossRef]
10. Elshabrawy, T.; Robert, J. Capacity planning of LoRa networks with joint noise-limited and interference-limited coverage considerations. *IEEE Sens. J.* **2019**, *19*, 4340–4348. [CrossRef]

11. Hoeller, A.; Souza, R.D.; López, O.L.A.; Alves, H.; de Noronha Neto, M.; Brante, G. Analysis and Performance Optimization of LoRa Networks With Time and Antenna Diversity. *IEEE Access* **2018**, *6*, 32820–32829. [CrossRef]
12. Sanchez-Iborra, R.; Sanchez-Gomez, J.; Ballesta-Viñas, J.; Cano, M.D.; Skarmeta, A.F. Performance Evaluation of LoRa Considering Scenario Conditions. *Sensors* **2018**, *18*, 772. [CrossRef]
13. Farhad, A.; Kim, D.H.; Pyun, J.Y. Resource Allocation to Massive Internet of Things in LoRaWANs. *Sensors* **2020**, *20*, 2645. [CrossRef] [PubMed]
14. Matni, N.; Moraes, J.; Oliveira, H.; Rosario, D.; Cerqueira, E. LoRaWAN Gateway Placement Model for Dynamic Internet of Things Scenarios. *Sensors* **2020**, *20*, 4336. [CrossRef] [PubMed]
15. Escobar, J.J.L.; Gil-Castineira, F.; Redondo, R.P.D. JMAC Protocol: A Cross-Layer Multi-Hop Protocol for LoRa. *Sensors* **2020**, *20*, 6893. [CrossRef]
16. Bravo-Arrabal, J.; Fernandez-Lozano, J.J.; Seron, J.; Gomez-Ruiz, J.A.; García-Cerezo, A. Development and Implementation of a Hybrid Wireless Sensor Network of Low Power and Long Range for Urban Environments. *Sensors* **2021**, *21*, 567. [CrossRef] [PubMed]
17. Novak, V.; Stoces, M.; Cízkova, T.; Jarolimek, J.; Kanska, E. Experimental Evaluation of the Availability of LoRaWAN Frequency Channels in the Czech Republic. *Sensors* **2021**, *21*, 940. [CrossRef] [PubMed]
18. Zhao, Y.; Cao, C.; Liu, Z.; Mu, E. Intelligent Control Method of Hoisting Prefabricated Components Based on Internet-of-Things. *Sensors* **2021**, *21*, 980. [CrossRef] [PubMed]
19. Chinchilla-Romero, N.; Navarro-Ortiz, J.; Muñoz, P.; Ameigeiras, P. Collision Avoidance Resource Allocation for LoRaWAN. *Sensors* **2021**, *21*, 1218. [CrossRef]
20. Lee, J.; Yoon, Y.S.; Oh, H.W.; Park, K.R. DG-LoRa: Deterministic Group Acknowledgment Transmissions in LoRa Networks for Industrial IoT Applications. *Sensors* **2021**, *21*, 1444. [CrossRef]
21. Elshabrawy, T.; Robert, J. Closed-form approximation of LoRa modulation BER performance. *IEEE Commun. Lett.* **2018**, *22*, 1778–1781. [CrossRef]
22. Dias, C.F.; Lima, E.R.D.; Fraidenraich, G. Bit error rate closed-form expressions for LoRa systems under Nakagami and Rice fading channels. *Sensors* **2019**, *19*, 4412. [CrossRef]
23. Courjault, J.; Vrigenau, B.; Berder, O.; Bhatnagar, M. A Computable Form for LoRa Performance Estimation: Application to Ricean and Nakagami Fading. *IEEE Access* **2021**, *9*, 81601–81611. [CrossRef]
24. Ma, H.; Cai, G.; Fang, Y.; Chen, P.; Han, G. Design and Performance Analysis of a New STBC-MIMO LoRa System. *IEEE Trans. Commun.* **2021**, *69*, 5744–5757. [CrossRef]
25. Xu, W.; Cai, G.; Chen, G. Performance analysis of a two-hop relaying LoRa system. In Proceedings of the IEEE/CIC International Conference on Communications in China (ICCC), Xiamen, China, 28–30 July 2021. [CrossRef]
26. Wolfram Research, Inc. *Mathematica*; Version 13.0.0; Wolfram Research, Inc.: Champaign, IL, USA, 2021.
27. Yacoub, M.D. The κ-μ and the η-μ distribution. *IEEE Antennas Propag. Mag.* **2007**, *49*, 68–81. [CrossRef]
28. Simon, M.K.; Alouini, M.S. *Digital Communication Over Fading Channels*, 2nd ed.; Wiley: New York, NY, USA, 2005.
29. Prudnikov, A.P.; Brychkov, Y.A.; Marichev, O.I. *Integrals and Series Volume 4: Direct Laplace Transforms*, 1st ed.; CRC Press: Boca Raton, FL, USA, 1992.
30. Gradshteyn, L.; Ryzhik, I.M. *Table of Integrals, Series, and Products*, 6th ed.; Academic Press: San Diago, CA, USA, 2000.
31. Cui, G.; Kong, L.; Yang, X.; Ran, D. Two useful integrals involving generalised Marcum Q-function. *Electron. Lett.* **2012**, *48*, 1017–1018. [CrossRef]
32. Abdi, A.; Lau, W.C.; Alouini, M.S.; Kaveh, M. A New Simple Model for Land Mobile Satellite Channels: First- and Second-Order Statistics . *IEEE Trans. Wirel. Commun.* **2003**, *2*, 519–528. [CrossRef]
33. Prudnikov, A.P.; Brychkov, Y.A.; Marichev, O.I. *Integrals and Series Volume 3: More Special Functions*, 1st ed.; Gordon and Breach Science Publishers: Langhorne, PA, USA, 1986.
34. Munkhammar, J.; Mattsson, L.; Rydén, J. Polynomial probability distribution estimation using the method of moments. *PLoS ONE* **2017**, *12*, e0174573. [CrossRef]
35. Nuttall, A.H. Some Integrals Involving the Q_M function *IEEE Trans. Inf. Theory* **1975**, *21*, 95–96. [CrossRef]
36. Ermolova, N.Y.; Tirkkonen, O. Laplace Transform of Product of Generalized Marcum Q, Bessel I, and Power Functions with Applications. *IEEE Trans. Signal Process.* **2014**, *62*, 2938–2944.
37. Morales-Jimenez, D.; Paris, J.F. Outage Probability Analysis for η-μ Fading Channels. *IEEE Commun. Lett.* **2010**, *14*, 521–523. [CrossRef]
38. Ermolova, N.Y.; Tirkkonen, O. The η-μ fading distribution with integer values of μ. *IEEE Trans. Wireless Commun.* **2011**, *10*, 1976–1982. [CrossRef]
39. Peppas, K.P.; Alexandropoulos, G.; Mathiopoulos, P.T. Performance Analysis of Dual-Hop AF Relaying Systems over Mixed $\eta-\mu$ and $\kappa-\mu$ Fading Channels. *IEEE Trans. Veh. Technol.* **2013**, *62*, 3149–3163. [CrossRef]
40. Atapattu, S.; Tellambura, C.; Jiang, H. A Mixture Gamma Distribution to Model the SNR of Wireless Channels. *IEEE Trans. Wirel. Commun.* **2011**, *60*, 4193–4203. [CrossRef]

41. Loukatos, D.; Fragkos, A.; Arvanitis, K., Experimental Performance Evaluation Techniques of LoRa Radio Modules and Exploitation for Agricultural Use. In *Information and Communication Technologies for Agriculture—Theme I: Sensors*; Bochtis, D.D., Lampridi, M., Petropoulos, G.P., Ampatzidis, Y., Pardalos, P., Eds.; Springer International Publishing: Cham, Switzerland, 2022; pp. 101–120. [CrossRef]
42. Zhu, S.; Ghazaany, T.S.; Jones, S.M.R.; Abd-Alhameed, R.A.; Noras, J.M.; Buren, T.V.; Wilson, J.; Suggett, T.; Marker, S. Probability Distribution of Rician-Factor in Urban, Suburban and Rural Areas Using Real-World Captured Data. *IEEE Trans. Antennas Propag.* **2014**, *62*, 3835–3839. [CrossRef]

sensors

MDPI

Article

Towards Double Defense Network Security Based on Multi-Identifier Network Architecture

Yunmin Wang [1], Abla Smahi [1], Huayu Zhang [2] and Hui Li [1,3,*]

1 School of Electronic and Computer Engineering, Peking University, Shenzhen 518055, China; wangyunmin@pku.edu.cn (Y.W.); smahi_abla@pku.edu.cn (A.S.)
2 Purple Mountain Laboratories, Nanjing 211111, China; 1301111205@sz.pku.edu.cn
3 Peking University Lab of China Environment for Network Innovation, Peking University, Shenzhen 518055, China
* Correspondence: lih64@pkusz.edu.cn

Abstract: Recently, more and more mobile devices have been connected to the Internet. The Internet environment is complicated, and network security incidents emerge endlessly. Traditional *blocking and killing* passive defense measures cannot fundamentally meet the network security requirements. Inspired by the heuristic establishment of multiple lines of defense in immunology, we designed and prototyped a Double Defense strategy with Endogenous Safety and Security (DDESS) based on multi-identifier network (MIN) architecture. DDESS adopts the idea of a zero-trust network, with identity authentication as the core for access control, which solves security problems of traditional IP networks. In addition, DDESS achieves individual static security defense through encryption and decryption, consortium blockchain, trusted computing whitelist, and remote attestation strategies. At the same time, with the dynamic collection of data traffic and access logs, as well as the understanding and prediction of the situation, DDESS can realize the situation awareness of network security and the cultivation of immune vaccines against unknown network attacks, thus achieving the active herd defense of network security.

Keywords: network security; double defense; zero trust; situation awareness; immunology

Citation: Wang, Y.; Smahi, A.; Zhang, H.; Li, H. Towards Double Defense Network Security Based on Multi-Identifier Network Architecture. *Sensors* **2022**, *22*, 747. https://doi.org/10.3390/s22030747

Academic Editors: Peter Han Joo Chong and Omprakash Kaiwartya

Received: 15 December 2021
Accepted: 17 January 2022
Published: 19 January 2022

1. Introduction

With the development of the Internet and its deep integration with human social life, more and more mobile devices are connected. They are heterogeneous and ubiquitous, and put forward higher requirements for network security. The predominant *best-effort* design of TCP/IP network architecture focuses on the end-to-end communication between non-commercially and mutually trusted users. The network data transmission is entirely transparent, and the service and the bearer are separated. However, it instead hands over the security control to the users. This helps to improve the network availability and flexibility but at the cost of network security [1]. The Internet appears to be incapable of responding to users' demands for security in obtaining massive content. In addition, the current internet IP addresses have positioning, identity, and forwarding functions that pose many challenges for supporting quality of service (QoS), especially for the Internet of things (IoT) [2], Internet of Vehicles (IoV) [3], and the authenticity and credibility of the fusion of the virtual and real world in the metaverse [4].

Currently, network security protection is mainly based on IP networks' characteristics and adopts passive defense measures with *blocking and killing* methods, such as firewalls, authentication technology, access control, vulnerability scanning, disaster recovery, and honeypot technology. The defense capabilities of these measures can be passive or static, depending on predetermined settings before accessing the system and updating the preset defense library during use. Therefore, they can only detect and defend against a number of predefined network security attacks. Moreover, although these methods require higher user permissions and privileges, they can be easily controlled and exploited by attackers [5]. It is worth mentioning that the network security vulnerabilities constantly emerge while the attack methods are persistently refreshed. As demonstrated in Figure 1, attackers usually scan the weakest link in security defense to launch network attacks. Traditional defense measures focus on improving the protection capabilities against attacks rather than identifying, tracking, and investigating the responsibility of the attackers. They passively receive every intrusion attack, which is difficult to detect, identify, and respond to emerging attack methods, and it is challenging to solve network security problems fundamentally.

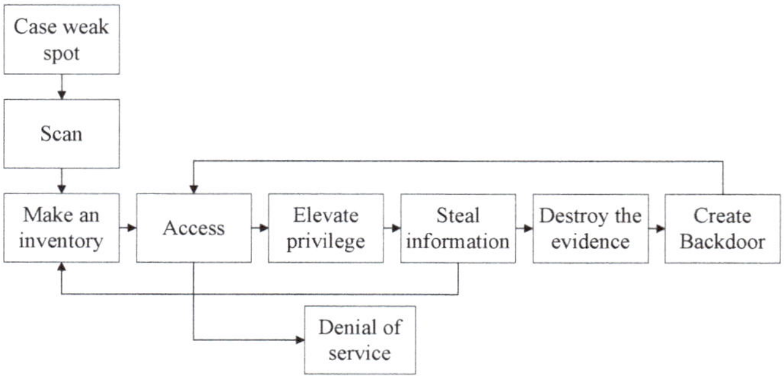

Figure 1. Network attack procedures.

While paying attention to network security, we found much inspiration from the biological immune system. The biological immune system has the advantages of feature extraction, distributed detection, self-tolerance, self-adaptation, robustness, and the capabilities of pattern recognition, learning, and memory [6]. It forms individual immunity to viral infections through its physical barrier, innate immune system, and adaptive immune system, and achieves the public herd immunity through collaboration between heterogeneous nodes and the cultivation and injection of vaccines [7]. Inspired by the multiple defense lines in immunology [8], this paper proposes a double defense strategy with endogenous safety and security (DDESS) as shown in Figure 2 based on the multi-identifier network (MIN) architecture [9]. We adopt identity authentication as the core access control method to solve traditional IP network security problems, and implement static network security defense through key encryption technology, blockchain technology, and trusted computing whitelist strategy. At the same time, through the dynamic collection of data traffic and access logs, DDESS can find the network vulnerabilities in the existing system, understand and predict the situation, realize the situation awareness of network attacks, and cultivate the immune vaccine of unknown network attacks, to complete the active dynamic defense of network security and the balance of network availability and security.

In summary, this paper makes the following contributions:

- This paper discusses the problem of the network's bottleneck resulting from the traditional IP-network carrying capacity and the poor network security. In this regard, we inspire our proposed system from the idea of zero-trust network [10]. In doing so, based on MIN architecture [9,11], we built a security strategy with an identity identifier as the core of network transmission. This strategy is built based on identity identifiers rather than network locations. Only after user authentication, authorization, and account verification can the user and the application communicate.
- Signature of network transmission packets protects data from theft. A variety of the data related to the identifiers is stored in the consortium blockchain of the MIN to ensure the identifier data's non-repudiation, tampering-resistance, and traceability. The independently developed voting consensus algorithm Proof of Vote (PoV) [12] improves the throughput of the system. At the same time, the encryption key and data

are stored in the trusted computing modules (TCM) to ensure that they are not read or disclosed [13], and protected by remote authentication [14]. Real-time monitoring of the trusted whitelist of accessing applications can promptly detect and respond to attacks.

- The network traffic and application access logs are collected and analyzed for further situation awareness. DDESS classifies and extracts appropriate attack information, including attack behavior and mode, and makes use of the convolutional neural networks (CNN) to evaluate the network security attack index of attack information. In addition, DDESS predicts the network security of the existing network system and provides the corresponding solutions. DDESS simulates network behaviors in the sandbox. It then analyzes these simulation results, cultivates a security immune vaccine, enriches the network attack behavior database, and prevents unknown network attack behaviors.

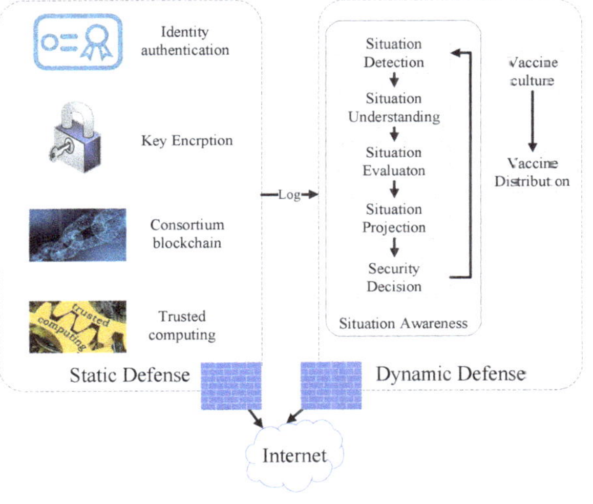

Figure 2. Double defense strategy with endogenous safety and security (DDESS).

The remainder of this paper is organized as follows: Section 2 reviews the related work of image anomaly detection. A detailed description of our proposed method is given in Section 3. In Section 4, we present experimental setups and comparisons. Section 5 carries out the corresponding experiments and gives out the simulations results. Section 6 concludes the paper and points out some of our future research.

2. Related Work

The traditional static defense of network security is divided into three types: reinforcement protection of the system, intrusion detection, and network deception [15].

Firewall (including packet filtering, proxy type, state inspection, in-depth inspection, web application) [16], encryption and decryption, data authentication [17], and access control [18] focus on protecting information and enhancing the security of the network system itself. They play a protective role in ensuring the normal access channels of the network system, authenticating legitimate user identities and rights management, and the security of confidential data and information.

Intrusion detection [19], vulnerability detection [20], traffic analysis [21], log auditing [22], and other attack methods for known characteristic information make use of methods such as characteristic scanning, pattern matching, and comprehensive data analysis to conduct dynamic monitoring, linkage alarms, and emergency response to prevent or eliminate attack threats.

Honeypot technology [23] deploys some hosts and network services as decoys to induce attackers to carry out attacks, thereby capturing and analyzing the attack behaviors, understanding the tools and methods used by the supplier, and speculating the intention and motivation of the attack, thereby enhancing its security protection capabilities [24]. However, it is difficult to deploy a honeypot environment that is not readily perceivable by intruders [25].

The above static defense methods can better defend against attacks with known characteristics and fixed patterns. However, it cannot defend against attacks based on unknown vulnerabilities backdoors, complicated and changeable multimode joint attacks, and attacks from within the network. With the continuous improvement of the automation and intelligence of countermeasures, the construction and strengthening of the network security defense system alone can no longer meet the actual needs of network security defense. Dynamic defense technology has gradually attracted widespread attention and is considered a revolutionary technology that changes the asymmetry of network security, such as Moving Target Defense (MTD) [26], Cyberspace Mimic Defense (CMD) [27], and Cyber Deception (CD) [28].

Let us dive into a specific scenario: vulnerabilities have constantly been discovered in the Internet of Vehicles (IoV), and the security problems of Vehicle to Everything (V2X) interactive communication are gradually emerging. The Internet of Vehicles requires real-time and timely resolution of network security issues. Therefore, many machine learning-based solutions in V2X scenarios have been proposed to provide dynamic defenses and address these problems. The surveillance of physical layer security (PLS) was explored in the field of connected vehicles [29]. A delimitated anti jammer scheme based on machine learning [30] secures the network and alleviates the traffic congestion simultaneously; in the meantime, it can reduce the computing delay.

Network dynamic defense is an innovative network defense technology system gradually developed to deal with the increasingly severe cyberspace security situation, which makes it possible to break the long-standing impregnable asymmetry and will balance the difficulty of network attack and defense in the future.

In this paper, we present a network defense strategy that integrates static and dynamic defenses. It adopts the ideas of the zero-trust network, and employs identity authentication, blockchain technology, and trusted computing technology, with situation awareness and dynamic immune functions.

3. Static Defense

We adopt the idea of a zero-trust network and build a multi-identifier network system with identity as the core, supplemented by data signature, blockchain technology, and trusted computing technology to realize the static defense of network security to improve the autoimmunity of network individuals.

3.1. Multi-Identifier Network System with Identity as the Core

The network identifier is the data carried in the network packet for addressing and forwarding by the intermediate router. In the traditional IP network, the target IP address of a network packet is its network identifier, forming a thin waist hourglass structure with IP as the core in the network layer, which identifies two network nodes for end-to-end communication. This network architecture is no longer suitable for the network communication requirements of the existing network to obtain content and services. At the same time, since the original design purpose of the IP network is to communicate between non-commercial users who trust each other, the network transmission is entirely transparent to the intermediate router, which makes data theft and monitoring easy. The IP network lacks top-level design and usually adopts a passive patching method for network security, making it increasingly challenging to implement [31].

The multi-identifier network (MIN) system [11] takes identity as the core, and supports the coexistence of multiple network identifiers such as identity, content [32], service, space,

space, and location information, and IP, to solve the depletion of IP addresses and the security problems existing in IP networks, as shown in Figure 3. The multi-identifier network is divided into a management plane (multi-identifier system, MIS) and data plane (multi-identifier routers, MIR), which supports simultaneous transmission of multiple network identifiers in the network, as depicted in Figure 4. MIS supports the management and resolution of network identifiers by multiple parties equally, and is responsible for the affairs related to identifiers combined with offline. Its main functions include identifier registration, identifier query, identifier generation, and identifier query. The multi-identifier router (MIR) is mainly responsible for the addressing, forwarding, and mutual translation functions of multiple identifier network packets. It supports the push transmission mode of identity and IP identifiers and the pull transmission mode of content identifier and service identifier.

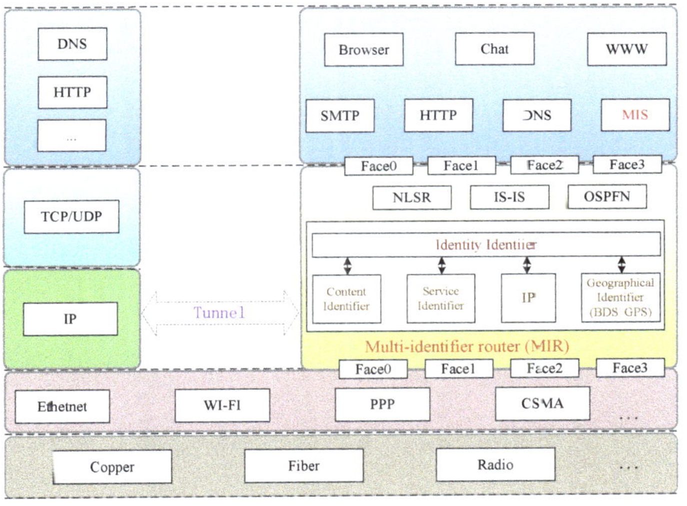

Figure 3. Multi-identifier network architecture.

Identity is not limited to users but also includes the unique identification of network communication physical entities (from now on referred to as users) such as devices, interfaces, applications, and business systems in the digital network world. It is the equivalent of physical network entities in the digital network world. The identity of a physical entity is unique in the digital network world, which is different from a username. A physical entity can have different usernames in different systems.

The identity identifier is the core network identifier in the MIN, and all multi-identifier routers need to support routing and forwarding of the identity identifier. Other types of network identifiers will be associated with a particular identity identifier. In an unsupported network domain, it can go back to the identity identifier for data forwarding. The hash value of the real identity information of the network communication entity (such as ID number, fingerprint, face, voiceprint, iris, and other biometric information, password) is included in the signature field in the data packet. It carries the sender's accurate identity signature information and time stamps, realizes dynamic access control based on identity, establishes a unified digital identity identifier and life cycle management for users and other physical entities participating in network communications, and saves them in the consortium blockchain of the MIS system.

After the communication entity is registered in the MIS, a unique identity is formed, corresponding permissions are assigned according to its role identification, and fine-grained permission management is carried out by monitoring its access behavior to realize the management and control of dynamic permissions. The principle of the least authority is implemented for communication entities, and their authority is allocated reasonably to ensure that each communication entity (including applications, services, users) can only access the required information or resources. When it is necessary to access sensitive resources, the identity recognition module will call other real identity information saved in the MIS for secondary verification, such as SMS verification code, dynamic password, face verification, to form access control and dynamic threat identification, privilege confirmation, alarm, and blocking at the regional boundary. The information transferred by users in the system will be recorded. The data package is highly bound with the user identity information to realize the traceability of data and behaviors.

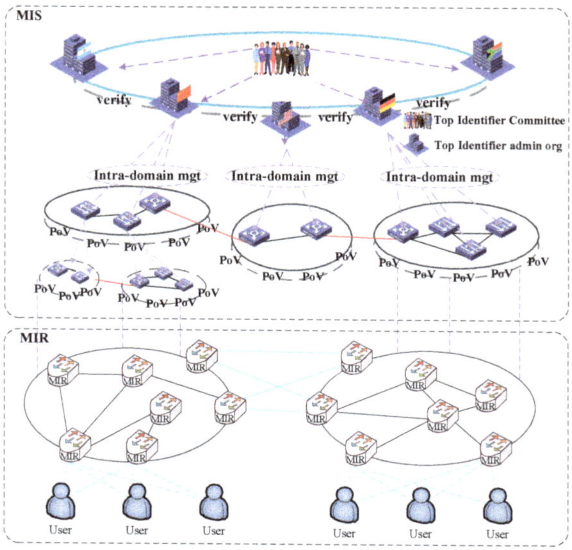

Figure 4. Separation of multi-identifier network management plane and data plane.

3.2. Encryption and Blockchain Protection

Identity authentication is based on Elliptic Curve Cryptography (ECC), rather than RSA. In addition to the registration phase, the subsequent authentication phase does not require the participation of a trusted third party, which ensures communication security and reduces the overhead of computing and communication. Data transmitted in the network will be protected by hash and asymmetric encryption at the packet level, rather than at the channel level, to prevent sensitive information from being eavesdropped, intercepted, or tampered with.

The nodes in the MIS consortium chain are divided into committee nodes, accounting nodes, and ordinary user nodes. The committee nodes are elected by members who volunteer to maintain the consortium chain and have equal voting rights jointly. The bookkeeping node is a node with the privilege to produce blocks voted by the committee nodes.

During user registration, the user will generate his public and private key pair, package the public key and the real identity information signed by the private key, and submit it to any blockchain node in the MIS system. The node will check the format of the registration request and find whether the user information already exists in the local database to avoid repeated registration. After that, this node will perform primary verification of the content of the registration information, package it into blockchain transaction information and submit it to the accounting node in the MIS consortium blockchain, and store the transaction information in the transaction pool.

We adopted the PoV consensus algorithm [33] to accelerate the process of consortium blockchain consensus, as depicted in Figure 5. At the beginning of the blockchain consensus, the accounting node will take out some transaction information from the transaction pool to generate a pre-block, and send it to the committee node in the consortium chain to request a signature. The committee node verifies the pre-block header and the content of each transaction, and sends it back to the accounting node after signing. If the billing node receives more than half of the signatures of the committee nodes within a given time threshold, it will store the signature information in the block header and set the timestamp to write this pre-block to the master of the consortium chain, and the pre-block becomes an official block. Otherwise, the pre-block will be deleted; the transaction will be taken out from the transaction pool again, and the pre-block will be generated and sent to all committee nodes for verification and signature.

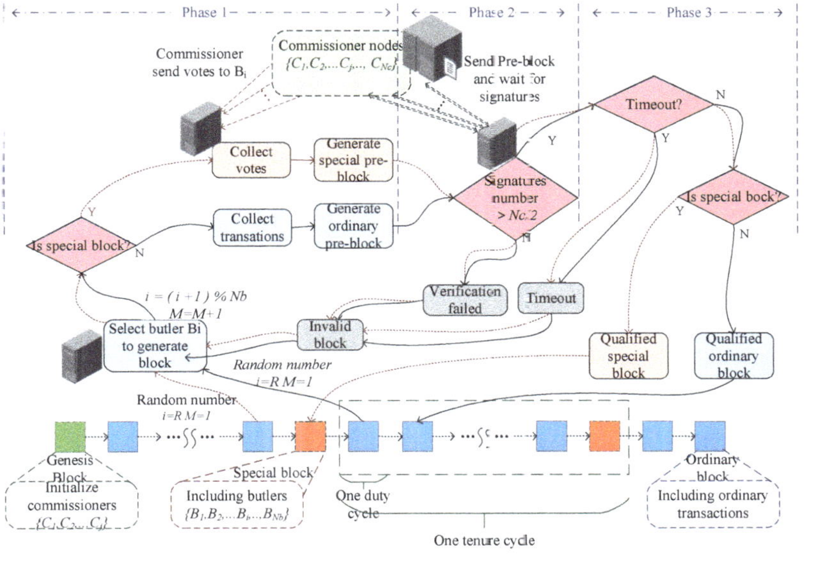

Figure 5. PoV consensus and block generation procedure.

When a user performs network data transmission, the user's signature information will be included in the data packet to record the user's access path in the multi-identifier network and realize the traceability of network security. When abnormal access is detected, the registered identity information reserved by the user will be retrieved from the consortium chain database, and the user will be warned or prohibited.

3.3. Trusted Computing Whitelist

Trustworthiness is the expectation that an entity can acquire the expected effect when realizing a given goal. It should be salable, adaptive, and auto-configurable, especially for the ubiquitous computing scenarios, such as Internet of Things and Internet of Vehicles [34]. We adopted the active immune trusted computing module (TCM) independently developed by China [13] in the client host hardware to save the encryption keys used in the consensus process of the blockchain nodes.

The private key is stored in the TCM chip and cannot be leaked and read. The public key can be used everywhere to identify the node identity and for signature verification. At the same time, all kinds of data generated by the node (such as blocks, transactions) are signed using the signature algorithm provided by TCM, to protect the private data such as encryption keys, and ensure data integrity, confidentiality, and security of the data through the transmission of the trusted chain. Identity authentication and encryption key mechanisms ensure the credibility of the computing environment.

Whitelist is adopted for application identification, and protection, such as NFD, PSYNC; only a few trusted applications are allowed to run in the current network system to realize the supervision and protection of the whole life cycle of applications from startup, loading, and operation, which reduces the load of the system and improves the availability of the system, as shown in Figure 6. The application developers use their private key to issue the application signatures and register them with the MIS, or the MIS trusted service to provide the verification benchmark and establish the application white list database. For applications on the whitelist, we conduct essential behavior analysis on their regular operations, and establish the whitelist behavior rule base of the applications. When an application starts, the hash measurement value of the application is obtained and compared with the expected benchmark value taken from the trusted module (TCM) to complete the integrity measurement. The execution of the application will be allowed only when the application is integrity and has not been tampered with. At the same time, during the operation of the application, the running status of its key behaviors is monitored in real-time, and the pre-established application behavior rule base is compared to detect abnormal behaviors in time, and record their operations in the log, to trade off between security and availability, and ensure the orderly operation of the application.

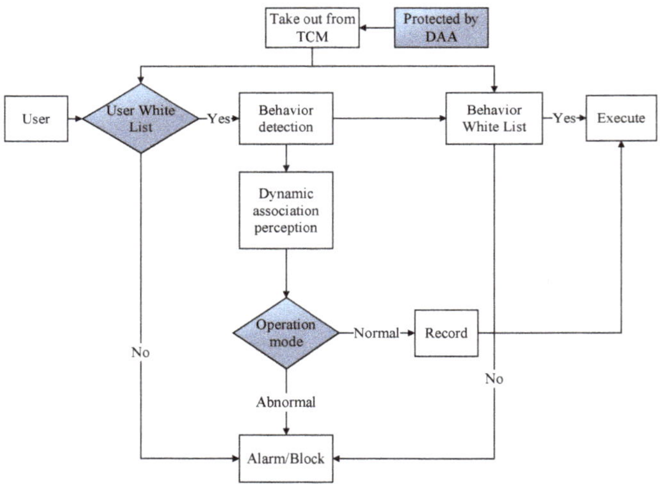

Figure 6. Trusted computing remote attestation and whitelisting.

At the same time, the security of TCMs needs to be guaranteed. [35] The feasibility of trusted computing storage and computing environment can be confirmed remotely through remote attestation Direct Anonymous Attestation (DAA) technology [14]. DAA uses the Carmenisch–Lysyanskaya signature mechanism [36] to sign certificates of the public key generated by the TCM to ensure the legitimacy of the TCM module. The TCM module uses the DAA certificate to interact with the DAA verifier, thereby identifying and distinguishing the compromised TCM module and ensuring the security of the trusted root.

4. Dynamic Defense

Static defense methods, such as identity authentication, data signatures, blockchain, and trusted computing, provide sufficient external barriers for network security to obtain balanced network confidentiality, integrity, and availability and hence protecting network nodes' security. They can effectively prevent and trace security problems. At the same time, nodes in the network are not simply isolated. They may face group network problems and new types of severe or unknown attacks. Therefore, it is necessary to build an effective dynamic defense system. The measures taken into consideration by DDESS include situation awareness and network security vaccine training and distribution.

4.1. Situation Awareness

Network situation awareness refers to the acquisition, understanding, and display of security elements that can bring about the network situation changes in a large-scale network environment, as well as the prediction of network development trends [37,38]. It is divided into network security situational extraction, understanding, and prediction.

When a network individual conducts a static defense, numerous original log files and backbone network traffic data will be generated, including user usage logs and alarm information, which provides a wealth of training materials for dynamic defense situation awareness of network security.

DDESS builds a decision tree containing all or part of the rules in the security policy rule set on the log file, which parses, filters, omissions, and normalizes the entries in the log file. Firstly, the raw data generated by the network monitoring equipment and management system are preprocessed, including data cleaning, noise reduction, dimensionality reduction, standardized merging or conversion, and data verification, removing duplicate and redundant information, merging similar information, and correcting error information, in order to obtain standardized asset data sets, threat data sets, and vulnerability data sets. Then, it carries out data correlation analysis and uses expert knowledge to model network activities and their regulations and characteristics, and identifies the existence and forms of various individuals in the network, and further identifies three different behaviors: malicious attack behavior, abnormal risk behavior (including weak password, account risk login, remote control), and normal access behavior. Behaviors that do not conform to the behavior baseline will trigger behavior alarms to detect risks in time and better find hidden attacks.

Based on the identified attack activities and their characteristics, the attacker's intention is inferred by further analyzing the semantics of these attack activities and the possible correlation among them. Its main tasks include identifying the source and type of these attack activities and judging the attacker's capability, opportunity, and the possibility of conducting a successful attack. Based on the network attack chain, network traffic abnormalities are found, such as external attacks, internal malicious scanning, ARP spoofing attack, and internal illegal accesses. The attacker's intentions can be mainly analyzed from the attack's behavior and target. The attack behavior prediction analyzes the logical relationship between attack behaviors and infers possible changes. In addition, we need to consider the function and importance of network assets to infer the attacker's attack intention and source. DDESS situational understanding draws portraits of user behaviors, including individuals and groups. Moreover, it constructs the network attack knowledge graph and draws the security graph from different perspectives, such as external attacks, internal horizontal penetration, and network data leakage. Therefore, DDESS can understand the current overall network security situation, detect and discover security events, and analyze and evaluate the network vulnerabilities and the impact of attacks.

DDESS first checks the network's security status to achieve network security situation awareness. Then, it understands in detail the various assets and participants within the networks and their potential vulnerabilities to attacks. It finally draws a graph of the network assets and the corresponding security vulnerabilities. After comprehensively acquiring network threat status data, convolutional neural networks (CNN) [39] are used to evaluate the potential network security risks and the behavioral patterns of different nodes. The resulting infliction of existing attack behaviors is based on the current network status and the identified attack activities, along with the vulnerabilities of network assets, as shown in Algorithm 1.

Algorithm 1 Training of security situational project model

Input:
The total number of layers L, the number of neurons in each hidden layer and output layer, activation function f, loss function, iteration step η,
maximum iteration times MAX and stop iteration threshold ϵ

Output:
linear relation coefficient matrix ω and bias vector b of each hidden layer and output layer.

```
Input the risk vector dataset R = {r_1, r_2, r_3, ···, r_n}.
for m=1 to L do
    Extract the m−layer features δ^m with a convolution kernel
    δ^m(r) ← f(Σ ω^m δ^{m−1} + b_m)
    Reduce network scale with max pooling:
    δ^m(r) = maxpooling(o^{m−1}(r) + b_m)
end for
update the weight ω with backpropagation:
for l = L to 2 do
    Compute δ^l based on δ^{l+1} and ω^{l+1} and z^l
    Compute the gradient ∇ω^{l+1} and ∇b^l
    Update ω^l and b^l in lth layer:
    ω^l ← ω^l − η Σ_{i=1}^{m} δ^{i,l} (a^{i,l−1})^T
    b^l ← b^l − η Σ_{i=1}^{m} δ^{i,l}
    if Δω < ε or L > MAX then
        break;
    end if
end for
```

The overall time complexity of Algorithm 1 is the cumulative time complexity of all convolution layers

$$\text{Time} : O\left(\sum_{l=1}^{D} m_l^2 \cdot f_l^2 \cdot C_{l-1} \cdot C_l\right), \tag{1}$$

and the spatial complexity is

$$\text{Space} : O\left(\sum_{l=1}^{D} f_l^2 \cdot C_{l-1} \cdot C_l + \sum_{l=1}^{D} m_l^2 \cdot C_l\right) \tag{2}$$

where D is the network depth, f_l is the filter size, C_l is the filter number and the output channels number of layer l, and C_{l-1} also represents the input channels number of layer l. m_l is the size of the output feature map, and $m_l = \lfloor (n_l - f_l + 2 * p_l)/s_l \rfloor + 1$, where n_l is the size of the input matrix, p_l is the padding, and s_l is the stride. In addition, we eliminate gradient vanishing problem with batch normalization [40] and introduce max pooling layer and 1×1 convolution kernel [41] for dimensionality reduction to reduce both the time and space complexity. DDESS can predict the future security status and the changing trend of the network and takes effective defensive measures.

4.2. Network Security Vaccine Training and Herd Immunity

The active immune trusted computing technology adopted in static defense can provide immune capabilities for network information systems. The vaccine culture used in organisms usually selects pathogens with strong immunogenicity, loses toxicity through biochemical inactivation or repeatedly immune iteration of animal cells, and retains its immunogenicity.

Biological vaccines are usually injected into organisms to produce immunity. However, network vaccines are different from biological vaccines in a way that they aim to provide a real-time and automatic defense. While defending against attacks, such vaccines improve the network's resilience and elasticity to maintain its availability and thus return to a normal state eventually. Therefore, the cultivation of network defense vaccines needs to be carried out dynamically in the network environment. The network vaccine can be described as:

$$V : (antigen, srcaddr, destaddr, timestamp, protocol). \quad (3)$$

The vaccine of network security's immunity is a string extracted from the characteristics of data packets transmitted over the network. Antibodies are the measures taken in static defenses and the rules generated in dynamic defenses. During the matching process of antigens and antibodies, the immune cells constantly defend against network attacks, form clonal regeneration, and evolve into memory cells. The memory cells record accurately the network intrusions that have occurred, and encapsulate them into a vaccine cell after stamping them with a timestamp according to the vaccine format as shown in Equation (3).

The encapsulated network security vaccine is transmitted to the vaccine defense center of the adjacent network to be finally distributed among the network nodes. The selection of such a network considers the local routing table, routing, and forwarding. When the antigenic match with the vaccine happens, it becomes activated and can detect network attacks. At the same time, network nodes can also be used for training and cultivating network vaccines as described by Equation (3). When new types of attacks are discovered, these nodes will continue training and learning to enhance the vitality of antibodies in vaccine cells to improve the autoimmunity and achieve adversarial collaboration.

5. Experiments

We evaluated the overall defense effectiveness, and intrusion detection performance of DDESS compared with related state-of-the-art schemes in this section.

5.1. Overall Performance Evaluation

Experiments were conducted on a testbed using the topology presented in Figure 7. The router *pkusz*11 plays the role of the edge router that is connected to the Internet. Experiments and simulations are based on the published KDD-99 dataset that consists of 5 million records [42]. These records are totally made up of 41 attributes and one attack category field which marks all observations as either "normal" or "attacked" with one of the following attacks: Denial of Service (DoS), Remote to Local (R2L), User to Root (U2R), and Probing/surveillance. We carry out attack experiments on DDESS combined with the multi-identifier network. The selected attack methods are conventional under IP networks, such as target detection, attack injection, ARP attacks, and so on. Experiments' results are demonstrated in Table 1, where ✓ indicates that the attack was successful, while ✗ represents an attack failure.

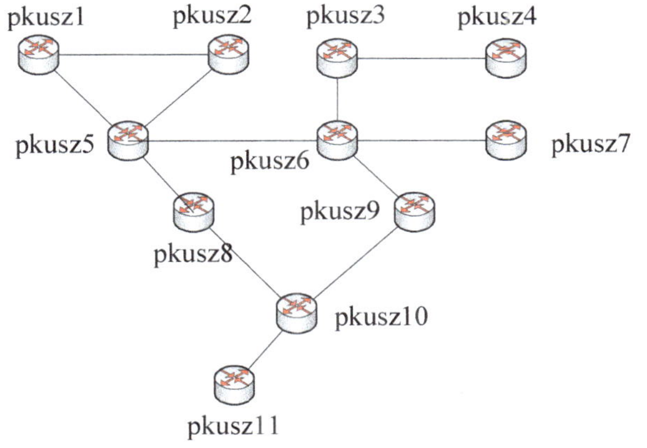

Figure 7. DDESS performance evaluation topology.

Table 1. Attack results of experiments.

Attack Phase	Description		IP-IP	IP-MIN
Target detection	Host discovery		✓	✗
	ping scan		✓	✗
	OS recognition		System fingerprints obtained	Host non-survival
	Port scan		All ports probed	The host is alive, but no port detected
Attack injection	Trojan	TCP trojan	✓	✗
		UDP trojan	✓	✗
		ICMP trojan	✓	✗
	One-sentence shell		✓	✗
Action	ARP Attack		Information sniffing	Target cannot be sniffed
			Network disconnection attack	Target not affected

MIN can effectively defend against target detection attacks such as host discovery, ping scanning, and operating system recognition. It can also prevent Trojan attack injections and one-sentence shell attacks. Attacks within the action phase were divided into ARP disconnection and ARP spoofing. We used "arpspoof" to send fake MAC-IP binding packets, hence poisoning the gateway's ARP cache. The attacker can therefore disconnect the target network or monitor its traffic. For the ARP disconnection attack, ARP spoofing was first initiated with an arpspoof tool to change the IP forwarding path from one target host to another within the LAN. Both types of ARP attacks were successfully blocked in MIN as shown in Table 1.

5.2. *Performance of Dynamic Defense*

We compared the intrusion detection performance of DDESS on the KDD-99 dataset [42] with that of state-of-the-art machine learning based methods with 10-fold cross-validation. Our comparison references included Gradient Boosted Machine (GBM) [43], k-Nearest Neighbor (kNN) [44], Classification and Regression Trees (CART) [45], Multi-Layer Perceptron (MLP) [46], and AdaBoost [47]. The evaluation results are depicted in Figure 8.

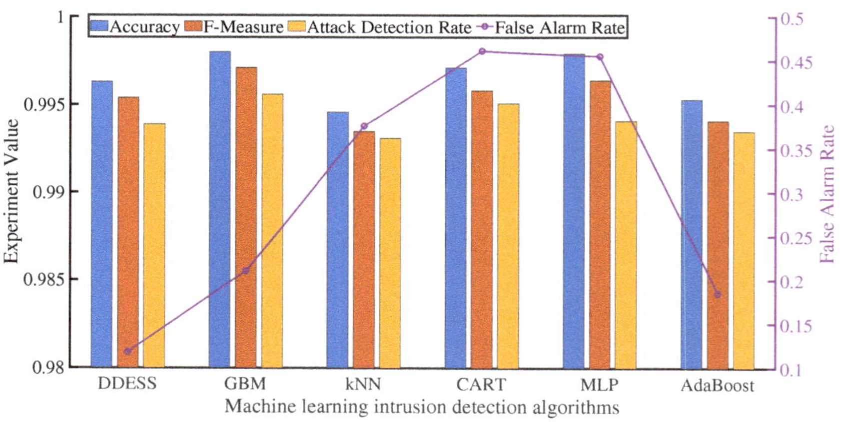

Figure 8. Performance comparison of intrusion detection.

It was observed that, compared with other algorithms, GBM, MLP, and CHAT have better performance in accuracy, reaching 99.81%, 99.79%, and 99.71%, respectively. In order to trade-off the precision rate and recall rate, F-Measure was usually applied [48],

$$\text{F-Measure} = \frac{(1+\beta^2)P \times R}{\beta^2 P + R}, \tag{4}$$

where P means the precision rate and R represents the recall rate. Here, we chose $\beta^2 = 1$ because we attached importance of both P and R. GBM still achieved the best F-Measure value (99.71%) across all the algorithms, whereas kNN obtains the worst one (99.35%).

As far as attack detection rate (ADR) is concerned, algorithms which are adaptive to continuous attacks have better effects. In Figure 8, GBM (99.56%) and CART (99.51%) can outperform other methods in ADR.

In addition, DDESS can acquire the best false alarm rate (0.118%) because the deep neural networks can reduce the opportunities for false positives. A low false alarm rate and low latency are important for time-sensitive networks (TSN) of Internet of Things (IoT) [31]. Therefore, in order to better monitor network security, we proposed DDESS to comprehensively measure these indicators and come to trade-offs.

We further evaluated the prediction performance of dynamic defense and compared it with TSA-AdaBoost [49], a situation awareness algorithm based on the AdaBoost machine learning method. DDESS adopts the batch normalization [40] to reduce the influence of gradient vanishing problem, and max-pooling layer and 1×1 convolutional kernel [41] for dimensionality reduction. Therefore, the overall fitness of the proposed DDESS is better than TSA-AdaBoost for the prediction of test samples, as shown in Figure 9a. We select mean absolute error (MAE), root mean square error (RMSE), and mean absolute percentage error (MAPE) as the prediction evaluation metrics to evaluate the proposed prediction model, where

$$\text{MAE} = \frac{1}{N}\sum_{i=1}^{N} \hat{y}_i - y_i| \tag{5}$$

$$\text{RMSE} = \sqrt{\frac{1}{N}\sum_{i=1}^{N} (\hat{y}_i - y_i)^2} \tag{6}$$

$$\text{MAPE} = \frac{100\%}{N}\sum_{i=1}^{N} \left| \frac{\hat{y}_i - y_i}{y_i} \right|, \tag{7}$$

where $\hat{y}_i$ means the prediction situation value. As shown in Figure 9b, the MAE and RMSE values of DDESS are 0.0306 and 0.035, respectively, while these of TSA-AdaBoost are 0.0417 and 0.0486, respectively. Compared with TSA AdaBoost, DDESS has less overall error precision in predicting network security situation value. In addition, the MAPE value of DDESS and TSA-AdaBoost is 6.67% and 8.95%, respectively, which means that DDESS has better accuracy than TSA AdaBoost.

Let us analyze forecast results in Figure 9a from a network administrators' perspective. Although the situational value on day 3 is relatively low, it is predicted that the network situational value will have a higher tendency on the next day (i.e., day 4), which will warn network administrators about the occurrence of network attacks. At the same time, based on the 11th-day forecast trend, this is likely to be the end phase of network attacks, which will make administrators pay special attention to the network behavior logs for the next two days to ensure that they are not deleted or destroyed by attackers. On the other hand, the forecasts on the 12th and 13th show that network attacks continue, and some vulnerabilities probably exist in the network system.

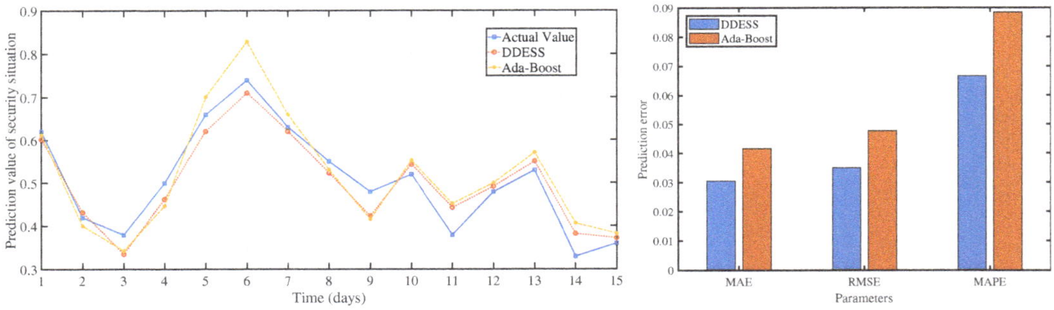

(a) Prediction performance (b) Prediction error analysis

Figure 9. Prediction performance comparison on situation awareness.

DDESS can train the attack model and make corresponding portraits of the attack and the attacker without requiring additional hardware, such as FPGA. To perceive the motivation behind the attack, DDESS will classify and mark the attack behaviors of the same source. This helps learn and predict potential attacks behaviors from the same source in the future.

5.3. Competition and Trial

The multi-identifier network system (MIN) combined with DDESS technology has withstood the "Network Security Challenge Competition" held by the Purple Mountain Laboratory [50]. Forty-eight teams continuously carried out remote online high-intensity network attacks for 72 h. During this period, the cumulative number of attacks reached 3.58 million. More precisely, the MIN was one of the few network security systems any team has not broken through. It also successfully prevented most competition-related qualification attacks such as Linux privilege escalation, virtual machine escape, session forging and hijacking, and simulation PWN. The DDESS-based system has been proved to be more robust when compared with traditional commercial workarounds.

6. Conclusions and Future Work

To overcome the current internet security problems, we propose a double defense strategy with endogenous safety and security (DDESS) based on the MIN system. DDESS provides both static and dynamic defense strategies against different network attacks. The proposed system preserves network security using static defensive measures such as identity verification, encryption protection, and trusted computing. It also uses dynamic defensive measures like situation awareness and vaccine cultivation and distribution. Experiments and performance analysis showed that DDESS provides sufficient availability and robust security compared to other existing network defense solutions. In the future work, we plan to introduce edge computing technology to reduce the pressure of central servers and better adapt to the network development, such as 6G, Internet of things (IoT), etc. In addition, we will study the intrusion detection algorithms, and put forward more effective strategies for network security detection and prediction in our future work.

Author Contributions: Conceptualization, H.L. and Y.W.; methodology, H.Z.; software, Y.W.; validation, A.S. and Y.W.; formal analysis, Y.W.; investigation, Y.W.; resources, H.L.; data curation, H.Z.; writing—original draft preparation, Y.W.; writing—review and editing, A.S.; visualization, A.S.; supervision, A.S.; project administration, H.L.; funding acquisition, H.L. All authors have read and agreed to the published version of the manuscript.

Funding: This work was partially supported by the Guangdong R&D Key Program (2019B010137001), Shenzhen Research Program (GXWD20201231165807007-20200807164903001, JCYJ20210324122013036, JCYJ20190808155607340), China Environment for Network Innovation (CENI) (GJFGW No.[2020]386, SZFGW [2019] 261), and the National Keystone R&D Program of China (2017YFB0803204).

Institutional Review Board Statement: Not applicable

Informed Consent Statement: Not applicable.

Data Availability Statement: The data used for this research is publicly available at https://kdd.org/kdd-cup/view/kdd-cup-1999/Data.

Conflicts of Interest: The authors declare no conflict of interest.

Abbreviations

The following abbreviations are used in this manuscript:

DDESS	Double Defense strategy with Endogenous Safety and Security
NDN	Named Data Networking
MIN	Multi-Identifier Network
MIS	Multi-Identifier System
MIR	Multi-Identifier Router
PoV	Proof of Vote
DAA	Direct Anonymous Attestation
TCM	Trusted Computing Module
ACC	Accuracy
ADR	Attack Detection Rate
FAR	False Alarm Rate

References

1. Blumenthal, M.S.; Clark, D.D. Rethinking the design of the internet: The end-to-end arguments vs. the brave new world. *ACM Trans. Internet Technol.* **2001**, *1*, 70–109. [CrossRef]
2. Kouicem, D.E.; Bouabdallah, A.; Lakhlef, H. Internet of things security: A top-down survey. *Comput. Netw.* **2018**, *141*, 199–221. [CrossRef]
3. Contreras-Castillo, J.; Zeadally, S.; Guerrero-Ibañez, J.A. Internet of vehicles: Architecture, protocols, and security. *IEEE Internet Things J.* **2017**, *5*, 3701–3709. [CrossRef]
4. Park, S.M.; Kim, Y.G. A Metaverse: Taxonomy, components, applications, and open challenges. *IEEE Access* **2022**, *10*, 4209–4251. [CrossRef]
5. Wu, J. *Cyberspace Endogenous Safety and Security: Mimic Defense and Generalized Robust Control*; Science Press: Beijing, China, 2020.
6. Forrest, S.; Hofmeyr, S.A.; Somayaji, A. Computer immunology. *Commun. ACM* **1997**, *40*, 88–96. [CrossRef]
7. Dasgupta, D. Advances in artificial immune systems. *IEEE Comput. Intell. Mag.* **2006**, *1*, 40–49. [CrossRef]
8. Yu, Q.; Ren, J.; Zhang, J.; Liu, S.; Fu, Y.; Li, Y.; Ma, L.; Jing, J.; Zhang, W. An Immunology-Inspired Network Security Architecture. *IEEE Wirel. Commun.* **2020**, *27*, 168–173. [CrossRef]
9. Li, H.; Yang, X. Architecture of Sovereignty Network. In *Co-Governed Sovereignty Network: Legal Basis and Its Prototype & Applications with MIN Architecture*; Springer: Singapore, 2021; pp. 61–94. [CrossRef]
10. Rose, S.W.; Borchert, O.; Mitchell, S.; Connelly, S. Zero Trust Architecture, Special Publication (NIST SP). 2020. Available online: https://www.nist.gov/publications/zero-trust-architecture (accessed on 29 November 2021).
11. Wang, Y.; Li, H.; Huang, T.; Zhang, X.; Bai, Y. Scalable Identifier System for Industrial Internet Based on Multi-Identifier Network Architecture. *IEEE Internet Things J.* **2021**, 1. [CrossRef]
12. Li, K.; Li, H.; Hou, H.; Li, K.; Chen, Y. Proof of vote: A high-performance consensus protocol based on vote mechanism & consortium blockchain. In Proceedings of the IEEE HPCC, Bangkok, Thailand, 18–20 December 2017.
13. Shen, C. To Create a Positive Cyberspace by Safeguarding Network Security with Active Immune Trusted Computing 3.0. *J. Inf. Secur. Res.* **2018**, *4*, 282–302.
14. Brickell, E.; Camenisch, J.; Chen, L. Direct anonymous attestation. In Proceedings of the 11th ACM conference on Computer and communications security, Washington, DC, USA, 25–28 October 2004; pp. 132–145.
15. Kaur, T.; Malhotra, V.; Singh, D. Comparison of network security tools-firewall, intrusion detection system and Honeypot. *Int. J. Enhanc. Res. Sci. Technol. Eng.* **2014**, *3*, 200–204.
16. Song, X. Firewall technology in computer network security in 5G environment. *J. Phys. Conf. Ser.* **2020**, *1544*, 012090. [CrossRef]
17. Aman, M.N.; Basheer, M.H.; Sikdar, B. Data provenance for IoT with light weight authentication and privacy preservation. *IEEE Internet Things J.* **2019**, *6*, 10441–10457. [CrossRef]
18. Sandhu, R.S.; Samarati, P. Access control: Principle and practice. *IEEE Commun. Mag.* **1994**, *32*, 40–48. [CrossRef]

19. Tidjon, L.N.; Frappier, M.; Mammar, A. Intrusion detection systems: A cross-domain overview. *IEEE Commun. Surv. Tutor.* **2019**, *21*, 3639–3681. [CrossRef]
20. Williams, L.; McGraw, G.; Migues, S. Engineering security vulnerability prevention, detection, and response. *IEEE Softw.* **2018**, *35*, 76–80. [CrossRef]
21. Kausar, F.; Aljumah, S.; Alzaydi, S.; Alroba, R. Traffic analysis attack for identifying users' online activities. *IT Prof.* **2019**, *21*, 50–57. [CrossRef]
22. Roger, M.; Goubault-Larrecq, J. Log auditing through model-checking. In Proceedings of the Proceedings of the 14th IEEE workshop on Computer Security Foundations, Cape Breton, NS, Canada, 11–13 June 2001; pp. 220–234.
23. Sun, Y.; Tian, Z.; Li, M.; Su, S.; Du, X.; Guizani, M. Honeypot Identification in Softwarized Industrial Cyber–Physical Systems. *IEEE Trans. Ind. Inform.* **2020**, *17*, 5542–5551. [CrossRef]
24. Fan, W.; Du, Z.; Fernández, D.; Villagrá, V.A. Enabling an anatomic view to investigate honeypot systems: A survey. *IEEE Syst. J.* **2017**, *12*, 3906–3919. [CrossRef]
25. Shrivastava, R.K.; Bashir, B.; Hota, C. Attack detection and forensics using honeypot in IoT environment. In Proceedings of the International Conference on Distributed Computing and Internet Technology, Bhubaneswar, India, 10–13 January 2019; pp. 402–409.
26. Zhuang, R.; DeLoach, S.A.; Ou, X. Towards a theory of moving target defense. In Proceedings of the ACM Workshop on Moving Target Defense, Scottsdale, AZ, USA, 7 November 2014; pp. 31–40.
27. Wu, J. *Cyberspace Mimic Defense*; Springer: Cham, Switzerland, 2020.
28. Wang, C.; Lu, Z. Cyber deception: Overview and the road ahead. *IEEE Secur. Priv.* **2018**, *16*, 80–85. [CrossRef]
29. Makarfi, A.U.; Rabie, K.M.; Kaiwartya, O.; Adhikari, K.; Nauryzbayev, G.; Li, X.; Kharel, R. Toward Physical-Layer Security for Internet of Vehicles: Interference-Aware Modeling. *IEEE Internet Things J.* **2020**, *8*, 443–457. [CrossRef]
30. Kumar, S.; Singh, K.; Kumar, S.; Kaiwartya, O.; Cao, Y.; Zhou, H. Delimitated anti jammer scheme for Internet of vehicle: Machine learning based security approach. *IEEE Access* **2019**, *7*, 113311–113323. [CrossRef]
31. Wu, J. Thoughts on the development of novel network technology. *Sci. China Inf. Sci.* **2018**, *61*, 101301. [CrossRef]
32. Zhang, L.; Afanasyev, A.; Burke, J.; Jacobson, V.; Claffy, K.; Crowley, P.; Papadopoulos, C.; Wang, L.; Zhang, B. Named data networking. *ACM SIGCOMM Comp. Commun. Rev.* **2014**, *44*, 66–73. [CrossRef]
33. Li, H.; Yang, X. Key Technologies of Sovereignty Network. In *Co-Governed Sovereignty Network*; Springer: Cham, Switzerland, 2021; pp. 95–182.
34. López, J.; Maña, A.; Muñoz, A. A secure and auto-configurable environment for mobile agents in ubiquitous computing scenarios. In Proceedings of the International Conference on Ubiquitous Intelligence and Computing, Wuhan, China, 3–6 September 2006; pp. 977–987.
35. Gürgens, S.; Rudolph, C.; Scheuermann, D.; Atts, M.; Plaga, R. Security evaluation of scenarios based on the TCG's TPM specification. In Proceedings of the European Symposium on Research in Computer Security, Oslo, Norway, 11–13 September 2007; pp. 438–453.
36. Camenisch, J.; Lysyanskaya, A. A signature scheme with efficient protocols. In Proceedings of the International Conference on Security in Communication Networks, Amalfi, Italy, 11–13 September 2002; pp. 268–289.
37. Endsley, M.R. Design and evaluation for situation awareness enhancement. In Proceedings of the Human Factors Society Annual Meeting, Anaheim, CA, USA, 24 October 1988; Volume 32, pp. 97–101.
38. Bass, T. Intrusion detection systems and multisensor data fusion. *Commun. ACM* **2000**, *43*, 99–105. [CrossRef]
39. LeCun, Y.; Bengio, Y.; Hinton, G. Deep learning. *Nature* **2015**, *521*, 436–444. [CrossRef]
40. Ioffe, S.; Szegedy, C. Batch normalization: Accelerating deep network training by reducing internal covariate shift. In Proceedings of the International Conference on Machine Learning, Lille, France, 6–11 July 2015; pp. 448–456.
41. Lin, M.; Chen, Q.; Yan, S. Network in network. *arXiv* **2013**, arXiv:1312.4400.
42. Tavallaee, M.; Bagheri, E.; Lu, W.; Ghorbani, A.A. A detailed analysis of the KDD CUP 99 data set. In Proceedings of the 2009 IEEE Symposium on Computational Intelligence for Security and Defense Applications, Ottawa, ON, Canada, 8–10 July 2009; pp. 1–6.
43. Tama, B.A.; Rhee, K.H. An in-depth experimental study of anomaly detection using gradient boosted machine. *Neural Comput. Appl.* **2019**, *31*, 955–965. [CrossRef]
44. Li, W.; Yi, P.; Wu, Y.; Pan, L.; Li, J. A new intrusion detection system based on KNN classification algorithm in wireless sensor network. *J. Electr. Comput. Eng.* **2014**, *2014*, 240217. [CrossRef]
45. Soylu, T.; Erdem, O.; Carus, A.; Güner, E.S. Simple CART based real-time traffic classification engine on FPGAs. In Proceedings of the 2017 International Conference on ReConFigurable Computing and FPGAs (ReConFig) (ReConFig), Cancun, Mexico, 4–6 December 2017; pp. 1–8.
46. Ahmad, I.; Abdullah, A.; Alghamdi, A.; Alnfajan, K.; Hussain, M. Intrusion detection using feature subset selection based on MLP. *Sci. Res. Essays* **2011**, *6*, 6804–6810.
47. Hu, W.; Hu, W. Network-based intrusion detection using Adaboost algorithm. In Proceedings of the The 2005 IEEE/WIC/ACM International Conference on Web Intelligence (WI'05), Compiegne, France, 19–22 September 2005; pp. 712–717.
48. Fawcett, T. An introduction to ROC analysis. *Pattern Recognit. Lett.* **2006**, *27*, 861–874. [CrossRef]

49. Zhao, Y.; Cheng, G.; Duan, Y.; Gu, Z.; Zhou, Y.; Tang, L. Secure IoT edge: Threat situation awareness based on network traffic. *Comput. Netw.* **2021**, *201*, 108525. [CrossRef]
50. Laboratories, P.M. Mimic Defense International Elite Challenge. 2021. Available online: http://www.pmlabs.com.cn/plus/view.php?aid=1464 (accessed on 29 November 2021).

sensors

MDPI

Review

Deep Learning in Diverse Intelligent Sensor Based Systems

Yanming Zhu [1], Min Wang [2], Xuefei Yin [2], Jue Zhang [2], Erik Meijering [1] and Jiankun Hu [2,*]

[1] School of Computer Science and Engineering, University of New South Wales, Sydney, NSW 2052, Australia
[2] School of Engineering and Information Technology, University of New South Wales, Canberra, ACT 2612, Australia
* Correspondence: j.hu@adfa.edu.au

Abstract: Deep learning has become a predominant method for solving data analysis problems in virtually all fields of science and engineering. The increasing complexity and the large volume of data collected by diverse sensor systems have spurred the development of deep learning methods and have fundamentally transformed the way the data are acquired, processed, analyzed, and interpreted. With the rapid development of deep learning technology and its ever-increasing range of successful applications across diverse sensor systems, there is an urgent need to provide a comprehensive investigation of deep learning in this domain from a holistic view. This survey paper aims to contribute to this by systematically investigating deep learning models/methods and their applications across diverse sensor systems. It also provides a comprehensive summary of deep learning implementation tips and links to tutorials, open-source codes, and pretrained models, which can serve as an excellent self-contained reference for deep learning practitioners and those seeking to innovate deep learning in this space. In addition, this paper provides insights into research topics in diverse sensor systems where deep learning has not yet been well-developed, and highlights challenges and future opportunities. This survey serves as a catalyst to accelerate the application and transformation of deep learning in diverse sensor systems.

Keywords: deep learning; computer vision; biomedical imaging; biometrics; remote sensing; cybersecurity; Internet of Things; natural language processing; audio and speech processing; control system and robotics; information system; food; agriculture; chemistry

Citation: Zhu, Y.; Wang, M.; Yin, X.; Zhang, J.; Meijering, E.; Hu, J. Deep Learning in Diverse Intelligent Sensor Based Systems. *Sensors* **2023**, *23*, 62. https://doi.org/10.3390/s23010062

Academic Editor: Yang Yue

Received: 2 November 2022
Revised: 6 December 2022
Accepted: 14 December 2022
Published: 21 December 2022

1. Introduction

In recent years, driven by the rapid increase in available data and computational resources, deep learning has achieved extraordinary advances and almost become the defacto standard approach in virtually all fields of science and engineering. Essentially, deep learning is a part of the field of machine learning, a subfield of artificial intelligence (AI) concerned with learning data representations using computational methods. In traditional machine learning algorithms, manually choosing features and a classifier is needed, while in a deep learning algorithm, the features are extracted automatically by the algorithm through learning from its own errors. It is this automatic feature extraction that distinguishes deep learning from the field of machine learning.

Neural networks make up the backbone of deep learning algorithms. A neural network aims to learn nonlinear maps between inputs and outputs through its elementary computational cells (also called "neurons"). It is the number of layers (also called depth) of neural networks that distinguishes a shallow network from a Deep Neural Network (DNN). Typically, a network must have more than three layers to be considered a DNN. Deep networks learn representations of the data in a hierarchical manner to simulate the mechanism of the human brain in extracting information from given data.

The increasing complexity and the large volume of data collected by diverse sensor systems have brought about significant developments in deep learning, which have fundamentally transformed the way the data are acquired, processed, analyzed, and interpreted. Therefore, in this paper, we provide a comprehensive investigation of deep learning in diverse intelligent sensor based systems, covering fundamentals of deep learning models and methods, deep learning techniques for fundamental tasks in individual sensor systems, insights of reformulation of these fundamental tasks for broader applications in diverse intelligent sensor based systems, and challenges of breaking through the bottleneck of current deep learning approaches in exploring the full potential of deep learning. We searched Google Scholar (GS) and Web of Science (WOS) with the keywords deep learning (DL) and sensor. This resulted in 16,100 articles from 2020. We further selected, based on the top journals and conferences, around 150 most relevant papers for careful inspection, and traced some further relevant references from there. From these, we observed that existing relevant surveys [1–4] have one or more of the following limitations: (1) touching only a small subset of topics in individual domains, (2) lacking an overview of common techniques/algorithms from different domains, and (3) lacking a holistic view based on the individual domains of diverse intelligent sensor based systems. This survey aims to be a catalyst for accelerating the application and transformation of deep learning across diverse intelligent sensor based systems.

The contributions of this paper can be summarized as follows

- This is the first paper to provide a comprehensive investigation of deep learning in diverse sensor systems from the perspective, in a holistic view, of different data modalities across different intelligent sensor based systems and application domains.
- This paper presents the fundamentals of deep learning and the most widely used deep learning models and methods in a concise and high-level way, which would be very useful for people to get a quick start in the field.
- This paper provides a comprehensive summary of deep learning implementation tips and links to tutorials, open-source codes, and pretrained models, which can serve as an excellent self-contained reference for deep learning practitioners and researchers. This is a unique feature that makes it distinguishable from existing literature survey papers.
- This paper identifies the fundamental tasks in individual intelligent sensor based systems and provides insights to reformulation of these task for broader applications for those seeking to innovate deep learning in diverse sensor systems.
- This paper provides insights into research topics where deep learning has not yet been well-developed, and highlights the challenges and future directions of deep learning in diverse intelligent sensor based systems.

2. Deep Learning Basics

2.1. History of Deep Neural Networks

The origin of DNNs can be traced back to 1943, when McCulloch and Pitts proposed the first artificial neural network [5]. Since then, deep learning has grown gradually and achieved a few significant milestones in its development. One of them worth mentioning is Rosenblatt's "perceptron" introduced in 1958. It demonstrated that a perceptron will converge when what they are trying to learn can be represented [6]. However, such a model has obvious limitations, and multilayer perceptrons are required by complex tasks, but at that time, it was not clear how to train these models. Subsequently, deep learning encountered its first winter.

Until 1985, Hinton et al. proposed the back-propagation algorithm, which has greatly stimulated the development of this field [7]. At almost the same period, the "neocogitron" which inspired the Convolutional Neural Networks (CNNs), the Recurrent Neural Networks (RNNs), and the DNNs were proposed [8–10]. However, due to the limitation of hardware, these models were hard to use for handling large data, and thus the development of deep learning was trapped again.

By 2006, Hinton and others solved the training problem of DNNs by using a layer-wise pretraining framework, which greatly revitalized the field [11,12]. At the same time, algorithms for training deep AutoEncoders (AEs), and other deep architectures were proposed [13], which allowed deep learning to develop at an exponential rate. From then, a variety of deep learning methods increasingly emerged, including Deep Belief Networks (DBNs), Restricted Boltzmann Machines (RBMs), CNNs, Generative Adversarial Networks (GANs), Graph Neural Networks (GNNs), and so on.

In recent years, two astounding deep learning applications made a global splash and shocked the world. One is AlphaGo, which defeated the world champion Go players using deep learning with the support of abundant hardware resources (https://www.deepmind.com/research/highlighted-research/alphago accessed on 2 November 2022). Another is AlphaFold, which solved the 50-year-old challenging protein folding problem. These further stimulated the rapid development of deep learning in various domains. Nowadays, with the advancements in Graphics Processing Units (GPUs) and High-Performance Computing (HPC), deep learning has become one of the most efficient tools with outstanding performance in almost every domain.

2.2. *Fundamentals of Deep Neural Networks*

DNNs try to mimic the way biological neurons send signals to each other through numerous neurons (also called nodes). Generally, the architecture of a DNN consists of multiple neuron layers including an input layer, an output layer, and one or many hidden layers [14] (Figure 1). Each neuron is connected to another neuron to pass information. The input to a DNN can be numbers, characters, audios, images, etc., which are broken down into bits of binary data that a computer can process. The output can be continuous values, binary values, or categorical values, depending on the tasks. A DNN relies on training data to learn and improve its accuracy over time. During the learning, if it cannot accurately recognize a particular pattern for a given task, an algorithm would adjust its weights until it determines the correct mathematical manipulation to fully process the data [13].

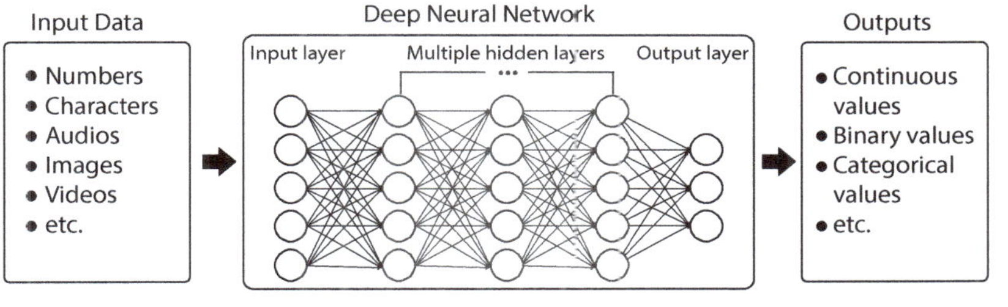

Figure 1. Diagram of a DNN.

2.2.1. Neuron Perception

A neuron multiplies each of its inputs by an associated weight and then sums these weighted inputs and adds a predetermined number called the bias (Figure 2). The neuron is activated if its output is above a specified threshold and will pass its output to the next layer of the DNN. That is, in a DNN, neurons in each layer get inputs from the previous layer, learn representations, and then pass the information to the next layer. Each successive layer of a DNN uses the output of the previous layer for its input. This way, a DNN produces an output at the end.

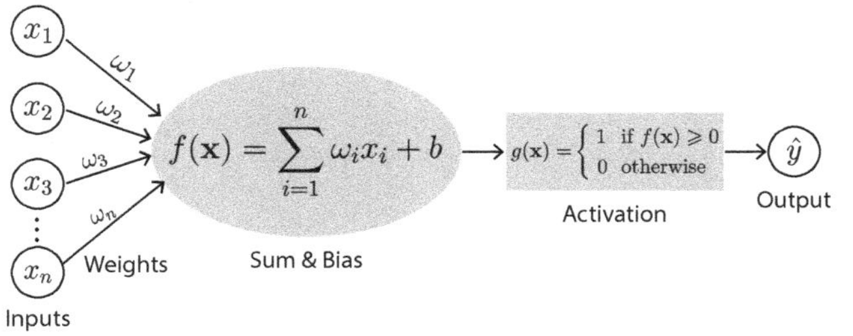

Figure 2. Diagram of the perception of a neuron.

2.2.2. Activation Functions

The activation function is an important aspect of a DNN. It defines the output of a node given inputs and is mainly used to generate a nonlinear relationships between the input and the output. Currently, there are 10 types of nonlinear activation functions (Table 1). Here, we elaborate on four most popular activation functions, namely sigmoid, tanh, ReLU, and leaky ReLu, describe their application scenarios, and analyze their pros and cons.

Table 1. Ten types of nonlinear activation functions.

Name	Definition
Sigmoid	$f(x) = \frac{1}{1+e^{-x}}$
Tanh	$f(x) = \frac{e^x - e^{-x}}{e^x + e^{-x}}$
ReLU	$f(x) = \max(0, x)$
LeakyReLU	$f(x) = \max(0.1x, x)$
Parametric ReLU	$f(x) = \max(ax, x)$
ELU	$\begin{cases} x & \text{for} \quad x \geq 0 \\ \alpha(e^x - 1) & \text{for} \quad x < 0 \end{cases}$
Probability	$\text{sig}(x) = \frac{1}{1+e^{-x}}$
Softmax	$\text{softmax}(x_i) = \frac{\exp(x_i)}{\sum_j \exp(x_j)}$
Swish	$\sigma(x) = \frac{x}{1+e^{-x}}$
GELU	$f(x) = 0.5x\left(1 + \tanh\left[\sqrt{\frac{2}{\pi}}(x + ax^3)\right]\right)$ with $a = 0.044715$

The sigmoid activation function is one of the most widely used activation functions. It takes an arbitrary value as input and outputs a value between 0 and 1. The larger the input, the closer the output value is to 1. This function is differentiable and provides a smooth gradient, and is suitable for tasks that require predicting probabilities as outputs. Its limitation is that it stops the DNN from learning and makes the DNN suffer from the vanishing gradient problem as the gradient value approaches zero.

The tanh activation function also has an S-shape like the sigmoid function, but with the difference in output range of -1 to 1. That is, with tanh, the larger the input, the closer the output value is to 1. This function is widely used for hidden layers of a DNN, because it can help to center the data and make the learning for the next layer easier. However, it also faces the same problem of vanishing gradients as the sigmoid function. However, in practice, the tanh activation function is more preferred than sigmoid due to its zero-centered nature.

The ReLU activation function, which stands for Rectified Linear Unit, is another most important and popular activation function. Its main feature is that it does not activate all the nodes at the same time, and only the nodes with an output larger than 0 will be activated. Therefore, this function is computationally efficient, compared to the sigmoid and tanh activation functions. In addition, it facilitates the convergence of gradient descent towards the global minimum of the loss function. Its limitation is that it may cause possible dead nodes due to the negative side of the curve making the gradient value zero.

Therefore, the leaky ReLU activation function, which is an improvement of ReLU, has been proposed to solve the dying ReLU problem. It has a small positive slope for the negative side, which enables back-propagation for negative inputs. This way, the gradient of the negative side of the curve will be a nonzero value, and the problem of dead nodes is solved. The limitation of this activation function is that it makes the learning of the DNN time-consuming.

2.2.3. Stochastic Gradient Descent (SGD)

The SGD is an efficient approach for fitting linear classifiers or regressors under convex loss functions, especially in high-dimensional optimization. Therefore, it has been widely and successfully used as an important optimization method for training a DNN [7,14–16]. It has the advantages of high efficiency and ease of implementation, but also the disadvantages of requiring some hyperparameters such as the regularization parameter and the number of iterations, and being sensitive to feature scaling.

2.2.4. Back-Propagation (BP)

The BP, which is short for "backward propagation of errors", is the most prominent algorithm to train a DNN. Strictly, it refers only to the algorithm for computing the gradient, not how to use the gradient. However, loosely and generally, it refers to using the mean squared error and the SGD to fine-tune the weights of a DNN. Specifically, it calculates the gradient of a loss function with respect to all the weights in the DNN by the chain rule, and employs the SGD to decide how to use the gradient to properly tune the weights. A DNN's weights are iteratively tuned until the desired output is achieved.

2.3. Learning Scenarios of Deep Learning

2.3.1. Supervised Learning

Supervised learning is a learning paradigm that uses a set of labeled examples as training data and makes predictions for all unseen points [17]. Supervised algorithms are expected to learn the mapping between pairs of inputs and output values, also called annotations or labels. This scenario includes two types of problem: classification and regression.

2.3.2. Semi-Supervised Learning

Semi-Supervised learning (SSL) aims to learn predictive models that make use of both labeled and unlabeled data. SSL provides a feasible solution in the setting where unlabeled data are easily accessible, but labels are difficult to obtain [18]. By exploring the pattern in additional unlabeled data, SSL methods can improve the learning performance. Deep SSL has dominated this research area in recent years [19–21].

2.3.3. Unsupervised Learning

In contrast to supervised learning, unsupervised learning constructs models where only unlabeled data are available [22]. The key of unsupervised methods is to discover hidden patterns and discriminative feature representations without human intervention. Clustering and dimensionality reduction are examples of unsupervised learning problems.

2.3.4. Reinforcement Learning

Different from supervised learning, reinforcement learning refers to the learning scenario where the learner receives rewards after a course of actions by interacting with the environment, and then determines the optimal actions by maximizing the rewards to achieve the goal [17]. With different states of the environment, the problem can be divided into two settings: the planning problem and learning problem.

2.4. Training Strategy and Performance

2.4.1. Learning Rate

Learning rate is one of the most important hyperparameters when configuring a neural network. It controls how much a model is changed based on the estimated error each time the model weights are updated [23]. Choosing an appropriate learning rate is very challenging, because a very small value may cause the training process to be too long and get stuck, while a very large value may result in learning a suboptimal set of weights or with an unstable training process. A typical solution to choose the appropriate learning rate is to reduce the learning rate during training. Currently, there are three kinds of popular ways to achieve this: constant, factored, and exponential decay.

2.4.2. Weight Decay

Weight decay is a regularization technique applied to the weights of a neural network for shrinking the weights during back-propagation. It works by adding a penalty term, which is usually the L2 norm of the weights, to the loss function. It can help to prevent overfitting and avoid exploding gradient.

2.4.3. Dropout

Dropout is widely used to prevent overfitting by randomly dropping out neural units in a neural network. It is a strong regularization to prevent complex co-adaptations on training data [24]. More technically, at each training stage, individual nodes are either dropped out of the network with probability $1 - p$ or kept with probability p, leading to a reduced network. Dropout forces a neural network to learn more robust features and roughly doubles the number of iterations required to converge, but the training time for each epoch is less.

2.4.4. Early Stopping

Early stopping is a training strategy used to reduce overfitting without compromising on model accuracy. The underlying idea behind early stopping is to stop training before a model starts to overfit. There are mainly three strategies for early stopping: training models on a preset number of epochs, stop when the loss function update becomes very small, and observing the changes of training and validation errors with the number of epochs.

2.4.5. Batch Normalization

Batch normalization is a technique to standardize the inputs to a neural network for stabilizing the learning process and reducing the number of training epochs required to train deep networks [25]. With batch normalization, a network can use higher learning rates, achieve better results, and the training can be faster. It also makes activation functions viable by regulating the inputs to them, and adds noise which reduces overfitting with a regularization effect.

2.4.6. Data Augmentation

Data augmentation refers to a set of techniques to artificially increase the amount of training data by generating new data from existing data. It is a low-cost and effective method to improve the performance and accuracy of deep learning models in data-constrained environments. For visual data, alterations such as cropping, rotating, scaling, flipping, contrast changing, adding noise are effective and popular data augmentation methods. For other kinds of data, data augmentation is not as popular as for visual data, due to the complexity of the data. Some advanced models such as GANs are popular for data augmentation [26–28].

2.5. Deep Learning Platforms and Resources

2.5.1. Deep Learning Platforms

The two currently most renowned end-to-end open source platforms for deep learning are TensorFlow [29] and PyTorch [30]. They provide comprehensive and flexible ecosystems of tools, libraries, and community resources that let engineers and researchers easily build and deploy deep learning powered applications.

TensorFlow (https://www.tensorflow.org/ accessed on 2 November 2022) is developed by researchers and engineers at Google and was released in 2015. It is a symbolic math library and is best suited for data flow programming across a wide variety of tasks. It provides multiple abstraction levels for building and training a DNN. In addition, it has adopted Keras (https://keras.io/ accessed on 2 November 2022), which is a functional API that extends TensorFlow and allows users to easily code some high-level functional sections. It provides system-specific functionality such as pipelining, estimators, and eager execution, and supports various topologies with different combinations of inputs, output, and layers.

PyTorch (https://pytorch.org/ accessed on 2 November 2022) is based on Torch and is relatively new compared to TensorFlow. It is developed by researchers at Facebook and was released in 2017. It is well known for its simplicity, ease of use, flexibility, efficient memory usage, and dynamic computational graphs. Due to its computation power and native programming feeling, PyTorch is emerging as a winner. Furthermore, it has a large community of developers and researchers who have built rich and powerful tools and libraries to extend PyTorch. Some popular libraries include GPyTorch, BoTorch, and Allen NLP.

Other frameworks include Caffe [31], Torch [32], DL4j (https://deeplearning4j.konduit.ai/ accessed on 2 November 2022, Neon (https://github.com/NervanaSystems/neon accessed on 2 November 2022, Theano [33], MXNet [34], and CNTK [35]. The choice of which platform is superior has always been controversial, but PyTorch and TensorFlow are undoubtedly the two most popular deep learning frameworks today.

2.5.2. Codes and Pretrained Models

While TensorFlow and PyTorch have provided official tutorials on how to use them, topic-specific tutorials for different levels are beneficial and complementary. There are many reputable courses online, for example, Practical Deep Learning for Coders (https://course.fast.ai/ accessed on 2 November 2022), which provides practical programming skills and an easy-to-use code library for most important deep learning techniques. Furthermore, it is free and without ads, and is designed for learners with various background levels. More useful courses can be found at the collection of AI Curriculum from top universities (https://github.com/Machine-Learning-Tokyo/AI_Curriculum accessed on 2 November 2022). A comprehensive collection of deep learning books, videos, lectures, workshops, datasets, tools, etc., is available on GitHub (https://github.com/ChristosChristofidis/awesome-deep-learning accessed on 2 November 2022).

Open source code can greatly help to learn deep learning and improve the efficiency of the learning. The distinguished Papers With Code website https://paperswithcode.com/accessed on 2 November 2022) collects new research papers and their corresponding open source codes, as well as the latest trending directions and state-of-the-art results across many standard benchmarks.

As we will describe in later sections, utilizing pretrained models is an important technique in transfer leaning and can greatly improve the efficiency of deep leaning. A collection of pretrained models is available for both TensorFlow (https://github.com/tensorflow/models accessed on 2 November 2022) and Pytorch (https://pytorch.org/vision/stable/models.html accessed on 2 November 2022). The AI community Hugging Face (https://huggingface.co/accessed on 2 November 2022) also provides a huge collection of pretrained models as well as the codes to train these models. The website Model

Zoo (https://modelzoo.co/ accessed on 2 November 2022) is also a great place to discover pretrained models and open source deep learning codes.

2.5.3. Computing Resources

Training deep learning models requires relatively high computing resources. Therefore, open source web-based development environments that run entirely in the cloud are very helpful for average researchers. The two currently popular web applications for interactive computing are Jupyter Notebook (https://jupyter.org/ accessed on 2 November 2022) and Colab (https://colab.research.google.com/ accessed on 2 November 2022). They are very similar, and both require zero configuration, provide access to GPUs free of charge, and support most popular machine learning libraries. They are easy to use and to create documents that contain live code, equations, visualizations, and text. Furthermore, their flexible interfaces allow users easily to configure, arrange, and share workflows for team work.

Tracking and visualizing metrics such as loss and accuracy during the model training is a vital process of training a DNN. A predominant toolkit for this purpose is Tensorboard (https://www.tensorflow.org/tensorboard accessed on 2 November 2022), which works for both TensorFlow and Pytorch. In addition to the above functions, it can visualize model graphs, views histograms of weights, biases, or other tensors as they change over time, project embeddings to a lower dimensional space, display images, text, and audio, and so on.

3. Deep Learning Models and Methods

3.1. Convolutional Neural Network (CNN)

3.1.1. Introduction of CNN

The design of convolutional networks was inspired by biological processes where the pattern of connections between neurons resembles the organization of the human visual cortex: individual cortical neurons respond only to stimuli in the receptive fields, which partially overlap to cover the entire field of view [36].

A typical CNN consists of several convolutional layers and pooling layers followed by fully connected layers at the end (Figure 3). The input of a CNN is a tensor arranged in four dimensions ($N \times h \times w \times c$), where N denotes the number of inputs, h and w are the height and width of the input, and c the depth or number of channels of the input ($c = 3$ for an RGB image). The convolutional layer convolves the input with k kernels/filters of size ($k_h \times k_w \times k_c$), where $k_h < h$, $k_w < w$, and $k_c \leq c$, and generates and passes k feature maps to the following layer. These kernels share the same parameters and form the base of local connections. The convolution operation performs a dot product (usually the Frobenius inner product) of the kernel with a small region of the layer's input matrix each time, then an activation function (usually the ReLU function) is applied. As the kernel slides along the input matrix, a feature map is generated. The pooling layers reduce the dimension of the feature maps by subsampling, thus decreasing the number of parameters for training. The pooling operation usually takes the maximum (max pooling) or average value (average pooling) of the local cluster of neurons (local pooling) or all neurons (global pooling) in the feature map. The last few layers of a CNN are fully connected layers, as in a multilayer perception that connect every neuron in one layer to every neuron in the following layer. Through these layers, the CNN extracts high-level representations from the input data, and its final layer outputs the probabilities that the instance belongs to each class.

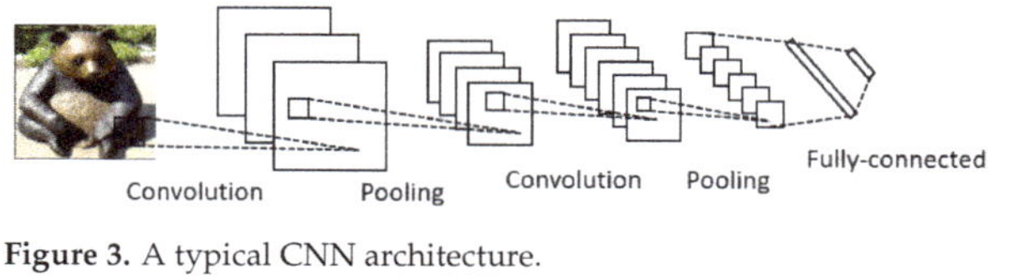

Figure 3. A typical CNN architecture.

CNNs improve the fully connected networks in three major aspects: (1) local connections, (2) weight sharing, and (3) subsampling. These mechanisms significantly reduce the number of parameters, speed up convergence, and make CNN an outstanding algorithm in the field of deep learning. CNNs are particularly popular in computer vision applications since they fully exploit the two-dimensional structure of the input image data [37].

Since its first introduction, the CNN design has received widespread attention from researchers, and various variant models and improvements have been proposed. Next, we introduce several representative CNN models and their main contributions. Table 2 summarizes these models and following works.

Table 2. Summary of popular CNN architectures.

Model	Usage	Main Contribution	Code	Year
AlexNet [37]	Recognition	Depth is essential	✓	2012
VGG [38]	Recognition	Small kernel size	✓	2013
GoogLeNet/Inception [39]	Recognition	Inception module (sparse connections)	✓	2013
ZfNet [40]	Visualisation	Understanding network activity	✓	2014
ResNet [41]	Recognition	Residual module (skip connections)	✓	2015
DenseNet [42]	Recognition	Dense concatenation	✓	2017
UNet [43]	Segmentation	U-shaped encoder-decoder architecture	✓	2015
Faster R-CNN [44]	Segmentation	Region proposal network	✓	2015
Highway Networks [45]	Recognition	Cross-layer connection	✓	2015
YOLO [46]	Detection	High efficiency 'only look once'	✓	2016
Mask R-CNN [47]	Segmentation	Object mask	✓	2017
MobileNet [48]	Recognition/Detection	Depthwise separable convolutions	✓	2017
Pyramidal Net [49]	Recognition	Pyramidal structure	✓	2017
Xception [50]	Recognition	Extreme version of Inception	✓	2017
Inception-ResNet [51]	Recognition	Inception with residual connections	✓	2017
PolyNet [52]	Training solution	Optimize networks	✓	2017

3.1.2. AlexNet

AlexNet [37] consists of eight layers: five convolutional layers, some of which followed by max-pooling layers, concatenated with three fully connected layers. It uses the ReLU activation function, which shows improved training performance over tanh and sigmoid which are prone to the vanishing gradient problem [53] (e.g., the derivative of sigmoid becomes very small in the saturating region, and therefore, the updates to the weights almost vanish). A dropout layer is used after every fully connected layer, reducing overfitting. AlexNet was one of the first deep neural networks to push ImageNet classification accuracy up by a significant amount (a top five accuracy of 80.2%) in comparison to traditional methods. The depth of the model was essential for its high performance, and while computationally expensive, training was made feasible by the utilization of GPUs.

3.1.3. VGG

VGG [39] improves over AlexNet by replacing large size kernels (11 and 5 in the first and second convolutional layer, respectively) with multiple 3×3 kernels one after another. The idea behind this is that with a given receptive field, stacking multiple kernels of smaller size is better than using one kernel of larger size. This is because multiple nonlinear layers increase the depth of the network, which enables it to learn more complex features at a lower cost. In addition, the 3×3 kernels help retain finer representations of the input. In VGG-D, blocks with the same kernel size are applied multiple times to extract more complex and representative features. This concept of blocks or modules became a common theme in the networks after VGG. It achieved top five accuracy of 91.2% on ImageNet.

3.1.4. GoogLeNet (Inception)

GoogLeNet [38] introduces the inception module to form a sparse architecture rather than the previous dense connection architecture to reduce the computation requirement of training deep networks such as VGG. It builds on the idea that most of the activations in a deep network are either unnecessary or redundant because of correlations between them. Therefore, the most efficient architecture of a deep network will have a sparse connection between the activations, rather than a dense connection architecture. Thus, the inception module (Figure 4) approximates a sparse CNN with a normal dense construction. Since only a few neurons are effective, the width and number of the convolutional filters of a particular kernel size is kept small. Convolutions of different sizes are used to capture features at varied scales (5×5, 3×3, 1×1). A bottleneck layer (1×1 convolutions) is introduced for massive reduction of the computational cost. All these changes allow the network to have a large width and depth. GoogLeNet is built on top of the inception blocks and it replaces the fully-connected layers at the end with a simple global average pooling which averages out the channel values across the 2D feature map. This drastically reduces the total number of parameters. It achieves 93.3% top five accuracy on ImageNet and is much faster to train than VGG.

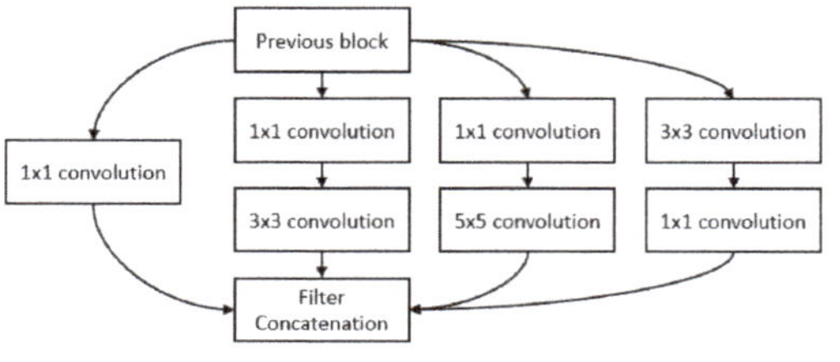

Figure 4. The inception module in GoogLeNet.

3.1.5. ResNet

ResNet [41] was proposed to solve the vanishing gradient problem [54] and degradation problem. The vanishing gradient prevents the update of the weights and hinders convergence from the beginning due to the increased depth. The degradation problem refers to the phenomenon that as the network depth increases, accuracy gets saturated and then degrades rapidly (this is not caused by overfitting but adding more layers leads to higher training error) [41]. Degradation of training accuracy indicates that not all systems are similarly easy to optimize. Hence, the residual learning framework is designed to recast the original mapping $\mathcal{H}(x)$ into a residual mapping which is easier to optimize than the original mapping. The residual module (Figure 5) creates a shortcut connection between the input and output to the module, implying an identity mapping, thus allowing the stacked nonlinear layers to fit a residual mapping $\mathcal{G}(x) := \mathcal{H}(x) - x$. With these shortcuts, the residual module helps to build deeper neural networks as large as a network depth of 152. In addition, ResNet adopts a global average pooling followed by the classification layer as in GoogLeNet. It achieves better accuracy (95.51% top five accuracy with ResNet-152) than VGGNet and GoogLeNet while being computationally more efficient than VGGNet.

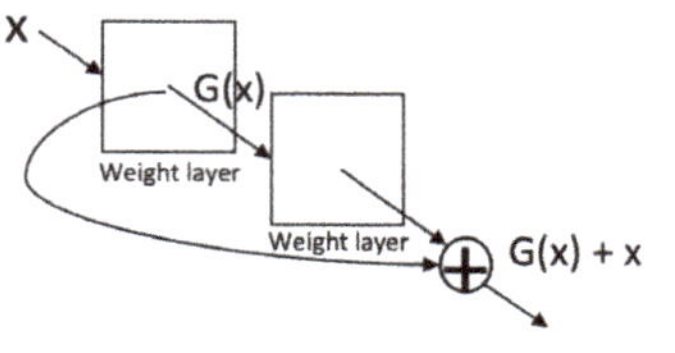

Figure 5. Illustration of a residual learning module.

3.1.6. DenseNet

DenseNet [42] is one of the new discoveries in neural networks for visual object recognition. DenseNet is quite similar to ResNet but with some fundamental differences: ResNet uses an additive method to merge the previous layer (identity) with the future layer, whereas DenseNet concatenates the output of the previous layer with the future layer. For ResNet, the identity shortcut that stabilizes training also limits its representation capacity, while DenseNet has a higher capacity with multilayer feature concatenation. In DenseNet, each layer obtains additional inputs from all preceding layers and passes on its own feature maps to all subsequent layers (Figure 6). With concatenation, each layer is receiving collective knowledge from all preceding layers. However, the dense concatenation requires higher GPU memory and more training time.

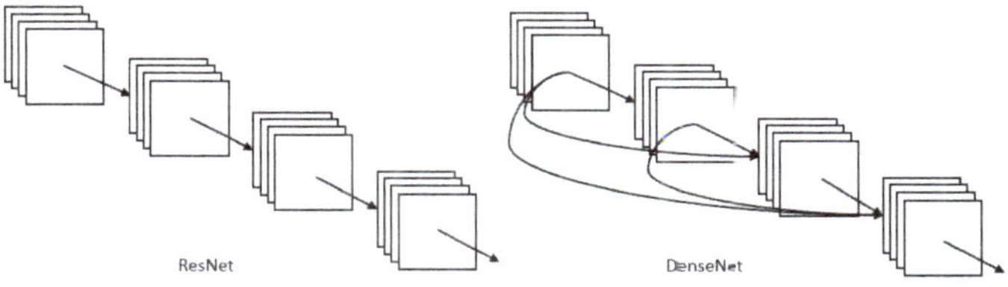

Figure 6. DenseNet block vs. ResNet block.

3.1.7. UNet

UNet [43] is an architecture originally developed for biomedical image segmentation and is now one of the most popular approaches in semantic segmentation tasks. UNet is a U-shaped encoder-decoder network architecture consisting of four encoder blocks and four decoder blocks that are connected via a bridge (Figure 7). The encoder network (contracting path) acts as the feature extractor and learns an abstract representation of the input image through a sequence of the encoder blocks. It halves the spatial dimensions and doubles the number of filters at each encoder block. The decoder network takes the abstract representation and generates a semantic segmentation mask. It doubles the spatial dimensions and half the number of feature channels.

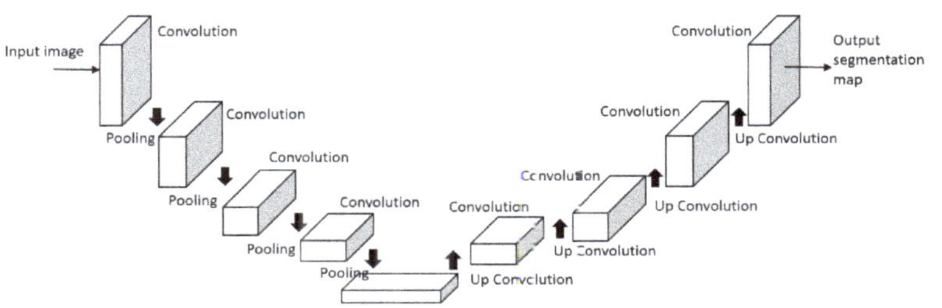

Figure 7. UNet architecture.

3.1.8. Mask R-CNN

Mask Region-based CNN (mask R-CNN) [47] is the state-of-the-art in terms of image segmentation. It detects objects in an image and generates a high-quality segmentation mask for each instance. Mask R-CNN can deal with two types of image segmentation: semantic segmentation separates the subjects of the image from the background without differentiating object instances; and instance segmentation accentuates the subjects by detecting all objects in the image while segmenting each instance. The R-CNN is a type of model that utilizes bounding boxes across the object regions and then evaluates CNNs independently on all the Regions of Interest (RoI) to classify multiple image regions into the proposed classes. An improved version of R-CNN is Fast R-CNN [55] which extracts features using RoI Pooling from each candidate box and performs classification and bounding-box regression. Faster R-CNN [44] was then designed to add the attention mechanism with a region proposal network to the Fast R-CNN architecture. Mask R-CNN is an extension of Faster R-CNN by adding a branch for predicting an object mask in parallel with the existing branch for bounding box recognition (Figure 8). It outputs a class label, a bounding-box offset, and the object mask, where the mask output requires the extraction of a fine spatial layout of an object. The key element of Mask R-CNN is the pixel-to-pixel alignment, which is the main missing piece of Fast/Faster R-CNN. Mask R-CNN is simple to implement and train given the Faster R-CNN framework, which facilitates a wide range of flexible architecture designs. Additionally, the mask branch only adds a small computational overhead, enabling a fast system and rapid experimentation.

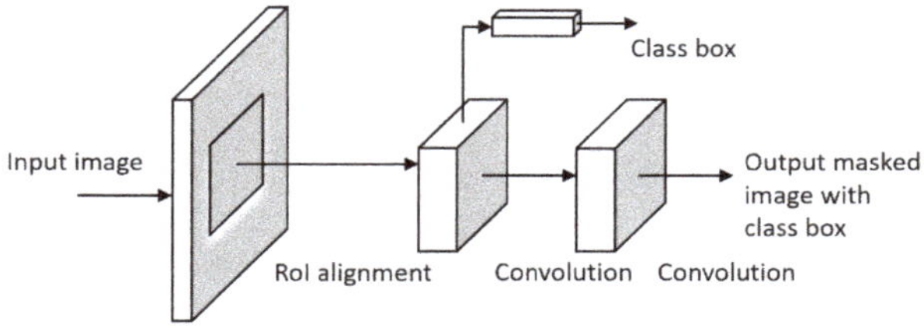

Figure 8. Mask R-CNN model.

3.1.9. YOLO

YOLO [46] is a popular model for real-time object detection, which concerns what and where objects are inside a given image. The algorithm applies a single neural network to the full image, and then divides the image into regions and predicts bounding boxes and probabilities for each region. These bounding boxes are weighted by the predicted probabilities. YOLO is popular because it achieves high accuracy while also being able to run in real-time. The algorithm 'only looks once' at the image in the sense that it requires only one forward propagation pass through the neural network to make predictions. After non-max suppression (which makes sure the object detection algorithm only detects each object once), it then outputs recognized objects together with the bounding boxes. With YOLO, a single CNN simultaneously predicts multiple bounding boxes and class probabilities for those boxes. It trains on full images and directly optimizes detection performance.

3.2. Recurrent Neural Network (RNN)

3.2.1. Introduction of RNN

The RNN is a type of artificial neural network that is especially suitable for processing sequential information such as natural languages or time series data such as videos [56,57]. Applications of RNNs include handwriting recognition [58], speech recognition [59], gesture recognition [60], image captioning [61], natural language processing [62] and understanding [63], sound event prediction [64], tracking and monitoring [65–69], etc.

Unlike traditional neural networks, the RNN can exploit sequential information by means of a connection that acts as feedback to prior layers (Figure 9). The most distinguished characteristic of an RNN is that it has memory, taking information from prior inputs to influence the current input and output. Because of this unique characteristic, an RNN can remember important information of the input, which allows it to predict with great precision what will happen next. That is why the RNN is the method of choice for processing sequential data. Another salient characteristic of the RNN is that it shares the same weight parameters within each layer of the network, whereas a normal feed-forward network has different weights on each node.

The RNN employs the back-propagation through time (BPTT) algorithm to adjust and fit the parameters of the model [70–72]. BPTT is almost the same as the standard BP, except that it sums errors at each time step, while BP does not need to sum errors because it does not share parameters between each layer. This also makes the RNN have two main issues of vanishing gradients and exploding gradients [73]. In other words, gradients may decay or explode exponentially due to the multiplications of a large number of small or large gradients during training over time. Therefore, the RNN tends to forget the previous inputs as the new inputs come in. One solution to these issues is to clip the gradient and scale the gradient. Long Short-Term Memory (LSTM) [63] (see below) is proposed to handle this issue by providing memory blocks in its recurrent connections.

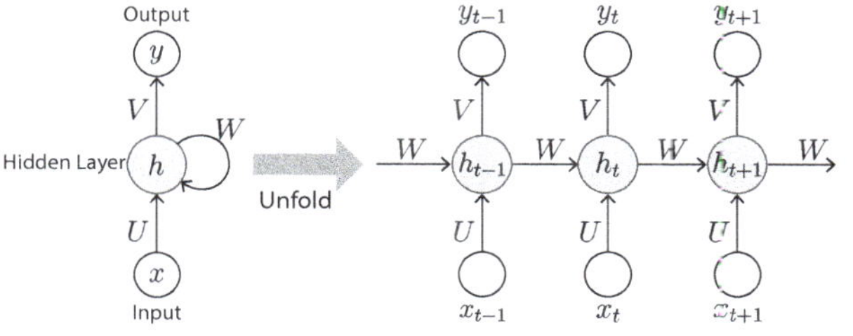

Figure 9. Diagram of an RNN. x, y represent the input and output, respectively, t is the time, h is the memory unit of the network, and U, V, and W are weight matrices.

3.2.2. Bidirectional Recurrent Neural Network (BRNN)

The BRNN was firstly invented in 1997 by Schuster and Paliwal for increasing the amount of input information available to the network [74]. It is a variant architecture of the RNN. While the classical RNN can learn only from previous layers to predict the current state, the BRNN learns from future data to improve its accuracy. This is achieved by a structure of connecting two hidden layers of opposite direction to the same output (Figure 10). BRNNs are especially beneficial in cases where the context of the input is required. For example, in handwriting recognition, performance can be improved by knowing the letters before and after the current letter [75]. The BRNN is more common in supervised learning rather than semi-supervised or unsupervised learning because it is difficult to compute a reliable probabilistic model.

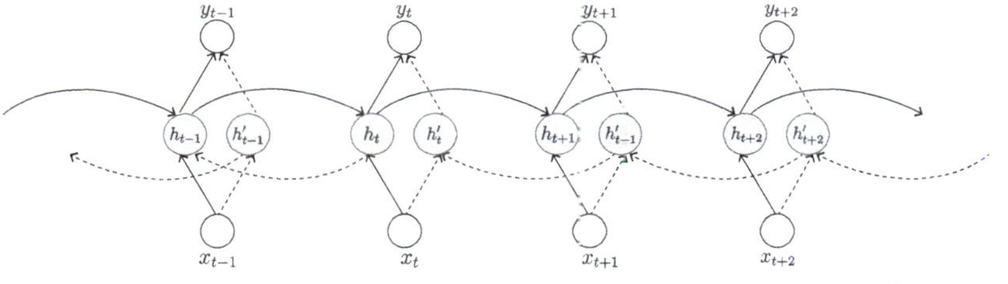

Figure 10. Diagram of a BRNN. x, y represent the input and output, respectively, h, h' represent the two bidirectional memory units, and t is the time. Solid arrows represent data forward propagation, and dashed arrows represent data back propagation.

The training of a BRNN is similar to the BPTT algorithm. However, since there are forward and backward passes, simultaneously updating the weights for the two processes leads to erroneous results. Therefore, to update forward and backward passes separately, the forward and backward states are firstly processed in the forward pass, and then the output values are passed. Subsequently, the reverse takes place for the backward pass; that is, the output values are processed first, and then the forward and backward states are processed. Finally, the weights are updated after the completion of both forward and backward passes.

3.2.3. Long Short-Term Memory (LSTM)

The LSTM was proposed by Hochreiter and Schmidhuber, and has been widely used for many applications [76]. It is an improved version of RNN, with the memory blocks (also called cells) able to let new information in, forget information, and give information enough importance to affect the output. It uses a mechanism of 'gates' for controlling its memory process (Figure 11). There are three gates: input gate, output gate, and forget gate. The input gate is responsible for accepting new information and information from the previous hidden state. The forget gate is responsible for deciding the storage or removal of information based on the learned weights. The output gate is responsible for determining the value of the next hidden state. This gate mechanism regulates the flow of information in the RNN and resolves the short-term memory issue, thus enabling an RNN to hold its value for a sufficient amount of time.

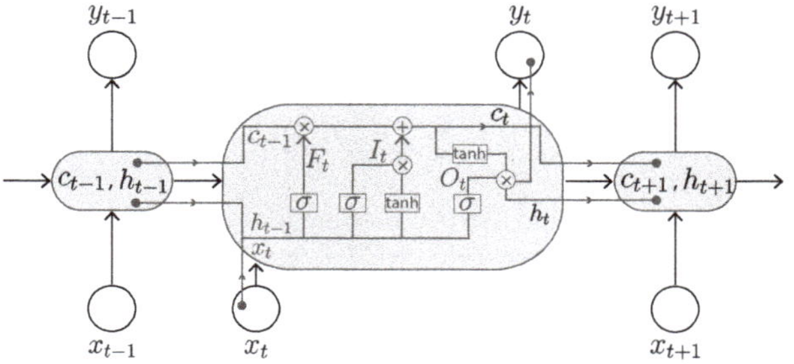

Figure 11. Diagram of a LSTM memory cell. c, h are the cell state and hidden state, respectively, t is the time, and F_t, I_t, O_t are the forget gate, input gate, and output gate, respectively.

The gates in the LSTM are modeled in the form of sigmoid function. To decide which information can pass through and what information can be discarded, the short-term memory and input pass through the sigmoid function, which transforms the values to be between 0 and 1, where 0 indicates the information is unimportant and 1 indicates the information is valuable. The use of the sigmoid function also guarantees that the gates can be back-propagated. The LSTM keeps the gradients steep enough and thus solves the issue of vanishing gradients in RNNs. This also makes its training comparatively short and its accuracy comparatively high.

3.2.4. Gated Recurrent Unit (GRU)

The GRU, proposed by Cho et al. in 2014 [56], is also a variant of RNN and is very similar to the LSTM and, in some cases, produces equally good results [77]. It has two gates, an update gate and a reset gate (Figure 12), rather than three gates as in LSTM. The reset gate is responsible for the short-term memory and controls what information goes out or is discarded. The update gate is responsible for long-term memory and regulates information to be retained from previous memory as well as the new memory to be added. In addition, the GRU uses hidden states rather than separate cell states in LSTM to regulate the flow of information. Therefore, due to the reduced number of parameters and its simpler architecture, GRU is faster to train with high effectiveness and accuracy. The GRU is also able to address the short-term memory problem of RNN and to effectively hold long-term dependencies in sequential data.

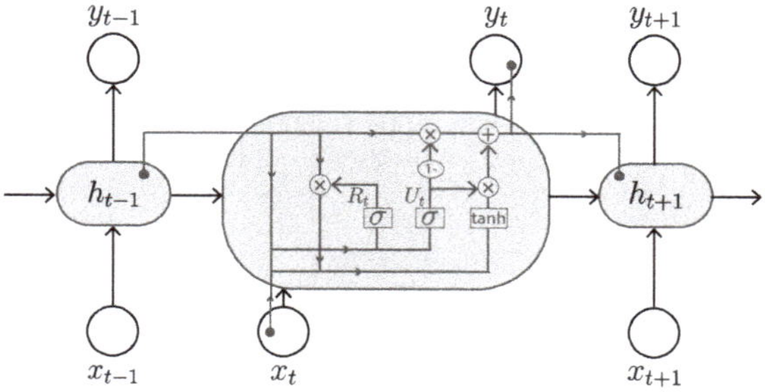

Figure 12. Diagram of a GRU memory cell. $xandy$ are the input and output, respectively, h is the hidden state, t is the time, and R_t and U_t are the reset gate and update gate.

3.2.5. RNN with Attention

Introducing attention to RNNs is probably the most significant innovation in sequential models in recent times. Attention refers to the ability of a model to focus on specific elements in the data. As mentioned, RNNs try to remember the entire input sequence through a hidden unit before predicting the output. However, compressing all information into one hidden unit may lead to information loss, especially for long sequences. To help the RNN focus on the most important elements of the input sequence, the attention mechanism assigns different attention weights to each input element. These attention weights designate how important or relevant a given input sequence element is at a given time step.

The first attention mechanism developed for RNNs was proposed by Bahdanau et al. [78] in 2014, who used it for language translation. Later, several RNN variants with attention mechanism were proposed. Examples include the dual state attention based RNN for time series prediction [79], the attention based GRU for visual question answering [80], and the outstanding attention-LSTM for Google's neural machine translation system [81]. The success of attention-LSTM has inspired more research of neural networks based on attention mechanism, and with more and more powerful computing resources becoming available, state-of-the-art models now typically use a memory-hungry architectural style called transformers (Section 3.7).

3.3. *AutoEncoder (AE)*

3.3.1. Introduction of AE

The AE is a type of artificial neural network that can learn data representation in an unsupervised manner [13]. It is a specific type of feed-forward neural network where the input is the same as the output. Its aim is to learn a low-dimensional representation (also called latent-space representation or encoding) of high-dimensional data by training the network to capture the most important elements of the inputs, usually for dimensionality reduction. By using it as an encoding and decoding technique, and combing it with other DNNs such as CNN and RNN, the AE concept has been extensively applied for data (images, audio, etc.) denoising [82,83], information retrieval [84,85], image inpainting and enhancement [86,87], and anomaly detection [88,89].

A classical AE consists of three components named encoder, code, and decoder (Figure 13). The encoder maps the input data to the feature space and produces the code, while the decoder then reconstructs the data by mapping this code back to the data space. The encoder is essentially a fully-connected neural network (though other types of networks such as CNNs can also be used), and the decoder has a similar mirror network structure as the encoder. The code is a compressed representation of the input and is important to prevent the AE from memorizing the input and overfitting on the data.

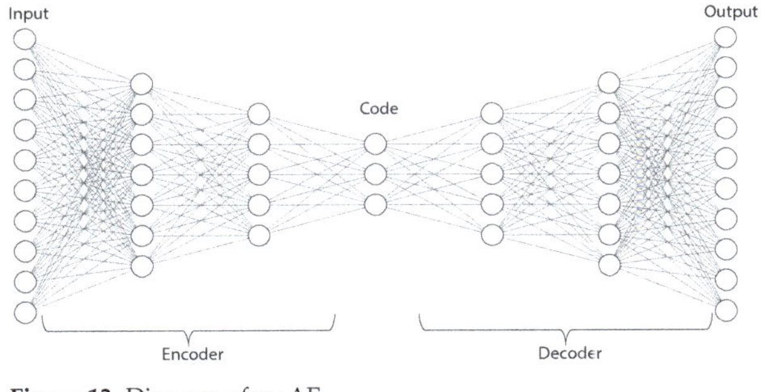

Figure 13. Diagram of an AE.

Since the goal of an AE is to get an output identical to the input, it can be trained by minimizing a reconstruction loss formulated as:

$$L_A(x, \hat{x}) = ||x - \hat{x}||^2, \quad (1)$$

where x is the input and $\hat{x}$ is the corresponding reconstruction by the AE. It is trained the same way as a DNN via BP, and also has the vanishing gradient problem because gradients may become too small as they go back through many layers of the AE.

3.3.2. Sparse AE (SAE)

The SAE is a regularized AE proposed by Ranzato et al. [90] to learn sparse representations. It is used to learn latent representations instead of redundant information of the input data, and has been shown to improve performance on classification tasks. A SAE selectively activates regions of the network, depending on the input data. As a result, it is restrained to memorize the input data but can effectively extract features from the data. More specifically, a SAE adds a nonlinear sparsity between its encoder and decoder to force the code vector into a quasi-binary sparse code. There are two ways to impose this sparsity regularization, and both are adding a constraint term to the loss function. By adding an L1 regularization as the constraint term, the loss function is formulated as:

$$L_S(x, \hat{x}) = L(x, \hat{x}) + \alpha \sum |a^h|, \quad (2)$$

where $L(x, \hat{x})$ is computed using Equation (1), α is the parameter to control the regularization strength, and a is the activation of the hidden layer h. By adding a KL-divergence as the constraint term, the loss function is formulated as:

$$L_S(x, \hat{x}) = L(x, \hat{x}) + \beta \mathrm{KL}(\rho||\hat{\rho}), \quad (3)$$

where $L(x, \hat{x})$ is computed using Equation (1), β is the parameter to control the regularization strength, $\hat{\rho}$ is the average activation of the code over the input data, ρ is a sparsity hyperparameter, and $\mathrm{KL}(\rho||\hat{\rho})$ is the KL divergence of $(\rho||\hat{\rho})$, with minimum at $\hat{\rho} = \rho$.

3.3.3. Contractive AE (CAE)

The CAE is another variant of the classical AE, which adds a contractive regularization to the code to improve its feature representation capability [91]. Its basic principle is that similar inputs should have similar encodings and similar latent space representations. To this end, CAE requires the derivative of the hidden layer activations to be small with respect to the input. Thus, the mapping from the input to the representation will converge with higher probability. The loss function of the CAE is defined as:

$$L_C(x, \hat{x}) = L(x, \hat{x}) + \gamma ||J(x)||_F^2, \quad (4)$$

where $L(x, \hat{x})$ is computed using Equation (1), γ is the parameter to control the regularization strength, $J(x)$ represents the Jacobian matrix of the encoder, and $||J(x)||_F^2$ is the square of the Frobenius norm of the Jacobian matrix. It is worth mentioning that these two terms in the CAE loss function contradict each other. While the reconstruction loss $L(x, \hat{x})$ aims to distinguish the difference between two inputs and observe changes in the data, the regularization $||J(x)||_F^2$ aims to allow the model to ignore changes in the input data. However, a loss function with these two terms enables the hidden layers of the CAE to capture only the most essential information.

3.3.4. Denoising AE (DAE)

The DAE was originally proposed by Vincent et al. [92,93] based on the AE for removing noise of the input. Now, it has become an important and essential tool for feature extraction and selection. Different from the above types of AEs, the DAE does not have the input image as its ground truth. Its basic idea is to slightly corrupt the input data but still use the uncorrupted data as target output. This way, it can force the DAE to recover a noise-free version of the input data. Furthermore, a DAE model cannot simply learn a map that memorizes the input and overfits the data because the input and target output are no longer the same. Essentially, a DAE gets rid of noise with the help of nonlinear dimensionality reduction. The loss function used by the DAE is expressed as:

$$L_D(x, \hat{x'}) = ||x - \hat{x'}||^2, \tag{5}$$

where x' is the corrupted version of input x, and $\hat{x'}$ is the reconstruction by the DAE. A DAE can exploit the statistical dependencies inherent in the input data and remove the detrimental effects of noisy inputs.

3.3.5. Variational AE (VAE)

While AEs can learn a representative code from the input data and reconstruct the data from this compressed code, the distribution of this compressed code remains unknown and cannot be expressed in a probabilistic fashion. The VAE [94] is designed to handle this issue and learn to format the code as a probability distribution. This way, the learned code can be easily sampled and interpolated to generate new unseen data. Therefore, the VAE is a kind of deep generative model. The VAE makes the code to be a Gaussian distribution, so that the encoder can be trained to return its mean μ and variance σ^2. The loss function for VAE training is defined as:

$$L_V(x, \hat{x}) = L(x, \hat{x}) + \mathrm{KL}(N(\mu, \sigma), N(0, 1)), \tag{6}$$

where $L(x, \hat{x})$ is computed using Equation (1), $\mathrm{KL}(N(\mu, \sigma), N(0, 1))$ is a regularization term on the learned code to force the distribution of the extracted code to be close to a standard normal distribution. The reason why an input is encoded as a distribution with some variance rather than a single point is that it expresses the latent space regularization very naturally. Sampling from this latent distribution and feeding it to the decoder can lead to new data being generated by the VAE.

3.4. Restricted Boltzmann Machine (RBM)

The RBM was invented by Hinton in 2007 for learning a probability distribution over its set of inputs [95]. It is a generative stochastic artificial neural network that has wide applications in different areas such as dimensionality reduction [96], classification [97], regression [98], collaborative filtering [99], feature learning [100], and topic modeling [101].

A classical RBM has two layers, named visible layer and hidden layer (Figure 14). The visible layer has input nodes to receive input data, while the hidden layer is formed by nodes that extract feature information from the data and output a weighted sum of the input data [102]. An important and unique characteristic of the RBM is that the output generated by the hidden layer is further processed to become a new input to the visible layer. This process is called reconstruction or backward pass, and is repeated until the regenerated input is aligned with the original input. This way, an RBM is able to learn a probability distribution over the input. In an RBM, there is no typical output layer. In addition, every node can be connected to every other node, and there are no connections from visible to visible or hidden to hidden nodes.

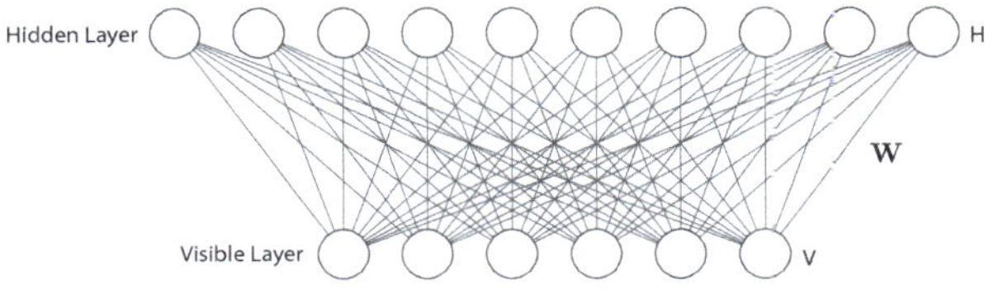

Figure 14. Diagram of an RBM. V, H, and $\mathbf{W}$ represent the state vector of the visible layer, the state vector of the hidden layer, and the weight matrix between hidden and visible layers, respectively.

An RBM is also a generative model. It represents a probability distribution by the connection weights learned from the data. Denote the m visible nodes as $V = (v_1, v_2, \ldots, v_m)$ and n hidden nodes as $H = (h_1, h_2, \ldots, h_n)$. In a binary RBM, the random variables (V, H) take values $(v, h) \in \{0, 1\}^{m+n}$, and the joint probability distribution is given by the Gibbs distribution $p(v, h) = 1/Ze^{-E(v,h)}$ with the energy function defined as [103]:

$$E(v,h) = -\sum_{i=1}^{n}\sum_{j=1}^{m} w_{ij}h_i v_j - \sum_{i=1}^{n} b_i h_i - \sum_{j=1}^{m} c_j v_j, \tag{7}$$

where $Z = \sum_{v,h} e^{-E(v,h)}$ is the normalization factor, $i \in 1, 2, \ldots, n$ and $j \in 1, 2, \ldots, m$, w_{ij} is a weight associated with the edge between nodes v_j and h_i, and b_i and c_j are biases associated with the ith visible and the jth hidden variable, respectively. The RBM has proven to be capable of achieving highly expressive marginal distributions [104].

3.5. Generative Adversarial Network (GAN)

3.5.1. Introduction of GAN

The GAN was firstly proposed by Goodfellow et al. [105] and has become one of the most popular generative adversarial models. Its purpose is to learn the distribution of input data and thus enable the network to generate new data from that same distribution. Since the GAN was proposed, it has gained much attention in various areas such as synthetic training data [106], image and audio style transfer [107], music generation [108], text to image generation [109], super-resolution [110], semantic segmentation [111], natural language processing [112], and predicting the next frame in a video [113].

A GAN is basically composed of two neural networks, named generator and discriminator (Figure 15). The generator takes a random vector sampled from a noise distribution as input and generates samples. The discriminator takes the generated samples and real samples as input and tries to distinguish them as real or fake. These two networks compete with each other. The goal of the generator is to generate fake samples that are hard for the discriminator to distinguish from real samples. The goal of the discriminator is to beat the generator by identifying whether its received samples are fake or real. This competition between the generator and discriminator goes on until the generator manages to generate fake samples that the discriminator cannot distinguish from real ones.

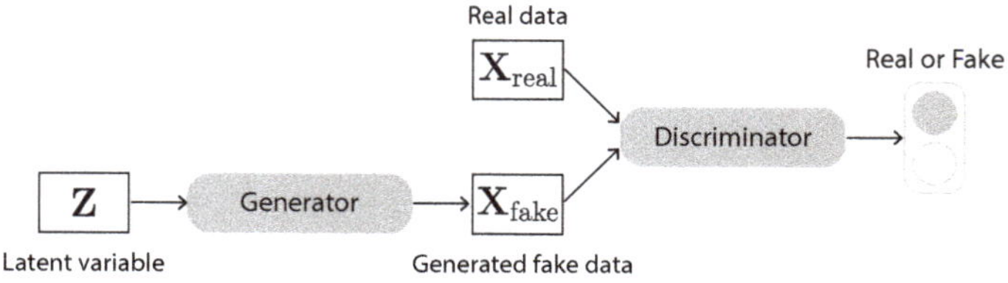

Figure 15. Diagram of a GAN.

This zero-sum game is modeled as an optimization problem by:

$$\min_G \max_D L(D, G), \tag{8}$$

where D and G denote the generator and discriminator, respectively, and

$$L(D, G) = \mathbb{E}_{x \sim p_{\text{data}}(x)}[\log(D(x))] - \mathbb{E}_{x \sim p_z(z)}[1 - \log(D(G(z)))], \tag{9}$$

where x is the input data, $p_{\text{data}}(x)$ is the distribution of input data, and z is noise from a distribution $p_z(z)$. The GAN is trained in an alternative way of firstly maximizing the discriminator loss and then minimizing the generator loss. Both generator and discriminator employ independent back-propagation procedures. In this way, GANs have the ability to learn the data distribution in an unsupervised manner.

3.5.2. Deep Convolutional GAN (DCGAN)

The DCGAN, proposed by Radford et al. [114], is a convolution-based GAN. It is one of the most powerful and successful types of GAN, and has been widely used in many convolution-based generation-based techniques. Compared to GAN, the DCGAN uses convolutional and convolutional-transpose layers to implement its generator and discriminator, and this is the origin of its name. Another interesting characteristic of DCGAN is that, unlike the typical neural networks to map input to a binary output, or a regression output, or even a categorical output, the generator of a DCGAN can map from random noise to images. For example, the generator of the DCGAN in [114] takes in a noise vector of size 100×1 and maps it into an output image of size $64 \times 64 \times 3$ (Figure 16). The DCGAN can be used to generate images as 'real' as possible from a distribution.

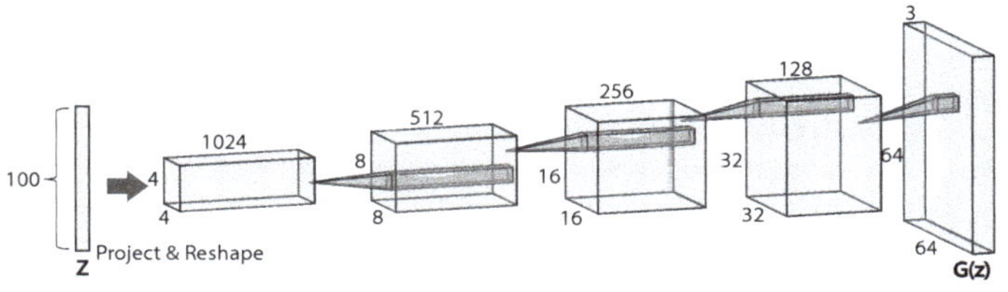

Figure 16. Diagram of a DCGAN.

3.5.3. Conditional GAN (cGAN)

The cGAN (Figure 17) is a type of GAN whose generator and discriminator are conditioned on some auxiliary information from other modalities [115]. As a result, it can learn multimodal mapping from inputs to outputs by feeding it with different contextual information. In other words, a cGAN allows us to guide the generator to generate the kind of fake samples we want. The input to the auxiliary layer can be class labels or some other properties we expect from the generated data. As the cGAN uses some kind of labels for it to work, it is not a strictly unsupervised learning algorithm. The advantages of using additional information are (1) the convergence will be faster and (2) the generator can generate specific output given a certain label.

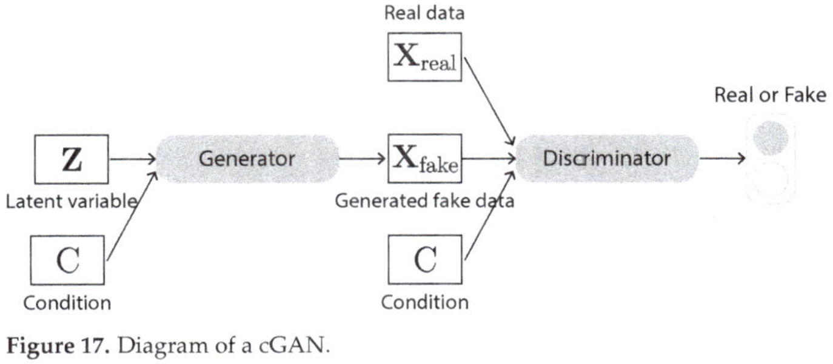

Figure 17. Diagram of a cGAN.

3.5.4. Other Types of GANs

Other well-known types of GANs include Info GAN (also called iGAN) [116], Auxilary Classifier GAN (ACGAN) [117], Stacked GAN [118], Wasserstein GAN [119], Cycle GAN [120], and Progressive GAN [121].

(1) The Info GAN is a modified GAN that aims to learn interpretable and meaningful representations. To this end, it splits the input of the generator into two parts: the typical noise and a new "latent code" which is composed of control variables. The code is then made meaningful by maximizing the mutual information between the code and the generated output. This way, the generator can be trained by using the control variables to affect specific properties of the generated outputs.

(2) The ACGAN is similar to the cGAN because both their generators take noise and labels as input. However, the ACGAN has an auxiliary class label output compared to the cGAN. Therefore, the ACGAN can be seen as an extension of the cGAN. It has the effect of stabilizing the training process and allowing the generation of large, high-quality images, while learning representations in a latent space independent of class labels.

(3) The Stacked GAN is an extension of the GAN for generating images from text by a hierarchical stack of cGANs. Its architecture is composed of a set of text-conditional and image-conditional GANs. More specifically, the first-level generator is conditioned on text and generates a low-resolution image. The second-level generator is conditioned on both the text and the low-resolution image and outputs a high-resolution image.

(4) The Wasserstein GAN is an advanced GAN that aims to better approximate the distribution of data observed in a given training dataset. To this end, it uses a critic rather than a discriminator to scores the realness or fakeness of a given image. Its underlying idea is to let the generator minimize the distance between the distribution of the data in the training dataset and the distribution of the generated samples. The advantage of Wasserstein GAN is that its training process is more stable and less sensitive to model architecture and hyperparameter configurations.

(5) The Cycle GAN is an advanced GAN proposed for image-to-image translation. Its outstanding characteristic is that it learns mapping between inputs and outputs using an unpaired dataset. The Cycle GAN simultaneously trains two generators and two discriminators. One generator is responsible for generating images for the resource domain learned from, and the other is responsible for generating images for the target domain. Each generator has a corresponding discriminator.

(6) The Progressive GAN is proposed for stable training and large-scale high-resolution image generation. Similar to a GAN, the Progressive GAN consists of a generator and a discriminator, which are symmetrical to each other. Its key feature is to progressively grow the generator and discriminator, starting from a low resolution, and then adding new layers to increase the model's fine details as training progresses. As a result, training is faster and more stable, producing images of unprecedented quality.

3.6. Graph Neural Network (GNN)

Graph neural networks are a class of neural networks that operate on the graph structure, where data are generated from non-Euclidean domains and represented as graphs with complex relationships and interdependencies between nodes [122]. Examples of graph data include social networks, citation networks, molecular structures, and many other types of data that are organized in a graph format.

A graph is represented as $G = (V, E)$, where V is the set of vertices or nodes, and E is the set of edges. Let $v_i \in V$ denote a node and $e_{ij} = (v_i, v_j) \in E$ denote an edge pointing from v_i to v_j. The neighborhood of a node v is defined as $N(v) = \{u \in V | (v, u) \in E\}$. The adjacency matrix $\boldsymbol{A}$ is an $n \times n$ matrix with $A_{ij} = 1$ if $e_{ij} \in E$ and $A_{ij} = 0$ if $e_{ij} \notin E$. A graph may have node attributes $\boldsymbol{X}$, where $\boldsymbol{X} \in \mathbb{R}^{n \times d}$ is a node feature matrix with $\boldsymbol{x}_v \in \mathbb{R}^d$ representing the feature vector of a node v. Furthermore, a graph may have edge attributes $\boldsymbol{X}^e$, where $\boldsymbol{X}^e \in \mathbb{R}^{m \times c}$ is an edge feature matrix with $\boldsymbol{x}^e_{v,u} \in \mathbb{R}^c$ representing the feature vector of an edge (v, u). A directed graph is a graph with all edges directed from one node to another. An undirected graph is considered as a special case of directed graphs, where there is a pair of edges with inverse directions if two nodes are connected. A graph is undirected if and only if the adjacency matrix is symmetric. A spatial–temporal graph is an attributed graph where the node attributes change dynamically over time. The spatial–temporal graph is defined as $G^{(t)} = (V, E, \boldsymbol{X}^{(t)})$ with $\boldsymbol{X}^{(t)} \in \mathbb{R}^{n \times d}$.

There are three general types of analytics tasks on graphs: graph-level, node-level, and edge-level. In a graph-level task, the goal is to predict a single property for an entire graph [123]. This is often referred to as a graph classification task, as the entire graph is associated with a label. To obtain a compact representation on the graph level, GNNs are often combined with pooling and readout operations [124–126]. Node-level tasks are concerned with predicting the identity or role of each node in a graph [127], and therefore, the model outputs relate to node regression and node classification tasks. Recurrent GNNs and convolutional GNNs can extract high-level node representations by information propagation and graph convolution. With a multiperceptron or a softmax layer as the output layer, GNNs are able to perform node-level tasks in an end-to-end manner. Similarly, an edge-level task predicts the property or presence of edges in a graph, hence the outputs relate to the edge classification and link prediction tasks. With two nodes' hidden representations from GNNs as inputs, a similarity function or a neural network can be utilized to predict the label/connection strength of an edge.

Based on the model architectures, GNNs can be categorized into recurrent graph neural networks, convolutional graph neural networks, graph autoencoders and generative graph neural networks, and spatial-temporal graph neural networks.

3.6.1. Recurrent Graph Neural Network (RecGNN)

RecGNNs aim to learn node representations with recurrent architectures. A representative model in this class is the GNN proposed by Scarselli et al. [128], which updates the states of nodes by exchanging neighborhood information recurrently until a stable equilibrium is researched, as in the following equation:

$$h_v^{(t)} = \sum_{u \in N(v)} f\left(x_v, x^e_{(v,u)}, x_u, h_u^{(t-1)}\right), \tag{10}$$

where $f(\cdot)$ is the parametric function and $h_v^{(0)}$ is the initial state randomly set. Other popular RecGNNs include the GraphESN [129] which extends echo state networks to improve the training efficiency of GNN, and the Gated GNN [130] which employs a gated recurrent unit as the recurrent function that reduces the recurrence to a fixed number of steps. RecGNNs are conceptually important and inspired later research on ConvGNNs. In particular, the idea of information passing is inherited by spatial-based ConvGNNs.

3.6.2. Convolutional Graph Neural Network (ConvGNN)

ConvGNNs generalize the operation of convolution from grid data to graph data. The main idea is to generate a representation of a node v by aggregating its own features x_v and neighbors' features x_u, where $u \in N(v)$. Different from RecGNNs, ConvGNNs stack multiple graph convolutional layers to extract high-level node representations. ConvGNNs play a central role in building up a great deal of other complex GNN models. ConvGNNs can be further divided into spectral-based methods and spatial-based methods: the first category defines graph convolutions by introducing filters from the perspective of graph signal processing [131], and the latter inherits ideas from RecGNNs to define graph convolutions by information propagation.

Spectral-based methods have a solid mathematical foundation in graph signal processing, and they are based on the normalized graph Laplacian matrix which is a mathematical representation of an undirected graph, defined as $L = I_n - D^{-1/2}AD^{-1/2}$, where D is a diagonal matrix of node degrees. This normalized Laplacian matrix can be factored as $L = U\Lambda U^T$, where Λ and U denote the ordered diagonal matrix of eigenvalues and the corresponding eigenvector matrix, respectively. The graph convolution of an input signal x with a filter $g \in \mathbb{R}^n$ is then defined as:

$$x *_G g = \mathscr{F}^{-1}(\mathscr{F}(x) \odot \mathscr{F}(g)) = U(U^T x \odot U^T g) \tag{11}$$

where $\odot$ denotes the element-wise product, and $\mathscr{F}(x)$ is the graph Fourier transform of the signal x. Let $g_\theta = \mathrm{diag}(U^T g)$ denote a filter, the spectral graph convolution is simplified as:

$$x *_G g_\theta = U g_\theta U^T x. \tag{12}$$

Popular spectral-based GNNs inlcude the Spectral CNN [132], ChebNet [125] and GCN [127], where the key difference lies in the design of the filter g_θ.

The spatial-based graph convolution is defined on the nodes' spatial relations, and it convolves a node's representation with its neighbors' representations to derive the updated representation, inheriting the idea of information propagation of RecGNNs. Representative spatial-based GNNs include the Diffusion CNN [133], message-passing neural network (MPNN) [134], GraphSage [135], and graph attention network (GAT) [136] (which brings in attention mechanisms), mixture model network (MoNet) [137], and FastGCN [138]. Since GCN [127] bridged the gap between spectral-based approaches and spatial-based approaches, spatial-based methods have developed rapidly recently due to their attractive efficiency, flexibility, and generality.

3.6.3. Graph Autoencoder (GAE) and Other Generative Graph Neural Networks

GAEs and generative GNNs are unsupervised learning frameworks that encode nodes into a latent vector space and decode graph information from the latent representations. GAEs are used to learn network embeddings and graph generative distributions. A network embedding is a low-dimensional vector representation of a node that preserves a node's topological information. For network embedding, GAEs learn latent node representations through reconstructing graph structural information, such as the graph adjacency matrix. Representative GAEs for network embedding include the DNGR [123], SDNE [139], GAE [140], Variational GAE [140], and GraphSage [135]. These models combine different AEs and other models such as ConvGNNs and LSTM. With multiple graphs, GAEs are able to learn the generative distribution of graphs by encoding graphs into hidden representations and decoding a graph structure given hidden representations. The majority of GAEs for graph generation are designed to solve the molecular graph generation problem [141], which has a high practical value in drug discovery. These methods either propose a new graph sequentially, such as DeepGMG [142] and GraphRNN [143], or in a global manner, such as GraphVAE [144]. GNNs are also integrated with the architecture and training strategy of GANs, resulting in MolGAN [145] and NetGAN [146].

3.6.4. Spatial–Temporal Graph Neural Network (STGNN)

Graphs in many real-world applications are dynamic, both in terms of graph structures and graph inputs. STGNNs occupy important positions in capturing the dynamics of graphs. The task of STGNNs can be forecasting future node values or labels, or predicting spatial–temporal graph labels. STGNNs capture spatial and temporal dependencies of a graph simultaneously. Current approaches integrate graph convolutions to capture spatial dependence with RNNs or CNNs to model temporal dependence. Most RNN-based approaches capture spatial–temporal dependencies by filtering inputs and hidden states passed to a recurrent unit using graph convolutions [147]. As alternative solutions, CNN-based approaches tackle spatial–temporal graphs in a non-recursive manner with the advantages of parallel computing, stable gradients, and low memory requirements. CNN-based approaches interleave 1-D-CNN layers with graph convolutional layers to learn temporal and spatial dependencies, respectively, as in the CGCN [148].

3.6.5. Training of GNNs

Given a single network with part of the nodes labeled and others unlabeled, ConvGNNs can be trained in a semi-supervised manner to learn a robust model that effectively identifies the class labels for the unlabeled nodes [127]. To this end, an end-to-end framework can be built by stacking a couple of graph convolutional layers followed by a softmax layer for multiclass classification. In addition, GNNs can be trained in a supervised manner for graph-level classification, which is achieved by applying the graph pooling layers and readout layers [123]. Finally, GNNs can learn graph embedding in a purely unsupervised manner in an end-to-end framework (e.g., an AE framework [140]).

3.7. Transformer

The transformer [149] is a prominent type of deep learning models that has achieved impressive advances on various tasks such as computer vision and audio processing. Originally proposed for natural language processing, the transformer mainly relies on deep neural networks and the self-attention mechanism, emphasizing the global dependencies between the input and output, thereby providing strong representation capability and state-of-the-art performance. Due to the significant improvement made by the transformer model, several variants have been proposed for either improving model performance or adapting the model to specific tasks in recent years.

3.7.1. Vanilla Transformer

The transformer follows the encoder-decoder structure (Figure 18). The encoder is composed of a stack of identical blocks with two modules: the multihead self-attention layers and the position-wise fully connected feed-forward network (FFN). A residual skip connection, followed by a batch normalization layer, is applied to each submodule. Besides the two modules in the encoder block, the decoder block inserts an additional masked multihead attention layer, which is specially modified to avoid positions from attending to subsequent positions. In the following, we introduce the two modules in more detail.

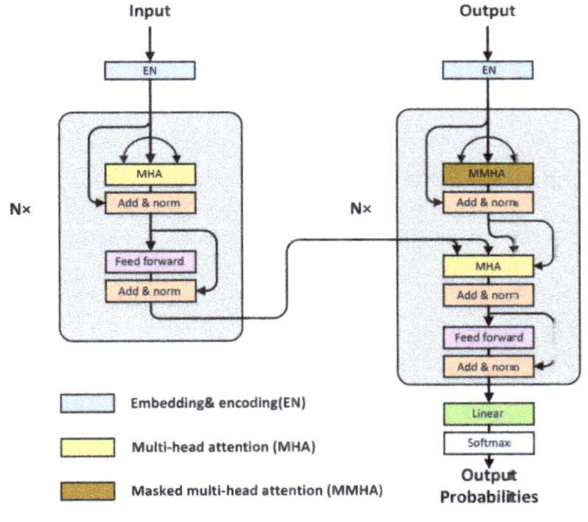

Figure 18. The model architecture of the Transformer.

(1) The multihead attention layer adopts the self-attention mechanism with the Query-Key-Value (Q-K-V) model. The inputs are first projected into three kinds of vectors: the query vector $\boldsymbol{q}$, the key vector $\boldsymbol{k}$ with dimension d_k, and the value vector $\boldsymbol{v}$ with dimension d_v. After packing a set of these vectors together into three matrices, namely queries $\boldsymbol{Q} \in \mathbb{R}^{N \times D_k}$, keys $\boldsymbol{K} \in \mathbb{R}^{N \times D_k}$, and values $\boldsymbol{V} \in \mathbb{R}^{N \times D_v}$, the scale dot-product attention can be computed as follows:

$$\text{Attention}(\boldsymbol{Q}, \boldsymbol{K}, \boldsymbol{V}) = \text{softmax}\left(\frac{\boldsymbol{Q} \cdot \boldsymbol{K}^{\top}}{\sqrt{d_k}}\right) \cdot \boldsymbol{V} = \boldsymbol{A}\boldsymbol{V}. \tag{13}$$

In this process, $\boldsymbol{Q} \cdot \boldsymbol{K}^{\top}$ computes a score between each pair of input vectors and yields the degree of attention. The produced scores are divided by $\sqrt{d_k}$ to avoid the vanishing gradient problem and improve the stability of training. The softmax operator transforms the divided scores into probabilities $\boldsymbol{A}$, which is also called the attention matrix. After multiplying values $\boldsymbol{V}$ with the attention matrix, vectors with higher probabilities receive more attention from the subsequent layers.

Rather than using a single self-attention operation, multihead attention learns h different linear projections and transforms the queries, keys, and values into h sets with D_k, D_k, D_v dimensions. Then, the self-attention operation can be implemented in parallel and produce different output values, which are subsequently concatenated and projected linearly back to D_m-dimension feature.

$$\begin{aligned} \text{Multihead}(\boldsymbol{Q}, \boldsymbol{K}, \boldsymbol{V}) &= \text{concat}(\text{head}_1, \ldots, \text{head}_h)\boldsymbol{W}^{\boldsymbol{O}} \\ \text{where head}_i &= \text{Attention}\left(\boldsymbol{Q}\boldsymbol{W}_i^{\boldsymbol{Q}}, \boldsymbol{K}\boldsymbol{W}_i^{\boldsymbol{K}}, \boldsymbol{V}\boldsymbol{W}_i^{\boldsymbol{V}}\right) \end{aligned} \tag{14}$$

where $\boldsymbol{W}_i^{\boldsymbol{Q}} \in \mathbb{R}^{D_{\text{model}} \times D_k}$, $\boldsymbol{W}_i^{\boldsymbol{K}} \in \mathbb{R}^{D_{\text{model}} \times D_k}$, $\boldsymbol{W}_i^{\boldsymbol{V}} \in \mathbb{R}^{D_{\text{model}} \times D_v}$ denote the parameters for linear projections for the Q, K, V branches, respectively. $\boldsymbol{W}_i^{\boldsymbol{O}} \in \mathbb{R}^{hD_v \times D_{\text{model}}}$ denote the parameters for linear projections after concatenation. In the vanilla transformer, $D_k = D_v = D_{\text{model}}/h = 64$ and $h = 8$.

(2) The fully connected feed-forward network consists of two linear transformations with a RelU activation function in between.

$$\text{FFN}(x) = \max(0, x\boldsymbol{W}_1 + b_1)\boldsymbol{W}_2 + b_2 \tag{15}$$

3.7.2. Transformer Variants

Motivated by the impressive success of the transformer, researchers have devoted numerous efforts to make further progress in a variety of tasks. Improvements have been achieved from three perspectives: using pretrained models (PTM), modifying the vanilla transformer architecture, and adapting to new tasks.

(1) Using pretrained models: Compared with training a model from scratch, using pretrained transformer models has been revealed to be beneficial for building up universal feature representations. Powerful PTMs help reduce the need for task-specific architectures by simple fine-tuning on the downstream datasets. Bidirectional Encoder Representations from Transformers (BERT) [150] is the first fine-tuning based model with transformer architecture for natural language understanding and pushed the performance frontier of 11 NLP tasks. Generative Pretrained Transformer (GPT) series [151,152] show that massive PTMs with large-scale parameters can help achieve strong universal representation ability and provide state-of-the-art performance on different types of tasks, even without the fine-tuning process. Bidirectional and Auto-Regressive Transformers (BART) [153] generalized the pretraining scheme and built a denoising auto-encoder model to further boost the capacity in language understanding.

(2) Modifying the vanilla transformer architecture: As self-attention is considered to be the fundamental component of the transformer, various architecture modifications have been proposed to address its limitations including computational complexity and ignorance of prior knowledge. Representative modifications including Low-rank based Sparse attention [154], linearized attention [155], improved multihead attention [156], and prior attention [157] have been designed to reduce complexity and make the most of the structural prior. Another branch of important modifications is adapting the architecture to be lightweight in terms of model size and computation, such as Lite Transformer [158], Funnel Transformer [159] and DelighT [160].

(3) Adopting to new tasks: Besides NLP, the transformer concept has been adapted in various fields, including computer vision [161–166] and multimodal data processing. For vision tasks, the transformer architecture has been extensively explored. ViT [161] is the first vanilla transformer architecture applied to image classification tasks without any alternation. It directly reshapes the image patches and flattens them into a sequence as the input. Experiments on large datasets such as ImageNet and JFT-300M show that the transformer has great potential in capturing long-range dependency and suits vision tasks well. Researchers also attempted to modify the network architecture and make it more feasible to vision tasks. Transformer in Transformer (TNT)[165], iGPT [162], and Swin Transformer [166] are representative models in this regard.

3.8. Bayesian Neural Network (BNN)

While DNNs have been shown to achieve great success in different applications, they are unable to deal with the uncertainty of a given task due to model uncertainty. This is due to their essence of using BP to approximate a minimal cost of point estimates of the network parameters, while discarding all other possible parametrizations of the network [167]. The BNN is proposed to mitigate this by providing a strict framework to train an uncertainty-aware neural network [168,169]. The application domains of BNN are very wide, including recommender systems [170], computer vision [171], natural language processing [172], speech recognition [173], biomedical applications [174], and so on.

The BNN is essentially a stochastic neural network trained using a Bayesian method [175,176]. A stochastic neural network is a type of DNN involving stochastic components into its network. The stochastic component is used to simulate multiple possible models with their associated probability distribution. The main aim of a stochastic neural network is to obtain a better idea of the uncertainty associated with the model. This is achieved by comparing the predictions of multiple models obtained by sampling the model parameterization. The uncertainty is low if these models generate consistent predictions, otherwise the uncertainty is high. This process can be formulated as:

$$y = \Phi_\theta(x) + \varepsilon, \tag{16}$$

where $\theta = (\boldsymbol{W}, \boldsymbol{b})$ are the parameters of the neural network which follow the probability distribution $p(\theta)$, and ε is the random noise used to ensure the function Φ represented by the network is only an approximation. This way, a BNN can be defined as a stochastic neural network trained using Bayesian inference [177].

The uncertainty of a neural network is a measure of how certain a model is with its prediction. With BNNs, there are two kinds of uncertainty: aleatoric uncertainty and epistemic uncertainty. The aleatoric uncertainty refers to the noise inherent in the observations, and cannot be reduced by collecting more data. The epistemic uncertainty is also known as model uncertainty and is caused by the model itself. It can be reduced by collecting more data. The BNN usually solves this issue by placing a probability distribution on the network weights or by learning a mapping from input to probabilistic outputs to derive the estimation of uncertainty. More specifically, the epistemic uncertainty is modeled by placing a prior distribution on the network weights and then capturing the degree of change of these weights over the data. The aleatoric uncertainty is modeled by placing a distribution on the outputs of the model.

One problem of BNNs is that they are hard to train. In practice, the Bayes by Backprop algorithm proposed by Blundell et al. [178] is used for learning a probability distribution on the network weights. Another problem of using BNNs is that they rely on prior knowledge, and it is challenging to derive insights about plausible parametrization for a given model before training. However, BNNs have become promising due to the following advantages. Firstly, thanks to its stochastic component, BNNs can quantify uncertainty, which means the uncertainty is more consistent with the observed errors. Moreover, BNNs are very data-efficient because they can learn from a small dataset without overfitting. This is due to the fact that they can distinguish the epistemic and aleatoric uncertainty. Finally, BNNs enable the analysis of learning methods, which is important for many fields such as traffic monitoring and medicine.

3.9. Fuzzy Deep Neural Networks (FDNN)

3.9.1. Introduction of FDNN

Typical DNNs are trained by minimizing the loss or error given an input through gradient descent-based weight update [179]. This is a calculus-based method that iteratively computes the minimum of the error function. However, obvious disadvantages of this method are that it is computationally intensive and may not find the global minimum [180]. To address this issue, multiple FDNNs have been proposed, for example, the fuzzy RBM [181] and the Takagi Sugeno fuzzy deep network [182]. As an emerging method, FDNNs have been applied in distributed systems [183], cloud computing [184], traffic control [185], healthcare [186], image processing [187], and various other areas.

A FDNN is a hybridization of DNNs and fuzzy logic methods, to solve various complex problems involving high-dimensional data. The key benefit of a DNN is its ability to learn from data, but it cannot clarify how its final output is achieved. Combined with fuzzy logic, a FDNN can interpret the results generated by the network [188]. More specifically, a FDNN introduces an additional fuzzy inference into a DNN to create an explainable rule-based structure. This way, through this rule-based structure, how a decision is made by the network is understandable.

3.9.2. Types of FDNN

A FDNN can be comprised by a broad category of DNNs and fuzzy inference systems in different architectures. Current architectures in the literature can be classified into three categories: sequential FDNN, parallel FDNN, and cooperative FDNN [189].

A sequential FDNN has a structure that passes the data through the DNN and the fuzzy inference system sequentially (Figure 19a). It is suitable for solving problems involving high linearity, such as text documents, time-series data, video classification, and speech recognition.

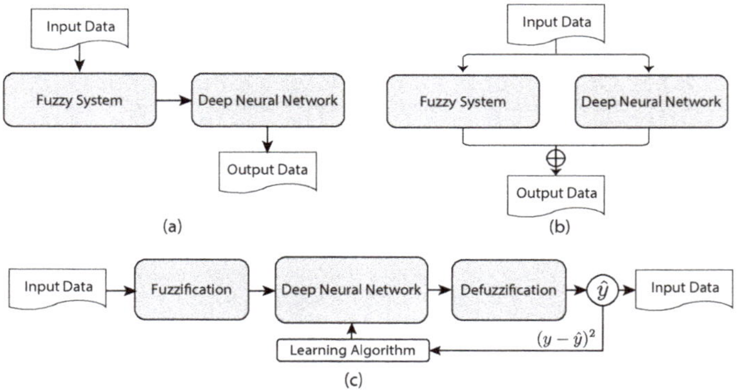

Figure 19. Diagram of three types of FDNN. (**a**) Sequential FDNN, (**b**) Parallel FDNN, and (**c**) Cooperative FDNN.

A parallel FDNN has a structure that passes the data separately through the DNN and the fuzzy inference system, and fuses the results to generate the output (Figure 19b). This kind of FDNN has been used for multiple classification tasks [190].

A cooperative FDNN has a structure where the input data are firstly passed through a fuzzy interface block to generate fuzzy values, which are subsequently input to a DNN followed by a defuzzification block to convert the fuzzy values into output data (Figure 19c). An example application of the cooperative FDNN is fuzzy classification [191].

3.10. Deep Reinforcement Learning (DRL)

A reinforcement learning (RL) agent executes a sequence of actions and observes states and rewards, with major components being the value function, policy and model. A RL problem may be formulated as a prediction, control or planning problem, and solution methods may be model-free or model-based, with value function and/or policy [192]. Exploration-exploitation is a fundamental trade-off in RL. Knowledge would be critical for RL. DRL, integrating deep learning and RL, represents a step forward in building autonomous systems with a higher-level understanding.

3.10.1. Deep Q-Network

Value function is a fundamental concept in reinforcement learning, and temporal difference learning [193] and its extension, Q-learning [194], are classic algorithms for learning state and action value functions respectively. Q-learning learns the action-value function $Q(s,a)$, i.e., how good it is to take an action a at a particular state s, to build a memory table $Q[s,a]$ that stores Q values for all possible combinations of s and a. However, if the combinations of states and actions are large, the memory and computation requirement for Q is very high. Deep Q-learning addresses this problem by generalizing the approximation of the Q-value function rather than remembering the solutions. The challenge in RL is that both the input and target change constantly during the process, which makes training unstable. Deep Q-Network (DQN) [195] ignited the field of DRL, making an important contribution in stabilizing the training of action value function approximation with DNNs using experience replay. In addition, it designs an end-to-end RL approach, with only the pixels and the game score as inputs, so that only minimal domain knowledge is required. Important extensions of DQN are the Double DQN [196] which addresses the over-estimate issue in Q-learning, and Dueling DQN [197] which uses two separate heads to compute the state value function $V(s)$ and associated advantage function $A(s,a)$ (Figure 20).

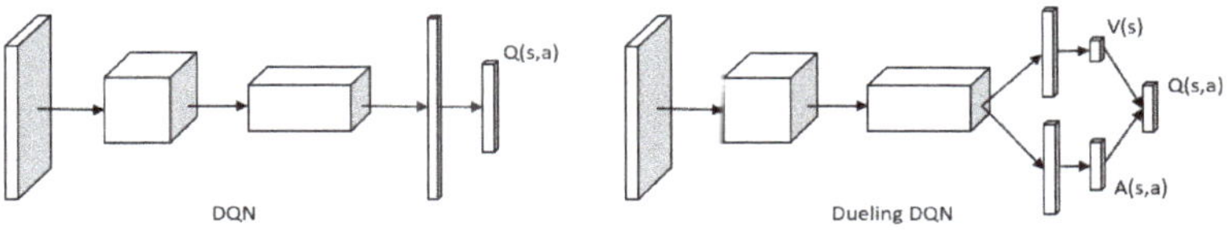

Figure 20. Deep Q-Network and Dueling Deep Q-Network architectures.

3.10.2. Asynchronous Advantage Actor-Critic (A3C)

A3C [198] uses multiple agents with each agent having its own network parameters and a copy of the environment. These agents interact with their respective environments asynchronously, learning with each interaction. Each agent is controlled by a global network. As each agent gains more knowledge, it contributes to the total knowledge of the global network. The presence of a global network allows each agent to have more diversified training data. An actor-critic algorithm predicts both the value function $V(s)$ and the optimal policy function $\pi(s)$. The learning agent uses the value of the value function (critic) to update the optimal policy function (actor). It determines the conditional probability $P(a|s;\theta)$, the parameterized probability that the agent chooses the action a when in state s. Different from most deep learning algorithms, asynchronous methods can run on a single multi-core CPU.

3.10.3. Trust Region Policy Optimization (TRPO)

A policy maps the state to action. Policy optimization is to find an optimal mapping from state to action. Policy gradient methods are popular in RL. The basic principle uses gradient ascent to follow policies with the steepest increase in rewards. However, large policy changes can destroy training, and it is not easy to map changes between policy and parameter space and to deal with the vanishing or exploding gradient problems and poor sample efficiency. The challenge is to have an accurate optimization method to limit the policy changes and guarantee any change will lead to improvement in rewards. A more commonly used method is to use a trust region, in which optimization steps are restricted to lie within a region where the approximation of the true cost function still holds. By preventing updated policies from deviating too wildly from previous policies, the chance of a catastrophically bad update is lessened, and many algorithms that use trust regions guarantee or practically result in monotonic improvement in policy performance. The idea of constraining each policy gradient update, as measured by the Kullback–Leibler (KL) divergence between the current and proposed policy, has a long history in RL [199]. TRPO [200] is an algorithm in this line of work that has been shown to be relatively robust and applicable to domains with high-dimensional inputs. To achieve this, TRPO optimizes a surrogate objective function—specifically, it optimizes an (importance sampled) advantage estimate, constrained using a quadratic approximation of the KL divergence. It avoids parameter updates that change the policy too much with a KL divergence constraint on the size of the policy update at each iteration. The generalized advantage estimation (GAE) proposed several more advanced variance reduction baselines [201]. The combination of TRPO and GAE remains one of the state-of-the-art RL techniques in continuous control.

3.11. Deep Transfer Learning (DTL)

Deep learning has a strong dependence on massive training data compared to traditional machine learning methods. Having sufficient training data is a prerequisite for a deep learning model to understand the latent patterns of the data. However, this is quite a challenge itself since the collection of data is time consuming and expensive. It is difficult to build a large-scale and high-quality annotated dataset in many fields. In addition, the training of deep learning models relies on intensive computation, which in practice can be challenging due to limited resources (e.g., high performance GPUs) and time constraints. Transfer learning is a concept of reusing a pretrained model on a new problem, which is an efficient way to tackle the insufficient training data problem and reduce the computational resource requirement and training time. It is very common in deep learning to use a pretrained model as a feature extractor in a new task or fine-tune the pretrained model (or some high-level parts of the model) to a new learning task.

Let D_s and D_t denote the source domain and target domain, and T_s and T_t denote two learning tasks ($D_s \neq D_t$ or $T_s \neq T_t$), respectively. Transfer learning can be defined as the process of enhancing the learning of the target predictive function $f_T(\cdot)$ in D_t using knowledge derived from D_s and T_s. It is a deep transfer learning task when $f_T(\cdot)$ is a nonlinear function that reflects a DNN. There are three forms of transfer learning: inductive transfer learning [202], transductive transfer learning [203], and unsupervised transfer learning [204]. In the first, T_s and T_t are different, and some labelled data in D_t are required to induce $f_T(\cdot)$ for use in D_t. In the second, we have the same T_s and T_t but different D_s and D_t, while no labelled data in D_t are available but labelled data in D_s are available. Finally, in the last setting, T_t is different from but related to T_s, and there are no labelled data in both D_s and D_t during training. The focus is on solving unsupervised learning tasks in D_t, such as clustering, dimensionality reduction, and density estimation.

According to the content to be transferred, transfer learning methods can be categorised into four cases: (1) instance-based approaches try to reweight the samples in D_s for learning in D_t [205]; (2) feature-based approaches encode knowledge into feature representations which are transferred across domains to help improve the performance of T_t [202]; (3) parameter-based approaches transfer knowledge across tasks through the shared parameters of the D_s and D_t learning models [206]; and (4) relational-based approaches, which transfer the knowledge through learning the common relationships between D_s and D_t. Recently, statistical relational learning techniques dominate this context [207].

3.12. Federated Learning (FL)

FL is applied in a situation where a group of clients wants to collaboratively train a global model without sharing their private local dataset [208]. Compared with conventional machine learning methods which require gathering different datasets, clients in FL collaboratively train a global model by exchanging local model weights/gradients without sharing their local dataset. There are typically two key players in FL: (1) the clients holding the local dataset and training the local model, and (2) the central server coordinating the training process and updating the global model. In general, FL contains three phases [209]:

Phase 1: FL initialization. The central server initializes the FL training model and sets the hyperparameters, including the number of FL training iterations, the total number of participating clients, the number of clients selected at each training

Phase 2: Local model training and updating. In each FL training iteration, clients first update the local model using the shared global model and train the local model using the local dataset. Then, clients send the local model weights or gradients to the central server for model aggregation.

Phase 3: Global model aggregation. The central server aggregates the model weights or gradients from the participating clients and shares the aggregated model to the clients for the next training iteration.

Algorithm 1 shows the pseudocode of an FL system proposed in [210]. According to the characteristics of training data, FL methods are usually classified into two categories: horizontal FL and vertical FL [208,211].

Algorithm 1 FedAvg [210]

Input:

N_{global}: Maximum number of global iterations, n: the total number of participating clients, m: the number of clients used in each global iteration, N_{local}: the number of local epochs, and η: the local learning rate.

Output:

Global model weight w_G

Processing:

[*Central Server*]
Initialize w_G^0
for each iteration t from 1 to N_{global} **do**
  $\mathcal{M}_t$ includes m clients randomly selected from the n clients
  for each client $i \in \mathcal{M}_t$ **in parallel do**
    $w_i^t, N_i \leftarrow$ **LocalTraining**(i, w_G^t)
  end for
  $w_G^{t+1} = \frac{1}{\sum_{j=1}^{m} N_j} \sum_{i=1}^{m} N_i w_i^t$
end for
[*Each Participating Client*]
LocalTraining(i, w):
$\mathcal{B}_i$ is the set of batches for the local dataset $\mathcal{D}_i$
for each epoch j from 1 to N_{local} **do**
  for each batch $b \in \mathcal{B}_i$ **do**
    $w \leftarrow w - \eta \nabla L(w; b)$
  end for
end for
return the weights w and $N_i = |\mathcal{D}_i|$

3.12.1. Horizontal FL (HFL)

HFL is used in scenarios where the datasets of the clients share the identical feature space but a different sample ID space [210,212]. For example, the electricity usage held by different electricity supplier companies may have the same feature space but different ID space. The communication protocols in FL can be divided into two classes: client-server protocol [210,213] and peer-to-peer protocol [212,214,215]. The client-server protocol deploys a central server to coordinate the training process, whereas the peer-to-peer protocol randomly selects a client as the server for the coordination work in each iteration.

In the client-server protocol, the clients are assumed honest and the server is assumed honest but curious. To avoid private information leakage, the exchanged model parameters are usually encrypted or masked by clients. The key steps are summarized as follows:

Step 1: The central server initializes the model and hyperparameters and allocates computation tasks to named clients.

Step 2: The participating clients train their local models on their local dataset, encrypt the model weights/gradients, and transmit them to the central server.

Step 3: The server conducts model aggregation, for example by averaging.

Step 4: The server broadcasts the updated model to all clients.
Step 5: The clients decrypt the model and update their local models.

In the peer-to-peer protocol, as there is no central server, two approaches are usually adopted to coordinate the training process:

(1) Cyclic Setting: All clients form a circular chain, denoted by $\{C_1, C_2, \ldots, C_n\}$. Client C_i transmits its local model to client C_{i+1}. Client C_{i+1} aggregates the received model with its local model which is trained on its local dataset and then transmits the updated model along the chain to client C_{i+2}. The training process stops once the termination condition is met.
(2) Random Setting: Client C_t randomly picks a client C_i from all participants with equal chance and sends its model information to another client C_i. C_i aggregates the received model with its local model which is trained on its local dataset, then randomly picks another client C_j with equal chance and sends the updated model to it. The training process stops once the termination condition is met.

3.12.2. Vertical FL (VFL)

VFL is used in scenarios where datasets between participating clients share the identical sample ID space but a different feature space. For example, a bank and an online shopping company may have the same customers but provide different services. The communication protocols for VFL can be divided into two classes: communication with a third-party coordinator [216] and communication without a third-party coordinator [217]. Assume that two clients, C_1 and C_2, plan to train a global model using their local datasets, and that samples from C_1 are labeled. In addition, C_1 and C_2 are assumed honest but curious to each other.

To protect the private data, the communication protocol with a third-party coordinator is designed as follows [216]:

Step 1: As the two datasets of C_1 and C_2 contain samples with different IDs, it is necessary to extract the common samples sharing the same IDs [218].
Step 2: The coordinator C_3 produces a pair of public and private keys and broadcasts the public key to C_1 and C_2.
Step 3: C_1 and C_2 compute encrypted gradients and add a mask. In addition, C_1 computes the encrypted loss. C_1 and C_2 then transmit the encrypted results to C_3.
Step 4: C_3 decrypts the received results and broadcasts them back to C_1 and C_2. C_1 and C_2 then update their local model using the received information.

To protect the private data, the communication protocol without a third-party coordinator is designed as follows [217]:

Step 1: A sample ID alignment process [219] is first employed to select the shared IDs between C_1 and C_2. Samples sharing the same IDs are confirmed to train a vertical FL model.
Step 2: C_1 produces an encryption key pair and transmits its public key to C_2.
Step 3: The two clients initialize their model weights and compute their partial prediction results. C_2 then transmits its result to C_1.
Step 4: C_1 computes the model residual, encrypts the residual, and transmits it to C_2.
Step 5: C_2 computes the encrypted gradient and transmits the masked gradient to C_1.
Step 6: C_1 decrypts the masked gradient and transmits it back to C_2. Then, C_1 and C_2 update their model locally.

3.13. *Multiple Instance Learning (MIL)*

3.13.1. Introduction of MIL

The concept of multiple instance learning was firstly proposed by Dietterich et al. [220] for investigating the problem of drug activity prediction. It is a type of weakly supervised learning where the training set is composed of many bags and each bag contains many instances, and a label is provided for the entire bag rather than each individual instance in it. This problem occurs when dealing with a lack of detailed annotation for large quantities of data. For example, it emerges when developing computer-aided diagnosis algorithms where medical images have only a patient-level diagnosis label rather than costly local labels annotated by experts [221]. Furthermore, it naturally occurs in a number of real-world learning scenarios, including image and video classification [222], document classification [223], and sound classification [224].

Generally, there are two assumptions in multiple instance learning: the standard and the collective assumption. The former assumes only negative instances are contained in negative bags, while one or more positive instances are contained in positive bags. This means that as long as there is one positive instance in the bag, the bag is positive. On the contrary, the collective assumption refers to cases where more than one positive instance is needed to identify a positive bag. These two assumptions are applied to different problem domains. For example, the standard assumption works well for drug activity prediction, while the collective assumption is more suitable for traffic jam detection.

3.13.2. Training Mechanism of MIL

The MIL problem under the standard assumption can be solved through alternate optimization. Specifically, the labels of all instances are assumed to be known at first, then a classification model can be obtained through supervised learning. Subsequently, this model is used to make predictions for each training instance, and the labels of the training instances are updated accordingly, and then this classification model can be retrained with the updated labels again, and this process repeats until convergence. Thus, the optimization process has mainly two parts: supervised learning and label updating.

When training the supervised learning model, only the predicted "most correct" (i.e., the highest classification score) is selected from the positive instance bag, and other instances in the positive instance bag are discarded, regardless of whether the prediction is positive. This is because, under the standard assumption, the MIL can only consider the "most correct" instance in the positive instance bag. Therefore, this selection strategy is exactly in line with the problem definition. In addition, if there are enough negative instances, only the instance that is predicted to be "most correct" in each negative instance bag can be used for training. Such a negative instance is also called hard instance or most violated sample. In practice, they are most effective for fast model convergence.

3.13.3. Challenges of Using MIL

The unique challenges of using MIL arise from four aspects: the level of prediction, the composition of bags, the ambiguity of instance labels, and the distribution of the data [225]. These factors affect the choice and the performance of MIL algorithms.

(1) The level of prediction refers to whether a network makes the prediction on a bag-level or an instance-level. These two kinds of tasks employ different loss functions, and thus algorithms designed for bag classification are not optimal for instance classification. Cheplygina et al. [226] details how to choose algorithms for different problems.

(2) The composition of bags refers to the ratio of instances from each class or the relation between instances. The proportion of positive instances in positive bags is generally defined as witness rate (WR). If the WR is very high, which means positive bags contain only a few negative instances, the problem can be solved in a regular supervised framework. However, if the WR is very low, which means a serious class imbalance problem because a few positive instances have a limited effect on training the network, many algorithms will have a poor performance. Several MIL algorithms have been proposed for this problem [227–229].

(3) The ambiguity of instance labels refers to label noise or instances not belonging to a class clearly. This is inherent to weakly supervised learning. Some MIL algorithms impose strict requirements on the correctness of bag labels, such as the DD

algorithm [230]. For practical problems where positive instances may be found in negative bags, algorithms working under the collective assumption are needed [231].

(4) The distributions of positive and negative instances also affect MIL algorithms. This has two sides. First, the positive instances can either be located in a single cluster in feature space or be corresponding to many clusters, which leads to different applicable MIL algorithms [230,232]. Second, the distribution of the training data can or cannot entirely represent the distribution of negative instances in the test data, which also leads to different applicable MIL algorithms [233,234].

4. Deep Learning in Diverse Intelligent Sensor Based Systems

4.1. General Computer Vision Sensor Systems

A well-known application domain of deep learning is general computer vision, where the processed data are images and videos acquired from camera-based sensor systems. The research in this domain focuses on enabling computers to gain an understanding like that of human vision from images or videos. Deep learning is used for a wide range of important tasks in this domain, as described next.

4.1.1. Image Classification

Conceptually, image classification is one of the simplest yet most fundamental problems in computer vision. It refers to the process of predicting information classes from an image. CNNs are the most commonly employed techniques for solving this problem. Specifically, the CNNs take an image as input and aim to output the class of the input image. Since AlexNet [37] achieved remarkable classification performance in the ImageNet challenge, many types of CNN models have been proposed for image classification, such as VGG [39], ResNet [41], and DenseNet [42]. In 2017, Xie et al. proposed ResNeXT [235], which is an extension of ResNet and VGG, and achieved the state-of-the-art performance of 3.03% top-five errors. Around the same time, the problem of supervised image classification was regarded as "solved", and the ImageNet classification challenge concluded. However, in many applications, the tasks cannot be formulated as plain vanilla image classification problems. Many object classes may be present in a single image. Therefore, more research efforts are being made toward object detection and segmentation.

4.1.2. Object Detection

Object detection is also a fundamental problem in computer vision. Its aim is to identify and localize different objects in an image. Therefore, deep learning models for this problem usually consist of two components: the backbone component, which is similar to an image classification model, and the region proposal component, for predicting bounding boxes. Region-proposal and Region-based CNN (R-CNN) is a pioneering work for object detection [236]. However, it requires much computing time and memory for training. Therefore, several improved variants of R-CNN have been proposed, such as the renowned Fast R-CNN [55] and Faster R-CNN [44]. Another kind of models for object detection is represented by YOLO (You Only Look Once), which achieved reasonable performance for real-time object detection [46]. Other advanced models include Region-based Fully Convolutional Networks (R-FCNs) [237], which use ResNet as an object detector and are faster than the Faster R-CNN, and the Single-Shot MultiBox Detector (SSD) [238], which is even faster than YOLO and has comparable accuracy to Faster R-CNN. Most object detection methods mentioned above incur a high computational cost due to their bounding box processing. More architectures addressing this issue and achieving higher accuracy can be found in recent overview articles [239,240].

4.1.3. Semantic Segmentation

Semantic segmentation refers to the process of dividing an image into meaningful types of regions. It is a vital step for many image processing and analysis tasks. Its aim is to label an image at the pixel level, or more accurately to assign each pixel to the class it most likely belongs to. A predominant and outstanding architecture particularly for the semantic segmentation problem is U-Net [43]. Due to the huge success of U-Net, many variants have been proposed to fit specific applications and achieved great results [241–245]. U-Net was originally proposed for addressing medical image segmentation. Similarly, other domain-oriented deep neural networks have been developed, including SegNet [246], PSPNet [247], and DeepLab [248]. More detailed discussions of semantic segmentation can be found in various surveys [249–251]. In a way, semantic segmentation can be seen as a rough classification, for example two different dog objects being segmented as one entity, but in many applications we need a finer segmentation, where each dog is segmented out separately. Hence, instance segmentation methods are important.

4.1.4. Instance Segmentation

Instance segmentation is essential for many real-world application scenarios such as autonomous driving, medical imaging, robot vision, and so on. It refers to detecting instances of objects and, more challenging, demarcating their boundaries. Because instance segmentation involves both the detection of different objects and their per-pixel segmentation, networks for solving it can be either R-CNN driven or FCN driven. Mask R-CNN [47] is one of the representative networks. Its overall structure is the two-stage object detection network Faster R-CNN, where the box head is used for detection and the mask head for segmentation. Fundamentally, these instance segmentation networks employ object detection networks in identifying the object bounding boxes, while an extra component mask head is used to further extract the foreground of the bounding boxes. Other reputable networks include YOLACT [252] and SOLO [253] inspired by YOLO, and PolarMask [254] and AdaptIS [255] inspired by the object detection network FCOS [256]. See several recent reviews of instance segmentation for more details [257–259].

4.1.5. Pose Estimation

Pose estimation refers to the problem of inferring the pose of a person or an object in an image or video. In other words, it concerns determining the position and orientation of the camera relative to a given person or object of interest. This is typically achieved by identifying, locating, and tracking a number of key points of the person or object. This problem is basic and important because it occurs in many real-world applications such as object/person tracking. Deep learning is often employed to detect and track these key points. There are many specific neural network architectures for this purpose, and the most robust and reliable ones that make good places to start include Stacked-Hourglass networks [260], Mask R-CNN, and PoseNet [261]. Key factors in designing DNNs for pose estimation include using dilation convolution, upsampling operations, and skip connections. This is because pose estimation requires a higher-resolution feature map and is more sensitive to the location of keypoints compared to classification/detection tasks. More advanced networks have been described in various reviews [262–264].

4.1.6. Style Transfer

Style transfer refers to the computer vision task of blending two input images, named content image and style reference image, and producing an output image that maintains the core elements of the content image while following the style of the style reference image. It can power practical applications such as photo and video editing, gaming, virtual reality, and so on. Neural networks have become the state-of-the-art method for style transfer. Generally, CNNs are the mainstream approaches for this problem, and a style transfer model usually consists of two networks, namely a pretrained feature extraction network and a transfer network. Significant networks that are good starting points include the model proposed by Johnson et al. [265] for single style transfer, the model proposed by Dumoulin et al. [266] at Google for multiple style transfer, and the model proposed by Huang et al. [267] for arbitrary style transfer. Detailed explorations and more advanced architectures can be found in recent surveys [107,268,269].

4.1.7. Video Analytics

Video analytics refers to generating descriptions of the content of, or events in the video, which involves tasks of object (persons, cars, or other objects) detection, tracking, as well as calculating their appearance and movements. It is also an important and essential computer vision technique and has significant practical benefits such as monitoring video for security incidents helps prevent crime, intelligent traffic systems, and more [270,271]. While its tasks overlap beyond image analysis tasks, they are more challenging because they involve both spatial and temporal information.

The object detection problem in video is associated with the object segmentation problem, because an accurate object segmentation facilitates object detection, and robust object detection in turn facilitates object segmentation. Recent neural networks for this problem are based on recurrent convolution neural networks (RCNN). For example, Donahue et al. [272] firstly proposed a long-term RCNN, where a set of CNNs is employed for visual understanding and then their outputs are fed to a set of RNNs for analyzing temporal information. Other representative RCNN based models include the one proposed by Ballas et al. [273], MaskRNN [274], and MoNet [275]. For comprehensive discussions of video analytics we refer to recent surveys [270,271,276,277].

4.1.8. Codes, Pretrained Models, and Benchmark Datasets

Various implementation codes and pretrained models of many of the above introduced methods can be found in the references provided in Section 2.5.2. Some renowned benchmark datasets that are widely used in general computer vision to evaluate different deep learning methods are listed as follows.

(1) MNIST: http://yann.lecun.com/exdb/mnist/ (accessed on 2 November 2022).
(2) CIFAR-10 and CIFAR-100: https://www.cs.toronto.edu/kriz/cifar.html (accessed on 2 November 2022).
(3) ImageNet: https://image-net.org/challenges/LSVRC/ (accessed on 2 November 2022).
(4) COCO: https://cocodataset.org/#home (accessed on 2 November 2022).
(5) PASCAL VOC: http://host.robots.ox.ac.uk/pascal/VOC/ (accessed on 2 November 2022).
(6) OpenImages: https://storage.googleapis.com/openimages/web/index.html (accessed on 2 November 2022).
(7) MIT pedestrian: http://cbcl.mit.edu/software-datasets/PedestrianData.html (accessed on 2 November 2022).
(8) Youtube-8M: https://research.google.com/youtube8m/ (accessed on 2 November 2022).
(9) SVHN: http://ufldl.stanford.edu/housenumbers/ (accessed on 2 November 2022).
(10) Caltech: http://www.vision.caltech.edu/datasets/ (accessed on 2 November 2022).

4.2. Biomedical Sensor Systems

Deep learning has fundamentally converted the way we process, analyze, and interpret data, including in biology and medicine. We discuss deep learning in biomedical sensor systems from the perspective of three different biomedical domains: biomedical imaging, omics data analysis, and prognostics and healthcare.

4.2.1. Biomedical Imaging

Biomedical image analysis is one of the most important and fundamental areas in biomedical science and has become a cornerstone of modern healthcare. Automatic biomedical image analysis involves a set of basic tasks introduced in Section 4.1, such as image reconstruction and registration, image or object classification, object detection, segmentation, and tracking. According to the different image types and their unique characteristics, we further divide biomedical imaging into four subareas and discuss the application of deep learning in each, as in [278].

(1) Medical Imaging. Medical images are typically acquired using devices such as X-ray CT (computed tomography), MRI (magnetic resonance imaging), and US (ultrasound). With the advancement of medical imaging devices, the quality of medical images has improved over the years, but their automated analysis is still a challenging task. DNNs can provide powerful solutions to this problem. For example, U-Net [43] and UNet++ [279] are two most reputable and popular architectures for medical image segmentation. In fact, U-Net has become the de facto standard method in medical image segmentation due to its huge success in the field. Various CNN-based architectures have achieved top performance for brain tumor analysis [280]. For a more in-depth discussion of DNN architectures in medical imaging, we refer to recent overview and survey papers [281–283].

(2) Pathological Imaging. Pathological images are generated from specimen slides by virtual microscopy, also called whole-slide imaging. Their visual interpretation is more challenging than for medical images due to the large size and high resolution of the images. As in medical imaging, deep learning brings great potential in providing reliable image interpretation in this subarea. For example, Zhu et al. [284] proposed a DeepConvSurv model based on CNN for survival analysis with pathological images. Li et al. [285] proposed a DenseNet based solution for pathological image classification. A recent trend in pathological image processing is to incorporate multiple instance learning to deal with the high resolution and weak labels of pathological images. More advanced models can be found in a recent survey paper [286].

(3) Preclinical Imaging. Preclinical imaging refers to the visualization of small living animals for conducting in-vivo studies for clinical translation. Preclinical images can be obtained by micro-US, MRI, and CT for anatomical imaging, or bioluminescence, PET (positron emission tomography), and SPECT (single photon emission computed tomography) for molecular visualization. Employing deep learning for interpreting these images is comparatively under-researched. A few related DNN-based methods are discussed in recent works [287,288].

(4) Biological Imaging. Biological images capture various aspects of organisms and biological systems that are not visible to the naked eye. Automated analysis and interpretation of these images is challenging, as they are typically very noisy and highly variable depending on experimental conditions, and they can be quite large. DNNs have proven to be very suitable for biological image analysis and have empowered biological research [289,290]. Moreover, to facilitate the design of DNN architectures for this purpose, neural architecture search-based solutions have been proposed for cell segmentation [291,292]. Architectures for deep learning-based biological image analysis have been discussed in several recent papers [293–295].

4.2.2. Omics Data Analysis

Omics data are complex, heterogeneous, and high-dimensional, and deep learning methods are specially suitable to analyze them. According to the different types of data, we introduce deep learning in omics data analysis from the following aspects.

(1) Genomics. Deep learning methods have been applied to genomics data analysis for several years, and have achieved impressive results. For example, CNNs have been employed for single-nucleotide polymorphisms and indels detection [296]. SAEs have been successful in predicting the effect of genetic variants on gene expression [297]. Both have achieved better results than traditional methods. A review of more architectures can be found in a recent survey paper [298].

(2) Transcriptomics. Analysis of transcriptomics data may yield an estimate of the expression level of each gene or transcript across several samples [299]. Therefore, it can be seen as a typical deep learning problem. Various deep learning methods have been proposed for addressing this problem. For example, a RAN-based solution for detecting long ncRNAs achieved a remarkable 99% accuracy [300]. For comprehensive introductions and discussions we refer to various survey papers [301,302].

(3) Proteomics. Protein data analysis mainly centers around two topics: protein structure prediction (PSP) and protein interaction prediction (PIP) [303]. For PSP, deep learning-based methods have been used to solve problems such as backbone angles prediction [304], protein secondary structure prediction [305], and protein loop modeling and disorder prediction [306]. Moreover, due to the success of deep learning in generating higher-level representations and ignoring irrelevant input changes, deep learning methods have become the technology of choice to help PSP. For PIP, deep learning-based methods have been used to analyze protein–protein interactions [307], drug–target interactions [308], and compound–protein interactions [309]. A latest trend in PSP is using GNNs to better learn complex relationships among protein interaction networks for PSP.

4.2.3. Prognostics and Healthcare

Clinical data and electronic medical records are vital for prognostics and healthcare management. Deep learning to handle these kinds of data is also rapidly growing [310,311]. For example, deep learning-based methods have been used for detecting cardiac arrhythmia from electrocardiograms [312] and for phenotype discovery using clinical data [313]. There are also examples of using DNNs and topic modeling techniques to learn effective representations from electronic health records [314,315]. A key challenge in this area is the efficient utilization of temporal information for achieving high performance [316]. Hybrid DNNs such as those incorporating RNN and CNN components are promising in addressing this challenge.

4.2.4. Codes, Pretrained Models, and Benchmark Datasets

Various implementation codes and pretrained models of many of the above introduced methods can be found in the references provided in Section 2.5.2. In addition, the implementation of nnU-net [317], which is a powerful self-adapting neural network framework that can automatically configure itself, including selecting the optimal preprocessing, architecture, training, and post-processing for any new task, is publicly available (https://github.com/MIC-DKFZ/batchgenerators accessed on 2 November 2022). Some renowned benchmark datasets that are widely used in the biomedical domain to evaluate different deep learning methods are listed as follows.

(1) Decathlon: http://medicaldecathlon.com/ (accessed on 2 November 2022)
(2) MedPix: https://medpix.nlm.nih.gov/home (accessed on 2 November 2022)
(3) NIH Pancreas-CT: https://academictorrents.com/details (accessed on 2 November 2022)
(4) AMRG Cardiac Atlas: http://www.cardiacatlas.org/studies/amrg-cardiac-atlas/ (accessed on 2 November 2022)
(5) Cancer Imaging Archive: https://wiki.cancerimagingarchive.net (accessed on 2 November 2022)
(6) OASIS Brains: http://www.oasis-brains.org/ (accessed on 2 November 2022)
(7) ADNI: https://adni.loni.usc.edu/data-samples/access-data/ (accessed on 2 November 2022)
(8) DDSM: http://www.eng.usf.edu/cvprg/ (accessed on 2 November 2022)
(9) CTC: http://celltrackingchallenge.net/ (accessed on 2 November 2022)
(10) ISIC Archive: https://www.isic-archive.com/#!/onlyHeaderTop/gallery (accessed on 2 November 2022)

4.3. Biometric Sensor Systems

Biometrics deals with recognizing people by using their physical and behavioral characteristics. Biometric recognition can be formulated as a verification or identification problem. The verification task aims to verify whether a person is who they claim to be by comparing the person's biometric template with the reference template of the claimed identity. The identification task compares a person's biometric template with references of all identities in the database to establish the person's identity. In either task, the system needs to collect the biometric data, extract features, and perform comparison or classification. Deep learning has a big impact on biometrics in terms of feature extraction and classification, which primarily involves supervised learning. Recent advances in this field also applied generative models with unsupervised learning to enhance the learning of features and improve recognition performance. In this section, we review deep learning approaches for biometric applications and discuss how the methods can benefit the field of biometrics and the open research questions.

4.3.1. Automatic Face Recognition

Faces are one of the most commonly used biometrics in surveillance, forensics, security, access control applications scenarios. Acquisition of face biometrics is based on cameras and the collected data are in the format of images or videos. While being noninvasive and convenient, face biometrics are subject to imaging conditions and physical factors related to illumination, pose, expression, aging, and other appearance changes.

Conventional methods for automatic face recognition can be categorized into feature-based approaches and appearance-based approaches, which extract local features and global representations, respectively. With a hierarchical structure, deep learning simultaneously extracts local and global representations while handling nuisance factors. Among different architectures, CNN-based models show the most significant impact in this field. For example, CNNs with different architectures and loss functions [318] were trained to learn DeepID features in joint identification-verification tasks. Verification essentially deals with the similarity between two faces, and therefore, metric learning such as joint Bayesian and triplet loss are adopted. Identification, on the other hand, is a multiclass classification problem, hence the cross-entropy is usually used in the loss function. Facebook proposed DeepFace [319], which is a nine-layer CNN trained on four million Facebook images from four thousand subjects. DeepFace addresses the alignment issue and learns effective face representations with high transferability. Google proposed FaceNet, a deep CNN with triplet loss [320] to learn direct embeddings of images, which are effective in face verification, identification and clustering tasks. In addition, various CNN frameworks are proposed to handle the pose and illumination variations. For example, the face identity-preserving framework [321] integrates feature extraction layers with a reconstruction layer to reconstruct face images in a canonical view. An ensemble of pose-aware CNNs [322] was proposed for face recognition, where each model is trained for a specific pose using pose-specific images generated by 3D rendering.

To improve the efficiency of training deep neural networks, researchers have proposed different learning strategies. For example, a sparse network can be trained iteratively from a denser model using correlations between neural activations of consecutive layers [323]. A face alignment network [324] trained jointly with the face recognition network can reduce the number of training samples needed. Furthermore, hybrid discriminative and generative models were proposed to learn identity-specific representations that are pose-invariant [325]. Generative models such as AEs and GANs are also used to generate identity-bearing facial images [326]. In addition, the recognition of facial attributes such as age and gender [327] is an important task because it helps narrow down candidate matches, which can then facilitate face recognition. Hierarchical representations from a CNN or an ensemble of CNNs have been used for this purpose via classification [327] or regression [328]. CNN with different constructs have been adopted, including the residual network [329]. Training such models requires crowdsourcing to get the age and gender labels, which usually results in small datasets not sufficient to train deep neural networks. Therefore, models such as VGGNet and GoogleNet pretrained on large datasets such as ImageNet are often adopted and fined-tuned for age and gender estimation [330].

4.3.2. Periocular Region and Iris

The periocular region presents salient traits for face and facial attribute recognition, which is helpful when the lower half of a face is occluded. Researchers have used CNN and

RBM models trained with unsupervised learning to learn representations from periocular image patches and transferred the representations to recognition tasks [331]. Deep learning models were also used with conventional handcrafted methods to enhance performance. For example, autoencoders were used to learn latent representations from the texture features extracted by handcrafted filters [332], and CNNs were trained on both face images and the SIFT features to gain higher recognition accuracy [333].

The iris is a highly distinctive biometric trait. However, the acquisition of iris images often suffers low user acceptance. Iris recognition relies on random texture information in the irises and the quality of the extracted information depends on the preprocessing steps, including iris segmentation, off-axis gaze correction and removal of eyelashes. Gabor filtering is the classic method widely used in real-world applications for capturing iris texture information [334]. Deep learning replaces the Gabor filters with neural network modules. For example, CNNs were used to learn source-specific filters for iris images from visible and near-infrared sources [335]. Deep CNNs integrating inception layers were proposed for iris recognition, providing robust performance in terms of segmentation and alignment. Sparse autoencoders were also trained for feature extraction in mobile applications where iris images were collected by mobile devices [336]. Moreover, representations learned by CNNs were fused with handcrafted features to improve recognition accuracy.

4.3.3. Fingerprint and Palmprint

Fingerprints are one of the most established biometric modalities. The acquisition of fingerprints uses cameras and the collected data are images. Two types of features are used, one is global features in terms of loop, delta, and whorl, and the other is local features in terms of ridge, valley, and minutiae. The major challenges of fingerprint recognition are the intra-subject variations caused by displacement, distortion, pressure, skin condition, and other noises. Applying deep learning to fingerprint recognition aims to extract deep global and local representations, as well as enhancing the fingerprint images.

CNNs are the most popular models in fingerprint biometric applications. With different designs in the neural network structures and training strategies, CNNs have been used for identification [337], authentication [338], liveness detection [339], double-identity detection [340], fingerprint alteration detection [341], spoofing detection [341], latent fingerprint recognition [342], cancellable recognition systems [343], and fingerprint segmentation [344], enhancement [345], and indexing [346]. Recent work also started to explore the use of CNNs for contactless and partial 3D fingerprint recognition [347–351]. DBNs are also used for fingerprint liveness detection, anti-spoofing, and enhancement [352]. One of the biggest challenges for most fingerprint recognition systems is the spoofing attack, which tries to circumvent a recognition system using artificial replicas of human characteristics similar to the legitimate enrolled trait. Models based on AEs such as stacked AEs and sparse AEs [353] were proposed to defend against spoofing attacks on fingerprint recognition systems and to perform liveness detection. Moreover, generative models based on GANs are widely used to generate fingerprint images [354]. The generation of high-quality fingerprints is used for fingerprint recovery [355] and presentation attack detection [356]. Furthermore, hybrid deep learning models or ensemble DL methods have been proposed to perform multiple tasks at once. For example, the Inception, MobileNet, and GAN are integrated in one framework [357] for localization and detection of altered fingerprints in order to address obfuscation presentation attack.

Palmprint and hand geometry share similar traits as fingerprints. Classic features include the hand/palm shape, principal lines, wrinkles, delta points, and minutiae features. Deep learning is used to learn these multiscale features from the palmprint and hand images. Various models based on CNN and RBM [358] have been used for palmprint recognition. The models are trained with either the whole palmprint images or regions of interest [359].

4.3.4. Voice-Based Speaker Recognition

Application scenarios of speaker recognition can be classified into speaker verification, speaker identification, speaker diarization (which is used for automated speech transcription systems where dialogue is generated along with the speaker's information), and speaker recognition in-the-wild (which refers to real-world scenarios where conditions are unknown or even corrupted with noise, echo and cross-talk). The in-the-world scenario is one of the major challenges targeted by researchers. The general types of speaker recognition are text-dependent and text-independent, and their difference lies in whether specific phrases are required or not. Deep learning methods are currently the state-of-the-art in the above-mentioned application scenarios and types. These methods process voice inputs in two patterns: raw sound waves and preprocessed data. Although some methods (e.g., the SincNet [360], RawNet [361], and AM-MobileNet [48]) are directly trained using raw speech data, most methods rely on signal preprocessing, which segments the signal into frames, performs normalization, converts signals to the frequency domain, and extracts spectrogram, mel-filterbank and mel-frequency cepstral coefficients (MFCC).

Based on the learning strategies, existing DL methods for speaker recognition can be categorized into stage-wise approaches and end-to-end systems. The stage-wise strategy involves two stages: speaker-specific feature extraction and classification of speakers. The i-vector [362] is a classic method for speaker recognition, consisting of a feature extractor based on Gaussian mixture models (GMM) and universal background models (UBM) and a classifier based on linear discriminant analysis. Inspired by i-vector, DL architectures are proposed to dig deeper representations, resulting in DL-based speaker embedding systems, d-vectors [363] (deep vector), x-vectors [364] (time-delay), and t-vectors [365] (triplet network). End-to-end systems do not require a multistage network. However, pretraining steps such as extraction of spectogram, MFCC, and mel-filterbank, or automatic feature learning with AEs, are employed to enhance recognition performance. Residual networks are widely used in feature extraction and end-to-end speaker recognition systems. Representative methods include the DeepSpeaker [192] (which integrates CNN with residual network), RawNet [361] (which consists of convolutional layers and gated recurrent unit layers with residual block constructs), and AM-MobileNet [48]. Some architectures adopted speech specific layers to facilitate speech signal processing. For example, the SincNet [360] uses a parameterized Sinc function to perform convolutions, which results in a smaller number of parameters and achieves better performance and faster convergence than standard CNNs. Autoencoders have also been widely used in speaker recognition for data encoding, feature dimension reduction, and data denoising [366]. Furthermore, generative models based on GANs are used for data augmentation and generation in speaker recognition systems to help extend short utterances into long speeches to enhance recognition performance [367]. An example is the SpeakerGAN [368] which is a variant of conditional GAN trained on inadequate speech data.

4.3.5. Behavioral Biometrics

Handwritten signature is the most popular behavioural biometrics that has been widely used in various applications in legal, medical, and banking sectors. Based on how the signature is acquired, signature verification can be operated in two scenarios, including offline methods that use static signature images as inputs and online methods that further take into account the dynamics of the signing process (such as the pressure and velocity). Various deep learning models have been proposed to extract deep representations from the signature images and the signing process to improve verification accuracy for both offline and online applications. Popular models include the RNNs [369] (LSTM, gated recurrent unit), CNNs [370], DBN [371], and the combination of these models with AEs [372]. Classic methods such as the length normalized path signature descriptors [373], direction features, and wavelet features [371] are also used as inputs to train deep nets, instead of raw images, to improve performance. A Siamese network structure with contrastive loss was used for writer-independent verification [374], which measures how likely two given signatures are written by the same writer without knowledge of the writer's identity.

Gait and keystrokes are two other popular behavioral biometrics which use the shape and motion cues of a person's walking style and the typing patterns respectively for person recognition. There are two ways to acquire gait data: one is to use cameras or motion sensors to capture image/video [375] during the gait phases and the transition periods between phases, the second is to use sensors such as accelerators [376] to capture the signal variations of the person during walking. Deep learning methods for image/video-based gait recognition share great similarities with those in computer vision applications. The major difference lies in the input images, where models for gait recognition are usually silhouette shape-based and are trained with gait energy images [375] or chrono-gait images [377]. In terms of models, CNNs, LSTM, AEs, and their combinations are popular. In particular, 3D CNNs with temporal information in gait sequences considered provide a significant improvement in performance [378].

4.3.6. Physiological Signals-Based Biometrics

Brain biometrics and heart biometrics are the major modalities in this category, which are an emerging branch of biometric technology. Brain biometrics are based on EEG (electroencephalogy) signals, which are recordings of the electrical pulses of the brain activity collected from a person's scalp using electrodes. Similarly, the heart signals are collected from the chest, finger, or arm using electrical, optical, acoustic and mechanical sensors. The resultant signals are referred to as ECG (electrocardiography), PPG (photoplethysmography), PCG (phonocardiogram), and SCG (seismocardiogram), respectively. Other physiological signals used for biometric applications include the EMG (electromyography), EDA (electrodermal activity), and EOG (electrooculogram). Deep learning contributes to brain, heart, and other physiological signals-based biometrics in two aspects: automatic representation learning and classification. Various models based on MLPs, LSTM, and CNNs [379] are proposed to directly learn deep representations from the physiological signals for biometric recognition for end-to-end systems. In addition, since the salient features of these signals are usually in the frequency domain, pretraining steps such as Fourier transform and wavelet package decomposition were adopted in many works to convert the signal into the frequency or time-frequency domain [380]. Other pretraining steps include constructing functional connectivity networks using multichannel EEG signals, followed by CNNs [381] or GCNNs [382] to learn structural representations from the networks. The acquisition of physiological data from human subjects is a difficult and time-consuming task, and therefore, the datasets are usually small. To address this issue, generative models based on AEs [383,384] and GANs [385] were proposed for data augmentation and incomplete data reconstruction. The results show a significant improvement in recognition performance with data augmentation. We refer to [386] for a comprehensive survey in this area.

4.3.7. Databases

Databases commonly used for biometric performance evaluation are summarized in Table 3. We separate the databases for different biometric modalities.

Table 3. Databases for biometric applications.

Modality	Database
Face	Labeled Faces in the Wild http://vis-www.cs.umass.edu/lfw/ (accessed on 2 November 2022)
Face	YouTube Faces http://www.cs.tau.ac.il/wolf/ytfaces/ (accessed on 2 November 2022)
Face	AR Face database [387]
Face	MORPH https://uncw.edu/oic/tech/morph.html (accessed on 2 November 2022)
Iris	VSSIRIS https://tsapps.nist.gov/BDbC/Search/Details/541 (accessed on 2 November 2022)
Iris	Mobile Iris Challenge Evaluation http://biplab.unisa.it/MICHE/ (accessed on 2 November 2022)
Iris	Q-FIRE [388]
Iris	LG2200 and LG4000 https://cvrl.nd.edu/projects/data/ (accessed on 2 November 2022)
Fingerprint	FVC-onGoing https://biolab.csr.unibo.it/FVCOnGoing/UI/Form/Home.aspx (accessed on 2 November 2022)
Fingerprint	NIST SD27 https://www.nist.gov/itl/iad/image-group/nist-special-database-2727a (accessed on 2 November 2022)
Palmprint	PolyU Palmprint database http://www4.comp.polyu.edu.hk/csajaykr/database.php (accessed on 2 November 2022)
Voice	Google Audioset https://research.google.com/audioset/ (accessed on 2 November 2022)
Voice	VoxCeleb https://www.robots.ox.ac.uk/vgg/data/voxceleb/ (accessed on 2 November 2022)
Signature	GPDS-960 corpus https://figshare.com/articles/dataset/GPDS960signature_database/1287360/1 (accessed on 2 November 2022)
Signature	Signature verification competition 2004 [389]
Gait	CASIA-B http://www.cbsr.ia.ac.cn/english/Gait20Databases.asp (accessed on 2 November 2022)
Gait	OU-ISIR LP dataset http://www.am.sanken.osaka-u.ac.jp/BiometricDB/GaitLPBag.html (accessed on 2 November 2022)
Keystroke	CMU Benchmark Dataset https://www.cs.cmu.edu/keystroke/ (accessed on 2 November 2022)
EEG	EEG Motor Movement/Imagery Dataset https://physionet.org/content/eegmmidb/1.0.0/ (accessed on 2 November 2022)
EEG	BED [390]
ECG	ECG-ID https://physionet.org/content/ecgiddb/1.0.0/ (accessed on 2 November 2022)
ECG	PTB https://www.physionet.org/content/ptbdb/1.0.0/ (accessed on 2 November 2022)

4.4. Remote Sensing Systems

In general, remote sensing refers to non-contact and long-distance detection technology, which uses remote sensors to capture the radiation and reflection characteristics of objects on the earth's surface [391]. Remote sensors, typically mounted on airborne and satellite platforms, are the core component in any remote sensing system, and can be classified into two types: passive and active sensors. Passive sensors measure energy that is naturally available, and are usually optical and camera-based, such as panchromatic and multispectral sensors, providing images in the visible range. Different from passive sensors, active sensors such as radar sensors receive the reflection of the impulse they emitted and are less influenced by the environment.

With the availability of remote sensing imagery, DL methods have seen a rapid surge of interest from the remote sensing community and made a remarkable breakthrough. Deep learning in remote sensing confronts some new challenges:

(1) Multiple image modalities. Multimodality remotely sensed datasets, such as multi- and hyperspectral data, light detection and ranging (LiDAR) data, and synthetic aperture radar (SAR) data differ from each other not only in the imaging mechanism but also in the imaging geometries and contents. Different data modalities are often complementary. The design of deep models is crucial in making the most of these data.

(2) Growing importance of prior knowledge. Remote sensing data presents the real geodetic measurements for the earth surface, with each data point containing geophysical or biochemical information. Hence, minimizing distortion and improving data quality are especially crucial to remote sensing tasks. Pure data-driven models, without any prior knowledge, will lead to possible misinterpretation or blind trust.

In the following sections, we will investigate how deep learning models are modified to cope with these two challenges from the perspective of image classification, scene classification, object detection and segmentation, and multimodal data fusion.

4.4.1. Image Classification

Image classification is one of the most active research topics in remote sensing, which aims to assign semantic labels to every pixel in the image. Various works used machine learning algorithms such as random forest (RF) and support vector machine (SVM) to improve the accuracy. The advent of deep learning pushed the boundary even further. Chen et al. [392] proposed the first deep learning-based classification model, which uses a stacked AE to extract hierarchical spectral information. Soon afterwards, DBN [393], and sparse SAEs [394] were introduced to learn stable and effective features for hyperspectral data classification. Makantasis et al. [395] proposed to use CNNs as feature extractor and a multiLayer perceptron (MLP) responsible for the classification task. Santara et al. [396] constructed an end-to-end CNN-based framework and generated band specific spectral-spatial features for classification. In [397], Li et al. proposed a pixel-pair strategy for CNN-based classification, and achieved state-of-the-art performance even with limited training samples. Recent improvements can be attributed to (1) specially designed architectures such as Siamese CNNs [398], the capsule network [399], and Transformer [400], and (2) improved feature representation [401,402].

4.4.2. Scene Classification

Scene classification, which aims to automatically classify the image into the category it belongs to, has become one of the most active areas of high-resolution remote sensing image understanding, and attracted growing attention in the past decade [403]. It is a relatively challenging task because even different scenes may contain objects with similar features. Such variations make scene classification considerably difficult. Compared with traditional approaches based on bag-of-visual-words (BoVW), deep models have distinct advantages in learning more abstract and discriminative features, thereby providing much better classification accuracy. Hence, most of the recent works paid much attention to building a robust and informative scene representation. Using PTMs [404,405] is a popular technique in scene classification, as it is difficult and time-consuming to train a CNN model from scratch with a limited number of training samples. Fine-tuning [406] also helps the PTMs adapt to the specific task and learn oriented feature representation for remote sensing images. Another family of methods focuses on feature selection [407], features aggregation [402,408,409], and fusion [410–412]. For example, Lu et al. [409] proposed a supervised feature encoding module and a progressive aggregation strategy to make full use of intermediate features. To cope with large intra-class variance caused by large resolution variance and confusing information, Zhao et al. [412] proposed a multigranularity multilevel feature fusion branch to extract structural information and fine-grained features.

4.4.3. Object Detection

With the rapid development of intelligent earth observation, automatic interpretation of remote sensing data has become increasingly important. Object detection in remote sensing aims to identify ground objects of interest such as vehicles, roads, buildings or airports from images and correctly classify them. In recent years, DL-based methods have been dominating this research area and made remarkable progress.

Preliminary work for object detection in remote sensing images [413–415] borrows the coarse-localization-fine-classification pipeline and CNN models from the computer vision community. Zhu et al. [416] introduced AlexNet CNN [37] to extract robust features, combined them with an image segmentation method for localization, and finally employed an SVM classifier for detection. Chen et al. [413] presented a hybrid DNN (HDNN) for vehicle detection, which used a DNN as feature extractor and a MLP as classifier. To further adapt CNN models to remote sensing object detection, researchers also take the rotation-invariant characteristic and context information into consideration. Cheng et al. [417] used and a newly proposed rotation-invariant layer to cope with object rotation variations. To cope with performance drop resulting from object appearance differences,

Zhang et al. [418] proposed to use attention-modulated features as well as global and local contexts to detect objects from remote sensing images.

The advent of two-stage models such as RCNN [236] and faster RCNN [44], and one-stage methods such as YOLO series [46,419,420], made another leap in detection accuracy. By adapting two-stage models, most work focuses on improving the quality of region proposals [421–423]. More recently, advanced deep architectures such as Transformer[424] have also been introduced to advance the performance.

4.4.4. Multimodal Data Fusion

Data fusion, as a fundamental task in the field of remote sensing, has been extensively studied for decades. With the availability of multimodal remote sensing data, data fusion techniques are expected to integrate complementary information and help boost the performance of downstream tasks. We briefly discuss two main topics in this area: (1) pansharpening, and (2) task-specific data fusion.

The goal of pansharpening is to integrate panchromatic (PAN) images and multispectral (MS) images, which are two types of optical remote sensing images with inevitable trade-off between spectral diversity and spatial resolution [425]. In general, PAN images provide high spatial resolution but contain limited spectral information, while MS images have much higher spectral resolution with less spatial details. The key point in pansharpening is that while ensuring the spatial increment, the detail injection implemented should preserve the unified spatial–spectral fidelity for fusion products [426]. The first DL-based pansharpening was proposed by Huang et al. [427], in which a modified sparse denoising autoencoder (MSDA) algorithm was used to learn the relationship between high-resolution (HR) and low-resolution (LR) image patches. Masi et al. [428] utilized a shallow CNN to upsample the intensity band after the intensity–hue–saturation (IHS) transform. As pansharpening aims to maximize the spatial injection and minimize spectral distortion, much effort has been devoted to making network architectures good at extracting spatial details while preserving spectral information [429,430]. To this end, Yuan et al. [431] proposed a multiscale and multidepth CNN to better fulfill spatial detail extraction and improve the fusion quality. Yang et al. [432] designed structural and spectral preservation modules and trained the network in the high-pass domain for more effective spatial injection. Zhang et al. [426] introduced saliency analysis as a measure to indicate the demand for spectral and spatial details, and treated them differently in the CNN based fusion process.

Unlike pansharpening aiming only at producing high-quality fusion products, task-specific data fusion usually leverages feature-level or decision-level fusion with specific downstream tasks such as land cover mapping and object detection in a unified framework [433]. A simple way of utilizing multimodal data for training a NN-based model is to concatenate them into an N-dimensional input. In [434], Lagrange et al. found that combining a digital surface model channel with RGB data in the training process can help retrieve some specific classes. For the image classification task, Hong et al. [433] designed an extraction Network (Ex-Net) and a fusion Network (Fu-Net) to learn from two different types of modality. Experiments on HS-LiDAR and MS-SAR data reveal the superiority of multimodal data fusion. Irwin et al. [435] combined SAR, optical imagery and airborne LiDAR data for surface water detection, in which a multilevel decision tree is developed to synthesize the results from a single data source.

4.4.5. Codes, Pretrained Models, and Benchmark Datasets

To fulfill the demand of training deep learning-based models, a number of datasets are proposed by research groups in the earth observation community. Details of the publicly available datasets are shown in Table 4.

Table 4. Databases for remote sensing applications.

Database	Task	Imagery	Resolution	Channels
UCMerced LandUse [436]	Image classification	Multispectral	-	115
University of Pavia [437]		Hyperspectral	1.3 m	11
Salinas [437]		Hyperspectral	3.7 m	224
WHU RS19 [438]	Scene classification	Aerial	up to 0.5 m	3
AID [439]		Aerial	-	3
NaSC-TG2 [440]		Multispectral	100 m	4
NWPU-RESISC45 [403]		Multispectral	30–0.2 m	3
NWPU VHR-10 [441]	Object detection	-	0.5–2 m	3
UCAS-AOD [416]		Aerial	-	3
HRSC2016 [442]		-	2–0.4 m	3
DOTA [443]		Aerial	-	3
DIOR [444]		Aerial	-	3
HRSID [445]		SAR	0.5–3 m	-

4.5. Intelligent Sensor Based Cybersecurity Systems

Cybersecurity is the practice of protecting critical systems and sensitive information from digital attacks, such as intrusion attacks and malware attacks. This section briefly reviews deep learning applications used in the detection of the four types of attacks: intrusion detection, malware detection, phishing detection, and spam detection.

4.5.1. Intrusion Detection

Intrusion detection has become an essential task in the cybersecurity field. The objective of an intrusion detection system (IDS) is to distinguish malicious activities in network traffic and protect sensitive information. The following is a summary of the common attack types used in intrusion attacks.

(1) Denial-of-Service (DoS) attacks, such as botnet and smurf, aim to crash a machine or network service by flooding it with traffic, rendering it inaccessible to its users.
(2) Distributed DoS (DDoS) attacks aim to interrupt the regular traffic of a targeted network by flooding the target or its surrounding infrastructure with huge quantities of network traffic.
(3) User-to-Root (U2R) attacks attempt to get root access as a normal user by exploiting system weaknesses.
(4) Remote-to-Local (R2L) attacks are attempts by a remote system to obtain unauthorized access to the root.
(5) Password-based attacks attempt to obtain access to a system by attempting to guess or crack passwords.
(6) Injection attacks use well-designed instructions or queries to steal sensitive information or obtain unauthorized access to a system.

Deep learning-based techniques have demonstrated exceptional performance for intrusion detection in complicated, large-scale traffic conditions. For example, several recent methods [446–448] have introduced neural networks based on DBNs to achieve improved detection accuracy on the NSL-KDD dataset [449]. However, DBNs-based methods have the drawback that they are computationally unfeasible to train in an end-to-end supervised manner. AEs are widely used as a preprocessing step in intrusion detection, followed by the application of a deep learning classifier. For example, Abolhasanzadeh et al. [450] proposed an AE-based model with seven layers to extract compact and discriminant representations of the input data, and achieved high detection accuracy on the NSL-KDD dataset. In addition, several recent methods based on AEs have considered using stacked AEs for intrusion detection [451–454]. Vu et al. [455] proposed combining variational AEs [94] with several classifiers, such as naive Bayes, SVM, decision tree, and random forest classifiers for intrusion detection, and achieved good results on the NSL-KDD and UNSW-NB15 datasets. These AEs-based methods have the drawback of requiring an additional model to perform classification in additional to the AEs. To address this drawback, recent deep learning-based methods increasingly use CNNs for intrusion detection systems [456,457]. Especially, the LSTM networks have proven very useful because they have the strong ability to process data in intrusion detection that is often structured as sequences of features evolving over time. Several intrusion detection methods in the literature are based on LSTM networks [458]. Among these methods, the one proposed in [459] adopts a three-layer LSTM network, which achieves high detection accuracies on the ADFA-LD and UNM datasets. Similarly, the method proposed in [460] adopts a cascade of three LSTM network modules, which achieve an impressive intrusion detection accuracy by combining them through a voting mechanism. In addition, to take full advantage of LSTMs in processing time series data and CNNs in extracting spatial patterns, several recent methods consider combinations of LSTMs and CNNs for intrusion detection. For example, the method described in [461] uses both a CNN and a hybrid LSTM-CNN to perform the intrusion detection. The method developed in [462] uses a hybrid LSTM-CNN model based on the LeNet. GANs have been used for intrusion detection because their advantage of learning in an unsupervised manner is very suitable for learning the characteristics of data distributions in specific situations (e.g., under normal conditions) in the IDS context. For example, Schlegl et al. proposed a CNN-based GAN [463] to learn the characteristics of data captured under normal conditions, which is then used to detect anomalies by computing the distance between freshly captured data and normal data. In addition, Zenati et al. [464] proposed to further improve the computational efficiency of the GAN in [463] to achieve a faster detection.

4.5.2. Malware Detection

Malware is a malicious software that is disseminated to compromise a system's security, integrity, and functioning. The types of malware include viruses, worms, trojans, backdoors, spyware, botnets, and so on. Deep learning in this field is mainly concentrated on malware detection and analysis. The developed techniques can be generally classified into two categories: PC-based and Android-based malware detection.

(1) *PC-based malware detection.* Deep learning can be used to learn the language of malware through the executed instructions, and thus to help extract resilient features. To achieve this goal, Pascanu et al. [465] firstly proposed a method based on the Echo State Network (ESN) and RNN to classify malware samples. Later, David et al. [466] proposed a DeepSign to automatically generate malware signatures, which does not rely on any specific aspect of the malware. This model uses stacked denoising AE (SDAE) and creates an invariant compact representation of the general behavior of the malware. In 2017, Yousefi-Azar et al. [467] proposed a generative feature learning-based method for malware classification and achieved a network-based anomaly detection using AE. Recently, two GAN-based methods for malware detection have been proposed [468,469]. Specifically, in [468], Kim et al. adopted a transferred deep convolutional GAN (tDCGAN) to generate the fake malware and learn to distinguish it from the real one, which achieves robust zero-day malware detection. In [469], latent semantic controlling GAN (LSCGAN) is proposed to detect obfuscated malware, where features are first extracted using a VAE and then transferred to a generator to generate virtual data from a Gaussian distribution.

(2) *Android-based malware detection.* Malicious Android apps detection is vital and highly demanded by app markets. Deep learning models can automatically learn features without any human interference. The first investigation of applying deep learning to Android malware detection was Droid-Sec [470], which learns more than 200 features from both the static and dynamic analysis of Android apps for malware detection. Later, Hou et al. [471] proposed DroidDelver to deal with Android malware threats, which firstly categorizes the API calls of the Smali code into a block and then applies a DBN for newly unknown Android malware detection. Su et al. [472] proposed the DroidDeep for Android malware detection, which is also a DBN-based model. In 2017, CNN was firstly applied to Android malware detection context

by McLaughlin et al. [473]. They used CNN to extract raw opcode sequences from disassembled code, with the purpose of removing the need to count the vast number of distinct n-grams. Later, Nix et al. [474] proposed a CNN-based framework for Android malware classification, which gets help from API-call sequences. Specifically, a pseudo-dynamic program analyzer is firstly used to generate a sequence of API calls along the program execution path. Then, the CNN learns sequential patterns for each location by performing convolution alongside the sequence and sliding the convolution window down the sequence. Recently, Jan et al. [475] employed a Deep Convolutional GAN (DCGAN) for investigating the dynamic behavior of Android applications.

4.5.3. Phishing Detection

Phishing is a form of fraud in which the attacker tries to learn sensitive information such as login credentials or account information by sending emails or other communication messages. Therefore, phishing detection is a vital task in cybersecurity. Deep learning has also been researched to facilitate the solving of this task. For example, Zhang et al. [476] proposed to detect phishing email attacks by using a 3-layer FCN which consists of one input layer, one hidden layer, and one output layer. In addition, in this network, tanh and sigmoid activation functions are used to better fit the data. Mohammad et al. [477] proposed a self-structuring neural network for detecting phishing website attacks. It can automate the process of structuring the network, which is important for extracting the dynamic phishing-related features. Benavides et al. [478] investigated a variety of networks for cyber-attacks classification and found that the most regularly utilized are DNN and CNN. Although diverse deep learning-based methods have been presented and analyzed, there is still a research gap in the application of deep learning in cyber-attacks recognition.

4.5.4. Spam Detection

The research of spam detection can be basically classed into text spam detection and multimedia spam detection.

(1) *Text Spam Detection.* Text-based spam content generally includes malicious URLs, hashtags, fake reviews/comments, posts, SMS, chat messages, etc. Wu et al. [479] developed a deep learning-based method to identify spam on Twitter, which employs MLP classifiers to learn the syntax of many tweets to perform pre-processing and create high-dimensional vectors. It outperforms the traditional feature-based machine learning methods such as random forest. Jain et al. [480] proposed a semantic CNN (SCNN) that employs a CNN with an additional semantic layer for malicious URL detection, where the semantic layer is a Word2Vec network used to map the word. Thejas et al. [481] proposed a hybrid deep network for click fraud detection, which involves an ANN and auto-encoders (AEs). The ANN is used to gain learning and pass knowledge to the other layers in the hybrid neural network, while the AEs are used to acquire the distribution of human clicks. The proposed hybrid network achieved high accuracy on a real-time dataset of ad-clicks data. Singh et al. [482] proposed using a CNN to classify the aggressive behavior on social networks, which achieved significant accuracy. Ban et al. [483] proposed using a Bi-LSTM network to extract features from Twitter text for spam detection.

(2) *Multimedia Spam Detection.* Deepfake is a currently famous technology that synthesizes media to create falsified content by replacing or synthesizing faces, speech, and manipulating emotions. It uses deep neural networks to learn from large and real samples to simulate human behavior, voices, expressions, variations, etc., and thus, its generated content seems genuine [484]. This technology can be valued in many applications such as movies, games, education, etc. However, it can seriously eradicate trust due to giving forged reality [485]. It also brings many challenges for the spam detection, as its synthetic media is generated by deep learning techniques. Therefore, an arms race between Deepfake techniques and spam detection algorithms

has begun. For example, Hasan et al. [485] proposed employing a blockchain-based Ethereum smart contract framework to deal with media content authenticity. This system can preserve all historical information related to the creator and publisher of the digital media, and then it checks the authenticity of video content by tracking whether it is from some reliable or trustworthy source or not. Fagni et al. [486] proposed a TweepFake to detect deepfake tweets, which involves CNN and bidirectional gate recurrent unit (GRU). For more advanced neural networks for multimedia spam detection we refer to the survey paper [487].

4.5.5. Codes, Pretrained Models, and Benchmark Datasets

Various implementation codes and pretrained models of many of the above introduced methods can be found in the references provided in Section 2.5.2. The popular benchmark datasets for intrusion detection are summarized in Table 5.

Table 5. Benchmarks for cybersecurity intrusion detection.

Dataset	Year	Main Attack Types
AWID3 [488]	2021	Flooding, injection, Botnet
CIC-IDS2017 [489]	2017	DoS/DDoS, port scan, web attacks
AWID2 [490]	2016	Flooding, injection, web attack
UNSW-NB15 [491]	2015	DoS, worms, back-doors, generic
ADFA-LD [492]	2013	Password, web attacks
NSL-KDD [449]	2009	DoS, Probe, U2R, R2L

4.6. *Internet of Things (IoT) Systems*

With the development of commodity sensors and increasingly powerful embedded systems, the research of IoT is rapidly emerging and developing. According to the different sensor systems in the IoT, we describe deep learning in this domain from four aspects: smart healthcare, smart home, smart transportation, and smart industry.

4.6.1. Smart Healthcare

Deep learning and IoT in smart healthcare systems can be researched in the following two aspects.

(1) *Health Monitoring*. Sensor-equipped mobile phones and wearable sensors enable a number of mobile applications for health monitoring. In these applications, human activity recognition is used to analyze health conditions [493]. However, extracting effective representative features from the massive raw health-related data to recognize human activity is one of the significant challenges. Deep learning is employed for this purpose in these applications. For example, Hammerla et al. [494] proposed to use CNNs and LSTM to analyze the movement data and then combine the analysis results to make a better freezing gaits prediction for Parkinson disease patients. Zhu et al. [495] proposed using a CNN model to predict energy expenditure from triaxial accelerometers and heart rate sensors, and achieved promising results to relieve chronic diseases. Hannun et al. [496] proposed using a CNN with 34 layers to map from a sequence of ECG records obtained by a single-lead wearable monitor to a sequence of rhythm classes, and achieved higher performance than that of board certified cardiologists in detecting heart arrhythmias. Gao et al. [497] proposed a novel recurrent 3D convolutional neural network (R3D), which can extract efficient and discriminating spatial-temporal features for action recognition through aggregating the R3D entries to serve as an input to the LSTM architecture. Therefore, with wearable devices, it can monitor health state and standardize the way of life at any time. Deploying deep learning-based methods on low-power wearable devices can be very challenging because of the limited resources of the wearable devices. Therefore, some research works employing deep learning for health monitoring focus on addressing this issue. For example, Ravi et al. [498] utilized a spectral domain

preprocessing for the data input to the deep learning framework to optimize the real-time on-node computation in resource-limited devices.

(2) *Disease Analysis.* Using the comparatively cheap and convenient mobile phone-based or wearable sensors for disease analysis is increasingly important for healthcare. Deep learning has been widely used in assisting this. For example, CNNs have been used to automatically segment cartilage and predict the risk of osteoarthritis by inferring hierarchical representations of low-field knee magnetic resonance imaging (MRI) scans [499]. Another work using CNNs is to identify diabetic retinopathy from retinal fundus photographs [500], which has achieved both high sensitivity and specificity over about 10,000 test images with respect to certified ophthalmologist annotations. Other examples of employing deep learning for disease analysis include the work of Zeng et al. [501], where a deep learning-based pill image recognition model is proposed to identify unknown prescription pills using mobile phones. In addition, Lopez et al. [502] proposed a deep learning-based method to classify whether a dermotropic image contains a malignant or benign skin lesion. Chen et al. [503] proposed a ubiquitous healthcare framework UbeHealth for addressing the challenges in terms of network latency, bandwidth, and reliability. Chang et al. [504] proposed a deep learning-based intelligent medicine recognition system ST-MedBox, which can help chronic patients take multiple medications correctly and avoid taking wrong medications.

4.6.2. Smart Home

Smart home enables the interconnection of smart home devices through home networking for better living. In recent years, a variety of systems has been developed with the application of deep learning techniques. Two main kinds of smart home applications are indoor localization and home robotics, described below.

(1) *Indoor Localization.* With the spread of mobile phones, indoor localization has become a critical research topic because it is not feasible to employ Global Positioning System (GPS) in an indoor environment. Indoor localization covers several tasks such as baby monitoring and intruder detection. However, there are a lot of challenges to achieve these task, e.g, the multi-path effect, the delay distortion, etc. In addition, high processing speed and accuracy are essential for indoor localization systems. Fingerprinting-based indoor localization is a powerful strategy to address these challenges. For example, Gu et al. [505] proposed a semisupervised deep extreme learning machine (SDELM), which takes advantage of semi-supervised learning, deep learning, and extreme learning machine, and achieves a satisfactory localization performance while reducing the calibration effort. Mohammadi et al. [506] proposed a semisupervised DRL model, which uses VAEs as the inference engine to generalize the optimal policies. Wang et al. [507] proposed using an RBM with four layers to process the raw CSI data to obtain the locations. One challenge of applying deep learning in this field is the lack of suitable databases for large indoor structures such as airports, shopping malls, and convention centers. In addition, DRL-based fingerprinting is another area that has not received much attention. However, DRL is gaining enormous momentum and may push the boundaries of performance.

(2) *Home Robotics.* Equipped with commodity sensors, home robots can perform a variety of tasks in home environments. For example, popular tasks include localization, navigation, map building, human–robot interaction, object recognition, and object handling. However, case-specific strategies are needed for guiding a mobile robot to any desired locations when GPS is not available. In [508], a deep learning-based method for autonomous navigation to identify markers or objects from images and videos is proposed, which uses pattern recognition and CNNs. Levine et al. [509] proposed to train a large CNN to achieve successful grasps of the robot gripper using only monocular camera images. This method can predict the probability of the task-space motion of the gripper, and is independent of the camera calibration or

the current robot pose. Therefore, it greatly improves the hand-eye coordination of a robot for object handling, and thus improve human–robot interaction. Reinforcement learning and unsupervised learning will be promising in this area because it is inefficient to manually label data that may change dramatically depending on the user and environment in a smart home.

4.6.3. Smart Transportation

Nowadays, intelligent transportation systems heavily depend on the historical and real-time traffic data collected from all kinds of sensors, such as inductive loops, cameras, crowd-sourcing, and social media. Deep learning in various smart transportation systems currently has the following focuses.

(1) *Traffic Flow Prediction.* Traffic flow prediction is a basic and essential problem for transportation modeling and management in intelligent transportation systems. Deep learning has been increasingly used in this area to exploit the rich amount of traffic data and thus extract highly representative features. For example, Huang et al. [510] proposed using a DBN to capture effective features from each part of road traffic networks, and then these features from related roads and stations are grouped to explore the nature of the whole road traffic network to predict traffic flow. Lv et al. [511] proposed a stack of AEs model to extract features from historical traffic data to make the prediction. In addition, there are a lot of works focused on using deep learning for traffic and crowd flow prediction [512,513] Most current methods to predict traffic flow are for short-term prediction while long-term prediction horizons can reduce costs and provide better intelligent transportation system management. Research in this field is very challenging due to the difficulty of achieving high accuracy of long-term prediction. A promising solution is using data-driven methods.

(2) *Traffic Monitoring.* Traffic monitoring is one of the most popular research fields in smart transportation. Its aim is to both reduce the workload of human operators and warn drivers of dangerous situations. Therefore, traffic video analytics is a key part of traffic monitoring. One of the key tasks in traffic monitoring is object detection, which includes pedestrian detection, on-road vehicle detection, unattended object detection, and so on. As in other tasks (Section 4.1), deep neural networks for object detection have also played an important role here, and have significantly improved the accuracy and speed of traffic monitoring. For example, Ren et al. [44] proposed using a region proposal network (RPN), which shares full-image convolutional features with the detection network and can achieve nearly cost-free region proposals. Redmon et al. [46] proposed to formulate frame object detection as a regression problem, which separates the processes of recognizing bounding boxes and computing class probabilities. Another important task in traffic monitoring is object tracking, which plays a significant role in surveillance systems, including tracking suspected people or target vehicles for safety monitoring, urban flow management, and autonomous driving. Deep learning has also been widely in this area. For example, Vincent et al. [451] proposed building deep networks based on stacking layers of denoising AEs for this purpose. Li et al. [514] proposed a robust tracking algorithm based on a single CNN to learn effective feature representations for the target object. Ondruska et al. [515] proposed an end-to-end object tracking approach, which uses RNN to directly map from raw sensor input to object tracks in sensor space.

(3) *Autonomous Driving.* Autonomous driving is crucial to city automation. Vision-based autonomous driving systems have two main paradigms: mediated perception-based and behavior reflex-based. The underlying idea of mediated perception-based methods is to recognize multiple driving-relevant objects, such as lanes, traffic signs, traffic lights, cars, and pedestrians. However, most of these systems rely on highly precise instruments and thus bring unnecessarily high complexity and cost. Therefore, current autonomous driving systems focus more on real-time inference speed, small model size, and energy efficiency [516]. Deep learning is adopted here to

learn a map from input images/videos to driving behaviors, or to construct a direct map from the sensory input to a driving action. For example, Bojarski et al. [517] trained a CNN to map raw pixels from a single front-facing camera directly to steering commands. Xu et al. [518] proposed using an end-to-end FCN-LSTM network to predict multimodal discrete and continuous driving behaviors. Readers interested in finding more deep learning-based methods for this topic are referred to the survey paper [519]. Currently, most papers on deep learning for self-driving cars focus on perception and end-to-end learning. Although deep learning has made great progress in the accuracy of object detection and recognition, the level of recognition detail still needs to be improved to perceive and track more objects in real time in the autonomous driving scene. In addition, the gap between image-based and 3D-based perception needs to be filled.

4.6.4. Smart Industry

Smart industry, also known as industry 4.0, represents the latest trend of the manufacturing revolution. In the era of smart industry, explosive data produced in manufacture can be analyzed by deep learning to empower the manipulators with human-like abilities. Deep learning in several main research topics are described as follows.

(1) *Manufacture Inspection.* Manufacture inspection refers to inspecting and assessing the quality of products. Various deep learning-based visual inspection methods have been proposed and become a powerful tool to extract representative features and thus to detect product defects in large scale production. For example, Li et al. [520] proposed a CNN-based classification model to implement a robust inspection system, which significantly improves the efficiency. Park et al. [521] proposed a generic CNN-based method to extract patch features and predict defect areas through thresholding and segmenting for surface integration inspection. Deep learning based methods have achieved the best experimental results so far in this domain, with accuracies ranging from 86.20% up to 99.00%.

(2) *Fault Assessment.* Fault assessment is crucial to building smart factories. Specific application tasks include machinery conditions monitoring, incipient defects identification, root cause of failures diagnosis, fault detection of rotating machines with vibration sensors, bearing diagnosis, tool wear diagnosis, and so on. This information can then be incorporated into manufacturing production and control. Deep learning has also been used here to solve these tasks. For example, Cinar [522] proposed using transfer learning models for equipment condition monitoring. Chen et al. [523] investigated the latest deep learning based methods for machinery fault diagnostics. Wang et al. [524] proposed a wavelet-based CNN to achieve automatic machinery fault diagnosis. Specifically, a wavelet transform is used to transfer a one-dimensional vibration signal into a two-dimensional one which is then fed into the CNN model. Wang et al. [525] proposed a continuous sparse auto-encoder (CSAE), which incorporates a Gaussian stochastic unit into its activation function to extract nonlinear features of the input data. Lei et al. [526] proposed a sparse filtering based two-layer neural network model, which is used to learn representative features from the mechanical vibration signals in an unsupervised manner. Generally, AE fits well with high-dimensional data and thus is a good technique of choice for fault assessment.

(3) Others. Deep learning has also been used in many sectors of renewable power systems. For example, Alassery et al. [527] proposed using neural networks for solar radiation prophesy models for green energy utilization in the energy management system. Another promising application of deep learning in the smart industry field is smart agriculture. For example, Khan et al. [528] proposed an optimized smart irrigation system for effective energy management, which overcomes the problems of transmitting data failure, energy consumption, and network lifetime reduction in the field of IoT-based agriculture. DNNs have also been applied in waste management.

For example, Kshirsagar et al. [529] proposed using a customized LeNet model to classify garbage into cartons and plastics.

4.6.5. Codes, Pretrained Models, and Benchmark Datasets

Implementation codes and pretrained models of many of the above introduced applications can be found in the references provided in Section 2.5.2. In addition, some commonly used datasets suitable for building deep learning applications in IoT are listed as follows.

(1) CGIAR Dataset: http://www.ccafs-climate.org/ (accessed on 2 November 2022)
(2) Educational Process Mining: https://archive.ics.uci.edu/ml/datasets/mining (accessed on 2 November 2022)
(3) Commercial Building Energy Dataset: https://combed.github.io/ (accessed on 2 November 2022)
(4) Electric Power Consumption: https://archive.ics.uci.edu/ml/datasets/power (accessed on 2 November 2022)
(5) AMPds Dataset: http://ampds.org/ (accessed on 2 November 2022)
(6) Uk-dale Dataset: https://jack-kelly.com/data/ (accessed on 2 November 2022)
(7) PhysioBank Databases: https://physionet.org/data/ (accessed on 2 November 2022)
(8) T-LESS: http://cmp.felk.cvut.cz/t-less/ (accessed on 2 November 2022)
(9) Malaga Datasets: http://datosabiertos.malaga.eu/dataset (accessed on 2 November 2022)
(10) ARAS Human Activity Datasets: https://www.cmpe.boun.edu.tr/aras/ (accessed on 2 November 2022)

4.7. *Natural Language Processing (NLP)*

NLP is a crucial and widely researched field. It is a subfield of AI that is concerned with enabling computers to understand text and spoken language in much the same way humans do. Due to the ambiguities of human language, NLP is a very challenging problem. Some involved popular tasks include speech recognition, sentiment analysis, machine translation, and question answering, introduced in the following.

4.7.1. Speech Recognition

Speech recognition, also called speech-to-text, refers to the task of enabling a computer to translate human speech into text. There are many algorithms for speech recognition, but deep learning provides more advanced solutions. This is because DNNs can combine several aspects of the voice signals such as grammar, syntax, structure, and composition to understand and process human speech. The initial success in speech recognition was achieved by Zweig et al. [530] on a small-scale dataset with an error rate of 34.8%. After that, more advanced neural networks were proposed to improve recognition accuracy such as the representative networks Segmental RNN, EdgeRNN, and Quanaum CNN [531–533]. Comprehensive introductions of architectures for speech recognition can be found in recent survey papers [534–536].

4.7.2. Sentiment Analysis

Sentiment analysis refers to the task of determining the attitude of reviewers, more specifically, the task of determining whether data are positive, negative, or neutral. It focuses on the polarity of a text but also aims to detect specific feelings and emotions such as happy and sad, and intentions such as interested and not interested. Popular types of sentiment analysis include graded sentiment analysis, emotion detection, and multilingual sentiment analysis. When applying deep learning to sentiment analysis, it is usually formulated as a classification problem, where DNN takes texts as input and outputs a category representing the sentiment class. For example, the Bag-of-Words (BoW) model is one of the most reputable methods for document level sentiment classification [537]. The Recursive AE (RAE) network proposed by Socher et al. is the first model for sentence level sentiment classification [538]. The Adaptive RNN (AdaRNN) is a renowned model for aspect-level sentiment classification [539]. More introductions and discussions of DNNs for sentiment analysis are given in other review papers [540,541].

4.7.3. Machine Translation

Machine translation refers to the process of automatically translating text from one language to another without human involvement. This is one of the first applications of computing power, starting in the 1950s. Deep learning is well suited for this problem because DNNs can consider the whole input sentence at each step for generating the output sentence. This way, it can address the limitations of traditional methods that need to break an input sentence into words and phrases, and thus provide better translation quality. Basically, DNNs for machine translation have an encoder-decoder structure, where the encoder learns to extract the important features from its input sentence, and the decoder processes the extracted features and outputs the target sentence. For example, Kalchbrenner et al. [542] proposed a model with a CNN encoder and RNN decoder, which is the most original and classic structure of machine translation. More advanced architectures for machine translation are discussed in recent survey papers [543,544].

4.7.4. Question Answering

Question answering refers to building systems that can answer questions posed in a natural language by humans. For this problem, a DNN takes a specific question and a paragraph of text as input and aims to output an answer to this question based on the given text. Such DNNs need to understand the structure of the language and have a semantic understanding of the context and the question, thus, attention-based DNNs are needed to handle the complex training. More specifially, attention-based RNNs are suitable for this task. One of the most famous networks for question answering is R-Net [545], which employs a gated attention-based RNN. Other renowned architectures include FusionNet [546] and the recently emerging Transformer. For comprehensive introductions and discussions on question answering we refer to recent surveys [547,548].

4.7.5. Codes, Pretrained Models, and Benchmark Datasets

In addition to the references provided in Section 2.5.2, we refer to a survey of pretrained models for NLP [549]. A collection of renowned benchmark datasets that are widely used in NLP to evaluate different deep learning methods can be found at https://github.com/niderhoff/nlp-datasets/blob/master/README.md accessed on 2 November 2022. In addition, we list the most advanced pretrained language models as below.

(1) BERT: https://github.com/google-research/bert (accessed on 2 November 2022)
(2) GPT2: https://github.com/openai/gpt-2 (accessed on 2 November 2022)
(3) XLNet: https://github.com/zihangdai/xlnet (accessed on 2 November 2022)
(4) RoBERTa: https://github.com/facebookresearch/roberta (accessed on 2 November 2022)
(5) ALBERT: https://github.com/google-research/albert (accessed on 2 November 2022)
(6) T5: https://github.com/google-research/T5 (accessed on 2 November 2022)
(7) GPT3: https://github.com/openai/gpt-3 (accessed on 2 November 2022)
(8) ELECTRA: https://github.com/google-research/electra (accessed on 2 November 2022)
(9) DeBERTa: https://github.com/microsoft/DeBERTa (accessed on 2 November 2022)
(10) PaLM: https://github.com/lucidrains/PaLM-pytorch (accessed on 2 November 2022)

4.8. Audio Signal Processing

Audio signal processing was an early application of deep learning and is still one of its major application domains. Before deep learning, conventional methods for audio signal processing relied on handcrafted feature extraction, including the mel frequency cepstral coefficients (MFCCs), discrete cosine transform, and mel filter bank. Deep learning improves the processing performance by learning hierarchical representations from the audio signal using various models such as CNNs, RNNs, and GANs. These models are either trained using raw audio signals or classical features extracted from audio signals. This section briefly reviews the application of deep learning in the main scenarios of audio signal processing, including speech recognition, music and environmental sound analysis, localization and tracking, source separation, audio enhancement, and synthesis.

4.8.1. Speech Recognition

Different from the speech recognition in Section 4.7.1, here speech recognition refers to converting speech into sequences of words in the context, which is the base for any speech-based interaction system. It is widely used in virtual assistance systems such as Google Home, Apple Siri, and Microsoft Cortana, and speech transcriptions such as the YouTube caption function. For a long time, the modeling of speech was dominated by methods based on Gaussian mixture models and hidden Markov models due to their mathematical elegance. However, deep learning models dramatically reduced the word error rate on various recognition tasks, and hence became mainstream [550]. Popular models for speech recognition include LSTMs GRUs [551], and a combination of LSTM layers with convolutional layers [552]. RNN blocks (including LSTM and GRU) are widely used to model the temporal correlations in speech sequences. Sequence-to-sequence models such as CTC (connectionist temporal classification) [553] and LAS (listen, attend and spell) [554] were also proposed. Transfer learning also plays an important role to enhance systems on low resource language with data from rich resources languages [555]. In addition to speech recognition, other applications related to speech are voice activity detection, speaker recognition (see Section 4.3.4), speech translation, and language detection [556].

4.8.2. Music and Environmental Sound Analysis

Music analysis involves low-level tasks such as onset/offset detection, fundamental frequency estimation, rhythm analysis, and harmonic analysis, and high-level tasks such as instrument detection, separation, transcription, segmentation, artist recognition, genre classification, discovery of repeated themes, music similarity estimation, and score alignment. These tasks were previously done by handcrafted features and conventional classifiers, and are now addressed by deep learning algorithms such as LSTMs, CNNs, and RNNs [557]. Modern systems integrate temporal modeling [558], applying 2D convolution on spectro-temporal inputs before doing 1D convolution to fuse representations across frequencies, followed by GRU to capture the sequence dependencies.

Environmental sound analysis is often used in multimedia indexing and retrieval, acoustic surveillance, and context-aware devices. In terms of recognition tasks, deep learning models are mainly used for acoustic scene classification [559] (the scene labels can be home and street), acoustic event detection [560] (detect the start and end time of an event and assign a label to the event) and tagging [561] (predict multiple sound classes at the same time).

4.8.3. Source Separation, Enhancement, Localization and Tracking

Source separation aims to recover one or several source signals from a given mixture signal. It is an important task in audio signal processing in real-world environments, and is often performed before speech recognition to improve the data quality. In single-channel setups (only one microphone is used), deep learning aims to model the single-channel spectrum or the separation mask of a target source [562]. Convolutional and recurrent layers are often used in such models. Furthermore, some methods integrate supervised learning and supervised learning for source separation. For example, deep clustering [563] performs supervised learning to estimate embedding vectors for each time-frequency point, then cluster them in an unsupervised manner for separation. In multi-channel setups (e.g., audio data are collected from multiple microphones), the separation can be improved by taking into account the spatial locations of sources or the mixing process. In this case, the input of DNNs contains spatial features as well as spectral features, and the DNNs are used to estimate the weights of a multi-channel mask [564].

Audio enhancement aims to reduce noise and improve the audio quality. It is a critical component for robust systems. Deep learning in audio enhancement is mainly designed for reconstructing clean speech [565] or estimating masks [566] from noisy signals. To this end, researchers have proposed various models based on GANs [112], denoising AEs [567], CNNs [567] and RNNs [568].

For localization and tracking, deep neural networks are often trained on the phase spectrum [569], magnitude spectrum [570], and cross-correlation between channels [567]. The key is to design an architecture, e.g., a CNN, in a way that can learn the inter-channel information while extracting within-channel representations.

4.8.4. Sound Synthesis

Sound synthesis can be used to generate realistic sound samples, speeches [571], music and art [572]. It is achieved by generative models which learn the characteristics of sound from a database and output desired sound samples. When the deep learning model is operated to generate fake speeches for a given person, it is often referred to as DeepFake. Popular deep learning models used for sound synthesis include VAEs and GANs [573], where the sound is synthesized and upsampled from a low-dimensional latent representation. Autoregressive approaches such as LSTM and GRU, on the other hand, generate new samples iteratively based on previous samples [574]. With multiple stacked layers, such methods are able to process sound at different temporal resolutions. The WaveNet [575] is a popular model in this regard. It stacks dilated convolutional layers, providing context windows of reasonable size to allow the model to learn context information (e.g., speaker identity). Furthermore, the problem of autoregressive sample prediction is cast into a classification problem. Follow-up models such as the parallel WaveNet [576] further improve the computational efficiency during the training stage.

4.9. Robotic Systems

Applications of deep learning in robotics are mainly aimed at addressing the challenges in learning complex and high-dimensional dynamics, learning control policies in dynamic environments, advanced motion manipulation, object recognition and localization, human action interpretation and prediction, sensor fusion, and task planning. In terms of deep learning architectures and strategies, existing methods for robotics can be classified into discriminative models, generative and unsupervised models, recurrent models, and policy learning models trained with reinforcement learning. This section briefly reviews how these models are used in different tasks.

4.9.1. Learning Complex Dynamics and Control Policies

Robots often need to cope with states with high-level uncertainty, which requires the system to be able to quickly and autonomously adapt to new dynamics. This is important in tasks such as grasping new objects, traveling over surfaces with unknown or uncertain properties, managing interactions with a new tool or environment, and adapting to system degradation. Discriminative models, such as CNNs, were trained to assess the possibility of a specific robot motion for successfully grasping common office objects from image data [509]. DeepMPC [577] is a recurrent conditional deep predictive dynamics model for robotic food-cutting which is a controlling task with complex nonlinear dynamics. Transforming recurrent units were adopted to handle the time-dependent dynamics by integrating long-term information while allowing transitions in dynamics. Generative models such as AEs and GANs were also used to model the nonlinear dynamics of simple physical systems [578] and inverse dynamics of a manipulator [579]. Reinforcement learning plays a significant role in robotic control tasks. It is useful in learning to operate dynamic systems from partial state information. For example, it has been used to learn deep control policies for autonomous aerial vehicles control [580].

4.9.2. Motion Manipulation

It remains elusive to find robust solutions for robotic motion tasks such as grasping deformable or complex geometries, using tools, and actuating in dynamic environments. The corresponding challenges approached with deep learning methods are grasp detection, path and trajectory planning, and motion control. Deep learning models based on recurrent units, CNNs [581,582], and deep spatial AEs [583] have been used for learning visuomotor and manipulation action plans.

4.9.3. Scene/Object Recognition and Localization

Scene and object recognition as well as localization are critical tasks for robot systems, since knowing what kind of objects are there in the environment and the locations of those objects is a prerequisite for performing other tasks. Deep learning methods have shown promising performance in recognizing and classifying objects for grasp detection [581,584], including advanced applications such as recognizing deformable objects and estimating their state and pose for grasping [585], semantic tasks [586], and path specification [587].

4.9.4. Human Action Interpretation and Prediction

Effective human–robot interaction requires the robot to have social skills, hence, the robot needs to be capable of inferring the intentions of humans and giving corresponding responses or actions accordingly. Such skills are critical in human–robot collaborative applications such as social robots, manufacturing, and autonomous vehicles. Interpreting and predicting human social behavior is a complex task, and it is difficult to formulate handcrafted solutions. Deep learning methods present great potential in this area. Learning by demonstration [581] is one way to solve the problem, where deep learning models are trained to learn manipulation action plans by watching unconstrained videos from the World Wide Web. In another study, a recurrent model was trained for the robot to learn grasping actions from a human collaborator [588].

4.9.5. Sensor Fusion

The use of multiple sources of information is necessary in robotic systems, as it provides a plethora of rich representations of the environment and brings proper redundancy to the system to deal with uncertainties. The challenge is how to construct meaningful and useful representations from the various data sources. Due to the hierarchical structures, deep learning models naturally support the processing and integration of high-level representations learned from different data streams. For example, generative models [589] and recurrent models [590] with unsupervised learning were proposed for integrating multimodal sensorimotor data, including video, audio, and joint angles, for robotic systems. The level of abstraction depends on the application specifics.

4.9.6. Knowledge Adaptation in Robotic Systems

Training deep learning models can be time-consuming and data demanding. A robotic system should be crafted in a way that is easy to adapt to a similar task. Transfer learning plays an important role in leveraging the knowledge gained by previous solutions of similar problems to solve new problems. To this end, researchers have used pretrained models [236] and proposed sim-to-real approaches [591] to facilitate the learning process and improve efficiency. For example, AlexNet, GoogleNet, and VGG models pretrained on the ImageNet dataset have been used for extracting high-level representations from image data for object recognition in robotic systems. The sim-to-real approaches offer a way to create the solution in a simulation environment and apply it to the real-world problem, which is a safer and more economical way than the traditional trial-and-error approaches. Furthermore, some works focused on extraction of domain invariant features [592] to transfer knowledge across domains. Other works proposed learning by imitation/demonstrations approaches to help robots to learn manipulation skills [593]

4.10. Information Systems

Deep learning has received increasing attention in information systems. Major applications include social network analysis, information retrieval, and recommendation.

4.10.1. Social Network Analysis

Social network analysis is an important problem in data mining. It targets social media networks such as Facebook, Twitter and Instagram to analyze their patterns and infer knowledge from them. Network representation learning is an important task in social network analysis. It encodes network data into low-dimensional representations, namely network embeddings, which effectively preserves network topology and other attribute information, facilitating subsequent tasks such as classification [594], link prediction [595], semantic evaluation [596], anomaly detection [597], and clustering [598].

Semantic evaluation helps machines understand the semantic meaning of users' posts in social networks and infer the users' opinions. Examples of sentiment classification are the SemEval [599] and Amazon purchase review [596]. Link prediction is widely used in recommendation and social ties prediction applications, where deep learning models are trained to learn robust representations to enhance prediction performance and deal with the scalability issue [595]. Popular models for link prediction include RBMs, DBNs [600], and GNNs [601]. In some studies, transfer learning with pretrained models (e.g., RBMs) was applied to improve the training efficiency and address the insufficient data issue [600]. Anomaly detection aims at spotting malicious activities in social networks, such as spamming and fraud. Such activities can be interpreted as outliers that deviate from the majority of normal activities. Deep learning approaches based on network embedding techniques are receiving increasing attention in this field [597]. Anomaly detection is also related to crisis response [602] which focuses on detecting natural and man-made disasters, where deep learning models are trained to identify information from the posts and classify them into classes such as bushfire and earthquake. It is worth noting that attention mechanisms have been widely adopted in sequence-based tasks to allow the deep learning models to focus on relevant parts of the input during the learning process. Attention layers are also used for aggregating important features from the local neighbors of nodes, as in graph attention networks [136].

4.10.2. Information Retrieval

Deep learning approaches are also employed in document retrieval and web search applications [603]. A representative work is the deep-structured semantic modeling (DSSM) [603] which adopts a DNN for latent semantic analysis. A following work improved DSSM by applying convolutional layers to integrate representations extracted from each word in the sequence in order to generate representations for a subset of words [604]. Moreover, deep stacking networks were proposed for general information retrieval tasks, where multiple network blocks were stacked on top of each other to extract high-level, low-dimensional abstractions in the final feature space.

4.10.3. Recommendation Systems and Others

Recommender systems play an important role in online shopping services by helping users discover items of interest from a large resource collection. A memory augmented graph neural network (MA-GNN) can capture both the long- and short-term user interests. Ma et al. [605] proposed memory augmented graph neural networks for sequential recommendation. Specifically, a GNN was used to model the item contextual information within a short-term period, a shared memory network was designed to capture the long-range dependencies between items, and co-occurrence patterns of related items were captured to model the user interests. Furthermore, a heterogeneous information network containing different types of nodes and links is a powerful information model in this field. Hence, researchers have proposed embedding methods to represent the network for recommender systems [606]. Other applications include bibliometric analysis such as citation prediction [607] and co-authorship network analysis. In such works, deep learning models, especially graph neural networks, were proposed to learn patterns from the citation networks, co-authorship networks, and heterogeneous bibliometric networks [608].

4.11. Other Applications

4.11.1. Deep Learning in Food

Deep learning has recently been introduced in food science and engineering and has proved to be an advanced technology. The research of deep learning in food mainly focuses on the following topics.

(1) *Food Recognition and Classification.* Food analysis is important for the health of human beings. As image sensing has become an easy and low-cost information acquisition tool, food analysis based on images of food has become popular. Food images contain important information of food characteristics, which can be used to recognize and classify food to help people record their daily diets. Currently, with the great success of CNN in various recognition and classification tasks, several CNN variants have been adopted for food recognition and classification [609–611]. These methods achieve relatively good results, yet there is still room for improvement in accuracy and efficiency.

(2) *Food Calorie Estimation.* Food calorie estimation is widely adopted in many mobile apps to help people monitor and control nutrition intake, lose weight, and improve dietary habits to stay healthy. An image-based food calorie estimation method has been proposed and become popular [612]. It uses a multitask CNN and outperforms the traditional search-based methods. Following this, more CNN-based methods have been proposed for this task and proved that CNNs are effective for image-based food calorie estimation [613,614].

(3) *Food Quality Detection.* Food quality is vital for the health of human beings. Food quality detection can be further divided into subtopics of vegetable quality detection, fruit quality detection, and meat and aquatic quality detection. Among them, vegetables and fruits quality detection are currently hot and challenging topics. Stacked sparse AE and CNN were adopted for detecting vegetable quality based on hyperspectral imaging [615], where the diversity of surface defects in size and color are problematic for traditional methods based on the average spectrum of the whole sample. DNNs coupled with spectral sensing methods have been proposed for addressing problems of varieties classification, nutrient content prediction, and disease and damage detection in fruit quality detection [616,617].

(4) *Food Contamination.* Food contamination is a serious threat to human health, and thus has received great attention from all over the world. Several deep learning based methods have been proposed for predicting, monitoring, and identifying food contamination. For example, Song et al. [618] proposed using DNNs to predict the morbidity of gastrointestinal infections by food contamination. Gorji et al. [619] proposed using deep learning to automatically identify fecal contamination on meat carcasses. We refer to the survey paper [620] for more works and discussions. Generally, CNNs and their variants are still the most widely used and effective methods in this field.

4.11.2. Deep Learning in Agriculture

Since the concept of precision farming was proposed, it has brought new problems and challenges. Deep learning has been adopted to develop agricultural intelligent machinery equipment due to its strong ability of extracting features from image and structured data.

(1) *Plant Diseases Detection.* Detecting diseases of crop is important for improving productivity. There are many types of disease species to be inspected. Deep learning technologies have been applied to crop disease classification or detection. For example, Ha et al. [621] proposed a deep learning based method to detect radish disease, where the radish was classified into diseased and healthy through a CNN.

Ma et al. [622] also proposed using a CNN to recognize the four types of cucumber diseases. Lu et al. [623] proposed using CNNs to identify ten types of rice diseases, which proved the superiority of CNN-based methods in identifying rice diseases.

(2) *Smart Animal Breeding Environment.* Deep learning technologies have been adopted for monitoring and improving animal breeding environment. The currently most popular research in this domain is face recognition and behavior analysis of pigs and cows. For example, Yang et al. [624] proposed using a CNN combined with spatial and temporal information to detect nursing behaviors in a pig farm. Qiao et al. [625] proposed using a Mask R-CNN to settle cattle contour extraction and instance segmentation in a sophisticated feedlot surrounding. These works demonstrated the effectiveness of CNNs in automatic recognition of nursing interactions for animal farms. In addition, Hansen et al. [626] proposed a CNN-based method to recognize pigs. Tian et al. [627] proposed using CNN to count pigs.

(3) *Land Cover Change Detection.* Land cover change is vital for the natural basis of human survival, the Earth's biochemical circle, and the energy and material circulation of the Earth system. One of the fundamental tasks in land cover change is cover classification. Deep learning techniques have been adopted for addressing this task. For example, Kussul et al. [628] proposed a multilevel deep learning architecture to classify the land cover and crop types using remote sensing data. Gaetano et al. [629] proposed a two-branch CNN for land cover classification. In addition, several CNN variants and transfer learning are adopted in the literature to validate land cover and classify wetland classes. See the survey papers [630,631] for details.

4.11.3. Deep Learning in Chemistry

Deep learning has been actively and widely used in computational chemistry in the past few years. Several hot and popular research topics are discussed as follows. To build a molecule with a particular property would first require developing methods to accurately correlate any given structure to the property. These can then be used to intelligently design a molecule that maximizes the desired property. The final step is to design an efficient synthesis from readily available starting materials.

(1) *Materials Design.* Advanced materials are fundamental for many modern technologies such as batteries and renewable energy. Deep learning in this field is comparatively new, but there has been a rapid growth in the past few years. Xie et al. [632] proposed using a crystal CGNN to capture the crystalline structure for accurate and interpretable prediction of material properties. In addition, CGNNs and several CGNN variants have been proposed to predict the properties of bulk materials [633], optimize polymer properties [634], and explore chemical materials space [635]. These works demonstrated great potential of deep learning in exploring properties of materials. In addition to this, deep learning has been used to optimize synthesis parameters [636] and perform defect detection [637].

(2) *Drug Design.* Drug design is one of the most important applications of chemistry. Its aim is to identify molecules that achieve a particular biological function with maximum efficacy. Deep learning has been used to optimize the properties of molecules to improve potency and specificity, while decrease side effects and production costs. Specifically, AEs, GANs, and RNNs have been used to generate potent drug molecules [638–640]. More deep learning based methods are reviewed and discussed in recent papers [641–643].

(3) *Retrosynthesis.* The underlying challenge of retrosynthesis is similar to that of board games such as Chess and Go [644]. It can be solved by formulating the retrosynthesis as a tree search, where the branching factor is how many possible steps can be taken from a particular point. Therefore, inspired by AlphaGo, one of the predominant retrosynthetic AI was proposed by Segler et al. [645], which adopted the AlphaGo methodology of Monte Carlo Tree Search with deep neural network. This method has shown great potential. However, assessing synthesis plans is a challenging task.

Other research has been using RNNs and AE to perform retrosynthetic analysis of small molecules [646].

(4) *Reaction Prediction*. Reaction prediction refers to taking a set of known reagents and conditions and predicting the products that will form. Deep learning has been used in this field to reduce the high computational cost in chemical space exploration. A representative work using DNNs to predict which products can be formed is presented by Wei et al. [647]. RNN variants and Siamese architectures have also been proposed for reaction prediction [648,649]. Emphasizing interpretability by using GCNN to predict reaction in a manner similar to human intuition is currently a hot research direction in this field [650].

5. Deep Learning Challenges and Future Directions

5.1. *Efficiency*

One of the growing problems of deep learning is computing efficiency. With the increasing volume of data and increasing complexity of DNNs, the requirement for computing power is increasingly high. This can be solved to some extent by advanced multicore GPUs, and tensor processing units (TPUs). However, more efficiency is often needed when optimizing deep learning architectures for embedded devices applications. This can be achieved through codesigning model architectures, training algorithms, software, and hardware to allow multimachine parallelism and scalable distributed deep learning [278]. For example, using compression techniques to compress the layers and thus optimize the model architecture; trimming the number of parameters to achieve a smaller footprint or a more efficient model; designing layers and architectures specifically with efficiency to save the number of parameters and avoid over-parameterization. Another challenging and promising direction is to design programmable computational arrays, bare-hardware implementation, and stochastic computation mechanisms [1].

5.2. *Explainability*

A major problem that affects the deployment of deep learning in various areas is the lack of transparency, which also called the "black box" problem. Deep learning algorithms learn from data to find patterns and correlations that human experts would not normally notice, and their decision-making processes often confuse even the engineers who created them. This might not be a problem when deep learning is performing a trivial task where a wrong result will cause little or no damage. However, when it comes to medical diagnosis or financial trades, a mistake can have very serious consequences. Therefore, the transparency issue is a potential liability when applying deep learning. Various visual analysis tools have been proposed to dissect DNNs and reveal what they actually learn, as investigated in the paper [651]. In addition, there are some techniques such as LIME [652] and Deep Lift [653] that can be used to explain the model using feature importance. However, the transparency issue has been well solved now. A promising way is to link neural networks to the existing well-known physical or biological phenomena [1]. This will help to develop a metaphysical relationship to demystify the DNN's "brain".

5.3. *Generalizability*

Generalizability is an important concern when applying a trained deep learning model in practice. It is challenging to demonstrate a deep learning model's generalizability before implementing the model. To address the problem of model generalizability, many researchers try to use as much and as diverse data as possible to train a deep learning model. However, this is very challenging for some applications such as clinical scenarios where obtaining sufficient training samples with labels is extremely expensive and labor-intensive. Some researchers work on optimization algorithms that minimize the training error to achieve generalization. For example, Neyshabur et al. [654] proposed Path-SGD for better generalization, which is invariant to rescaling of weights. Hardt et al. [655] proposed to use stochastic gradient descent to ensure uniform stability, and thus to improve generalization for convex objectives. However, these are based on the assumption of having a "closed set" where the possible conditions in the test data are exactly the same as those in the training data. For many practical applications, the scenario is that "incomplete knowledge of the world is present at training time, and unknown classes can be submitted to an algorithm during testing" [656]. Therefore, a possible and promising research direction is using more generalizable or "open set" approaches to develop and evaluate deep learning models, such as open-set recognition [657].

5.4. Ethical and Legal Issues

Though deep learning has been widely deployed in many fields and has gained great success, some ethical and legal issues are emerging. There are two prominent issues. The first is the biased learning issue, where the model will provide a biased and prejudiced prediction/recommendation. One typical real-life example is the COMPAS (Correctional Offender Management Profiling for Alternative Sanctions) algorithm, which is used in US court systems to predict the probability that a defendant would become a recidivist [658]. This algorithm produced two times as many false positives for recidivism for black criminals (45%) than white criminals (23%). Another critical issue is the privacy of the deep learning training data. For example, social network face images could be used for training the deep learning model without the prior consent of the subjects. The lack of relevant governing frameworks on the regulation of these ethical and legal issues affects the wide application of deep learning in sensitive areas such as healthcare, finance, security, and law enforcement. The community is in desperate need of developing a relevant code of ethics and legal frameworks for addressing those issues.

5.5. Automated Learning

Deep learning has achieved great success in automatically learning representative features and performing recognition of these learned features. Although this has greatly eliminated the cumbersome process of handcrafting features, the development of deep learning models is resource-intensive, requiring significant domain knowledge and time to produce and compare dozens of models. Various software tool kits have been developed for getting production-ready deep learning models with great ease and efficiency [659,660]. However, these are not satisfactory enough for developing high-level and user-friendly platforms that are easy also for non-experts to adopt existing DNNs or to design their own solutions. Automated machine learning (AutoML) is a research field for this purpose. For deep learning, a variety of neural architecture search (NAS) methods have been proposed to automate the network designing process [291,661], which will be a promising way to solve the automated learning problem.

5.6. Distributed Learning

With the development of IoT and smart-world applications, massive numbers of smart mobiles and embedded devices are incorporated into the computing, resulting in network congestion and latency. Recent research in edge computing and in-device computing has provided solutions to this problem by utilizing IoT devices and some novel mechanisms within centralized and distributed computing frameworks [662,663]. Despite the achievements, several critical issues have yet to be well solved, and significant work still needs to be done. For example, the training of deep learning models in IoT devices is a problem that needs to be further solved. A possible way is to locally train the distributed and partial neural network input in IoT devices through offloading pretrained feature output for additional training at higher layers. In addition, developing appropriate paradigms to analyze data in a timely manner is another challenging problem requiring further research. Possible and promising research can be undertaken in the following directions: (1) distributed deep learning at the network edge, and more specifically, developing and optimizing parallel simultaneous edge network architectures for self-organization and runtime; and (2) in-device deep learning, and more specifically, implementing deep networks in IoT devices by considering the limited hardware and computational capabilities.

5.7. Privacy-Preserving Federated Learning

Nowadays, increasing privacy concerns have emerged along with the aggregation of distributed computing results. Privacy-preserving federated learning has become a solution for privacy-preserving deep learning [208,664]. By training deep learning models on separate datasets that are distributed across different devices or parties, it can preserve the local data privacy to a certain extent. However, despite the achievements, the challenge of protecting data privacy while maintaining the data utility through deep learning still remains. Potential and promising research problems and directions are: (1) how to effectively apply the privacy-preserving mechanisms [665,666] to federated learning frameworks for better privacy preservation; (2) develop efficient solutions to defend the final model against inference attacks extracting sensitive information from it; and (3) how to efficiently handle data memorization in federated learning to prevent privacy leakage.

5.8. Multimodal Learning

With the development of various sensor system, increasing numbers of data modalities can be obtained. Different modalities are characterized by different statistical properties, and thus it is important to discover the relationship between different modalities. Research in many application areas needs to be based on data from multiple modalities to achieve a more complete picture of the task, for example, biomedical studies typically involve both image and "omics" data. Therefore, multimodal learning, which can represent the joint representations of different modalities, is required for taking full advantage of all available data in such studies. This has been well recognized in several works [667,668] but deserves more attention. Potential research problems and directions are: (1) designing new learning frameworks with more powerful computing architectures to effectively learn feature structures of the multimodal data of increasing volume; (2) developing new deep learning models for multimodal data that take semantic relationships into consideration to mine the intermodality and crossmodality knowledge; and (3) designing online and incremental multimodal deep learning models for data fusion to learn new knowledge from new data without much loss of historical knowledge.

6. Conclusions

Deep learning has become a predominant method for solving data analysis problems in virtually all fields of science and engineering. The increasing complexity and the large volume of data collected by diverse sensor systems have brought about a significant development of deep learning, which has also fundamentally transformed the way data are acquired, processed, analyzed, and interpreted. In this paper we have provided a comprehensive investigation of deep learning in diverse sensor systems, starting from the fundamentals of deep learning models and methods, to mapping specific deep learning methods with individual suitable sensor systems. This paper also provides a comprehensive summary of implementation tips and links to tutorials, open-sourced codes, and pretrained models for new deep learning practitioners and those seeking to innovate deep learning in diverse sensor systems. In addition, this paper provides insights into research topics where deep learning has not yet been well-developed, but may have potential, and highlights the challenges and future of deep learning in diverse sensor systems. We hope this survey will provide an excellent self-contained and comprehensive reference for industry practitioners and researchers in the field.

Author Contributions: Conceptualization, Y.Z., M.W., X.Y., J.Z, E.M. and J.H.; methodology, Y.Z., M.W., X.Y., J.Z, E.M. and J.H.; investigation, Y.Z., M.W. and X.Y.; writing—original draft preparation, Y.Z., M.W., X.Y. and J.Z; writing—review and editing, Y.Z., M.W., X.Y., J.Z., E.M. and J.H.; visualization, Y.Z., M.W., X.Y., and J.Z.; supervision, E.M. and J.H.; project administration, Y.Z. and J.H.; funding acquisition, J.H. All authors have read and agreed to the published version of the manuscript.

Funding: This work was supported by the Australian Research Council through the discovery grant DP200103207.

Institutional Review Board Statement: Not applicable.

Informed Consent Statement: Not applicable.

Data Availability Statement: Not applicable.

Conflicts of Interest: The authors declare no conflict of interest.

References

1. Pouyanfar, S.; Sadiq, S.; Yan, Y.; Tian, H.; Tao, Y.; Reyes, M.P.; Shyu, M.L.; Chen, S.C.; Iyengar, S.S. A survey on deep learning Algorithms, techniques, and applications. *ACM Comput. Surv. (CSUR)* **2018**, *51*, 1–36. [CrossRef]
2. Dong, S.; Wang, P.; Abbas, K. A survey on deep learning and its applications. *Comput. Sci. Rev.* **2021**, *40*, 100379. [CrossRef]
3. Raghu, M.; Schmidt, E. A survey of deep learning for scientific discovery. *arXiv* **2020**, arXiv:2003.11755.
4. Dargan, S.; Kumar, M.; Ayyagari, M.R.; Kumar, G. A survey of deep learning and its applications: A new paradigm to machine learning. *Arch. Comput. Methods Eng.* **2020**, *27*, 1071–1092. [CrossRef]
5. McCulloch, W.S.; Pitts, W. A logical calculus of the ideas immanent in nervous activity. *Bull. Math. Biophys.* **1943**, *5*, 115–133 [CrossRef]
6. Rosenblatt, F. The perceptron: A probabilistic model for information storage and organization in the brain. *Psychol. Rev.* **1958**, *65*, 386 [CrossRef]
7. Ackley, D.H.; Hinton, G.E.; Sejnowski, T.J. A learning algorithm for Boltzmann machines. *Cogn. Sci.* **1985**, *9*, 147–169. [CrossRef]
8. Fukushima, K.; Miyake, S. Neocognitron: A self-organizing neural network model for a mechanism of visual pattern recognition In *Competition and Cooperation in Neural Nets*; Springer: Berlin/Heidelberg, Germany, 1982; pp. 267–285.
9. Jordan, M.I. Serial order: A parallel distributed processing approach. In *Advances in Psychology*; Elsevier: Amsterdam, The Netherlands, 1997; Volume 121, pp. 471–495.
10. LeCun, Y.; Bottou, L.; Bengio, Y.; Haffner, P. Gradient-based learning applied to document recognition. *Proc. IEEE* **1998**, *86*, 2278–2324. [CrossRef]
11. Hinton, G.E.; Osindero, S.; Teh, Y.W. A fast learning algorithm for deep belief nets. *Neural Comput.* **2006**, *18*, 1527–1554. [CrossRef]
12. Hinton, G.E.; Salakhutdinov, R.R. Reducing the dimensionality of data with neural networks. *Science* **2006**, *313*, 504–507 [CrossRef]
13. Bengio, Y. Learning deep architectures for AI. *Found. Trends Mach. Learn.* **2009**, *2*, 1–127. [CrossRef]
14. Schmidhuber, J. Deep learning in neural networks: An overview. *Neural Netw.* **2015**, *61*, 85–117. [PubMed]
15. Bengio, Y.; Courville, A.; Vincent, P. Representation learning: A review and new perspectives. *IEEE Trans. Pattern Anal. Mach. Intell.* **2013**, *35*, 1798–1828. [CrossRef] [PubMed]
16. Bottou, L. Stochastic gradient descent tricks. In *Neural Networks: Tricks of the Trade*; Springer: Berlin, Germany, 2012; pp. 421–436
17. Mohri, M.; Rostamizadeh, A.; Talwalkar, A. *Foundations of Machine Learning*; MIT Press: Cambridge, MA, USA, 2018.
18. Yang, X.; Song, Z.; King, I.; Xu, Z. A survey on deep semi-supervised learning. *arXiv* **2021**, arXiv:2103.00550.
19. Oliver, A.; Odena, A.; Raffel, C.A.; Cubuk, E.D.; Goodfellow, I. Realistic evaluation of deep semi-supervised learning algorithms *Adv. Neural Inf. Process. Syst.* **2018**, *31*, 1–12.
20. Sajjadi, M.; Javanmardi, M.; Tasdizen, T. Regularization with stochastic transformations and perturbations for deep semi-supervised learning. *Adv. Neural Inf. Process. Syst.* **2016**, *29*, 1–9.
21. Yan, P.; Li, G.; Xie, Y.; Li, Z.; Wang, C.; Chen, T.; Lin, L. Semi-supervised video salient object detection using pseudo-labels In Proceedings of the IEEE/CVF International Conference on Computer Vision, Seoul, Republic of Korea, 27 October–2 November 2019; pp. 7284–7293.
22. Schmarje, L.; Santarossa, M.; Schröder, S.M.; Koch, R. A survey on semi-, self-and unsupervised learning for image classification *IEEE Access* **2021**, *9*, 82146–82168. [CrossRef]
23. Jacobs, R.A. Increased rates of convergence through learning rate adaptation. *Neural Netw.* **1988**, *1*, 295–307. [CrossRef]
24. Srivastava, N.; Hinton, G.; Krizhevsky, A.; Sutskever, I.; Salakhutdinov, R. Dropout: A simple way to prevent neural networks from overfitting. *J. Mach. Learn. Res.* **2014**, *15*, 1929–1958.
25. Ioffe, S.; Szegedy, C. Batch normalization: Accelerating deep network training by reducing internal covariate shift. In Proceedings of the International Conference on Machine Learning, Miami, FL, USA, 9–11 December 2015; pp. 448–456.
26. Shorten, C.; Khoshgoftaar, T.M. A survey on image data augmentation for deep learning. *J. Big Data* **2019**, *6*, 1–48. [CrossRef]
27. Wen, Q.; Sun, L.; Yang, F.; Song, X.; Gao, J.; Wang, X.; Xu, H. Time series data augmentation for deep learning: A survey. *arXiv* **2020**, arXiv:2002.12478.
28. Feng, S.Y.; Gangal, V.; Wei, J.; Chandar, S.; Vosoughi, S.; Mitamura, T.; Hovy, E. A survey of data augmentation approaches for NLP. *arXiv* **2021**, arXiv:2105.03075.
29. Abadi, M.; Agarwal, A.; Barham, P.; Brevdo, E.; Chen, Z.; Citro, C.; Corrado, G.S.; Davis, A.; Dean, J.; Devin, M.; et al. Tensorflow: Large-scale machine learning on heterogeneous distributed systems. *arXiv* **2016**, arXiv:1603.04467.

30. Paszke, A.; Gross, S.; Massa, F.; Lerer, A.; Bradbury, J.; Chanan, G.; Killeen, T.; Lin, Z.; Gimelshein, N.; Antiga, L.; et al. Pytorch: An imperative style, high-performance deep learning library. *Adv. Neural Inf. Process. Syst.* **2019**, *32*, 1–12.
31. Jia, Y.; Shelhamer, E.; Donahue, J.; Karayev, S.; Long, J.; Girshick, R.; Guadarrama, S.; Darrell, T. Caffe: Convolutional architecture for fast feature embedding. In Proceedings of the 22nd ACM International Conference on Multimedia, New York, NY, USA, 14 October 2014; pp. 675–678.
32. Collobert, R.; Bengio, S.; Mariéthoz, J. *Torch: A Modular Machine Learning Software Library*; Technical Report 02-46; Idiap: Lausanne, Switzerland, 2002; pp. 1–9.
33. Al-Rfou, R.; Alain, G.; Almahairi, A.; Angermueller, C.; Bahdanau, D.; Ballas, N.; Bastien, F.; Bayer, J.; Belikov, A.; Belopolsky, A.; et al. Theano: A Python framework for fast computation of mathematical expressions. *arXiv* **2016**, arXiv:1605.02688.
34. Chen, T.; Li, M.; Li, Y.; Lin, M.; Wang, N.; Wang, M.; Xiao, T.; Xu, B.; Zhang, C.; Zhang, Z. Mxnet: A flexible and efficient machine learning library for heterogeneous distributed systems. *arXiv* **2015**, arXiv:1512.01274.
35. Yu, D.; Eversole, A.; Seltzer, M.; Yao, K.; Huang, Z.; Guenter, B.; Kuchaiev, O.; Zhang, Y.; Seide, F.; Wang, H.; et al. *An Introduction to Computational Networks and the Computational Network Toolkit*; Technical Report MSR-TR-2014-112; Microsoft Research: Washington, DC, USA, 2014; pp. 1–150.
36. LeCun, Y.; Bengio, Y.; Hinton, G. Deep learning. *Nature* **2015**, *521*, 436–444. [CrossRef]
37. Krizhevsky, A.; Sutskever, I.; Hinton, G.E. Imagenet classification with deep convolutional neural networks. *Commun. ACM* **2017**, *60*, 84–90. [CrossRef]
38. Szegedy, C.; Liu, W.; Jia, Y.; Sermanet, P.; Reed, S.; Anguelov, D.; Erhan, D.; Vanhoucke, V.; Rabinovich, A. Going deeper with convolutions. In Proceedings of the IEEE Conference on Computer Vision and Pattern Recognition, Boston, MA, USA, 7–12 June 2015; pp. 1–9.
39. Simonyan, K.; Zisserman, A. Very deep convolutional networks for large-scale image recognition. In Proceedings of the International Conference on Learning Representations, San Diego, CA, USA, 7–9 May 2015.
40. Zeiler, M.D.; Fergus, R. Visualizing and understanding convolutional networks. In Proceedings of the European Conference on Computer Vision, Zurich, Switzerland, 6–12 September 2014; pp. 818–833.
41. He, K.; Zhang, X.; Ren, S.; Sun, J. Deep residual learning for image recognition. In Proceedings of the IEEE Conference on Computer Vision and Pattern Recognition, Las Vegas, NV, USA, 27–30 June 2016; pp. 770–778.
42. Huang, G.; Liu, Z.; Van Der Maaten, L.; Weinberger, K.Q. Densely connected convolutional networks. In Proceedings of the IEEE Conference on Computer Vision and Pattern Recognition, Honolulu, HI, USA, 21–26 July 2017; pp. 4700–4708.
43. Ronneberger, O.; Fischer, P.; Brox, T. U-Net: Convolutional networks for biomedical image segmentation. In Proceedings of the International Conference on Medical Image Computing and Computer-Assisted Intervention, Munich, Germany, 5–9 October 2015; pp. 234–241.
44. Ren, S.; He, K.; Girshick, R.; Sun, J. Faster R-CNN: Towards real-time object detection with region proposal networks. *Adv. Neural Inf. Process. Syst.* **2015**, *28*, 1–9. [CrossRef]
45. Srivastava, R.K.; Greff, K.; Schmidhuber, J. Highway networks. *arXiv* **2015**, arXiv:1505.00387.
46. Redmon, J.; Divvala, S.; Girshick, R.; Farhadi, A. You only look once: Unified, real-time object detection. In Proceedings of the IEEE Conference on Computer Vision and Pattern Recognition, Las Vegas, NV, USA, 27–30 June 2016; pp. 779–788.
47. He, K.; Gkioxari, G.; Dollár, P.; Girshick, R. Mask R-CNN. In Proceedings of the IEEE International Conference on Computer Vision, Venice, Italy, 22–29 October 2017; pp. 2961–2969.
48. Howard, A.G.; Zhu, M.; Chen, B.; Kalenichenko, D.; Wang, W.; Weyand, T.; Andreetto, M.; Adam, H. Mobilenets: Efficient convolutional neural networks for mobile vision applications. *arXiv* **2017**, arXiv:1704.04861.
49. Han, D.; Kim, J.; Kim, J. Deep pyramidal residual networks. In Proceedings of the IEEE Conference on Computer Vision and Pattern Recognition, Honolulu, HI, USA, 21–26 July 2017; pp. 5927–5935.
50. Chollet, F. Xception: Deep learning with depthwise separable convolutions. In Proceedings of the IEEE Conference on Computer Vision and Pattern Recognition, Honolulu, HI, USA, 21–26 July 2017; pp. 1251–1258.
51. Szegedy, C.; Ioffe, S.; Vanhoucke, V.; Alemi, A.A. Inception-v4, inception-resnet and the impact of residual connections on learning. In Proceedings of the AAAI Conference on Artificial Intelligence, San Francisco, CA, USA, 4–9 February 2017.
52. Zhang, X.; Li, Z.; Change Loy, C.; Lin, D. Polynet: A pursuit of structural diversity in very deep networks. In Proceedings of the IEEE Conference on Computer Vision and Pattern Recognition, Honolulu, HI, USA, 21–26 July 2017; pp. 718–726.
53. Bengio, Y.; Simard, P.; Frasconi, P. Learning long-term dependencies with gradient descent is difficult. *IEEE Trans. Neural Netw.* **1994**, *5*, 157–166. [CrossRef] [PubMed]
54. Glorot, X.; Bengio, Y. Understanding the difficulty of training deep feedforward neural networks. In Proceedings of the Thirteenth International Conference on Artificial Intelligence and Statistics, Sardinia, Italy, 3-15 May 2010; pp. 249–256.
55. Girshick, R. Fast R-CNN. In Proceedings of the IEEE International Conference on Computer Vision, Santiago, Chile, 7–13 December 2015; pp. 1440–1448.
56. Cho, K.; Van Merriënboer, B.; Gulcehre, C.; Bahdanau, D.; Bougares, F.; Schwenk, H.; Bengio, Y. Learning phrase representations using RNN encoder-decoder for statistical machine translation. *arXiv* **2014**, arXiv:1406.1078.
57. Pascanu, R.; Gulcehre, C.; Cho, K.; Bengio, Y. How to construct deep recurrent neural networks. *arXiv* **2013**, arXiv:1312.6026.
58. Graves, A.; Liwicki, M.; Fernández, S.; Bertolami, R.; Bunke, H.; Schmidhuber, J. A novel connectionist system for unconstrained handwriting recognition. *IEEE Trans. Pattern Anal. Mach. Intell.* **2008**, *31*, 855–868. [CrossRef]

59. Graves, A.; Mohamed, A.r.; Hinton, G. Speech recognition with deep recurrent neural networks. In Proceedings of the 2013 IEEE International Conference on Acoustics, Speech and Signal Processing, Vancouver, BC, Canada, 26–31 May 2013; pp. 6645–6649.
60. Lefebvre, G.; Berlemont, S.; Mamalet, F.; Garcia, C. Inertial gesture recognition with BLSTM-RNN. In *Artificial Neural Networks*; Springer: Berlin, Germany, 2015; pp. 393–410.
61. You, Q.; Jin, H.; Wang, Z.; Fang, C.; Luo, J. Image captioning with semantic attention. In Proceedings of the IEEE Conference on Computer Vision and Pattern Recognition, Las Vegas, NV, USA, 27–30 June 2016; pp. 4651–4659.
62. Yao, L.; Guan, Y. An improved LSTM structure for natural language processing. In Proceedings of the 2018 IEEE International Conference of Safety Produce Informatization (IICSPI), Chongqing, China, 10–12 December 2018; pp. 565–569.
63. Li, X.; Wu, X. Constructing long short-term memory based deep recurrent neural networks for large vocabulary speech recognition. In Proceedings of the 2015 IEEE International Conference on Acoustics, Speech and Signal Processing, South Brisbane, QSD, Australia, 19–24 April 2015; pp. 4520–4524.
64. Chatterjee, C.C.; Mulimani, M.; Koolagudi, S.G. Polyphonic sound event detection using transposed convolutional recurrent neural network. In Proceedings of the 2020 IEEE International Conference on Acoustics, Speech and Signal Processing, Barcelona, Spain, 4–8 May 2020; pp. 661–665.
65. Li, Y.; Yu, R.; Shahabi, C.; Liu, Y. Graph convolutional recurrent neural network: Data-driven traffic forecasting. *arXiv* **2017**, arXiv:1707.01926.
66. Azzouni, A.; Pujolle, G. A long short-term memory recurrent neural network framework for network traffic matrix prediction. *arXiv* **2017**, arXiv:1705.05690.
67. Altché, F.; de La Fortelle, A. An LSTM network for highway trajectory prediction. In Proceedings of the 2017 IEEE 20th International Conference on Intelligent Transportation Systems (ITSC), Yokohama, Japan, 16–19 October 2017; pp. 353–359.
68. Khairdoost, N.; Shirpour, M.; Bauer, M.A.; Beauchemin, S.S. Real-time driver maneuver prediction using LSTM. *IEEE Trans. Intell. Veh.* **2020**, *5*, 714–724. [CrossRef]
69. Li, L.; Zhao, W.; Xu, C.; Wang, C.; Chen, Q.; Dai, S. Lane-change intention inference based on RNN for autonomous driving on highways. *IEEE Trans. Veh. Technol.* **2021**, *70*, 5499–5510. [CrossRef]
70. Robinson, A.; Fallside, F. *The Utility Driven Dynamic Error Propagation Network*; University of Cambridge Department of Engineering Cambridge: Cambridge, UK, 1987.
71. Werbos, P.J. Generalization of backpropagation with application to a recurrent gas market model. *Neural Netw.* **1988**, *1*, 339–356. [CrossRef]
72. Mozer, M.C. A focused backpropagation algorithm for temporal. *Backpropag. Theory Archit. Appl.* **1995**, *137*, 137–170.
73. Hochreiter, S.; Bengio, Y.; Frasconi, P.; Schmidhuber, J. Gradient flow in recurrent nets: The difficulty of learning long-term dependencies. In *A Field Guide to Dynamical Recurrent Networks*; Wiley-IEEE Press: New York, NY, USA, 2001; pp. 237–243.
74. Schuster, M.; Paliwal, K.K. Bidirectional recurrent neural networks. *IEEE Trans. Signal Process.* **1997**, *45*, 2673–2681. [CrossRef]
75. Liwicki, M.; Graves, A.; Fernàndez, S.; Bunke, H.; Schmidhuber, J. A novel approach to on-line handwriting recognition based on bidirectional long short-term memory networks. In Proceedings of the 9th International Conference on Document Analysis and Recognition, Curitiba, Paraná, Brazil, 23–26 September 2007.
76. Hochreiter, S.; Schmidhuber, J. Long short-term memory. *Neural Comput.* **1997**, *9*, 1735–1780. [CrossRef] [PubMed]
77. Karpathy, A.; Fei-Fei, L. Deep visual-semantic alignments for generating image descriptions. In Proceedings of the IEEE Conference on Computer Vision and Pattern Recognition, Boston, MA, USA, 7–12 June 2015; pp. 3128–3137.
78. Bahdanau, D.; Cho, K.; Bengio, Y. Neural machine translation by jointly learning to align and translate. *arXiv* **2014**, arXiv:1409.0473.
79. Qin, Y.; Song, D.; Chen, H.; Cheng, W.; Jiang, G.; Cottrell, G. A dual-stage attention-based recurrent neural network for time series prediction. *arXiv* **2017**, arXiv:1704.02971.
80. Lu, J.; Yang, J.; Batra, D.; Parikh, D. Hierarchical question-image co-attention for visual question answering. *Adv. Neural Inf. Process. Syst.* **2016**, *29*, 1–9.
81. Wu, Y.; Schuster, M.; Chen, Z.; Le, Q.V.; Norouzi, M.; Macherey, W.; Krikun, M.; Cao, Y.; Gao, Q.; Macherey, K.; et al. Google's neural machine translation system: Bridging the gap between human and machine translation. *arXiv* **2016**, arXiv:1609.08144.
82. Lu, X.; Tsao, Y.; Matsuda, S.; Hori, C. Speech enhancement based on deep denoising autoencoder. In Proceedings of the Interspeech, Lyon, France, 25–29 August 2013; pp. 436–440.
83. Saad, O.M.; Chen, Y. Deep denoising autoencoder for seismic random noise attenuation. *Geophysics* **2020**, *85*, V367–V376. [CrossRef]
84. Krizhevsky, A.; Hinton, G.E. Using very deep autoencoders for content-based image retrieval. In Proceedings of the European Symposium on Artificial Neural Networks, Bruges, Belgium, 27–29 April 2011.
85. Feng, F.; Wang, X.; Li, R. Cross-modal retrieval with correspondence autoencoder. In Proceedings of the 22nd ACM International Conference on Multimedia, Orlando, FL, USA, 3–7 November 2014; pp. 7–16.
86. Shcherbakov, O.; Batishcheva, V. Image inpainting based on stacked autoencoders. *J. Phys. Conf. Ser.* **2014**, *536*, 012020. [CrossRef]
87. Zhu, Y.; Yin, X.; Hu, J. FingerGAN: A Constrained Fingerprint Generation Scheme for Latent Fingerprint Enhancement. *arXiv* **2022**, arXiv:2206.12885.
88. Tagawa, T.; Tadokoro, Y.; Yairi, T. Structured denoising autoencoder for fault detection and analysis. In Proceedings of the Asian Conference on Machine Learning, Kuala Lumpur, Malaysia, 3–6 November 2015; pp. 96–111.

89. Zhou, C.; Paffenroth, R.C. Anomaly detection with robust deep autoencoders. In Proceedings of the 23rd ACM SIGKDD International Conference on Knowledge Discovery and Data Mining, Halifax, NS, Canada, 13–17 August 2017; pp. 665–674.
90. Ranzato, M.; Poultney, C.; Chopra, S.; LeCun, Y. Efficient learning of sparse representations with an energy-based model. *Adv. Neural Inf. Process. Syst.* **2007**, *19*, 1137.
91. Rifai, S.; Vincent, P.; Muller, X.; Glorot, X.; Bengio, Y. Contractive auto-encoders: Explicit invariance during feature extraction. In Proceedings of the International Conference on Machine Learning, Fort Lauderdale, FL, USA, 11–13 April 2011.
92. Vincent, P.; Larochelle, H.; Bengio, Y.; Manzagol, P.A. Extracting and composing robust features with denoising autoencoders. In Proceedings of the 25th International Conference on Machine Learning, Helsinki, Finland, 5–9 July 2008; pp. 1096–1103.
93. Vincent, P. A connection between score matching and denoising autoencoders. *Neural Comput.* **2011**, *23*, 1661–1674. [CrossRef] [PubMed]
94. Kingma, D.P.; Welling, M. Auto-encoding variational bayes. In Proceedings of the International Conference on Learning Representations, Banff, AB, Canada, 14–16 April 2014; pp. 1–14.
95. Hinton, G.E. Boltzmann machine. *Scholarpedia* **2007**, *2*, 1668. [CrossRef]
96. Zhang, K.; Liu, J.; Chai, Y.; Qian, K. An optimized dimensionality reduction model for high-dimensional data based on restricted Boltzmann machines. In Proceedings of the The 27th Chinese Control and Decision Conference, Qingdao, China, 23–25 May 2015; pp. 2939–2944.
97. Larochelle, H.; Mandel, M.; Pascanu, R.; Bengio, Y. Learning algorithms for the classification restricted Boltzmann machine. *J. Mach. Learn. Res.* **2012**, *13*, 643–669.
98. Elaiwat, S.; Bennamoun, M.; Boussaïd, F. A spatio-temporal RBM-based model for facial expression recognition. *Pattern Recognit.* **2016**, *49*, 152–161. [CrossRef]
99. Salakhutdinov, R.; Mnih, A.; Hinton, G. Restricted Boltzmann machines for collaborative filtering. In Proceedings of the 24th International Conference on Machine learning, Corvalis, OR, USA, 20–24 June 2007; pp. 791–798.
100. Längkvist, M.; Karlsson, L.; Loutfi, A. A review of unsupervised feature learning and deep learning for time-series modeling. *Pattern Recognit. Lett.* **2014**, *42*, 11–24. [CrossRef]
101. Hinton, G.E.; Salakhutdinov, R.R. Replicated softmax: An undirected topic model. *Adv. Neural Inf. Process. Syst.* **2009**, *22*, 1–8.
102. Fischer, A.; Igel, C. An introduction to restricted Boltzmann machines. In Proceedings of the Iberoamerican Congress on Pattern Recognition, Havana, Cuba, 28–31 October 2012; pp. 14–36.
103. Fischer, A.; Igel, C. Training restricted Boltzmann machines: An introduction. *Pattern Recognit.* **2014**, *47*, 25–39. [CrossRef]
104. Smolensky, P. *Information Processing in Dynamical Systems: Foundations of Harmony Theory*; Technical Report; Colorado University at Boulder Department of Computer Science: Boulder, CO, USA, 1986; pp. 1–56.
105. Goodfellow, I.; Pouget-Abadie, J.; Mirza, M.; Xu, B.; Warde-Farley, D.; Ozair, S.; Courville, A.; Bengio, Y. Generative adversarial nets. *Adv. Neural Inf. Process. Syst.* **2014**, *27*, 2672–2680.
106. Bowles, C.; Chen, L.; Guerrero, R.; Bentley, P.; Gunn, R.; Hammers, A.; Dickie, D.A.; Hernández, M.V.; Wardlaw, J.; Rueckert, D. GAN augmentation: Augmenting training data using generative adversarial networks. *arXiv* **2018**, arXiv:1810.10863.
107. Jing, Y.; Yang, Y.; Feng, Z.; Ye, J.; Yu, Y.; Song, M. Neural style transfer: A review. *IEEE Trans. Vis. Comput. Graph.* **2019**, *26*, 3365–3385. [CrossRef] [PubMed]
108. Dong, H.W.; Hsiao, W.Y.; Yang, L.C.; Yang, Y.H. MuseGAN: Multi-track sequential generative adversarial networks for symbolic music generation and accompaniment. In Proceedings of the AAAI Conference on Artificial Intelligence, New Orleans, LA, USA, 2–7 February 2018; pp. 34–41.
109. Reed, S.; Akata, Z.; Yan, X.; Logeswaran, L.; Schiele, B.; Lee, H. Generative adversarial text to image synthesis. In Proceedings of the International Conference on Machine Learning, New York, NY, USA, 20–22 June 2016; pp. 1060–1069.
110. Dahl, R.; Norouzi, M.; Shlens, J. Pixel recursive super resolution. In Proceedings of the IEEE International Conference on Computer Vision, Venice, Italy, 22–29 October 2017; pp. 5439–5448.
111. Souly, N.; Spampinato, C.; Shah, M. Semi supervised semantic segmentation using generative adversarial network. In Proceedings of the IEEE International Conference on Computer Vision, Venice, Italy, 22–29 October 2017; pp. 5688–5696.
112. Pascual, S.; Bonafonte, A.; Serra, J. SEGAN: Speech enhancement generative adversarial network. *arXiv* **2017**, arXiv:1703.09452.
113. Kwon, Y.H.; Park, M.G. Predicting future frames using retrospective cycle GAN. In Proceedings of the IEEE/CVF Conference on Computer Vision and Pattern Recognition, Long Beach, CA, USA, 15–20 June 2019; pp. 1811–1820.
114. Radford, A.; Metz, L.; Chintala, S. Unsupervised representation learning with deep convolutional generative adversarial networks. *arXiv* **2015**, arXiv:1511.06434.
115. Dai, B.; Fidler, S.; Urtasun, R.; Lin, D. Towards diverse and natural image descriptions via a conditional GAN. In Proceedings of the IEEE International Conference on Computer Vision, Venice, Italy, 22–29 October 2017; pp. 2970–2979.
116. Isola, P.; Zhu, J.Y.; Zhou, T.; Efros, A.A. Image-to-image translation with conditional adversarial networks. In Proceedings of the IEEE Conference on Computer Vision and Pattern Recognition, Venice, Italy, 22–29 October 2017; pp. 1125–1134.
117. Odena, A.; Olah, C.; Shlens, J. Conditional image synthesis with auxiliary classifier GANs. In Proceedings of the International Conference on Machine Learning, Sydney, Australia, 6–11 August 2017; pp. 2642–2651.
118. Huang, X.; Li, Y.; Poursaeed, O.; Hopcroft, J.; Belongie, S. Stacked generative adversarial networks. In Proceedings of the IEEE Conference on Computer Vision and Pattern Recognition, Honolulu, HI, USA, 21–26 July 2017; pp. 5077–5086.
119. Adler, J.; Lunz, S. Banach wasserstein GAN. *Adv. Neural Inf. Process. Syst.* **2018**, *31*, 1–10.

120. Zhu, J.Y.; Park, T.; Isola, P.; Efros, A.A. Unpaired image-to-image translation using cycle-consistent adversarial networks In Proceedings of the IEEE International Conference on Computer Vision, Venice, Italy, 22–29 October 2017; pp. 2223–2232.
121. Karras, T.; Aila, T.; Laine, S.; Lehtinen, J. Progressive growing of GANs for improved quality, stability, and variation. *arXiv* **2017**, arXiv:1710.10196.
122. Wu, Z.; Pan, S.; Chen, F.; Long, G.; Zhang, C.; Philip, S.Y. A comprehensive survey on graph neural networks. *IEEE Trans. Neural Networks Learn. Syst.* **2020**, *32*, 4–24. [CrossRef]
123. Zhang, M.; Cui, Z.; Neumann, M.; Chen, Y. An end-to-end deep learning architecture for graph classification. In Proceedings of the AAAI Conference on Artificial Intelligence, New Orleans, LA, USA, 2–7 February 2018.
124. Henaff, M.; Bruna, J.; LeCun, Y. Deep convolutional networks on graph-structured data. *arXiv* **2015**, arXiv:1506.05163.
125. Defferrard, M.; Bresson, X.; Vandergheynst, P. Convolutional neural networks on graphs with fast localized spectral filtering *Adv. Neural Inf. Process. Syst.* **2016**, *29*, 1–9.
126. Lee, J.; Lee, I.; Kang, J. Self-attention graph pooling. In Proceedings of the International Conference on Machine Learning, Long Beach, CA, USA, 9–15 June 2019; pp. 3734–3743.
127. Kipf, T.N.; Welling, M. Semi-supervised classification with graph convolutional networks. In Proceedings of the International Conference on Learning Representations, Toulon, France, 24–26 April 2017.
128. Scarselli, F.; Gori, M.; Tsoi, A.C.; Hagenbuchner, M.; Monfardini, G. The graph neural network model. *IEEE Trans. Neural Networks* **2008**, *20*, 61–80. [CrossRef]
129. Gallicchio, C.; Micheli, A. Graph echo state networks. In Proceedings of the The 2010 International Joint Conference on Neural Networks (IJCNN), Barcelona, Spain, 18–23 July 2010; pp. 1–8.
130. Li, Y.; Tarlow, D.; Brockschmidt, M.; Zemel, R. Gated graph sequence neural networks. In Proceedings of the International Conference on Learning Representations, San Juan, Puerto Rico, 2–4 May 2016.
131. Shuman, D.I.; Narang, S.K.; Frossard, P.; Ortega, A.; Vandergheynst, P. The emerging field of signal processing on graphs: Extending high-dimensional data analysis to networks and other irregular domains. *IEEE Signal Process. Mag.* **2013**, *30*, 83–98 [CrossRef]
132. Bruna, J.; Zaremba, W.; Szlam, A.; LeCun, Y. Spectral networks and deep locally connected networks on graphs. In Proceedings of the International Conference on Learning Representations, Banff, AB, Canada, 14–16 April 2014.
133. Atwood, J.; Towsley, D. Diffusion-convolutional neural networks. *Adv. Neural Inf. Process. Syst.* **2016**, *29*, 1993–2001.
134. Gilmer, J.; Schoenholz, S.S.; Riley, P.F.; Vinyals, O.; Dahl, G.E. Neural message passing for quantum chemistry. In Proceedings of the International Conference on Machine Learning, Sydney, Australia, 6–10 August 2017; pp. 1263–1272.
135. Hamilton, W.; Ying, Z.; Leskovec, J. Inductive representation learning on large graphs. *Adv. Neural Inf. Process. Syst.* **2017**, *30*, 1024–1034.
136. Veličković, P.; Cucurull, G.; Casanova, A.; Romero, A.; Lio, P.; Bengio, Y. Graph attention networks. In Proceedings of the International Conference on Learning Representations, Vancouver, BC, Canada, 30 April–3 May 2018.
137. Monti, F.; Boscaini, D.; Masci, J.; Rodola, E.; Svoboda, J.; Bronstein, M.M. Geometric deep learning on graphs and manifolds using mixture model CNNs. In Proceedings of the IEEE Conference on Computer Vision and Pattern Recognition, Honolulu, HI, USA, 21–26 July 2017; pp. 5115–5124.
138. Chen, J.; Ma, T.; Xiao, C. FastGCN: Fast learning with graph convolutional networks via importance sampling. In Proceedings of the International Conference on Learning Representations, Vancouver, BC, Canada, 30 April–3 May 2018.
139. Wang, D.; Cui, P.; Zhu, W. Structural deep network embedding. In Proceedings of the 22nd ACM SIGKDD International Conference on Knowledge Discovery and Data Mining, San Francisco, CA, USA, 13–17 August 2016; pp. 1225–1234.
140. Kipf, T.N.; Welling, M. Variational graph auto-encoders. In Proceedings of the Neural Information Processing Systems Workshop, Barcelona, Spain, 5–10 December 2016.
141. Gómez-Bombarelli, R.; Wei, J.N.; Duvenaud, D.; Hernández-Lobato, J.M.; Sánchez-Lengeling, B.; Sheberla, D.; Aguilera-Iparraguirre, J.; Hirzel, T.D.; Adams, R.P.; Aspuru-Guzik, A. Automatic chemical design using a data-driven continuous representation of molecules. *ACS Cent. Sci.* **2018**, *4*, 268–276. [CrossRef] [PubMed]
142. Li, Y.; Vinyals, O.; Dyer, C.; Pascanu, R.; Battaglia, P. Learning deep generative models of graphs. In Proceedings of the International Conference on Learning Representations Workshop, Vancouver, BC, Canada, 30 April–3 May 2018.
143. You, J.; Ying, R.; Ren, X.; Hamilton, W.; Leskovec, J. GraphRNN: Generating realistic graphs with deep auto-regressive models. In Proceedings of the International Conference on Machine Learning, Stockholm, Sweden, 10–15 July 2018; pp. 5708–5717.
144. Simonovsky, M.; Komodakis, N. Graphvae: Towards generation of small graphs using variational autoencoders. In Proceedings of the International Conference on Artificial Neural Networks, Rhodes, Greece, 4–7 October 2018; pp. 412–422.
145. De Cao, N.; Kipf, T. MolGAN: An implicit generative model for small molecular graphs. *arXiv* **2018**, arXiv:1805.11973.
146. Bojchevski, A.; Shchur, O.; Zügner, D.; Günnemann, S. Netgan: Generating graphs via random walks. In Proceedings of the International Conference on Machine Learning, Stockholm, Sweden, 10–15 July 2018; pp. 610–619.
147. Seo, Y.; Defferrard, M.; Vandergheynst, P.; Bresson, X. Structured sequence modeling with graph convolutional recurrent networks. In Proceedings of the International Conference on Neural Information Processing, Siem Reap, Cambodia, 13–16 December 2018; pp. 362–373.
148. Yu, B.; Yin, H.; Zhu, Z. Spatio-temporal graph convolutional networks: A deep learning framework for traffic forecasting. In Proceedings of the 27th International Joint Conference on Artificial Intelligence, Stockholm, Sweden, 13–19 July 2018; pp. 3634–3640.

149. Vaswani, A.; Shazeer, N.; Parmar, N.; Uszkoreit, J.; Jones, L.; Gomez, A.N.; Kaiser, Ł.; Polosukhin, I. Attention is all you need. *Adv. Neural Inf. Process. Syst.* **2017**, *30*, 5998–6008.
150. Devlin, J.; Chang, M.W.; Lee, K.; Toutanova, K. BERT: Pre-training of deep bidirectional transformers for language understanding. In Proceedings of the 2019 Conference of the North American Chapter of the Association for Computational Linguistics: Human Language Technologies, Minneapolis, MN, USA, 2–7 June 2019; pp. 4171–4186.
151. Rae, J.W.; Potapenko, A.; Jayakumar, S.M.; Lillicrap, T.P. Compressive transformers for long-range sequence modelling. In Proceedings of the International Conference on Learning Representations, Addis Ababa, Ethiopia, 26–30 April 2020.
152. Brown, T.; Mann, B.; Ryder, N.; Subbiah, M.; Kaplan, J.D.; Dhariwal, P.; Neelakantan, A.; Shyam, P.; Sastry, G.; Askell, A.; et al. Language models are few-shot learners. *Adv. Neural Inf. Process. Syst.* **2020**, *33*, 1877–1901.
153. Lewis, M.; Liu, Y.; Goyal, N.; Ghazvininejad, M.; Mohamed, A.; Levy, O.; Stoyanov, V.; Zettlemoyer, L. BART: Denoising sequence-to-sequence pre-training for natural language generation, translation, and comprehension. In Proceedings of the 58th Annual Meeting of the Association for Computational Linguistics, Online, 5–10 July 2020; pp. 7871–7880.
154. Guo, Q.; Qiu, X.; Xue, X.; Zhang, Z. Low-rank and locality constrained self-attention for sequence modeling. *IEEE/ACM Trans. Audio Speech Lang. Process.* **2019**, *27*, 2213–2222. [CrossRef]
155. Katharopoulos, A.; Vyas, A.; Pappas, N.; Fleuret, F. Transformers are RNNs: Fast autoregressive transformers with linear attention. In Proceedings of the International Conference on Machine Learning, Virtual Event, 13–18 July 2020; pp. 5156–5165.
156. Guo, Q.; Qiu, X.; Liu, P.; Xue, X.; Zhang, Z. Multi-scale self-attention for text classification. In Proceedings of the AAAI Conference on Artificial Intelligence, New York, NY, USA, 7–12 February 2020; pp. 7847–7854.
157. Guo, M.; Zhang, Y.; Liu, T. Gaussian transformer: A lightweight approach for natural language inference. In Proceedings of the AAAI Conference on Artificial Intelligence, HI, USA, 27 January–1 February 2019; pp. 6489–6496.
158. Wu, Z.; Liu, Z.; Lin, J.; Lin, Y.; Han, S. Lite transformer with long-short range attention. In Proceedings of the International Conference on Learning Representations, Addis Ababa, Ethiopia, 26–30 April 2020.
159. Dai, Z.; Lai, G.; Yang, Y.; Le, Q. Funnel-transformer: Filtering out sequential redundancy for efficient language processing. *Adv. Neural Inf. Process. Syst.* **2020**, *33*, 4271–4282.
160. Mehta, S.; Ghazvininejad, M.; Iyer, S.; Zettlemoyer, L.; Hajishirzi, H. DeLighT: Very deep and light-weight transformer. In Proceedings of the International Conference on Learning Representations, Vienna, Austria, 3–7 May 2021; pp. 1–19.
161. Dosovitskiy, A.; Beyer, L.; Kolesnikov, A.; Weissenborn, D.; Zhai, X.; Unterthiner, T.; Dehghani, M.; Minderer, M.; Heigold, G.; Gelly, S.; et al. An image is worth 16x16 words: Transformers for image recognition at scale. In Proceedings of the International Conference on Learning Representations, Addis Ababa, Ethiopia, 26–30 April 2020.
162. Chen, M.; Radford, A.; Child, R.; Wu, J.; Jun, H.; Luan, D.; Sutskever, I. Generative pretraining from pixels. In Proceedings of the International Conference on Machine Learning, Virtual Event, 13–18 July 2020; pp. 1691–1703.
163. Zeng, Y.; Fu, J.; Chao, H. Learning joint spatial-temporal transformations for video inpainting. In Proceedings of the European Conference on Computer Vision, Glasgow, UK, 23–28 August 2020; pp. 528–543.
164. Zhou, L.; Zhou, Y.; Corso, J.J.; Socher, R.; Xiong, C. End-to-end dense video captioning with masked transformer. In Proceedings of the IEEE Conference on Computer Vision and Pattern Recognition, Salt Lake City, UT, USA, 18–23 June 2018; pp. 8739–8748.
165. Han, K.; Xiao, A.; Wu. E.; Guo, J.; XU, C.; Wang, Y. Transformer in transformer. *Adv. Neural Inf. Process. Syst.* **2021**, *34*, 15908–15919.
166. Cao, H.; Wang, Y.; Chen, J.; Jiang, D.; Zhang, X.; Tian, Q.; Wang, M. Swin-unet: Unet-like pure transformer for medical image segmentation. *arXiv* **2021**, arXiv:2105.05537.
167. Jospin, L.V.; Laga, H.; Boussaid, F.; Buntine, W.; Bennamoun, M. Hands-on Bayesian neural networks—A tutorial for deep learning users. *IEEE Comput. Intell. Mag.* **2022**, *17*, 29–48. [CrossRef]
168. Neal, R.M. *Bayesian Learning for Neural Networks*; Springer Science & Business Media: Berlin, Germany, 2012; Volume 118.
169. Denker, J.; LeCun, Y. Transforming neural-net output levels to probability distributions. *Adv. Neural Inf. Process. Syst.* **1990**, *3*, 853–859.
170. He, J.; Liu, R.; Zhuang, F.; Lin, F.; Niu, C.; He, Q. A general cross-domain recommendation framework via Bayesian neural network. In Proceedings of the 2018 IEEE International Conference on Data Mining (ICDM), Singapore, 17–20 November 2018; pp. 1001–1006.
171. Nie, S.; Zheng, M.; Ji, Q. The deep regression bayesian network and its applications: Probabilistic deep learning for computer vision. *IEEE Signal Process. Mag.* **2018**, *35*, 101–111. [CrossRef]
172. Chien, J.T. Deep Bayesian natural language processing. In Proceedings of the 57th Annual Meeting of the Association for Computational Linguistics: Tutorial Abstracts, Florence, Italy, 28 July–2 August 2019; pp. 25–30.
173. Xue, B.; Hu, S.; Xu, J.; Geng, M.; Liu, X.; Meng, H. Bayesian Neural Network Language Modeling for Speech Recognition. *IEEE/ACM Trans. Audio Speech Lang. Process.* **2022**, *30*, 2900–2917. [CrossRef]
174. Kwon, Y.; Won, J.H.; Kim, B.J.; Paik, M.C. Uncertainty quantification using Bayesian neural networks in classification: Application to biomedical image segmentation. *Comput. Stat. Data Anal.* **2020**, *142*, 106816. [CrossRef]
175. Lampinen, J.; Vehtari, A. Bayesian approach for neural networks—Review and case studies. *Neural Netw.* **2001**, *14*, 257–274. [CrossRef]
176. Titterington, D. Bayesian methods for neural networks and related models. *Stat. Sci.* **2004**, 128–139. [CrossRef]
177. MacKay, D.J. A practical Bayesian framework for backpropagation networks. *Neural Comput.* **1992**, *4*, 448–472. [CrossRef]

178. Blundell, C.; Cornebise, J.; Kavukcuoglu, K.; Wierstra, D. Weight uncertainty in neural network. In Proceedings of the International Conference on Machine Learning, Lille, France, 6–11 July 2015; pp. 1613–1622.
179. Mitchell, T.M.; Mitchell, T.M. *Machine Learning*; McGraw-Hill: New York, NY, USA, 1997.
180. Ruder, S. An overview of gradient descent optimization algorithms. *arXiv* **2016**, arXiv:1609.04747.
181. Chen, C.P.; Zhang, C.Y.; Chen, L.; Gan, M. Fuzzy restricted Boltzmann machine for the enhancement of deep learning. *IEEE Trans. Fuzzy Syst.* **2015**, *23*, 2163–2173. [CrossRef]
182. Rajurkar, S.; Verma, N.K. Developing deep fuzzy network with Takagi Sugeno fuzzy inference system. In Proceedings of the 2017 IEEE International Conference on Fuzzy Systems (FUZZ-IEEE), Naples, Italy, 9–12 July 2017; pp. 1–6.
183. Bhatia, V.; Rani, R. Dfuzzy: A deep learning-based fuzzy clustering model for large graphs. *Knowl. Inf. Syst.* **2018**, *57*, 159–181. [CrossRef]
184. Chen, D.; Zhang, X.; Wang, L.; Han, Z. Prediction of cloud resources demand based on fuzzy deep neural network. In Proceedings of the 2018 IEEE Global Communications Conference (GLOBECOM), Abu Dhabi, United Arab Emirates, 9–13 December 2018; pp. 1–5.
185. Hernandez-Potiomkin, Y.; Saifuzzaman, M.; Bert, E.; Mena-Yedra, R.; Djukic, T.; Casas, J. Unsupervised incident detection model in urban and freeway networks. In Proceedings of the 2018 21st International Conference on Intelligent Transportation Systems (ITSC), Maui, HI, USA, 4–7 November 2018; pp. 1763–1769.
186. Ali, F.; El-Sappagh, S.; Islam, S.R.; Kwak, D.; Ali, A.; Imran, M.; Kwak, K.S. A smart healthcare monitoring system for heart disease prediction based on ensemble deep learning and feature fusion. *Inf. Fusion* **2020**, *63*, 208–222. [CrossRef]
187. Al-Dmour, H.; Al-Ani, A. A clustering fusion technique for MR brain tissue segmentation. *Neurocomputing* **2018**, *275*, 546–559. [CrossRef]
188. An, J.; Fu, L.; Hu, M.; Chen, W.; Zhan, J. A novel fuzzy-based convolutional neural network method to traffic flow prediction with uncertain traffic accident information. *IEEE Access* **2019**, *7*, 20708–20722. [CrossRef]
189. Talpur, N.; Abdulkadir, S.J.; Alhussian, H.; Hasan, M.H.; Aziz, N.; Bamhdi, A. Deep Neuro-Fuzzy System application trends, challenges, and future perspectives: A systematic survey. *Artif. Intell. Rev.* **2022**, 1–49. [CrossRef]
190. Chen, D.; Zhang, X.; Wang, L.; Han, Z. Prediction of cloud resources demand based on hierarchical pythagorean fuzzy deep neural network. *IEEE Trans. Serv. Comput.* **2019**, *14*, 1890–1901. [CrossRef]
191. Yeganejou, M.; Dick, S.; Miller, J. Interpretable deep convolutional fuzzy classifier. *IEEE Trans. Fuzzy Syst.* **2019**, *28*, 1407–1419. [CrossRef]
192. Li, Y. Deep reinforcement learning: An overview. *arXiv* **2017**, arXiv:1701.07274.
193. Sutton, R.S. Learning to predict by the methods of temporal differences. *Mach. Learn.* **1988**, *3*, 9–44. [CrossRef]
194. Watkins, C.J.; Dayan, P. Q-learning. *Mach. Learn.* **1992**, *8*, 279–292. [CrossRef]
195. Mnih, V.; Kavukcuoglu, K.; Silver, D.; Rusu, A.A.; Veness, J.; Bellemare, M.G.; Graves, A.; Riedmiller, M.; Fidjeland, A.K.; Ostrovski, G.; et al. Human-level control through deep reinforcement learning. *Nature* **2015**, *518*, 529–533. [CrossRef]
196. Van Hasselt, H.; Guez, A.; Silver, D. Deep reinforcement learning with double Q-learning. In Proceedings of the AAAI Conference on Artificial Intelligence, Phoenix, AZ, USA, 12–17 February 2016; Volume 30.
197. Wang, Z.; Schaul, T.; Hessel, M.; Hasselt, H.; Lanctot, M.; Freitas, N. Dueling network architectures for deep reinforcement learning. In Proceedings of the International Conference on Machine Learning, New York, NY, USA, 20–22 June 2016; pp. 1995–2003.
198. Mnih, V.; Badia, A.P.; Mirza, M.; Graves, A.; Lillicrap, T.; Harley, T.; Silver, D.; Kavukcuoglu, K. Asynchronous methods for deep reinforcement learning. In Proceedings of the International Conference on Machine Learning, New York, NY, USA, 20–22 June 2016; pp. 1928–1937.
199. Kakade, S.M. A natural policy gradient. *Adv. Neural Inf. Process. Syst.* **2001**, *14*, 1531–1538.
200. Schulman, J.; Levine, S.; Abbeel, P.; Jordan, M.; Moritz, P. Trust region policy optimization. In Proceedings of the International Conference on Machine Learning, Lille, France, 6–11 July 2015; pp. 1889–1897.
201. Schulman, J.; Moritz, P.; Levine, S.; Jordan, M.; Abbeel, P. High-dimensional continuous control using generalized advantage estimation. *arXiv* **2015**, arXiv:1506.02438.
202. Raina, R.; Battle, A.; Lee, H.; Packer, B.; Ng, A.Y. Self-taught learning: Transfer learning from unlabeled data. In Proceedings of the 24th International Conference on Machine Learning, Corvalis, OR, USA, 20–24 June 2007; pp. 759–766.
203. Daume III, H.; Marcu, D. Domain adaptation for statistical classifiers. *J. Artif. Intell. Res.* **2006**, *26*, 101–126. [CrossRef]
204. Dai, W.; Yang, Q.; Xue, G.R.; Yu, Y. Self-taught clustering. In Proceedings of the 25th International Conference on Machine Learning, New York, NY, USA, 5–9 July 2008; pp. 200–207.
205. Yao, Y.; Doretto, G. Boosting for transfer learning with multiple sources. In Proceedings of the 2010 IEEE Computer Society Conference on Computer Vision and Pattern Recognition, San Francisco, CA, USA, 13–18 June 2010; pp. 1855–1862.
206. Lawrence, N.D.; Platt, J.C. Learning to learn with the informative vector machine. In Proceedings of the 21st International Conference on Machine Learning, Banff, AB, Canada, 4–8 July 2004; p. 65.
207. Mihalkova, L.; Mooney, R.J. Transfer learning by mapping with minimal target data. In Proceedings of the AAAI Workshop on Transfer Learning for Complex Tasks, Chicago, IL, USA, 13–17 July 2008.
208. Yin, X.; Zhu, Y.; Hu, J. A comprehensive survey of privacy-preserving federated learning: A taxonomy, review, and future directions. *ACM Comput. Surv. (CSUR)* **2021**, *54*, 1–36. [CrossRef]

209. Lim, W.Y.B.; Luong, N.C.; Hoang, D.T.; Jiao, Y.; Liang, Y.C.; Yang, Q.; Niyato, D.; Miao, C. Federated learning in mobile edge networks: A comprehensive survey. *IEEE Commun. Surv. Tutor.* **2020**, *22*, 2031–2063. [CrossRef]
210. McMahan, B.; Moore, E.; Ramage, D.; Hampson, S.; Arcas, B.A. Communication-efficient learning of deep networks from decentralized data. In Proceedings of the Artificial Intelligence and Statistics, Fort Lauderdale, FL, USA, 20–22 April 2017; pp. 1273–1282.
211. Zhu, H.; Zhang, H.; Jin, Y. From federated learning to federated neural architecture search: A survey. *Complex Intell. Syst.* **2021**, *7*, 639–657. [CrossRef]
212. Zantedeschi, V.; Bellet, A.; Tommasi, M. Fully decentralized joint learning of personalized models and collaboration graphs. In Proceedings of the International Conference on Artificial Intelligence and Statistics, Online, 26–28 August 2020; pp. 864–874.
213. Charles, Z.; Garrett, Z.; Huo, Z.; Shmulyian, S.; Smith, V. On large-cohort training for federated learning. *Adv. Neural Inf. Process. Syst.* **2021**, *34*, 20461–20475.
214. Wang, H.; Mu noz-González, L.; Eklund, D.; Raza, S. Non-IID data re-balancing at IoT edge with peer-to-peer federated learning for anomaly detection. In Proceedings of the 14th ACM Conference on Security and Privacy in Wireless and Mobile Networks, Abu Dhabi, United Arab Emirates, 28 June–2 July 2021; pp. 153–163.
215. Wink, T.; Nochta, Z. An approach for peer-to-peer federated learning. In Proceedings of the 2021 51st Annual IEEE/IFIP International Conference on Dependable Systems and Networks Workshops (DSN-W), Taipei, Taiwan, 21–24 June 2021; pp. 150–157.
216. Yang, Q.; Liu, Y.; Chen, T.; Tong, Y. Federated machine learning: Concept and applications. *ACM Trans. Intell. Syst. Technol. (TIST)* **2019**, *10*, 1–19. [CrossRef]
217. Yang, S.; Ren, B.; Zhou, X.; Liu, L. Parallel distributed logistic regression for vertical federated learning without third-party coordinator. In Proceedings of the IJCAI Workshop on Federated Machine Learning for User Privacy and Data Confidentiality, Macao, China, 10–16 August 2019.
218. Scannapieco, M.; Figotin, I.; Bertino, E.; Elmagarmid, A.K. Privacy preserving schema and data matching. In Proceedings of the 2007 ACM SIGMOD International Conference on Management of Data, Beijing China, 11–14 June 2007; pp. 653–664.
219. Liang, G.; Chawathe, S.S. Privacy-preserving inter-database operations. In Proceedings of the International Conference on Intelligence and Security Informatics, Atlanta, GA, USA, 19–20 May 2004; pp. 66–82.
220. Dietterich, T.G.; Lathrop, R.H.; Lozano-Pérez, T. Solving the multiple instance problem with axis-parallel rectangles. *Artif. Intell.* **1997**, *89*, 31–71. [CrossRef]
221. Zhu, W.; Sun, L.; Huang, J.; Han, L.; Zhang, D. Dual attention multi-instance deep learning for Alzheimer's disease diagnosis with structural MRI. *IEEE Trans. Med Imaging* **2021**, *40*, 2354–2366. [CrossRef]
222. Chen, Y.; Bi, J.; Wang, J.Z. MILES: Multiple-instance learning via embedded instance selection. *IEEE Trans. Pattern Anal. Mach. Intell.* **2006**, *28*, 1931–1947. [CrossRef]
223. Zhou, Z.H.; Sun, Y.Y.; Li, Y.F. Multi-instance learning by treating instances as non-iid samples. In Proceedings of the 26th Annual International Conference on Machine Learning, Montreal, QC, Canada, 14–18 June 2009; pp. 1249–1256.
224. Briggs, F.; Fern, X.Z.; Raich, R. Rank-loss support instance machines for MIML instance annotation. In Proceedings of the 18th ACM SIGKDD International Conference on Knowledge Discovery and Data Mining, Beijing, China, 12–16 August 2012; pp. 534–542.
225. Carbonneau, M.A.; Cheplygina, V.; Granger, E.; Gagnon, G. Multiple instance learning: A survey of problem characteristics and applications. *Pattern Recognit.* **2018**, *77*, 329–353. [CrossRef]
226. Cheplygina, V.; Tax, D.M.; Loog, M. On classification with bags, groups and sets. *Pattern Recognit. Lett.* **2015**, *59*, 11–17. [CrossRef]
227. Bunescu, R.C.; Mooney, R.J. Multiple instance learning for sparse positive bags. In Proceedings of the 24th International Conference on Machine Learning, New York, NY, USA, 20–24 June 2007; pp. 105–112.
228. Gärtner, T.; Flach, P.A.; Kowalczyk, A.; Smola, A.J. Multi-instance kernels. In Proceedings of the International Conference on Machine Learning, Las Vegas, NV, USA, 24–27 June 2002; Volume 2.
229. Gehler, P.V.; Chapelle, O. Deterministic annealing for multiple-instance learning. In Proceedings of the Artificial Intelligence and Statistics, San Juan, Puerto Rico, 21–24 March 2007; pp. 123–130.
230. Venkatesan, R.; Chandakkar, P.; Li, B. Simpler non-parametric methods provide as good or better results to multiple-instance learning. In Proceedings of the IEEE International Conference on Computer Vision, Santiago, Chile, 7–13 December 2015; pp. 2605–2613.
231. Amores, J. Vocabulary-based approaches for multiple-instance data: A comparative study. In Proceedings of the 20th International Conference on Pattern Recognition, Istanbul, Turkey, 23–26 August 2010; pp. 4246–4250.
232. Cheplygina, V.; Tax, D.M.; Loog, M. Multiple instance learning with bag dissimilarities. *Pattern Recognit.* **2015**, *48*, 264–275. [CrossRef]
233. Wang, Z.; Zhao, Z.; Zhang, C. Learning with only multiple instance positive bags. In Proceedings of the 2016 International Joint Conference on Neural Networks (IJCNN), Vancouver, BC, Canada. 24–29 July 2016; pp. 334–341.
234. Xiao, Y.; Liu, B.; Hao, Z. A sphere-description-based approach for multiple-instance learning. *IEEE Trans. Pattern Anal. Mach. Intell.* **2016**, *39*, 242–257. [CrossRef]
235. Xie, S.; Girshick, R.; Dollár, P.; Tu, Z.; He, K. Aggregated residual transformations for deep neural networks. In Proceedings of the IEEE Conference on Computer Vision and Pattern Recognition, Honolulu, HI, USA, 21–26 July 2017; pp. 1492–1500.

236. Girshick, R.; Donahue, J.; Darrell, T.; Malik, J. Rich feature hierarchies for accurate object detection and semantic segmentation. In Proceedings of the IEEE Conference on Computer Vision and Pattern Recognition, Columbus, OH, USA, 23–28 June 2014; pp. 580–587.
237. Dai, J.; Li, Y.; He, K.; Sun, J. R-FCN: Object detection via region-based fully convolutional networks. *Adv. Neural Inf. Process. Syst.* **2016**, *29*.
238. Liu, W.; Anguelov, D.; Erhan, D.; Szegedy, C.; Reed, S.; Fu, C.Y.; Berg, A.C. SSD: Single shot multibox detector. In Proceedings of the European Conference on Computer Vision, Amsterdam, The Netherlands, 11–14 October 2016; pp. 21–37.
239. Liu, L.; Ouyang, W.; Wang, X.; Fieguth, P.; Chen, J.; Liu, X.; Pietikäinen, M. Deep learning for generic object detection: A survey. *Int. J. Comput. Vis.* **2020**, *128*, 261–318. [CrossRef]
240. Zhao, Z.Q.; Zheng, P.; Xu, S.t.; Wu, X. Object detection with deep learning: A review. *IEEE Trans. Neural Networks Learn. Syst.* **2019**, *30*, 3212–3232. [CrossRef]
241. Oktay, O.; Schlemper, J.; Folgoc, L.L.; Lee, M.; Heinrich, M.; Misawa, K.; Mori, K.; McDonagh, S.; Hammerla, N.Y.; Kainz, B.; et al. Attention u-net: Learning where to look for the pancreas. *arXiv* **2018**, arXiv:1804.03999.
242. Punn, N.S.; Agarwal, S. Inception u-net architecture for semantic segmentation to identify nuclei in microscopy cell images. *ACM Trans. Multimed. Comput. Commun. Appl. (TOMM)* **2020**, *16*, 1–15. [CrossRef]
243. Li, D.; Dharmawan, D.A.; Ng, B.P.; Rahardja, S. Residual u-net for retinal vessel segmentation. In Proceedings of the IEEE International Conference on Image Processing (ICIP), Taipei, Taiwan, 22–25 September 2019; pp. 1425–1429.
244. Wang, C.; Zhao, Z.; Ren, Q.; Xu, Y.; Yu, Y. Dense U-net based on patch-based learning for retinal vessel segmentation. *Entropy* **2019**, *21*, 168. [CrossRef]
245. Zhou, Z.; Siddiquee, M.M.R.; Tajbakhsh, N.; Liang, J. Unet++: Redesigning skip connections to exploit multiscale features in image segmentation. *IEEE Trans. Med Imaging* **2019**, *39*, 1856–1867. [CrossRef]
246. Badrinarayanan, V.; Kendall, A.; Cipolla, R. SegNet: A deep convolutional encoder-decoder architecture for image segmentation. *IEEE Trans. Pattern Anal. Mach. Intell.* **2017**, *39*, 2481–2495. [CrossRef]
247. Zhao, H.; Shi, J.; Qi, X.; Wang, X.; Jia, J. Pyramid scene parsing network. In Proceedings of the IEEE Conference on Computer Vision and Pattern Recognition, Honolulu, HI, USA, 21–26 July 2017; pp. 2881–2890.
248. Chen, L.C.; Papandreou, G.; Kokkinos, I.; Murphy, K.; Yuille, A.L. Deeplab: Semantic image segmentation with deep convolutional nets, atrous convolution, and fully connected CRFs. *IEEE Trans. Pattern Anal. Mach. Intell.* **2017**, *40*, 834–848. [CrossRef]
249. Hao, S.; Zhou, Y.; Guo, Y. A brief survey on semantic segmentation with deep learning. *Neurocomputing* **2020**, *406*, 302–321. [CrossRef]
250. Lateef, F.; Ruichek, Y. Survey on semantic segmentation using deep learning techniques. *Neurocomputing* **2019**, *338*, 321–348. [CrossRef]
251. Yu, H.; Yang, Z.; Tan, L.; Wang, Y.; Sun, W.; Sun, M.; Tang, Y. Methods and datasets on semantic segmentation: A review. *Neurocomputing* **2018**, *304*, 82–103. [CrossRef]
252. Bolya, D.; Zhou, C.; Xiao, F.; Lee, Y.J. Yolact: Real-time instance segmentation. In Proceedings of the IEEE/CVF International Conference on Computer Vision, Seoul, Republic of Korea, 27 October–2 November 2019; pp. 9157–9166.
253. Wang, X.; Zhang, R.; Shen, C.; Kong, T.; Li, L. Solo: A simple framework for instance segmentation. *IEEE Trans. Pattern Anal. Mach. Intell.* **2021**, 44, 8587–8601. [CrossRef]
254. Xie, E.; Sun, P.; Song, X.; Wang, W.; Liu, X.; Liang, D.; Shen, C.; Luo, P. Polarmask: Single shot instance segmentation with polar representation. In Proceedings of the IEEE/CVF Conference on Computer Vision and Pattern Recognition, Seattle, WA, USA, 13–19 June 2020; pp. 12193–12202.
255. Sofiiuk, K.; Barinova, O.; Konushin, A. Adaptis: Adaptive instance selection network. In Proceedings of the IEEE/CVF International Conference on Computer Vision, Seoul, Republic of Korea, 27 October–2 November 2019; pp. 7355–7363.
256. Tian, Z.; Shen, C.; Chen, H.; He, T. Fcos: Fully convolutional one-stage object detection. In Proceedings of the IEEE/CVF International Conference on Computer Vision, Seoul, Republic of Korea, 27 October–2 November 2019; pp. 9627–9636.
257. Hafiz, A.M.; Bhat, G.M. A survey on instance segmentation: State of the art. *Int. J. Multimed. Inf. Retr.* **2020**, *9*, 171–189. [CrossRef]
258. Zhang, H.; Sun, H.; Ao, W.; Dimirovski, G. A survey on instance segmentation: Recent advances and challenges. *Int. J. Innov. Comput. Inf. Control* **2021**, *17*, 1041–1053.
259. Anoob, N.; Ebey, S.J.; Praveen, P.; Prabudhan, P.; Augustine, P. A Comparison on Instance Segmentation Models. In Proceedings of the 2021 International Conference on Advances in Computing and Communications (ICACC), Kochi, Kakkanad, India, 21–23 October 2021; pp. 1–5.
260. Newell, A.; Yang, K.; Deng, J. Stacked hourglass networks for human pose estimation. In Proceedings of the European Conference on Computer Vision, Amsterdam, The Netherlands, 11–14 October 2016; pp. 483–499.
261. Kendall, A.; Grimes, M.; Cipolla, R. Posenet: A convolutional network for real-time 6-dof camera relocalization. In Proceedings of the IEEE International Conference on Computer Vision, Santiago, Chile, 7–13 December 2015; pp. 2938–2946.
262. Marchand, E.; Uchiyama, H.; Spindler, F. Pose estimation for augmented reality: A hands-on survey. *IEEE Trans. Vis. Comput. Graph.* **2015**, *22*, 2633–2651. [CrossRef]
263. Liu, Z.; Zhu, J.; Bu, J.; Chen, C. A survey of human pose estimation: The body parts parsing based methods. *J. Vis. Commun. Image Represent.* **2015**, *32*, 10–19. [CrossRef]

264. Sarafianos, N.; Boteanu, B.; Ionescu, B.; Kakadiaris, I.A. 3D human pose estimation: A review of the literature and analysis of covariates. *Comput. Vis. Image Underst.* **2016**, *152*, 1–20. [CrossRef]
265. Johnson, J.; Alahi, A.; Fei-Fei, L. Perceptual losses for real-time style transfer and super-resolution. In Proceedings of the European Conference on Computer Vision, Amsterdam, The Netherlands, 11–14 October 2016; pp. 694–711.
266. Dumoulin, V.; Shlens, J.; Kudlur, M. A learned representation for artistic style. *arXiv* **2016**, arXiv:1610.07629.
267. Huang, X.; Belongie, S. Arbitrary style transfer in real-time with adaptive instance normalization. In Proceedings of the IEEE International Conference on Computer Vision, Venice, Italy, 22–29 October 2017; pp. 1501–1510.
268. Jin, D.; Jin, Z.; Hu, Z.; Vechtomova, O.; Mihalcea, R. Deep learning for text style transfer: A survey. *Comput. Linguist.* **2022**, *48*, 155–205. [CrossRef]
269. Zhao, C. A survey on image style transfer approaches using deep learning. In Proceedings of the Journal of Physics: Conference Series, Xi'an, China, 18–19 October 2020; Volume 1453, p. 012129.
270. Olatunji, I.E.; Cheng, C.H. Video analytics for visual surveillance and applications: An overview and survey. *Mach. Learn. Paradig.* **2019**, 475–515.
271. Bhuiyan, M.R.; Abdullah, J.; Hashim, N.; Al Farid, F. Video analytics using deep learning for crowd analysis: A review. *Multimed. Tools Appl.* **2022**, *81*, 27895–27922. [CrossRef]
272. Donahue, J.; Anne Hendricks, L.; Guadarrama, S.; Rohrbach, M.; Venugopalan, S.; Saenko, K.; Darrell, T. Long-term recurrent convolutional networks for visual recognition and description. In Proceedings of the IEEE Conference on Computer Vision and Pattern Recognition, Boston, MA, USA, 7–12 June 2015; pp. 2625–2634.
273. Ballas, N.; Yao, L.; Pal, C.; Courville, A. Delving deeper into convolutional networks for learning video representations. *arXiv* **2015**, arXiv:1511.06432.
274. Hu, Y.T.; Huang, J.B.; Schwing, A. Maskrnn: Instance level video object segmentation. *Adv. Neural Inf. Process. Syst.* **2017**, *30*, 325–334.
275. Xiao, H.; Feng, J.; Lin, G.; Liu, Y.; Zhang, M. Monet: Deep motion exploitation for video object segmentation. In Proceedings of the IEEE Conference on Computer Vision and Pattern Recognition, Salt Lake City, UT. USA, 18–23 June 2018; pp. 1140–1148.
276. Zhang, T.; Aftab, W.; Mihaylova, L.; Langran-Wheeler, C.; Rigby, S.; Fletcher, D.; Maddock, S.; Bosworth, G. Recent advances in video analytics for rail network surveillance for security, trespass and suicide prevention—A survey. *Sensors* **2022**, *22*, 4324. [CrossRef]
277. Sánchez, F.L.; Hupont, I.; Tabik, S.; Herrera, F. Revisiting crowd behaviour analysis through deep learning: Taxonomy, anomaly detection, crowd emotions, datasets, opportunities and prospects. *Inf. Fusion* **2020**, *64*, 318–335. [CrossRef]
278. Meijering, E. A bird's-eye view of deep learning in bioimage analysis. *Comput. Struct. Biotechnol. J.* **2020**, *18*, 2312–2325. [CrossRef]
279. Zhou, Z.; Rahman Siddiquee, M.M.; Tajbakhsh, N.; Liang, J. Unet++: A nested U-Net architecture for medical image segmentation. In *Deep learning in Medical Image Analysis and Multimodal Learning for Clinical Decision Support*; Springer: Berlin, Germany, 2018; pp. 3–11.
280. Nadeem, M.W.; Ghamdi, M.A.A.; Hussain, M.; Khan, M.A.; Khan, K.M.; Almotiri, S.H.; Butt, S.A. Brain tumor analysis empowered with deep learning: A review, taxonomy, and future challenges. *Brain Sci.* **2020**, *10*, 118. [CrossRef]
281. Litjens, G.; Kooi, T.; Bejnordi, B.E.; Setio, A.A.A.; Ciompi, F.; Ghafoorian, M.; Van Der Laak, J.A.; Van Ginneken, B.; Sánchez, C.I. A survey on deep learning in medical image analysis. *Med Image Anal.* **2017**, *42*, 60–88. [CrossRef]
282. Haskins, G.; Kruger, U.; Yan, P. Deep learning in medical image registration: A survey. *Mach. Vis. Appl.* **2020**, *31*, 1–18. [CrossRef]
283. Liu, X.; Song, L.; Liu, S.; Zhang, Y. A review of deep-learning-based medical image segmentation methods. *Sustainability* **2021**, *13*, 1224. [CrossRef]
284. Zhu, X.; Yao, J.; Huang, J. Deep convolutional neural network for survival analysis with pathological images. In Proceedings of the 2016 IEEE International Conference on Bioinformatics and Biomedicine (BIBM), Shenzhen, China, 15–18 December 2016; pp. 544–547.
285. Li, Y.; Xie, X.; Shen, L.; Liu, S. Reverse active learning based atrous DenseNet for pathological image classification. *BMC Bioinform.* **2019**, *20*, 1–15. [CrossRef]
286. Deng, S.; Zhang, X.; Yan, W.; Chang, E.I.; Fan, Y.; Lai, M.; Xu, Y. Deep learning in digital pathology image analysis: A survey. *Front. Med.* **2020**, *14*, 470–487. [CrossRef]
287. Rogers, M.A.; Aikawa, E. Cardiovascular calcification: Artificial intelligence and big data accelerate mechanistic discovery. *Nat. Rev. Cardiol.* **2019**, *16*, 261–274. [CrossRef]
288. Choi, H. Deep learning in nuclear medicine and molecular imaging: Current perspectives and future directions. *Nucl. Med. Mol. Imaging* **2018**, *52*, 109–118. [CrossRef] [PubMed]
289. Moen, E.; Bannon, D.; Kudo, T.; Graf, W.; Covert, M.; Van Valen, D. Deep learning for cellular image analysis. *Nat. Methods* **2019**, *16*, 1233–1246. [CrossRef] [PubMed]
290. Cheng, H.J.; Hsu, C.H.; Hung, C.L.; Lin, C.Y. A review for cell and particle tracking on microscopy images using algorithms and deep learning technologies. *Biomed. J.* **2021**, *21*, S2319–S4170. [CrossRef] [PubMed]
291. Zhu, Y.; Meijering, E. Automatic improvement of deep learning-based cell segmentation in time-lapse microscopy by neural architecture search. *Bioinformatics* **2021**, *37*, 4844–4850. [CrossRef]
292. Zhu, Y.; Meijering, E. Neural architecture search for microscopy cell segmentation. In Proceedings of the International Workshop on Machine Learning in Medical Imaging, Lima, Peru, 4 October 2020; pp. 542–551.

293. de Haan, K.; Rivenson, Y.; Wu, Y.; Ozcan, A. Deep-learning-based image reconstruction and enhancement in optical microscopy. *Proc. IEEE* **2019**, *108*, 30–50. [CrossRef]
294. Wu, Y.; Rivenson, Y.; Wang, H.; Luo, Y.; Ben-David, E.; Bentolila, L.A.; Pritz, C.; Ozcan, A. Three-dimensional virtual refocusing of fluorescence microscopy images using deep learning. *Nat. Methods* **2019**, *16*, 1323–1331. [CrossRef]
295. Liu, Z.; Jin, L.; Chen, J.; Fang, Q.; Ablameyko, S.; Yin, Z.; Xu, Y. A survey on applications of deep learning in microscopy image analysis. *Comput. Biol. Med.* **2021**, *134*, 104523. [CrossRef]
296. Poplin, R.; Chang, P.C.; Alexander, D.; Schwartz, S.; Colthurst, T.; Ku, A.; Newburger, D.; Dijamco, J.; Nguyen, N.; Afshar, P.T.; et al. A universal SNP and small-indel variant caller using deep neural networks. *Nat. Biotechnol.* **2018**, *36*, 983–987. [CrossRef]
297. Xie, R.; Wen, J.; Quitadamo, A.; Cheng, J.; Shi, X. A deep auto-encoder model for gene expression prediction. *BMC Genomics* **2017**, *18*, 39–49. [CrossRef]
298. Zou, J.; Huss, M.; Abid, A.; Mohammadi, P.; Torkamani, A.; Telenti, A. A primer on deep learning in genomics. *Nat. Genet.* **2019**, *51*, 12–18. [CrossRef]
299. Martorell-Marugán, J.; Tabik, S.; Benhammou, Y.; del Val, C.; Zwir, I.; Herrera, F.; Carmona-Sáez, P. Deep learning in omics data analysis and precision medicine. *Exon Publ.* **2019**, 37–53.
300. Tripathi, R.; Patel, S.; Kumari, V.; Chakraborty, P.; Varadwaj, P.K. DeepLNC, a long non-coding RNA prediction tool using deep neural network. *Netw. Model. Anal. Health Inform. Bioinform.* **2016**, *5*, 1–14. [CrossRef]
301. Heydari, A.A.; Sindi, S.S. Deep learning in spatial transcriptomics: Learning from the next next-generation sequencing. *BioRxiv* **2022**. [CrossRef]
302. Zhang, Z.; Zhao, Y.; Liao, X.; Shi, W.; Li, K.; Zou, Q.; Peng, S. Deep learning in omics: A survey and guideline. *Briefings Funct. Genom.* **2019**, *18*, 41–57. [CrossRef]
303. Zemouri, R.; Zerhouni, N.; Racoceanu, D. Deep learning in the biomedical applications: Recent and future status. *Appl. Sci.* **2019**, *9*, 1526. [CrossRef]
304. Gao, Y.; Wang, S.; Deng, M.; Xu, J. RaptorX-Angle: Real-value prediction of protein backbone dihedral angles through a hybrid method of clustering and deep learning. *BMC Bioinform.* **2018**, *19*, 73–84. [CrossRef]
305. Hu, Y.; Nie, T.; Shen, D.; Yu, G. Sequence translating model using deep neural block cascade network: Taking protein secondary structure prediction as an example. In Proceedings of the 2018 IEEE International Conference on Big Data and Smart Computing (BigComp), Shanghai, China, 15–17 January 2018; pp. 58–65.
306. Nguyen, S.P.; Li, Z.; Xu, D.; Shang, Y. New deep learning methods for protein loop modeling. *IEEE/ACM Trans. Comput. Biol. Bioinform.* **2017**, *16*, 596–606. [CrossRef]
307. Lei, H.; Wen, Y.; You, Z.; Elazab, A.; Tan, E.L.; Zhao, Y.; Lei, B. Protein–protein interactions prediction via multimodal deep polynomial network and regularized extreme learning machine. *IEEE J. Biomed. Health Informatics* **2018**, *23*, 1290–1303. [CrossRef]
308. Bahi, M.; Batouche, M. Drug-target interaction prediction in drug repositioning based on deep semi-supervised learning. In Proceedings of the IFIP International Conference on Computational Intelligence and Its Applications, Oran, Algeria, 8–10 May 2018; pp. 302–313.
309. Li, S.; Wan, F.; Shu, H.; Jiang, T.; Zhao, D.; Zeng, J. MONN: A multi-objective neural network for predicting compound-protein interactions and affinities. *Cell Syst.* **2020**, *10*, 308–322. [CrossRef]
310. Baldi, P. Deep learning in biomedical data science. *Annu. Rev. Biomed. Data Sci.* **2018**, *1*, 181–205. [CrossRef]
311. Fink, O.; Wang, Q.; Svensen, M.; Dersin, P.; Lee, W.J.; Ducoffe, M. Potential, challenges and future directions for deep learning in prognostics and health management applications. *Eng. Appl. Artif. Intell.* **2020**, *92*, 103678. [CrossRef]
312. Sahoo, S.; Dash, M.; Behera, S.; Sabut, S. Machine learning approach to detect cardiac arrhythmias in ECG signals: A survey. *Innov. Res. Biomed. Eng.* **2020**, *41*, 185–194. [CrossRef]
313. Xiao, C.; Choi, E.; Sun, J. Opportunities and challenges in developing deep learning models using electronic health records data: A systematic review. *J. Am. Med. Inform. Assoc.* **2018**, *25*, 1419–1428. [CrossRef]
314. Miotto, R.; Li, L.; Kidd, B.A.; Dudley, J.T. Deep patient: An unsupervised representation to predict the future of patients from the electronic health records. *Sci. Rep.* **2016**, *6*, 1–10. [CrossRef]
315. Li, Y.; Rao, S.; Solares, J.R.A.; Hassaine, A.; Ramakrishnan, R.; Canoy, D.; Zhu, Y.; Rahimi, K.; Salimi-Khorshidi, G. BEHRT: Transformer for electronic health records. *Sci. Rep.* **2020**, *10*, 1–12. [CrossRef]
316. Pham, T.; Tran, T.; Phung, D.; Venkatesh, S. Predicting healthcare trajectories from medical records: A deep learning approach. *J. Biomed. Inform.* **2017**, *69*, 218–229. [CrossRef]
317. Isensee, F.; Jaeger, P.F.; Kohl, S.A.; Petersen, J.; Maier-Hein, K.H. nnU-Net: A self-configuring method for deep learning-based biomedical image segmentation. *Nat. Methods* **2021**, *18*, 203–211. [CrossRef]
318. Sun, Y.; Chen, Y.; Wang, X.; Tang, X. Deep learning face representation by joint identification-verification. *Adv. Neural Inf. Process. Syst.* **2014**, *27*, 1988–1996.
319. Taigman, Y.; Yang, M.; Ranzato, M.; Wolf, L. Web-scale training for face identification. In Proceedings of the IEEE Conference on Computer Vision and Pattern Recognition, Boston, MA, USA, 7–12 June 2015; pp. 2746–2754.
320. Schroff, F.; Kalenichenko, D.; Philbin, J. Facenet: A unified embedding for face recognition and clustering. In Proceedings of the IEEE Conference on Computer Vision and Pattern Recognition, Boston, MA, USA, 7–12 June 2015; pp. 815–823.
321. Zhu, Z.; Luo, P.; Wang, X.; Tang, X. Deep learning identity-preserving face space. In Proceedings of the IEEE International Conference on Computer Vision, Sydney, Australia, 1–8 December 2013; pp. 113–120.

322. Masi, I.; Rawls, S.; Medioni, G.; Natarajan, P. Pose-aware face recognition in the wild. In Proceedings of the IEEE Conference on Computer Vision and Pattern Recognition, Las Vegas, NV, USA, 27–30 June 2016; pp. 4838–4846.
323. Sun, Y.; Wang, X.; Tang, X. Sparsifying neural network connections for face recognition. In Proceedings of the IEEE Conference on Computer Vision and Pattern Recognition, Las Vegas, NV, USA, 27–30 June 2016; pp. 4856–4864.
324. Peng, X.; Ratha, N.; Pankanti, S. Learning face recognition from limited training data using deep neural networks. In Proceedings of the 2016 23rd International Conference on Pattern Recognition (ICPR), Cancún, Mexico, 4–8 December 2016; pp. 1442–1447.
325. Tran, L.; Yin, X.; Liu, X. Representation learning by rotating your faces. *IEEE Trans. Pattern Anal. Mach. Intell.* **2018**, *41*, 3007–3021. [CrossRef]
326. Yin, W.; Fu, Y.; Sigal, L.; Xue, X. Semi-latent GAN: Learning to generate and modify facial images from attributes. *arXiv* **2017**, arXiv:1704.02166.
327. Huerta, I.; Fernández, C.; Segura, C.; Hernando, J.; Prati, A. A deep analysis on age estimation. *Pattern Recognit. Lett.* **2015**, *68*, 239–249. [CrossRef]
328. Wang, X.; Guo, R.; Kambhamettu, C. Deeply-learned feature for age estimation. In Proceedings of the 2015 IEEE Winter Conference on Applications of Computer Vision, Waikoloa, HI, USA, 5–9 January 2015; pp. 534–541.
329. Liu, H.; Lu, J.; Feng, J.; Zhou, J. Label-sensitive deep metric learning for facial age estimation. *IEEE Trans. Inf. Forensics Secur.* **2017**, *13*, 292–305. [CrossRef]
330. Rothe, R.; Timofte, R.; Van Gool, L. Deep expectation of real and apparent age from a single image without facial landmarks. *Int. J. Comput. Vis.* **2018**, *126*, 144–157. [CrossRef]
331. Nie, L.; Kumar, A.; Zhan, S. Periocular recognition using unsupervised convolutional RBM feature learning. In Proceedings of the 2014 22nd International Conference on Pattern Recognition, Stockholm, Sweden, 24–28 August 2014; pp. 399–404.
332. Raghavendra, R.; Busch, C. Learning deeply coupled autoencoders for smartphone based robust periocular verification. In Proceedings of the 2016 IEEE International Conference on Image Processing (ICIP), Phoenix, AZ, USA, 25–28 September 2016; pp. 325–329.
333. Ahuja, K.; Islam, R.; Barbhuiya, F.A.; Dey, K. A preliminary study of CNNs for iris and periocular verification in the visible spectrum. In Proceedings of the 2016 23rd International Conference on Pattern Recognition (ICPR), Cancun, Mexico, 4–8 December 2016; pp. 181–186.
334. Daugman, J. New methods in iris recognition. *IEEE Trans. Syst. Man Cybern. Part B (Cybern.)* **2007**, *37*, 1167–1175. [CrossRef]
335. Liu, N.; Zhang, M.; Li, H.; Sun, Z.; Tan, T. DeepIris: Learning pairwise filter bank for heterogeneous iris verification. *Pattern Recognit. Lett.* **2016**, *82*, 154–161. [CrossRef]
336. Raja, K.B.; Raghavendra, R.; Vemuri, V.K.; Busch, C. Smartphone based visible iris recognition using deep sparse filtering. *Pattern Recognit. Lett.* **2015**, *57*, 33–42. [CrossRef]
337. Minaee, S.; Azimi, E.; Abdolrashidi, A. Fingernet: Pushing the limits of fingerprint recognition using convolutional neural network. *arXiv* **2019**, arXiv:1907.12956.
338. Sajjad, M.; Khan, S.; Hussain, T.; Muhammad, K.; Sangaiah, A.K.; Castiglione, A.; Esposito, C.; Baik, S.W. CNN-based anti-spoofing two-tier multi-factor authentication system. *Pattern Recognit. Lett.* **2019**, *126*, 123–131. [CrossRef]
339. Nogueira, R.F.; de Alencar Lotufo, R.; Machado, R.C. Fingerprint liveness detection using convolutional neural networks. *IEEE Trans. Inf. Forensics Secur.* **2016**, *11*, 1206–1213. [CrossRef]
340. Goel, I.; Puhan, N.B.; Mandal, B. Deep convolutional neural network for double-identity fingerprint detection. *IEEE Sensors Lett.* **2020**, *4*, 1–4. [CrossRef]
341. Chugh, T.; Cao, K.; Jain, A.K. Fingerprint spoof buster: Use of minutiae-centered patches. *IEEE Trans. Inf. Forensics Secur.* **2018**, *13*, 2190–2202. [CrossRef]
342. Cao, K.; Jain, A.K. Automated latent fingerprint recognition. *IEEE Trans. Pattern Anal. Mach. Intell.* **2018**, *41*, 788–800. [CrossRef] [PubMed]
343. Abdellatef, E.; Omran, E.M.; Soliman, R.F.; Ismail, N.A.; Abd Elrahman, S.E.S.; Ismail, K.N.; Rihan, M.; El-Samie, A.; Fathi, E.; Eisa, A.A. Fusion of deep-learned and hand-crafted features for cancelable recognition systems. *Soft Comput.* **2020**, *24*, 15189–15208. [CrossRef]
344. Zhu, Y.; Yin, X.; Jia, X.; Hu, J. Latent fingerprint segmentation based on convolutional neural networks. In Proceedings of the 2017 IEEE Workshop on Information Forensics and Security (WIFS), Rennes, France, 4–7 December 2017; pp. 1–6.
345. Liu, M.; Qian, P. Automatic segmentation and enhancement of latent fingerprints using deep nested unets. *IEEE Trans. Inf. Forensics Secur.* **2020**, *16*, 1709–1719. [CrossRef]
346. Song, D.; Tang, Y.; Feng, J. Aggregating minutia-centred deep convolutional features for fingerprint indexing. *Pattern Recognit.* **2019**, *88*, 397–408. [CrossRef]
347. Yin, X.; Hu, J.; Xu, J. Contactless fingerprint enhancement via intrinsic image decomposition and guided image filtering. In Proceedings of the 2016 IEEE 11th Conference on Industrial Electronics and Applications (ICIEA), Hefei, China, 5–7 June 2016; pp. 144–149.
348. Yin, X.; Zhu, Y.; Hu, J. A robust contactless fingerprint enhancement algorithm. In Proceedings of the International Conference on Mobile Networks and Management, Melbourne, VIC, Australia, 13–15 December 2017; pp. 127–136.
349. Lin, C.; Kumar, A. Contactless and partial 3D fingerprint recognition using multi-view deep representation. *Pattern Recognit.* **2018**, *83*, 314–327. [CrossRef]

350. Yin, X.; Zhu, Y.; Hu, J. Contactless fingerprint recognition based on global minutia topology and loose genetic algorithm. *IEEE Trans. Inf. Forensics Secur.* **2019**, *15*, 28–41. [CrossRef]
351. Yin, X.; Zhu, Y.; Hu, J. A survey on 2D and 3D contactless fingerprint biometrics: A taxonomy, review, and future directions. *IEEE Open J. Comput. Soc.* **2021**, *2*, 370–381. [CrossRef]
352. Kim, S.; Park, B.; Song, B.S.; Yang, S. Deep belief network based statistical feature learning for fingerprint liveness detection. *Pattern Recognit. Lett.* **2016**, *77*, 58–65. [CrossRef]
353. Yuan, C.; Chen, X.; Yu, P.; Meng, R.; Cheng, W.; Wu, Q.; Sun, X. Semi-supervised stacked autoencoder-based deep hierarchical semantic feature for real-time fingerprint liveness detection. *J. Real-Time Image Process.* **2020**, *17*, 55–71. [CrossRef]
354. Minaee, S.; Abdolrashidi, A. Finger-GAN: Generating realistic fingerprint images using connectivity imposed GAN. *arXiv* **2018**, arXiv:1812.10482.
355. Lee, S.; Jang, S.W.; Kim, D.; Hahn, H.; Kim, G.Y. A novel fingerprint recovery scheme using deep neural network-based learning. *Multimed. Tools Appl.* **2021**, *80*, 34121–34135. [CrossRef]
356. Kim, H.; Cui, X.; Kim, M.G.; Nguyen, T.H.B. Fingerprint generation and presentation attack detection using deep neural networks. In Proceedings of the 2019 IEEE Conference on Multimedia Information Processing and Retrieval, San Jose, CA, USA, 28–30 March 2019; pp. 375–378.
357. Tabassi, E.; Chugh, T.; Deb, D.; Jain, A.K. Altered fingerprints: Detection and localization. In Proceedings of the 2018 IEEE 9th International Conference on Biometrics Theory, Applications and Systems, Redondo Beach, CA, USA, 22–25 October 2018; pp. 1–9.
358. Jalali, A.; Mallipeddi, R.; Lee, M. Deformation invariant and contactless palmprint recognition using convolutional neural network. In Proceedings of the 3rd International Conference on Human-agent Interaction, Daegu, Republic of Korea, 21–24 October 2015; pp. 209–212.
359. Svoboda, J.; Masci, J.; Bronstein, M.M. Palmprint recognition via discriminative index learning. In Proceedings of the 2016 23rd International Conference on Pattern Recognition, Cancun, Mexico, 4–8 December 2016; pp. 4232–4237.
360. Ravanelli, M.; Bengio, Y. Speaker recognition from raw waveform with SincNet. In Proceedings of the 2018 IEEE Spoken Language Technology Workshop (SLT), Athens, Greece, 18–21 December 2018; pp. 1021–1028.
361. Jung, J.w.; Heo, H.S.; Kim, J.h.; Shim, H.j.; Yu, H.J. RawNet: Advanced end-to-end deep neural network using raw waveforms for text-independent speaker verification. *arXiv* **2019**, arXiv:1904.08104.
362. Dehak, N.; Kenny, P.J.; Dehak, R.; Dumouchel, P.; Ouellet, P. Front-end factor analysis for speaker verification. *IEEE Trans. Audio Speech, Lang. Process.* **2010**, *19*, 788–798. [CrossRef]
363. Variani, E.; Lei, X.; McDermott, E.; Moreno, I.L.; Gonzalez-Dominguez, J. Deep neural networks for small footprint text-dependent speaker verification. In Proceedings of the 2014 IEEE International Conference on Acoustics, Speech and Signal Processing (ICASSP), lorence, Italy, 4–9 May 2014; pp. 4052–4056.
364. Snyder, D.; Garcia-Romero, D.; Sell, G.; Povey, D.; Khudanpur, S. X-vectors: Robust DNN embeddings for speaker recognition. In Proceedings of the 2018 IEEE International Conference on Acoustics, Speech and Signal Processing (ICASSP), Calgary, AB, Canada, 15–20 April 2018; pp. 5329–5333.
365. Zhang, C.; Bahmaninezhad, F.; Ranjan, S.; Dubey, H.; Xia, W.; Hansen, J.H. UTD-CRSS systems for 2018 NIST speaker recognition evaluation. In Proceedings of the 2019 IEEE International Conference on Acoustics, Speech and Signal Processing (ICASSP), Brighton, UK, 12–17 May 2019; pp. 5776–5780.
366. Zhang, Z.; Wang, L.; Kai, A.; Yamada, T.; Li, W.; Iwahashi, M. Deep neural network-based bottleneck feature and denoising autoencoder-based dereverberation for distant-talking speaker identification. *EURASIP J. Audio Speech Music. Process.* **2015**, *2015*, 1–13. [CrossRef]
367. Hu, Z.; Fu, Y.; Luo, Y.; Xu, X.; Xia, Z.; Zhang, H. Speaker recognition based on short utterance compensation method of generative adversarial networks. *Int. J. Speech Technol.* **2020**, *23*, 443–450. [CrossRef]
368. Chen, L.; Liu, Y.; Xiao, W.; Wang, Y.; Xie, H. SpeakerGAN: Speaker identification with conditional generative adversarial network. *Neurocomputing* **2020**, *418*, 211–220. [CrossRef]
369. Nathwani, C. Online signature verification using bidirectional recurrent neural network. In Proceedings of the 2020 4th International Conference on Intelligent Computing and Control Systems (ICICCS), Madurai, India, 13–15 May 2020; pp. 1076–1078.
370. Lai, S.; Jin, L.; Lin, L.; Zhu, Y.; Mao, H. SynSig2Vec: Learning representations from synthetic dynamic signatures for real-world verification. In Proceedings of the AAAI Conference on Artificial Intelligence, New York, NY, USA, 7–12 February 2020, Volume 34, pp. 735–742.
371. Ribeiro, B.; Gonçalves, I.; Santos, S.; Kovacec, A. Deep learning networks for off-line handwritten signature recognition. In Proceedings of the Iberoamerican Congress on Pattern Recognition,Havana, Cuba, 28–31 October 2011; pp. 523–532.
372. Ahrabian, K.; BabaAli, B. Usage of autoencoders and Siamese networks for online handwritten signature verification. *Neural Comput. Appl.* **2019**, *31*, 9321–9334. [CrossRef]
373. Lai, S.; Jin, L.; Yang, W. Online signature verification using recurrent neural network and length-normalized path signature descriptor. In Proceedings of the 2017 14th IAPR International Conference on Document Analysis and Recognition (ICDAR), Kyoto, Japan, 9–15 November 2017; Volume 1, pp. 400–405.
374. Dey, S.; Dutta, A.; Toledo, J.I.; Ghosh, S.K.; Lladós, J.; Pal, U. Signet: Convolutional siamese network for writer independent offline signature verification. *arXiv* **2017**, arXiv:1707.02131.

375. Han, J.; Bhanu, B. Individual recognition using gait energy image. *IEEE Trans. Pattern Anal. Mach. Intell.* **2005**, *28*, 316–322. [CrossRef]
376. Zou, Q.; Wang, Y.; Wang, Q.; Zhao, Y.; Li, Q. Deep Learning-Based Gait Recognition Using Smartphones in the Wild. *IEEE Trans. Inf. Forensics Secur.* **2020**, *15*, 3197–3212. [CrossRef]
377. Wang, C.; Zhang, J.; Pu, J.; Yuan, X.; Wang, L. Chrono-gait image: A novel temporal template for gait recognition. In Proceedings of the European Conference on Computer Vision, Heraklion, Greece, 5–11 September 2010; pp. 257–270.
378. Lin, B.; Zhang, S.; Bao, F. Gait recognition with multiple-temporal-scale 3D convolutional neural network. In Proceedings of the 28th ACM International Conference on Multimedia, New York, NY, USA, 12–16 October 2020; pp. 3054–3062.
379. El-Fiqi, H.; Wang, M.; Salimi, N.; Kasmarik, K.; Barlow, M.; Abbass, H. Convolution neural networks for person identification and verification using steady state visual evoked potential. In Proceedings of the 2018 IEEE International Conference on Systems, Man, and Cybernetics (SMC), Miyazaki, Japan, 7–10 October 2018; pp. 1062–1069.
380. Yang, S.; Deravi, F.; Hoque, S. Task sensitivity in EEG biometric recognition. *Pattern Anal. Appl.* **2018**, *21*, 105–117. [CrossRef]
381. Wang, M.; Kasmarik, K.; Bezerianos, A.; Tan, K.C.; Abbass, H. On the channel density of EEG signals for reliable biometric recognition. *Pattern Recognit. Lett.* **2021**, *147*, 134–141. [CrossRef]
382. Wang, M.; El-Fiqi, H.; Hu, J.; Abbass, H.A. Convolutional neural networks using dynamic functional connectivity for EEG-based person identification in diverse human states. *IEEE Trans. Inf. Forensics Secur.* **2019**, *14*, 3259–3272. [CrossRef]
383. El-Fiqi, H.; Wang, M.; Kasmarik, K.; Bezerianos, A.; Tan, K.C.; Abbass, H.A. Weighted gate layer autoencoders. *IEEE Trans. Cybern.* **2021**, *52*, 7242–7253. [CrossRef]
384. Wang, M.; Abdelfattah, S.; Moustafa, N.; Hu, J. Deep gaussian mixture-hidden markov model for classification of EEG signals. *IEEE Trans. Emerg. Top. Comput. Intell.* **2018**, *2*, 278–287. [CrossRef]
385. Abdelfattah, S.M.; Abdelrahman, G.M.; Wang, M. Augmenting the size of EEG datasets using generative adversarial networks. In Proceedings of the 2018 International Joint Conference on Neural Networks (IJCNN), Rio de Janeiro, Brazil, 8–13 July 2018; pp. 1–6.
386. Wang, M.; Yin, X.; Zhu, Y.; Hu, J. Representation Learning and Pattern Recognition in Cognitive Biometrics: A Survey. *Sensors* **2022**, *22*, 5111. [CrossRef]
387. Martinez, A.; Benavente, R. *The AR Face Database*; Technical Report 24; CVC Technical Report; Elsevier: Amsterdam, The Netherlands, 1998; p. 8.
388. Johnson, P.A.; Lopez-Meyer, P.; Sazonova, N.; Hua, F.; Schuckers, S. Quality in face and iris research ensemble (Q-FIRE). In Proceedings of the 2010 Fourth IEEE International Conference on Biometrics: Theory, Applications and Systems (BTAS), Washington, DC, USA, 27–29 September 2010; pp. 1–6.
389. Yeung, D.Y.; Chang, H.; Xiong, Y.; George, S.; Kashi, R.; Matsumoto, T.; Rigoll, G. SVC2004: First international signature verification competition. In Proceedings of the International Conference on Biometric Authentication, Hong Kong, China, 15–17 July 2004; pp. 16–22.
390. Arnau-González, P.; Katsigiannis, S.; Arevalillo-Herráez, M.; Ramzan, N. BED: A new data set for EEG-based biometrics. *IEEE Internet Things J.* **2021**, *8*, 12219–12230. [CrossRef]
391. Toth, C.; Jóźków, G. Remote sensing platforms and sensors: A survey. *ISPRS J. Photogramm. Remote. Sens.* **2016**, *115*, 22–36. [CrossRef]
392. Chen, Y.; Lin, Z.; Zhao, X.; Wang, G.; Gu, Y. Deep learning-based classification of hyperspectral data. *IEEE J. Sel. Top. Appl. Earth Obs. Remote. Sens.* **2014**, *7*, 2094–2107. [CrossRef]
393. Chen, Y.; Zhao, X.; Jia, X. Spectral–spatial classification of hyperspectral data based on deep belief network. *IEEE J. Sel. Top. Appl. Earth Obs. Remote. Sens.* **2015**, *8*, 2381–2392. [CrossRef]
394. Tao, C.; Pan, H.; Li, Y.; Zou, Z. Unsupervised spectral–spatial feature learning with stacked sparse autoencoder for hyperspectral imagery classification. *IEEE Geosci. Remote. Sens. Lett.* **2015**, *12*, 2438–2442.
395. Makantasis, K.; Karantzalos, K.; Doulamis, A.; Doulamis, N. Deep supervised learning for hyperspectral data classification through convolutional neural networks. In Proceedings of the 2015 IEEE International Geoscience and Remote Sensing Symposium (IGARSS), Milan, Italy, 26–31 July 2015; pp. 4959–4962.
396. Santara, A.; Mani, K.; Hatwar, P.; Singh, A.; Garg, A.; Padia, K.; Mitra, P. BASS net: Band-adaptive spectral-spatial feature learning neural network for hyperspectral image classification. *IEEE Trans. Geosci. Remote. Sens.* **2017**, *55*, 5293–5301. [CrossRef]
397. Li, W.; Wu, G.; Zhang, F.; Du, Q. Hyperspectral image classification using deep pixel-pair features. *IEEE Trans. Geosci. Remote. Sens.* **2017**, *55*, 844–853. [CrossRef]
398. Liu, X.; Zhou, Y.; Zhao, J.; Yao, R.; Liu, B.; Zheng, Y. Siamese convolutional neural networks for remote sensing scene classification. *IEEE Geosci. Remote. Sens. Lett.* **2019**, *16*, 1200–1204. [CrossRef]
399. Zhang, W.; Tang, P.; Zhao, L. Remote sensing image scene classification using CNN-CapsNet. *Remote Sens.* **2019**, *11*, 494. [CrossRef]
400. Bazi, Y.; Bashmal, L.; Rahhal, M.M.A.; Dayil, R.A.; Ajlan, N.A. Vision transformers for remote sensing image classification. *Remote Sens.* **2021**, *13*, 516. [CrossRef]
401. Zhang, C.; Li, G.; Du, S. Multi-scale dense networks for hyperspectral remote sensing image classification. *IEEE Trans. Geosci. Remote. Sens.* **2019**, *57*, 9201–9222. [CrossRef]

402. Sun, H.; Li, S.; Zheng, X.; Lu, X. Remote sensing scene classification by gated bidirectional network. *IEEE Trans. Geosci. Remote. Sens.* **2019**, *58*, 82–96. [CrossRef]
403. Cheng, G.; Han, J.; Lu, X. Remote sensing image scene classification: Benchmark and state of the art. *Proc. IEEE* **2017**, *105*, 1865–1883. [CrossRef]
404. Othman, E.; Bazi, Y.; Alajlan, N.; Alhichri, H.; Melgani, F. Using convolutional features and a sparse autoencoder for land-use scene classification. *Int. J. Remote. Sens.* **2016**, *37*, 2149–2167. [CrossRef]
405. Penatti, O.A.; Nogueira, K.; Dos Santos, J.A. Do deep features generalize from everyday objects to remote sensing and aerial scenes domains? In Proceedings of the IEEE Conference on Computer Vision and Pattern Recognition Workshops, Boston, MA, USA, 7–12 June 2015; pp. 44–51.
406. Hu, F.; Xia, G.S.; Hu, J.; Zhang, L. Transferring deep convolutional neural networks for the scene classification of high-resolution remote sensing imagery. *Remote Sens.* **2015**, *7*, 14680–14707. [CrossRef]
407. Zou, Q.; Ni, L.; Zhang, T.; Wang, Q. Deep learning based feature selection for remote sensing scene classification. *IEEE Geosci. Remote. Sens. Lett.* **2015**, *12*, 2321–2325. [CrossRef]
408. He, N.; Fang, L.; Li, S.; Plaza, A.; Plaza, J. Remote sensing scene classification using multilayer stacked covariance pooling. *IEEE Trans. Geosci. Remote. Sens.* **2018**, *56*, 6899–6910. [CrossRef]
409. Lu, X.; Sun, H.; Zheng, X. A feature aggregation convolutional neural network for remote sensing scene classification. *IEEE Trans. Geosci. Remote. Sens.* **2019**, *57*, 7894–7906. [CrossRef]
410. Li, E.; Xia, J.; Du, P.; Lin, C.; Samat, A. Integrating multilayer features of convolutional neural networks for remote sensing scene classification. *IEEE Trans. Geosci. Remote. Sens.* **2017**, *55*, 5653–5665. [CrossRef]
411. Mei, S.; Yan, K.; Ma, M.; Chen, X.; Zhang, S.; Du, Q. Remote sensing scene classification using sparse representation-based framework with deep feature fusion. *IEEE J. Sel. Top. Appl. Earth Obs. Remote. Sens.* **2021**, *14*, 5867–5878. [CrossRef]
412. Zhao, Q.; Lyu, S.; Li, Y.; Ma, Y.; Chen, L. MGML: Multigranularity multilevel feature ensemble network for remote sensing scene classification. *IEEE Trans. Neural Networks Learn. Syst.* **2021**, *1*, 1–15. [CrossRef]
413. Chen, X.; Xiang, S.; Liu, C.L.; Pan, C.H. Vehicle detection in satellite images by hybrid deep convolutional neural networks. *IEEE Geosci. Remote. Sens. Lett.* **2014**, *11*, 1797–1801. [CrossRef]
414. Ševo, I.; Avramović, A. Convolutional neural network based automatic object detection on aerial images. *IEEE Geosci. Remote. Sens. Lett.* **2016**, *13*, 740–744. [CrossRef]
415. Tang, J.; Deng, C.; Huang, G.B.; Zhao, B. Compressed-domain ship detection on spaceborne optical image using deep neural network and extreme learning machine. *IEEE Trans. Geosci. Remote. Sens.* **2014**, *53*, 1174–1185. [CrossRef]
416. Zhu, H.; Chen, X.; Dai, W.; Fu, K.; Ye, Q.; Jiao, J. Orientation robust object detection in aerial images using deep convolutional neural network. In Proceedings of the 2015 IEEE International Conference on Image Processing (ICIP), Quebec City, QC, Canada, 27–30 September 2015; pp. 3735–3739.
417. Cheng, G.; Zhou, P.; Han, J. Learning rotation-invariant convolutional neural networks for object detection in VHR optical remote sensing images. *IEEE Trans. Geosci. Remote. Sens.* **2016**, *54*, 7405–7415. [CrossRef]
418. Zhang, G.; Lu, S.; Zhang, W. CAD-Net: A context-aware detection network for objects in remote sensing imagery. *IEEE Trans. Geosci. Remote. Sens.* **2019**, *57*, 10015–10024. [CrossRef]
419. Xu, D.; Wu, Y. Improved YOLO-V3 with DenseNet for multi-scale remote sensing target detection. *Sensors* **2020**, *20*, 4276. [CrossRef]
420. Liu, Y.; He, G.; Wang, Z.; Li, W.; Huang, H. NRT-YOLO: Improved YOLOv5 based on nested residual transformer for tiny remote sensing object detection. *Sensors* **2022**, *22*, 4953. [CrossRef]
421. Zhang, S.; He, G.; Chen, H.B.; Jing, N.; Wang, Q. Scale adaptive proposal network for object detection in remote sensing images. *IEEE Geosci. Remote. Sens. Lett.* **2019**, *16*, 864–868. [CrossRef]
422. Zhang, Z.; Guo, W.; Zhu, S.; Yu, W. Toward arbitrary-oriented ship detection with rotated region proposal and discrimination networks. *IEEE Geosci. Remote. Sens. Lett.* **2018**, *15*, 1745–1749. [CrossRef]
423. Feng, X.; Han, J.; Yao, X.; Cheng, G. Progressive contextual instance refinement for weakly supervised object detection in remote sensing images. *IEEE Trans. Geosci. Remote. Sens.* **2020**, *58*, 8002–8012. [CrossRef]
424. Xu, X.; Feng, Z.; Cao, C.; Li, M.; Wu, J.; Wu, Z.; Shang, Y.; Ye, S. An improved swin transformer-based model for remote sensing object detection and instance segmentation. *Remote Sens.* **2021**, *13*, 4779. [CrossRef]
425. Zhang, L.; Zhang, J. A new saliency-driven fusion method based on complex wavelet transform for remote sensing images. *IEEE Geosci. Remote. Sens. Lett.* **2017**, *14*, 2433–2437. [CrossRef]
426. Zhang, L.; Zhang, J.; Ma, J.; Jia, X. SC-PNN: Saliency cascade convolutional neural network for pansharpening. *IEEE Trans. Geosci. Remote. Sens.* **2021**, *59*, 9697–9715. [CrossRef]
427. Huang, W.; Xiao, L.; Wei, Z.; Liu, H.; Tang, S. A new pan-sharpening method with deep neural networks. *IEEE Geosci. Remote. Sens. Lett.* **2015**, *12*, 1037–1041. [CrossRef]
428. Masi, G.; Cozzolino, D.; Verdoliva, L.; Scarpa, G. Pansharpening by convolutional neural networks. *Remote Sens.* **2016**, *8*, 594. [CrossRef]
429. Hu, J.; Hu, P.; Kang, X.; Zhang, H.; Fan, S. Pan-sharpening via multiscale dynamic convolutional neural network. *IEEE Trans. Geosci. Remote. Sens.* **2021**, *59*, 2231–2244. [CrossRef]

130. He, L.; Rao, Y.; Li, J.; Chanussot, J.; Plaza, A.; Zhu, J.; Li, B. Pansharpening via detail injection based convolutional neural networks. *IEEE J. Sel. Top. Appl. Earth Obs. Remote. Sens.* **2019**, *12*, 1188–1204. [CrossRef]
131. Yuan, Q.; Wei, Y.; Meng, X.; Shen, H.; Zhang, L. A multiscale and multidepth convolutional neural network for remote sensing imagery pan-sharpening. *IEEE J. Sel. Top. Appl. Earth Obs. Remote. Sens.* **2018**, *11*, 978–989. [CrossRef]
132. Yang, J.; Fu, X.; Hu, Y.; Huang, Y.; Ding, X.; Paisley, J. PanNet: A deep network architecture for pan-sharpening. In Proceedings of the IEEE International Conference on Computer Vision (ICCV), Venice, Italy, 22–29 October 2017; pp. 1753–1761.
133. Hong, D.; Gao, L.; Yokoya, N.; Yao, J.; Chanussot, J.; Du, Q.; Zhang, B. More diverse means better: Multimodal deep learning meets remote-sensing imagery classification. *IEEE Trans. Geosci. Remote. Sens.* **2021**, *59*, 4340–4354. [CrossRef]
134. Lagrange, A.; Le Saux, B.; Beaupère, A.; Boulch, A.; Chan-Hon-Tong, A.; Herbin, S.; Randrianarivo, H.; Ferecatu, M. Benchmarking classification of earth-observation data: From learning explicit features to convolutional networks. In Proceedings of the 2015 IEEE International Geoscience and Remote Sensing Symposium (IGARSS), Milan, Italy, 26–31 July 2015; pp. 4173–4176. [CrossRef]
135. Irwin, K.; Beaulne, D.; Braun, A.; Fotopoulos, G. Fusion of SAR, optical imagery and airborne LiDAR for surface water detection. *Remote Sens.* **2017**, *9*, 890. [CrossRef]
136. Yang, Y.; Newsam, S. Bag-of-visual-words and spatial extensions for land-use classification. In Proceedings of the 18th SIGSPATIAL International Conference on Advances in Geographic Information Systems, San Jose, CA, USA, 2–5 November 2010; pp. 270–279.
137. Zuo, X. Hyperspectral Data. 2022. Available online: https://ieee-dataport.org/documents/hyperspectral-data (accessed on 2 November 2022).
138. Dai, D.; Yang, W. Satellite image classification via two-layer sparse coding with biased image representation. *IEEE Geosci. Remote. Sens. Lett.* **2010**, *8*, 173–176. [CrossRef]
139. Xia, G.S.; Hu, J.; Hu, F.; Shi, B.; Bai, X.; Zhong, Y.; Zhang, L.; Lu, X. AID: A benchmark data set for performance evaluation of aerial scene classification. *IEEE Trans. Geosci. Remote. Sens.* **2017**, *55*, 3965–3981. [CrossRef]
140. Zhou, Z.; Li, S.; Wu, W.; Guo, W.; Li, X.; Xia, G.; Zhao, Z. NaSC-TG2: Natural scene classification with Tiangong-2 remotely sensed imagery. *IEEE J. Sel. Top. Appl. Earth Obs. Remote. Sens.* **2021**, *14*, 3228–3242. [CrossRef]
141. Cheng, G.; Han, J. A survey on object detection in optical remote sensing images. *ISPRS J. Photogramm. Remote Sens.* **2016**, *117*, 11–28. [CrossRef]
142. Liu, Z.; Wang, H.; Weng, L.; Yang, Y. Ship rotated bounding box space for ship extraction from high-resolution optical satellite images with complex backgrounds. *IEEE Geosci. Remote. Sens. Lett.* **2016**, *13*, 1074–1078. [CrossRef]
143. Xia, G.S.; Bai, X.; Ding, J.; Zhu, Z.; Belongie, S.; Luo, J.; Datcu, M.; Pelillo, M.; Zhang, L. DOTA: A large-scale dataset for object detection in aerial images. In Proceedings of the IEEE Conference on Computer Vision and Pattern Recognition, Salt Lake City, UT, USA, 18–22 June 2018; pp. 3974–3983.
144. Li, K.; Wan, G.; Cheng, G.; Meng, L.; Han, J. Object detection in optical remote sensing images: A survey and a new benchmark. *ISPRS J. Photogramm. Remote. Sens.* **2020**, *159*, 296–307. [CrossRef]
145. Wei, S.; Zeng, X.; Qu, Q.; Wang, M.; Su, H.; Shi, J. HRSID: A high-resolution SAR images dataset for ship detection and instance segmentation. *IEEE Access* **2020**, *8*, 120234–120254. [CrossRef]
146. Gao, N.; Gao, L.; Gao, Q.; Wang, H. An intrusion detection model based on deep belief networks. In Proceedings of the 2014 Second International Conference on Advanced Cloud and Big Data, Huangshan, China, 20–22 November 2014; pp. 247–252.
147. Alom, M.Z.; Bontupalli, V.; Taha, T.M. Intrusion detection using deep belief networks. In Proceedings of the 2015 National Aerospace and Electronics Conference (NAECON), New York, NY, USA, 15–19 June 2015; pp. 339–344.
148. Alrawashdeh, K.; Purdy, C. Toward an online anomaly intrusion detection system based on deep learning. In Proceedings of the 2016 15th IEEE international Conference on Machine Learning and Applications (ICMLA), Anaheim, CA, USA, 18–20 December 2016; pp. 195–200.
149. Tavallaee, M.; Bagheri, E.; Lu, W.; Ghorbani, A.A. A detailed analysis of the KDD CUP 99 data set. In Proceedings of the 2009 IEEE Symposium on Computational Intelligence for Security and Defense Applications, Ottawa, ON, Canada, 8–10 July 2009; pp. 1–6.
150. Abolhasanzadeh, B. Nonlinear dimensionality reduction for intrusion detection using auto-encoder bottleneck features. In Proceedings of the 2015 7th Conference on Information and Knowledge Technology (IKT), Urmia, Iran, 26–28 May 2015; pp. 1–5.
151. Vincent, P.; Larochelle, H.; Lajoie, I.; Bengio, Y.; Manzagol, P.A.; Bottou, L. Stacked denoising autoencoders: Learning useful representations in a deep network with a local denoising criterion. *J. Mach. Learn. Res.* **2010**, *11*.
152. Niyaz, Q.; Sun, W.; Javaid, A.Y. A deep learning based DDoS detection system in software-defined networking (SDN). *EAI Endorsed Trans. Secur. Saf.* **2017**, *4*, e2. [CrossRef]
153. Shone, N.; Ngoc, T.N.; Phai, V.D.; Shi, Q. A deep learning approach to network intrusion detection. *IEEE Trans. Emerg. Top. Comput. Intell.* **2018**, *2*, 41–50. [CrossRef]
154. Parker, L.R.; Yoo, P.D.; Asyhari, T.A.; Chermak, L.; Jhi, Y.; Taha, K. DEMISe: Interpretable deep extraction and mutual information selection techniques for IoT intrusion detection. In Proceedings of the 14th International Conference on Availability, Reliability and Security, Canterbury, UK, 26–29 August 2019; pp. 1–10.
155. Vu, L.; Nguyen, Q.U.; Nguyen, D.N.; Hoang, D.T.; Dutkiewicz, E. Learning latent distribution for distinguishing network traffic in intrusion detection system. In Proceedings of the 2019 IEEE International Conference on Communications (ICC), Shanghai, China, 20–24 May 2019; pp. 1–6.

456. Yin, X.; Zhu, Y.; Hu, J. A subgrid-oriented privacy-preserving microservice framework based on deep neural network for false data injection attack detection in smart grids. *IEEE Trans. Ind. Inform.* **2021**, *18*, 1957–1967. [CrossRef]
457. Yin, X.; Zhu, Y.; Xie, Y.; Hu, J. PowerFDNet: Deep learning-based stealthy false data injection attack detection for AC-model transmission systems. *IEEE Open J. Comput. Soc.* **2022**, *3*, 149–161. [CrossRef]
458. Brown, A.; Tuor, A.; Hutchinson, B.; Nichols, N. Recurrent neural network attention mechanisms for interpretable system log anomaly detection. In Proceedings of the 1st Workshop on Machine Learning for Computing Systems, Tempe, AZ, USA, 12 June 2018; pp. 1–8.
459. Kim, G.; Yi, H.; Lee, J.; Paek, Y.; Yoon, S. LSTM-based system-call language modeling and robust ensemble method for designing host-based intrusion detection systems. *arXiv* **2016**, arXiv:1611.01726. .
460. Jiang, F.; Fu, Y.; Gupta, B.B.; Liang, Y.; Rho, S.; Lou, F.; Meng, F.; Tian, Z. Deep learning based multi-channel intelligent attack detection for data security. *IEEE Trans. Sustain. Comput.* **2018**, *5*, 204–212. [CrossRef]
461. Wang, W.; Sheng, Y.; Wang, J.; Zeng, X.; Ye, X.; Huang, Y.; Zhu, M. HAST-IDS: Learning hierarchical spatial-temporal features using deep neural networks to improve intrusion detection. *IEEE Access* **2017**, *6*, 1792–1806. [CrossRef]
462. Zhang, Y.; Chen, X.; Jin, L.; Wang, X.; Guo, D. Network intrusion detection: Based on deep hierarchical network and original flow data. *IEEE Access* **2019**, *7*, 37004–37016. [CrossRef]
463. Schlegl, T.; Seeböck, P.; Waldstein, S.M.; Schmidt-Erfurth, U.; Langs, G. Unsupervised anomaly detection with generative adversarial networks to guide marker discovery. In Proceedings of the International Conference on Information Processing in Medical Imaging, Boone, NC, USA, 25–30 June 2017; pp. 146–157.
464. Zenati, H.; Foo, C.S.; Lecouat, B.; Manek, G.; Chandrasekhar, V.R. Efficient GAN-based anomaly detection. In Proceedings of the 20th IEEE International Conference on Data Mining, Sorrento, Italy, 17–20 November 2018; pp. 1–11.
465. Pascanu, R.; Stokes, J.W.; Sanossian, H.; Marinescu, M.; Thomas, A. Malware classification with recurrent networks. In Proceedings of the 2015 IEEE International Conference on Acoustics, Speech and Signal Processing (ICASSP), South Brisbane, QSD, Australia, 19–24 April 2015; pp. 1916–1920.
466. David, O.E.; Netanyahu, N.S. Deepsign: Deep learning for automatic malware signature generation and classification. In Proceedings of the 2015 International Joint Conference on Neural Networks (IJCNN), Killarney, Ireland, 12–17 July 2015; pp. 1–8.
467. Yousefi-Azar, M.; Varadharajan, V.; Hamey, L.; Tupakula, U. Autoencoder-based feature learning for cyber security applications. In Proceedings of the 2017 International Joint Conference on Neural Networks (IJCNN), Anchorage, AK, USA, 14–19 May 2017; pp. 3854–3861.
468. Kim, J.Y.; Bu, S.J.; Cho, S.B. Zero-day malware detection using transferred generative adversarial networks based on deep autoencoders. *Inf. Sci.* **2018**, *460*, 83–102. [CrossRef]
469. Kim, J.Y.; Cho, S.B. Detecting intrusive malware with a hybrid generative deep learning model. In Proceedings of the International Conference on Intelligent Data Engineering and Automated Learning, Madrid, Spain, 21–23 November 2018; pp. 499–507.
470. Yuan, Z.; Lu, Y.; Wang, Z.; Xue, Y. Droid-Sec: Deep learning in Android malware detection. In Proceedings of the 2014 ACM Conference on SIGCOMM, Chicago, IL, USA, 17–22 August 2014; pp. 371–372.
471. Hou, S.; Saas, A.; Ye, Y.; Chen, L. Droiddelver: An android malware detection system using deep belief network based on api call blocks. In Proceedings of the International Conference on Web-Age Information Management, Nanchang, China, 3–5 June 2016; pp. 54–66.
472. Su, X.; Zhang, D.; Li, W.; Zhao, K. A deep learning approach to android malware feature learning and detection. In Proceedings of the 2016 IEEE Trustcom/BigDataSE/ISPA, Tianjin, China, 23–16 August 2016; pp. 244–251.
473. McLaughlin, N.; Martinez del Rincon, J.; Kang, B.; Yerima, S.; Miller, P.; Sezer, S.; Safaei, Y.; Trickel, E.; Zhao, Z.; Doupé, A.; et al. Deep android malware detection. In Proceedings of the 7th ACM on Conference on Data and Application Security and Privacy, Scottsdale, AZ, USA, 22–24 March 2017; pp. 301–308.
474. Nix, R.; Zhang, J. Classification of Android apps and malware using deep neural networks. In Proceedings of the 2017 International Joint Conference on Neural Networks (IJCNN), Anchorage, AK, USA, 14–19 May 2017; pp. 1871–1878.
475. Jan, S.; Ali, T.; Alzahrani, A.; Musa, S. Deep convolutional generative adversarial networks for intent-based dynamic behavior capture. *Int. J. Eng. Technol.* **2018**, *7*, 101–103.
476. Zhang, N.; Yuan, Y. *Phishing Detection Using Neural Network*; CS229 Lecture Notes; Stanford University: Stanford, CA, USA, 2012; pp. 1–5.
477. Mohammad, R.M.; Thabtah, F.; McCluskey, L. Predicting phishing websites based on self-structuring neural network. *Neural Comput. Appl.* **2014**, *25*, 443–458. [CrossRef]
478. Benavides, E.; Fuertes, W.; Sanchez, S.; Sanchez, M. Classification of phishing attack solutions by employing deep learning techniques: A systematic literature review. *Dev. Adv. Def. Secur.* **2020**, 51–64.
479. Wu, T.; Liu, S.; Zhang, J.; Xiang, Y. Twitter spam detection based on deep learning. In Proceedings of the Australasian Computer Science Week Multiconference, Geelong, Australia, 31 January–3 February 2017; pp. 1–8.
480. Jain, G.; Sharma, M.; Agarwal, B. Spam detection on social media using semantic convolutional neural network. *Int. J. Knowl. Discov. Bioinform. (IJKDB)* **2018**, *8*, 12–26. [CrossRef]
481. Thejas, G.; Boroojeni, K.G.; Chandna, K.; Bhatia, I.; Iyengar, S.; Sunitha, N. Deep learning-based model to fight against ad click fraud. In Proceedings of the 2019 ACM Southeast Conference, Kennesaw, GA, USA, 18–20 April 2019; pp. 176–181.

482. Singh, V.; Varshney, A.; Akhtar, S.S.; Vijay, D.; Shrivastava, M. Aggression detection on social media text using deep neural networks. In Proceedings of the 2nd Workshop on Abusive Language Online (ALW2), Brussels, Belgium, 21 October 2018; pp. 43–50.
483. Ban, X.; Chen, C.; Liu, S.; Wang, Y.; Zhang, J. Deep-learnt features for Twitter spam detection. In Proceedings of the 2018 International Symposium on Security and Privacy in Social Networks and Big Data (SocialSec), Santa Clara, CA, USA, 10–12 December 2018; pp. 208–212.
484. Tolosana, R.; Vera-Rodriguez, R.; Fierrez, J.; Morales, A.; Ortega-Garcia, J. Deepfakes and beyond: A survey of face manipulation and fake detection. *Inf. Fusion* **2020**, *64*, 131–148. [CrossRef]
485. Hasan, H.R.; Salah, K. Combating deepfake videos using blockchain and smart contracts. *IEEE Access* **2019**, *7*, 41596–41606. [CrossRef]
486. Fagni, T.; Falchi, F.; Gambini, M.; Martella, A.; Tesconi, M. TweepFake: About detecting deepfake tweets. *PLoS ONE* **2021**, *16*, e0251415. [CrossRef]
487. Verdoliva, L. Media forensics and deepfakes: An overview. *IEEE J. Sel. Top. Signal Process.* **2020**, *14*, 910–932. [CrossRef]
488. Chatzoglou, E.; Kambourakis, G.; Kolias, C. Empirical evaluation of attacks against IEEE 802.11 enterprise networks: The AWID3 dataset. *IEEE Access* **2021**, *9*, 34188–34205. [CrossRef]
489. Sharafaldin, I.; Lashkari, A.H.; Ghorbani, A.A. Toward generating a new intrusion detection dataset and intrusion traffic characterization. In Proceedings of the 4th International Conference on Information Systems Security and Privacy, Madeira, Portugal, 22–24 January 2018; pp. 108–116.
490. Kolias, C.; Kambourakis, G.; Stavrou, A.; Gritzalis, S. Intrusion detection in 802.11 networks: Empirical evaluation of threats and a public dataset. *IEEE Commun. Surv. Tutorials* **2016**, *18*, 184–208. [CrossRef]
491. Moustafa, N.; Slay, J. UNSW-NB15: A comprehensive data set for network intrusion detection systems (UNSW-NB15 network data set). In Proceedings of the 2015 Military Communications and Information Systems Conference (MilCIS), Canberra, Australia, 10–12 November 2015; pp. 1–6.
492. Creech, G.; Hu, J. Generation of a new IDS test dataset: Time to retire the KDD collection. In Proceedings of the 2013 IEEE Wireless Communications and Networking Conference (WCNC), Shanghai, China, 7–10 April 2013; pp. 4487–4492.
493. Miotto, R.; Wang, F.; Wang, S.; Jiang, X.; Dudley, J.T. Deep learning for healthcare: Review, opportunities and challenges. *Briefings Bioinform.* **2018**, *19*, 1236–1246. [CrossRef]
494. Hammerla, N.Y.; Halloran, S.; Plötz, T. Deep, convolutional, and recurrent models for human activity recognition using wearables. In Proceedings of the Twenty-Fifth International Joint Conference on Artificial Intelligence, New York, NY, USA, 9–15 July 2016; pp. 1533–1540.
495. Zhu, J.; Pande, A.; Mohapatra, P.; Han, J.J. Using deep learning for energy expenditure estimation with wearable sensors. In Proceedings of the 17th International Conference on E-health Networking, Application & Services (HealthCom), Boston, MA, USA, 14–17 October 2015; pp. 501–506.
496. Hannun, A.Y.; Rajpurkar, P.; Haghpanahi, M.; Tison, G.H.; Bourn, C.; Turakhia, M.P.; Ng, A.Y. Cardiologist-level arrhythmia detection and classification in ambulatory electrocardiograms using a deep neural network. *Nat. Med.* **2019**, *25*, 65–69. [CrossRef]
497. Gao, Y.; Xiang, X.; Xiong, N.; Huang, B.; Lee, H.J.; Alrifai, R.; Jiang, X.; Fang, Z. Human action monitoring for healthcare based on deep learning. *IEEE Access* **2018**, *6*, 52277–52285. [CrossRef]
498. Ravi, D.; Wong, C.; Lo, B.; Yang, G.Z. A deep learning approach to on-node sensor data analytics for mobile or wearable devices. *IEEE J. Biomed. Health Inform.* **2016**, *21*, 56–64. [CrossRef]
499. Prasoon, A.; Petersen, K.; Igel, C.; Lauze, F.; Dam, E.; Nielsen, M. Deep feature learning for knee cartilage segmentation using a triplanar convolutional neural network. In Proceedings of the International Conference on Medical Image Computing and Computer-Assisted Intervention, Nagoya, Japan, 22–26 September 2013; pp. 246–253.
500. Gulshan, V.; Peng, L.; Coram, M.; Stumpe, M.C.; Wu, D.; Narayanaswamy, A.; Venugopalan, S.; Widner, K.; Madams, T.; Cuadros, J.; et al. Development and validation of a deep learning algorithm for detection of diabetic retinopathy in retinal fundus photographs. *JAMA* **2016**, *316*, 2402–2410. [CrossRef]
501. Zeng, X.; Cao, K.; Zhang, M. MobileDeepPill: A small-footprint mobile deep learning system for recognizing unconstrained pill images. In Proceedings of the 15th Annual International Conference on Mobile Systems, Applications, and Services, Niagara Falls, NY, USA, 19–23 June 2017; pp. 56–67.
502. Lopez, A.R.; Giro-i Nieto, X.; Burdick, J.; Marques, O. Skin lesion classification from dermoscopic images using deep learning techniques. In Proceedings of the 13th IASTED International Conference on Biomedical Engineering (BioMed), Innsbruck, Austria, 20–21 February 2017; pp. 49–54.
503. Chen, M.; Yang, J.; Zhou, J.; Hao, Y.; Zhang, J.; Youn, C.H. 5G-smart diabetes: Toward personalized diabetes diagnosis with healthcare big data clouds. *IEEE Commun. Mag.* **2018**, *56*, 16–23. [CrossRef]
504. Chang, W.J.; Chen, L.B.; Hsu, C.H.; Lin, C.P.; Yang, T.C. A deep learning-based intelligent medicine recognition system for chronic patients. *IEEE Access* **2019**, *7*, 44441–44458. [CrossRef]
505. Gu, Y.; Chen, Y.; Liu, J.; Jiang, X. Semi-supervised deep extreme learning machine for Wi-Fi based localization. *Neurocomputing* **2015**, *166*, 282–293. [CrossRef]
506. Mohammadi, M.; Al-Fuqaha, A.; Guizani, M.; Oh, J.S. Semisupervised deep reinforcement learning in support of IoT and smart city services. *IEEE Internet Things J.* **2017**, *5*, 624–635. [CrossRef]

507. Wang, X.; Gao, L.; Mao, S.; Pandey, S. CSI-based fingerprinting for indoor localization: A deep learning approach. *IEEE Trans. Veh. Technol.* **2016**, *66*, 763–776. [CrossRef]
508. Erol, B.A.; Majumdar, A.; Lwowski, J.; Benavidez, P.; Rad, P.; Jamshidi, M. Improved deep neural network object tracking system for applications in home robotics. In *Computational Intelligence for Pattern Recognition*; Springer: Berlin, Germany, 2018; pp. 369–395.
509. Levine, S.; Pastor, P.; Krizhevsky, A.; Ibarz, J.; Quillen, D. Learning hand-eye coordination for robotic grasping with deep learning and large-scale data collection. *Int. J. Robot. Res.* **2018**, *37*, 421–436. [CrossRef]
510. Huang, W.; Song, G.; Hong, H.; Xie, K. Deep architecture for traffic flow prediction: Deep belief networks with multitask learning. *IEEE Trans. Intell. Transp. Syst.* **2014**, *15*, 2191–2201. [CrossRef]
511. Lv, Y.; Duan, Y.; Kang, W.; Li, Z.; Wang, F.Y. Traffic flow prediction with big data: A deep learning approach. *IEEE Trans. Intell. Transp. Syst.* **2014**, *16*, 865–873. [CrossRef]
512. Zhao, Z.; Chen, W.; Wu, X.; Chen, P.C.; Liu, J. LSTM network: A deep learning approach for short-term traffic forecast. *IET Intell. Transp. Syst.* **2017**, *11*, 68–75. [CrossRef]
513. Polson, N.G.; Sokolov, V.O. Deep learning for short-term traffic flow prediction. *Transp. Res. Part C Emerg. Technol.* **2017**, *79*, 1–17. [CrossRef]
514. Li, H.; Li, Y.; Porikli, F. Deeptrack: Learning discriminative feature representations online for robust visual tracking. *IEEE Trans. Image Process.* **2015**, *25*, 1834–1848. [CrossRef]
515. Ondrúška, P.; Posner, I. Deep tracking: Seeing beyond seeing using recurrent neural networks. In Proceedings of the Thirtieth AAAI Conference on Artificial Intelligence, Phoenix, AZ, USA, 12–17 February 2016; pp. 3361–3367.
516. Wu, B.; Iandola, F.; Jin, P.H.; Keutzer, K. Squeezedet: Unified, small, low power fully convolutional neural networks for real-time object detection for autonomous driving. In Proceedings of the IEEE Conference on Computer Vision and Pattern Recognition Workshops, Honolulu, HI, USA, 21–26 July 2017; pp. 129–137.
517. Bojarski, M.; Del Testa, D.; Dworakowski, D.; Firner, B.; Flepp, B.; Goyal, P.; Jackel, L.D.; Monfort, M.; Muller, U.; Zhang, J.; et al. End to end learning for self-driving cars. *arXiv* **2016**, arXiv:1604.07316.
518. Xu, H.; Gao, Y.; Yu, F.; Darrell, T. End-to-end learning of driving models from large-scale video datasets. In Proceedings of the IEEE Conference on Computer Vision and Pattern Recognition, Honolulu, HI, USA, 21–26 July 2017; pp. 2174–2182.
519. Grigorescu, S.; Trasnea, B.; Cocias, T.; Macesanu, G. A survey of deep learning techniques for autonomous driving. *J. Field Robot.* **2020**, *37*, 362–386. [CrossRef]
520. Li, L.; Ota, K.; Dong, M. Deep learning for smart industry: Efficient manufacture inspection system with fog computing. *IEEE Trans. Ind. Inform.* **2018**, *14*, 4665–4673. [CrossRef]
521. Park, J.K.; Kwon, B.K.; Park, J.H.; Kang, D.J. Machine learning-based imaging system for surface defect inspection. *Int. J. Precis. Eng. Manuf. Green Technol.* **2016**, *3*, 303–310. [CrossRef]
522. Cinar, E. A Sensor Fusion Method Using Transfer Learning Models for Equipment Condition Monitoring. *Sensors* **2022**, *22*, 6791. [CrossRef]
523. Chen, H.; Zhong, K.; Ran, G.; Cheng, C. Deep Learning-Based Machinery Fault Diagnostics. In *Machine*; MDPI: Basel, Switzerland, 2022; Volume 10, p. 690.
524. Wang, J.; Zhuang, J.; Duan, L.; Cheng, W. A multi-scale convolution neural network for featureless fault diagnosis. In Proceedings of the 2016 International Symposium on Flexible Automation (ISFA), Cleveland, Ohio, 1–3 August 2016; pp. 65–70.
525. Wang, L.; Zhao, X.; Pei, J.; Tang, G. Transformer fault diagnosis using continuous sparse autoencoder. *SpringerPlus* **2016**, *5*, 1–13. [CrossRef]
526. Lei, Y.; Jia, F.; Lin, J.; Xing, S.; Ding, S.X. An intelligent fault diagnosis method using unsupervised feature learning towards mechanical big data. *IEEE Trans. Ind. Electron.* **2016**, *63*, 3137–3147. [CrossRef]
527. Alassery, F.; Alzahrani, A.; Khan, A.; Irshad, K.; Kshirsagar, S.R. An artificial intelligence-based solar radiation prophesy model for green energy utilization in energy management system. *Sustain. Energy Technol. Assess.* **2022**, *52*, 102060. [CrossRef]
528. Khan, A.I.; Alsolami, F.; Alqurashi, F.; Abushark, Y.B.; Sarker, I.H. Novel energy management scheme in IoT enabled smart irrigation system using optimized intelligence methods. *Eng. Appl. Artif. Intell.* **2022**, *114*, 104996. [CrossRef]
529. Kshirsagar, P.R.; Kumar, N.; Almulihi, A.H.; Alassery, F.; Khan, A.I.; Islam, S.; Rothe, J.P.; Jagannadham, D.; Dekeba, K. Artificial Intelligence-Based Robotic Technique for Reusable Waste Materials. *Comput. Intell. Neurosci.* **2022**, *2022*, 2073482. [CrossRef]
530. Zweig, G. Classification and recognition with direct segment models. In Proceedings of the 2012 IEEE International Conference on Acoustics, Speech and Signal Processing (ICASSP), Kyoto, Japan, 25–30 March 2012; pp. 4161–4164.
531. Lu, L.; Kong, L.; Dyer, C.; Smith, N.A.; Renals, S. Segmental recurrent neural networks for end-to-end speech recognition. *arXiv* **2016**, arXiv:1603.00223.
532. Yang, S.; Gong, Z.; Ye, K.; Wei, Y.; Huang, Z.; Huang, Z. EdgeRNN: A compact speech recognition network with spatio-temporal features for edge computing. *IEEE Access* **2020**, *8*, 81468–81478. [CrossRef]
533. Yang, C.H.H.; Qi, J.; Chen, S.Y.C.; Chen, P.Y.; Siniscalchi, S.M.; Ma, X.; Lee, C.H. Decentralizing feature extraction with quantum convolutional neural network for automatic speech recognition. In Proceedings of the ICASSP 2021–2021 IEEE International Conference on Acoustics, Speech and Signal Processing (ICASSP), Toronto, ON, Canada, 6–11 June 2021; pp. 6523–6527.
534. Bell, P.; Fainberg, J.; Klejch, O.; Li, J.; Renals, S.; Swietojanski, P. Adaptation algorithms for neural network-based speech recognition: An overview. *IEEE Open J. Signal Process.* **2020**, *2*, 33–66. [CrossRef]

35. Wang, D.; Wang, X.; Lv, S. An overview of end-to-end automatic speech recognition. *Symmetry* **2019**, *11*, 1018. [CrossRef]
36. Malik, M.; Malik, M.K.; Mehmood, K.; Makhdoom, I. Automatic speech recognition: A survey. *Multimed. Tools Appl.* **2021**, *80*, 9411–9457. [CrossRef]
37. Moraes, R.; Valiati, J.F.; Neto, W.P.G. Document-level sentiment classification: An empirical comparison between SVM and ANN. *Expert Syst. Appl.* **2013**, *40*, 621–633. [CrossRef]
38. Socher, R.; Pennington, J.; Huang, E.H.; Ng, A.Y.; Manning, C.D. Semi-supervised recursive autoencoders for predicting sentiment distributions. In Proceedings of the Conference on Empirical Methods in Natural Language Processing, Scotland, UK, 27–31 July 2011; pp. 151–161.
39. Dong, L.; Wei, F.; Tan, C.; Tang, D.; Zhou, M.; Xu, K. Adaptive recursive neural network for target-dependent Twitter sentiment classification. In Proceedings of the 52nd Annual Meeting of the Association for Computational Linguistics, Baltimore, MD, USA, 23–24 June 2014; pp. 49–54.
40. Zhang, L.; Wang, S.; Liu, B. Deep learning for sentiment analysis: A survey. *Wiley Interdiscip. Rev. Data Min. Knowl. Discov.* **2018**, *8*, e1253. [CrossRef]
41. Yadav, A.; Vishwakarma, D.K. Sentiment analysis using deep learning architectures: A review. *Artif. Intell. Rev.* **2020**, *53*, 4335–4385. [CrossRef]
42. Kalchbrenner, N.; Blunsom, P. Recurrent continuous translation models. In Proceedings of the 2013 Conference on Empirical Methods in Natural Language Processing, Seattle, WA, USA, 19–21 October 2013; pp. 1700–1709.
43. Singh, S.P.; Kumar, A.; Darbari, H.; Singh, L.; Rastogi, A.; Jain, S. Machine translation using deep learning: An overview. In Proceedings of the 2017 International Conference on Computer, Communications and Electronics, Jaipur, India, 1–2 July 2017; pp. 162–167.
44. Yang, S.; Wang, Y.; Chu, X. A survey of deep learning techniques for neural machine translation. *arXiv* **2020**, arXiv:2002.07526.
45. Natural Language Computing Group *R-NET: Machine Reading Comprehension with Self-Matching Networks*; Microsoft Research Lab-Asia: Beijing, China, 2017; pp. 1–11.
46. Huang, H.Y.; Zhu, C.; Shen, Y.; Chen, W. Fusionnet: Fusing via fully-aware attention with application to machine comprehension. *arXiv* **2017**, arXiv:1711.07341.
47. Abbasiantaeb, Z.; Momtazi, S. Text-based question answering from information retrieval and deep neural network perspectives: A survey. *Wiley Interdiscip. Rev. Data Min. Knowl. Discov.* **2021**, *11*, e1412. [CrossRef]
48. Srivastava, Y.; Murali, V.; Dubey, S.R.; Mukherjee, S. Visual question answering using deep learning: A survey and performance analysis. In Proceedings of the International Conference on Computer Vision and Image Processing, Prayagraj, India, 4–6 December 2020; pp. 75–86.
49. Qiu, X.; Sun, T.; Xu, Y.; Shao, Y.; Dai, N.; Huang, X. Pre-trained models for natural language processing: A survey. *Sci. China Technol. Sci.* **2020**, *63*, 1872–1897. [CrossRef]
50. Hinton, G.; Deng, L.; Yu, D.; Dahl, G.E.; Mohamed, A.r.; Jaitly, N.; Senior, A.; Vanhoucke, V.; Nguyen, P.; Sainath, T.N.; et al. Deep neural networks for acoustic modeling in speech recognition: The shared views of four research groups. *IEEE Signal Process. Mag.* **2012**, *29*, 82–97. [CrossRef]
51. Sak, H.; Vinyals, O.; Heigold, G.; Senior, A.; McDermott, E.; Monga, R.; Mao, M. Sequence discriminative distributed training of long short-term memory recurrent neural networks. In Proceedings of the Interspeech, Singapore, 14–18 September 2014; pp. 17–18.
52. Sainath, T.N.; Vinyals, O.; Senior, A.; Sak, H. Convolutional, long short-term memory, fully connected deep neural networks. In Proceedings of the 2015 IEEE International Conference on Acoustics, Speech and Signal Processing (ICASSP), South Brisbane, QSD, Australia, 19–24 April 2015; pp. 4580–4584.
53. Soltau, H.; Liao, H.; Sak, H. Neural speech recognizer: Acoustic-to-word LSTM model for large vocabulary speech recognition. *arXiv* **2016**, arXiv:1610.09975.
54. Prabhavalkar, R.; Rao, K.; Sainath, T.N.; Li, B.; Johnson, L.; Jaitly, N. A Comparison of sequence-to-sequence models for speech recognition. In Proceedings of the Interspeech, Stockholm, Sweden, 20–24 August 2017; pp. 939–943.
55. Li, B.; Zhang, Y.; Sainath, T.; Wu, Y.; Chan, W. Bytes are all you need: End-to-end multilingual speech recognition and synthesis with bytes. In Proceedings of the 2019 IEEE International Conference on Acoustics, Speech and Signal Processing (ICASSP), Brighton, UK, 12–17 May 2019; pp. 5621–5625.
56. Lopez-Moreno, I.; Gonzalez-Dominguez, J.; Plchot, O.; Martinez, D.; Gonzalez-Rodriguez, J.; Moreno, P. Automatic language identification using deep neural networks. In Proceedings of the 2014 IEEE International Conference on Acoustics, Speech and Signal Processing (ICASSP), Florence, Italy, 4–9 May 2014; pp. 5337–5341.
57. Durand, S.; Bello, J.P.; David, B.; Richard, G. Robust downbeat tracking using an ensemble of convolutional networks. *IEEE/ACM Trans. Audio, Speech, Lang. Process.* **2016**, *25*, 76–89. [CrossRef]
58. McFee, B.; Bello, J.P. Structured training for large-vocabulary chord recognition. In Proceedings of the 18th International Society for Music Information Retrieval Conference, Suzhou, China, 23–27 October 2017; pp. 188–194.
59. Vivek, V.; Vidhya, S.; Madhanmohan, P. Acoustic scene classification in hearing aid using deep learning. In Proceedings of the 2020 International Conference on Communication and Signal Processing (ICCSP), Chennai, India, 28–30 July 2020; pp. 0695–0699.

560. Mesaros, A.; Heittola, T.; Benetos, E.; Foster, P.; Lagrange, M.; Virtanen, T.; Plumbley, M.D. Detection and classification of acoustic scenes and events: Outcome of the DCASE 2016 challenge. *IEEE/ACM Trans. Audio, Speech, Lang. Process.* **2017**, *26*, 379–393. [CrossRef]
561. Purwins, H.; Li, B.; Virtanen, T.; Schlüter, J.; Chang, S.Y.; Sainath, T. Deep learning for audio signal processing. *IEEE J. Sel. Top. Signal Process.* **2019**, *13*, 206–219. [CrossRef]
562. Wang, Y.; Narayanan, A.; Wang, D. On training targets for supervised speech separation. *IEEE/ACM Trans. Audio Speech Lang. Process.* **2014**, *22*, 1849–1858. [CrossRef]
563. Isik, Y.; Roux, J.L.; Chen, Z.; Watanabe, S.; Hershey, J.R. Single-channel multi-speaker separation using deep clustering. *arXiv* **2016**, arXiv:1607.02173.
564. Xiao, X.; Watanabe, S.; Erdogan, H.; Lu, L.; Hershey, J.; Seltzer, M.L.; Chen, G.; Zhang, Y.; Mandel, M.; Yu, D. Deep beamforming networks for multi-channel speech recognition. In Proceedings of the 2016 IEEE International Conference on Acoustics, Speech and Signal Processing (ICASSP), Shanghai, China, 20–25 March 2016; pp. 5745–5749.
565. Feng, X.; Zhang, Y.; Glass, J. Speech feature denoising and dereverberation via deep autoencoders for noisy reverberant speech recognition. In Proceedings of the 2014 IEEE International Conference on Acoustics, Speech and Signal Processing (ICASSP), Florence, Italy, 4–9 May 2014; pp. 1759–1763.
566. Li, B.; Sim, K.C. A spectral masking approach to noise-robust speech recognition using deep neural networks. *IEEE/ACM Trans. Audio, Speech, Lang. Process.* **2014**, *22*, 1296–1305. [CrossRef]
567. Vesperini, F.; Vecchiotti, P.; Principi, E.; Squartini, S.; Piazza, F. A neural network based algorithm for speaker localization in a multi-room environment. In Proceedings of the 2016 IEEE 26th International Workshop on Machine Learning for Signal Processing (MLSP), Salerno, Italy, 13–16 September 2016; pp. 1–6.
568. Weninger, F.; Erdogan, H.; Watanabe, S.; Vincent, E.; Roux, J.L.; Hershey, J.R.; Schuller, B. Speech enhancement with LSTM recurrent neural networks and its application to noise-robust ASR. In Proceedings of the International Conference on Latent Variable Analysis and Signal Separation, Liberec, Czech Republic, 25–28 August 2015; pp. 91–99.
569. Chakrabarty, S.; Habets, E.A. Multi-speaker localization using convolutional neural network trained with noise. *arXiv* **2017**, arXiv:1712.04276.
570. Adavanne, S.; Politis, A.; Virtanen, T. Direction of arrival estimation for multiple sound sources using convolutional recurrent neural network. In Proceedings of the 2018 26th European Signal Processing Conference (EUSIPCO), Roma, Italy, 3–7 September 2018; pp. 1462–1466.
571. Jia, Y.; Zhang, Y.; Weiss, R.; Wang, Q.; Shen, J.; Ren, F.; Nguyen, P.; Pang, R.; Lopez Moreno, I.; Wu, Y.; et al. Transfer learning from speaker verification to multispeaker text-to-speech synthesis. *Adv. Neural Inf. Process. Syst.* **2018**, *31*, 4485–4495.
572. Ghose, S.; Prevost, J.J. Autofoley: Artificial synthesis of synchronized sound tracks for silent videos with deep learning. *IEEE Trans. Multimed.* **2020**, *23*, 1895–1907. [CrossRef]
573. Donahue, C.; McAuley, J.; Puckette, M. Adversarial audio synthesis. *arXiv* **2018**, arXiv:1802.04208.
574. Kalchbrenner, N.; Elsen, E.; Simonyan, K.; Noury, S.; Casagrande, N.; Lockhart, E.; Stimberg, F.; Oord, A.; Dieleman, S.; Kavukcuoglu, K. Efficient neural audio synthesis. In Proceedings of the International Conference on Machine Learning, Stockholm, Sweden, 10–15 July 2018; pp. 2410–2419.
575. Oord, A.v.d.; Dieleman, S.; Zen, H.; Simonyan, K.; Vinyals, O.; Graves, A.; Kalchbrenner, N.; Senior, A.; Kavukcuoglu, K. Wavenet: A generative model for raw audio. *arXiv* **2016**, arXiv:1609.03499.
576. Oord, A.; Li, Y.; Babuschkin, I.; Simonyan, K.; Vinyals, O.; Kavukcuoglu, K.; Driessche, G.; Lockhart, E.; Cobo, L.; Stimberg, F.; et al. Parallel wavenet: Fast high-fidelity speech synthesis. In Proceedings of the International Conference on Machine Learning, Stockholm, Sweden, 10–15 July 2018; pp. 3918–3926.
577. Lenz, I.; Knepper, R.A.; Saxena, A. DeepMPC: Learning deep latent features for model predictive control. In Proceedings of the Robotics: Science and Systems, Rome, Italy, 13–17 July 2015; Volume 10, pp. 1–9.
578. Watter, M.; Springenberg, J.; Boedecker, J.; Riedmiller, M. Embed to control: A locally linear latent dynamics model for control from raw images. *Adv. Neural Inf. Process. Syst.* **2015**, *28*, 2746-2754.
579. Polydoros, A.S.; Nalpantidis, L.; Krüger, V. Real-time deep learning of robotic manipulator inverse dynamics. In Proceedings of the 2015 IEEE/RSJ International Conference on Intelligent Robots and Systems (IROS), Hamburg, Germany, 28 September–3 October 2015; pp. 3442–3448.
580. Zhang, T.; Kahn, G.; Levine, S.; Abbeel, P. Learning deep control policies for autonomous aerial vehicles with MPC-guided policy search. In Proceedings of the 2016 IEEE International Conference on Robotics and Automation (ICRA), Stockholm, Sweden, 16–21 May 2016; pp. 528–535.
581. Yang, Y.; Li, Y.; Fermuller, C.; Aloimonos, Y. Robot learning manipulation action plans by "watching" unconstrained videos from the world wide web. In Proceedings of the AAAI Conference on Artificial Intelligence, Austin, TX, USA, 25–30 January 2015; Volume 29, pp. 3686–3692.
582. Levine, S.; Finn, C.; Darrell, T.; Abbeel, P. End-to-end training of deep visuomotor policies. *J. Mach. Learn. Res.* **2016**, *17*, 1334–1373.
583. Finn, C.; Tan, X.Y.; Duan, Y.; Darrell, T.; Levine, S.; Abbeel, P. Deep spatial autoencoders for visuomotor learning. In Proceedings of the 2016 IEEE International Conference on Robotics and Automation (ICRA), Stockholm, Sweden, 16–21 May 2016; pp. 512–519.

584. Redmon, J.; Angelova, A. Real-time grasp detection using convolutional neural networks. In Proceedings of the 2015 IEEE International Conference on Robotics and Automation (ICRA), Seattle, WA, USA, 26–30 May 2015; pp. 1316–1322.
585. Mariolis, I.; Peleka, G.; Kargakos, A.; Malassiotis, S. Pose and category recognition of highly deformable objects using deep learning. In Proceedings of the 2015 International Conference on Advanced Robotics (ICAR), Taipei, Taiwan, 29-31 May 2015; pp. 655–662.
586. Crespo, J.; Barber, R.; Mozos, O. Relational model for robotic semantic navigation in indoor environments. *J. Intell. Robot. Syst.* **2017**, *86*, 617–639. [CrossRef]
587. Neverova, N.; Wolf, C.; Taylor, G.W.; Nebout, F. Multi-scale deep learning for gesture detection and localization. In Proceedings of the European Conference on Computer Vision, Zurich, Switzerland, 6–12 September 2014; pp. 474–490.
588. Hwang, J.; Jung, M.; Madapana, N.; Kim, J.; Choi, M.; Tani, J. Achieving "synergy" in cognitive behavior of humanoids via deep learning of dynamic visuo-motor-attentional coordination. In Proceedings of the 2015 IEEE-RAS 15th International Conference on Humanoid Robots (Humanoids), Seoul, Republic of Korea, 3–5 November 2015; pp. 817–824.
589. Wu, J.; Yildirim, I.; Lim, J.J.; Freeman, B.; Tenenbaum, J. Galileo: Perceiving physical object properties by integrating a physics engine with deep learning. *Adv. Neural Inf. Process. Syst.* **2015**, *28*, 127–135.
590. Noda, K.; Arie, H.; Suga, Y.; Ogata, T. Multimodal integration learning of robot behavior using deep neural networks. *Robot. Auton. Syst.* **2014**, *62*, 721–736. [CrossRef]
591. Peng, X.B.; Andrychowicz, M.; Zaremba, W.; Abbeel, P. Sim-to-real transfer of robotic control with dynamics randomization. In Proceedings of the 2018 IEEE International Conference on Robotics and Automation (ICRA), Brisbane, Australia, 21–25 May 2018; pp. 3803–3810.
592. Zhuang, F.; Cheng, X.; Luo, P.; Pan, S.J.; He, Q. Supervised representation learning: Transfer learning with deep autoencoders. In Proceedings of the Twenty-Fourth International Joint Conference on Artificial Intelligence, Buenos Aires, Argentina, 25–31 July 2015.
593. Nair, A.; McGrew, B.; Andrychowicz, M.; Zaremba, W.; Abbeel, P. Overcoming exploration in reinforcement learning with demonstrations. In Proceedings of the 2018 IEEE International Conference on Robotics and Automation (ICRA), Brisbane, Australia, 21–25 May 2018; pp. 6292–6299.
594. Liao, L.; He, X.; Zhang, H.; Chua, T.S. Attributed social network embedding. *IEEE Trans. Knowl. Data Eng.* **2018**, *30*, 2257–2270. [CrossRef]
595. Wang, P.; Xu, B.; Wu, Y.; Zhou, X. Link prediction in social networks: The state-of-the-art. *Sci. China Inf. Sci.* **2015**, *58*, 1–38. [CrossRef]
596. Zhang, X.; Zhao, J.; LeCun, Y. Character-level convolutional networks for text classification. *Adv. Neural Inf. Process. Syst.* **2015**, *28*, 649–657.
597. Peng, Z.; Luo, M.; Li, J.; Liu, H.; Zheng, Q. ANOMALOUS: A Joint Modeling Approach for Anomaly Detection on Attributed Networks. In Proceedings of the 27th International Joint Conference on Artificial Intelligence, Stockholm, Sweden, 13–19 July 2018; pp. 3513–3519.
598. Wang, X.; Cui, P.; Wang, J.; Pei, J.; Zhu, W.; Yang, S. Community preserving network embedding. In Proceedings of the Thirty-first AAAI Conference on Artificial Intelligence, San Francisco, CA, USA, 4–9 February 2017; Volume 31, pp. 203–209.
599. Rosenthal, S.; Farra, N.; Nakov, P. SemEval-2017 task 4: Sentiment analysis in Twitter. *arXiv* **2019**, arXiv:1912.00741.
600. Liu, F.; Liu, B.; Sun, C.; Liu, M.; Wang, X. Deep belief network-based approaches for link prediction in signed social networks. *Entropy* **2015**, *17*, 2140–2169. [CrossRef]
601. Liu, Y.; Zeng, K.; Wang, H.; Song, X.; Zhou, B. Content matters: A GNN-based model combined with text semantics for social network cascade prediction. In Proceedings of the Pacific-Asia Conference on Knowledge Discovery and Data Mining, Virtual Event, 11–14 May 2021; pp. 728–740.
602. Nguyen, D.T.; Joty, S.; Imran, M.; Sajjad, H.; Mitra, P. Applications of online deep learning for crisis response using social media information. *arXiv* **2016**, arXiv:1610.01030.
603. Huang, P.S.; He, X.; Gao, J.; Deng, L.; Acero, A.; Heck, L. Learning deep structured semantic models for web search using clickthrough data. In Proceedings of the 22nd ACM International Conference on Information & Knowledge Management, San Francisco, CA, USA, 27 October–1 November 2013; pp. 2333–2338.
604. Shen, Y.; He, X.; Gao, J.; Deng, L.; Mesnil, G. Learning semantic representations using convolutional neural networks for web search. In Proceedings of the 23rd International Conference on World Wide Web, Seoul, Republic of Korea, 7–11 April 2014; pp. 373–374.
605. Ma, C.; Ma, L.; Zhang, Y.; Sun, J.; Liu, X.; Coates, M. Memory augmented graph neural networks for sequential recommendation. In Proceedings of the AAAI Conference on Artificial Intelligence, New York, NY, USA, 7–12 February 2020; Volume 34, pp. 5045–5052.
606. Shi, C.; Hu, B.; Zhao, W.X.; Philip, S.Y. Heterogeneous information network embedding for recommendation. *IEEE Trans. Knowl. Data Eng.* **2018**, *31*, 357–370. [CrossRef]
607. Holm, A.N.; Plank, B.; Wright, D.; Augenstein, I. Longitudinal citation prediction using temporal graph neural networks. *arXiv* **2020**, arXiv:2012.05742.
608. Lu, H.; Zhu, Y.; Lin, Q.; Wang, T.; Niu, Z.; Herrera-Viedma, E. Heterogeneous knowledge learning of predictive academic intelligence in transportation. *IEEE Trans. Intell. Transp. Syst.* **2022**, *23*, 3737–3755. [CrossRef]

609. Ciocca, G.; Napoletano, P.; Schettini, R. CNN-based features for retrieval and classification of food images. *Comput. Vis. Image Underst.* **2018**, *176*, 70–77. [CrossRef]
610. Zhou, L.; Zhang, C.; Liu, F.; Qiu, Z.; He, Y. Application of deep learning in food: A review. *Compr. Rev. Food Sci. Food Saf.* **2019**, *18*, 1793–1811. [CrossRef]
611. Kiourt, C.; Pavlidis, G.; Markantonatou, S. Deep learning approaches in food recognition. In *Machine Learning Paradigms*; Springer: Berlin, Germany, 2020; pp. 83–108.
612. Ege, T.; Yanai, K. Image-based food calorie estimation using recipe information. *IEICE Trans. Inf. Syst.* **2018**, *101*, 1333–1341. [CrossRef]
613. Yunus, R.; Arif, O.; Afzal, H.; Amjad, M.F.; Abbas, H.; Bokhari, H.N.; Haider, S.T.; Zafar, N.; Nawaz, R. A framework to estimate the nutritional value of food in real time using deep learning techniques. *IEEE Access* **2018**, *7*, 2643–2652. [CrossRef]
614. Naritomi, S.; Yanai, K. CalorieCaptorGlass: Food calorie estimation based on actual size using hololens and deep learning. In Proceedings of the 2020 IEEE Conference on Virtual Reality and 3D User Interfaces Abstracts and Workshops (VRW), Atlanta, GA, USA, 2–26 March 2020; pp. 818–819.
615. Liu, C.; Cao, Y.; Luo, Y.; Chen, G.; Vokkarane, V.; Yunsheng, M.; Chen, S.; Hou, P. A new deep learning-based food recognition system for dietary assessment on an edge computing service infrastructure. *IEEE Trans. Serv. Comput.* **2017**, *11*, 249–261. [CrossRef]
616. Rodríguez, F.J.; García, A.; Pardo, P.J.; Chávez, F.; Luque-Baena, R.M. Study and classification of plum varieties using image analysis and deep learning techniques. *Prog. Artif. Intell.* **2018**, *7*, 119–127. [CrossRef]
617. Koirala, A.; Walsh, K.B.; Wang, Z.; McCarthy, C. Deep learning–Method overview and review of use for fruit detection and yield estimation. *Comput. Electron. Agric.* **2019**, *162*, 219–234. [CrossRef]
618. Song, Q.; Zheng, Y.J.; Xue, Y.; Sheng, W.G.; Zhao, M.R. An evolutionary deep neural network for predicting morbidity of gastrointestinal infections by food contamination. *Neurocomputing* **2017**, *226*, 16–22. [CrossRef]
619. Gorji, H.T.; Shahabi, S.M.; Sharma, A.; Tande, L.Q.; Husarik, K.; Qin, J.; Chan, D.E.; Baek, I.; Kim, M.S.; MacKinnon, N.; et al. Combining deep learning and fluorescence imaging to automatically identify fecal contamination on meat carcasses. *Sci. Rep.* **2022**, *12*, 2392. [CrossRef]
620. Song, Q.; Zheng, Y.J.; Yang, J. Effects of food contamination on gastrointestinal morbidity: Comparison of different machine-learning methods. *Int. J. Environ. Res. Public Health* **2019**, *16*, 838. [CrossRef] [PubMed]
621. Ha, J.G.; Moon, H.; Kwak, J.T.; Hassan, S.I.; Dang, M.; Lee, O.N.; Park, H.Y. Deep convolutional neural network for classifying Fusarium wilt of radish from unmanned aerial vehicles. *J. Appl. Remote. Sens.* **2017**, *11*, 042621. [CrossRef]
622. Ma, J.; Du, K.; Zheng, F.; Zhang, L.; Gong, Z.; Sun, Z. A recognition method for cucumber diseases using leaf symptom images based on deep convolutional neural network. *Comput. Electron. Agric.* **2018**, *154*, 18–24. [CrossRef]
623. Lu, Y.; Yi, S.; Zeng, N.; Liu, Y.; Zhang, Y. Identification of rice diseases using deep convolutional neural networks. *Neurocomputing* **2017**, *267*, 378–384. [CrossRef]
624. Yang, A.; Huang, H.; Zhu, X.; Yang, X.; Chen, P.; Li, S.; Xue, Y. Automatic recognition of sow nursing behaviour using deep learning-based segmentation and spatial and temporal features. *Biosyst. Eng.* **2018**, *175*, 133–145. [CrossRef]
625. Qiao, Y.; Truman, M.; Sukkarieh, S. Cattle segmentation and contour extraction based on Mask R-CNN for precision livestock farming. *Comput. Electron. Agric.* **2019**, *165*, 104958. [CrossRef]
626. Hansen, M.F.; Smith, M.L.; Smith, L.N.; Salter, M.G.; Baxter, E.M.; Farish, M.; Grieve, B. Towards on-farm pig face recognition using convolutional neural networks. *Comput. Ind.* **2018**, *98*, 145–152. [CrossRef]
627. Tian, M.; Guo, H.; Chen, H.; Wang, Q.; Long, C.; Ma, Y. Automated pig counting using deep learning. *Comput. Electron. Agric.* **2019**, *163*, 104840. [CrossRef]
628. Kussul, N.; Lavreniuk, M.; Skakun, S.; Shelestov, A. Deep learning classification of land cover and crop types using remote sensing data. *IEEE Geosci. Remote. Sens. Lett.* **2017**, *14*, 778–782. [CrossRef]
629. Gaetano, R.; Ienco, D.; Ose, K.; Cresson, R. A two-branch CNN architecture for land cover classification of PAN and MS imagery. *Remote Sens.* **2018**, *10*, 1746. [CrossRef]
630. Ren, C.; Kim, D.K.; Jeong, D. A survey of deep learning in agriculture: Techniques and their applications. *J. Inf. Process. Syst.* **2020**, *16*, 1015–1033.
631. Vali, A.; Comai, S.; Matteucci, M. Deep learning for land use and land cover classification based on hyperspectral and multispectral earth observation data: A review. *Remote Sens.* **2020**, *12*, 2495. [CrossRef]
632. Xie, T.; Grossman, J.C. Crystal graph convolutional neural networks for an accurate and interpretable prediction of material properties. *Phys. Rev. Lett.* **2018**, *120*, 145301. [CrossRef]
633. Jain, A.; Bligaard, T. Atomic-position independent descriptor for machine learning of material properties. *Phys. Rev. B* **2018**, *98*, 214112. [CrossRef]
634. Li, H.; Collins, C.R.; Ribelli, T.G.; Matyjaszewski, K.; Gordon, G.J.; Kowalewski, T.; Yaron, D.J. Tuning the molecular weight distribution from atom transfer radical polymerization using deep reinforcement learning. *Mol. Syst. Des. Eng.* **2018**, *3*, 496–508. [CrossRef]
635. Xie, T.; Grossman, J.C. Hierarchical visualization of materials space with graph convolutional neural networks. *J. Chem. Phys.* **2018**, *149*, 174111. [CrossRef] [PubMed]

36. Kim, E.; Huang, K.; Jegelka, S.; Olivetti, E. Virtual screening of inorganic materials synthesis parameters with deep learning. *NPJ Comput. Mater.* **2017**, *3*, 1–9. [CrossRef]
37. Feng, S.; Zhou, H.; Dong, H. Using deep neural network with small dataset to predict material defects. *Mater. Des.* **2019**, *162*, 300–310. [CrossRef]
38. Polykovskiy, D.; Zhebrak, A.; Vetrov, D.; Ivanenkov, Y.; Aladinskiy, V.; Mamoshina, P.; Bozdaganyan, M.; Aliper, A.; Zhavoronkov, A.; Kadurin, A. Entangled conditional adversarial autoencoder for de novo drug discovery. *Mol. Pharm.* **2018**, *15*, 4398–4405. [CrossRef]
39. Kadurin, A.; Nikolenko, S.; Khrabrov, K.; Aliper, A.; Zhavoronkov, A. druGAN: An advanced generative adversarial autoencoder model for de novo generation of new molecules with desired molecular properties in silico. *Mol. Pharm.* **2017**, *14*, 3098–3104. [CrossRef]
40. Segler, M.H.; Kogej, T.; Tyrchan, C.; Waller, M.P. Generating focused molecule libraries for drug discovery with recurrent neural networks. *ACS Cent. Sci.* **2018**, *4*, 120–131. [CrossRef] [PubMed]
41. Zhang, L.; Tan, J.; Han, D.; Zhu, H. From machine learning to deep learning: Progress in machine intelligence for rational drug discovery. *Drug Discov. Today* **2017**, *22*, 1680–1685. [CrossRef] [PubMed]
42. Walters, W.P.; Barzilay, R. Applications of deep learning in molecule generation and molecular property prediction. *Accounts Chem. Res.* **2020**, *54*, 263–270. [CrossRef] [PubMed]
43. Gupta, R.; Srivastava, D.; Sahu, M.; Tiwari, S.; Ambasta, R.K.; Kumar, P. Artificial intelligence to deep learning: Machine intelligence approach for drug discovery. *Mol. Divers.* **2021**, *25*, 1315–1360. [CrossRef]
44. Mater, A.C.; Coote, M.L. Deep learning in chemistry. *J. Chem. Inf. Model.* **2019**, *59*, 2545–2559. [CrossRef]
45. Segler, M.H.; Preuss, M.; Waller, M.P. Planning chemical syntheses with deep neural networks and symbolic AI. *Nature* **2018**, *555*, 604–610. [CrossRef]
46. Dong, J.; Zhao, M.; Liu, Y.; Su, Y.; Zeng, X. Deep learning in retrosynthesis planning: Datasets, models and tools. *Briefings Bioinform.* **2022**, *23*, bbab391. [CrossRef]
47. Wei, J.N.; Duvenaud, D.; Aspuru-Guzik, A. Neural networks for the prediction of organic chemistry reactions. *ACS Cent. Sci.* **2016**, *2*, 725–732. [CrossRef]
48. Schwaller, P.; Gaudin, T.; Lanyi, D.; Bekas, C.; Laino, T. "Found in Translation": Predicting outcomes of complex organic chemistry reactions using neural sequence-to-sequence models. *Chem. Sci.* **2018**, *9*, 6091–6098. [CrossRef]
49. Fooshee, D.; Mood, A.; Gutman, E.; Tavakoli, M.; Urban, G.; Liu, F.; Huynh, N.; Van Vranken, D.; Baldi, P. Deep learning for chemical reaction prediction. *Mol. Syst. Des. Eng.* **2018**, *3*, 442–452. [CrossRef]
50. Coley, C.W.; Jin, W.; Rogers, L.; Jamison, T.F.; Jaakkola, T.S.; Green, W.H.; Barzilay, R.; Jensen, K.F. A graph-convolutional neural network model for the prediction of chemical reactivity. *Chem. Sci.* **2019**, *10*, 370–377. [CrossRef] [PubMed]
51. Chatzimparmpas, A.; Martins, R.M.; Jusufi, I.; Kerren, A. A survey of surveys on the use of visualization for interpreting machine learning models. *Inf. Vis.* **2020**, *19*, 207–233. [CrossRef]
52. Ribeiro, M.T.; Singh, S.; Guestrin, C. "Why should i trust you?" Explaining the predictions of any classifier. In Proceedings of the 22nd ACM SIGKDD International Conference on Knowledge Discovery and Data Mining, San Francisco, CA, USA, 13–17 August 2016; pp. 1135–1144.
53. Li, J.; Zhang, C.; Zhou, J.T.; Fu, H.; Xia, S.; Hu, Q. Deep-LIFT: Deep label-specific feature learning for image annotation. *IEEE Trans. Cybern.* **2021**, *52*, 7732–7741. [CrossRef] [PubMed]
54. Neyshabur, B.; Salakhutdinov, R.R.; Srebro, N. Path-sgd: Path-normalized optimization in deep neural networks. *Adv. Neural Inf. Process. Syst.* **2015**, *28*, 2422-2430..
55. Hardt, M.; Recht, B.; Singer, Y. Train faster, generalize better: Stability of stochastic gradient descent. In Proceedings of the International Conference on Machine Learning, New York, NY, USA, 19–24 June 2016; pp. 1225–1234.
56. Scheirer, W.J.; de Rezende Rocha, A.; Sapkota, A.; Boult, T.E. Toward open set recognition. *IEEE Trans. Pattern Anal. Mach. Intell.* **2012**, *35*, 1757–1772. [CrossRef]
57. Geng, C.; Huang, S.j.; Chen, S. Recent advances in open set recognition: A survey. *IEEE Trans. Pattern Anal. Mach. Intell.* **2020**, *43*, 3614–3631. [CrossRef]
58. Skeem, J.; Eno Louden, J. Assessment of Evidence on the Quality of the Correctional Offender Management Profiling for Alternative Sanctions (COMPAS). Unpublished Report Prepared for the California Department of Corrections and Rehabilitation. 2007. Available online: https://webfiles.uci.edu/skeem/Downloads.html (accessed on 2 November 2022).
59. Erickson, B.J.; Korfiatis, P.; Akkus, Z.; Kline, T.; Philbrick, K. Toolkits and libraries for deep learning. *J. Digit. Imaging* **2017**, *30*, 400–405. [CrossRef]
60. Nguyen, G.; Dlugolinsky, S.; Bobák, M.; Tran, V.; López García, Á.; Heredia, I.; Malík, P.; Hluchý, L. Machine learning and deep learning frameworks and libraries for large-scale data mining: A survey. *Artif. Intell. Rev.* **2019**, *52*, 77–124. [CrossRef]
61. Elsken, T.; Metzen, J.H.; Hutter, F. Neural architecture search: A survey. *J. Mach. Learn. Res.* **2019**, *20*, 1997–2017.
62. Hatcher, W.G.; Yu, W. A survey of deep learning: Platforms, applications and emerging research trends. *IEEE Access* **2018**, *6*, 24411–24432. [CrossRef]
63. Yin, X.; Wang, S.; Zhu, Y.; Hu, J. A novel lLength-flexible lightweight cancelable fingerprint template for privacy-preserving authentication systems in resource-constrained IoT applications. *IEEE Internet Things J.* **2022**. [CrossRef]

664. Yin, X.; Wang, S.; Shahzad, M.; Hu, J. An IoT-oriented privacy-preserving fingerprint authentication system. *IEEE Internet Things J.* **2022**, *9*, 11760–11771. [CrossRef]
665. Jiang, H.; Li, J.; Zhao, P.; Zeng, F.; Xiao, Z.; Iyengar, A. Location privacy-preserving mechanisms in location-based services: A comprehensive survey. *ACM Comput. Surv.* **2021**, *54*, 1–36. [CrossRef]
666. Cunha, M.; Mendes, R.; Vilela, J.P. A survey of privacy-preserving mechanisms for heterogeneous data types. *Comput. Sci. Rev* **2021**, *41*, 100403. [CrossRef]
667. Guo, W.; Wang, J.; Wang, S. Deep multimodal representation learning: A survey. *IEEE Access* **2019**, *7*, 63373–63394. [CrossRef]
668. Gao, J.; Li, P.; Chen, Z.; Zhang, J. A survey on deep learning for multimodal data fusion. *Neural Comput.* **2020**, *32*, 829–864 [CrossRef]

sensors

MDPI

Review

A Survey of 3D Indoor Localization Systems and Technologies

Andrey Sesyuk [1,*,†], Stelios Ioannou [2,†] and Marios Raspopoulos [2,†]

1 School of Engineering, University of Central Lancashire, Preston PR12HE, UK
2 School of Sciences, University of Central Lancashire Cyprus, Larnaca 7080, Cyprus
* Correspondence: asesyuk@uclan.ac.uk
† These authors contributed equally to this work.

Abstract: Indoor localization has recently and significantly attracted the interest of the research community mainly due to the fact that Global Navigation Satellite Systems (GNSSs) typically fail in indoor environments. In the last couple of decades, there have been several works reported in the literature that attempt to tackle the indoor localization problem. However, most of this work is focused solely on two-dimensional (2D) localization, while very few papers consider three dimensions (3D). There is also a noticeable lack of survey papers focusing on 3D indoor localization; hence, in this paper, we aim to carry out a survey and provide a detailed critical review of the current state of the art concerning 3D indoor localization including geometric approaches such as angle of arrival (AoA), time of arrival (ToA), time difference of arrival (TDoA), fingerprinting approaches based on Received Signal Strength (RSS), Channel State Information (CSI), Magnetic Field (MF) and Fine Time Measurement (FTM), as well as fusion-based and hybrid-positioning techniques. We provide a variety of technologies, with a focus on wireless technologies that may be utilized for 3D indoor localization such as WiFi, Bluetooth, UWB, mmWave, visible light and sound-based technologies. We critically analyze the advantages and disadvantages of each approach/technology in 3D localization.

Keywords: 3D indoor localization; location-based services; Internet of Things

Citation: Sesyuk, A.; Ioannou, S.; Raspopoulos, M. A Survey of 3D Indoor Localization Systems and Technologies. *Sensors* **2022**, *22*, 9380. https://doi.org/10.3390/s22239380

Academic Editor: Peter Han Joo Chong

Received: 7 November 2022
Accepted: 27 November 2022
Published: 1 December 2022

1. Introduction

For centuries, scientists have been fascinated by the idea of determining position. The first positioning systems appeared several millennia ago, when people driven by their need to know their position when travelling typically used natural landmarks to orientate themselves before establishing their own landmarks (trails, lighthouses, etc). Down the line, other approaches were introduced such as celestial and astronomic methods as well as dead reckoning for ocean navigation. Most of them, however, were extremely limited in range while all of them relied on visual observations, at least to some extent, and hence required clear lines of sight between the light source and the user to be positioned. This restricted their use to specific times of day or to specific weather conditions. The late-nineteenth-century discovery of radio waves paved the way for radio-based navigation/positioning. Radio frequency signals have a greater transmission range than visible light while the can be transmitted through clouds or fog or even propagate as ground waves over vast distances, depending on the frequency of transmission overcoming the range issue for ground-based and satellite-based navigation systems [1].

For many years, location-based services (LBSs), applications and systems have been playing an important role in our lives. Outdoor localization has been very successfully implemented using Global Navigation Satellite Systems (GNSSs) which was typically the de facto approach in wireless positioning. Various GNSSs have been established over the years such as the American Global Positioning System (GPS), Russia's Global Navigation Satellite System (GLONASS) and the European GALILEO. GNSSs require at least three satellites to determine the specific location on the globe as well as one more satellite for time synchronization. Therefore, it is imperative that these satellites have an unobstructed path between them and the receiving device being positioned. Due to this, heavily shadowed urban areas (areas of dense and tall buildings, usually referred to as 'urban canyons') or indoor areas cannot be reliably supported by GNSSs. Therefore, there has been significant work reported in the literature [2] over the last 20-30 years which includes many solutions and approaches for solving the localization problem in satellite-denied environments using—over the years—the current available radio technologies. However, none of these solutions have been standardized as the universal solution (like GNSSs for outdoors) for this kind of environment. Various reasons could be found for this, such as the incremental need for more and more accuracy, the rapid evolution of wireless (and other) technologies that facilitate the support of this higher accuracy which makes the adoption of one system unreasonable if it is going to become obsolete in the near future, the cost and maturity of the underlying technologies to be integrated in mobile devices, etc. Several attempts have been proposed in the literature for improving GNSS localization by fusing the data with IMU sensors and although the accuracy as shown in [3] was indeed improved by 20% or as shown in [4] by 38%, no considerable efforts have been identified that present accurate enough results for indoor environments. Moreover, localization accuracy is relatively subject to the application used. For instance, typical GPS-level accuracy (3-10m) would be sufficient for automobile navigation while room-level accuracy (2-4m) would be enough to identify the presence of someone in a room or area of an indoor environment [5,6].

The global indoor positioning and indoor navigation market was valued at USD 6.1 billion in 2020 and is expected to increase at a compound annual growth rate (CAGR) of 22.9 percent from 2020 to 2025, reaching USD 17.0 billion by 2025 [7]. The growing integration of beacons in cameras, LED lightings, Point of Sale (PoS) devices and digital signage; the proliferation of smartphones, connected devices and location-based apps among customers; and the inefficiency of GPS technology in the indoor environment are driving the global adoption of the indoor location market [8]. The COVID-19 pandemic, which started in 2019 has had an impact on the indoor location market; however, businesses are now using it for facility management, virus monitoring, personnel tracking and management and smart quarantining. Indoor location solutions are being adopted by governments and private organizations across industries to keep residents indoors and track them. For example, Inpixon is providing its location-based technology applications and services free of charge or at a reduced rate (depending on the solution) to healthcare providers and other organizations looking for solutions to help control the spread of COVID-19 or manage the impact of the pandemic to ensure citizens' safety and well being [9,10].

In the last couple of decades, there have been several positioning systems proposed and implemented using different techniques and approaches in an attempt to tackle the indoor localization problem. Most such systems solve the problem only in two dimensions, meaning that the position is estimated only on a horizontal (x-y) plane, ignoring the vertical (z) dimension. One practical implication of this could be the inability to recognize if a device is located in a pocket or is held up high or whether a user is located on the first or ground floor of a shopping mall (see Figure 1). This additional localization data in some cases might be crucial. Examples may include a drone used for seeding and fertilizing crops in a greenhouse, where knowing the altitude of the drone with respect to the crops is important or a drone used in search and rescue operations to rescue climbers in canyons or miners in mines, where GNSSs might fail. In most of these cases, accuracy better than sub-meter level is required to avoid crashing the UAV on obstacles. Precise 3D positioning can also find applications in supporting wireless communication and effectiveness in antenna orientation and beamforming [11], pilot assignment [12], channel prediction and resource allocation [13]. Furthermore, due to the rapid increase in the world's population, not only are buildings nowadays built upwards (skyscrapers), but also road traffic in cities is increasing, which will eventually lead to development of self-driving underground cities in the form of tunnels where GPS will no longer be able to provide localization and navigation. The fact that 3D positioning methods enable the identification of the accurate position of UAVs in space, for example, in urban canyon scenarios could also be extended to perform accurate positioning of a device underwater by utilizing more appropriate ranging technologies (e.g., acoustic) [14–16]. In the past decade, there has been a tremendous technical development in indoor positioning/navigation; however, there is yet to be technology that is affordable enough for general market adoption, as opposed to outdoor, well established GNSSs. There are so many factors that could play a role in improving localization accuracy, such as signal attenuation, NLOS conditions and even corporal shade, that the precision of the indoor positioning systems is highly vital in order to reach the most accurate results. While there are several papers on 2D indoor positioning in the literature, to the best of our knowledge, no comprehensive survey on 3D indoor positioning has been conducted. Therefore, in this paper, we discuss existing techniques and technologies for 3D indoor localization and establish a precedent for the need of 3D positioning in the said domain. Furthermore, our work follows an intuitive flow by highlighting the challenges and issues in indoor localization and outlining the existing solutions. The utilization of 5G-related technology has become the development trend of the future 3D indoor positioning. 5G operates through MIMO (Multi-user Multiple Input Multiple Output) antennas, which provide a precise orientation of the signal in one specific direction instead of a multi-directional broadcast. 5G technologies can achieve centimeter-level accuracy for 3D indoor positioning; however, they have not yet reached the necessary global implementation levels. With the rapid rise of more 5G-supported devices, this is soon to be changed. Already, the discussions for the next generation (6G) of wireless systems have begun, envisioning precise localization and sensing systems, as it is believed that 6G systems will accelerate the transition to even higher frequency operation, such as mmWave and THz ranges, as well as significantly wider bandwidths. It is evident that 6G communication opens up a new range of challenges and opportunities in localization and sensing which the authors of [17] summarize in five key research questions: (1) How can cm-level 3D positioning/sensing accuracy be achieved by utilizing the range of technologies used in 6G? (2) How can novel waveform designs be devised to better facilitate localization and sensing in addition to providing the fundamental communication benefits? (3) How can energy efficiency, high positioning/sensing accuracy and (we also say) low cost be supported in very high frequency and very highly mobile and dynamic environments in 6G systems? (4) Can real-time energy efficient AI/ML algorithms be used to further facilitate and support the localization and sensing process? (5) How can the quality and accuracy between active and passive sensing be bridged?

This survey paper focused on studies performed specifically on three-dimensional indoor positioning systems as well as the techniques and technologies to facilitate them by studying various books, articles and papers published by various reputable journals. Although this paper focuses mainly on indoor positioning, some sections may include outdoor positioning examples such as drone navigation as similar principles apply to both cases as both cannot accurately predict the vertical positioning using satellites. Furthermore, this paper excludes two-dimensional indoor positioning studies as there are plenty already reported in the literature.

The remainder of this paper is organized as follows:

- Section 2: We discuss different 3D localization techniques such as geometric approaches like AoA, ToA and TDoA. Moreover, we discuss fingerprinting approaches as they are one of the widely used methods based on metrics such as RSS, CSI, MF and FTM. Furthermore, we discuss the principles of sensor fusion and specifically filtering approaches such as Kalman and Particle Filtering as well as cooperative positioning and PDR. To conclude this section, we discuss the fusion of positioning approaches, also known as hybrid positioning systems and existing systems found in the literature.
- Section 3: We provide a variety of technologies, with a focus on wireless technologies that may be utilized for 3D indoor localization such as WiFi, Bluetooth, UWB, mmWave, visible light and sound-based technologies such as acoustic signals and ultrasound. We analyze the advantages and disadvantages of each technology item primarily focusing the discussion on their applicability for 3D localization.

- Section 4: We discuss the principles of machine learning for 3D indoor localization and provide various existing systems reported to date in the literature.
- Section 5: We provide a critical discussion and conclusions concerning the survey.

Figure 1. 3D indoor positioning application example in a multi-storey mall (by Unknown Author (https://meet.bnext.com.tw/blog/view/3442? accessed on 26 November 2022)— is licensed under CC BY-NC-ND)

A summary of the various notations and symbols used in this paper is shown in Table 1.

Table 1. Notations and Symbols Used Throughout the Paper.

2D	2-Dimensional	3D	3-Dimensional
BLE	Bluetooth Low Energy	CSI	Channel State Information
FP	Fingerprinting	FT	Fixed Terminal
FTM	Fine Time Measurement	GNSS	Global Navigation Satellite System
GPS	Global Positioning System	IMU	Inertial Measurement Unit
IoT	Internet of Things	KF	Kalman Filter
LBS	Location Based System	LOS	Line of Sight
MF	Magnetic Field	ML	Machine Learning
mmWave	Millimeter Wave	NLOS	Non-Line of Sight
PDoA	Phase Difference of Arrival	PDR	Pedestrian Dead Reckoning
PF	Partice Filter	RAT	Radio Access Technology
RSS	Received Signal Strength	TDoA	Time Difference of Arrival
ToA	Time of Arrival	ToF	Time of Flight
TWTF	Two Way Time of Flight	UAV	Unmanned Aerial Vehicle
UWB	Ultra Wideband	VLC	Visible Light Communication

2. 3D Localization Techniques

In this section, the current state of the art on 3D localization is reviewed. This review covers geometric, fingerprinting-based, sensor-fusion-based as well as hybrid approaches, critically evaluating and quantifying their 3D positioning performance based on the work reported in the literature. At the end of this section in Table 2, a summary of all geometric approaches can be found, describing their advantages and disadvantages, as well as their accuracies found in the literature.

2.1. Geometric Approach

Among the many indoor positioning techniques (2D and 3D) that have been reported in literature, the most widely used and recognized are the ones based on a geometric approach. This approach suggests that localization is generally carried out in two steps. The mobile device first records one or more signal parameters that are dependent on the mobile user's location from an adequate number of transmitters and then computes the relative location coordinates in a 2D or 3D plane using standard geometry. In this method, there are three approaches: angle-distance, timing-based techniques (ToA and TDoA) and angular-based techniques (AoA). In timing-based techniques the approximate distance to each transmitter is computed by determining the time required for the signal to reach the terminal when transmitted from a specific access point. The latter technique relies on the ability of the terminal to record the angle of arrival of a signal from a given access point or base station. Modern radio technologies such as UWB and even more millimeter-wave (mmWave) radio create opportunities for very accurately estimating the time and angle of arrival (using phased antenna arrays) [18–20]. There are three prevalent terminologies that describe the geometric approach to determine position, based on distance or angle of arrival measurements: triangulation, trilateration and angulation (see Figure 2). Triangulation is the estimation of a 2D or 3D location using unilateral or multilateral measurements (the position is determined from the measured lengths of three sides of a triangle). Trilateration is the estimation of location using several distance measurements, whereas angulation uses angles relative to known positions. In this subsection, we will describe techniques which utilize all these approaches [21].

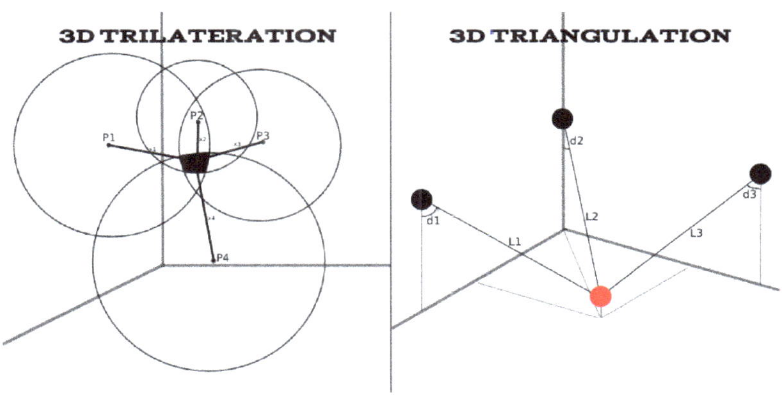

Figure 2. 3D Trilateration and 3D Triangulation.

2.1.1. Angle-Distance

The simplest geometrical method for estimating the device's position is one that uses the distance and angle of arrival from a single transmitter. This appears to be dependent on the device's ability to execute direction finding and distance measurement. Direction finding on the terminal can be achieved through the use of a rotating directional antenna installed on the mobile terminal or through the use of specific procedures if the system is

Multiple Input Multiple Output (MIMO). Ranging can be estimated by either translating the recorded time of arrival into distance (multiplying by the speed of light) or by applying the free space path loss formulation to the recorded received signal strength.

2.1.2. Angle of Arrival—AoA

AoA utilizes the triangulation concept described earlier, where a mobile terminal (MT) obtains the angles of arrival of two signals from two fixed transmitting locations. The main advantage of AoA is that, with it, it is possible to establish a position with as low as two sensors for 2D or three for 3D localization, as there is no need for an extra sensor for time synchronization (as is the case in time-based approaches which lead to distance estimation) [22]. Although AoA can give accurate estimates when the distance between the transmitter and the receiver is modest as compared to RSS approaches, it requires more sophisticated equipment and much more careful calibration and as the transmitter-receiver distance increases, its accuracy decays, meaning that a small error in the angle of arrival calculation translates into a large error in the actual location estimation. Furthermore, because of multi-path effects in indoor environments, the AoA may be sometimes difficult to measure [2].

The estimation of the angle of arrival (AoA) has received a lot of attention from researchers mainly due to the advances in phased-array antenna technology that facilitate the accurate estimation of angles of arrivals. mmWave technologies further enhance this ability as these arrays need to be relatively small to be implemented in microcontroller boards and handheld devices; however, the range is limited [20,23]. In order to achieve accurate results, several existing AoA estimate algorithms examine the entire angle space. Although the existing methods may be reasonable in a variety of different scenarios, for future commercial small and low power wireless devices such as Bluetooth-based Internet of Things (IoT) devices, they may not be practical. This problem is exacerbated in 3D systems, as the elevation angles must also be considered [24]. In terms of work reported concerning this topic, researchers in [25] present an AoA-based algorithm for tracking the position of an anonymous target in 3D space. The placement of dispersed sensors allows for the measurement of the azimuth and elevation angles of the AoAs. The extended Kalman filter (EKF) (see Section 2.3.1) is used to create a unified factor graph (FG) framework, presuming the target movement is non-linear. The observation procedure is carried out using a practical AoA-based position detector. RMSEs were produced in order to assess the suggested tracking technique's accuracy. The observer attained RMSE = 1.6 m using the 3D location detector, while the suggested EKF reduces this value to 1.4 m.

2.1.3. Time of Arrival—ToA

AoA techniques are typically impractical in more typical everyday scenarios, as it is typically difficult to obtain the angular information using current conventional mobile devices such as smartphones. In this respect, distance can be calculated instead by either the Received Signal Strength (RSS) readings [26,27] or the ToA (sometimes referred to as Time of Flight—ToF) measurements [28,29]. A limiting factor for the ToA case is that the receiving and transmitting clocks must be synchronized in order to accurately estimate the ToA and produce more precise distance estimates. This is usually achieved by introducing an extra synchronizing node (as in GNSS) [22]. For three-dimensional positioning, at least four fixed nodes are required.

At least four non-coplanar anchor nodes (ANs) are required for the ToA-based 3D positioning to enable unique position estimation. However, direct method (DM) and particle filter (PF) (see Section 2.3.1) algorithms were developed to address the three-anchor ToA-based 3D positioning problem in [30]. The proposed DM reduces this problem to the solution of a quadratic equation, exploiting the knowledge about the workspace, to first estimate the x- or z-coordinate and then the remaining two coordinates. The implemented PF uses 1000 particles to represent the posterior probability density function (PDF) of the AN's 3D position. The prediction step generates new particles by a resampling procedure.

The ToA measurements determine the importance of these particles to enable updating the posterior PDF and estimating the 3D position of the AN. The DM achieved a horizontal accuracy of 10 cm and a vertical accuracy of 5 cm, while the PF achieved 9 cm and 5 cm, respectively. To reduce the impact of the non-line of sight (NLOS) error, which significantly reduces the localization accuracy, a ToA-based 3D indoor localization algorithm named LMR (LLS Minimum Residual) is proposed in [31]. Firstly, the NLOS error is estimated and used to correct the measurement distances and then to calculate the target location with the linear least squares (LLS) solution. The final node location can be obtained accurately by NLOS error mitigation. The average accuracy achieved was around 0.8 m.

A system based entirely on ToF sensors is proposed in [32]. A major contribution is a new distance measuring method, enabling Time-of-Flight sensors to sense the 3D positions of fast moving reflective markers. ToF sensors are tiny depth sensing systems that are becoming more common in augmented reality smartphones and embedded systems. ToF sensors measure the amount of time it takes for light to travel from the camera to the scene and return to the sensor. This generates photos in which each pixel represents the distance between the camera and the related objects. The sensors are able to be placed on a device and are capable of determining a position with low latency and at rapid update rates. A ToF camera emits light and records three-dimensional images of reflective markers. Distance measurements may be used by a device equipped with a ToF imaging sensor to estimate the relative 3D location of each visible marker. While the final precision of the proposed positioning system depends on the geometry of the captured scene, this evaluation shows that it is possible to use ToF 3D imaging systems for centimeter-level (0.9–1.4 cm) indoor positioning. Along with the high achievable update rates and the simple implementation with a single sensor, it is believed that these results prove the feasibility of this positioning solution for a wide range of applications.

2.1.4. Time Difference of Arrival—TDoA

ToA approaches are a fairly simple position finding approach that use ranging measurements; nevertheless, as mentioned previously, they are susceptible to proper synchronization between transmitter and receiver clocks, as well as the fact that the receiving entity must issue a notification that the transmission has occurred. Time Difference of Arrival (TDoA) is a modified version of the ToA technique that solves this constraint; all it requires is that the transmission has a distinct and unambiguous starting point [33]. The advantage of using ToA and TDoA techniques is the fact that the distances between reference node and target node when increased do not affect the accuracy unless the transmitters in the area outside the ToA and TDoA sites are used. Weak synchronization of time, multipath propagation and low SNR, however, will reduce the resolution of ToA/TDoA measurements [34].

The authors in [35] present a novel TDoA-based approach suitable for single-anchor positioning systems, implemented by phase wrapping-impaired array antenna, with the latter being a typical occurrence in large Switched Beam Antenna (SBA) operating in the low microwave range. The proposed method takes advantage of the large bandwidth of radio link, established between the anchor and the positioning target by generating an unambiguous equivalent phase relationship between antenna array elements. The technique is validated by adopting a relatively large SBA antenna operating in the 4.75–6.25 GHz bandwidth and capable of positioning a target in a 3D domain. Combining range and angle errors, the associated cumulative distribution function error in 90% of cases shows an error of 0.13 m.

Table 2. Geometric Approach 3D Positioning Existing Systems.

Technique	Advantages	Disadvantages	Accuracy	Ref.
AoA	-Do not require clock synchronization	-Accurate angle measurements may require additional equipment such as directional antenna to support the system which will increase the cost.	1.4 m	[25]
ToA	-The distances between reference node and target node when increased do not affect the accuracy	-Weak synchronization of time -Multipath propagation -Low SNR will reduce the resolution of ToA measurements	0.05 m 0.8 m	[30] [31]
TDoA	-Similar to ToA	-Similar to ToA	0.13 m	[35]

2.2. *Fingerprinting Approaches*

Another approach is fingerprinting (FP). The FP process consists of an online and an offline phase. During the offline one (also known as the data collection phase), the received signal strength (RSS) measurements are obtained at multiple different locations across a known environment. These measured fingerprints are pre-stored and are then used as reference when comparing them to the measured signals collected during the online stage to estimate the user location [36,37]. Fingerprinting techniques have gained popularity due to their ability to enable positioning estimation without additional hardware knowledge of the space layout or AP positions. An advantage of such techniques is that they may be used in a variety of indoor environments, even including underground [38]. Fingerprinting offers a discrete rather than a continuous estimate of the user location. Technically, the precision of position estimate may be enhanced by decreasing the distance between offline measurement locations, which would increase the density of the fingerprint field, until nearly continuous location estimation is achieved. However, due to channel statistics and measurement noise, the difference in signal intensity between two neighbour points will become considerably different, making an estimate of the right location nearly impossible [2]. The RSS fingerprint's values may often fluctuate due to signal interference such as objects being moved, doors opening/closing and the amount of people within the given environment. Because of this, there is a need to constantly update and calibrate the "fingerprinting map" [38]. This causes a massive disadvantage as it requires a lot of effort and time to renew the fingerprints especially in large buildings. One solution is to use channel models to construct the fingerprinting map. For instance, in [39], an FP map is constructed using 3D Ray Tracing and this map has been calibrated with a small set of manually collected data across the environment to calibrate it for multiple types of devices. Another way to address this problem is crowd-sourcing mapping which has been proposed in [40]. In other words this is called cooperative positioning technique. The radio map is constructed and maintained in those systems using fingerprints acquired and expressly annotated by users. However, the quality of the users' input might have an impact on cooperative systems, resulting in low position accuracy. For these reasons, several alternative techniques explore inertial sensors and interfaces incorporated in mobile phones in order to generate the radio map using user motion patterns [38]. The principles of cooperative positioning will further be discussed in the next section.

At the end of this section in Table 3, a summary of all fingerprinting approaches can be found, describing their advantages and disadvantages, as well as their accuracies found in the literature.

2.2.1. RSS-Based Fingerprinting

Received Signal Strength (RSS) is obtained by measuring the power of the signal at the receiver. It is either used directly as a fingerprint or is plugged into signal model equations to determine the distance between the transmitting and the receiving device. The strength of the signal is proportional to the distance between the devices—the closer the transmitter and receiver are to each other, the greater the RSS value. RSS is typically used in conjunction with other techniques and technologies such as Wi-Fi, ultrasound, ZigBee, UWB and fingerprinting approaches [36]. Due to its simplicity and low cost, RSS-based approaches are the most common and widely used localization techniques. However, in some scenarios (such as NLOS conditions), it suffers from poor localization accuracy due to increased signal attenuation caused by transmission through walls and other possible obstructions such as movement of humans inside a building, as well as excessive RSS fluctuation caused by multipath fading and noise. To counteract these issues, several filter or averaging mechanisms can be applied; however, in most cases, to achieve high localization accuracy, a relatively complex algorithm must be employed [2]. For example, in [41], a Kalman filter is applied to eliminate a large part of the noise from the RSS data and therefore enhance the accuracy. A multilateration problem is formulated via Singular Value Decomposition (SVD), which is extensively applied in numerous fields such as control systems, in order to estimate the location of target nodes in three-dimensional settings. The distance between the reference nodes and the target nodes is approximated using RSS for a given set of reference nodes and the position of the target node, meaning the 3D coordinate, may then be computed.

Woodman and Harle [42] describe another method for obtaining relatively good positioning accuracy as well as accurate continuous information about the current location on the z-axis. The entire system was evaluated experimentally, using an independent tracking system for ground truth. The results show that it can track a user throughout an 8725 m^2 building spanning three floors to within 0.5 m 75% of the time and to within 0.73 m 95% of the time. In [43], the expansion of the 2D RSS-based WLAN fingerprinting localization technique to 3D is presented by implementing and extending the Isolines and Euclidian Distance Algorithms. The third dimension is regarded discretely as the floor level. Both algorithms were tested in two different environments of university and a museum. Within the university test bed, the floor level (z-position) could be estimated correctly in 86.67% of cases for the Isolines Algorithm and 93.33% of cases for the Euclidean Distance Algorithm. The results in the museum test bed reached 96.84% with the Isolines Algorithm and 100% with the Euclidean Distance Algorithm.

2.2.2. CSI-Based Fingerprinting

Channel State Information (CSI) refers to known channel properties of a communication link when establishing a wireless communication. Through this information, it is possible to identify the propagation characteristics of the channel between the transmitter and the receiver. This gives access to information such as scattering, fading and power decay with distance which is typically not available with conventional RSS measurements.

Most of the Wi-Fi-based indoor positioning technology can be divided into two main categories: RSS-based and CSI-based. However, in the indoor environment, the RSS signal, as a kind of coarse-grained information, is highly susceptible to interference from other signals and the indoor multipath effect, so it cannot provide sufficient accuracy and reliability [44,45]. For Wi-Fi signals using IEEE 802.11n [46] communications protocol, it can obtain CSI in Orthogonal Frequency Division Multiplexing (OFDM) subcarriers by modifying the wireless network card driver [47].

CSI is classified into two types: (a) channel impulse response (CIR) and (b) channel frequency response (CFR). CIR is a time-domain representation of the complex channel and describes the channel's amplitude and phase in time bins, whereas CFR is its frequency-domain equivalent, which displays the complex channel in frequency sub-carriers. CIR requires an impulse signal to be generated, whereas CFR may be simply retrieved using orthogonal frequency division multiplexing (OFDM) devices. Due to insufficient synchronization, CIR is also more prone to error. Phase compensation techniques for CFR help to solve synchronization problems. CSI fingerprinting is recommended over RSS fingerprinting because it can work with a single AP in both LOS and NLOS scenarios. Because many RF systems use OFDM, it can achieve high positioning resolution up to the centimeter level and can be readily supported by existing infrastructure [48].

In terms of work reported on this topic, the researchers in [49] designed a positioning system based on CSI for the tracking and navigation of UAVs. As the authors state, with UAV technology, due to a common issue and a phenomenon called "black flying" which involves acts such as illegally intruding into certain areas such as airports, gas stations, nuclear power plants, petrochemical plants, detention centres and others [50], it is necessary to introduce necessary countermeasures for such scenarios. The system operates by firstly monitoring the communication information between the UAV and the controller and analyzing the CSI. Secondly, the angle of azimuth (AoA) and angle of elevation (EOA) is estimated for the direct LOS signal and then utilizes the positioning model to calculate the position of the UAV. Eventually, Wireless Insite (WI) is applied to verify the system which is a simulation software that applies and analyzes the operating aspects of radio transmission and wireless communication systems using Ray Tracing model methods. The testing results show that the 2D position error is around 1.1 m and the 3D position error is around 2.02 m.

Machine Learning (ML) (see Section 3) has also been used in conjunction with CSI. For instance in [51], WiCluster is introduced, which uses a novel ML technique for passive indoor positioning. WiCluster can predict both zone-level and exact 2D or 3D positions without the need for accurate location labels during training. As stated in this paper, initially CSI-based indoor positioning studies focused on non-parametric digital signal-processing (DSP) techniques. More recently, however, the focus has been shifted to parametric approaches (e.g., fully supervised ML methods). However, these methods do not handle the complexity of real-world environments well and do not meet the requirements for large-scale commercial deployments: the accuracy of DSP-based methods degrades significantly in non-line of sight conditions, whereas supervised ML methods require large amounts of difficult-to-obtain centimeter accuracy position labels. WiCluster, on the other hand, is precise and it requires lower label information that is easily acquired and works well in non-line of sight settings. This system demonstrates meter-level accuracy in three separate realistic environments: two offices and one multi-story building. The average accuracy was around 0.97 m. The positioning system performs effectively even in rooms with no direct line of sight to the transmitter or receiver.

2.2.3. Magnetic Field-Based Fingerprinting

Despite that most of the fingerprinting techniques are based on Wi-Fi RSS measurements, recently there have been major advancements in Magnetic Field (MF)-based location fingerprinting techniques for indoor positioning that take advantage of MF anomalies. The Earth's Magnetic Field (EMF) is a ubiquitous and location-specific signal. Due to the fact that the local MFs in steel-frame buildings can be influenced by both natural and man-made sources (e.g., steel and reinforced concrete structures, electric current and electronic appliances), causing anomalies in the local MF inside the building, it is a promising resource that can be used in accurate global self-localization. When compared to other existing indoor localization systems, the MF system is more cost- and energy-efficient while maintaining the same precision and it relies on built-in EMF sensors on smartphones without the need for additional equipment [37]. MF anomalies, on the other hand, can only affect specific types of regions. Because of the sensitivity limits of the smartphone built-in sensors, the limited discernibility of received local MF signals may result in multiple positions having the same MF-location information in regions away from disturbances. This makes distinguishing between different positions of the same local MF value extremely challenging.

The authors of [52] introduce a 3D MF-based tracking system where the recorded data is analyzed using the Kalman Filter, which removes the overlays caused by kinematic effects in order to obtain reliable distance and elevation measurements between mobile stations and reference points. The results reveal that in a typical indoor environment, good positioning accuracy may be achieved in the range of 0.5–1.5 m with regards to the horizontal plane as well as to the z-value. To further improve and assist the MF localization system, a visual-based camera-assisted indoor positioning system is introduced in [37]. This vision-based approach, similarly to previous fingerprint-based positioning systems, employs image feature points as the matching resource. By comparing the query image to the pre-built image database, the location where an image was captured by a user may then be determined. Unlike previous systems, this methodology may display the user's location on a visual 3D map of the indoor environment, allowing people to identify their position more precisely. Unlike most original MF-based indoor location systems, which rely just on MF fingerprinting to find individuals, this multi-pronged approach is significantly improved by using a camera-based visual positioning technique in places with less disturbances. All results reveal that the camera-aided MF indoor positioning system outperforms others in both accuracy and reliability when compared to two competing systems evaluated using smart mobile devices in different indoor conditions. Compared with results using MF alone, the camera-aided MF solution achieves more than a 50% improvement in average error distance in both cases of fewer and abundant disturbance environments.

Table 3. Fingerprinting Approach 3D Positioning Existing Systems.

Technique	Advantages	Disadvantages	Accuracy	Ref.
RSS	-Simple to set up and use -Low cost as it does not require additional hardware	-Suffers from poor accuracy in NLOS conditions -Very laborious	0.73 m 2.2 m	[42] [43]
CSI	-Immune to noises and fading	-Insufficient synchronization which may lead to error	0.97 m 2.02 m	[49] [51]
MF	-Cost- and energy-efficient while maintaining similar precision -Relies on built-in EMF sensors on smartphones without the need for additional equipment	-MF anomalies can only affect specific types of environments	0.5–1.5 m	[52]
FTM	-Does not require offline training, which saves significant labour	-Performs poorly in NLOS and multipath propagation scenarios	1.11 m	[53]

2.2.4. Fine Time Measurement-Based Fingerprinting

The fine timing measurement (FTM) protocol which was standardized in IEEE 802.11 [54] can achieve meter-level positioning accuracy with time of flight (TOF) echo technology. One of the major issues for positioning, as with many other ranging measurements, is the mitigation of NLOS effects [55]. If the direct path between a fixed terminal (FT) and a mobile terminal (MT) is obstructed, the signal's time of arrival (ToA) at the FT is delayed, introducing a positive bias. The use of such ToA estimations may considerably reduce positioning accuracy [56]. The fingerprint-based Wi-Fi location approach is mostly implemented using received signal strength (RSS) or channel status information (CSI). In comparison to RSS-based solutions, this new technology does not require offline training, which saves significant labour [57].

In this context, ref. [53] proposes a real-time 3D indoor localization algorithm based on Wi-Fi FTM together with built-in sensors. The received signal strength indicator and round-trip duration acquired from Wi-Fi Access Points (APs) are combined for proximity recognition and provide more precise range results. The adaptive extended Kalman filter (AEKF) is utilized to estimate the pedestrian's real-time direction and walking speed. Additionally, the AEKF, proximity detection and Wi-Fi ranging findings are combined using the unscented particle filter. The combination of the Wi-Fi FTM-based method and built-in sensor-based method effectively improves the positioning accuracy and stability. The final CDF error of 2D positioning is within 1.11 m at 67.5% and the altitude error is within 0.28 m at 67.5%.

2.3. Sensor Fusion

Sensor fusion is the technique of combining data from various sensors in an attempt to minimize the amount of error in a positioning system. Despite the fact that many traditional localization frameworks not utilizing sensor fusion have been enhanced in various other ways to decrease the uncertainty or improve accuracy, sensor fusion frameworks often provide a further improvement in the positioning accuracy [58]. Sensor fusion networks are usually categorized based on the type of sensor configuration. There are three main types [59]:

- Complementary: Sensors give independent types of information about the environment. Sensors are not directly reliant on each other, but can be combined to provide a more comprehensive image of the area of interest. This fixes the issue of sensor data inadequacy. In general, fusing complementary data is simple since data from different sensors may be added to one other. A complementary configuration would be the use of numerous cameras, each watching different sections of a room.
- Competitive/redundant: Sensors are designed competitively if each sensor provides independent measurements of the same property. Competitive configuration is often distinguished by either fusion of the data from different sensors or the fusion of measurements from a single sensor obtained at different instants.
- Cooperative: A cooperative sensor network leverages information from two (or more) independent sensors to extract information that would not be obtainable from a single sensor. Stereoscopic vision is an example of a cooperative sensor configuration—by integrating two-dimensional images from two cameras at slightly different angles to form a three-dimensional image of the scene.

The three fundamental sensor communication methods are as follows [60]:

- Distributed: Information is sent between nodes at a set communication rate (e.g., every five scans)
- Decentralized: There is no communication between the sensor nodes. In decentralized systems, every node makes its own decision. The final behavior of the system is the aggregate of the decisions of the individual nodes.
- Centralized: All sensors send data to a single node. The centralized system is a subset of the distributed scheme in which the sensors interact with each other every scan.

Current indoor positioning technologies can be divided into two types: infrastructure-based approaches and infrastructure-free approaches. Infrastructure-based techniques achieve indoor positioning using data gathered from external infrastructure or equipment such as network nodes, WiFi signals, Bluetooth signals, radio frequency (RF) signals, magnetic signals and video signals. Infrastructure-free techniques are able to achieve indoor positioning without any external signals. The majority of these techniques rely on inertial sensors, such as accelerometers, magnetometers and gyroscopes. These sensors are able to achieve accurate results even in complex indoor environments. However, these sensors' drift and bias flaws present major issues. Infrastructure-based solutions demand the installation of different equipment which in most cases is quite costly, whereas infrastructure-free solutions are more flexible and cost-effective, as they involve the sensors that are already built into the smart devices. The trend in recent years has been toward infrastructure-free solutions; however, the accuracy is too insufficient to be employed in real-world applications [61]. Smartphones offer various types of measures that can be used to achieve indoor positioning using simply smartphone-based data. By adding relative height information, such as from barometer data and using suitable filtering, 3D positioning may be easily achieved.

At the end of this section in Table 4, a summary of all sensor fusion approaches can be found, describing their advantages and disadvantages, as well as their accuracies found in the literature.

2.3.1. Filtering Approaches

In many environments, the measurements from positioning systems still contain unwanted noise and the quality of the measurement data can be enhanced using filters [62]. Filtering is a typical example of sensor fusion. The two most common used filtering approaches are kalman filters (KFs) and particle filters (PFs). KFs and PFs, which represent location probability as a set of samples (particles), are among the most-efficient methods due to their ability to accommodate non-linear state and measurement models, handle multiple hypotheses and seamlessly combine different types of information.

Kalman Filter

The Kalman Filter (KF) [63] is one of the most common implementations of Bayesian filters [1]. Kalman filtering is an algorithm which uses a series of measurements observed over time, including statistical noise and other inaccuracies, to produce estimates of unknown variables that are more accurate than those based on a single measurement alone, by estimating a joint probability distribution over the variables for each timeframe [41]. One of the key advantages of KFs is their computational efficiency in implementing Gaussian process mean and covariances using just matrix and vector operations. The algorithm operates in two stages. The KF generates estimates of the current state variables, together with their uncertainty, for the prediction phase. Once the result of the next measurement is seen (which is unavoidably corrupted with some error, including random noise), these estimates are updated using a weighted average [64]. The algorithm is able to work in real time with only the present input measurements and the previously determined state, as well as its uncertainty matrix. Extensions and modifications of the filter, such as the extended Kalman filter (EKF) [65] and the unscented Kalman filter (UKF) [66], which operate on nonlinear systems, have also been developed. Furthermore, Kalman filtering has been effectively applied in multi-sensor fusion and distributed sensor networks to produce more distributed Kalman filtering [67]. In most cases, Kalman filtering is used to eliminate systematic errors of different systems.

Work reported in [61] describes a methodology for attaining 3D indoor position using foot-mounted sensors by extending an existing 2D model to 3D. The Zero Velocity Potential Update (ZUPT) algorithm was utilized to detect when a pedestrian has stopped moving and this information was used in the Kalman filter to eliminate systematic errors. To acquire correct height information, a 3D indoor positioning barometer was added and merged with an accelerometer using a Kalman filter. The particle filter was removed due to its high processing time cost and difficulties in implementing wearable devices. The suggested approach has been tested in a number of real-world and simulated settings. The distance errors are around 1% and the positioning errors are less than 1% of the total travelled distance. Results demonstrate that the suggested system outperforms other comparable systems that make use of the same low-cost IMUs. In [41], RSS is used to estimate the distance between reference nodes and the target nodes for 3D position estimation. However, due to RSS fluctuations, which lead to rather inaccurate distance estimations, a Kalman filter is applied to the measurements to reduce these fluctuations. The experiment findings demonstrate that increasing the number of reference nodes (used in the computation of multilateration localization) improves accuracy, but only up to six nodes. The estimation error increases as the number of reference nodes goes beyond six (i.e., seven and eight nodes). This differs with the theoretical notion that increasing the number of nodes leads to increased location accuracy. The average accuracy achieved was around 0.6 m. Ref. [68] proposes a high-scale 3D indoor positioning system that uses EKF for real-time 3D pose estimation (position and orientation) by integrating IMU relative motion data with camera measurements to fixed LED landmarks with known absolute positions. The findings demonstrated that by observing one LED on average in each camera frame, this technique can confidently predict the global 3D position of the sensor pair with less than 0.4 m accuracy. Some other existing works which utilize Kalman filtering have already been discussed previously (see Sections 2.1.2, 2.2.1 and 2.2.4).

Particle Filter

Another important type of Bayesian filter is based on estimation of integrals by numerical integration. These approaches, known as particle filters (PFs) [69], have grown in popularity to be used in position tracking applications. Particle Filtering's underlying idea is the representation of the state Probability Density Function (PDF) by a predefined number of hypotheses; hence, it does not implement an analytical function. In comparison to KFs, PFs often have a substantially higher complexity depending on the amount of particles that must be created to model the PDF. Furthermore, PFs are subject to inconsistent behaviour, due to phenomena such as sample degeneracy or sample impoverishment.

PFs have recently been used in some works on 3D positioning. For instance, ref. [42] outlines a smartphone-targeted positioning system that employs numerous sensors such as accelerometers, gyroscopes and barometers, as well as technologies such as PDR, WiFi positioning and PF (which is able to work in three-dimensional space). The research reported in this paper aims to provide solutions to three main problems: real-time indoor localization in multi-story buildings, re-sampling in 3D Particle Filter (PF) related to transition between floors and determining final position from a cloud of particles. According to the test findings, the accuracy of the 3D algorithm is higher for all final location estimators. The mean error for 2D PF reached roughly 1.7 m, whereas 3D PF reached about 1.4 m. The most significant advantage of a 3D particle filter is that particles maintain their XY locations and headings when travelling across levels. In the 2D version of the algorithm, the particles were generated afresh after floor change. In this situation, the global heading needs to stabilize once again, even though on the previous floor the majority of the particles had steady heading. The authors of [70] use a PF with three states (XYZ) to estimate the 3D position of a moving node. Due to the fact that no movement information is available, the PF uses a measurement model to produce some random particle motion once every second. When a range measurement to a beacon is obtained, the distance between all particles and that beacon is estimated. The moving object's location is determined by computing the weighted mean position of all particles. The experiments were conducted with the help of Bespoon (https://bespoon.xyz, accessed on 26 November 2022) and Decawave (https://www.decawave.com, accessed on 26 November 2022) equipment [71], reaching mean positioning accuracies in NLOS conditions of 0.51 m and 0.24 m, respectively.

Pedestrian Dead Reckoning—PDR

Similar to maps for outdoor localization, building structure information is required as basic data for many indoor positioning services; however, this information may not always apply to all buildings. As a result, research is being carried out on determining the building environment using pedestrian sensing data acquired from multiple people moving within a building. To obtain precise building structural data, high-accuracy pedestrian trajectories must first be estimated from sensor data. Dead reckoning is the technique of computing the current position of a moving object by utilizing a previously established position as well as integrating estimations of speed, heading direction and course over elapsed time. Due to the rapid advancements of smartphone capabilities because of the increase in the variety of different built-in sensors, such as accelerometers which may be utilized as pedometers, and because magnetometers can be used as compass heading providers, these can be used to estimate the direction in which a person is walking and to estimate movement relative to initial location [72]. Pedestrian dead reckoning (PDR) can be used to enhance other positioning techniques by expanding the range into places where other positioning systems are inaccessible [73]. One of the biggest issues when employing PDR to determine position and velocity is due to sensor inaccuracy; the system in most cases eventually diverges. However, the extended Kalman filter calculation is adopted to tackle this problem. The extended Kalman filter calculates system state errors for altitude, angular velocity, position, velocity and acceleration, as presented in [67,74]. Along the same lines, ref. [75] presents a Cascade Pedestrian Dead Reckoning (C-PDR) approach. C-PDR is a 3D pedestrian dead reckoning method that does not require any infrastructure and is based on a waist-worn inertial system. This system utilizes data from a triaxial accelerometer, gyroscope and magnetometer. The ability to wear the inertial platform around the user's waist allows the system to be implemented in a wider range of different applications. Wearing the tracking device on one or more limbs may limit the agility of the users' movements, for example, in defense and rescue services. In order to track the walking path in 3D space, C-PDR combines a pedestrian activity classifier with a position estimation mechanism.

2.3.2. Cooperative Positioning

Due to the rapid evolution of smart devices and the Internet of Things (IoT) concept, another promising terrestrial positioning method known as cooperative positioning has sparked the attention of the research community in recent years. Since modern wireless smartphones are capable of establishing peer-to-peer (P2P) connections and performing a variety of ranging measurements, the cooperative positioning approach is based on the fact that smart devices can communicate and share data with one another in order to form location estimations within a given indoor environment. Cooperative positioning is split into two categories: deterministic and probabilistic methods. Deterministic approaches consider the location of users as undefined variables to be calculated from measured ranges, angles, signal strengths, etc. Although deterministic localization strategies perform well in wireless sensor networks, they are not suited for situations where the tracking of people is required, as they are unable to accurately process the prior user positioning data [76]. Probabilistic methods (also known as Bayesian), alternatively, consider the users' locations as random variables in space whose distribution of probability must be derived from measurements of intrinsic uncertainties [77]. The Bayesian method is a statistical approach that is based on the Bayesian understanding of probability and reflects a degree of confidence in the occurrence of an event. Bayesian techniques are implemented as recursive filters (such as KFs and PFs as described above) that incorporate previous position information as well as motion data produced by inertial sensors, making them ideal for the location and monitoring of moving objects in indoor environments.

Some 3D cooperative positioning works exist in the literature. In [78], a 3D universal cooperative localizer (3D UCL) is proposed for VANETs in 3D space under several forms of ranging data such as ToA, RSS, AoA and Doppler frequency. One ranging measurement contains three hybrid variables, which are derived by subtracting the node receiving this ranging measurement's x, y and z positions from the position of its pairing neighbor. Testing results show accuracy of around 0.8 m. In [79], a 3D cooperative approach is developed that outperforms non-cooperative algorithms in accuracy and robustness to anchors; nonetheless, the fundamental issue of a lack of height reference data remains unsolved. In order to address the underlying problem of a lack of reference data at altitude in the 3D cooperative localization approach, previous location information is provided to limit the initial position when a severe fall in accuracy arises. The testing was carried out at a university in a room of approximately 6 m by 15 m. The test was performed using 15 test points which are set on two lines, whose angles with the central axis of the door are 90 and 30 degrees, respectively. At each test point, 10 testers were asked to estimate the distance with the central axis of the door by themselves. The average error of the 10 testers at 15 points was achieved at about 0.3 m.

Table 4. Sensor Fusion 3D Positioning Existing Systems.

Technique	Advantages	Disadvantages	Accuracy	Ref.
KF	-Capable of handling Non-linear models -Low computational complexity	-Designed for Gaussian noises	0.6 m 0.4 m	[41] [68]
PF	-Capable of handling non-Gaussian and non-linear estimations -Methodologically simple and flexible	-Number of particles is a trade-off between computational complexity and accuracy -Issue of filter initialization [80]	1.4 m	[42]
Cooperative	-Incorporates the sensors within the smart devices to communicate and share data with one another -Cost-efficient as no additional hardware is required	-Computational complexity, communication bottlenecks, scalability and lack of robustness against failure [80]	0.3–0.8 m	[79]
PDR	-Can be used to enhance other positioning techniques by expanding the range into places where other positioning systems are inaccessible	-Possible IMU sensor errors -Estimation errors increase with the distance to the known initial position	1.24 m	[75]

2.4. Hybrid 3D Positioning Systems

To further leverage the benefits of various sensors and approaches, different types of measurements from different networks or technologies can be combined to enhance positioning. Different networks or technologies that provide various types of metrics can be combined to produce hybrid positioning systems. Different types of measures obtained from different networks/technologies are more effective than the same type of measurements provided within the same network/technology because they combine the benefits of different technology [81]. Some examples of existing hybrid positioning systems are presented below.

2.4.1. ToA/AoA

One example of a hybrid positioning system is proposed in [82], where the authors take advantage of the ToA and AoA approaches for dealing with access node (AN) location uncertainty without increasing computing complexity. First, the 2D positioning approach is expanded to geometry-based 3D, in which a robot's location is determined by using both the ToA and the AoA measurements in the robot positioning algorithms. Second, an EKF-based (see Section 2.3.1) positioning algorithm is developed and implemented with the AN location uncertainty mapped to the measurement noise statistics. Aside from 3D positioning accuracy, vertical accuracy is also used as a performance parameter, as vertical accuracy is important in certain applications. The 3D and vertical RMSE as well as the number of operations needed to implement the considered algorithms (at one time instant) were utilized as the metrics for comparison. The numerical findings demonstrated that the EKF-based algorithms used remained a preferable choice in terms of both 3D and vertical RMSE performances as long as the error in AN placement was kept under 0.5 m (i.e., standard deviation along the x-direction). On the other hand, the proposed geometry-based approach, namely weighted centroid geometric (WCG), was capable of maintaining a higher positioning accuracy than EKF-based approaches when exposed to AN locations uncertainty larger than 0.5 m (standard deviation error), thus yielding a higher robustness. The 3D RMSE averaged about 1.9 m, while vertical RMSE averaged about 0.4 m.

2.4.2. PDR/Fingerprinting

The authors in [83] propose an indoor navigation algorithm by combining both PDR and fingerprinting approaches. It employs a variety of sensors and technologies, including nine-axis sensors (such as 3D gyros, accelerometers and magnetometers), WiFi and magnetic matching. PDR is utilized to provide continuous position solutions as well as to detect errors in both WiFi fingerprinting and magnetic matching. Meanwhile, WiFi fingerprinting uses point-by-point matching technology, whereas magnetic matching focuses on profile matching. Finally, the position-tracking module receives updates from the WiFi and magnetic matching results. This algorithm was tested with Samsung Galaxy S4 and Xiaomi 4 smartphones in different indoor environments (i.e., Environment 1 with abundant WiFi APs and significant magnetic changes and Environment 2 with less WiFi and magnetic information). In these conditions, the hybrid PDR/WiFi/MM algorithm provided RMS accuracy of 2.8 m and 2.9 m in the two test environments. Another work [40] outlines the software navigation engine for indoor positioning by utilizing the already existing data from smartphone sensors and communications modules such as IMU (3D accelerometer, gyroscope), a magnetic field sensor (magnetometer), WiFi and BLE modules, together with the floor premises plan. Indoor navigation software uses such technologies as PDR, Wi-Fi fingerprinting, geomagnetic fingerprinting and map matching. Being blended in the particle filter, dissimilar measurements allow solving a set of principal tasks. Positioning results given for different indoor environments in a shopping mall and in a big exhibition hall show fast TTFF indoors and accurate and reliable real-time indoor positioning with accuracy of about 1–2 m.

As can be seen, to the best of the authors' knowledge, only two existing 3D hybrid positioning works have been found in the literature. Table 5 showcases these systems and it can be observed that the accuracy varies from 1 to 2 m.

Table 5. Hybrid 3D Positioning Existing Systems.

Technique	Accuracy	Ref.
ToA/AoA	1.9 m	[82]
PDR/Fingerprinting	1–2 m	[83]

3. Machine Learning for 3D Indoor Positioning

For widespread deployment of indoor positioning, accuracy, dependability, scalability and environmental adaptation remain the main challenges, in particular unpredictable radio propagation characteristics in constantly changing indoor environments as well as access technology constraints. Indoor environments, in contrast to outdoor, are extremely complex, with various shapes and sizes, as well as the presence or absence of stationary and moving objects (e.g., furniture and people). These variables drastically change both LOS and NLOS radio signal propagation, resulting in unpredictable attenuation, scattering, shadowing and blind spots that significantly reduce indoor positioning accuracy. To solve these issues, artificial intelligence (AI) and machine learning (ML) techniques have recently been extensively researched and have achieved reasonable success [84]. The fundamental benefit of using AI/ML techniques is their ability to make effective decisions based on observed data without the need for precise mathematical formulation. Moreover, ML has also proven to be a useful tool for fusing multidimensional data acquired from various location sensors, technologies and techniques. Due to rapid advancements in machine learning in recent years, several computer vision-based positioning systems exist in the literature such as [85,86]; however, these papers focus on 2D positioning. Such papers would be great benchmarks for further research and 3D expansion. Three-dimensional point cloud classification could also be considered, for example, [87].

One work utilizing ML has already been mentioned previously in the paper (see Section 2.2.2). Another work is proposed where the researchers in [88] have designed a miniaturized indoor positioning device while considering several machine learning optimization algorithms and using a hybrid method of Levenberg–Marquardt and ToA positioning algorithm to achieve 3D positioning in space. The purpose of this system and the utilization of machine learning is the ability to precisely locate the height of the target in the absence of height difference between base stations. The hybrid method was able to achieve more accurate results of 19.19 cm RMSE compared to the traditional ToA and

TDoA methods of 2.7 m RMSE with no significant degradation in efficiency. The reason for such differences in the RMSE measurements are in fact the errors measured in the z-axis, with 271.85 cm using the traditional method and 12.20 cm using the machine learning hybrid method, which justifies the contribution of this algorithm. The authors in [89] propose a 3D positioning approach for navigation within a hospital building. This system is designed particularly for multiple-story buildings. It aims to obtain the building level, longitude and latitude for a specific location. This system can recognize the horizontal information of the plane space, as well as the vertical information of different floors. In order to estimate the positions of mobile stations, it employs deep learning algorithms to analyze the received signal strength from cellular networks and Wi-Fi access points. In order to determine the precise position information (building level, longitude and latitude) in multiple-level buildings, a two-stage deep learning process (level classification and location determination) has been developed. A deep learning neural network was trained for the first stage of level classification. Three deep learning neural networks were trained to obtain the distinct location coordinates (longitude and latitude) for three different building levels. The average distance error of the location determination for different floors was 0.28 m.

4. Technologies for 3D Localization

Due to the uniqueness of each indoor environment and the immaturity and cost of various technologies (e.g., UWB, mmWave), there are no established standards for indoor positioning systems yet. In practice, each installation is adapted to spatial dimensions, structural materials, accuracy specifications and budget restrictions. Therefore, several different wireless positioning techniques and algorithms are currently being utilized and several more have been reported in the literature, which take advantage of Radio Access Technologies (RATs) such as Wi-Fi, Bluetooth, Ultrawideband (UWB), mmWave, cellular (2G–6G), etc. The importance of such technologies is their integration in modern smart devices. Alternative non-radio technologies applied in modern systems are ultrasound, inertial sensors and Visible Light Communication (VLC). At the end of this section in Table 6, a summary of all 3D positioning technologies can be found, describing their advantages and disadvantages, as well as their accuracies found in the literature. As well as Table 7, compares these technologies from the perspective of reception range, availability, energy efficiency, cost and scalability.

4.1. Wi-Fi

Nowadays, smartphones have become one of the most common technologies in everyday society and they are mostly used indoors. Ref. [40] states that "80% of smartphone usage happens inside buildings." The majority of modern smart devices are WiFi capable, making WiFi a great choice for indoor localization as well as one of the most thoroughly researched localization technologies in the literature. Because existing Wi-Fi access points may also be utilized as signal collection reference points, modest localization systems (with reasonable localization accuracy) can be created without the requirement for additional infrastructure [2]. Wi-Fi positioning systems have been in the lead for commercialized indoor localization, due to the massive deployment of Wi-Fi access points by mobile network carriers. Unfortunately, WPSs majorly depend on the density and distribution of Wi-Fi access points (APs) in the known environment, which directly affects the accuracy and the availability of the systems. Unfortunately, WPS accuracy and availability degrades as a result of its reliance on the number and distribution of Wi-Fi APs in its unique indoor service region. Although unsupervised as well as supervised Wi-Fi APs have been used to improve the location databases (DBs), such as fingerprinting DB or AP location DB, to increase the localization performance, taking environmental factors into account has little or no effect on improving location effectiveness in Wi-Fi dead zones. While the installation of additional APs will improve the system performance, the mobile network carriers usually are not willing, as they make the systems less time-efficient and more costly [74]. As mentioned previously in the paper, Wi-Fi is also the technology used for fingerprinting approaches such as RSS, CSI and FTM (see Sections 2.2.1, 2.2.2 and 2.2.4).

Ref. [74] proposes and implements a highly scalable 3D indoor positioning system based on loosely linked Wi-Fi/Sensor integration. Location database, which is derived using dynamic surveying data, is used to estimate Wi-Fi location. PDR is utilized as a time update model to compensate for the limitations of pedestrian motion modeling. The test findings suggest that providing a stable and accurate 3D indoor location in a scaled indoor environment is doable by using the basic yet complimentary loosely coupled Kalman filtering.

The researchers in [36] propose a robust 3D indoor positioning system appropriate for an indoor IoT application. This system is based on a Bayesian network that operates by determining the intensity of Wi-Fi signals. Using just four APs and a modest number of RPs, the suggested 3D Bayesian Graphical Model (3D-BGM) obtained an overall localization accuracy of 2.9 m.

WiFi round-trip time (RTT) was utilized in [90] for a 3D indoor localization algorithm for smartphones. In the proposed algorithm, the weighted centroid (WC) algorithm is utilized to estimate the rough two-dimensional (2D) position due to its easy implementation and low complexity. The coarse target altitude is acquired according to pedestrian activity. Then, the coarse altitude and 2D position combine into a rough 3D position, which is regarded as the initial position of the standard particle swarm optimization (SPSO) algorithm. The SPSO algorithm aims to estimate a more accurate 3D location on the basis of the cursory 3D position of the smartphone. To reduce computation, the density-based spatial clustering of applications with noise (DBSCAN) algorithm was used to assist in updating the SPSO particles. Experimental results show that the proposed positioning algorithm has better 3D accuracy than WC and least-squares (LS) algorithms, with a 2D accuracy of 1.147 m and an altitude precision of 0.305 m.

In [91], a smartphone-based 3D indoor positioning method is proposed which takes into account information from a WiFi interface and from the barometer sensor. Several experiments have been performed in two real scenarios and measurements have been made over commercial mobile devices. When tested in two different environments, it was distinguished that this method allows obtaining a lower positioning error even if few APs are available: when more than five Access Points (APs) are used, the proposed 3D positioning system is able to accurately localize the user with an error below 2 m and 1.2 m, respectively.

4.2. Bluetooth

Bluetooth was established as an open specification with low power, short range wireless data and voice connections and has long been used in the communication and proximity markets. It is used to transmit data over short ranges between devices via ultra high frequency (UHF) radio waves, ranging from 2.402 GHz to 2.48 GHz. Initially, it was developed as a wireless replacement for the RS-232 data cable. Similarly to WiFi, due to its broad availability in smart devices, it also seems like a great option for indoor localization. There are currently two main types of Bluetooth indoor positioning solutions: connection-based and inquiry-based [92].

While Bluetooth Low Energy (BLE) may be utilized with many localization approaches such as RSS, AoA and ToF, the majority of existing BLE-based localization solutions rely on RSS-based inputs since RSS-based systems are believed to be much simpler. However, due to the fact that it is strongly dependent on RSS-based inputs, the localization accuracy is limited. Despite the fact that BLE in its original form can be used for localization (due to its range, low cost and energy consumption), two BLE-based protocols, iBeacons (by Apple Inc., California, U.S.) and Eddystone (by Google Inc , California, U.S.), have recently been proposed, primarily for context aware proximity-based services [2].

The research in [92] presents an inquiry-based Bluetooth indoor positioning method using RSS probability distributions. The results suggest that the RSS probabilistic technique is a viable option for Bluetooth positioning. On the other hand, Bluetooth positioning has a substantial bottleneck owing to the low power consumption protocol: the updating frequency. Considering the accuracy of position determination is not very high, the test results show that the technique suggested in this study performs rather well. When compared to WLAN positioning, however, the Bluetooth signal characteristics and the number of access points result in lower accuracy.

The authors in [93] discuss low-cost 3D indoor positioning with Bluetooth smart device and least square (LS) methods. Nonlinear least square (NLS) method is adopted for parameter estimation of Bluetooth signal propagation model and various linear least square methods are used for 3D location estimation of the target Bluetooth device. Simulation and hardware experiment results illustrate that the nonlinear least square method is suitable for parameter estimation of Bluetooth signal propagation and the generalized least square (GLS) method has better performance than total least square methods. The proposed method also has the merits of low cost, low power consumption, high usability and high location precision. The hardware experiments have achieved a 3D positioning accuracy of 2.27 m and this was lowered to 1.97 m when combined with a barometer.

4.3. Cellular (2G–6G)

In cellular-based localization, downlink transmissions from the Base Station (BS) to the mobile device and uplink transmissions from the mobile device to the BS can be used to facilitate user positioning. The cellular-positioning techniques can be divided into two types based on the entity that computes the position: (1) mobile-based, in which the user device calculates its own location, and (2) network-based, in which the network location server computes the user device's position. Most cellular-based positioning systems are network-based due to their centralized design, which provides the network operator complete control of the location service, as well as their support for older devices. After an extensive literature review, no relevant 3D positioning works have been identified utilizing 2G–4G cellular technology. This is mainly due to the fact that at the time that these technologies were developed the need for 3D positioning was not as high as it is now. Therefore, the majority of existing systems using cellular technology are 5G-based. Ref. [94] suggested a 3D positioning method in a simulated indoor 5G ultra-dense network. The paper suggests a 3D dynamic reconstruction fingerprint matching technique, with the first step being to rebuild the entire fingerprint matrix from partial data. The sub-optimal service base stations are then removed from the dataset to simplify the fingerprint data. Finally, the 3D coordinates are estimated using the k-nearest neighbor matching approach. Positioning errors are assessed at various Signal-to-Noise Ratio (SNR) levels. The mean error is 0.31 m at SNR = 2 dB and 0.16 m at SNR = 20 dB. Ref. [95] focused their research on positioning a single cell (base station) equipped with a wideband 5G signal and a vector antenna (VA). This technique avoids the problems of multi-cell systems, such as base station synchronization and greater deployment costs due to system complexity. They employed statistics-based expectation maximization and the subspace-spaced technique to estimate position. The results which were obtained using sounding reference signals in a line of sight scenario demonstrate that VA is capable of providing 3D positioning with sub-meter accuracy in 5G networks without the need for numerous cells or antennas. The researchers in [96] discuss various ways of utilizing space detection to achieve more accurate and precise results for indoor localization. The designed and developed 5G simulation as well as the 5G-based particle filter fusion resulted in a reliable localization performance. For this, two approaches were proposed, the first one being the map data out of computer aided design (CAD) plans and the second the accuracy clarification of the positioning technique performance followed by a simple 5G-based PF which uses map information and geospatial analysis, smartphone sensor values and 5G simulation as input to provide a 3D trajectory for a long term robust performance in both online and offline environments. The results of this investigation show that map and routing graph preparation can be carried out efficiently, which ensures the accuracy and precision of indoor localization.

The approaches in map generation, simulation and localization were developed using available data sources as well as common algorithms with new usages in the 5G-based fusion domain. Moreover, a novel interpretation in accuracy and precision analyses has been discussed and tested with the simulated 5G measurements, based on the desired 3GPP standards. For a complex building design, errors below 3 m can be considered as the target accuracy of the 5G campus network. In the 4G era, cellular positioning was used for emergency services and services associated with lawful interception. Commercial use cases have gained significant interest concerning 5G and use cases such as factory automation, transportation and logistics are included in 5G alongside regulatory use cases. Positioning and location services are expected to be a critical components of the system requested by most commercial applications, such as AR/VR/XR, gaming, sensing, low-cost tracking and new industrial applications requiring exceptionally high precision as we move closer to 6G. This could also be enhanced by fusing with artificial intelligence powered mobile networks as suggested in [97]. As a result, location accuracy and latency requirements are expected to tighten even more with 5G [98]. The fifth generation (5G) new radio (NR) had a successful worldwide release in 2020. After a few years, the majority of the world has already adapted to this new communication standard and there is now a need to aim for new potential technologies while finding substantial use cases for the next generation of wireless systems, termed 6G communication systems. Wireless networks are frequently praised only for their communication capabilities, while their inherent positioning and sensing benefits are disregarded. In this sense, the 5G NR access interface, with its high carrier frequency, large bandwidth and massive antenna array, provides excellent prospects for precise localization and sensing systems. Furthermore, 6G systems will accelerate the transition to even higher frequency operation, such as millimeter wave (mmWave) and THz ranges, as well as significantly wider bandwidths. Furthermore, the THz frequency range provides several opportunities, including not just precise localization but also high-quality imaging and frequency spectroscopy [17]. In the 5G evolution to 6G, connectivity remains one of the most significant enablers of new services, but monetization of private networks requires more than simply a wireless connection. Beyond connectivity, for example, in industrial automation, high-accuracy positioning and sensing must be smoothly integrated into a single communication system [99]. 6G systems built for communication, sensing and location will enable new applications while improving traditional connectivity [98,100]. Future trends in wireless communication indicate that 6G radios are likely to use signals at the mmWave range and have channel bandwidths which are at least five times wider than 5G. From a localization and sensing perspective this has multiple benefits: (1) there is a more direct relation between the propagation paths and the environment as the signals on these frequencies do not typically penetrate walls; (2) the very fine time resolution of the power delay in these wide channels facilitates the resolvability of multi-path components and especially the LoS ones to more accurately estimate ranges; (3) smaller wavelengths that mean smaller antennas, especially phased array antennas that facilitate the good estimation of azimuth and elevation angles and hence enable accurate 3D positioning [17]. In addition to these, the high frequencies to be used in 6G systems open up a new potential in terms of sensing and imaging based on the radar-like technology that arises. The fact that multi-path components are highly resolvable in terms of time, angle and Doppler in the the power delay profile or impulse response enables the acquisition of spatial knowledge about the physical environment (known as imaging). The availability of this environment spatial information will better facilitate the use of Simultaneous Localization and Mapping (SLAM) approaches.

4.4. Ultra-Wideband

Ultra-wideband (UWB) is a short-range wireless technology which uses much wider bandwidths compared to the narrow-band transmissions typically used in Wi-Fi systems. UWB systems typically use frequencies ranging from 3.1 to 10.6 GHz but the bandwidth needs to be at least 20% of the central frequency. In addition, instead of measuring the signal strengths (RSS), the positioning is achieved by using the transit time methodology (ToA). The advantage of UWB technology compared to other Radio Access Technologies is that it offers "spatial awareness" since the wide bandwidth allows for better resolution in the time domain allowing for more accurate time and thereafter distance estimates to be measured. The localization accuracy could reach a centimeter level of approximately 10–30 cm, in comparison to GPS (1–3 m) or Wi-Fi (2–10 m) [101]. However, the issue with using UWB is that it is extremely short-ranged and requires a direct line of sight between receiver and transmitter due to high losses experienced when signals propagate through obstructions. This requires a greater number of transmitters within an indoor environment, which subsequently increases the cost. Even though it is not as widespread or cost-efficient as other RATs, utilizing the "spatial awareness" of this technology and especially combining it with the cooperative positioning approach, makes UWB a technology to consider in the future. The world's largest smartphone manufacturers, such as Apple, Samsung and Huawei, are all currently capitalizing on the UWB projects, specifically the manufacturing of the chips and antennas. However, Apple is the first to actually deploy it in a phone, with the others expecting to shortly follow.

In recent years, UWB technology has received a lot of interest for indoor positioning. Several systems have already been implemented commercially, while many others are being utilized as experimental testbeds such as those provided by Decawave and Bespoon companies. These systems have been thoroughly researched and validated for specific purposes. Other activities have focused on modelling the LOS and NLOS circumstances in order to develop NLOS identification metrics that will allow some NLOS mitigation methods to be implemented. The NLOS problem, which is the primary source of inaccuracy in UWB range and positioning, is still an open research topic [70].

Ref. [102] proposes a UWB positioning system which utilizes two way time of flight (TWTF) to compute range measurements. These readings are employed in the multilateration approach to determine the trans-receiver location (TAG). The authors of this paper state that this type of system has the advantage of providing high accuracy positioning (about 10 cm from the state of the art), as well as low power consumption, high multipath resolution, high data rate and other benefits. The system's testing has statically analyzed the system's positioning and range capabilities in an indoor office environment. The test yielded an average 3D accuracy of 100 ± 25 mm.

The authors in [103] propose a 3D ToA positioning algorithm while utilizing the UWB technology. The main idea of the proposed algorithm is to replace the quadratic term in the positioning estimation with a new variable and the usage of the weighted least squares linear estimation followed by the combination with Kalman filter to reduce the interference error in the transmission process. The simulation results show that the positioning accuracy can reach about 5–10 cm.

Another example is proposed in [104], where a high resolution UWB positioning radar system based on TDoA was developed. The UWB radar system provides millimeter accuracy in dense multipath indoor environments for 1D, 2D and 3D localization. The system is fully compliant with the FCC UWB regulations and utilizes time domain measurements to suppress both multipath signals and NLOS errors and has a potential for even sub-mm accuracy. Specifically, a 3 mm maximum error was achieved for the x, y dimensions with a 7 mm maximum error in the z-dimension.

The authors in [105] present a novel approach to a self-localizing anchor-system calibration that uses a calibration unit (CU) for improved localization accuracy. This study confirmed that the use of the CU decreases the average positional error of the anchors in 3D UWB localization systems. In addition, the simulations were confirmed to be a valid tool for determining the best position of the CU. Finally, the first demonstration of an anchor calibration with a CU and anchors localized in the working coordinate system in 3D was presented. It had an error of 0.32 m.

Mobile laser scanning (MLS) has been widely used in 3D city modelling data collection, such as Google cars for Google Map/Earth. Building Information Modelling (BIM) has recently emerged and become prominent. Three-dimensional models of buildings are essential for BIM. Static laser scanning is usually used to generate 3D models for BIM, but this method is inefficient if a building is very large or it has many turns and narrow corridors. Therefore, the researchers in [106] propose using MLS for BIM 3D data collection. The positions and attitudes of the mobile laser scanner are important for the correct geo-referencing of the 3D models. This paper proposes using three high-precision ultra-wide band (UWB) tags to determine the positions and attitudes of the mobile laser scanner. The accuracy of UWB-based MLS 3D models is assessed by comparing the coordinates of target points, as measured by static laser scanning and a total station survey. The UWB system can achieve centimeter positioning accuracy on the horizontal plane (around 8 cm), but decimeter accuracy in height (around 19 cm).

4.5. mmWave

Millimeter-wave (mmWave) technology is defining a new era in wireless communication by providing very wide bandwidths. This technology is currently used in some Wi-Fi systems (e.g., IEEE802.11ad) and is planned to be used in 5G communications in the near future as it offers much more flexibility to use wider bandwidths and hence have the strong potential to achieve much higher data rates and capacity. mmWave communication systems typically operate in the frequency range between 30 and 300 GHz. The first standardized consumer radios were in the 60 GHz unlicensed band, i.e., 57–64 GHz, where 2 GHz signal bandwidth is typical in applications. The very large availability of bandwidth, together with the use of massive phase array antennas that allow the estimation of the phase can be used for achieving cm-level accuracy or better [18]. Additionally, mmWave systems have higher transmit power allowance compared to UWB systems which compensates partly the high path losses that are typically experienced on those very high frequencies. Another way to alleviate those loss is by using beamforming. Directional beamforming is a challenging task as it requires good knowledge of the propagating channel and also imposes an extra difficulty and challenge in mmWave-based positioning as the exact orientation (azimuth, elevation) angle of the user equipment (UE) should be well known. In [19], the authors derived theoretically the Cramér–Rao bound (CRB) on position and rotation angle estimation uncertainty from mm-wave signals from a single transmitter, in the presence of scatterers. They demonstrated that in open Line of Sight (LoS) conditions, it is possible to estimate the target's position and orientation angle by exploiting the information coming from the multipath, though with a significant performance penalty. Moreover, the authors of [20] demonstrated the benefits of array antennas towards identifying the orientation of the device. Finally, due to this high sensitivity of the mmWave technology, positioning accuracy seems to be strongly correlated with the distance away from the target to be positioned. For instance, the authors of [23] conducted AoA and signal measurements in a 35 m by 65.5 m open space and achieved a position accuracy ranging from 16 cm to 3.25 m. Positioning research using this mmWave technology is still in very early stages but early theoretical findings and some practical experiments demonstrate its strong potential to achieve the very high accuracy required by modern smart applications. The authors in [107] propose a multipath-assisted localization (MAL) model based on millimeter-wave radar to achieve the localization of indoor electronic devices. The model fully considers the help of the multipath effect when describing the characteristics of the reflected signal and precisely locates the target position by using the MAL area formed by the reflected signal. At the same time, for the situation where the radar in the traditional Single-Input Single-Output (SISO) mode cannot obtain the 3D spatial position information of the target, the advantage of the MAL model is that the 3D information of the target can be obtained after the mining process of the multipath effect. Experiments show that the proposed MAL model enables the millimeter-wave multipath positioning model to achieve a 3D positioning error within 15 cm. A virtualized indoor office scenario with only one mmWave base station (BS) is considered in [108]. User equipment (UE) motion feature, mmWave line of sight (LoS) and first order reflection paths' AoA-ToA are fused for indoor positioning.

Firstly, an improved least mean square (LMS) algorithm that combines motion message is proposed to refine the multi-path AoA estimation. Furthermore, a modified multi-path unscented Kalman filter (UKF) is proposed to track UE's position in the scenario. The information exchanges of the two stages not only consists of estimates (position, AoA) but also variance of position. Based on the simulation results, the proposed methods provide two times LoS-AoA estimation gains and centimeter 3D positioning accuracy, respectively of around 60 cm. In addition, this strategy is capable of positioning task with insufficient anchor nodes (ANs).

4.6. Visible Light

Indoor localization based on visible light communication (VLC) has gained a lot of attention in recent years. One of its main advantages is its ability to provide high-accuracy positioning by utilizing the ubiquitous LED lights found in modern buildings without the need for any additional specialized infrastructure for location services [68]. According to the optical receiver in use, VLC-based positioning methods in the literature may be divided into two types, camera-based [109] and photodiode-based [110]. Camera-based solutions in particular have proven popular with both academics and industry, for example because of the high positioning precision achieved by imaging geometry and the strong interoperability of user devices. On a standard smartphone with a front-facing camera state-of-the-art commercial solutions may provide centimeter-level precision. Despite already existing systems' promising performance, there are still several practical challenges to be solved [68].

A large majority of VLP solutions rely on multilateration or triangulation to obtain location estimations. However, because of the physical field-of-view limits of both the luminaire (transmitter) and photodiode (receiver) in 3D, performance qualifications of these approaches in 3D positioning are limited and often unattainable. The limitations of FOV have an influence on line of sight (LOS) access to luminaires, which is problematic when several luminaires are required for positioning. Recently, several researchers have been trying to enhance lighting with other peripherals such as more PDs, a steerable laser and even a rotating receiver to eliminate the requirement to position with more than one luminaire while still enabling 3D positioning in the most recent literature. These additional peripherals improve positioning accuracy, especially if they have angular diversity. The developers in [111] introduce the notion of Ray-Surface Positioning (RSP). This method combines angular information from a steerable beam with range information obtained from an isointense envelope measured at a receiver. The first implementation of RSP is discussed to test theoretical and simulated predictions on 3D positioning accuracy and was averaged at around 30 cm.

The authors of [112] describe an RSS-based VLP as a "possible competitor" to UWB positioning. The paper also describes some approaches already developed by other researchers; for example, in [113], a three-dimensional VLP approach is proposed which is based on Artificial Neural Networks (ANN) utilizing the hybrid between phase PDoA and RSS approach. The approach is believed to minimize the distortions caused by inaccurate modeling as well as improve the overall robustness of conventional VLP systems. In [114] an LED-based 3D IPS is proposed which is aimed at both lighting and communications. The system is based on experimentally measured RSS with less than 3 cm of error. Another efficient 3D VLP algorithm is [115], with the intention of utilizing it for drone navigation. The receiver module did not require any extra height sensors; therefore, a four-LED arrangement was studied. However, simulations revealed that a traditional design of four Light Emitting Diodes (LEDs) arranged in a square form is incapable of solving the 3D position properly achieving accuracy of around 50 cm.

4.7. Sound-Based Technologies

A sound is a mechanical wave-like vibration that propagates or travels across any medium. The medium through which the waves propagate or travel can be either solid liquid or gas. A sound wave is also the pattern of disturbances caused by the movement of energy away from the source of the sound. Sound waves are sometimes known as longitudinal waves which means the propagation of particle vibration is somewhat parallel to the propagation of energy waves [116]. A source is necessary for the generation of sound. A speaker is an example of a sound source as its diaphragm is able to vibrate in order to produce sound. When a sound source vibrates, the particles in the medium around it vibrate as well. As the medium continues to vibrate because of the vibrating particles, the vibrating particles travel further away from the source of sound. The propagation of vibrating particles away from the source occurs at the speed of sound, therefore creating a sound wave [117].

Sound signals, which are pressure waves moving in the air, benefit from the fact that sound travels at a significantly slower pace than electromagnetic signals, making it much easier to measure the time between signal generation and arrival. Because the radio signal arrives at the sensor almost instantly and the sound signal arrives later, the difference between these two times can be used to calculate distance [118].

4.7.1. Ultrasound

Ultrasonic sounds have a high frequency that cannot be heard or identified by the human ear, greater than $20KHz$. Humans are unable to hear or recognize ultrasonic sounds nor can they generate them. Ultrasonic systems have generally been recognized as a captivating technology for indoor applications, due to some of its advantages such as low power consumption, adequate centimeter level accuracy under line of sight (LoS) conditions and even low cost, especially when considering the hardware devices and equipment required for practical real-time implementations [2,119]. Some 3D ultrasonic positioning systems have previously been created utilizing two major approaches: emitters are fixed in place while receivers move in the environment and vice versa. In most cases, a trilateration method is used to estimate the positions of the receivers, which is often based on the determination of time differences of arrival (TDoA), times of arrival (ToA), angle of arrival (AoA) or even hybrid techniques, to measure the distances between emitters and receivers. Some examples of existing ultrasonic-based localization systems based on trilateration include Active Bat, Cricket, Dolphin and Millibots [120].

Active Bat [121] and Cricket [122] are two of the most well known systems that utilize ultrasonic signals. The architecture of Active Bat requires mobile users to wear ultrasonic tags. Ceiling-mounted ultrasound receivers pick up the signal from the tag and send it to a central server. Active Bat employs ultrasonic time-of-flight lateration, in which the user delivers both an ultrasound and a radio signal and the system computes the difference in arrival times between the two signals to establish the user's position. Cricket improves Active Bat by narrowing the time frame in which arriving signals are processed by employing radio signal arrival time. Dolphin is another distributed ultrasonic positioning system. Only a few nodes' locations are known in Dolphin, while the rest of the nodes can infer their own locations based on the locations of reference nodes [123].

The researchers in [124] describe a 3D positioning system that uses broadband ultrasonic chirp pulses to acquire high-precision distance measurements. The higher bandwidth solves most of the difficulties associated with narrowband signals often employed with conventional piezo-ultrasonic transducers (typically with a bandwidth of 2 KHz), such as poor resolution, low ambient noise immunity, limited range and low robustness to the Doppler effect. A set of experiments were performed to evaluate the proposed system. Very stable 3D position estimates were obtained (absolute standard deviation less than 2.3 cm) and a position refresh rate of 350 ms was achieved.

Ultrasonic advantages include high accuracy at close range distances. The disadvantage is that they are highly prone to NLOS propagation and multipath effects.

4.7.2. Audible Sound

The human ear can effortlessly identify or sense frequencies ranging from 20 Hz to 20 kHz. As a result, sound waves with frequencies ranging from 20 Hz to 20 kHz are referred to as audible sounds. Human ears are sensitive to every minute pressure difference occurring in the air if it lies in the audible frequency ranges. They can detect pressure fluctuations as small as one billionth of atmospheric pressure [116].

It is also feasible to encode information for positioning systems using audible sound signals. Obviously, the simple concept of just generating an artificial audible sound has too many problems, the most significant of which is that it would irritate persons nearby. However, there are more complex systems to solve this issue, which works by watermarking an already available sound, like music in malls and other public locations in a way that the human ear cannot detect [118].

Table 6. Comparison of 3D Indoor Positioning Technologies.

Technology	Approach	Advantages	Disadvantages	Accuracy	Ref.
Wi-Fi	-RSS FP -CSI FP -RTT -FP + Barometer -FTM	-Simple to set up and use -Low cost as it does not require additional hardware	-Suffers from poor accuracy in NLOS conditions -Low accuracy when compared to other technologies	2.90 m 0.97 m 1.15 m 1.20 m 0.5–1.5 m	[36] [51] [90] [91] [52]
Bluetooth	-GLS + Barometer	-Easy to set up -Easy to operate -Inexpensive -Low energy consumption	-Difficult to calibrate each BLE beacon -Need extra hardware, medium accuracy -Prone to radio interference	1.97 m	[93]
Cellular	-FP -ToA/AoA -PDR	-Can be implemented with existing hardware in smart devices -No interference with other devices which operate at same frequency	-Low reliability due to varying signal propagation conditions -Requires synchronized base stations	0.16 m 1 m 3 m	[94] [95] [96]
Magnetic Field	-FP	-Cost- and energy-efficient while maintaining similar precision -Relies on built-in EMF sensors on smartphones without the need for additional equipment	-MF anomalies can only affect specific types of environments	0.5–1.5 m	[52]
UWB	-TWTF -ToA -TDoA -TOF	-High accuracy positioning even in the presence of severe multipath -Does not interfere with existing RF systems	-Need extra hardware -Expensive compared to other technologies	0.1 m 0.05–0.1 m 0.07 m 0.32 m	[102] [103] [104] [105]
mmWave	-ToA/AoA	-Higher transmission rate -Large bandwidth -Low interference	-More expensive -Compatibility issue, not all devices are able to support mmWave -Higher power consumption	0.15 m 0.6 m	[108] [111]
VLC	-RSP -PDOA/RSS -Trilateration	-Not affected due to EM radiations from RF systems -Easy to install -Performs well in LOS conditions	-Performs poor in NLOS conditions -Has interference issues from other ambient light sources -Short range	0.3 m 0.03 m 0.5 m	[111] [113] [115]
Ultrasound		-High accuracy at close range distances	-Highly prone to NLOS propagation and multipath effects -Receiver and transmitter need to see each other directly	0.02 m	[53]
Audible		-Widely supported -Works well in a wide variety of environments	-Can be heard by humans -Position is computed only when the user requests it -Performs poorly in NLOS conditions	0.6 m	[123]

Ultrasound systems, because of their reliance on technologies, require the user to acquire additional hardware such as badges or tags. As a result, many positioning approaches have been suggested by researchers that utilize the hardware already present in the users' smart devices, such as audible sound positioning systems like Beep [123]. Beep is a 3D localization system that uses audible sound for positioning. Existing devices (cell phones, PDAs, PCs, etc.) support audible sound, making it the foundation for a low-cost and widespread location system. Audible sound removes the requirement of additional user infrastructure. Beep provides on-demand positioning, which means that position is computed only when the user requests it, which saves power by avoiding continual communication between the user's device and the sensors. The testing results show that in more than 97% of cases, the measured location is accurate to within 0.6 m.

Table 7. Comparison of 3D Indoor Positioning Technology Attributes.

Technology	Reception Range	Availability	Energy Efficiency	Cost	Scalability	Ref.
Wi-Fi	45 m	High	Low	Low	High	[2,81,118]
Bluetooth	100 m	High	High	Low	Low	
Cellular	1 km	High	Low	Low	High	
Magnetic Field	~	High	High	Low	Medium	
UWB	10–20 m	Medium	Low	High	Medium	
mmWave	10–20 m	Low	Low	High	Medium	
VLC	1.4 km	Low	Low	Medium	High	
Ultrasound	20 m	Low	Medium	High	High	
Audible	2 m	High	Medium	Medium	High	

5. Critical Discussion and Conclusions

This paper reviews and discusses the current state of the art on 3D indoor positioning. This review includes the different techniques/approaches and technologies which can be used and/or combined to achieve the high 3D accuracy requirements of modern smart applications while maintaining cost efficiency. Table 1 showcases the various technologies that have been utilized for 3D indoor positioning, indicating their potential advantages and disadvantages. For instance, Wi-Fi, a technology that has been extensively utilized by either adopting fingerprinting approaches (RSS, CSI or FTM-based) as well as various geometric approaches is considered a technology that can be fairly easy to set up at a relatively low cost; however, it demonstrates poor accuracy in NLOS conditions compared to technologies like UWB and mmWave. Likewise, Bluetooth, given its simplicity and inexpensiveness, is similar to Wi-Fi; however, it is prone to radio interference; therefore, it is typically linked with low positioning accuracy. VLC and Ultrasound, despite the fact that they demonstrate relatively good accuracy compared to other technologies, are both extremely short-ranged and applicable only in line of sight situations. Moreover, audible sound, considering the fact that it is widely supported in various types of environments and able to achieve sub-meter level accuracy, cannot be utilized in common positioning scenarios mainly due to the disturbing noise it causes. Finally, UWB and mmWave technologies demonstrate the most promising results compared to other technologies, reaching centimeter-level accuracy even in multipath scenarios and are relatively insensitive to interference. Nevertheless, their main disadvantage is the fact that as of today there is a lack of supporting devices (mostly mmWave), making them a relatively expensive or infeasible option. However, the global technology evolution trend demonstrates that this is likely to change in the near future. In terms of approaches, the geometric ones which use angular (AoA) or timing information (ToA, TDoA) and base their principle of operation on the utilization of signals collected by a receiver from a dispersed collection of transmitters constitute a fundamental and relatively accurate way of estimating 3D position. Positioning accuracy obviously relies on the accuracy of the measured distances or angles and this accuracy seems to be strongly correlated with the underlying technology used. For instance, UWB and mmWave technologies demonstrate a high accuracy in estimating distance (based on timing measurements), while the introduction of phased arrays in these modern system facilitates the accurate estimation of angular information. The number of dispersed nodes also plays an important role in 3D localization estimation. The greater the number of transmitters the higher the accuracy; however, this imposes an additional financial, implementation and administrative cost when implementing such systems, especially in more complex and crowded areas, as well as considering scenarios where objects or people are continually moving. Nevertheless, the rapid evolution of the Internet of Things and the availability of many moving nodes facilitate the 3D localization process especially by using sensor fusion filtering or even cooperative positioning strategies. Cooperative positioning appears to be a promising solution, as the devices within space are interconnected and can determine location relative to one another. Fingerprinting also constitutes a candidate approach for 3D positioning typically complemented or combined with other approaches or technologies (e.g., barometers) to calculate the z-dimension or improve the accuracy (by using filtering). The problem is that the data collection process is typically extremely laborious and extra challenges emerge when dynamism appears in the environment either when people are moving around or when geometric or morphological changes happen to the environment itself or even when users use different devices and hold them in various different ways. The literature reports that magnetic field-based positioning could be less laborious; however, it only works in specific types of indoor environments.

Author Contributions: Conceptualization, M.R. and A.S.; methodology, A.S. and S.I. and M.R. investigation, A.S. and M.R.; writing—original draft preparation, A.S.; writing—review and editing M.R. and S.I.; supervision, M.R.; project administration, S.I. All authors have read and agreed to the published version of the manuscript.

Funding: This research received no external funding.

Conflicts of Interest: The Authors declare no conflict of interest.

References

1. Sand, S.; Dammann, A.; Mensing, C. References. In *Positioning in Wireless Communications Systems*; John Wiley & Sons: Hoboken NJ, USA, 2013; pp. 233–244. [CrossRef]
2. Zafari, F.; Gkelias, A.; Leung, K.K. A Survey of Indoor Localization Systems and Technologies. *IEEE Commun. Surv. Tutor.* **2019** *21*, 2568–2599. [CrossRef]
3. Sun, R.; Wang, J.; Cheng, Q.; Mao, Y.; Ochieng, W.Y. A New IMU-Aided Multiple GNSS Fault Detection and Exclusion Algorithm for Integrated Navigation in Urban Environments. *GPS Solut.* **2021**, *25*, 147. [CrossRef]
4. Mao, Y.; Sun, R.; Wang, J.; Cheng, Q.; Kiong, L.C.; Ochieng, W.Y. New Time-Differenced Carrier Phase Approach to GNSS/INS Integration. *GPS Solut.* **2022**, *26*, 122. [CrossRef]
5. Laoudias, C.; Moreira, A.; Kim, S.; Lee, S.; Wirola, L.; Fischione, C. A Survey of Enabling Technologies for Network Localization Tracking, and Navigation. *IEEE Commun. Surv. Tutor.* **2018**, *20*, 3607–3644. [CrossRef]
6. Shi, G.; Ming, Y. Survey of Indoor Positioning Systems Based on Ultra-wideband (UWB) Technology. In *Wireless Communications Networking and Applications*; Springer: New Delhi, India, 2016; pp. 1269–1278. [CrossRef]
7. Radaelli, L.; Jensen, C.S. Towards fully organic indoor positioning. In Proceedings of the Fifth ACM SIGSPATIAL International Workshop on Indoor Spatial Awareness, Orlando, FL, USA, 5 November 2013; pp. 16–20.
8. Pankaj, L. Indoor Positioning and Indoor Navigation (IPIN) Market Outlook: 2025. Available online: alliedmarketresearch.com (accessed on 26 November 2022).
9. Indoor Location Market. Available online: marketsandmarkets.com (accessed on 26 November 2022).
10. Laoudias, C.; Raspopoulos, M.; Christoforou, S.; Kamilaris, A. Privacy-Preserving Presence Tracing for Pandemics Via Machine-to-Machine Exposure Notifications. In Proceedings of the 2022 23rd IEEE International Conference on Mobile Data Management (MDM), Paphos, Cyprus, 6–9 June 2022; pp. 355–360. [CrossRef]
11. Han, C.; Zhu, X.; Doufexi, A.; Kocak, T. Location-Aided Multi-User Beamforming for 60 GHz WPAN Systems. In Proceedings of the 2012 IEEE 75th Vehicular Technology Conference (VTC Spring), Yokohama, Japan, 6–9 May 2012; pp. 1–5. [CrossRef]
12. Akbar, N.; Yan, S.; Yang, N.; Yuan, J. Mitigating Pilot Contamination through Location-Aware Pilot Assignment in Massive MIMO Networks. In Proceedings of the 2016 IEEE Globecom Workshops (GC Wkshps), Washington, DC, USA, 4–8 December 2016; pp. 1–6. [CrossRef]

13. Muppirisetty, L.S.; Svensson, T.; Wymeersch, H. Spatial Wireless Channel Prediction under Location Uncertainty. *IEEE Trans. Wirel. Commun.* **2016**, *15*, 1031–1044. [CrossRef]
14. Luo, J.; Han, Y.; Fan, L. Underwater Acoustic Target Tracking: A Review. *Sensors* **2018**, *18*, 112. [CrossRef] [PubMed]
15. Farr, N.; Bowen, A.; Ware, J.; Pontbriand, C.; Tivey, M. An integrated, underwater optical /acoustic communications system. In Proceedings of the OCEANS'10 IEEE SYDNEY, Sydney, NSW, Australia, 24–27 May 2010; pp. 1–6. [CrossRef]
16. Zhang, T.; Yan, L.; Han, G.; Peng, Y. Fast and Accurate Underwater Acoustic Horizontal Ranging Algorithm for an Arbitrary Sound-Speed Profile in the Deep Sea. *IEEE Internet Things J.* **2022**, *9*, 755–769. [CrossRef]
17. Bourdoux, A.; Barreto, A.N.; van Liempd, B.; de Lima, C.; Dardari, D.; Belot, D.; Lohan, E.S.; Seco-Granados, G.; Sarieddeen, H.; Wymeersch, H.; et al. 6G White Paper on Localization and Sensing. *arXiv* **2020**, arXiv 2006.01779. [CrossRef]
18. Wang, D.; Fattouche, M.; Zhan, X. Pursuance of mm-Level Accuracy: Ranging and Positioning in mmWave Systems. *IEEE Syst. J.* **2019**, *13*, 1169–1180. [CrossRef]
19. Shahmansoori, A.; Garcia, G.E.; Destino, G.; Seco-Granados, G.; Wymeersch, H. Position and Orientation Estimation Through Millimeter-Wave MIMO in 5G Systems. *IEEE Trans. Wirel. Commun.* **2018**, *17*, 1822–1835. [CrossRef]
20. Han, Y.; Shen, Y.; Zhang, X.P.; Win, M.Z.; Meng, H. Performance Limits and Geometric Properties of Array Localization. *IEEE Trans. Inf. Theory* **2016**, *62*, 1054–1075. [CrossRef]
21. Bensky, A. *Wireless Positioning Technologies and Applications*; Artech House, Inc.: Norwood, MA, USA, 2007.
22. Brás, L.; Carvalho, N.; Pinho, P.; Kulas, L.; Nyka, K. A Review of Antennas for Indoor Positioning Systems. *Int. J. Antennas Propag.* **2012**, *2012*, 953269. [CrossRef]
23. Kanhere, O.; Rappaport, T.S. Position Locationing for Millimeter Wave Systems. *arXiv* **2018**, arXiv:1808.07094. [CrossRef]
24. Zhu, Z.; Bocus, M.Z. A Computationally Efficient Method for Direction Finding with Known Transmit Sequence. In Proceedings of the 2018 International Conference on Indoor Positioning and Indoor Navigation (IPIN), Nantes, France, 24–27 September 2018; pp. 1–6. [CrossRef]
25. Zhang, H.; Zhang, Z. AoA-Based Three-Dimensional Positioning and Tracking Using the Factor Graph Technique. *Symmetry* **2020**, *12*, 1400. [CrossRef]
26. Hacioglu, G.; Sesli, E. Improved RSS Based Distance Estimation for Autonomous Vehicles. *Wirel. Pers. Commun.* **2022**, *125*, 325–350. [CrossRef]
27. Gonendik, E.; Gezici, S. Fundamental Limits on RSS Based Range Estimation in Visible Light Positioning Systems. *IEEE Commun. Lett.* **2015**, *19*, 2138–2141. [CrossRef]
28. Coluccia, A.; Fascista, A. On the Hybrid ToA/RSS Range Estimation in Wireless Sensor Networks. *IEEE Trans. Wirel. Commun.* **2018**, *17*, 361–371. [CrossRef]
29. Obeidat, H.; Ahmad, I.; Rawashdeh, M.R.; Abdullah, A.A.; Shuaieb, W.; Obeidat, O.; Abd-Alhameed, R.A. Enhanced ToA Estimation Using OFDM over Wide-Band Transmission Based on a Simulated Model. *Wirel. Pers. Commun.* **2022**, *123*, 3449–3461. [CrossRef]
30. Khalaf-Allah, M. Novel Solutions to the Three-Anchor ToA-Based Three-Dimensional Positioning Problem. *Sensors* **2021**, *21*, 7325. [CrossRef]
31. Wang, W.; Zhang, Y.; Tian, L. ToA-based NLOS error mitigation algorithm for 3D indoor localization. *China Commun.* **2020**, *17*, 63–72. [CrossRef]
32. Plank, H.; Egger, T.; Steffan, C.; Steger, C.; Holweg, G.; Druml, N. High-performance indoor positioning and pose estimation with time-of-flight 3D imaging. In Proceedings of the 2017 International Conference on Indoor Positioning and Indoor Navigation (IPIN), Sapporo, Japan, 18–21 September 2017; pp. 1–8. [CrossRef]
33. Kaune, R. Accuracy studies for TDoA and ToA localization. In Proceedings of the 2012 15th International Conference on Information Fusion, Singapore, 9–12 July 2012; pp. 408–415.
34. *Comparison of Time-Difference-of-Arrival and Angle-of-Arrival Methods of Signal Geolocation*; ITU: Geneva, Switzerland, 2011.
35. Passafiume, M.; Collodi, G.; Ciervo, E.; Cidronali, A. A Novel TDoA-Based Method for 3D Combined Localization Techniques Using an Ultra-Wideband Phase Wrapping-Impaired Switched Beam Antenna. *Electronics* **2021**, *10*, 2137. [CrossRef]
36. Alhammadi, A.; Alraih, S.; Hashim, F.; Rasid, M.F.A. Robust 3D Indoor Positioning System Based on Radio Map Using Bayesian Network. In Proceedings of the 2019 IEEE 5th World Forum on Internet of Things (WF-IoT), Limerick, Ireland, 15–18 April 2019; pp. 107–110. [CrossRef]
37. Du, Y.; Arslan, T.; Juri, A. Camera-aided region-based magnetic field indoor positioning. In Proceedings of the 2016 International Conference on Indoor Positioning and Indoor Navigation (IPIN), Alcala de Henares, Spain, 4–7 October 2016; pp. 1–7. [CrossRef]
38. Pendão, C.; Moreira, A. FastGraph—Organic 3D Graph for Unsupervised Location and Mapping. In Proceedings of the 2018 International Conference on Indoor Positioning and Indoor Navigation (IPIN), Nantes, France, 24–27 September 2018; pp. 206–212. [CrossRef]
39. Raspopoulos, M. Multidevice Map-Constrained Fingerprint-Based Indoor Positioning Using 3-D Ray Tracing. *IEEE Trans. Instrum. Meas.* **2018**, *67*, 466–476. [CrossRef]
40. Berkovich, G. Accurate and reliable real-time indoor positioning on commercial smartphones. In Proceedings of the 2014 International Conference on Indoor Positioning and Indoor Navigation (IPIN), Busan, Korea, 27–30 October 2014; pp. 670–677. [CrossRef]

41. Yang, J.; Lee, H.; Moessner, K. Multilateration localization based on Singular Value Decomposition for 3D indoor positioning. In Proceedings of the 2016 International Conference on Indoor Positioning and Indoor Navigation (IPIN), Alcala de Henares, Spain, 4–7 October 2016; pp. 1–8. [CrossRef]
42. Jaworski, W.; Wilk, P.; Zborowski, P.; Chmielowiec, W.; Lee, A.Y.; Kumar, A. Real-time 3D indoor localization. In Proceedings of the 2017 International Conference on Indoor Positioning and Indoor Navigation (IPIN), Sapporo, Japan, 18–21 September 2017; pp. 1–8. [CrossRef]
43. Gansemer, S.; Hakobyan, S.; Püschel, S.; Großmann, U. 3D WLAN indoor positioning in multi-storey buildings. In Proceedings of the 2009 IEEE International Workshop on Intelligent Data Acquisition and Advanced Computing Systems: Technology and Applications, Rende, Italy, 21–23 September 2009; pp. 669–672. [CrossRef]
44. AlShamaa, D.; Mourad-Chehade, F.; Honeine, P. Localization of sensors in indoor wireless networks: An observation model using WiFi RSS. In Proceedings of the 2018 9th IFIP International Conference on New Technologies, Mobility and Security (NTMS), Paris, France, 26–28 February 2018; pp. 1–5.
45. Wu, F.; Xing, J.; Dong, B. An indoor localization method based on rssi of adjustable power WiFi router. In Proceedings of the 2015 Fifth International Conference on Instrumentation and Measurement, Computer, Communication and Control (IMCCC), Qinhuangdao, China, 18–20 September 2015; pp. 1481–1484.
46. *IEEE Std 802.11n-2009*; IEEE Standard for Information technology—Local and metropolitan area networks—Specific requirements—Part 11: Wireless LAN Medium Access Control (MAC)and Physical Layer (PHY) Specifications Amendment 5: Enhancements for Higher Throughput. IEEE: Piscataway, NJ, USA, 2009; pp. 1–565. [CrossRef]
47. Dang, X.; Tang, X.; Hao, Z.; Liu, Y. A device-free indoor localization method using CSI with Wi-Fi signals. *Sensors* **2019**, *19*, 3233. [CrossRef] [PubMed]
48. Rocamora, J.M.; Ho, I.W.H.; Mak, W.; Lau, A. Survey of CSI Fingerprinting-based indoor positioning and mobility tracking systems. *IET Signal Process.* **2020**, *14*, 407–419. [CrossRef]
49. Li, Y.; Nie, W.; He, W.; Wang, Y.; Yang, X. UAV 3D Localization System Using CSI. In Proceedings of the 2021 International Conference on Microwave and Millimeter Wave Technology (ICMMT), Nanjing, China, 23–26 May 2021; pp. 1–3. [CrossRef]
50. Stop the Occurrence of "Black Flight" through UAV Reaction Technology. 1999. Available online: sma818.com (accessed on 26 November 2022).
51. Karmanov, I.; Zanjani, F.G.; Merlin, S.; Kadampot, I.; Dijkman, D. WiCluster: Passive Indoor 2D/3D Positioning using WiFi without Precise Labels. In Proceedings of the 2021 IEEE Global Communications Conference (GLOBECOM), Madrid, Spain, 7–11 December 2021.
52. Hellmers, H.; Eichhorn, A.; Norrdine, A.; Blankenbach, J. IMU/magnetometer based 3D indoor positioning for wheeled platforms in NLoS scenarios. In Proceedings of the 2016 International Conference on Indoor Positioning and Indoor Navigation (IPIN), Alcala de Henares, Spain, 4–7 October 2016; pp. 1–8. [CrossRef]
53. Yu, Y.; Chen, R.; Chen, L.; Xu, S.; Li, W.; Wu, Y.; Zhou, H. Precise 3-D Indoor Localization Based on Wi-Fi FTM and Built-In Sensors. *IEEE Internet Things J.* **2020**, *7*, 11753–11765. [CrossRef]
54. Hiertz, G.R.; Denteneer, D.; Stibor, L.; Zang, Y.; Costa, X.P.; Walke, B. The IEEE 802.11 universe. *IEEE Commun. Mag.* **2010**, *48*, 62–70. [CrossRef]
55. Han, K.; Yu, S.; Kim, S.L. Smartphone-based Indoor Localization Using Wi-Fi Fine Timing Measurement. In Proceedings of the 2019 International Conference on Indoor Positioning and Indoor Navigation (IPIN), Pisa, Italy, 28 November 2019; pp. 1–5. [CrossRef]
56. Si, M.; Wang, Y.; Xu, S.; Sun, M.; Cao, H. A Wi-Fi FTM-Based Indoor Positioning Method with LOS/NLOS Identification. *Appl. Sci.* **2020**, *10*, 956. [CrossRef]
57. Si, M.; Wang, Y.; Seow, C.K.; Cao, H.; Liu, H.; Huang, L. An Adaptive Weighted Wi-Fi FTM-Based Positioning Method in an NLOS Environment. *IEEE Sens. J.* **2022**, *22*, 472–480. [CrossRef]
58. Poulose, A.; Kim, J.; Han, D. A Sensor Fusion Framework for Indoor Localization Using Smartphone Sensors and Wi-Fi RSSI Measurements. *Appl. Sci.* **2019**, *9*, 4379. [CrossRef]
59. Chauhan, K.; Chauhan, R.K.; Saini, A. Chapter 11—Medical image fusion methods: Review and application in cardiac diagnosis. In *Image Processing for Automated Diagnosis of Cardiac Diseases*; Chauhan, K., Chauhan, R.K., Eds.; Academic Press: Cambridge, MA, USA, 2021; pp. 195–215. [CrossRef]
60. Se, S.; Lowe, D.; Little, J. Mobile Robot Localization and Mapping with Uncertainty using Scale-Invariant Visual Landmarks. *Int. J. Robot. Res.* **2002**, *21*, 735–760. [CrossRef]
61. Zheng, L.; Zhou, W.; Tang, W.; Zheng, X.; Peng, A.; Zheng, H. A 3D indoor positioning system based on low-cost MEMS sensors. *Simul. Model. Pract. Theory* **2016**, *65*, 45–56. [CrossRef]
62. Pastell, M.; Frondelius, L.; Järvinen, M.; Backman, J. Filtering methods to improve the accuracy of indoor positioning data for dairy cows. *Biosyst. Eng.* **2018**, *169*, 22–31. [CrossRef]
63. Kalman, R.E. A New Approach to Linear Filtering and Prediction Problems. *Trans. ASME–J. Basic Eng.* **1960**, *82*, 35–45. [CrossRef]
64. Ebner, F.; Fetzer, T.; Deinzer, F.; Köping, L.; Grzegorzek, M. Multi sensor 3D indoor localisation. In Proceedings of the 2015 International Conference on Indoor Positioning and Indoor Navigation (IPIN), Banff, AB, Canada, 13–16 October 2015; pp. 1–11. [CrossRef]

5. Julier, S.J.; Uhlmann, J.K. New extension of the Kalman filter to nonlinear systems. In Proceedings of the Signal Processing, Sensor Fusion and Target Recognition VI, Orlando, FL, USA, 21–24 April 1997; Kadar, I., Ed.; International Society for Optics and Photonics, SPIE: Bellingham, WA, USA, 1997; Volume 3068, pp. 182–193. [CrossRef]
6. Wan, E.; Van Der Merwe, R. The unscented Kalman filter for nonlinear estimation. In Proceedings of the IEEE 2000 Adaptive Systems for Signal Processing, Communications and Control Symposium (Cat. No.00EX373), Lake Louise, AB, Canada, 4 October 2000; pp. 153–158. [CrossRef]
7. Wang, R.; Zheng, L.; Wu, D.; Peng, A.; Tang, B.; Lu, H.; Shi, H.; Zheng, H. Research on multiple gait and 3D indoor positioning system. In Proceedings of the 2017 International Conference on Indoor Positioning and Indoor Navigation (IPIN), Sapporo, Japan, 18–21 September 2017; pp. 1–7. [CrossRef]
8. Liang, Q.; Lin, J.; Liu, M. Towards Robust Visible Light Positioning Under LED Shortage by Visual-inertial Fusion. In Proceedings of the 2019 International Conference on Indoor Positioning and Indoor Navigation (IPIN), Pisa, Italy, 30 September–3 October 2019; pp. 1–8. [CrossRef]
9. Ristic, B. *Particle Filters for Random Set Models*; Springer: Berlin/Heidelberg, Germany, 2013; Volume 798.
0. Jiménez, A.; Seco, F. Comparing Decawave and Bespoon UWB location systems: Indoor/outdoor performance analysis. In Proceedings of the 2016 International Conference on Indoor Positioning and Indoor Navigation (IPIN), Alcala de Henares, Spain, 4–7 October 2016; pp. 1–8. [CrossRef]
1. Ruiz, A.R.J.; Granja, F.S. Comparing ubisense, bespoon and decawave uwb location systems: Indoor performance analysis. *IEEE Trans. Instrum. Meas.* **2017**, *66*, 2106–2117. [CrossRef]
2. Kaji, K.; Kawaguchi, N. Estimating 3D pedestrian trajectories using stability of sensing signal. In Proceedings of the 2016 International Conference on Indoor Positioning and Indoor Navigation (IPIN), Alcala de Henares, Spain, 4–7 October 2016; pp. 1–8. [CrossRef]
3. Yu, N.; Zhan, X.; Zhao, S.; Wu, Y.; Feng, R. A Precise Dead Reckoning Algorithm Based on Bluetooth and Multiple Sensors. *IEEE Internet Things J.* **2018**, *5*, 336–351. [CrossRef]
4. Cho, Y.S.; Ji, M.I.; Kim, J.Y.; Jeon, J.I. High-scalable 3D indoor positioning algorithm using loosely-coupled Wi-Fi/sensor integration. In Proceedings of the 2015 17th International Conference on Advanced Communication Technology (ICACT), PyeongChang, Republic of Korea, 1–3 July 2015; pp. 96–99. [CrossRef]
5. Inderst, F.; Pascucci, F.; Santoni, M. 3D pedestrian dead reckoning and activity classification using waist-mounted inertial measurement unit. In Proceedings of the 2015 International Conference on Indoor Positioning and Indoor Navigation (IPIN), Banff, AB, Canada, 13–16 October 2015; pp. 1–9. [CrossRef]
6. Seco, F.; Jiménez, A.R. Smartphone-Based Cooperative Indoor Localization with RFID Technology. *Sensors* **2018**, *18*, 266. [CrossRef] [PubMed]
7. Ristic, B.; Arulampalam, S.; Gordon, N. *Beyond the Kalman Filter: Particle Filters for Tracking Applications*; Artech House: Norwood, MA, USA, 2003.
8. Wang, S.; Jiang, X. Three-Dimensional Cooperative Positioning in Vehicular Ad-hoc Networks. *IEEE Trans. Intell. Transp. Syst.* **2021**, *22*, 937–950. [CrossRef]
9. Xiaoxuan, W.; Peng, S.; Minlin, C.; Hucheng, W.; Zhi, W. A 3-D Cooperative Base Station Localization Method Applied in Large Complex Indoor Environment. Available online: https://ceur-ws.org/Vol-2498/short49.pdf (accessed on 26 November 2022) *Sensors* **2021**, *21*, 1002.
0. Pascacio, P.; Casteleyn, S.; Torres-Sospedra, J.; Lohan, E.S.; Nurmi, J. Collaborative Indoor Positioning Systems: A Systematic Review. *Sensors* **2021**, *21*, 1002. [CrossRef]
1. Guo, X.; Ansari, N.; Hu, F.; Shao, Y.; Nkrow, R.; Li, L. A Survey on Fusion-Based Indoor Positioning. *IEEE Commun. Surv. Tutor.* **2019**, *22*, 566–594. [CrossRef]
2. Lu, Y.; Koivisto, M.; Talvitie, J.; Valkama, M.; Lohan, E.S. EKF-based and Geometry-based Positioning under Location Uncertainty of Access Nodes in Indoor Environment. In Proceedings of the 2019 International Conference on Indoor Positioning and Indoor Navigation (IPIN), Pisa, Italy, 30 September–3 October 2019; pp. 1–7. [CrossRef]
3. Li, Y.; Zhang, P.; Lan, H.; Zhuang, Y.; Niu, X.; El-Sheimy, N. A modularized real-time indoor navigation algorithm on smartphones. In Proceedings of the 2015 International Conference on Indoor Positioning and Indoor Navigation (IPIN), Banff, AB, Canada, 13–16 October 2015; pp. 1–7. [CrossRef]
4. Nessa, A.; Adhikari, B.; Hussain, F.; Fernando, X.N. A Survey of Machine Learning for Indoor Positioning. *IEEE Access* **2020**, *8*, 214945–214965. [CrossRef]
5. Clark, R.; Trigoni, N.; Markham, A. Robust Vision-Based Indoor Localization. In Proceedings of the 14th International Conference on Information Processing in Sensor Networks (IPSN '15), Seattle, WA, USA, 14–16 April 2015; Association for Computing Machinery: New York, NY, USA, 2015; pp. 378–379. [CrossRef]
6. Peng, P.; Yu, C.; Xia, Q.; Zheng, Z.; Zhao, K.; Chen, W. An Indoor Positioning Method Based on UWB and Visual Fusion. *Sensors* **2022**, *22*, 1394. [CrossRef] [PubMed]
7. Huang, C.Q.; Jiang, F.; Huang, Q.H.; Wang, X.Z.; Han, Z.M.; Huang, W.Y. Dual-Graph Attention Convolution Network for 3-D Point Cloud Classification. *IEEE Trans. Neural Netw. Learn. Syst.* **2022**, 1–13. [CrossRef] [PubMed]

88. Liu, X.; Zhang, Z.; Cai, R.; Du, C.; Yu, B.; Yang, D. UWB-based Machine Learning Optimized 3D Positioning Algorithm. In Proceedings of the 2022 IEEE 6th Information Technology and Mechatronics Engineering Conference (ITOEC), Chongqing, China, 4–6 March 2022; Volume 6, pp. 1799–1803. [CrossRef]
89. Zhang, Q.; Wang, Y. A 3D mobile positioning method based on deep learning for hospital applications. *EURASIP J. Wirel. Commun. Netw.* **2020**, *2020*, 170. [CrossRef]
90. Cao, H.; Wang, Y.; Bi, J. Smartphones: 3D Indoor Localization Using Wi-Fi RTT. *IEEE Commun. Lett.* **2021**, *25*, 1201–1205. [CrossRef]
91. Bisio, I.; Sciarrone, A.; Bedogni, L.; Bononi, L. WiFi Meets Barometer: Smartphone-Based 3D Indoor Positioning Method. In Proceedings of the 2018 IEEE International Conference on Communications (ICC), Kansas City, MO, USA, 20–24 May 2018; pp. 1–6. [CrossRef]
92. Pei, L.; Chen, R.; Liu, J.; Tenhunen, T.; Kuusniemi, H.; Chen, Y. An Inquiry-based Bluetooth indoor positioning approach for the Finnish pavilion at Shanghai World Expo 2010. In Proceedings of the IEEE/ION Position, Location and Navigation Symposium, Indian Wells, CA, USA, 4–6 May 2010; pp. 1002–1009. [CrossRef]
93. Li, H. Low-Cost 3D Bluetooth Indoor Positioning with Least Square. *Wirel. Pers. Commun.* **2014**, *78*, 1331–1344. [CrossRef]
94. Zhang, Y.; Jin, J.; Liu, C.; Jia, P. Indoor 3D Dynamic Reconstruction Fingerprint Matching Algorithm in 5G Ultra-Dense Network. *KSII Trans. Internet Inf. Syst.* **2021**, *15*, 343–364. [CrossRef]
95. Sun, B.; Tan, B.; Wang, W.; Valkama, M.; Morlaas, C.; Lohan, E.S. 5G Positioning Based on the Wideband Electromagnetic Vector Antenna. In Proceedings of the WiP Proceedings of the International Conference on Localization and GNSS (ICL-GNSS 2021), Tampere, Finland, 1–3 June 2021.
96. Shoushtari, H.; Askar, C.; Harder, D.; Willemsen, T.; Sternberg, H. 3D Indoor Localization using 5G-based Particle Filtering and CAD Plans. In Proceedings of the 2021 International Conference on Indoor Positioning and Indoor Navigation (IPIN), Lloret de Mar, Spain, 29 November–2 December 2021; pp. 1–8. [CrossRef]
97. Luo, G.; Yuan, Q.; Li, J.; Wang, S.; Yang, F. Artificial Intelligence Powered Mobile Networks: From Cognition to Decision. *arXiv* **2021**, arXiv:2112.04263. [CrossRef]
98. Säily, M.; Yilmaz, O.N.C.; Michalopoulos, D.S.; Pérez, E.; Keating, R.; Schaepperle, J. Positioning Technology Trends and Solutions Toward 6G. In Proceedings of the 2021 IEEE 32nd Annual International Symposium on Personal, Indoor and Mobile Radio Communications (PIMRC), Helsinki, Finland, 13–16 September 2021; pp. 1–7. [CrossRef]
99. Viswanathan, H.; Mogensen, P.E. Communications in the 6G Era. *IEEE Access* **2020**, *8*, 57063–57074. [CrossRef]
100. Wild, T.; Braun, V.; Viswanathan, H. Joint Design of Communication and Sensing for Beyond 5G and 6G Systems. *IEEE Access* **2021**, *9*, 30845–30857. [CrossRef]
101. Alarifi, A.; Al-Salman, A.; Alsaleh, M.; Alnafessah, A.; Al-Hadhrami, S.; Al-Ammar, M.A.; Al-Khalifa, H.S. Ultra Wideband Indoor Positioning Technologies: Analysis and Recent Advances. *Sensors* **2016**, *16*, 707. [CrossRef] [PubMed]
102. Dabove, P.; Di Pietra, V.; Piras, M.; Jabbar, A.A.; Kazim, S.A. Indoor positioning using Ultra-wide band (UWB) technologies: Positioning accuracies and sensors' performances. In Proceedings of the 2018 IEEE/ION Position, Location and Navigation Symposium (PLANS), Monterey, CA, USA, 23–26 April 2018; pp. 175–184. [CrossRef]
103. Ni, D.; Postolache, O.A.; Mi, C.; Zhong, M.; Wang, Y. UWB Indoor Positioning Application Based on Kalman Filter and 3-D ToA Localization Algorithm. In Proceedings of the 2019 11th International Symposium on Advanced Topics in Electrical Engineering (ATEE), Bucharest, Romania, 28–30 March 2019; pp. 1–6. [CrossRef]
104. Zhang, C.; Kuhn, M.; Merkl, B.; Mahfouz, M.; Fathy, A.E. Development of an UWB Indoor 3D Positioning Radar with Millimeter Accuracy. In Proceedings of the 2006 IEEE MTT-S International Microwave Symposium Digest, San Francisco, CA, USA, 11–16 June 2006; pp. 106–109. [CrossRef]
105. Krapež, P.; Munih, M. Anchor Calibration for Real-Time-Measurement Localization Systems. *IEEE Trans. Instrum. Meas.* **2020**, *69*, 9907–9917. [CrossRef]
106. Lau, L.; Quan, Y.; Wan, J.; Zhou, N.; Wen, C.; Qian, N.; Jing, F. An Autonomous Ultra-Wide Band-Based Attitude and Position Determination Technique for Indoor Mobile Laser Scanning. *ISPRS Int. J. Geo-Inf.* **2018**, *7*, 155. [CrossRef]
107. Hao, Z.; Yan, H.; Dang, X.; Ma, Z.; Jin, P.; Ke, W. Millimeter-Wave Radar Localization Using Indoor Multipath Effect. *Sensors* **2022**, *22*, 5671. [CrossRef] [PubMed]
108. Jia, Y.; Tian, H.; Fan, S.; Liu, B. Motion Feature and Millimeter Wave Multi-path AoA-ToA Based 3D Indoor Positioning. In Proceedings of the 2018 IEEE 29th Annual International Symposium on Personal, Indoor and Mobile Radio Communications (PIMRC), Bologna, Italy, 9–12 September 2018; pp. 1–7. [CrossRef]
109. Li, Y.; Ghassemlooy, Z.; Tang, X.; Lin, B.; Zhang, Y. A VLC smartphone camera based indoor positioning system. *IEEE Photonics Technol. Lett.* **2018**, *30*, 1171–1174. [CrossRef]
110. Bai, B.; Chen, G.; Xu, Z.; Fan, Y. Visible Light positioning based on LED traffic light and photodiode. In Proceedings of the 2011 IEEE Vehicular Technology Conference (VTC Fall), San Francisco, CA, USA, 5–8 September 2011; pp. 1–5.
111. Lam, E.; Little, T. Indoor 3D Localization with Low-Cost LiFi Components. In Proceedings of the 2019 Global LIFI Congress (GLC), Paris, France, 12–13 June 2019; pp. 1–6. [CrossRef]
112. Plets, D.; Bastiaens, S.; Ijaz, M.; Almadani, Y.; Martens, L.; Raes, W.; Stevens, N.; Joseph, W. Three-dimensional Visible Light Positioning: An Experimental Assessment of the Importance of the LEDs' Locations. In Proceedings of the 2019 International Conference on Indoor Positioning and Indoor Navigation (IPIN), Pisa, Italy, 30 September–3 October 2019; pp. 1–6. [CrossRef]

113. Zhang, S.; Du, P.; Chen, C.; Zhong, W.D.; Alphones, A. Robust 3D Indoor VLP System Based on ANN Using Hybrid RSS/PDOA. *IEEE Access* **2019**, *7*, 47769–47780. [CrossRef]
114. Yang, S.; Jeong, E.; Kim, D.; Kim, H.; Son, Y.; Han, S.K. Indoor three-dimensional location estimation based on LED visible light communication. *Electron. Lett.* **2013**, *49*, 54–56. [CrossRef]
115. Joseph, D.; Ijaz, M. Efficient 3D trilateration algorithm for Visible Light positioning. *J. Opt.* **2019**, *21*, 05LT01.
116. Begault, D.R. Audible and inaudible early reflections: Thresholds for auralization system design. In Proceedings of the Audio Engineering Society Convention 100, Copenhagen, Denmark, 11–14 May 1996; Audio Engineering Society: New York, NY, USA, 1996.
117. Embleton, T.F. Tutorial on sound propagation outdoors. *J. Acoust. Soc. Am.* **1996**, *100*, 31–48. [CrossRef]
118. Brena, R.; García-Vázquez, J.; Galván Tejada, C.; Munoz, D.; Vargas-Rosales, C.; Fangmeyer, J., Jr.; Palma, A. Evolution of Indoor Positioning Technologies: A Survey. *J. Sens.* **2017**, *2017*, 2630413. [CrossRef]
119. Mannay, K.; Ureña, J.; Hernández, A.; Machhout, M.; Aguili, T. Characterization of an Ultrasonic Local Positioning System for 3D Measurements. *Sensors* **2020**, *20*, 2794. [CrossRef] [PubMed]
120. Kapoor, R.; Ramasamy, S.; Gardi, A.; Bieber, C.; Silverberg, L.; Sabatini, R. A Novel 3D Multilateration Sensor Using Distributed Ultrasonic Beacons for Indoor Navigation. *Sensors* **2016**, *16*, 1637. [CrossRef] [PubMed]
121. Woodman, O.J.; Harle, R.K. Concurrent scheduling in the active bat location system. In Proceedings of the 2010 8th IEEE International Conference on Pervasive Computing and Communications Workshops (PERCOM Workshops), Mannheim, Germany, 29 March–2 April 2010; pp. 431–437.
122. Priyantha, N.B. The Cricket Indoor Location System. Ph.D. Thesis, Massachusetts Institute of Technology, Cambridge, MA, USA, 2005.
123. Mandal, A.; Lopes, C.; Givargis, T.; Haghighat, A.; Jurdak, R.; Baldi, P. Beep: 3D indoor positioning using audible sound. In Proceedings of the Second IEEE Consumer Communications and Networking Conference (CCNC 2005), Las Vegas, NV, USA, 6 January 2005; pp. 348–353. [CrossRef]
124. Lopes, S.I.; Vieira, J.M.N.; Albuquerque, D. High Accuracy 3D Indoor Positioning Using Broadband Ultrasonic Signals. In Proceedings of the 2012 IEEE 11th International Conference on Trust, Security and Privacy in Computing and Communications, Liverpool, UK, 25–27 June 2012; pp. 2008–2014. [CrossRef]

sensors

MDPI

Review

A Comprehensive Review on Time Sensitive Networks with a Special Focus on Its Applicability to Industrial Smart and Distributed Measurement Systems

Tommaso Fedullo [1,2], Alberto Morato [3], Federico Tramarin [1,*], Luigi Rovati [1] and Stefano Vitturi [3]

1 Department of Engineering "Enzo Ferrari", University of Modena and Reggio Emilia, 41125 Modena, Italy; tommaso.fedullo@unimore.it (T.F.); luigi.rovati@unimore.it (L.R.)
2 Department of Management and Engineering, University of Padova, S. S. Nicola 3, 36100 Vicenza, Italy
3 National Research Council of Italy, CNR–IEIIT, Via Gradenigo 6/B, 35131 Padova, Italy; alberto.morato@ieiit.cnr.it (A.M.); stefano.vitturi@ieiit.cnr.it (S.V.)
* Correspondence: federico.tramarin@unimore.it

Abstract: The groundbreaking transformations triggered by the Industry 4.0 paradigm have dramatically reshaped the requirements for control and communication systems within the factory systems of the future. The aforementioned technological revolution strongly affects industrial smart and distributed measurement systems as well, pointing to ever more integrated and intelligent equipment devoted to derive accurate measurements. Moreover, as factory automation uses ever wider and complex smart distributed measurement systems, the well-known Internet of Things (IoT) paradigm finds its viability also in the industrial context, namely Industrial IoT (IIoT). In this context, communication networks and protocols play a key role, directly impacting on the measurement accuracy, causality, reliability and safety. The requirements coming both from Industry 4.0 and the IIoT, such as the coexistence of time-sensitive and best effort traffic, the need for enhanced horizontal and vertical integration, and interoperability between Information Technology (IT) and Operational Technology (OT), fostered the development of enhanced communication subsystems. Indeed, established technologies, such as Ethernet and Wi-Fi, widespread in the consumer and office fields, are intrinsically non-deterministic and unable to support critical traffic. In the last years, the IEEE 802.1 Working Group defined an extensive set of standards, comprehensively known as Time Sensitive Networking (TSN), aiming at reshaping the Ethernet standard to support for time-, mission- and safety-critical traffic. In this paper, a comprehensive overview of the TSN Working Group standardization activity is provided, while contextualizing TSN within the complex existing industrial technological panorama, particularly focusing on industrial distributed measurement systems. In particular, this paper has to be considered a technical review of the most important features of TSN, while underlining its applicability to the measurement field. Furthermore, the adoption of TSN within the Wi-Fi technology is addressed in the last part of the survey, since wireless communication represents an appealing opportunity in the industrial measurement context. In this respect, a test case is presented, to point out the need for wirelessly connected sensors networks. In particular, by reviewing some literature contributions it has been possible to show how wireless technologies offer the flexibility necessary to support advanced mobile IIoT applications.

Keywords: TSN; Industry 4.0; smart distributed measurement systems; IIoT; IoT; Ethernet; Wi-Fi

Citation: Fedullo, T.; Morato, A.; Tramarin, F.; Rovati, L.; Vitturi, S. A Comprehensive Review on Time Sensitive Networks with a Special Focus on Its Applicability to Industrial Smart and Distributed Measurement Systems. *Sensors* **2022**, 22, 1638. https://doi.org/10.3390/s22041638

Academic Editor: Peter Han Joo Chong

Received: 26 January 2022
Accepted: 14 February 2022
Published: 19 February 2022

1. Introduction

The need to communicate information has driven human activities over the years, adapting to and impacting on the technological and economical growth of the society. Notably, the communication of data between sensors, controllers and actuators becomes of critical importance, thus impacting on the measurement accuracy and the possibility to stably control an industrial process. Moreover, the novel smart factory requires ever integrated measurement systems, able to communicate data from and to the field with the management areas of the industrial plant. Nowadays, the Time Sensitive Networking (TSN) project is capturing much research interest as a promising set of standards, able to cope with strict requirements coming from different application areas. Although this paper focuses on industrial measurement networks, in fact, TSN is the outcome of the interweaving in the recent history of the industrial field and the consumer one. This is the reason why a brief dive into the past is needed, to better understand the importance and the power of TSN.

The year was 1999 when the term Internet of Things (IoT) was coined, by Kevin Ashton during a famous presentation [1]. Objects have always been fundamental, as they allow people to interface with (and even to modify) the physical world, but in the IoT context they acquire the capability to use some of the five functional senses through a specific sensor network [2]. In this context, objects acquire *computational* and *communication* capabilities all being interconnected, thus allowing access to information and data *anywhere and at any time* [3]. Additionally, the famous "click" is becoming obsolete: it is possible to directly "ask" your house to close the shutters or to turn off the light. This *smart* approach has enormous advantages in various application fields, for example, smart buildings [4], smart cities [5], e-health [6], transportation [7] and even smart farming [8]. Moreover, using IoT to develop smart and distributed Industrial measurement systems brings several advantages, thus giving the possibility to take continuous, thorough and real time measurements on wide areas [9]. In this context, the development of smart, distributed and IoT-based measurement systems definitely foresees the design of high-performing and real time communication networks, able to accurately transfer sensor and control data. Indeed, the transmission delay uncertainty of measurement data between several sensors placed in a wide and challenging industrial area, has an impact on the measurement quality. Furthermore, during last years, thanks to the advanced technologies derived from the IoT and Cyber Physical Systems (CPSs) [10], the Industry 4.0 plan mandates the integration of these networks comprising accurate measurement systems and smart actuators, within the whole industrial system. At present, the usage of IoT technologies to develop smart industrial systems namely Industrial Internet of Things (IIoT) [3], acquires a fundamental importance. Indeed *integration* must be guaranteed both *horizontally* and *vertically*, respectively, within the same level and between levels of the automation pyramid. An effective way to provide vertical integration is the usage of the OPC-UA (Unified Architecture) protocol [11], developed in 2008 by the Open Platform Communications Foundation. In this context, measurement systems may also communicate with higher levels of the automation pyramid and even exchange data between different plants, paving the way for a fully integrated smart factory. The rise of attention towards Industry 4.0 and IIoT made them evolving into strategic technologies, and a considerable pressure towards effective implementations comes from diverse scenarios [12–17]. Moreover, suitable communication and computation technologies devoted to measurement activities are needed, to guarantee specific accuracy levels. In this context, reasonably, the industrial network must handle not only an increased amount of measurement traffic in a deterministic way, but also the coexistence of *time-critical* and *normal* data exchange [9,18–21]. Indeed, in time-critical applications, the concept of *time* assumes a greater importance and a more refined meaning, and the *best effort* behavior is no longer sufficient. The communication network has to guarantee bounded latency and jitter, as well as a real-time behavior, defined as the ability to deliver the useful data before a specified instant of time, namely the *deadline*. From a metrological perspective measurement data must be sent before a specific deadline, while managing also best effort traffic, like network configuration, and higher priority traffic like alarms. Moreover handling time-critical and accurate measurements is of fundamental importance not only to provide enough stability to the controlled system, but also to handle safety-critical applications. Safety operation of an industrial system is unavoidable due to the strict interaction between humans and machines, thus underlining the need for deterministic and accurate measurement systems devoted to safety [22,23].

Currently, industrial communication systems are based on several established technologies, namely Fieldbuses and Real Time Ethernet (RTE) networks. However, novel approaches had to be pursued to accomplish the crucial requirements of the foreseen advanced applications, with strict time-criticality being one of them, along with high reliability, fault tolerance, and security. Furthermore, there are additional issues to consider, such as dense networks, sensors-to-cloud data exchange, seamless reconfiguration, support for a big data approach and convergence [24,25]. This requires a strict and seamless coexistence of IT and OT, that is, of best efforts and deterministic/time-critical data and protocols to the field level devices.

This new set of requirements poses several challenges that may be difficult to address, even by the best performing industrial communication systems, such as RTE networks. This has driven the interest of the industrial community towards a complete redesign of the whole architecture of the communication system. An important opportunity in this direction has been represented by the IEEE 802.1 Time-Sensitive Networking (TSN) standardization activity, which is currently recognized as the future *de facto* standard for industrial communications, as it will be better explained in the next paragraphs [26,27].

This review paper aims to provide a comprehensive analysis of the state-of-the art on TSN, providing appropriate bibliographic references to allow the reader to go in deep with the specific topics. Indeed, the TSN standardization project comprises many standards, and this survey can become a compass to, hopefully, guide the reader into the TSN world. Moreover, the applicability of TSN to the Instrumentation and Measurement field is analyzed, by demonstrating the impact and benefits of deterministic communication in the measurement uncertainty. In detail, the paper is organized as follows. Section 2 collocates TSN in the context of the scientific literature and provides the motivations to address the adoption of TSN in both the Industrial Automation and Instrumentation and Measurement fields. Section 3 provides an introduction of fieldbus and RTE technologies, pointing out the limits of the established industrial panorama that stimulated the introduction of TSN. Section 4 presents the TSN family of standards. Section 5 briefly describes the TSN Industrial Automation profile. Section 6 addresses the usage of TSN networks to design smart measurement systems, possibly based on wireless communication. This is achieved by the introduction of a meaningful test case, as well as by a discussion about the implementation of TSN over Wi-Fi. Finally, Section 7 concludes the paper and provides some future research directions.

2. Background and Motivation

The TSN project aims to provide all the features needed to handle time-critical traffic in different scenarios, and this resulted in a set of protocols that can be properly adopted and configured to design networks able to cope with the specific requirements for the application at hand. Given the intrinsic potentialities and the disrupting changes in the networking architecture the project introduced, the research interest towards the TSN development has been steadily growing in the last years, as can be evinced from an analysis of the recently published research articles in this field. To this regard, Figure 1 reports at a glance the outcomes of such an analysis. In detail, Figure 1a on the left reports the number of articles indexed by Scopus related to TSN topics over the last ten years, showing a marked increase. The same plot also reports the number of published surveys and reviews about TSN. Even more importantly, Figure 1b reports the percentages, among such articles, of those published in journals and conferences belonging to either the Industrial Automation or the Instrumentation and Measurement (I&M) fields. As it can be observed, such percentages are rather low, and this has been also confirmed by a further analysis relevant to the papers specifically concerned with the Industrial Automation Profile of TSN, that will be discussed later in Section 5. As a matter of fact, the data in Figure 2 clearly shows that the number of contributions concerned with the Industrial Automation profile is still limited, revealing that such field of application of TSN, and the strictly related ones like I&M, need to be further addressed.

Moving from the above considerations, this paper investigates the adoption of TSN in the industrial scenario focusing, in particular, on its possible usage to develop smart, distributed and IoT based measurement systems.

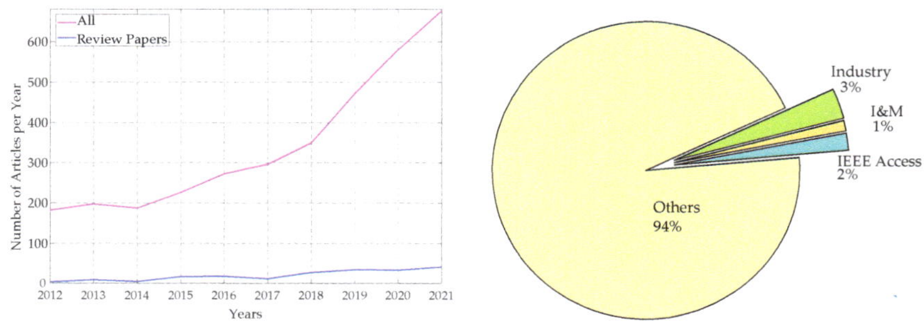

(**a**) (**b**)

Figure 1. An analysis of the recently published research articles in the field of TSN. (**a**) Number of articles per year on Scopus. (**b**) Number of articles published on conferences or journals belonging to the I&M and industrial fields on Scopus.

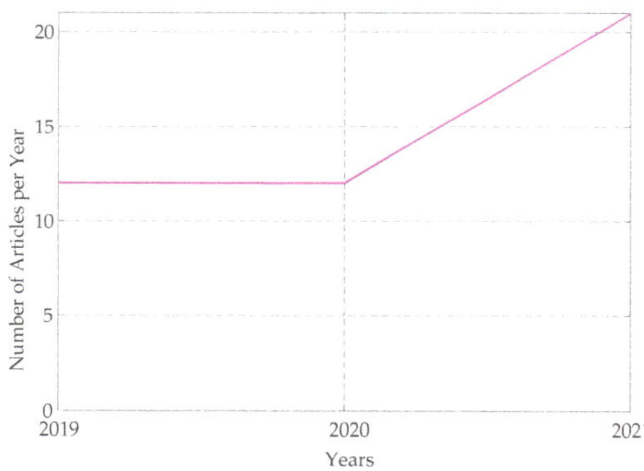

Figure 2. Number of articles per year on Scopus. Research key: IEEE/IEC 60802 in all the article fields.

Despite this, from the observations made above, it is important to underline that an Industrial communication network needs to handle a variety of data flows, with different requirements. This topic is even more critical considering the need for IoT smart measurement systems [9], and wireless connectivity, as also underlined by the Physikalisch-Technische Bundesanstalt (PTB) [28]. In this context, the measurements coming from sensors cover a widespread importance, impacting on several data flows. In particular, on cyclic real-time flows, alarms and events. In this context, sensors data need to guarantee certain levels of measurement precision and accuracy, thus allowing suitable handling of critical situations and/or stably controlling a process. Furthermore, in the harsh industrial environment, the analysis of the impact of complex, distributed and IIoT measurement networks, even wirelessly connected, on the measurement accuracy is of fundamental importance. In particular, there is the need for a precise analysis (by using also new measurement metrics) of the impact of such new intelligent and smart systems on the measurement activity. This problem is even more critical if the measurement system uses Artificial Intelligence techniques, or vision systems, the latter dramatically increasing the amount of time-critical data to send and process. In this context, a significant example is the impact of the transmission delay on the measurement process. Indeed, assuming that at a specific instant of time t_s the sensor sends a measure x_s, the data will be received at an instant of time $t_r = t_s + t_d$, where t_d is a random variable describing the delay introduced by the communication network. For this reason, the communication network has an impact on the measurement uncertainty, as the real value of the measured variable is $x_r \neq x_s$. From a measurement point of view, this issue seems to present different simple solutions, especially if the t_d uncertainty can be neglected. Indeed, considering only the measurement aspects, it is both possible to timestamp the data coming from sensors or adjust the deterministic error after data reception. In this context, the measurement error linearly increases with the network delay, as experimented by the authors of [29]. Unfortunately, in an Industrial scenario, measurements need to be used in real-time to control a process or handle critical situations, such as alarms or sporadic events. In this context, several works focus on the possibility, from a control perspective, to model and take into account the network delay in the control design stage. For example, authors of [30] try to compensate both network delay and packet loss by suitably designing the control stage. For example, the delay could be taken into account by using a e^{-st_d} term in the control model. In this context, the problem is that network delay is not even deterministic, as in general, t_d follows a specific probability function. From a measurement point of view, according to the ISO Guide to the Expression of Uncertainty in Measurement (GUM) [31], it is possible to evaluate the uncertainty (type B) introduced by the communication network delay t_d, as per Equation (1).

$$u(x) = \left| \frac{\delta x(t, t_d)}{\delta t_d} \right| \cdot u(t_d) \tag{1}$$

where x is the signal received from the sensor, which depends on both t and t_d, and $u(t_d)$ is the uncertainty on the knowledge of t_d. From the latter observation, it is possible to conclude that lowering $u(t_d)$, by using a deterministic network, lowers the measurement uncertainty. In particular, measurement data need to be handled with a certain priority, given by the critical level of the specific operation, that in turn reflects on the measurement uncertainty. It is worth observing that, practically, the calculation of $\frac{\delta x(t,t_d)}{\delta t_d}$ can be approximated by evaluating the dynamics of the specific sensors employed, thus deriving the $\frac{\Delta x}{\Delta t}$ of the sensor. This is possible as $x(t, t_d) = x(t - t_d)$, thus involving in $\left|\frac{\delta x(t,t_d)}{\delta t_d}\right| = \left|\frac{\delta x(t,t_d)}{\delta t}\right|$. If the measurement system has been well designed, the sensor dynamics needs to be fast enough to capture the measurand variations, thus being the latter approach a worst-case analysis. In the next Section, the widespread used communication networks for industrial applications are presented, underlining why they are not applicable to handle the requirements coming from the Industry 4.0 paradigm and to limit the measurement uncertainty.

3. The Long Journey from Fieldbus to the RTEs Technologies

The main important events that had an impact on today's technological panorama are presented in Figure 3.

In the early days of industrial automation systems, the need for data sharing among different parts of a machine soon led to the design of dedicated communication systems, targeted at the industrial scenario, universally known as *fieldbuses* [32]. First installations of fieldbuses date back to the early 1970s, but the number of available solutions quickly diverged, to such an extent that it was referred to as a "fieldbus war" [33], where several manufacturers have proposed proprietary industrial communication protocols, often with similar but completely non-interoperable functionality. To overcome this fragmentation, many research energies were spent in standardization processes. the project was shelved to develop a unique communication system, in 1999 the first version of the IEC 61158 international standard was released, which comprised several fieldbuses [34]. During the years, the IEC 61158 standard became a huge project collecting a lot of different fieldbuses, the majority of the total, for example Profibus, ControlNet, and Interbus (only to cite a few).

Significant limitations characterized these networks: low data rates, low number of connected nodes, as well as significantly reduced interoperability capabilities. Indeed, the integration of heterogeneous technologies and the sharing of data among different solutions were severely limited and internetworking capabilities were substantially absent [35].

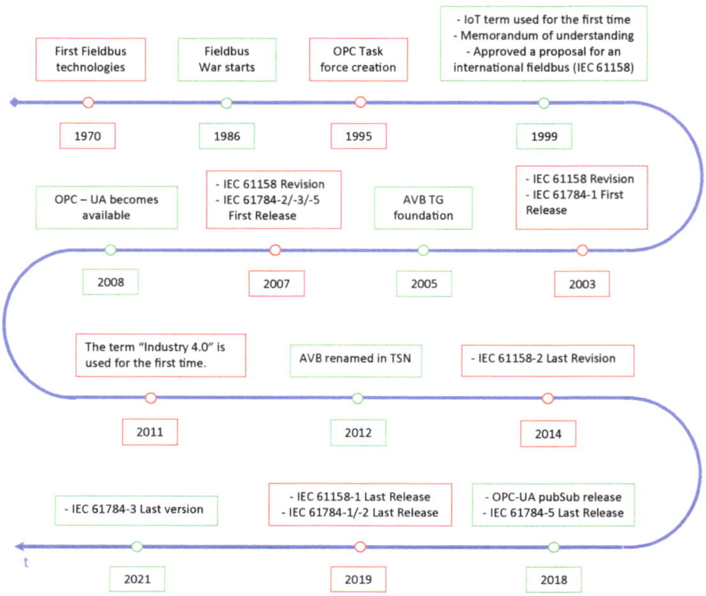

Figure 3. Main players involved in the fieldbus war.

With the subsequent proliferation of Ethernet technologies and the widespread availability of Internet connections, the automation world started to develop a new set of Ethernet–based systems, using the IEEE 802.1/802.3 specifications for the lowest communication layers. However, unless strict traffic and access controls are implemented, legacy Ethernet was unable to guarantee the required network latency, reliability and determinism. This intrinsic lack of real-time capabilities gave rise to the development of several dedicated (and proprietary) solutions, collectively referred to as real-time Ethernet (RTE), or Industrial Ethernet, networks [36]. The IEC 61158 and IEC 61784 international standards gathered several of them, e.g., PROFINET, Ethernet/IP, Modbus/TCP, and Ethernet POWERLINK, to name a few. Unfortunately, again the number of available RTEs rapidly increased, impairing interoperability, convergence, integration/implementation costs, and substantially replicating the former fieldbus battle [37,38].

Several shortcomings led to this situation. Indeed, one of the major barriers to the realization of a "one fits all" solution was that different standardization bodies were involved in the design of a new RTE protocol, as well as *consortia* (e.g., Profibus, ODVA, etc.) has been formed to protect relevant market shares and brands. This resulted in a widespread adoption of RTE solutions in the last years, with a large industrial pervasiveness, but also in different approaches to obtain the desired performance. Indeed, irrespective to the market share, these consortia had no control over the standardization process of the underlying Ethernet (IEEE 802.1/802.3) standard, and often an RTE solution has been obtained introducing some protocol "hack" over the legacy Ethernet. Particularly, a well-accepted classification of different RTEs systems follows the sketch in Figure 4, which identifies three different RTE classes with respect to different real-time performance [39].

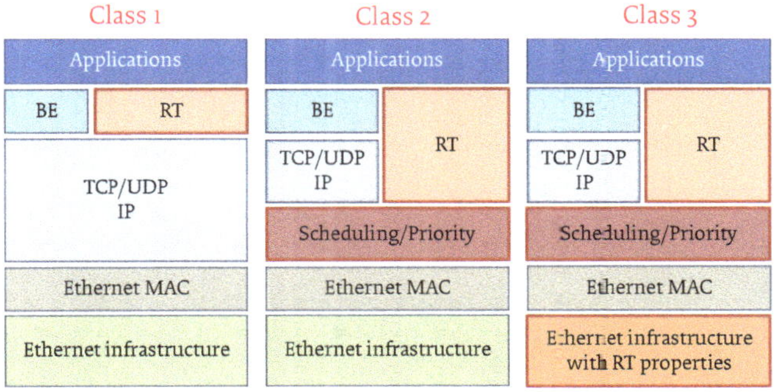

Figure 4. A widespread classification of RTE industrial networks.

For the aim to provide interoperation, several studies were made to connect different fieldbuses to each others or with different technologies using specific hardware or middleware protocol structure such as [40–43]. The latter one is also an example of a hybrid wired and wireless network, being a mixed network, a key solution to develop smart measurement systems. Nowadays, how to adapt the widespread used fieldbus and RTE systems to the requirements of Industry 4.0 is still challenging. Several research activities have been made in this direction [44,45]. Despite all this, the complex technological panorama is so broad that the development of a plethora of "adapters" to interconnect different fieldbus and RTE systems is practically infeasible. In this scenario, the development of new systems to use the CPS architecture and enact the Industry 4.0 revolution is undoubtedly required [46].

4. The Time Sensitive Networking Project

The Industry 4.0 paradigm highlights the need for increasingly standardized and integrated networks [47]. In this context, Time Sensitive Networking (TSN) standards offer a viable solution, pointing to the development of a novel smart factory paradigm. The idea underlying the whole TSN project is to deeply modify the Ethernet standard at its roots (by the development of a new Ethernet MAC layer and a new Ethernet infrastructure), to introduce all those intrinsic mechanisms required to support a broad range of time-, mission-, and safety-critical applications. Indeed, on the contrary, all the available RTE networks build upon the legacy features of Ethernet, use protocol strategies (as a clever use of Virtual LAN [VLAN] prioritization) or even out-of-standard data-link layers to introduce real-time capabilities over a network support that is intrinsically non-real-time [48,49].

Nevertheless, the first efforts in the stated direction have been pursued by the consumer electronics industry, and specifically for targeting the needs for deterministic Ethernet connections for professional audio and video streaming. This pushed towards introducing the needed modifications directly within the IEEE related standards. For this reason, in 2005 the Audio Video Bridging (AVB) Task Group (TG) was formed within the IEEE 802.1 standard committee. In parallel, the AVnu Alliance has been formed, an associated group of manufacturers and vendors to support the compliance and marketing activities. The activities of the AVB TG allowed to strongly enhance the real-time capabilities of Ethernet with four new IEEE standards: 802.1AS-2011, 802.1Qat-2010, 802.1Qav-2009 and 802.1BA-2011. The new potentialities of Ethernet AVB were soon deemed suitable also for the industrial scenario [50]. For this reason, it was rapidly evident that the AVB name was not appropriate to cover all the potential use cases that the achievable performance attracted.

In 2012, AVB was renamed in TSN Task Group, a subgroup of IEEE 802.1 Working Group [51]. The suitability of these set of standards to different fields of application, has led to the definition of different *profiles*, that represent one of the most powerful characteristic of TSN and have been presented in Section 5. The IEEE 802.1 defines Data Link Layer (DLL) protocols, as can be noticed from Table 1.

Table 1. IEEE 802.1 Contribution within IEEE 802.

ISO/OSI Layer	IEEE 802 Standard	
Data Link Layer	802.2 Logical Link Layer	
	802.1 Bridging	
	802.3 MAC	802.11 MAC
Physical	802.3 PHY	802.11 PHY

As it is possible to notice from Table 1, a network-specific Medium Access Control (MAC) layer is located right under the 802.1 bridging layer. In this article, two different LANs are considered: the IEEE 802.3 (Ethernet) and the IEEE 802.11 (Wi-Fi) one. TSN, traditionally, aims to enhance the performances of the IEEE 802.3 networks, but could also be applied to IEEE 802.11 networks, to reduce both delay and jitter [52]. The TSN over Wi-Fi networks will be analyzed in Section 6.2. The TSN standardization project focuses mainly on the IEEE 802.1Q (*IEEE Standard for Local and Metropolitan Area Networks–Bridges and Bridged Networks*) [53], with the development of several amendments to the standard. Indeed, time-sensitive traffic in different scenarios may have different QoS requirements, involving in the need of a set of configurable mechanism and protocols. Standards and amendments within the TSN project [54] are listed in Table 2.

Table 2. The TSN standardization project.

Standard	Description	Reference
IEEE 802.1AB	Station and Media Access Control Connectivity Discovery	[55]
IEEE 802.1AS	Timings & Syncronization	[56]
IEEE 802.1AX	Link Aggregation	[57]
IEEE 802.1CB	Frame Replication & Elimination	[58]
IEEE 802.1CS	Link Local Registration Protocol	[59]
	Ongoing Projects	
IEEE P802.1CQ	Multicast and Local Address Assignment	[60]
IEEE P802.1DC	Quality of Service Provision by Network Systems	[61]
IEEE P802f	YANG Data Model for EtherTypes (amending IEEE 802-2014 [62])	[63]
IEEE P802.1ABcu	LLDP YANG Data Model (amending IEEE 802.1AB [55])	[64]
IEEE P802.1ABdh	Support for Multiframe PDUs (amending IEEE 802.1AB [55])	[65]
IEEE P802.1ASdm	Hot Standby (amending IEEE 802.1AS [56])	[66]
IEEE P802.1ASdn	YANG Data Model (amending IEEE 802.1AS [56])	[67]
IEEE P802.1CBcv	FRER YANG Data Model (amending IEEE 802.1CB [58])	[68]
IEEE P802.1CBdb	FRER Extended Stream Identification Funs (amending IEEE 802.1CB [58])	[69]
	Amendments to the IEEE 802.1Q standard	
Amendment	**Description**	**Reference**
802.1Qat	Stream Reservation Protocol (SRP)	[70]
802.1Qav	Credit based Shaper	[71]
802.1Qaz	Stream Resv. Pot.	[72]
802.1Qbu	Frame Preemption	[73]
802.1Qbv	Enhancements for Scheduled Traffic	[74]
802.1Qca	Path Control	[75]
802.1Qcc	TSN Configuration	[76]
802.1Qch	Cyclic Queuing	[77]
802.1Qci	Per–stream Filtering	[78]

Table 2. *Cont.*

Standard	Description	Reference
802.1Qcp	Yang Data Model	[79]
802.1Qcr	Asynchronous Shaping	[80]
802.1Qcx	YANG Data Model for Connectivity Fault Management	[81]
	Ongoing Projects	
P802.1Qcj	Automatic Attachment to Provider Backbone Bridging (PBB) services	[82]
P802.1Qcw	YANG Data Models	[83]
P802.1Qcz	Congestion Isolation	[84]
P802.1Qdd	Resource Allocation Protocol	[85]
P802.1Qdj	Configuration Enhancements for Time-Sensitive Networking	[86]
	Amendments to the IEEE 802.3 standard	
Amendment	**Description**	**Reference**
802.3br	Interspersing Express Traffic	[87]

In the Table, the IEEE 802.3br amendment to the IEEE 802.3 standard is also reported, as the TSN preemption support requires a slight modification of the Ethernet standard. Moreover, in Table 2 are listed, among the others, several 802.1 ongoing projects, thus underlining that the TSN task group is still performing a ceaseless standardization activity. For this reason, Table 2 has not been considered exhaustive and definitive. Moreover, it is worth observing that this paper focuses on the most important standards for industrial measurement applications, and does not address all the aforementioned standards. This wide range of mechanisms and protocols offered by TSN, comprehensively aiming to reduce frame loss, synchronize stations among each other, provide bounded latency and high reliability [76], and need to be precisely configured in each bridge of the considered network, to meet specific QoS requirements.

4.1. Network Architecture and Configuration

Smart and distributed measurement systems foresee to send measurement data from a *talker* to several *Listeners*, through a proper network. IEEE 802.1Q [53] standard defines the *Bridged Network* providing structures, protocols and services to connect different LANs by means of *bridges*. Several unidirectional flows of frames called *streams*, are transferred between *end-stations*, such that the role of "talker" and "listener" is assigned to an end-station basing on the specific stream. Indeed, a specific end-station could be a talker for the i-th stream and a listener for the j-th one. A network structure example is provided in Figure 5.

In Figure 5, two data streams are considered, the red and the light blue one. It is worth observing that End Station ES_3 receives frames within the light blue stream and transmit data by means of the red one, being, respectively, both a *Listener* and a *Talker*. Furthermore, the standard comprises both MAC and VLAN bridges, the latter one allowing, by means of meaningful tags, to logically split the whole network into different Virtual LANs. This logical partition enhances the capability of the network, giving the possibility to properly limit and filter the traffic between different VLANs while allowing an unrestricted data flow within a specific VLAN. As TSN is composed by several mechanisms to handle time-critical traffic, each bridge in the network must be properly configured, basing on the Quality of Service (QoS) requirements of the specific stream. The IEEE 802.1 Qcc amendment [76] (Otherwise noted, this document has to be considered the reference for this section), in Clause 46, addresses the configuration process of a Time-Sensitive Network. This amendment, by providing mechanisms to specifically configure the TSN network, gives for the first time a vision of TSN as a well-structured and defined network. Indeed, it may be of interest to observe that the acronym "TSN" is introduced in IEEE 802.1Q by the Qcc amendment. Meaningful configuration information, containing requirements for a specific stream, are conveyed from talkers and listeners (in this context generally referred as *users* of a stream) to the bridges forming the *network* in charge of transmitting the frames within the stream. An interface, namely the User/Network Interface (UNI), manages the transmission of configuration information between users and network, introducing a certain degree of abstraction between the two parts. This data exchange is bidirectional: the join or leave requests from users, respectively configuring and releasing communication resources for the stream, are followed by the status responses from the network. There are different ways to manage the configuration information, correspondingly to three different models: the *fully distributed*, *centralized network/distributed user* and *fully centralized* ones. The first two methods foresee that the talker and listeners convey configuration information to the network, in the first case directly to the bridges, and in the second one through the nearest bridge, to a Centralized Network Configuration (CNC) device. Conversely, in the fully centralized approach, a Centralized User Configuration (CUC) entity establishes the time-sensitive requirements based on user's information, and communicates them to the CNC. A complete schematic representing a *fully centralized* architecture is shown in Figure 5. As can be seen, using this architecture, both talkers and listeners convey the stream's management information to the Centralized Unit CUC through the orange dashed lines (the purple and green dashed lines must not be considered in the fully centralized architecture) and the CUC properly informs the CNC. On the other hand, by removing all the orange elements, it allows us to obtain the *centralized network/distributed user* model, where the user's information is conveyed to the CNC by means of the purple and green dashed lines. The CNC, where present, properly manages the streams, scheduling frames in all the bridges of the network, basing on the UNI information. Centralized configuration model allows to run computationally complex configuration mechanisms in centralized entities rather than in all the bridges and to handle single streams requirements with a comprehensive vision of the network and the user's requirements. The latter feature covers a fundamental importance considering time-critical applications. The *fully distributed* model is obtained in Figure 5 removing both the orange and green elements: the management information are conveyed by users to the bridges placed at the network boundaries (purple dashed lines) and from there to the whole network. Within the talker parameters set, besides identification, stream, data frame and management information, the traffic parameter set contains QoS indications such as the maximum allowed jitter (that has an impact on the needed synchronization performances), latency and redundancy (that specifies the number of trees to generate for the specific stream) to cite only a few.

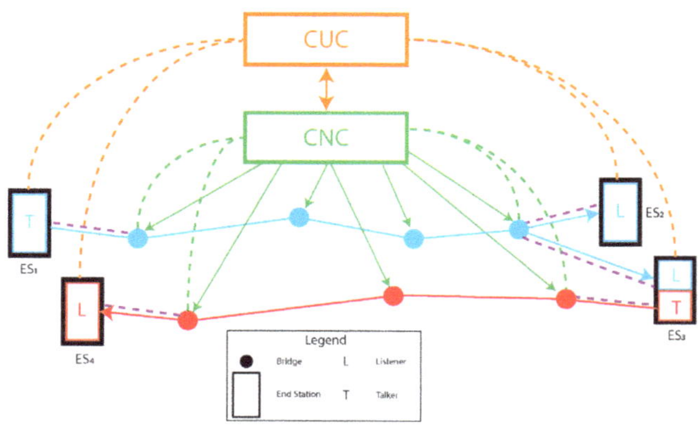

Figure 5. A simple network example.

The configuration capabilities of TSN are attracting much research interest, with several solutions already offered in the literature. Both the authors of [88,89], taking advantage from the freedom given by the standard on the choice of the communication protocol between end station and the CUC, suggest the usage of OPC-UA solutions. In particular, they propose the usage of a fully centralized model where end stations communicate the *join* message to the CUC through a OPC-UA network, the CUC conveys the stream's requirements to the CNC that manages the bridges and then transmits back the status information through the CUC to the end stations. Furthermore, in [89], an interesting TSN architecture is used to enable fog computing. Additionally, authors of [90] propose a solution to configure a multiple-domain TSN network and in [91] a learning-based self-configuration mechanism is developed to automatically reconfigure a TSN network basing on proper traffic measurements. In this regard, recently, automated configuration mechanisms and tools seem to attract interest from the research community, since they allow a seamless on-the-fly reconfiguration of dynamic TSN networks. For example, the authors in [92] retrieve the optimal network configuration by analyzing traffic in the edge switches. In this way, traffic requirements are extracted and forwarded to the CNC, which in turn properly configure the network, allowing a fast response to varying demands. Similarly, in [93], a "knowledge base entity" directly communicates with network entities using the NETCONF Event Notifications protocol obtaining devices' configuration and capabilities. In case of network changes, the knowledge base entity is automatically notified. Based on the stored information, the CNC elaborates the appropriate configuration. As a matter of fact, this standard covers a fundamental importance to suitably configure the sensor network, under both time and measurement strict requirements.

4.2. *Synchronization*

The aforementioned needs for a deterministic communication in modern distributed systems requires an accurate time measurement carried out with subsequent timestamps. For this reason, all the devices in the network need to share a *common notion of time* [94], in other terms they need to be accurately synchronized, especially when carrying measurement data [95]. The TSN synchronization standard, IEEE 802.1AS [56] (In this section when referring to IEEE 802.1AS capabilities, otherwise noted, this document has to be considered as the reference), specifies different media-dependent features, in Clause 10, 11, 12, and 13. In this section, Full-Duplex Ethernet LANs are considered (Clause 11), while in Section 6.2, Wi-Fi LANs are addressed (Clause 12). In this context, the synchronization protocol is based on the IEEE 1588, which is generally also known as Precision Time Protocol (PTP) [96] (In this section when referring to IEEE 1588 characteristics, otherwise noted, this document has to be considered as the reference). In particular, PTP comprises several protocols and parameters that can be used to compose flexible configurations (the so-called *profiles*) able to cope with different requirements and applications and to provide a synchronization accuracy in the order of microseconds. IEEE 802.1AS synchronization protocol, namely generalized PTP (gPTP), can be considered the *TSN profile* of PTP [97].

4.2.1. Network Time–Aware Devices

The gPTP protocol considers a network comprising several so-called *time-aware systems*, connected by a proper IEEE 802.3 full-duplex LAN. *End stations* and *bridges* forming the bridged network discussed in Section 4.1 may be considered as time-aware stations in the 802.1AS standard, and they correspond, respectively, to IEEE 1588 ordinary and boundary clocks. Stations that are not able to run the gPTP algorithm, called *ordinary stations*, are not involved in the synchronization process. The network presents a hierarchical logical structure where a root station namely GranMaster (GM) is used as a clock reference. The timing information is then communicated from the GM to the whole Time Sensitive Network. The so-called *synchronization spanning tree* is generated using the *Best Master Clock Algorithm* (BMCA), since the commonly used IEEE 802.1D [98] spanning tree generated by the Rapid Spanning Tree Protocol (RSTP), which is also encompassed by the IEEE 802.1Q [53] specification, is often considered sub-optimal for synchronization purposes. Indeed, RSTP is used to both provide redundancy while avoiding *logical loops* in the network.

It logically defines an *active topology* to be used as the default one, and an alternative path when a fault is detected [99]. Redundancy is then discussed in Section 4.7, but bases its behavior on the specific Spanning Tree Protocol employed. The Spanning Tree, generated by BMCA, avoids the *cyclic forwarding* of the *timing* messages, in agreement with the IEEE 1588 specification. End stations, for example, sensors and actuators in an industrial network, are modeled as the so-called *ordinary clocks* in the IEEE 1588-2008 standard. Ordinary clocks are devices characterized by a single port from which they will receive both the timing and regular messages, respectively, from an *event interface* and a *general interface*. A *local clock*, whose characteristics are addressed in Appendix B of the standard, is used as a source of time and, accordingly to the PTP protocol, has to be synchronized with the GM clock. Finally, some blocks built to run specific functions need to be mentioned, such as the *Timestamp Generation block* (linked only with the *event interface*) and the *PTP protocol engine*. PTP boundary clocks, instead, may be used to properly model the gPTP bridges. The latter device typology differs from the first one only for the presence of multiple ports, each of them comprising both the event and general interface. Obviously, one port is used for input message and the others for output ones. In the following, the synchronization process is analyzed.

4.2.2. The Synchronization Process

Each time-aware station in the network comprises a local clock, to properly timestamp the needed timing information. Unfortunately, different clocks may present both syntonization and synchronization problems, i.e., the associated square waves may have different frequencies and phases, respectively. The PTP aim is to communicate to all the stations a meaningful timing information, from which it is possible to syntonize and synchronize all the attached clocks. Consider the *i*-th station in the spanning tree. The timing information is communicated from the *i-1*-th to the considered station by a *Sync* and eventually a *Follow-Up* message, as represented in Figure 6.

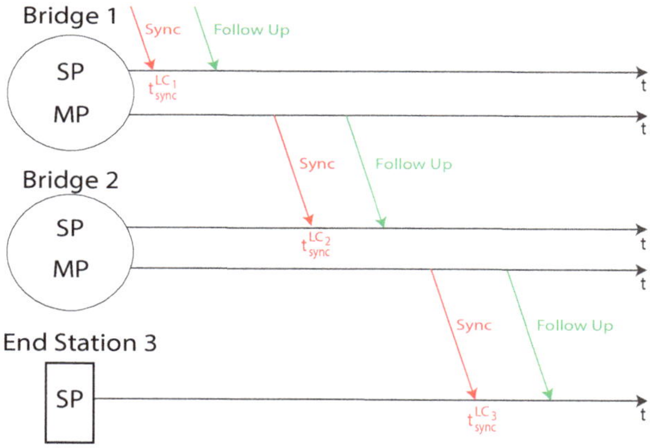

Figure 6. The gPTP synchronization activity.

The BMCE algorithm gives to each port within a time-aware device a specific role, namely Master Port (MP), Slave Port (SP), Passive Port (PP) and Disabled Port (DP). As can be seen in Figure 6, a MP is a port within a bridge or the GrandMaster enabled to send or forward timing information. In contrast, a SP is a port within bridges and end stations enabled only to receive timing information. PPs are ports that can potentially be elected GrandMaster, but that has been set to a wait state because in the network there is a better quality or higher priority master. Finally, DPs are ports that do not participate to the synchronization process. They discard all PTP messages, except for management ones Both the syntonization and synchronization activities can be carried out by means of two parameters, as specified by the IEEE 1588 document. When the *i-th* station receives the *Sync* message, a timestamp is generated and the $t_{sync}^{LC_i}$ time in the local clock time base is measured. Then, from the timing information contained in both the *Sync* and *Follow-Up* messages, it is possible to calculate the exact t_{sync}^{GM} in the GM clock time base. The synchronization offset could be calculated as per Equation (2).

$$syncOffset_{LC_i} = t_{sync}^{GM} - t_{sync}^{LC_i} \tag{2}$$

Considering N different timing transmissions, from 1 to N, it is also possible to calculate the ratio between the GM and local clock frequencies, as per Equation (3).

$$freqRatio_{LC} = \frac{t_{sync,N}^{LC_i} - t_{sync,1}^{LC_i}}{t_{sync,N}^{GM} - t_{sync,1}^{GM}} \tag{3}$$

In accordance with the IEEE 1588 standard, from the two aforementioned parameters, it is then possible to correctly synchronize the clock (the practical mechanism to perform this operation is out of the scope of the standard). The GrandMaster timing information has to be communicated, through the *spanning tree*, to all the time-aware devices within the gPTP domain. All the stations performs the aforementioned synchronization and all the *bridges* transmit the timing information to the subsequent stations. The communication of the timing information through the *spanning tree* introduces two kind of delays, the *propagation* and *residence* one. The first one is related to the time needed to send a message between a station through all the links, while the second one is the latency introduced by each bridge on the network. Each station is going to evaluate the propagation delay on all the links connecting the considered device to other ones. In such a way, for each link L connecting two stations A and B, the propagation delay is measured twice and both A and B are aware of the propagation delay. In this way, the synchronization algorithm can be run in both directions. The propagation delay measurement is carried out with the usage of the peer delay measurement mechanism, specified by IEEE 1588–2008. Consider a station A measuring the propagation delay in the link $L_{A \to B}$ connecting A with B, the synchronization messages exchange is represented in Figure 7.

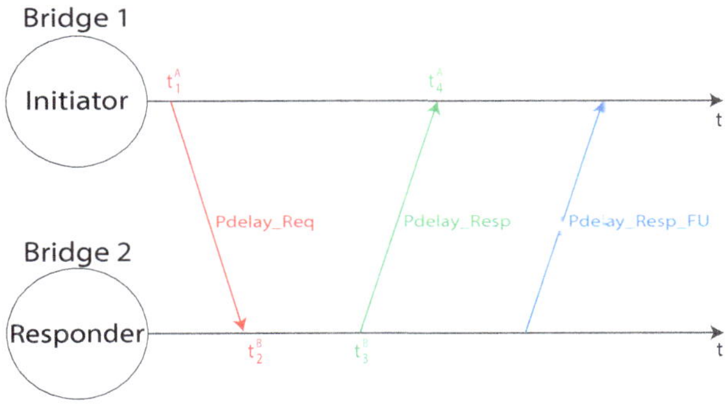

Figure 7. The propagation delay measurement: messages exchanged between two stations, A and B.

Station A starts the communication, sending a *Pdelay_Req* message at a specific timestamped time t_1^A, that is received by station B at the timestamped instant of time t_2^B. Station B sends a response message to A, *Pdelay_Resp*, at the timestamped time t_3^B, received by A at the time t_4^A. Subsequently, a *Pdelay_Resp_Follow_Up* message is sent from B to A, containing the t_3^B time stamp. Station A is now aware of all four timestamps taken: under

the assumption that the local clock frequencies, namely $f_A = f_B$, of the two stations is th
same, it is possible to calculate the propagation delay between A and B using Equation (4

$$d_{prop,AB} = \frac{(t_2^B - t_1^A) + (t_4^A - t_3^B)}{2} = \frac{(t_4^A - t_1^A) - (t_3^B - t_2^B)}{2} \quad (4$$

It is worth observing that, as the two stations have to still be considered not synchrc nized, the correspondent clocks may have different frequencies and phases. The phas issue is already solved, as Equation (4) foresees to calculate differences in the same tim base. For this reason, the phase shifts cancel each other out. Furthermore, as in genera $f_A \neq f_B$, Equation (4) needs to be properly modified by converting the timestamps taker by station B in the device' A local time base, as per Equation (5).

$$t_3^A - t_2^A = (t_3^B - t_2^B) * RR_{A \to B} \quad (5$$

It is worth noting that in Equation (5), $RR_{A\to B}$ represents the ratio between frequenc of station B local clock and station A one. As a last consideration, in general the transmissior time is not symmetrical, i.e., the delay from A to B is not exactly equal to the B to A one. In such a situation, the obtained value needs to be properly modified with the sc called *delayAssimetry* value. Both IEEE 802.1AS and IEEE 1588-2008 standards include a non-mandatory procedure to handle this issue, which is described in Clause 8.3 of [56 Furthermore, the residence delay is simply calculated by a *bridge*, time stamping both th reception of the timing message from the previous station and the transmission of th synchronization message from the specific Master Port.

The Follow-Up message contains several parameters useful to calculate the t_{sync}^{GM} ir Equations (2) and (3). Referring to a generic *i-1*-th station transmitting to the *i*-th device th timings information, the Follow-up message contains:

1. The preciseOriginTimeStamp, t_{origin}^{GM}, expressed in the GM timebase containing th timestamp originally created by the GM.
2. The correctionfield$_{i-1}$, d_{i-1}^{GM}, containing the total delay introduced from the generatior of t_{origin}^{GM}. This field is the sum of all the propagation delays introduced by the link used to convey the message before the considered stations and of all the residenc times introduced by the bridges used to convey the timing information before th considered station. This parameter is expressed in the GrandMaster time base.
3. The rate ratio RR_{i-1} between the the GM frequency and the *i-1*-th device.

After the reception of the timing messages each station can compute the t_{sync}^{GM} value t be used in Equations (2) and (3) as in Equation (6).

$$t_{sync}^{GM} = t_{origin}^{GM} + d_{i-1}^{GM} \quad (6$$

In Equation (6), for simplicity, the time bases in which the measurements are taker are not considered. It is worth noting that the transformation of a timing measuremen in a different time base can be carried out by multiplying or dividing timestamps by Rat Ratios between neighbors clock frequencies. If the current station *i* is a bridge, it compute the d_i^{GM} for each Master Port adding the residence time and the MP-specific propagatior time to the d_{i-1}^{GM} value.

The synchronization protocol performances have been evaluated in several work For example, Ref. [100] offers a comprehensive analysis targeted for an industrial scenaric carried out by a meaningful simulation assessment, that also take into account the PHYsica Jitter. Authors identified as a key parameter the *synchronization precision* (*SP*), defined as th maximum time difference between the time-aware systems local clocks and the GrandMas ter's one. Furthermore, moving from the assumption that within the industrial scenaric $SP \leq 1$ µs, they demonstrate that this condition can be surely met considering time-awar systems approximately placed between 1 and 30 hops away from the GrandMaster. Othe relevant works, targeted for different scenarios, are for example [101,102]. In conclusior as the measurement process is time-critical, the synchronization standard of TSN covers a great importance. In particular, as already stated in Section 2, the network must handle deterministic communication, thus reviling the need for precisely synchronized stations.

4.3. The Resource Reservation Capabilities of TSN

The early days of TSN within the IEEE 802.1Q standards date back to the 2009, when the IEEE 802.1Qav [71] amendment was approved. A peculiarity of this document is the introduction of the notions of Latency, Time-Sensitive Stream, Stream Reservation (SR) and Audio Video Traffic within the list of definitions at the beginning of the IEEE 802.1Q standard. According to [71], latency is defined as the propagation delay between two points of a network, where it is possible to take proper time-stamps. Time-sensitive streams are groups of frames for which the experienced latency needs to be bounded. An efficient mechanism to handle such time-aware data transmission is to split the streams in different traffic classes and provide bandwidth reservation for the time-critical ones, namely Stream Reservation (SR) classes. In conclusion, a bridge port supports from 1 to 8 queues, referring to different traffic classes and the standard defines as *forwarding process* as the ordered sequence of operations necessary to choose the frame to send in a specific instant. Figure 8 represents the queuing and forwarding process of IEEE 802.1Q, where it is possible to understand the relation between the different standards and mechanisms addressed in this section.

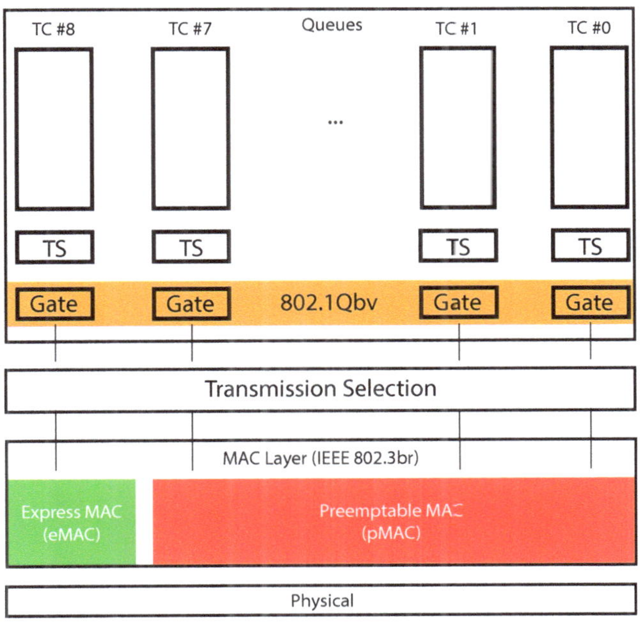

Figure 8. The queuing and forwarding process within IEEE 802.1Q.

The TSN working group gives a great importance to the forwarding process, that is comprehensively addressed in different standards. Indeed, it gives the possibility to design smart measurement systems with different data flows, characterized by different deadlines and priorities, thus allowing us to handle different uncertainty levels depending on the specific application. From Figure 8, it is worth observing that within the different Traffic Classes (TCs) a per-queue Transmission Selection (TS) algorithm is run to choose the specific frame to send. Afterwards, the IEEE802.1Qbv [74] standard defines what set of Traffic Queues can send data, by suitably opening a specific group of gates. Afterwards, an inter-queue TS algorithm allows to choose what frame to send. The last queue data can also preempt the transmission of the lower-priority queues by a specific mechanism described in Section 4.4. By using such a complex scheduling policy, it is possible to handle different traffic classes with a large variety of different requirements, thus allowing us to give to critical measurements a higher priority.

4.3.1. The Stream Reservation Protocol

The calculation of the amount of bandwidth reserved for each class can be performed by the Stream Reservation Protocol, now part of the IEEE 802.1Q standard [53] in clause 35. The original version dates back to 2010 and was outlined in the 802.1Qat [70] amendment, but some modifications are introduced by the Qcc [76] standard to enhance the performances of the algorithm and to adapt the Stream Reservation Protocol (SRP) to the new centralized approaches. The SRP protocol, basing on the *talker* and *listener* requirements, provide resource reservation in each bridge within the network path of the specific stream, with the aim to meet the QoS requirements. Afterwards, proper messages are sent to the end stations (both talkers and listeners) to inform on the result of the reservation activity, either successful or failed. It is worth observing that if required resources are correctly assigned to a stream, the transmission of its frame is guaranteed by each bridge within the network. As a last consideration, in order to handle emergency communication, various relevance levels are associated to the streams so that a bridge is allowed to give major priority to the most relevant streams.

4.3.2. The Transmission Selection Algorithms

The standard defines three different transmission policies: the strict priority algorithm, the Credit Based Shaper (CBS) and the Enhanced Transmission Protocol (ETS). CBS and ETS are described, respectively, in the Qav [71] and Qaz [72] amendments. The strict priority algorithm is the default scheduling algorithm since its implementation in bridges is mandatory. Furthermore, different algorithms can be used to generate the schedule on the condition that they are able to guarantee 802.1Q priorities requirements.

The *Credit Based Shaper* was introduced by the Qav amendment to properly provide to the SR classes the bandwidth previously determined, for example, with the usage of SRP or, in case of a fully centralized configuration model, also directly by the CNC [76]. Indeed, the frame selection following a pre-determined value of priorities (i.e., strict priority schedule), reviles the unsuitability to provide different bandwidth allocation to different traffic-classes. The CBS bases his foundation on a typical credit and debit system, where the currency are bits. Some examples, contained in the standard, are suitably represented in Figure 9.

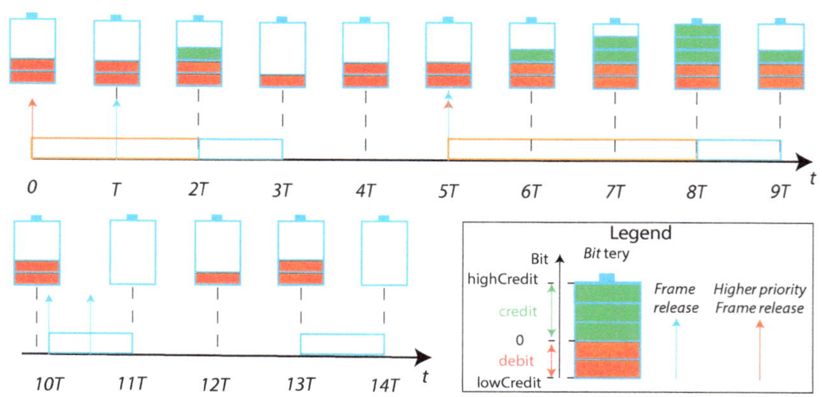

Figure 9. Credit Based Shaper principle.

The operations carried out by the CBS algorithm in Figure 9 are listed below:

1. For $0 \leq t \leq T$, the credit of a specific queue starts from 0 and maintains that level until a frame enters the related queue;
2. In $t = T$ a frame is queued but, due to the presence of the higher-priority frame, can not be transmitted immediately. For $T \leq t \leq 2T$, as the transmission of the queued frame is blocked by the higher priority one, the queue accumulates credit;
3. For $2T \leq t \leq 3T$, the frame is transmitted and the credit level decreases;
4. For $3T \leq t \leq 5T$, as the queue is indebted, also if no frame is queued the credit increases until reaches the null value;
5. For $5T \leq t \leq 8T$, as a frame is blocked by a higher priority transmission, the credit level reaches the maximum value;
6. For $8T \leq t \leq 9T$, the transmission of a frame decreases the credit. The remaining credit is positive, but no frame is queued so exactly after the instant $t = 9T$ the credit is restored to zero;
7. For $10T \leq t \leq 14T$, it is possible to notice that if the queue is indebted (i.e., the credit is negative) it is not possible to start a new frame transmission, and it is needed to wait until credit becomes non-negative.

A maximum indebtedness level is fixed, to give the possibility to the queue to send an entire frame also starting from a null credit. Vice versa, the queue stores credit when a higher priority class queue prevents the frame transmission, to be used for more than one consecutive frame transmission when the line becomes free. The algorithm need to be properly configured by tuning the rates at which the credit decreases during transmission and increases when blocked by higher priorities queues, respectively, denoted as *sendslope* and *idleslop*. Generally specking, these two values are different, as it possible to see in Figure 9 by comparing the time line with the relative *bittery* levels. It is possible to prove that the idleslope divided by the total transmission rate of the port, is the bandwidth fraction used by the queue [71]. For this reason, the *idleslope* has to be previously determined, for each supported queue, for example by means of the aforementioned SRP protocol.

As a last consideration, the IEEE 802.1Qcr [80] standard, needs to be mentioned. This standard foresees the inclusion of a different shaper, the Asynchronous Traffic Shaper (ATS). An interesting work addressing the shaping activity of TSN, can be found in [103]. Indeed, the authors firstly perform a theoretical evaluation of the delay bounds and secondly, by means of a meaningful case study, they demonstrate the tightness of the delay bounds already introduced.

4.4. Frame Preemption and Interspersing Express Traffic (IET)

The IEEE 802.1Qbu [73] is an amendment to the IEEE 802.1Q [53] standard, whose last version was developed in 2016 and it was received by IEEE 802.1Q in 2018. The amendment's aim is to support the IEEE 802.3br [87] (The original version of the standard [104] was developed in 2016 and was included in the Ethernet Standard [105] in 2018) Interspersing Express Traffic, that allows the *preemption* (i.e., the suspension of the transmission of) the ordinary traffic, to transmit the time-critical frames. This feature is surely important, since it allows us to give an higher priority to the time-sensitive frames, while guaranteeing the transmission of both time-critical and non-time-critical traffic. IEEE 802.3br comprises two different typologies of frames, the time critical (namely, Express) traffic and the preemptable one. The provision of IET allow a further step forward: a new MAC layer mechanism is introduced to temporary mark the completion of a frame that has been forcibly preempted. In this way, preempted frames are not lost, since the transmission of the remaining part can be completed in a later moment when the transmission medium is free from express traffic. A meaningful example is presented in Figure 10.

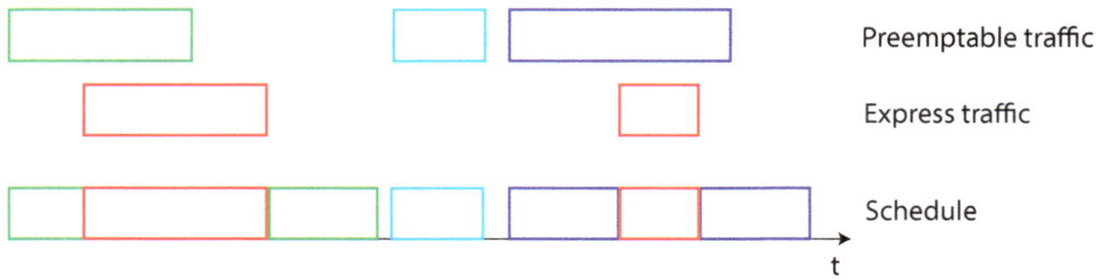

Figure 10. Express and preemptable traffic: an example.

The time-critical (or IET) frames, represented by red squares in Figure 10, are scheduled exactly when the transmission request is made (the bridge knows in advance their activation instant) and the communication activity is not subjected to interruptions. In such a way the real-time behavior of this kind of traffic is enhanced. Conversely, the preemptable traffic during a time-critical transmission need to suspended from the communication. The preemptable frame is then resumed when no express traffic is present, as illustrated in Figure 10 for both the green and purple frames. When frame preemption is not available the non-time-critical frames (for example the green one) will be delayed because the empty spaces between IET frames are not sufficient to accommodate for their transmission. Conversely, if frame preemption is available both at bridge (802.1Qbu) and at devices (802.3br), non-IET frames can be preempted, and the different chunks can fill the gaps. The relation between the IEEE 802.1Qbu and IEEE 802.3br amendments is represented in Figure 8, where it is possible to notice that two different MAC layers are introduced, eMAC and pMAC, to handle, respectively, express and preemptable traffic. The effect of the preemption capability was evaluated in several works [106–110]. Conversely, authors of [111] underlined the importance to study the impact of the preemption activity also on the delay introduced in the Best Effort (BE) traffic communication. Indeed, also the Best Effort traffic, conveying for example diagnostic or configuration messages, need to be properly exchanged. The results obtained in such a work reviled interesting, as the preemption activity allowed to exchange messages also for low ST traffic periods. Clearly when the ST traffic period increases the delay introduced in the BE traffic communication becomes lower.

4.5. *Enhancements for Scheduled Traffic*

The IEEE 802.1Qbv [74], developed in 2015, provides a mechanism to improve determinism in Time-Sensitive Networks. A system of queue-specific gates regulates the possibility to selectively send frames ready for the transmission from specific queues. In particular, a gate can be in two different states, namely *opened* and *closed*, respectively, allowing or denying the possibility to transmit a frame belonging to the specific queue. Within each queue with an opened gate, a specific scheduling algorithm is run to decide which frame of the queue will be sent. Furthermore, a precise scheduling of the time instants when to change the gate states must be performed. The latter problem can be formalized by a set of linear inequalities [112], which lead, especially on large networks, to computationally heavy problems as addressed by the authors in [113]. Some meaningful simulation results can be derived from [114], which show that using the enhancements for scheduled traffic it is possible to effectively bound the latency of time-critical classes.

4.6. *Cycling Queuing and Forwarding*

The aforementioned mechanisms used to manage the *forwarding process*, such as the Credit Based Shaper, the preemption and the Enhancements for scheduled traffic contributes to reduce and bound the latency. Furthermore, Cyclic Queuing and Forwarding (CQF), addressed in the IEEE 802.1Qch amendment [77], contributes to make the latency bounded and predictable. The main contribution of this document within the 802.1Q standard [53] is given by annex T, where CQF is explained. The basic principle of CQF is illustrated in Figure 11.

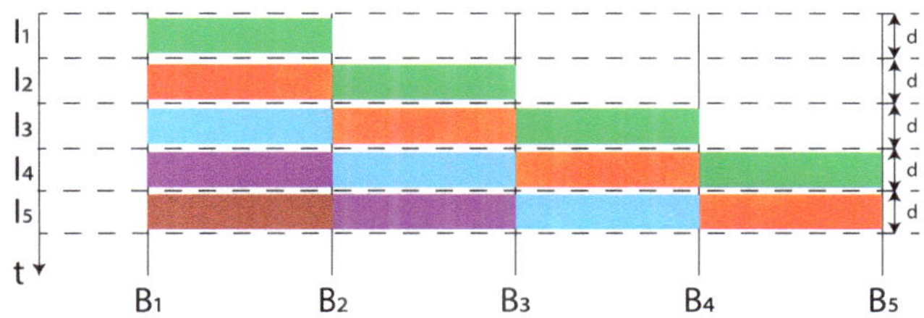

Figure 11. CQF principle.

The time is divided in intervals of duration d, namely $I_1, I_2, I_3, \ldots$ and each bridge B_j in the network sends frames received from B_{j-1} during I_i to B_{j+1} during I_{i+1}. For example, consider the green frames communication between B_1 and B_4. In a worst-case situation a frame is sent by B_1 at the beginning of I_1 and received by B_4 at the end of I_3. In the best situation, a frame is sent by B_1 at the end of I_1 and received by B_4 at the beginning of I_3. In conclusion, the latency introduced by CQF between B_1 and B_4 is expressed by Equation (7).

$$d \leq L_{1\to4} \leq 3 \cdot d \tag{7}$$

As a further example, latency introduced by a B_1 to B_5 frame transmission is expressed by Equation (8).

$$2 \cdot d \leq L_{1\to5} \leq 4 \cdot d \tag{8}$$

Summarizing, in consideration of the number of hops in the two previous examples, respectively $h_{1\to4} = 2$ and $h_{1\to5} = 3$, it is possible to generalize the previous relations, obtaining the result in Equation (9).

$$(h-1)d \leq L_h \leq (h+1)d \tag{9}$$

In Equation (9) h is the number of hops and L_h is the latency introduced for the transmission of the frames when the path is characterized by h hops. It is worth observing that the latency calculated by Equation (9) permits to pre-determine the h and d dependent latency value introduced by CQF, so that it is proved that CQF is deterministic.

4.7. Frame Replication and Elimination for Reliability (FRER)

Redundancy is traditionally considered a good methodology to increase the reliability of the communication. Several algorithms were developed over the years, such as the Rapid Spanning Tree Protocol [98], or the Media Redundancy Protocol (MRP), commonly based on the usage of an alternative path if a failure is detected on the default one [115]. Unfortunately, the latter ones foresee the introduction of a delay between the fault detection and the sending instant of the packet, so that others algorithms were developed to provide *seamless redundancy*. With the aim of standardization, TSN introduces the IEEE 802.1CB [58] (This document has to be considered as the reference for this section, otherwise noted.) standard, that comprises several functions, also known as Frame Replication and Elimination for Reliability (FRER), cooperating to replicate the packets and send them through different paths to the receiver. After the reception of the packets, extra copies are eliminated, introducing a *seamless redundancy*. Such an approach is considered fundamental for the Time Sensitive Networks in order to guarantee the reception of critical data also in case of equipment failure, providing low packet loss. Several *member streams*, conveying duplicated packets through different paths, are then created for aim of redundancy, whose combination forms the so-called *Compound Stream*. An example is provided in Figure 12, where it is supposed that multiple paths can be used for the red stream of Figure 5 transmission.

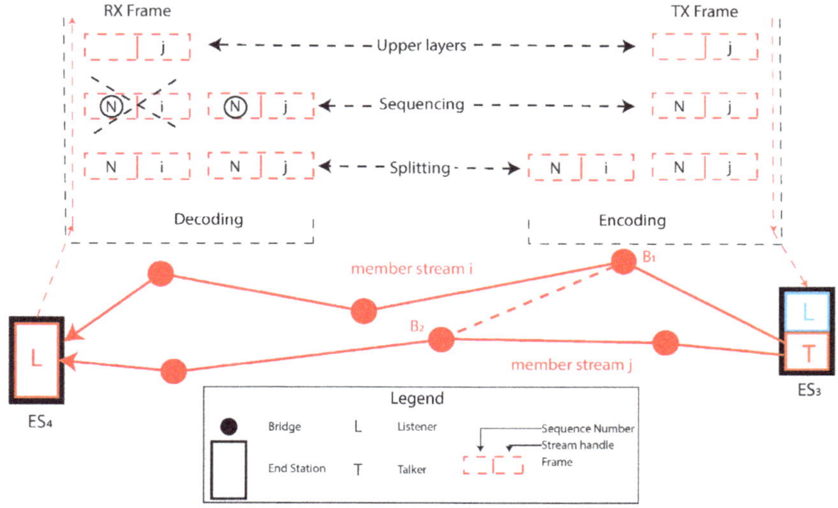

Figure 12. FRER example.

In such a situation, two member streams, i and j, are created between ES_3 and ES_4. FRER makes use of paths already created, for example, by means of the IEEE 802.1Qca [75] standard. Authors of [115] particularly describe the TSN approach, where features of the stream reservation (IEEE 802.1Qca [75]), configuration (IEEE 802.1Qcc [76]) and FRER standards strongly cooperates to provide redundancy. Furthermore, they qualitative compare the TSN approach with a total different methodology, based on the decoupling of the stream reservation and redundancy protocols. The main conclusion they draw, among others, is that FRER introduce advantages on the protocol overhead and bandwidth utilization, while introduces a lower flexibility. Furthermore, the algorithm can also replicate and eliminate frames in bridges within the network between Talker and Listeners. Consider the possibility to connect bridges B_1 and B_2 in Figure 12. In this situation, it is possible to also split frames in bridge B_1 and eliminate copies in bridge B_2, in order to make the packet loss even lower. Besides that, the FRER activity is managed with the usage of several functions, deeply analyzed in clause 7 of [58]. How each function behaves is clearly out of the scope of this article, but some topics useful to understand the general behavior of the FRER are now analyzed. Some of the activities of these functions are summarized on the top part of Figure 12. The so-called Stream Identification Function (SIF), addressed by Clause 6 of the standard, performs a key activity in the FRER context. This function is built on top of the MAC Layer, using one Service Access Point (SAP) to communicate packets to the lower layers (i.e., MAC and Physical) and several SAPs, serving different packet streams, to transmit packets through the layers above. In particular, the function uses the Internal Sublayer Service (ISS) specified by the IEEE 802.1AC [116] layering standard. ISS comprises two different primitives offered by the MAC layer, the *indication* and the *request* one, respectively, referring to the reception of a frame from the lower layers and the request of a frame transmission from the upper layers. In each primitive data set is present a *connection_identifier* parameter which, in turn, comprises two parameters, namely *stream_handle* and *sequence number*. The first one identifies the packet stream, while the second one identifies the packet sequence order. Both parameters are encapsulated by the FRER into the connection_identifier for internal use (they are not directly transmitted to the lower layers).

When the SIF function receives a packet, it identifies the stream and forwards the packet to the upper layer via the specific SAP if the stream is known. Otherwise, if the stream is unknown, the packet is handled by a specific SAP that serves the unknown streams. An interesting usage of this function is in the bridged network [53]. It is worth observing that the identification function comprises a Lower Identification and an upper one. In the first stage, the packets not belonging to a known stream are identified and transmitted to the peer device through a Non-Stream Transfer Function (NSTF). In the second stage, the proper SAP is identified and the FRER algorithm is used to convey the packets. On the top of the SIF is built the Sequence Encode/Decode Function (SEDF), that, with the usage of the *connection_identifier*'s *sequence_number* sub-parameter, decodes an incoming packet from the lower layer allowing the *Recovery Function* to discard extra packets. Conversely, when a transmission request is generated, SEDF encodes the packet sequence number in a frame to be transmitted through the underlying LAN. The latter activity is of fundamental importance to allow the peer station decoding operation and usually it is done adding an R-TAG in the transmitted frame containing both the stream and the packet number. Considering the frames to be transmitted, before the encapsulation activity, they are managed by the Sequencing Function, that assigns them a specific *sequence_number* and the Stream Splitting Function that replicates the packet assigning to each copy a specific stream_handle value. Additionally, it is to recall the presence of the so-called Latent Error Detection Function. The aim of this function is to trigger an event when some extra packets are not received, in order to signal an equipment failure on a specific path, that can be opportunely managed. For simplicity, some functions are not represented in Figure 12, such as SIF (that is placed right under the SEDF function) and the individual recovery function that performs a per-stream elimination activity. One drawback of FRER is the limited amount of available parameters, which are also strictly tied with the specific upper layer protocol, provided to identifies a stream. The ongoing project P802.1CBdb [69], known as FRER Extended Stream Identification Functions, overcome these problems by introducing a new set of parameters which are independent from the upper layer protocol in use.

Several articles in the literature highlight some of the limitations of the FRER algorithm. Authors in [117] pointed out that the arbitrary replication of all the packets may result in an inefficient network, suggesting the usage of a Machine Learning based algorithm for fault detection. In this way, a failure can be predicted and redundancy established just before the fault occurs. Furthermore, an interesting article [118] provides a critical overview of FRER, underlying some relevant limitations and challenges. Among others, it is worth observing that usually the Shortest Path Tree (SPT) is used as the default one. Then, the IEEE 802.1Qca standard [75] allows the usage of longer paths for aim of redundancy. This leads to different communication times through different paths, and a possible out-of-order communication. Indeed, the authors pointed out the need for a worst-case analysis of the algorithm. Moreover, authors of [26] demonstrated with a simple example, the non-composability of FRER with End-to-End (E2E) mechanisms. Finally, authors of [119] carried out an analysis on the performances of the FRER algorithm, evaluating the interval of time between the reception of the packet and its first copy.

5. The TSN Profile for Industrial Automation

Within the TSN standards, it is possible to create several configurations, called *profiles*, to adapt the network behavior to requirements coming from different fields of applications. The TSN working group, at present, is working on several profiles, listed in Table 3. Among them, the industrial automation profile is the most relevant to this analysis. It also targets the needs of the Instrumentation and Measurement field, since it has major knock-on effects in every aspect of the industrial scenario, not only revolutionizing real-time communications, but also the way of conceiving industrial devices and distributed measurement systems.

Table 3. TSN profiles.

Description	Standard	Reference
Audio Video Bridging (AVB) systems	IEEE Std 802.1BA	[120]
Time-Sensitive Networking for Fronthaul	IEEE 802.1CM	[121]
Ongoing Projects		
Industrial Automation	IEEE/IEC 60802	[122]
TSN Profile for Service Provider Networks	IEEE P802.1DF	[123]
TSN Profile for Automotive	IEEE P802.1 DG	[124]
TSN for Aerospace Onboard Ethernet Communications	IEEE P802.1 DP	[125]

The TSN profile for Industrial Automation (TSN-IA) aims to provide guidelines fo the configuration of TSN to meet Industrial Automation requirements. The Industria Automation *use cases* are analyzed in a specific document [126] and the IEEE/IEC join project 60802 is currently working on the aforementioned profile to cope with the specifie use cases. While the draft standard is not publicly available, some information can b inferred from the documents found on the WG website [122]. For instance, significan attention is given to synchronization and timing issues related to the IEEE 802.1AS standar to Energy Efficient Ethernet (EEE) capabilities [127] and to the new queuing and fram preemption options. The [126] document makes a list of the industrial traffic typologie that are briefly summarized in Table 4.

Table 4. Industrial traffic typologies.

Traffic Typology	Characteristics					
	Periodic	Sporadic	Deadline	Bandwidth	Bounded Latency	Priority
Isochronous cyclic real-time	X		X	X	X	
Cyclic real-time	X		X	X	X	
Network Control		X				X
Audio/Video	X			X	X	
Brownfield	X			X	X	
Alarms/Events		X		X	X	
Configuration/Diagnostic		X		X		
Internal/pass-through		X		X		
Best-Effort		X				

In the last part of the TSN-IA profile specifications, it is also possible to find a detaile analysis of the required functions for an industrial network. Here, the standard take into account some of the protocol features specified above (either mandatory or optional and specifies a fine tuning of their parameters. As a final confirmation that the standard ization activity is currently in progress, at the moment of writing, the standard covers in details the clock synchronization issues, whereas other sections have yet to be complete as for instance, the requirements for security, bridge delay, network access, etc.

6. TSN in Time–Critical, Possibly Wireless–Based, Measurement Systems

6.1. *A Representative Test Case*

The scheduling, bandwidth reservation, real-time behavior, Wi-Fi capabilities and other features of TSN, open up to interesting and advanced time–critical application wher a constant flow of information, often coming from heterogeneous sensors, is of vital im portance. An example is the scenario proposed by [128] where a swarm of quadcopters i controlled to perform maneuvers at high speed. In this application, measurements fro cameras and onboard sensors are used by a centralized control system to determine th references of each individual agent so that they can move in a coordinated way. Specificall a system consisting of eight cameras acquires the position and attitude of each vehicle wit a frequency of 200 Hz. The camera frames are sent via a UDP stream to a central processing unit. Furthermore, each quadrotor is equipped with on-board sensors (accelerometer and gyroscope), the measurements are sent via an XBee–UDP bridge to the central processing unit. Here, they are processed, and each vehicle receives setpoints for coordinated motion via a PPM analog transceiver with a 50Hz refresh rate. Another communication channel is a low priority downlink for the purpose of data logging. The real-time requirements are evident since the failure to comply with a deadline or delays in the communication chain could lead to unexpected and catastrophic results. The use of different types of traffic, such as real-time and best effort, is also evident, with the separation achieved through the use of physically separate communication channels. However, the communication architecture has some limitations. To maintain a sufficiently low latency and high bandwidth, the data flow from the cameras uses UDP, which does not provide any QoS mechanism, exposing the system to potential packet losses. Using bridges to switch from UDP to other communication systems can represent an additional bottleneck. Both of these downsides are destined to become critical if the number of agents, and therefore the data flow, increases. In this context, some of TSN's features can bring benefits. For example, bandwidth reservation and traffic scheduling can be used to prioritize video streams and cyclic data for the control system. The Frame Replication and Elimination for Reliability (FRER) can be used to increase the reliability of the communication. The use of these features allow us to lower the network latency and jitter, mitigating the effects discussed in Section 2. Additionally, the intrinsic clock synchronization required by TSN brings some advantages. Often in distributed autonomous systems GPS is used for clock synchronization in agents. TSN provides further improvements by providing a shared sub-microsecond time reference to the network's nodes, which can overcome GPS's existing constraints [129]. In addition to the decrease in latency, communication times, and improve synchronization, a precise time-stamping of measured data can also be used to compensate for further delays introduced by the measurement, processing, and control chain.

6.2. TSN over Wi-Fi

The smart interconnection of several objects of the everyday life within the Internet of Things vision, envisages a massive usage of wireless communications. The test case analyzed in the previous Section represents an iconic example of time-critical application that employs wireless communication. The development of increasingly efficient wireless technologies is also becoming of fundamental importance in the factory automation scenario, to provide enhanced mobility and to provide seamless connectivity to area which are difficult to cable. Indeed, wireless communication becomes a key player in the Industry 4.0 deployment process [130], introducing several benefits such as flexibility, reduction of maintenance and installation costs, and the reduction of network failures. The aforementioned advantages also reflect in the possibility for typical industrial controllers to acquire information from sensors and send control signals to the actuators via a wireless communication system, building up the so-called Wireless Networked Control Systems (WNCS) [131–133]. Some of the research activity, in the past, focused on IEEE 802.15.4 based-networks, such as WirelessHART ones [134]. These networks, by means of the Time Division Multiple Access protocol together with a proper scheduling algorithm, (for example the *rhythmic model* suggested by the authors of [135]), are characterized by enhanced real-time capabilities. In the last years, Wi-Fi was also revealed to be promising to be applied in factory automation as, compared with the IEEE 802.15.4 solutions, it gives the possibility to cope with the timing requirements of the modern control systems and to perform a useful Rate Adaptation activity [136]. Indeed, for example, authors of [137] underlined the necessity of a minimum control frequency of 1 kHz for some specific application, not achievable by wirelessHART since it is characterized by a time slot of at least 10 ms. How to adapt emergent wireless technologies, such as 5G and Wi-Fi, to the strict requirements of the factory automation is an open research field [138–141], together with recent works concerning industrial LoRa networks [142]. Some works suggest the usage of hybrid wired/wireless networks, integrating ethernet TSN networks with both Wi-Fi [52] and 5G [143]. Actually, TSN over Wi-Fi networks are promising to adapt Wi-Fi to the stringent requirements of the industrial context. At present, the IEEE 802.11AS standard [56] specifically refers also to IEEE 802.11 LANs, providing a synchronization mechanism similar to the one analyzed in Section 4.2. Indeed, the synchronization activity over Wi-Fi is performed exactly as presented in Section 4.2, with the exception of some media-dependent activities specified in IEEE 802.1AS [56], Clause 12. In particular, how to communicate the timing messages between a Master Port and the attached Slave Port in the generated *spanning tree* is quite different with the respect to the full-duplex Point to Point links. In this case, in fact, the IEEE 802.11 [144] Timing Measurement (TM) procedure is used to calculate the propagation time. The last version of the IEEE 802.11 standard allows also to use the Fine Timing Measurement (FTM) mechanism [144]. The transposition of the other TSN features in WiFi is still an open research field.

7. Conclusions

This article provided an assessment of TSN, aimed at investigating the adoption of such a wide family of standards in the context of Instrumentation and Measurements and Industrial Automation systems. As a first achievement, a careful bibliographic analysis showed that the aforementioned fields of applications are still not adequately addressed, as clearly indicated by the limited number of scientific contributions. Moving from this consideration, the paper provided a detailed description of the TSN features that are supposed to be more suitable for the targeted applications. Then, the impact of the ever performing TSN networks and protocols on the data exchange between sensors, actuators, controllers and measurement equipment was studied.

The analysis clearly evidenced the possible benefits deriving from the adoption of TSN, with respect to the state of the art communication systems. Nonetheless, it also showed the need for a better estimation of the effect of TSN networks on the measurement uncertainty. Moreover, the possible introduction of TSN on distributed Instrumentation and Measurement systems, based on wireless communication, was addressed. Although the analysis referred to specific examples, the benefits brought by TSN appear evident, thanks to its traffic prioritizing and synchronization features, that result in more precise time-stamping of the acquired sensor data, with the consequent performance improvement of the (wireless) distributed measuring system. Finally, the assessment carried out in this paper clearly outlines some future activities. Indeed, substantial efforts are expected in the development of theoretical and/or simulation analyses to improve awareness as well as knowledge in the relevant scientific community. Furthermore, practical experiments on prototype testbeds have to be carried out. This, on the one hand, allows us to check the quality of the theoretical/simulation models, to eventually validate them. On the other hand, experimental sessions allow us to practically assess some specific issues like the effects of TSN on the measurement accuracy, as well as the impact of the TSN protocol stack on limited-resource devices such as those often used in distributed measurement systems.

Author Contributions: Conceptualization, F.T. and S.V.; methodology and investigation, T.F., A.M and F.T.; writing original draft preparation, T.F. and A.M.; writing review and editing, T.F., A.M., F.T., L.R. and S.V.; supervision, F.T., L.R. and S.V. All authors have read and agreed to the published version of the manuscript.

Funding: This research received no external funding.

Conflicts of Interest: The authors declare no conflict of interest.

References

1. Ashton, K. That 'Internet of Things' Thing. *RFID J.* **2009**, *22*, 97–114.
2. Trappey, A.J.; Trappey, C.V.; Govindarajan, U.H.; Chuang, A.C.; Sun, J.J. A review of essential standards and patent landscapes for the Internet of Things: A key enabler for Industry 4.0. *Adv. Eng. Inform.* **2017**, *33*, 208–229. [CrossRef]

. Xu, H.; Yu, W.; Griffith, D.; Golmie, N. A Survey on Industrial Internet of Things: A Cyber-Physical Systems Perspective. *IEEE Access* **2018**, *6*, 78238–78259. [CrossRef]

. Včelák, J.; Vodička, A.; Maška, M.; Mrňa, J. Smart building monitoring from structure to indoor environment. In Proceedings of the 2017 Smart City Symposium Prague (SCSP), Prague, Czech Republic, 25–26 May 2017; pp. 1–5.

. Pai, P.; Shashikala, K.L. Smart City Services—Challenges and Approach. In Proceedings of the 2019 International Conference on Machine Learning, Big Data, Cloud and Parallel Computing (COMITCon), Faridabad, India, 14–16 February 2019; pp. 553–558.

. Monteiro, K.; Rocha, E.; Silva, E.; Santos, G.L.; Santos, W.; Endo, P.T. Developing an e-Health System Based on IoT, Fog and Cloud Computing. In Proceedings of the 2018 IEEE/ACM International Conference on Utility and Cloud Computing Companion (UCC Companion), Zurich, Switzerland, 17–20 December 2018; pp. 17–18.

. Ma, J.; Feng, S.; Li, X.; Zhang, X.; Zhang, D. Research on the Internet of Things Architecture for Intelligent Passenger Transportation Services and its Application. In Proceedings of the 2019 4th International Conference on Electromechanical Control Technology and Transportation (ICECTT), Guilin, China, 26–28 April 2019; pp. 194–197.

. Trilles, S.; González-Pérez, A.; Huerta, J. An IoT Platform Based on Microservices and Serverless Paradigms for Smart Farming Purposes. *Sensors* **2020**, *20*, 2418. [CrossRef] [PubMed]

. Ooi, B.Y.; Shirmohammadi, S. The potential of IoT for instrumentation and measurement. *IEEE Instrum. Meas. Mag.* **2020**, *23*, 21–26. [CrossRef]

0. Lu, Y. Industry 4.0: A survey on technologies, applications and open research issues. *J. Ind. Inf. Integr.* **2017**, *6*, 1–10. [CrossRef]

1. Ghazivakili, M.; Demartini, C.; Zunino, C. Industrial data-collector by enabling OPC-UA standard for Industry 4.0. In Proceedings of the 2018 14th IEEE International Workshop on Factory Communication Systems (WFCS), Imperia, Italy, 13–15 June 2018; pp. 1–8.

2. Jeong, S.; Na, W.; Kim, J.; Cho, S. Internet of Things for Smart Manufacturing System: Trust Issues in Resource Allocation. *IEEE Internet Things J.* **2018**, *5*, 4418–4427. [CrossRef]

3. Lin, J.; Yu, W.; Zhang, N.; Yang, X.; Zhang, H.; Zhao, W. A Survey on Internet of Things: Architecture, Enabling Technologies, Security and Privacy, and Applications. *IEEE Internet Things J.* **2017**, *4*, 1125–1142. [CrossRef]

4. Xu, G.; Yu, W.; Griffith, D.; Golmie, N.; Moulema, P. Toward Integrating Distributed Energy Resources and Storage Devices in Smart Grid. *IEEE Internet Things J.* **2017**, *4*, 192–204. [CrossRef]

5. Ahmed, N.; De, D.; Hussain, I. Internet of Things (IoT) for Smart Precision Agriculture and Farming in Rural Areas. *IEEE Internet Things J.* **2018**, *5*, 4890–4899. [CrossRef]

6. World Economic Forum. *Fourth Industrial Revolution, Beacons of Technology and Innovation in Manufacturing*; World Economic Forum: Cologny, Switzerland, 2019.

7. World Economic Forum. *Shaping the Sustainability of Production Systems: Fourth Industrial Revolution Technologies for Competitiveness and Sustainable Growth*; World Economic Forum: Cologny, Switzerland, 2019.

8. Yavari, A.; Jayaraman, P.P.; Georgakopoulos, D.; Nepal, S. ConTaaS: An Approach to Internet-Scale Contextualisation for Developing Efficient Internet of Things Applications. In Proceedings of the Hawaii International Conference on System Sciences, Hawaii County, HI, USA, 4–7 January 2017. [CrossRef]

9. Bhadoria, R.S.; Bajpai, D. Stabilizing Sensor Data Collection for Control of Environment-Friendly Clean Technologies Using Internet of Things. *Wirel. Pers. Commun.* **2019**, *108*, 493–510. [CrossRef]

0. Daponte, P.; Lamonaca, F.; Picariello, F.; De Vito, L.; Mazzilli, G.; Tudosa, I. A Survey of Measurement Applications Based on IoT. In Proceedings of the 2018 Workshop on Metrology for Industry 4.0 and IoT, Brescia, Italy, 16–18 April 2018; pp. 1–6. [CrossRef]

1. Gao, R.X.; Wang, L.; Helu, M.; Teti, R. Big Data Analytics for Smart Factories of the Future. *CIRP Ann.* **2020**, *69*, 668–692. [CrossRef]

2. Morato, A.; Vitturi, S.; Cenedese, A.; Fadel, G.; Tramarin, F. The Fail Safe over EtherCAT (FSoE) Protocol Implemented on the IEEE 802.11 WLAN. In Proceedings of the 2019 24th IEEE International Conference on Emerging Technologies and Factory Automation (ETFA), Zaragoza, Spain, 10–13 September 2019; pp. 1163–1170. [CrossRef]

3. Peserico, G.; Morato, A.; Tramarin, F.; Vitturi, S. Functional Safety Networks and Protocols in the Industrial Internet of Things Era. *Sensors* **2021**, *21*, 6073. [CrossRef] [PubMed]

4. Heymann, S.; Stojanovci, L.; Watson, K.; Nam, S.; Song, B.; Gschossmann, H.; Schriegel, S.; Jasperneite, J. Cloud-based Plug and Work architecture of the IIC Testbed Smart Factory Web. In Proceedings of the 2018 IEEE 23rd International Conference on Emerging Technologies and Factory Automation (ETFA), Turin, Italy, 4–7 September 2018; Volume 1, pp. 187–194. [CrossRef]

5. Kobzan, T.; Schriegel, S.; Althoff, S.; Boschmann, A.; Otto, J.; Jasperneite, J. Secure and Time-sensitive Communication for Remote Process Control and Monitoring. In Proceedings of the 2018 IEEE 23rd International Conference on Emerging Technologies and Factory Automation (ETFA), Turin, Italy, 4–7 September 2018; Volume 1, pp. 1105–1108. [CrossRef]

6. Lo Bello, L.; Steiner, W. A Perspective on IEEE Time-Sensitive Networking for Industrial Communication and Automation Systems. *Proc. IEEE* **2019**, *107*, 1094–1120. [CrossRef]

7. Bruckner, D.; Stanica, M.P.; Blair, R.; Schriegel, S.; Kehrer, S.; Seewald, M.; Sauter, T. An Introduction to OPC UA TSN for Industrial Communication Systems. *Proc. IEEE* **2019**, *107*, 1121–1131. [CrossRef]

8. PTB. Metrology for the Digitalization of the Economy and Society. Available online: https://www.ptb.de/cms/fileadmin/internet/forschung_entwicklung/digitalisierung/PTB-Digitalisierungsstudie_2018_EN.pdf (accessed on 11 February 2022).

29. Cristaldi, L.; Ferrero, A.; Muscas, C.; Salicone, S.; Tinarelli, R. The effect of net latency on the uncertainty in distributed measurement systems. In Proceedings of the 19th IEEE Instrumentation and Measurement Technology Conference (IEEE Cat. No.00CH37276), Anchorage, AK, USA, 21–23 May 2002; Volume 2, pp. 1265–1269. [CrossRef]
30. Branz, F.; Antonello, R.; Pezzutto, M.; Vitturi, S.; Tramarin, F.; Schenato, L. Drive-by-Wi-Fi: Model-Based Control Over Wireless at 1 kHz. *IEEE Trans. Control. Syst. Technol.* **2021**, 1–12. [CrossRef]
31. ISO Guide to the Expression of Uncertainty in Measurement (GUM). Available online: https://www.iso.org/standard/50461.html (accessed on 11 February 2022).
32. Sauter, T. The Three Generations of Field-Level Networks—Evolution and Compatibility Issues. *IEEE Trans. Ind. Electron.* **2010**, *57*, 3585–3595. [CrossRef]
33. Felser, M.; Sauter, T. The fieldbus war: History or short break between battles? In Proceedings of the 4th IEEE International Workshop on Factory Communication Systems, Vasteras, Sweden, 28–30 August 2002; pp. 73–80. [CrossRef]
34. Felser, M. The Fieldbus Standards: History and Structures. Available online: https://www.profilab.ch/papers/FE-TR-0205.pdf (accessed on 11 February 2022).
35. Wollschlaeger, M.; Sauter, T.; Jasperneite, J. The Future of Industrial Communication: Automation Networks in the Era of the Internet of Things and Industry 4.0. *IEEE Ind. Electron. Mag.* **2017**, *11*, 17–27. [CrossRef]
36. Danielis, P.; Skodzik, J.; Altmann, V.; Schweissguth, E.B.; Golatowski, F.; Timmermann, D.; Schacht, J. Survey on real-time communication via ethernet in industrial automation environments. In Proceedings of the 2014 IEEE Emerging Technology and Factory Automation (ETFA), Barcelona, Spain, 16–19 September 2014; pp. 1–8. [CrossRef]
37. Felser, M.; Sauter, T. Standardization of industrial Ethernet-the next battlefield? In Proceedings of the IEEE International Workshop on Factory Communication Systems, Vienna, Austria, 22–24 September 2004; pp. 413–420. [CrossRef]
38. Jasperneite, J.; Imtiaz, J.; Schumacher, M.; Weber, K. A Proposal for a Generic Real-Time Ethernet System. *IEEE Trans. Ind. Inform.* **2009**, *5*, 75–85. [CrossRef]
39. Jasperneite, J.; Schumacher, M.; Weber, K. Limits of increasing the performance of Industrial Ethernet protocols. In Proceedings of the 2007 IEEE Conference on Emerging Technologies and Factory Automation (EFTA 2007), Patras, Greece, 25–28 September 2007; pp. 17–24. [CrossRef]
40. Lv, Y.; Yu, H.; Wang, T.; Yang, Z. Fieldbus interoperation technologies. In Proceedings of the Fifth World Congress on Intelligent Control and Automation (IEEE Cat. No.04EX788), Hangzhou, China, 15–19 June 2004; Voume 4, pp. 3620–3623.
41. Yanjun, F.; Jun, X. An approach for interoperation between heterogeneous fieldbus systems. In Proceedings of the 2005 IEEE Conference on Emerging Technologies and Factory Automation, Catania, Italy, 19–22 September 2005; Voume 2, pp. 5–243.
42. Arjmandi, F.; Moshiri, B. Fieldbus Interoperability on Ethernet. In Proceedings of the 2007 5th IEEE International Conference on Industrial Informatics, Vienna, Austria, 23–27 June 2007; Voume 1, pp. 213–218.
43. Zhong, T.; Zhan, M.; Peng, Z.; Hong, W. Industrial wireless communication protocol WIA-PA and its interoperation with Foundation Fieldbus. In Proceedings of the 2010 International Conference on Computer Design and Applications, Qinhuangdao, China, 25–27 June 2010; Volume 4, pp. V4–370–V4–374.
44. Dang, T.; Merieux, C.; Pizel, J.; Deulet, N. On the Road to Industry 4.0: A Fieldbus Architecture to Acquire Specific Smart Instrumentation Data in Existing Industrial Plant for Predictive Maintenance. In Proceedings of the 2018 IEEE 27th International Symposium on Industrial Electronics (ISIE), Cairns, Australia, 13–15 June 2018; pp. 854–859.
45. Bellagente, P.; Ferrari, P.; Flammini, A.; Rinaldi, S.; Sisinni, E. Enabling PROFINET devices to work in IoT: Characterization and requirements. In Proceedings of the 2016 IEEE International Instrumentation and Measurement Technology Conference Proceedings, Taipei, Taiwan, 23–26 May 2016; pp. 1–6.
46. Vitturi, S.; Zunino, C.; Sauter, T. Industrial Communication Systems and Their Future Challenges: Next-Generation Ethernet, IIoT, and 5G. *Proc. IEEE* **2019**, *107*, 944–961. [CrossRef]
47. Li, Q.; Tang, Q.; Chan, I.; Wei, H.; Pu, Y.; Jiang, H.; Li, J.; Zhou, J. Smart Manufacturing Standardization: Architectures, Reference Models and Standards Framework. *Comput. Ind.* **2018**, *101*, 91–106. [CrossRef]
48. Schlesinger, R.; Springer, A.; Sauter, T. Concept for the coexistence of standard and Real-time Ethernet. In Proceedings of the 2018 14th IEEE International Workshop on Factory Communication Systems (WFCS), Imperia, Italy, 13–15 June 2018; pp. 1–10. [CrossRef]
49. Dietrich, D.; Bruckner, D.; Zucker, G.; Palensky, P. Communication and Computation in Buildings: A Short Introduction and Overview. *IEEE Trans. Ind. Electron.* **2010**, *57*, 3577–3584. [CrossRef]
50. Imtiaz, J.; Jasperneite, J.; Schriegel, S. A proposal to integrate process data communication to IEEE 802.1 Audio Video Bridging (AVB). In Proceedings of the ETFA2011, Toulouse, France, 5–9 September 2011; pp. 1–8. [CrossRef]
51. Zezulka, F.; Marcon, P.; Bradac, Z.; Arm, J.; Benesl, T. Time-Sensitive Networking as the Communication Future of Industry 4.0. *IFAC-PapersOnLine* **2019**, *52*, 133–138. [CrossRef]
52. Genc, E.; Del Carpio, L.F. Wi-Fi QoS Enhancements for Downlink Operations in Industrial Automation Using TSN. In Proceedings of the 2019 15th IEEE International Workshop on Factory Communication Systems (WFCS), Sundsvall, Sweden, 27–29 May 2019; pp. 1–6.
53. *IEEE Std 802.1Q-2018 (Revision of IEEE Std 802.1Q-2014)*; IEEE Standard for Local and Metropolitan Area Network–Bridges and Bridged Networks. IEEE: New York, NY, USA, 6 July 2018; pp. 1–1993.

54. Time-Sensitive Networking (TSN) Task Group Official Website. Available online: https://1.ieee802.org/tsn/ (accessed on 11 February 2022).
55. *IEEE Std 802.1AB-2016 (Revision of IEEE Std 802.1AB-2009)*; IEEE Standard for Local and Metropolitan Area Networks-Station and Media Access Control Connectivity Discovery. IEEE: New York, NY, USA, 11 March 2016; pp. 1–146. [CrossRef]
56. *IEEE Std 802.1AS-2020 (Revision of IEEE Std 802.1AS-2011)*; IEEE Standard for Local and Metropolitan Area Networks–Timing and Synchronization for Time-Sensitive Applications. IEEE: New York, NY, USA. 19 June 2020; pp. 1–421. [CrossRef]
57. *IEEE Std 802.1AX-2020 (Revision of IEEE Std 802.1AX-2014)*; IEEE Standard for Local and Metropolitan Area Networks–Link Aggregation. IEEE: New York, NY, USA, 29 May 2020, pp. 1–333. [CrossRef]
58. *IEEE Std 802.1CB-2017*; IEEE Standard for Local and Metropolitan Area Networks–Frame Replication and Elimination for Reliability. IEEE: New York, NY, USA, 27 October2017; pp. 1–102.
59. *IEEE Std 802.1CS-2020*; IEEE Standard for Local and Metropolitan Area Networks–Link-local Registration Protocol. IEEE: New York, NY, USA, 23 April 2021; pp. 1–151. [CrossRef]
60. P802.1CQ–Multicast and Local Address Assignment. Available online: https://1.ieee802.org/tsn/802-1cq/ (accessed on 11 February 2022).
61. P802.1DC–Quality of Service Provision by Network Systems. Available online: https://1.ieee802.org/tsn/802-1dc/ (accessed on 11 February 2022).
62. *IEEE Std 802-2014 (Revision to IEEE Std 802-2001)*; IEEE Standard for Local and Metropolitan Area Networks: Overview and Architecture. IEEE: New York, NY, USA, 30 June 2014; pp. 1–74. [CrossRef]
63. P802.1f–YANG Data Model for EtherTypes. Available online: https://1.ieee802.org/tsn/802f/ (accessed on 11 February 2022).
64. P802.1ABcu–LLDP YANG Data Model. Available online: https://1.ieee802.org/tsn/802-1abcu/ (accessed on 11 February 2022).
65. P802.1ABdh–Support for Multiframe Protocol Data Units. Available online: https://1.ieee802.org/tsn/802-1abdh/ (accessed on 11 February 2022).
66. P802.1ASdm–Hot Standby. Available online: https://1.ieee802.org/tsn/802-1asdm/ (accessed on 11 February 2022).
67. P802.1ASdn–YANG Data Model. Available online: https://1.ieee802.org/tsn/802-1asdn/ (accessed on 11 February 2022).
68. P802.1CBcv–FRER YANG Data Model and Management Information Base Module. Available online: https://1.ieee802.org/tsn/802-1cbcv/ (accessed on 11 February 2022).
69. P802.1CBdb–FRER Extended Stream Identification Functions. Available online: https://1.ieee802.org/tsn/802-1cbdb/ (accessed on 11 February 2022).
70. *IEEE Std 802.1Qat-2010 (Revision of IEEE Std 802.1Q-2005)*; IEEE Standard for Local and Metropolitan Area Networks-Virtual Bridged Local Area Networks Amendment 14: Stream Reservation Protocol (SRP). IEEE: New York, NY, USA, 30 September 2010, pp. 1–119.
71. *IEEE Std 802.1Qav-2009 (Amendment to IEEE Std 802.1Q-2005)*; IEEE Standard for Local and Metropolitan Area Networks-Virtual Bridged Local Area Networks Amendment 12: Forwarding and Queuing Enhancements for Time-Sensitive Streams. IEEE: New York, NY, USA, 5 January 2010; pp. C1–72.
72. *IEEE Std 802.1Qaz-2011 (Amendment to IEEE Std 802.1Q-2011 as Amended by IEEE Std 802.1Qbe-2011, IEEE Std 802.1Qbc-2011, and IEEE Std 802.1Qbb-2011)*; IEEE Standard for Local and Metropolitan Area Networks–Media Access Control (MAC) Bridges and Virtual Bridged Local Area Networks–Amendment 18: Enhanced Transmission Selection for Bandwidth Sharing Between Traffic Classes. IEEE: New York, NY, USA, 30 September 2011; pp. 1–110.
73. *IEEE Std 802.1Qbu-2016 (Amendment to IEEE Std 802.1Q-2014)*; IEEE Standard for Local and Metropolitan Area Networks—Bridges and Bridged Networks—Amendment 26: Frame Preemption. IEEE: New York, NY, USA, 30 August 2016; pp. 1–52.
74. *IEEE Std 802.1Qbv-2015 (Amendment to IEEE Std 802.1Q-2014 as Amended by IEEE Std 802.1Qca-2015, IEEE Std 802.1Qcd-2015, and IEEE Std 802.1Q-2014/Cor 1-2015)*; IEEE Standard for Local and Metropolitan Area Networks—Bridges and Bridged Networks—Amendment 25: Enhancements for Scheduled Traffic. IEEE: New York, NY, USA. 18 March 2016; pp. 1–57.
75. *IEEE Std 802.1Qca-2015 (Amendment to IEEE Std 802.1Q-2014 as Amended by IEEE Std 802.1Qcd-2015 and IEEE Std 802.1Q-2014/Cor 1-2015)*; IEEE Standard for Local and Metropolitan Area Networks—Bridges and Bridged Networks—Amendment 24: Path Control and Reservation. IEEE: New York, NY, USA, 11 March 2016; pp. 1–120.
76. *IEEE Std 802.1Qcc-2018 (Amendment to IEEE Std 802.1Q-2018 as Amended by IEEE Std 802.1Qcp-2018)*; IEEE Standard for Local and Metropolitan Area Networks—Bridges and Bridged Networks—Amendment 31: Stream Reservation Protocol (SRP) Enhancements and Performance Improvements. IEEE: New York, NY, USA, 31 October 2018; pp. 1–208.
77. *IEEE 802.1Qch-2017 (Amendment to IEEE Std 802.1Q-2014 as amended by IEEE Std 802.1Qca-2015, IEEE Std 802.1Qcd(TM)-2015, IEEE Std 802.1Q-2014/Cor 1-2015, IEEE Std 802.1Qbv-2015, IEEE Std 802.1Qbu-2016, IEEE Std 802.1Qbz-2016, and IEEE Std 802.1Qci-2017)*; IEEE Standard for Local and Metropolitan Area Networks—Bridges and Bridged Networks—Amendment 29: Cyclic Queuing and Forwarding. IEEE: New York, NY, USA, 28 June 2017; pp. 1–30.
78. *IEEE Std 802.1Qci-2017 (Amendment to IEEE Std 802.1Q-2014 as Amended by IEEE Std 802.1Qca-2015, IEEE Std 802.1Qcd-2015, IEEE Std 802.1Q-2014/Cor 1-2015, IEEE Std 802.1Qbv-2015, IEEE Std 802.1Qbu-2016, and IEEE Std 802.1Qbz-2016)*; IEEE Standard for Local and Metropolitan Area Networks–Bridges and Bridged Networks–Amendment 28: Per-Stream Filtering and Policing. IEEE: New York, NY, USA, 28 September 2017; pp. 1–65.
79. *IEEE Std 802.1Qcp-2018 (Amendment to IEEE Std 802.1Q-2018)*; IEEE Standard for Local and Metropolitan Area Networks—Bridges and Bridged Networks—Amendment 30: YANG Data Model. IEEE: New York, NY, USA, 14 September 2018; pp. 1–93.

80. *IEEE Std 802.1Qcr-2020 (Amendment to IEEE Std 802.1Q-2018 as Amended by IEEE Std 802.1Qcp-2018, IEEE Std 802.1Qcc-2018, IEEE Std 802.1Qcy-2019, and IEEE Std 802.1Qcx-2020)*; IEEE Standard for Local and Metropolitan Area Networks—Bridges and Bridged Networks—Amendment 34:Asynchronous Traffic Shaping. IEEE: New York, NY, USA, 6 November 2020; pp. 1–151. [CrossRef]
81. *IEEE Std 802.1Qcx-2020 (Amendment to IEEE Std 802.1Q-2018 as Amended by IEEE Std 802.1Qcp-2018, IEEE Std 802.1Qcc-2018, and IEEE Std 802.1Qcy-2019)*; IEEE Standard for Local and Metropolitan Area Networks—Bridges and Bridged Networks Amendment 33: YANG Data Model for Connectivity Fault Management. IEEE: New York, NY, USA, 5 October 2020; pp. 1–123. [CrossRef]
82. P802.1Qcj–Automatic Attachment to Provider Backbone Bridging (PBB) Services. Available online: https://1.ieee802.org/tsn/802-1qcj/ (accessed on 11 February 2022).
83. P802.1Qcw–YANG Data Models for Scheduled Traffic, Frame Preemption, and Per-Stream Filtering and Policing. Available online: https://1.ieee802.org/tsn/802-1qcw/ (accessed on 11 February 2022).
84. P802.1Qcz–Congestion Isolation. Available online: https://1.ieee802.org/tsn/802-1qcz/ (accessed on 11 February 2022).
85. P802.1Qdd–Resource Allocation Protocol. Available online: https://1.ieee802.org/tsn/802-1qdd/ (accessed on 11 February 2022).
86. P802.1Qdj–Configuration Enhancements for Time-Sensitive Networking. Available online: https://1.ieee802.org/tsn/802-1qdj/ (accessed on 11 February 2022).
87. *ISO/IEC/IEEE 8802-3:2017/Amd.5:2017(E)*; ISO/IEC/IEEE International Standard-Amendment 5: Specification and Management Parameters for Interspersing Express Traffic. IEEE: New York, NY, USA, 16 March 2018; pp. 1–62.
88. Zhou, Z.; Shou, G. An Efficient Configuration Scheme of OPC UA TSN in Industrial Internet. In Proceedings of the 2019 Chinese Automation Congress (CAC), Hangzhou, China, 22–24 November 2019; pp. 1548–1551.
89. Pop, P.; Raagaard, M.L.; Gutierrez, M.; Steiner, W. Enabling Fog Computing for Industrial Automation Through Time-Sensitive Networking (TSN). *IEEE Commun. Stand. Mag.* **2018**, *2*, 55–61. [CrossRef]
90. Böhm, M.; Ohms, J.; Wermser, D. Multi-Domain Time-Sensitive Networks—An East-Westbound Protocol for Dynamic TSN-Stream Configuration Across Domains. In Proceedings of the 2019 24th IEEE International Conference on Emerging Technologies and Factory Automation (ETFA), Zaragoza, Spain, 10–13 September 2019; pp. 1363–1366.
91. Gutiérrez, M.; Ademaj, A.; Steiner, W.; Dobrin, R.; Punnekkat, S. Self-configuration of IEEE 802.1 TSN networks. In Proceedings of the 2017 22nd IEEE International Conference on Emerging Technologies and Factory Automation (ETFA), Limassol, Cyprus, 12–15 September 2017; pp. 1–8.
92. Bülbül, N.S.; Ergenç, D.; Fischer, M. SDN-Based Self-Configuration for Time-Sensitive IoT Networks. *arXiv* **2021**, arXiv:2103.01282.
93. Garbugli, A.; Bujari, A.; Bellavista, P. End-to-End QoS Management in Self-Configuring TSN Networks. In Proceedings of the 2021 17th IEEE International Conference on Factory Communication Systems (WFCS), Linz, Austria, 9–11 June 2021; pp. 131–134 [CrossRef]
94. Anwar, F.; D'Souza, S.; Symington, A.; Dongare, A.; Rajkumar, R.; Rowe, A.; Srivastava, M. Timeline: An Operating System Abstraction for Time-Aware Applications. In Proceedings of the 2016 IEEE Real-Time Systems Symposium (RTSS), Porto, Portugal, 29 November–2 December 2016; pp. 191–202.
95. Skiadopoulos, K.; Tsipis, A.; Giannakis, K.; Koufoudakis, G.; Christopoulou, E.; Oikonomou, K.; Kormentzas, G.; Stavrakakis, I Synchronization of data measurements in wireless sensor networks for IoT applications. *Ad Hoc Netw.* **2019**, *89*, 47–57. [CrossRef]
96. *IEEE Std 1588-2019 (Revision ofIEEE Std 1588-2008)*; IEEE Standard for a Precision Clock Synchronization Protocol for Networked Measurement and Control Systems. IEEE: New York, NY, USA, 16 June 2020; pp. 1–499. [CrossRef]
97. Stanton, K.B. Distributing Deterministic, Accurate Time for Tightly Coordinated Network and Software Applications: IEEE 802.1AS, the TSN profile of PTP. *IEEE Commun. Stand. Mag.* **2018**, *2*, 34–40. [CrossRef]
98. *IEEE Std 802.1D-2004 (Revision of IEEE Std 802.1D-1998)*; IEEE Standard for Local and Metropolitan Area Networks: Media Access Control (MAC) Bridges. IEEE: New York, NY, USA, 9 June 2004; pp. 1–281.
99. Pallos, R.; Farkas, J.; Moldovan, I.; Lukovszki, C. Performance of rapid spanning tree protocol in access and metro networks. In Proceedings of the 2007 Second International Conference on Access Networks Workshops, Ottawa, ON, Canada, 22–24 August 2007; pp. 1–8.
100. Gutiérrez, M.; Steiner, W.; Dobrin, R.; Punnekkat, S. Synchronization Quality of IEEE 802.1AS in Large-Scale Industrial Automation Networks. In Proceedings of the 2017 IEEE Real-Time and Embedded Technology and Applications Symposium (RTAS), Pittsburgh, PA, USA, 18–21 April 2017; pp. 273–282.
101. Garner, G.M.; Gelter, A.; Teener, M.J. New simulation and test results for IEEE 802.1AS timing performance. In Proceedings of the 2009 International Symposium on Precision Clock Synchronization for Measurement, Control and Communication, Brescia, Italy, 12–16 October 2009; pp. 1–7.
102. Garner, G.M.; Ryu, H. Synchronization of audio/video bridging networks using IEEE 802.1AS. *IEEE Commun. Mag.* **2011**, *49*, 140–147. [CrossRef]
103. Mohammadpour, E.; Stai, E.; Mohiuddin, M.; Le Boudec, J. Latency and Backlog Bounds in Time-Sensitive Networking with Credit Based Shapers and Asynchronous Traffic Shaping. In Proceedings of the 2018 30th International Teletraffic Congress (ITC 30), Vienna, Austria, 3–7 September 2018; Volume 2, pp. 1–6.
104. *IEEE Std 802.3br-2016 (Amendment to IEEE Std 802.3-2015 as Amended by IEEE St802.3bw-2015, IEEE Std 802.3by-2016, IEEE Std 802.3bq-2016, and IEEE Std 802.3bp-2016)*; IEEE Standard for Ethernet Amendment 5: Specification and Management Parameters for Interspersing Express Traffic. IEEE: New York, NY, USA, 14 October 2016; pp. 1–58.

05. *IEEE Std 802.3-2018 (Revision of IEEE Std 802.3-2015)*; IEEE Standard for Ethernet. IEEE: New York, NY, USA, 31 August 2018; pp. 1–5600.
06. Hellmanns, D.; Falk, J.; Glavackij, A.; Hummen, R.; Kehrer, S.; Dürr, F. On the Performance of Stream-Based, Class-Based Time-Aware Shaping and Frame Preemption in TSN. In Proceedings of the 2020 IEEE International Conference on Industrial Technology (ICIT), Buenos Aires, Argentina, 26–28 February 2020; pp. 298–303. [CrossRef]
07. Lee, J.; Park, S. Time-Sensitive Network (TSN) Experiment in Sensor-Based Integrated Environment for Autonomous Driving. *Sensors* **2019**, *19*, 1111. [CrossRef]
08. Ojewale, M.A.; Yomsi, P.M.; Nikolić, B. Worst-Case Traversal Time Analysis of TSN with Multi-Level Preemption. *J. Syst. Archit.* **2021**, *116*, 102079. [CrossRef]
09. Zhao, L.; Pop, P.; Zheng, Z.; Daigmorte, H.; Boyer, M. Latency Analysis of Multiple Classes of AVB Traffic in TSN with Standard Credit Behavior Using Network Calculus. *IEEE Trans. Ind. Electron.* **2020**, *68*, 10291–10302. [CrossRef]
10. Zhou, Z.; Yan, Y.; Ruepp, S.; Berger, M. Analysis and Implementation of Packet Preemption for Time Sensitive Networks. In Proceedings of the 2017 IEEE 18th International Conference on High Performance Switching and Routing (HPSR), Campinas, Brazil, 18–21 June 2017; pp. 1–6. [CrossRef]
11. Houtan, B.; Ashjaei, M.; Daneshtalab, M.; Sjödin, M.; Mubeen, S. Work in Progress: Investigating the Effects of High Priority Traffic on the Best Effort Traffic in TSN Networks. In Proceedings of the 2019 IEEE Real-Time Systems Symposium (RTSS), Hong Kong, China, 3–6 December 2019; pp. 556–559.
12. Steiner, W.; Craciunas, S.S.; Oliver, R.S. Traffic Planning for Time-Sensitive Communication. *IEEE Commun. Stand. Mag.* **2018**, *2*, 42–47. [CrossRef]
13. Craciunas, S.S.; Oliver, R.S.; Steiner, W. Formal Scheduling Constraints for Time-Sensitive Networks. *arXiv* **2017**, arXiv:1712.02246.
14. Jiang, J.; Li, Y.; Hong, S.H.; Xu, A.; Wang, K. A Time-sensitive Networking (TSN) Simulation Model Based on OMNET++. In Proceedings of the 2018 IEEE International Conference on Mechatronics and Automation (ICMA), Changchun, China, 5–8 August 2018; pp. 643–648.
15. Kehrer, S.; Kleineberg, O.; Heffernan, D. A comparison of fault-tolerance concepts for IEEE 802.1 Time Sensitive Networks (TSN). In Proceedings of the 2014 IEEE Emerging Technology and Factory Automation (ETFA), Barcelona, Spain, 16–19 September 2014; pp. 1–8.
16. *ISO/IEC/IEEE 8802-1AC-2018(E)*; ISO/IEC/IEEE International Standard-Information Technology—Telecommunications and Information Exchange between Systems—Local and Metropolitan Area Networks—Part 1AC: Media Access Control (MAC) Service Definition. IEEE: New York, NY, USA, 30 April 2018; pp. 1–56.
17. Desai, N.; Punnekkat, S. Enhancing Fault Detection in Time Sensitive Networks using Machine Learning. In Proceedings of the 2020 International Conference on COMmunication Systems NETworkS (COMSNETS), Bangalore, India, 7–11 January 2020; pp. 714–719.
18. Hofmann, R.; Nikolić, B.; Ernst, R. Challenges and Limitations of IEEE 802.1CB-2017. *IEEE Embed. Syst. Lett.* **2019**, *12*, 105–108. [CrossRef]
19. Prinz, F.; Schoeffler, M.; Lechler, A.; Verl, A. End-to-end Redundancy between Real-time I4.0 Components based on Time-Sensitive Networking. In Proceedings of the 2018 IEEE 23rd International Conference on Emerging Technologies and Factory Automation (ETFA), Torino, Italy, 4–7 September 2018; Volume 1, pp. 1083–1086.
20. *IEEE Std 802.1BA-2021 (Revision of IEEE Std 802.1BA-2011)*; IEEE Standard for Local and Metropolitan Area Networks—Audio Video Bridging (AVB) Systems. IEEE: New York, NY, USA, 17 December 2021; pp. 1–45. [CrossRef]
21. *IEEE Std 802.1CM-2018*; IEEE Standard for Local and Metropolitan Area Networks—Time-Sensitive Networking for Fronthaul. IEEE: New York, NY, USA, 8 June 2018; pp. 1–62.
22. IEC/IEEE 60802 TSN Profile for Industrial Automation-WG Website. 2021. Available online: https://1.ieee802.org/tsn/iec-ieee-60802/ (accessed on 11 February 2022).
23. P802.1DF—TSN Profile for Service Provider Networks. Available online: https://1.ieee802.org/tsn/802-1df/ (accessed on 11 February 2022).
24. P802.1DG—TSN Profile for Automotive In-Vehicle Ethernet Communications, Draft 1.4. 2021. Available online: https://1.ieee802.org/tsn/802-1dg/ (accessed on 11 February 2022).
25. P802.1DP—TSN for Aerospace Onboard Ethernet Communications. Available online: https://1.ieee802.org/tsn/802-1dp/ (accessed on 11 February 2022).
26. Use Cases IEC/IEEE 60802 (V1.3). Available online: http://www.ieee802.org/1/files/public/docs2018/60802-industrial-use-cases-0918-v13.pdf (accessed on 11 February 2022).
27. Tramarin, F.; Vitturi, S. Strategies and Services for Energy Efficiency in Real-Time Ethernet Networks. *IEEE Trans. Ind. Inform.* **2015**, *11*, 841–852. [CrossRef]
28. Lupashin, S.; Schöllig, A.; Sherback, M.; D'Andrea, R. A simple learning strategy for high-speed quadrocopter multi-flips. In Proceedings of the 2010 IEEE International Conference on Robotics and Automation, Anchorage, AK, USA, 3–7 May 2010; pp. 1642–1648. [CrossRef]
29. Guo, M.; Wang, F.; Peng, F.; Lin, S.C. Design of Distributed Network Clock-Synchronization for Swarm UAV. In Proceedings of the 2020 International Conference on Computing and Data Science (CDS), Stanford, CA, USA, 1–2 August 2020; pp. 194–197. [CrossRef]

130. Ahmadi, A.; Moradi, M.; Cherifi, C.; Cheutet, V.; Ouzrout, Y. Wireless Connectivity of CPS for Smart Manufacturing: A Survey. In Proceedings of the 2018 12th International Conference on Software, Knowledge, Information Management Applications (SKIMA), Phnom Penh, Cambodia, 3–5 December 2018; pp. 1–8.
131. Park, P.; Coleri Ergen, S.; Fischione, C.; Lu, C.; Johansson, K.H. Wireless Network Design for Control Systems: A Survey. *IEEE Commun. Surv. Tutor.* **2018**, *20*, 978–1013. [CrossRef]
132. Sudhakaran, S.; Montgomery, K.; Kashef, M.; Cavalcanti, D.; Candell, R. Wireless Time Sensitive Networking for Industrial Collaborative Robotic Workcells. In Proceedings of the 2021 17th IEEE International Conference on Factory Communication Systems (WFCS), Linz, Austria, 9–11 June 2021; pp. 91–94. [CrossRef]
133. Cavalcanti, D.; Bush, S.; Illouz, M.; Kronauer, G.; Regev, A.; Venkatesan, G. Wireless TSN–Definitions, Use Cases & Standards Roadmap. *Avnu Alliance* **2020**, 1–16.
134. Song, J.; Han, S.; Mok, A.; Chen, D.; Lucas, M.; Nixon, M.; Pratt, W. WirelessHART: Applying Wireless Technology in Real-Time Industrial Process Control. In Proceedings of the 2008 IEEE Real-Time and Embedded Technology and Applications Symposium, St. Louis, MO, USA, 22–24 April 2008; pp. 377–386. [CrossRef]
135. Hong, S.; Hu, X.S.; Gong, T.; Han, S. On-Line Data Link Layer Scheduling in Wireless Networked Control Systems. In Proceedings of the 2015 27th Euromicro Conference on Real-Time Systems, Lund, Sweden, 8–10 July 2015; pp. 57–66. [CrossRef]
136. Luvisotto, M.; Tramarin, F.; Vitturi, S. A learning algorithm for rate selection in real-time wireless LANs. *Comput. Netw.* **2017**, *126*, 114–124. [CrossRef]
137. Branz, F.; Pezzutto, M.; Antonello, R.; Tramarin, F.; Schenato, L. Drive-by-Wi-Fi: Testing 1 kHz control experiments over wireless. In Proceedings of the 2019 18th European Control Conference (ECC), Naples, Italy, 25–28 June 2019; pp. 2990–2995.
138. Fedullo, T.; Tramarin, F.; Vitturi, S. The Impact of Rate Adaptation Algorithms on Wi-Fi-Based Factory Automation Systems. *Sensors* **2020**, *20*, 5195. [CrossRef]
139. Li, S.; Xu, L.D.; Zhao, S. 5G Internet of Things: A Survey. *J. Ind. Inf. Integr.* **2018**, *10*, 1–9. [CrossRef]
140. Maldonado, R.; Karstensen, A.; Pocovi, G.; Esswie, A.A.; Rosa, C.; Alanen, O.; Kasslin, M.; Kolding, T. Comparing Wi-Fi 6 and 5G Downlink Performance for Industrial IoT. *IEEE Access* **2021**, *9*, 86928–86937. [CrossRef]
141. Wijethilaka, S.; Liyanage, M. Survey on Network Slicing for Internet of Things Realization in 5G Networks. *IEEE Commun. Surv. Tutor.* **2021**, *23*, 957–994. [CrossRef]
142. Fedullo, T.; Morato, A.; Tramarin, F.; Bellagente, P.; Ferrari, P.; Sisinni, E. Adaptive LoRaWAN Transmission exploiting Reinforcement Learning: The Industrial Case. In Proceedings of the 2021 IEEE International Workshop on Metrology for Industry 4.0 IoT (MetroInd4.0 IoT), Rome, Italy, 7–9 June 2021; pp. 671–676. [CrossRef]
143. Neumann, A.; Wisniewski, L.; Ganesan, R.S.; Rost, P.; Jasperneite, J. Towards integration of Industrial Ethernet with 5G mobile networks. In Proceedings of the 2018 14th IEEE International Workshop on Factory Communication Systems (WFCS), Imperia, Italy, 13–15 June 2018; pp. 1–4.
144. *IEEE Std 802.11-2020 (Revision of IEEE Std 802.11-2016)*; IEEE Standard for Information Technology—Telecommunications and Information Exchange between Systems—Local and Metropolitan Area Networks—Specific Requirements—Part 11: Wireless LAN Medium Access Control (MAC) and Physical Layer (PHY) Specifications. IEEE: New York, NY, USA, 26 February 2021; pp. 1–4379. [CrossRef]

sensors

MDPI

Review

Drone Detection and Defense Systems: Survey and a Software-Defined Radio-Based Solution

Florin-Lucian Chiper, Alexandru Martian *, Calin Vladeanu, Ion Marghescu, Razvan Craciunescu and Octavian Fratu

Telecommunications Department, University Politehnica of Bucharest, 060042 Bucharest, Romania; florin_lucian.chiper@upb.ro (F.-L.C.); calin.vladeanu@upb.ro (C.V.); ion.marghescu@upb.ro (I.M.); razvan.craciunescu@upb.ro (R.C.); octavian.fratu@upb.ro (O.F.)

* Correspondence: alexandru.martian@upb.ro

Abstract: With the decrease in the cost and size of drones in recent years, their number has also increased exponentially. As such, the concerns regarding security aspects that are raised by their presence are also becoming more serious. The necessity of designing and implementing systems that are able to detect and provide defense actions against such threats has become apparent. In this paper, we perform a survey regarding the different drone detection and defense systems that were proposed in the literature, based on different types of methods (i.e., radio frequency (RF), acoustical, optical, radar, etc.), with an emphasis on RF-based systems implemented using software-defined radio (SDR) platforms. We have followed the preferred reporting items for systematic reviews and meta-analyses (PRISMA) guidelines in order to provide a concise and thorough presentation of the current status of the subject. In the final part, we also describe our own solution that was designed and implemented in the framework of the DronEnd research project. The DronEnd system is based on RF methods and uses SDR platforms as the main hardware elements.

Keywords: drone; UAV; RF methods; software-defined radio; detection system; defense system

Citation: Chiper, F.-L.; Martian, A.; Vladeanu, C.; Marghescu, I.; Craciunescu, R.; Fratu, O. Drone Detection and Defense Systems: Survey and a Software-Defined Radio-Based Solution. *Sensors* **2022**, *22*, 1453. https://doi.org/10.3390/s22041453

Academic Editor: Peter Han Joo Chong

Received: 14 January 2022
Accepted: 10 February 2022
Published: 14 February 2022

1. Introduction

Technical innovations continue to manifest at an ever-increasing speed, causing fast and drastic changes to modern society. These changes, driven by the possibilities offered by new technologies, affect citizens, governments, and all public and private industry sectors.

As a result, the development of small, low-cost unmanned aerial vehicles (UAVs), commonly known as drones, has resulted in an ever-increasing number of these devices being utilized in a variety of applications [1]. UAVs have introduced new participants in aviation, quickly evolving beyond their military origin to become powerful business tools [2,3].

Applications of UAVs range from recreation to commercial and military applications, including enjoyment, hobbies, games with drones, homemade entertainment videos, recreational movies [4–6], low altitude flying base stations [7], and the operation of UAVs for military purposes [8–13].

The following research questions were developed for this project:

- What functions should a drone detection and defense system (DDDS) have in order to prove its functionality?
- Which are the most popular methods used in the implementation of DDDSs?
- Which are the main parameters that should be taken into consideration in research?
- What gaps are in the current research of DDDSs?

A widely-used methodology was utilized to conduct a systematic literature review based on preferred reporting items for systematic reviews and meta-analyses (PRISMA) [14] in order to obtain the answers to our study questions. We conducted a literature search in scientific databases that encompass prominent computer science journals and conferences, such as IEEE Xplore, ACM digital library, ScienceDirect, SAGE Journals Online, and Springer Link, to discover key articles on the drone detection and defense systems topic. We used the following search string to discover the relevant publications and papers for our research: ('Drone' OR 'UAV') AND (Counter) in the domains of electrical engineering, applied physics, telecommunications, defense, and computer information systems, for the previous six years (2016–2021). In total, we gathered a set of 7349 potentially relevant publications, excluding grey literature and pre-prints.

We next looked at the titles, keywords, and abstracts of the publications in order to find the papers and articles that described at least one distributed ledger modeling or simulation approach. We chose a total of 99 publications in the process. We examined the references of the selected publications for other papers that were relevant to our inquiry in order to expand our literature collection. Figure 1 shows the overall number of articles produced as a result of this approach.

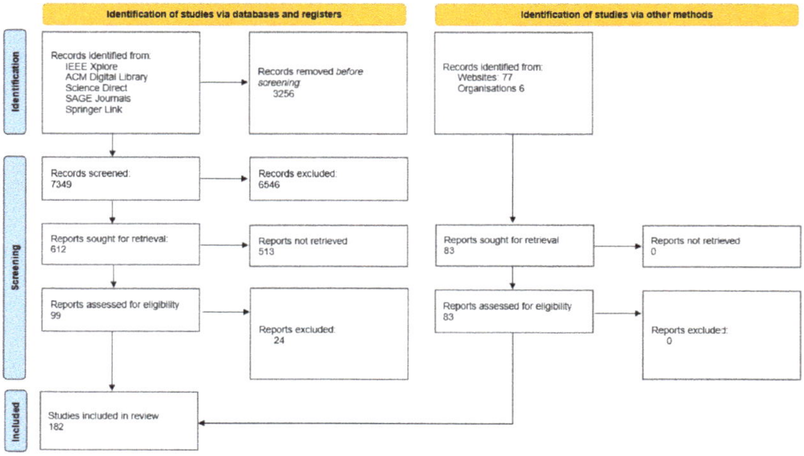

Figure 1. PRISMA 2020 flow diagram for systematic reviews.

The additional references that were identified in the bodies of the selected publications, or referencing those, were added to the literature list. We carefully studied the selected publications once the literature selection procedure was completed in order to determine the described applications and problems. The results of our analysis are reported in the following sections, which represent the core of the topical literature review.

The main contributions of our paper can be summarized as follows:

- We provide a detailed review regarding the drone defense systems based on RF methods, focusing on the solutions that are based on software-defined radio (SDR) platforms. To our best knowledge, other reviews that were performed concerning drone defense systems have not detailed that particular category of solutions;
- We discuss the current worldwide status of the legal issues regarding the jamming function, that enables the systems to annihilate the drones after they are detected;
- We present our own solution for an RF-based drone defense system that was designed and implemented within the framework of the DronEnd research project. The system was developed using several SDR platforms and a custom-made mount for dynamically adjusting the orientation of the jamming antenna, which enables the detection, localization, and annihilation of drones in a given monitored area.

The rest of this article is organized as follows: Section 2 reviews the most recent incidents that involved the reckless flying of UAV systems and the regulations taken by different governments and agencies around the world.

Section 3 describes the system requirements of a drone defense system in correlation with their basic mechanism/sensing technologies, considering their advantages and drawbacks. Also, this section highlights the specific models and architectures used in research for drone detection and defense systems. Section 4 details aspects regarding RF-based DDDSs and the use of SDR platforms in such systems. Section 5 contains a discussion regarding the challenges and the future research directions related to DDDSs. In Section 6, a solution for a drone detection and defense system based on SDR platforms, developed by the authors, is proposed and detailed, also highlighting the novel elements that are brought about, compared to the other existing solutions. The last section concludes the paper and includes the future perspectives of this work.

2. The Necessity of Drone Detection and Defense Systems: Incidents and Regulations

The drone industry's rapid rise has outpaced the rules for safe and secure drone operation, making them a symbol of illegal and destructive terror and crimes [15].

Drones have gained attention as a threat to safety and security since their entrance into civilian technology, which has fueled the development of anti-drone (or counter-drone) technologies. Anti-drone systems are designed to protect against drone accidents or terrorism, but they will need to evolve in order to deal with future drone flight systems [16].

UAVs have been used in a variety of military actions. Non-military UAVs have been accused of endangering airplanes, as well as persons and property on the ground. Due to the potential of an ingested drone to quickly damage an aircraft engine [17], safety concerns have been raised. Multiple near-misses and verified collisions have occurred involving hobbyist UAV pilots operating when violating the aviation safety standards [18].

2.1. Recently Reported Incidents

The necessity of anti-drone defense systems has gained importance, considering the large number of dangerous occurrences that are mentioned in Table 1.

Table 1. List of the recent UAV-related incidents.

Incident Type	Time and Place of the Event	Short Description of the Incident	Aftermaths	References
Aircraft collisions	17 April 2016/UK, London, Heathrow International Airport	An Airbus A320 collided with a Metropolitan Police UAV as it approached landing	There were no serious issues reported.	[19]
	21 September 2017/USA, Staten Island, New York City	A civilian UAV collided with a Black Hawk helicopter	The helicopter was able to continue flying and landed in a safe manner.	[20]
	12 October 2017/Canada, Jean Lesage Airport, Quebec City	A Skyjet Aviation Beech King Air A100 collided with a UAV	The plane landed safely, with only minor damage to its wings.	[21]
	13 December 2018/Mexico, Tijuana International Airport	On a Boeing 737–800 operating as Flight 773, a "quite loud noise" was heard	After a safe landing, the aircraft's nose was discovered to be damaged. The reason for the incident has not been identified; however, it was examined as a drone strike by the airline.	[22]
	10 August 2021/UK, Buttonville Municipal Airport	A Cessna 172 registered C-GKWL collided with a drone operated by the York Regional Police	The Cessna landed safely but with significant damage.	[23]

Table 1. *Cont.*

Incident Type	Time and Place of the Event	Short Description of the Incident	Aftermaths	References
Near-miss incidents	January 2017/P.R. China, Hangzhou Xiaoshan International Airport	A 23-year-old Xiaoshan UAV operator was arrested after taking footage with a drone that flew too close to planes landing	DJI, China's biggest drone manufacturer and the producer of the Mavic Pro drone (which was discovered to have been used in the event), issued a statement expressing its "strong condemnation" of the illegal filming.	[24]
	25 March 2018/New Zeeland, Auckland Airport	A UAV approached within 5 m of an Air New Zealand Boeing 777–200 on final approach to airport	The pilots spotted the UAV as the plane was approaching a position when evasive action was impossible, and they initially worried it would be pulled into an engine.	[25]
	19 December 2018/UK, Gatwick	A repeated deliberate intrusion of UAVs of "industrial standards" occurred	The suspension of all takeoffs and landings began at 9:03 p.m. on 19 December due to UAV sightings over the runway. Flights were briefly restarted the next morning but were banned again after more UAV sightings.	[26]
Other incidents that targeted officials and strategic objectives	April 2015/Japan	A small drone carrying radioactive materials was dropped on the roof of Japan's Prime Minister's mansion	The drone was not only able to fly to the Prime Minister's home, but it was also left unattended for over two weeks. Due to the characteristics of the area, notably privacy, it may have been difficult to deploy intensive detecting technology.	[27]
	October 2016/Syria	ISIL used two ultra-small drones purchased from Amazon to assassinate two Iranians in Syria	The first incidence of commercial drone terrorism, significant since commercial off-the-shelf drones were employed, demonstrating that a wide variety of drone terrorism was achievable because the drones could be cheaply bought without having the expert-level skill to fly.	[28]
	August 2018/Venezuela	Two bomb-carrying drones had a failed attempt to assassinate Venezuelan President Nicolas Maduro during a national outdoor celebration	The first time a drone was used to try to assassinate the country's leader. This incident emphasizes the importance of anti-drone technology for avoiding a traumatic event. Temporary anti-drone systems require rapid installation and deployment.	[29]

In addition to the highlighted incidents, the number of small mishaps caused by unauthorized or illegal drones invading restricted regions is increasing by the day [30]. This is another reason for anti-drone technology becoming increasingly important. As the regulations concerning drone usage are also a significant aspect to be considered when designing a DDDS, we review in the following subsection several aspects in this matter.

2.2. Regulations Regarding the Use of Drones

The most important agencies that regulate the use of drones (e.g., European Union Aviation Safety Agency (EASA), Federal Communication Commission (FCC), Australian Communication and Media Authority (ACMA), Civil Aviation Authority (CAA), etc.) have adopted action plans in order to ensure critical objectives against the illegal usage of UAVs [30–32].

For example, in order to address the hazards and threats posed by drones, European Union members in EASA have endorsed a counter-unmanned aerial systems (counter-UAS) action plan, proposed by the agency in 2019, which has subsequently been included in the European Plan for Aviation Safety (EPAS) [32].

The EASA's EPAS is applicable to all of the national and appropriate agencies, and it has resulted in the effective control of UAV hazards.

Furthermore, the EU has approved EASA's standard European guidelines in order to enable UAV integration and safe operation in the aviation system. The rules that apply to drones are outlined in Regulation (EU) 2019/94735 on the rules and procedures for the operation of unmanned aerial vehicles (UAVs) and Regulation (EU) 2019/945 on unmanned aerial vehicles and third-country operators of unmanned aerial vehicles (UAVs).

According to the document, there are three primary types of drone incident offenders that endanger civil aviation, as follows: non-criminal motivation, gross negligence, and criminal/terrorist motivation [30]. They relate to the drone's remote pilot's intention, as described in Table 2.

Table 2. EASA categorization of intention/motivation of pilots of unauthorized drones.

Negligence	Individuals Who Are Oblivious to or Are Unaware of the Appropriate Regulations and Constraints. As a Result, They Fly Their Drones across Sensitive or Forbidden Terrain. They Have a "Clueless" Mentality and Have No Intention of Disrupting Regular Aviation.
Gross negligence	Individuals who are reckless because they are aware of the appropriate regulations and constraints yet choose to break them for personal or professional advantage (e.g., aggressive spotters). Their actions can be described as "reckless", as they disrupt civil aviation while completely ignoring the implications of their conduct.
	Individuals who intentionally strive to use drones to disrupt aerodromes and flight operations, regardless of whether they are aware of the applicable legislation and limits. These individuals may even act as a group to maximize their impact. While their actions may have unexpected repercussions for aviation safety, they do not seek to put human lives in jeopardy.
Criminal/terrorist motivation	Criminals and terrorists are persons who intentionally strive to utilize drones to interfere with the safety and security of civil aviation, regardless of whether they are aware of the applicable legislation and limits. These persons should be considered criminally motivated or even terrorists because their actions are purposeful and show no concern for human lives and property.

3. Drone Detection and Defense Systems: Classification, Sensors, Countermeasures

In this section, we focus on the classification of drone detection and defense systems depending on different criteria, on the comparison of the different sensor types that can be used in order to detect the presence of the drones in the monitored area, on the classification of the countermeasures that can be adopted in order to annihilate the detected drones, and on the regulations regarding the use of jamming as countermeasure.

3.1. Classification of Drone Detection and Defense Systems

Firstly, it is necessary to classify DDDSs in order to understand their capabilities, as it is summarized in Table 3.

Table 3. Classification of DDDSs.

Category	Definition
Ground-based: fixed	Systems designed for usage in fixed locations [33]
Ground-based: mobile	Systems designed to be installed on automobiles and operated while they are in motion [33]
Hand-held	Systems designed to be operated by a single person using their hands; the majority of these systems resemble rifles [34]
UAV-based	Systems designed to be mounted on unmanned aerial vehicles (UAVs) [34]
UAV-swarm-based	Systems designed to use multiple drones [35]

A DDDS implies different available technologies for detection, tracking, and classification, in addition to neutralization techniques. The most essential elements recommended for the DDDS are considered to be detection, tracking, and classification of the target drones [30,34]. The different technologies that are used for allowing drone detection are summarized in Table 4.

Table 4. Technologies used for drone detection in DDDSs.

Technology	Description	References
Acoustic	UAVs are detected and tracked by using an array of microphones	[36–53]
Imaging (EO/IR)	UAVs are detected and tracked by using EO/IR cameras	[54–72]
Radar	UAVs are detected and tracked using their radar signature	[73–102]
Radio frequency (RF)	UAVs are detected, tracked, and identified by monitoring the radio frequencies used for communications; this technology could localize the UAV and the pilot	[103–113]
Hybrid	Combination of two or more of the above-mentioned technologies	[104,114]

3.2. Classification of Detection Sensors

All of the types of sensors that are currently used in DDDS present specific advantages and limitations and, as a direct consequence, such a system must incorporate more sensors of different types in order to achieve a higher detection rate [33].

A brief description of each category of sensors is given below and the different pros and cons for each category are summarized in Table 5.

Table 5. Pros and cons of sensors used in DDDSs.

Type	Pros	Cons	References
Acoustic	• Covers the spectrum of 20 Hz–20 kHz; • Acoustic signature library could be updated easily from flight to flight; • Lightweight and can be easily associated with other types of sensors.	• Limited range; • Vulnerable to ambient noise; • Susceptible to decoys.	[36–53]
Imaging	• Covers all of the visible and IR spectrum (3 MHz–300 GHz); • IR cameras could operate in cloudy weather and in day or night; • Could be assisted by computer-vision technologies.	• Provides 2D images; • Limited performances by weather conditions and background temperature; • Dependent of georeference data • LoS is required	[54–72]
Radar	• Bandwidth used: 3 MHz–300 GHz; • Could operate in all weather and day/night conditions; • Offers information regarding the velocity of the target; • Can recognize micro-Doppler signatures (MDS) • Offers high coverage; • Good accuracy; • Compact and high mobile, required for tactical applications; • High reliability.	• Large radar cross-section is desired; • Difficult to differentiate UAVs from birds; • Limited performance for low altitudes and speeds (death cone); • Could interfere easily with small objects, especially birds; • LoS is required; • High cost.	[73–102]
RF	• Capturing the communication spectrum and signals UAV and operators; • Low complexity and easy to implement; • Could operate in all weather and day/night conditions; • Easier to improve due to modular implementation of receivers and digital signal processing units used in implementation; • Possibility to localize the pilot.	• Knowledge regarding UAV communication specifications (e.g., frequency bands, modulations, etc.) is required; • Difficult to accurately determine AoA; • Difficult to use in urban areas due to fading and multipath phenomena; • Vulnerable to malicious or illegal modified RF that will exceed receiver capabilities.	[103–113]

3.2.1. Radio Frequency Sensors (RF)

UAV RF detection is a technique that involves the interception and analysis of the signals transmitted (Tx, Rx) between the UAV and the ground station. Usually, these signals consist of uplink (from the ground station) control signals and downlink (from the drone) data signals (position and video data) [103]. A detailed analysis of the DDDSs that are based on RF methods are presented in Section 4.

3.2.2. Radar

The Radar solution for drone defense systems represents an active method to identify and localize a potential UAV threat. In order to determine the range, angle, or velocity of a UAV, radar is widely used as an active sensor in sensing systems in a DDDS. A radar system consists of a transmitter, a receiver, and a processor [73].

3.2.3. Imaging Sensors

This technology involves the use of cameras that take images from a designated area in order to determine the presence of a target drone.

Electro-Optical (EO) Cameras

Some DDDS use imaging sensors (EO/IR), which could be led by other sensors (such as radar and RF) in order to obtain images of the drone and its main characteristics (e.g. payload). These images can be recorded and analyzed by specialists in order to determine the threat level [55].

The main disadvantage of this method is its low performance under dark and foggy conditions. Moreover, the quality of the images depends on the quality of the lenses and the angle of the photography (LoS is mandatory).

Infrared (IR) Cameras/Thermal

This method employs thermal IR cameras that are able to detect the heat produced by a UAV's hardware components, such as the motors, batteries, and processors.

This detection method presents disadvantages related to detection range and environment caused by the sensibility of the sensors that measure the thermal difference between the drone and the background. In consequence, the detection of drone presence depends on the drone's motor temperature, angle (LoS is mandatory), distance, and the temperature of the IR sensors [58].

3.2.4. Acoustic Sensors

This technology involves the use of a microphone array that captures the noise generated by the propellers and rotors of a UAV and compares it with an intern acoustic signature database [42].

Table 5 summarizes the advantages and limitations of each of the different technologies that were mentioned above.

3.3. Classification of Countermeasures

The necessity of DDDS arose for the first time in military applications under special regulations that exceed other governmental or structure capabilities and responsibilities. In consequence, the neutralization techniques are more numerous than the detection techniques [30].

The most important DDDS countermeasures are as follows:

- *Electromagnetic pulse (EMP)*—a beam generated with the goal to damage the internal electronics of the target drone [115–117];
- *Interceptor drone/Collision Drone*—a drone used to force the target drone to land or return home [118–123];
- *Lasers*—directed rays used to destroy the target or blind the camera (dazzler) [124–129];
- *Magnetic*—use powerful magnets in order to create a magnetic field around a protected area [130];
- *Prey birds*—eagles or falcons specially trained to attack the enemy's drone [131];
- *Shooting nets*—a net is launched towards the target drone to prevent the propellers from rotating [132];
- *Projectiles*—large-caliber ammunition used to destroy the target [133];
- *Missiles*—conventional ammunition, could be guided or unguided [133];
- *Guns*—conventional weapons and ammunition [133];
- *Water cannons*—a stream of water is directed towards the target drone [134];
- *RF/GNSS jamming*—disrupt the communication of the target drone with the control station and/or global navigation satellite system (GNSS) [135–139];
- *Spoofing*—decoys the drone by using imitation GNSS and control signals in order to take over the command [140–145];

- *Mixed countermeasure techniques*—use two or more countermeasures in order to maximize the neutralization rate.

The main advantages and drawbacks of each of the different countermeasure technique are presented in Table 6.

Table 6. Characteristics and limitations of countermeasure techniques.

Type	Pros	Cons	References
Electromagnetic pulse (EMP)	• Could burn or interfere with the internal electronics of the drone, disrupting its operation; • Could operate in both narrowband and wideband domains.	• Accurate direction of jamming is necessary; • Difficult to know the effectiveness of jamming.	[115–117]
Interceptor drones	• Searching and tracking capabilities; • Could carry weapons and ammunition.	• Requires a relatively close approach to the target; • Have a considerable delay.	[118–123]
Lasers	• Could operate at low powers (dazzlers) to blind the UAVs cameras or high power, which could burn/destroy the target; • Easy to track the target; • Cheaper and safer than projectiles or another physical countermeasure.	• Sensitive to weather conditions; • It is necessary to have an accurate measurement of the target's position; • High power lasers could interfere with other systems.	[124–129]
Magnetic	• Cost effective; • Could respond to multiple threats.	• Small protected area; • Could interfere with other systems.	[130]
Prey birds	• Does not require complex technology; • Fewer humans are required.	• Applicable only to slower and small UAVs. • Could harm the falcons.	[130]
Projectiles/ shooting nets/ water cannons	• Effective against any type of UAV; • Work in all weather conditions; • Quick reaction method.	• Might cause collateral damage; • High costs; • Requires professional operators.	[131–134]
RF/GNSS jamming	• Could neutralize grouped targets simultaneously, degrading their received signal-to-noise ratio (SNR); • GNSS frequencies and bands are widely known and relatively easy to jam; • The directivity diagram of the jamming signal can be oriented and directed as desired.	• Ineffective against autonomous UAVs; • Ineffective against drones that use inertial navigation systems/sensors (INS); • Ineffective against UAVs that use encrypted communications; • Effective only for short distances; • The jamming could interfere with other sensible equipment.	[135–139]
Spoofing	• DSP and AI algorithms could copy and reproduce the control communication signal with high accuracy in a relatively short time; • Could exploit the vulnerabilities of various systems of UAVs.	• It is necessary to have a consistent analysis of the targeted UAVs regarding their operation frequencies; • Spectrum sensing systems are desirable.	[140–145]

However, as pointed out in [35], destroying the drone does not mean that the problem is solved. Even if a drone is destroyed using one of the methods listed above, it is just half of the answer. It is critical to discover and detain the operator of the illegally flying UAV in order to resolve the problem completely. Without this, a motivated operator will almost certainly return with a newer and better UAV capable of causing even more disruption and damage.

3.4. Regulations Regarding the Use of Jamming in DDDSs

For most of the above-mentioned categories of countermeasures, there are not currently any regulations in force. However, in the case of RF jamming, several existing regulations apply, which will be detailed in the following paragraphs.

The neutralization of drones using jammers is still (in most countries) not legally permitted and is currently the subject of numerous regulatory and legal discussions.

The EU authorities were among the first organizations that took a position regarding the use of jamming devices. The Directive 2014/53/EU prohibits the use of such devices that could cause harmful interferences to the authorized radiocommunications and prevent the normal operation of the communications using radio frequencies [146]. This directive was transposed in all of the member state's legislations.

The Directive 2014/53/EU was transposed into Romanian legislation by Government Decision no.740/2016. According to this decision, the manufacture, importation, possession, advertising, placing on the market, making available on the market, putting into service and/or use of radio equipment or devices designed to cause harmful interference (jammers) are all prohibited and sanctioned with contravention [147].

In the UK, there were a lot of concerns regarding the collateral damage and the safety risks that must be taken into consideration when using jamming, because of the radio signal interference and the impact on other airspace users. However, only a few regulations have stated that such technology should not be used in any circumstances [148].

The FCC (Federal Communications Commission) in the United States does not merely state that the manufacture, sale, importation, and operation of jammers are all forbidden (Communications Act of 1934, Section 301), but that there are some exceptions, such as institutions under the US government. There is always the risk of a drone losing control, crashing, and causing property damage, or personal harm, when a drone jammer is deployed. This means that anyone using a drone jammer, even government-authorized workers, could be held liable. As a result, the deployment of drone jammers by private entities, such as power companies or airports, is still sporadic but strictly regulated. Only the federal government has the ability to approve the use of drone jammers, and this rigorous restriction extends to their manufacture, importation, and sales [149].

In the Russian Federation, flying a drone is legal. However, most Russian cities are equipped with GPS jammers, which create radio interference, preventing electronics, such as drones, from operating normally. As a consequence, drone users have to keep a safe distance from them because all of the major cities have integrated GPS jammers that can interfere with their drone positioning [150]. Also, there are some regulations that prohibit flying a drone within 500 m of a military installation.

In P. R. China, only the local authorities can use jammer "guns" and other RF DDDSs [151]

Despite of the lack of regulations regarding the use of RF jamming signals against drones, and some risks that should be taken into consideration, this method has to be considered to be among the most efficient.

4. Drone Detection and Defense Systems Based on RF Methods

As was mentioned in Section 3, one of the most used methods for drone detection is the identification of the RF signals that are exchanged by the drones with another entity (ground station/operator). Moreover, the annihilation of the detected drones can also be obtained by RF methods, by means of transmitting strong enough jamming signals that can interrupt the communication between the drone and its operator (as mentioned in Section 3.3).

Usually, drones operate on different frequencies, but most commercial drones operate in Industrial, Scientific, and Medical (ISM) frequency bands of 433 MHz and 2.4/5.8 GHz. The simple power detection in these bands will not work due to the presence of other legitimate users in the same geographical area. Therefore, most of the modern RF detection systems provide the detection and identification of the special and unique signals that are generated by the UAV or the data protocols implemented in a UAV.

There are two main functions that are necessary for the detection of the drones, as follows: The *identification* of the presence of the drones by scanning the frequency spectrum and *localization* of the drones. The *annihilation* function, which is necessary in order to allow the defense against the detected drones, can be performed by means of RF jamming, in order to interrupt the communication between the drones and their operators. Table 7 summarizes the main elements regarding the implementation of such systems. In the following paragraphs, each of the below mentioned categories will be detailed.

Table 7. RF-based drone detection and defense systems.

References	Implemented Functions	Methods	SDR Platform Used (Including Manufacturer, City and Country)
[152]	Identification Localization	RF fingerprinting (SFS, WEE, PSE) AoA (MUSIC, RAP MUSIC)	USRP-X310 (Ettus Research, Santa Clara, CA, USA)
[153]	Identification	RF fingerprinting (DRNN)	USRP-X310 (Ettus Research, Santa Clara, CA, USA)
[154]	Identification	RF fingerprinting (CNN)	USRP-X310 (Ettus Research, Santa Clara, CA, USA)
[155]	Identification	RF fingerprinting (KNN)	USRP-B210 (Ettus Research, Santa Clara, CA, USA)
[156]	Identification	RF fingerprinting (KNN, XGBoost)	-
[157]	Identification	RF fingerprinting (Wi-Fi)	-
[158]	Identification	RF fingerprinting	LimeSDR (Lime Microsystems, Guilford, UK)(customized)
[159]	Identification	RF fingerprinting	-
[160]	Localization	Received-signal strength (RSS)	USRP N210 (Ettus Research, Santa Clara, CA, USA)
[161]	Localization	RSS	AD-FMCOMMS5-EBZ Evaluation Board (Analog Devices, Wilmington, DC, USA)
[162–164]	Annihilation	RF jamming	BladeRF (Nuand, San Francisco, CA, USA)
[165]	Annihilation	RF jamming	Great Scott Gadgets HackRF One

Most of the RF-based solutions that are described in the literature focus only on the detection of the drones and do not propose countermeasures for the annihilation of the detected drones. One of the reasons behind this might be the increase in the complexity and price of the system that will be generated by the inclusion of such countermeasures in the designed system. A second reason might be related to the fact that most of the references that will be commented on in this section include academic research, in which the target was not the design of a complete commercial system. A third reason could be the fact that jamming equipment is not legal in many areas worldwide, as discussed in Sections 2 and 3. However, as mentioned previously, the jamming solution can be used in most of the countries if the system that generates it is used for national security or public order purposes.

Almost all of the implementations that were used for validating the solutions that are proposed in the literature are based on SDR platforms because of some of the significant advantages that are offered by this category of platforms, such as the following:

- Low to moderate cost;
- Extended frequency range, which can usually cover all of the frequency bands that are used by commercial drones;
- Scalability, allowing the extension of the platform, depending on the functions that are foreseen, to be implemented;

- Flexibility, allowing the processing of RF signals corresponding to different communication standards.

Only a few of the existing works include aspects that are related to both of the functions that were mentioned previously as necessary for the detection of the drones, *identification* and *localization*. Such an example is [152], where the authors proposed a drone detection system based on multi-dimensional signal feature identification. After identifying the channel on which the drone communicates with the controller, features, such as signal frequency spectrum (SFS), wavelet energy entropy (WEE), and power spectral entropy (PSE), are extracted in order to allow a precise identification of the drone. In a subsequent step, MUSIC and RAP-MUSIC algorithms are used for performing the localization of the drone, by using information, such as azimuth and elevation. The proposed solution is implemented and tested using USRP X310 SDR platforms and a circular antenna array, obtaining an average detection rate of more than 95%.

In most of the papers that are concerned with the *identification* of the drones RF fingerprinting techniques are used, which rely on the unique characteristics of the RF signal waveforms captured from different drones [153–157]. In [153], a classification of the detected drones is made using a deep residual neural network (DRNN), the results being validated using a USRP X310 SDR platform as a receiver and nine different drones as targets. The authors of [154] separate Wi-Fi and Bluetooth signals from UAV transmitted signals based on their bandwidth and modulation features and classify the UAV signals using machine learning (ML) techniques. In [155], the detection of multiple drones is performed using the k-nearest neighbor (KNN) algorithm after performing a short-time Fourier transform (STFT) on the received signal. A real-time testbed based on the USRP B210 SDR platform is also used for evaluating the performance of the proposed method. A combination of RF fingerprints and hierarchical learning is used in [156] for the classification of the detected drone signals. A Wi-Fi statistical fingerprint approach is proposed in [157], which accounts for the particular characteristics of the Wi-Fi control traffic produced by drones and their remote controllers.

In [158], a solution that is based on the low cost LimeSDR platform is developed for detecting the presence of drones in the 2.4–2.5 GHz ISM band. The authors use the LMS7002M RF chip from the LimeSDR module but customize the firmware of the FPGA located on the same SDR platform in order to implement the signal processing steps that are necessary for the identification of the RF signals that are transmitted by different drones.

The authors of [159] apply a STFT on the RF signals that are collected using a spectrum analyzer and calculate the time guards associated with the different hopping sequences using the autocorrelation function (ACF) in order to obtain an accurate differentiation of the different UAV remote control (RC) signals.

The following paragraphs will detail the different approaches that were proposed for the implementation of the *localization* function.

A received-signal strength- (RSS-) based 3D localization system utilizing a software defined radio is proposed in [160], using the recursive least squares (RLS) algorithm in order to numerically estimate the drone's 3D position.

The authors of [161] propose a localization approach based on the arrays of directional antennas, for obtaining the direction of arrival (DoA) of the NTSC signal that is transmitted by the drones.

Although the articles that were mentioned above only focused on the detection of drones based on RF methods, there are also papers that present implementations of the annihilation function using RF jamming as a countermeasure against drones [163–165].

In [162,163], a low-cost SDR platform, BladeRF X40 (Nuand, San Francisco, CA, USA) was used as hardware to implement a jamming system against unauthorized UAVs. The GNU Radio toolkit was used as a software environment for performing the necessary signal processing tasks. In [162], the communication of the drone with the remote control in the 2.4 GHz ISM band was targeted, whereas in [164] the GPS navigation system was targeted

The authors of [164] implemented a protocol-aware jammer using the BladeRF SDR platform as hardware. Tests were made to target the Futaba Advanced Spread Spectrum Technology (FASST) and the Advanced Continuous Channel Shifting Technology (ACCST) UAV remote control systems.

In [165], a portable jammer is proposed, based on the HackRF One SDR platform and a Raspberry Pi as a host computer. Multiple tests were made in order to validate the proposed solution, in both the 2.4 GHz and 5.8 GHz ISM bands and in the GPS L1 band.

5. Challenges and Future Perspectives for Drone Detection and Defense Systems

The previous sections contained a review of the different approaches that can be used for implementing a DDDS. In this section, the challenges that currently have to be faced when developing such a system will be detailed, together with a discussion regarding the future perspectives of this domain.

One of the challenges that is faced when implementing a DDDS is the ability to identify and, in a further step, to annihilate not only one, but several different target drones. In recent years, many applications have used multiple drones [166], therefore, such a feature becomes an important characteristic for a DDDS. Depending on the sensors that are used in the system, the possibility of detecting several target drones may or may not exist. A few examples of systems that include such a feature exist in the literature. In [167,168], algorithms are developed in order to allow multi-UAV detection using video streams. In [169], an RF-based deep learning (DL) algorithm is proposed for performing multiple drone detection. The possibility of a simultaneous annihilation of several drones is an even more challenging task. Electromagnetic pulses (EMP) have been proposed as a possible solution for defense against drone swarms [170]. RF jamming performed using antenna arrays could also generate, by means of signal processing methods (beamforming), multiple beams that could be targeted towards multiple target drones.

Another challenge that a DDDS would have to face, especially if the area in which the system is installed is a residential area, and there are several households in the close neighborhood, is to avoid interference or damage to nearby equipment (in the case of RF jamming and EMP) and to respect the privacy of the nearby neighbors (in the case of imaging sensors). In the case of RF jamming, this could be solved if the antennas that are used or the beams, in the case of using a beamforming approach, are very directive and targeted directly towards the target drone(s).

When referring to a DDDSs based on RF methods, one of the main challenges that has to be addressed is related to the legal issues around the use of jamming as a countermeasure, as was also commented on in Section 3.4. For the time being, in most of the regions worldwide, such a countermeasure can only be legally used when it is integrated into a system that is used for the defense of national security or for public order objectives. However, as the number of situations when such a system would be necessary also applies to the defense of private areas that cannot be included in the above mentioned categories, it is to be expected that the legislation in this domain might be modified in the near future in order to include the possibility of private users also legally using such a system, as long as the interference caused to the nearby areas is kept below certain well-defined thresholds.

An important limitation of RF-based DDDSs is related to the impossibility of detecting and annihilating autonomous drones in cases when they have a predefined flying path and do not have any active data communication with an operator located on the ground.

As mentioned in Table 5, each of the different types of sensors (RF, radar, imaging, and acoustic) has its own drawbacks and limitations. As such, the performance of a DDDS that is implemented using a single type of sensor is directly affected by the disadvantages and limitations of that particular category. By combining several different sensor types in a single *hybrid DDDS*, the system could benefit from the advantages of each different category of sensors. A first benefit would be the increase in accuracy that such a hybrid system could achieve, when the information regarding the identification and localization of the drone would be obtained from multiple different sensors. A second benefit would be related to the possibility of detecting the target drone in situations when one of the sensor types would not allow the detection on its own. For example, if we consider a hybrid DSSS that is implemented using both RF and imaging sensors, the imaging sensors could be used for detecting autonomous drones (that cannot be identified using the RF sensors) and the RF sensors could be used for detecting drones in low visibility conditions (when the imaging sensors could not provide the detection). Very few implementations of such hybrid systems are described in the literature (for example those in [105,115]), and we consider that such an approach is a promising future research and development direction for DDDSs.

6. DronEnd Detection and Defense System

In the current section, a drone detection and defense system, designed and implemented by the authors, together with a research team from the cybersecurity company Cyberwall [171], will be presented. The system was developed within the framework of the DronEnd research project [172]. The preliminary details regarding the project were given in [173].

The goal of the DronEnd ground defense system is to secure a certain area against the unauthorized presence of drones. In order to achieve this goal, the DronEnd system scans the RF spectrum in order to detect the presence of the drones in the supervised area, identifies the location of the drone by means of AoA algorithms, and annihilates the drone by using RF jamming methods. The block diagram of the implemented DronEnd ground defense system is presented in Figure 2.

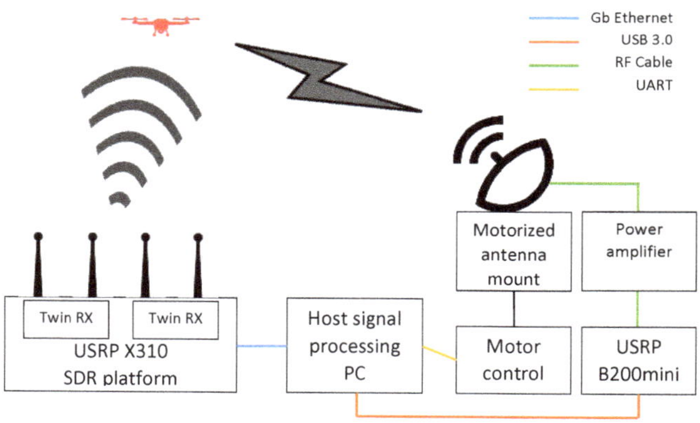

Figure 2. Block diagram of the DronEnd ground defense system.

In the following subsections, all of the elements of the system will be detailed, highlighting the steps that are necessary in order to perform the functions of detection, localization, and annihilation of the drone through jamming.

6.1. Detecting the Presence of the Drone Using Spectrum Sensing Algorithms

A first step required for detecting the presence of a drone in the case of RF-based drone defense systems is to monitor the radio spectrum through a spectrum sensing process in order to identify the signals that are transmitted by the drone. For the implementation of the spectrum sensing process in the DronEnd system, spectrum sensing algorithms based on the energy detection method have been used. Algorithms, such as 3EED [174] and 3EED with an adaptive threshold [175], that were previously developed, provide improved performance compared to the classical energy detection (CED) [176] algorithm and were used to identify the presence of the drones in the monitored area. The above-mentioned algorithms were implemented on SDR platforms from the USRP family (USRP X310 (Ettus Research, Santa Clara, CA, USA) [177] equipped with Twin-RX RF Daugterboards (Ettus Research,

Santa Clara, CA, USA) [178], 10–6000 MHz frequency range). The frequency bands that are used by the drones that were used to test the DronEnd system (DJI Mavic Air (SZ DJI Technology Co., Ltd., Shenzhen, China) [179], DJI Phantom 4 Pro v2.0 (SZ DJI Technology Co., Ltd., Shenzhen, China) [180], and DJI Mini 2 (SZ DJI Technology Co., Ltd., Shenzhen, China) [181]) were the 2.4 GHz (2400–2500 MHz) and the 5 GHz (5730–5830 MHz) ISM bands, which can be covered using the above-mentioned SDR platforms that can receive signals on frequencies up to 6 GHz. Because the position of the target drones was not initially known, omnidirectional antennas were used in this step.

Figure 3 shows the graphical user interface that was implemented in order to view the results of the spectrum sensing. The signal that was transmitted by the DJI Mavic Air drone on channel four of the ISM 2.4 GHz band can be seen as captured using the USRP X310 SDR platform. In the following subsections, the other elements of the DronEnd system will be detailed, highlighting the steps that are necessary in order to perform the functions of localization and annihilation of the drone through jamming. The capture of the RF data was performed using a GNU Radio python script. As the instantaneous bandwidth captured using the Twin-RX RF daughterboard is smaller than 100 MHz, in order to cover the 100 MHz bandwidth of the 2.4 GHz and 5 GHz ISM bands, several sub-bands were concatenated.

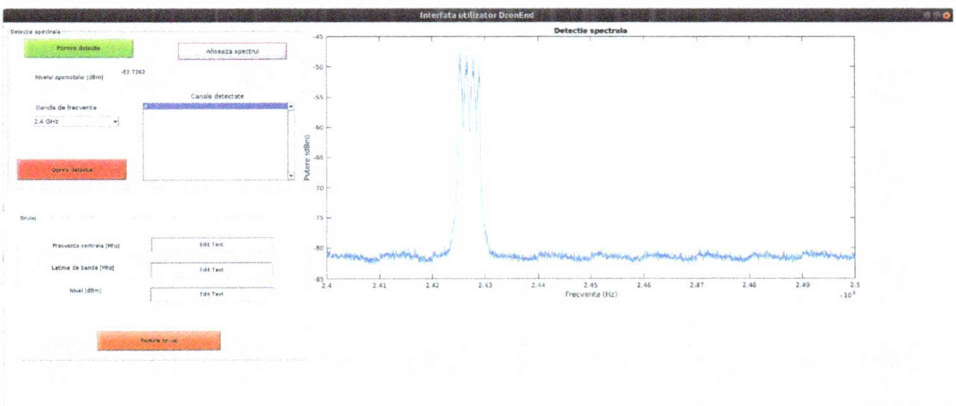

Figure 3. Graphical user interface of the spectrum sensing process implemented in the DronEnd system, showing the signal transmitted by the DJI Mavic Air drone in the 4th channel of the 2.4 GHz ISM band.

Once the signal that is transmitted by the target drone is detected, the next step is triggered, which is to localize the angle of arrival of the received signal, as will be discussed in the next subsection.

6.2. Localization of the Drone Using AoA Algorithms

Once the frequency that is used by the drone to communicate has been identified, a second necessary step is to obtain information about the position of the drone. This step was performed using AoA algorithms for detecting the angle of incidence of the detected RF signal. Such algorithms exploit the phase difference of the signals that are received from the drone using a multi-antenna system. The SDR platform that was used as the hardware for providing the RF receive front-end was the USRP X310 [177], on which two Twin-RX RF modules [178] were mounted (covered frequency range of 10–6000 MHz, instantaneous bandwidth 80 MHz). Each of the Twin-RX modules offers two coherent reception channels, and the local oscillator that was used can be shared by the two boards, so that in the end, a total of four coherent reception channels are obtained and are aligned in phase. The antenna system that was used was a linear system of four antennas, spaced at a distance equal to half the wavelength of the minimum frequency that the drones used for testin could transmit (2.4 GHz). In order to estimate the initial phase difference between the fou reception channels, a calibration step was required after each system restart, which involve the transmission of a test signal that will be received through the RF cables of equal lengt on all four of the reception channels. A 5-port RF splitter (Mini-Circuits ZN4PD1-63HP-S (Mini-Circuits, New York, USA) [182]) was used in order to distribute the signals. Figure shows both the antenna system that was used and the USRP X310 SDR platform during th calibration stage.

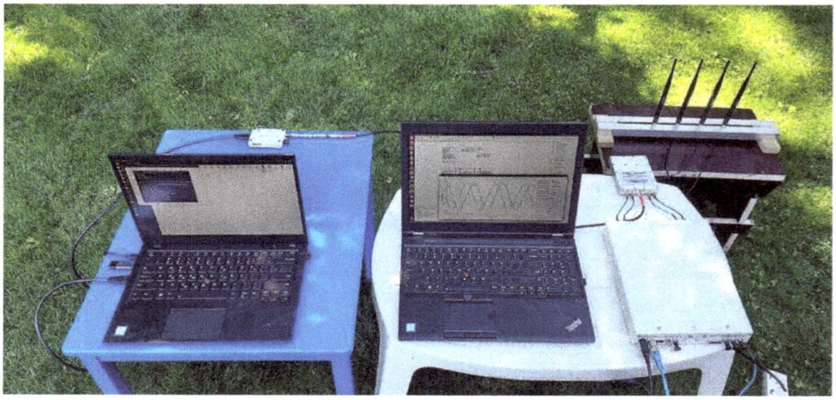

Figure 4. The linear antenna system that was used and the USRP X310 SDR platform during th calibration procedure of the Twin-RX RF modules.

Once the calibration step was completed, the four dipole antennas (VERT2450 [183] that make up the antenna system were connected to the four receive channels of the USR X310 SDR platform and, based on the phase difference of the signals that were receive the angle of incidence that corresponds to the drone location could be identified by usin AoA algorithms. We used one of the classical AoA algorithms, the MUSIC algorithm, an the result was both displayed on a graphical user interface, as shown in Figure 5, an forwarded as an input to the software module that is responsible for setting the orientatio of the jamming antenna, which will be detailed in the next subsection.

Figure 5. Tests performed using the DronEnd ground system (DJI Mavic AIR target drone). Th estimated angle of incidence can be noticed.

The positioning that was thus obtained was one in azimuth, as the antenna system that was used was placed horizontally. By using a second system that was located in a vertical plane, the elevation of the drone could be also estimated.

6.3. Annihilation of the Drone Using RF Jamming

A final step is to transmit a jamming signal to the identified target drone in order to disrupt the communication between the drone and its operator. As the jamming signal should only be transmitted in the direction of the target drone, in order to avoid interference with other equipment in the area, a directional antenna was used for the jamming operation. Figure 6 shows the following components that were used to implement this step: the transmitting antenna, the motorized antenna mount, the stepper motor control module that was used to move the antenna mount, and the power amplifier.

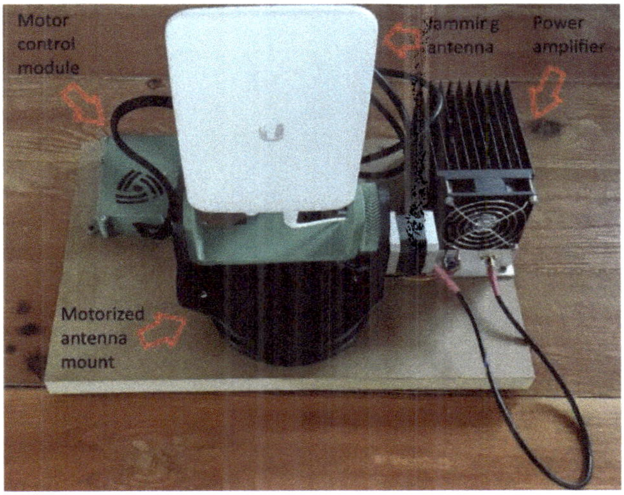

Figure 6. Components used for transmitting the jamming signal.

The angle of incidence that was detected by the AoA algorithm was processed (filtered) using a script implemented in the Matlab environment in order to remove any erroneous indications related to the position of the drone and was subsequently transmitted using a serial interface (UART) to the motor control module (MCM), which controls the stepper motors that are used to move the motorized support for positioning the jamming antenna. The MCM was based on a Microchip ATMega328p processor, which, using the angle information that is obtained using the AoA algorithm, controls the stepper motors. Two Nema 17 stepper motors, controlled using Texas instruments DRV8825 drivers, were used; one to adjust the azimuth and one to adjust the elevation of the jamming signal antenna. In the current configuration, given that the drone's position was estimated only in the azimuth plane, the commands were transmitted only to one of the two motors (the one that was responsible for the azimuth movement).

The SDR platform that was used to generate the jamming signal was a USRP B200mini platform (70–6000 MHz frequency range) (Ettus Research, Santa Clara, CA, USA) [184]. Given that the maximum power that can be obtained at the output of the SDR platform is 10 dBm, a power amplifier (Mini-Circuits ZHL-2W-63-S+ (Mini-Circuits, New York, NY, USA) [185]) was used to amplify the jamming signal in order to extend the range of the system, which offers a 42 dB gain and a maximum output power of 2 W. The antenna that was used to transmit the jamming signal was a Ubiquiti UMA-D (Ubiquiti Inc., New York, NY, USA) directional antenna [186], which covers the 2.4–2.5 GHz and 5.1–5.9 GHz bands and offers a 10 dBi gain in the 2.4 GHz band and a 15 dBi gain in the band of 5.8 GHz. By using a directional antenna that targets the location of the drone for the transmission of the jamming signal, the interferences that are caused to other communication systems that are operating in the neighborhood are minimized. Moreover, the transmit gain can be adjusted depending on the size of the area that has to be protected. Figure 7 shows the jamming signal with a 10 MHz bandwidth emitted in channel four of the 2.4–2.5 GHz ISM band, captured using an Anritsu MS2690A (Anritsu Corporation, Atsugi, Japan) spectrum analyzer.

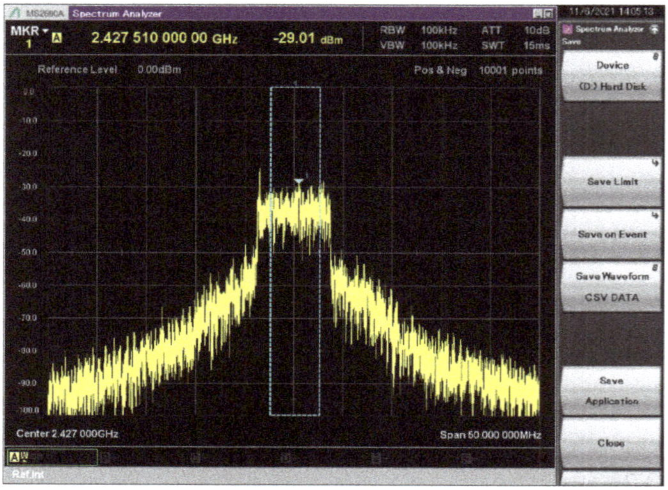

Figure 7. Jamming signal transmitted by the DronEnd ground system on channel 4 in the 2.4–2.5 GHz ISM band.

The tests were performed in an outdoor suburban scenario using the DJI Mavic Air, the DJI Phantom 4 Pro v2.0, and the DJI Mini 2 drones as targets and the annihilation of the drone, which resulted in a forced landing on the position where the drone was located when the jamming signal was turned on, was possible for distances of 40 m from the area where the DronEnd ground system was located.

6.4. Conclusion and Future Research Directions

To conclude, the main novel elements that are introduced by the DronEnd system, when compared to other drone detection and defense systems based on RF methods that were mentioned in Section 4, can be summarized as follows:

- Incorporates all of the three functions (identification, localization, and annihilation) that are necessary for a drone detection and defense system in an integrated and scalable platform, which can be reconfigured depending on the requirements of different use cases;
- Includes an agile and accurate identification subsystem, based on improved spectrum sensing algorithms, which performs a real-time identification of the signals that are transmitted by the drone and, moreover, allows a dynamic tracking of the signal transmitted by the drone, even when the transmit frequency is changed;
- Annihilates the detected drone by means of jamming, avoiding at the same time significant interference with nearby devices, as a directional antenna, targeted directly towards the target drone using a motorized antenna mount, is used.

Several aspects are considered as future research directions, in order to improve the performance of the proposed system.

The first direction is related to the possibility of replacing the mechanical motorized antenna mount that was used for targeting the directional jamming antenna with an equivalent static planar antenna array. By using such an approach, the orientation of the resulting antenna beam that was necessary for following the target drone would not involve any moving parts, as the steering would be obtained only by using signal processing methods. The advantages of such an approach would include a smaller delay, the possibility of adjusting the beamwidth by signal processing means, depending on the application necessity, and the absence of aging effects that might affect mechanical parts. However, as the transmit power level that is needed in order to obtain a large enough range for the system might be high, a challenge that would have to be addressed is the design of a power amplification stage for supplying the planar antenna array.

The second direction is related to the addition of a second antenna array, in an orthogonal plane, compared to the one in which the current antenna array is located. By using such a setup, the identification of the target drone could be performed both in azimuth and elevation, allowing for a more precise steering of the directional antenna that is used for transmitting the jamming signal.

The third research direction is related to a subject that was also commented on in Section 5, which is the implementation of a hybrid DDDS in order to improve the overall performance of the system. The addition of imaging sensors is considered, as such an approach would have a twofold contribution; it would improve the accuracy of the detected targets for the situations in which the target drone would be detected by both types of sensors, and it would allow the detection of the target drones also in the situations when only one type of sensor would be able to identify them.

7. Conclusions and Future Work

In this paper, a survey related to the current status of drone detection and defense systems was performed and our own solution for a drone defense system based on SDR platforms (DronEnd) was presented. Different aspects, such as regulatory issues and reported incidents that involved drones, were included in the survey. A classification of the drone detection systems that were based on the type of sensors that are used was performed. A detailed description of the RF-based drone detection and defense systems was made, with an emphasis on the use of SDR platforms for the implementation of such systems. The drone defense system that was developed by the authors within the framework of the DronEnd research project is presented in the final part of the paper. As future work, we intend to conduct a detailed testing of the DronEnd ground system, in order to verify the performance of our solution from the detection, localization, and annihilation points of view and we also plan to develop a flying version of the DronEnd system, by mounting an embedded SDR platform on a support drone and approaching the target drones from the air.

Author Contributions: Conceptualization, F.-L.C., A.M., C.V., I.M., R.C. and O.F.; methodology, F.-L.C. and A.M.; software, A.M.; validation, A.M., C.V., I.M. and O.F.; formal analysis, C.V., I.M. and O.F.; investigation, F.-L.C.; resources, F.-L.C. and A.M.; data curation, F.-L.C. and A.M.; writing—original draft preparation, F.-L.C. and A.M.; writing—review and editing, F.-L.C., A.M., C.V., I.M., R.C. and O.F.; visualization, F.-L.C. and R.C.; supervision, C.V., I.M. and O.F.; project administration, A.M.; funding acquisition, A.M. All authors have read and agreed to the published version of the manuscript.

Funding: This work was supported by a grant of the Romanian Ministry of Education and Research, CCCDI-UEFISCDI, project number PN-III-P2-2.1-PED-2019-1951, within PNCDI III.

Acknowledgments: The authors would like to acknowledge the editor and the anonymous reviewers for their comments and suggestions that have improved the presentation of this work.

Conflicts of Interest: The authors declare no conflict of interest.

Abbreviations

ACCST	Advanced Continuous Channel Shifting Technology
ACF	Autocorrelation Function
ACMA	Australian Communication and Media Authority
AI	Artificial Intelligence
AoA	Angle of Arrival
CNN	Convolutional Neural Network
DDDS	Drone Detection and Defense Systems
DoA	Direction of Arrival
DRNN	Deep Residual Neural Network
DSP	Digital Signal Processing
DL	Deep Learning
EASA	European Union Aviation Safety
EMP	Electromagnetic Pulses
EO	Electro-optical
FASST	Futaba Advanced Spread Spectrum Technology
FCC	Federal Communications Commission
FPGA	Field-Programmable Gate Array
GNSS	Global Navigation Satellite System
GPS	Global Positioning System
INS	Inertial Navigation Systems/Sensors
IR	Infrared
IS	Islamic State
ISM	Industrial, Scientific, and Medical
KNN	K-Nearest Neighbor
LoS	Line of Sight
MDS	Micro-Doppler Signatures
MCM	Motor Control Module
ML	Machine Learning
MUSIC	MUltiple SIgnal Classification
NTSC	National Television Standards Committee
PSE	Power Spectral Entropy
RC	Remote Control
RF	Radio Frequency
RLS	Recursive Least Squares
RSS	Received-Signal Strength
SDR	Software-Defined Radio
SFS	Signal Frequency Spectrum
SNR	Signal-to-Noise Ratio
STFT	Short-Time Fourier Transform
UAS	Unmanned Aerial Systems
UAV	Unmanned Air Vehicle
WEE	Wavelet Energy Entropy

References

1. Zeng, Y.; Zhang, R.; Lim, T.J. Wireless Communications with Unmanned Aerial Vehicles: Opportunities and Challenges. *IEE Commun. Mag.* **2016**, *54*, 36–42. [CrossRef]
2. World Economic Forum. Drones and Tomorrow's Airspace. 2020. Available online: https://www.weforum.org/communities drones-and-tomorrow-s-airspace (accessed on 13 January 2022).
3. Scott, G.; Smith, T. Disruptive Technology: What Is Disruptive Technology? *Investopedia* **2020**. Available online: https://www investopedia.com/terms/d/disruptive-technology.asp/ (accessed on 13 January 2022).
4. Germen, M. Alternative cityscape visualisation: Drone shooting as a new dimension in urban photography. In Proceedings of th Electronic Visualisation and the Arts (EVA), London, UK, 12–14 July 2016; pp. 150–157.
5. Kaufmann, E.; Gehrig, M.; Foehn, P.; Ranftl, R.; Dosovitskiy, A.; Koltun, V.; Scaramuzza, D. Beauty and the beast: Optima methods meet learning for drone racing. In Proceedings of the IEEE International Conference on Robotics and Automatio Montreal, QC, Canada, 20–24 May 2019; pp. 690–696.

6. Kaufmann, E.; Loquercio, A.; Ranftl, R.; Dosovitskiy, A.; Koltun, V.; Scaramuzza, D. Deep drone racing: Learning agile flight in dynamic environments. In Proceedings of the Conference on Robot Learning (CoRL), Zürich, Switzerland, 29–31 October 2018; pp. 133–145.
7. Ahmad, A.; Cheema, A.A.; Finlay, D. A survey of radio propagation channel modelling for low altitude flying base stations. *Comput. Netw.* **2020**, *171*, 107122. [CrossRef]
8. Tozer, T.; Grace, D.; Thompson, J.; Baynham, P. UAVs and HAPspotential convergence for military communications. In Proceedings of the IEEE Colloquium on Military Satellite Communications, London, UK, 6 June 2000; pp. 10-1–10-6.
9. Schneiderman, R. Unmanned drones are flying high in the military/aerospace sector. *IEEE Signal Process. Mag.* **2012**, *29*, 8–11. [CrossRef]
10. Chen, J.Y.C. UAV-guided navigation for ground robot tele-operation in a military reconnaissance environment. *Ergonomics* **2010**, *53*, 940–950. [CrossRef]
11. Coffey, T.; Montgomery, J.A. The emergence of mini UAVs for military applications. *Def. Horiz.* **2002**, *22*, 1.
12. Quigley, M.; Goodrich, M.A.; Griffiths, S.; Eldredge, A.; Beard, R.W. Target acquisition, localization, and surveillance using a fixed-wing mini-UAV and gimbaled camera. In Proceedings of the IEEE International Conference on Robotics and Automation, Barcelona, Spain, 18–22 April 2005; pp. 2600–2605.
13. Iscold, P.; Pereira, G.A.S.; Torres, L.A.B. Development of a hand launched small UAV for ground reconnaissance. *IEEE Trans. Aerosp. Electron. Syst.* **2010**, *46*, 335348. [CrossRef]
14. Page, M.J.; McKenzie, J.E.; Bossuyt, P.M.; Boutron, I.; Hoffmann, T.C.; Mulrow, C.D.; Shamseer, L.; Tezlaff, J.M.; Akl, E.A.; Brennan, S.E.; et al. The PRISMA 2020 statement: An updated guideline for reporting systematic reviews. *BMJ* **2021**, *372*, n71. Available online: http://www.prisma-statement.org/ (accessed on 13 January 2022). [CrossRef]
15. Butt, A.; Shah, S.I.A.; Zaheer, Q. Weapon launch system design of anti-terrorist UAV. In Proceedings of the International Conference on Engineering Technology (ICEET), Lahore, Pakistan, 21–22 February 2019; pp. 1–8.
16. EY India. What's the Right Strategy to Counter Rogue Drones? Available online: https://www.ey.com/en_in/emergingtechnologies/whats-the-right-strategy-to-counter-rogue-drones (accessed on 12 January 2022).
17. Ritchie, M.; Fioranelli, F.; Borrion, H. Micro UAV crime prevention: Can we help princess Leia? In *Crime Prevention 21st Century*; Springer: New York, NY, USA, 2017; pp. 359–376.
18. Van Voorst, B.R. Counter Drone System. U.S. Patent 15,443,143, 14 September 2017.
19. Drone Hits Plane at Heathrow Airport, Says Pilot. Available online: https://www.theguardian.com/uk-news/2016/apr/17/drone-plane-heathrow-airport-british-airways (accessed on 12 January 2022).
20. Alex Silverman, Drone Hits Army Helicopter Flying Over Staten Island. Available online: https://newyork.cbslocal.com/2017/09/22/drone-hits-army-helicopter/ (accessed on 12 January 2022).
21. Drone Collides with Commercial Aeroplane in Canada. Available online: https://www.bbc.com/news/technology-41635518 (accessed on 12 January 2022).
22. Boeing 737 Passenger Jet Damaged in Possible Mid-Air Jet. Available online: https://www.bloomberg.com/news/articles/2018-12-13/aeromexico-737-jetliner-damaged-in-possible-midair-drone-strike (accessed on 12 January 2022).
23. Plane Damaged after Being Hit by York Police Drone at Buttonville Airport. Available online: https://toronto.ctvnews.ca/plane-damaged-after-being-hit-by-york-police-drone-at-buttonville-airport-1.5554617 (accessed on 12 January 2022).
24. Drone's Operator Detained for Flying Near Chinese Airplane. Available online: https://edition.cnn.com/2017/01/17/asia/china-drone-passenger-plane-near-miss/ (accessed on 12 January 2022).
25. Air New Zealand Calls for Drone Legislation after Near Miss. Available online: https://www.bbc.com/news/world-asia-43551373.amp (accessed on 12 January 2022).
26. Gatwick Drones: As it Happened. Available online: https://www.bbc.com/news/live/uk-england-sussex-46564814 (accessed on 12 January 2022).
27. Ripley, C.W. Drone with Radioactive Material Found on Japanese Prime Minister's Roof. 2015. Available online: https://edition.cnn.com/2015/04/22/asia/japan-primeminister-rooftop-drone/index.html (accessed on 12 January 2022).
28. Gibbons-Neff, T. ISIS Used an Armed Drone to Kill Two Kurdish Fighters and Wound French Troops, Report Says. 2016. Available online: https://www.washingtonpost.com/news/checkpoint/wp/2016/10/11/isis-used-an-armed-drone-to-kill-twokurdish-ghters-and-wound-french-troops-report-says/ (accessed on 12 January 2022).
29. BBC. Venezuela President Maduro Survives Drone Assassination Attempt. 2018. Available online: https://www.bbc.com/news/world-latin-america-45073385 (accessed on 12 January 2022).
30. Drone Incident Management at Aerodromes. Available online: https://www.easa.europa.eu/sites/default/files/dfu/drone_incident_management_at_aerodromes_part1_website_suitable.pdf (accessed on 12 January 2022).
31. *FCC Enforcement Advisory, Cell Jammers, GPS Jammers, and Other Jamming Devices, document FCC RCD 1329(2)*; FCC: Washington, DC, USA, 2011.
32. Radiocommunications Exemption Arrangements for Drone Jamming Devices. Available online: https://www.acma.gov.au/sites/default/files/2019-08/IFC-4-2019-Consultation%20Paper%20-%20Radiocommunications%20exemption%20arrangements%20for%20drone%20jamming%20devices.docx (accessed on 12 January 2022).
33. Markarian, G.; Staniforth, A. *Countermeasures for Aerial Drones*; ARTECH HOUSE: Norwood, MA, USA, 2021; ISBN 13: 978-1-63081-801-2.

34. Michel, A.H. *Counter-Drones Systems*, 2nd ed.; Report from the Center of the Study of the Drone at Bard College: Annandale-On-Hudson, NY, USA, 2019; Available online: https://dronecenter.bard.edu/files/2018/02/CSD-Counter-Drone-Systems-Report.pdf (accessed on 13 January 2022).
35. Brust, M.R.; Danoy, G.; Stolfi, D.H.; Pascal, B. Swarm-based counter UAV defense system. *Discov Internet Things* **2021**, *1*, 2. [CrossRef]
36. Alsok. 2020. Available online: https://www.alsok.co.jp/en/ (accessed on 12 January 2022).
37. Ottoy, G.; de Strycker, L. An improved 2D triangulation algorithm for use with linear arrays. *IEEE Sens. J.* **2016**, *16*, 8238–8243. [CrossRef]
38. Shi, Z.; Chang, X.; Yang, C.; Wu, Z.; Wu, J. An Acoustic-Based Surveillance System for Amateur Drones Detection and Localization. *IEEE Trans. Veh. Technol.* **2020**, *69*, 2731–2739. [CrossRef]
39. Mezei, J.; Fiaska, V.; Molnar, A. Drone sound detection. In Proceedings of the 16th IEEE International Symposium on Computational Intelligence and Informatics (CINTI), Budapest, Hungary, 19–21 November 2015; pp. 333–338.
40. Hilal, A.A.; Mismar, T. Drone Positioning System Based on Sound Signals Detection for Tracking and Photography. In Proceedings of the 11th IEEE Annual Information Technology, Electronics and Mobile Communication Conference (IEMCON), Vancouver, BC, Canada, 4–7 November 2020; pp. 8–11. [CrossRef]
41. Kim, J.; Kim, D. Neural network based real-time UAV detection and analysis by sound. *J. Adv. Inf. Technol. Converg.* **2018**, *8*, 43–52. [CrossRef]
42. Busset, J.; Perrodin, F.; Wellig, P.; Ott, B.; Heutschi, K.; Rühl, T.; Nussbaumer, T. Detection and tracking of drones using advanced acoustic cameras. *Unmanned/Unattended Sens. Sens. Netw. XI Adv. Free. Space Opt. Commun. Tech. Appl.* **2015**, *9647*, 96470F.
43. Christnacher, F.; Hengy, S.; Laurenzis, M.; Matwyschuk, A.; Naz, P.; Schertzer, S.; Schmitt, G. Optical and acoustical UAV detection. *Electro-Opt. Remote Sens. X* **2016**, *9988*, 99880B.
44. Seo, Y.; Jang, B.; Im, S. Drone detection using convolutional neural networks with acoustic STFT features. In Proceedings of the 15th IEEE International Conference on Advanced Video and Signals-based Surveillance (AVSS), Auckland, New Zealand, 27–30 November 2018; pp. 1–6.
45. Bernardini, A.; Mangiatordi, F.; Pallotti, E.; Capodiferro, L. Drone detection by acoustic signature identication. *Electron. Imaging* **2017**, *2017*, 60–64. [CrossRef]
46. Hauzenberger, L.; Ohlsson, E.H. Drone Detection Using Audio Analysis. Master's Thesis, Department of Electrical and Information Technology, Faculty of Engineering, LTH, Lund University, Lund, Sweden, 2015.
47. Harvey, B.; O'Young, S. Acoustic detection of a xed-wing UAV. *Drones* **2018**, *2*, 4. [CrossRef]
48. Yang, C.; Wu, Z.; Chang, X.; Shi, X.; Wo, J.; Shi, Z. DOA Estimation Using Amateur Drones Harmonic Acoustic Signals. In Proceedings of the 2018 IEEE 10th Sensor Array and Multichannel Signal Processing Workshop (SAM), Sheffield, South Yorkshire, 8–11 July 2018; pp. 587–591. [CrossRef]
49. Kim, J.; Park, C.; Ahn, J.; Ko, Y.; Park, J.; Gallagher, J.C. Real-time UAV sound detection and analysis system. In Proceedings of the 2017 IEEE Sensors Applications Symposium (SAS), Glassboro, NJ, USA, 13–15 March 2017; pp. 1–5.
50. Siriphun, N.; Kashihara, S.; Fall, D.; Khurat, A. Distinguishing Drone Types Based on Acoustic Wave by IoT Device. In Proceedings of the 2018 22nd International Computer Science and Engineering Conference (ICSEC), Chiang Mai, Thailand, 21–24 November 2018; pp. 1–4. [CrossRef]
51. Droneshield. Dronesentry. 2020. Available online: https://www.droneshield.com/sentry (accessed on 13 January 2022).
52. Chang, X.; Yang, C.; Wu, J.; Shi, X.; Shi, Z. A surveillance system for drone localization and tracking using acoustic arrays. In Proceedings of the IEEE 10th Sensor Array Multichannel Signal Process Workshop (SAM), Sheffield, UK, 8–11 July 2018; pp. 573–577.
53. Al-Emadi, S.; Al-Ali, A.; Mohammad, A.; Al-Ali, A. Audio based drone detection and identication using deep learning. In Proceedings of the 15th International Wireless Communications & Mobile Computing Conference (IWCMC), Tangier, Morocco, 24–28 June 2019; pp. 459–464.
54. Opromolla, R.; Fasano, G.; Accardo, D. A vision-based approach to UAV detection and tracking in cooperative applications. *Sensors* **2018**, *18*, 3391. [CrossRef]
55. Rozantsev, A.; Lepetit, V.; Fua, P. Detecting flying objects using a single moving camera. *IEEE Trans. Pattern Anal. Mach. Intell.* **2017**, *39*, 879–892. [CrossRef]
56. Park, J.; Kim, D.H.; Shin, Y.S.; Lee, S. A comparison of convolutional object detectors for real-time drone tracking using a PTZ camera. In Proceedings of the 2017 17th International Conference on Control, Automation and Systems (ICCAS), Jeju, Korea, 18–21 October 2017; pp. 696–699. [CrossRef]
57. Nalamati, M.; Kapoor, A.; Saqib, M.; Sharma, N.; Blumenstein, M. Drone Detection in Long-Range Surveillance Videos. In Proceedings of the 2019 16th IEEE International Conference on Advanced Video and Signal Based Surveillance (AVSS), Taipei, Taiwan, 18–21 September 2019; pp. 1–6. [CrossRef]
58. Müller, T. Robust drone detection for day/night counter-UAV with static VIS and SWIR cameras. *Proc. SPIE* **2017**, *10190*, 302–313.
59. Magoulianitis, V.; Ataloglou, D.; Dimou, A.; Zarpalas, D.; Daras, P. Does deep super-resolution enhance UAV detection. In Proceedings of the 2019 16th IEEE International Conference on Advanced Video and Signal Based Surveillance (AVSS), Taipei, Taiwan, 18–21 September 2019; pp. 1–6.

0. Birch, G.C.; Woo, B.L. *Counter Unmanned Aerial Systems Testing: Evaluation of VIS SWIR MWIR and LWIR Passive Imagers*; SNL-NM: Albuquerque, NM, USA, 2017; Tech. Rep.; SAND2017-0921 650791.
1. Chen, H.; Wang, Z.; Zhang, L. Collaborative spectrum sensing for illegal drone detection: A deep learning-based image classification perspective. *China Commun.* **2020**, *17*, 81–92. [CrossRef]
2. Ringwald, T.; Sommer, L.; Schumann, A.; Beyerer, J.; Stiefelhagen, R. UAV-Net: A fast aerial vehicle detector for mobile platforms. In Proceedings of the IEE?/CVF Conference on Computer Vision and Pattern Recognit?on, CVPR Workshops, Long Beach, CA, USA, 14–19 June 2019; pp. 1–9.
3. Craye, C.; Ardjoune, S. Spatio-temporal semantic segmentation for drone detection. In Proceedings of the 16th IEEE International Conference on Advanced Video and Signal Based Surveillance (AVSS), Taipei, Taiwan, 18–21 September 2019; pp. 1–5.
4. Sapkota, K.R.; Roelofsen, S.; Rozantsev, A.; Lepetit, V.; Gillet, D.; Fua, P.; Martinoli, A. Vision-based unmanned aerial vehicle detection and tracking for sense and avoid systems. In Proceedings of the 2016 IEEE/RSJ International Conference on Intelligent Robots and Systems (IROS 2016), Daejeon, Korea, 9–14 October 2016; pp. 1556–1561.
5. Aker, C.; Kalkan, S. Using deep networks for drone detection. In Proceedings of the 14th IEEE International Conference on Advanced Video and Signal based Surveillance (AVSS 2017), Lecce, Italy, 29 August–1 September 2017; pp. 1–6.
6. Zhu, P.; Wen, L.; Du, D.; Bian, X.; Ling, H.; Hu, Q.; Nie, Q.; Cheng, H.; Liu, C.; Chenfeng, L.; et al. VisDrone-DET2018: The vision meets drone object detection in image challenge results. In Proceedings of the 15th European Conference, Munich, Germany, 8–14 September 2018; pp. 1–30.
7. Schumann, A.; Sommer, L.; Klatte, J.; Schuchert, T.; Beyerer, J. Deep cross-domain ying object classication for robust UAV detection. In Proceedings of the 14th IEEE International Conference on Advanced Video and Signal based Surveillance (AVSS 2017), Lecce, Italy, 29 August–1 September 2017; pp. 1–6.
8. Wang, L.; Ai, J.; Zhang, L.; Xing, Z. Design of airport obstacle free zone monitoring UAV system based on computer vision. *Sensors* **2020**, *20*, 2475. [CrossRef]
9. Saqib, M.; Khan, S.D.; Sharma, N.; Blumenstein, M. A study on detecting drones using deep convolutional neural networks. In Proceedings of the 14th IEEE Int. Conf. Adv. Video Signal Based Surveill. (AVSS), Lecce, Italy, 29 August–1 September 2017; pp. 1–5.
0. Cigla, C.; Thakker, R.; Matthies, L. Onboard stereo vision for drone pursuit or sense and avoid. In Proceedings of the 2018 IEEE/CVF Conference on Computer Vision and Pattern Recognition Workshops (CVPRW), Salt Lake City, UT, USA, 18–22 June 2018; pp. 738–746.
1. Crivellaro, A.; Rad, M.; Verdie, Y.; Yi, K.M.; Fua, P.; Lepetit, V. A novel representation of parts for accurate 3D object detection and tracking in monocular images. In Proceedings of the IEEE International Conference on Computer Vision (ICCV), Santiago, Chile, 7–13 December 2015; pp. 4391–4399.
2. Liu, H.; Qu, F.; Liu, Y.; Zhao, W.; Chen, Y. A drone detection with aircraft classication based on a camera array. In Proceedings of the 4th International Conference on Structure, Processing and Properties of Materials (SPPM 2018), Dhaka, Bangladesh, 1–3 March 2018; Volume 322. no. 5, Art. no. 052005.
3. Wellig, P.; Speirs, P.; Schupbach, C.; Oeschlin, R.; Renker, M.; Boeniger, U.; Pratisto, H. Radar Systems and Challenges for C-UAV. In Proceedings of the 19th International Radar Symposium IRS 2018, Bonn, Germany, 20–22 June 2018.
4. Torvik, B.; Olsen, K.E.; Griffiths, H. Classification of birds and UAVs based on radar polarimetry. *IEEE Geosci. Remote Sens. Lett.* **2016**, *13*, 13051309. [CrossRef]
5. Ren, J.; Jiang, X. Regularized 2-D complex-log spectral analysis and subspace reliability analysis of micro-Doppler signature for UAV detection. *Pattern Recognit.* **2017**, *69*, 225–237. [CrossRef]
6. Kim, B.K.; Kang, H.-S.; Park, S.-O. Drone classication using convolutional neural networks with merged Doppler images. *IEEE Geosci. Remote Sens. Lett.* **2017**, *14*, 38–42. [CrossRef]
7. Mahafza, B.R. *Radar Systems Analysis and Design Using MATLAB*; CRC Press: Boca Raton, FL, USA, 2013.
8. Li, C.J.; Ling, H. An investigation on the radar signatures of small consumer drones. *IEEE Antennas Wirel. Propag. Lett.* **2017**, *16*, 649–652. [CrossRef]
9. Shin, D.-H.; Jung, D.-H.; Kim, D.-C.; Ham, J.-W.; Park, S.-O. A distributed FMCW radar system based on fiber-optic links for small drone detection. *IEEE Trans. Instrum. Meas.* **2017**, *66*, 340–347. [CrossRef]
0. Mizushima, T.; Nakamura, R.; Hadama, H. Reection characteristics of ultra-wideband radar echoes from various drones in flight. In Proceedings of the IEEE Topical Conference on Wireless Sensors and Sensor Networks (WiSNeT), San Antonio, TX, USA, 17–20 January 2020.
1. Torvik, B.; Knapskog, A.; Lie-Svendsen, O.; Olsen, K.E.; Griffiths, H.D. Amplitude modulation on echoes from large birds. In Proceedings of the 11th European Radar Conference, Rome, Italy, 8–10 October 2014; pp. 177–180.
2. Guay, R.; Drolet, G.; Bray, J.R. Measurement and modelling of the dynamic radar cross-section of an unmanned aerial vehicle. *IET Radar Sonar Navigat.* **2017**, *11*, 1155–1160. [CrossRef]
3. Stateczny, A.; Lubczonek, J. FMCW radar implementation in river information services in poland. In Proceedings of the 16th International Radar Symposium (IRS), Dresden, Germany, 24–26 June 2015; pp. 852–857.
4. Farlik, J.; Kratky, M.; Casar, J.; Stary, V. Multispectral detection of commercial unmanned aerial vehicles. *Sensors* **2019**, *19*, 1517. [CrossRef]

85. Eriksson, N. Conceptual Study of a Future Drone Detection System-Countering a Threat Posed by a Disruptive Technolog Master's Thesis, Chalmers University Technology, Gothenburg, Sweden, 2018.
86. Chen, V.C. *The Micro-Doppler Effect in Radar*; Artech House: Boston, MA, USA, 2019.
87. Kim, B.K.; Kang, H.-S.; Park, S.-O. Experimental analysis of small drone polarimetry based on micro-Doppler signature. *IEE Geosci. Remote Sens. Lett.* **2017**, *14*, 1670–1674. [CrossRef]
88. Fang, G.; Yi, J.; Wan, X.; Liu, Y.; Ke, H. Experimental research of multistatic passive radar with a single antenna for drone detectio *IEEE Access* **2018**, *6*, 33542–33551. [CrossRef]
89. Colorado, J.; Perez, M.; Mondragon, I.; Mendez, D.; Parra, C.; Devia, C.; Martinez-Moritz, J.; Neira, L. An integrated aerial syster for landmine detection: SDR-based ground penetrating radar onboard an autonomous drone. *Adv. Robot.* **2017**, *31*, 791–80 [CrossRef]
90. Rahman, S.; Robertson, D.A. Millimeter-wave micro-Doppler measurements of small UAVs. *Proc. SPIE* **2017**, *10188*, 101880T.
91. Drozdowicz, J.; Wielgo, M.; Samczynski, P.; Kulpa, K.; Krzonkalla, J.; Mordzonek, M.; Bryl, M.; Jakielaszek, Z. 35 GHz FMCV drone detection system. In Proceedings of the 17th International Radar Symposium (IRS 2016), Krakow, Poland, 10–12 May 201 pp. 1–4.
92. Fontana, R.J.; Richley, E.A.; Marzullo, A.J.; Beard, L.C.; Mulloy, R.W.T.; Knight, E.J. An ultra wideband radar for micro air vehicl applications. In Proceedings of the 2002 IEEE Conference on Ultra Wideband Systems and Technologies (IEEE Cat. No.02EX580 Baltimore, MD, USA, 21–23 May 2002; pp. 187–191.
93. Liu, Y.; Wan, X.; Tang, H.; Yi, J.; Cheng, Y.; Zhang, X. Digital television based passive bistatic radar system for drone detection. I Proceedings of the IEEE Radar Conference (RadarConf), 8–12 May 2017; pp. 1493–1497.
94. Aldowesh, A.; BinKhamis, T.; Alnuaim, T.; Alzogaiby, A. Low Power Digital Array Radar for Drone Detection and Micro-Dopple Classification. In Proceedings of the 2019 Signal Processing Symposium (SPSympo), Krakow, Polan, 17–19 September 201 pp. 203–206. [CrossRef]
95. Jian, M.; Lu, Z.; Chen, V.C. Drone detection and tracking based on phase-interferometric Doppler radar. In Proceedings of th 2018 IEEE Radar Conference (RadarConf18), Oklahoma City, OK, USA, 23–27 April 2018; pp. 1146–1149. [CrossRef]
96. Semkin, V.; Yin, M.; Hu, Y.; Mezzavilla, M.; Rangan, S. Drone Detection and Classification Based on Radar Cross Sectio Signatures. In Proceedings of the 2020 International Symposium on Antennas and Propagation (ISAP), Osaka, Japan, 25–2 January 2021; pp. 223–224. [CrossRef]
97. Jarabo-Amores, M.P.; Mata-Moya, D.; Hoyo, P.J.G.; Bárcena-Humanes, J.; Rosado-Sanz, J.; Rey-Maestre, N.; Rosa-Zurera, M Drone detection feasibility with passive radars. In Proceedings of the 15th European Radar Conference (EuRAD), Madrid, Spai 26–28 September 2018; pp. 313–316.
98. Robin Radar Systems. Elvira. 2020. Available online: https://www.robinradar.com/elvira-anti-drone-system (accessed o 13 January 2022).
99. Björklund, S. Target Detection and Classification of Small Drones by Boosting on Radar Micro-Doppler. In Proceedings of th 2018 15th European Radar Conference (EuRAD), Madrid, Spain, 26–28 September 2018; pp. 182–185. [CrossRef]
100. Güvenç, I.; Ozdemir, O.; Yapici, Y.; Mehrpouyan, H.; Matolak, D. Detection, localization, and tracking of unauthorized UAS an jammers. In Proceedings of the 2017 IEEE/AIAA 36th Digital Avionics Systems Conference (DASC), St. Petersburg, FL, US 17–21 September2017; pp. 1–10.
101. Balleri, A. Measurements of the Radar Cross Section of a nano-drone at K-band. In Proceedings of the 2021 IEEE 8th Internationa Workshop on Metrology for AeroSpace (MetroAeroSpace), Naples, Italy, 23–25 June 2021; pp. 283–287. [CrossRef]
102. Al-Nuaim, T.; Alam, M.; Aldowesh, A. Low-Cost Implementation of a Multiple-Input Multiple-Output Radar Prototype for Dron Detection. In Proceedings of the 2019 International Symposium ELMAR, Zadar, Croatia, 23–25 September 2019; pp. 183–18 [CrossRef]
103. CRFS. Drone Detection: Myths and Reality. 2018. Available online: https://www.crfs.com/blog/drone-detection-myths-and reality/ (accessed on 13 January 2022).
104. Shi, X.; Yang, C.; Xie, W.; Liang, C.; Shi, Z.; Chen, J. Anti-drone system with multiple surveillance technologies: Architectur implementation, and challenges. *IEEE Commun. Mag.* **2018**, *56*, 68–74. [CrossRef]
105. Ezuma, M.; Erden, F.; Anjinappa, C.K.; Ozdemir, O.; Guvenc, I. Micro-UAV detection and classication from RF fingerprint using machine learning techniques. In Proceedings of the 2019 IEEE Aerospace Conference, Big Sky, MT, USA, 2–9 March 201 pp. 1–13.
106. CRFS. DroneDefense. 2020. Available online: https://pages.crfs.com/hubfs/CR-002800-GD-2-DroneDefense%20Brochure.pd (accessed on 12 January 2022).
107. Allahham, M.S.; Khattab, T.; Mohamed, A. Deep learning for RFbased drone detection and identication: A multi-channel 1-I convolutional neural networks approach. In Proceedings of the 2020 IEEE International Conference on Information Technolog (ICIoT), Doha, Qatar, 2–5 February 2020; pp. 112–117.
108. Al-Sa'd, M.F.; Al-Ali, A.; Mohamed, A.; Khattab, T.; Erbad, A. RF based drone detection and identication using deep learnin approaches: An initiative towards a large open source drone database. *Future Gener. Comput. Syst.* **2019**, *100*, 86–97. [CrossRef]
109. Nguyen, P.; Truong, H.; Ravindranathan, M.; Nguyen, A.; Han, R.; Vu, T. Matthan: Drone presence detection by identifyin physical signatures in the drone's RF communication. In Proceedings of the 15th ACM International Conference on Mobil Systems, Applications, and Services, Niagara Falls, NY, USA, 19–23 June 2017; pp. 211–224.

110. Rodhe and Schwarz. R&S Ardonis. 2020. Available online: https://scdn.rohde-schwarz.com/ur/pws/dl_downloads/dl_common_library/dl_brochures_and_datasheets/pdf_1/ARDRONIS_bro_en_5214-7035-12_v0600.pdf (accessed on 13 January 2022).
111. Nguyen, P.; Ravindranatha, M.; Nguyen, A.; Han, R.; Vu, T. Investigating cost-effective RF-based detection of drones. In Proceedings of the 2nd Workshop on Micro Aerial Vehicle Networks, Systems, and Applications for Civilian Use, Singapore, 26 June 2016; pp. 17–22.
112. DeDrone. RF-300 Data Sheet. 2020. Available online: https://assets.website-files.com/58fa92311759990d60953cd2/5d1e14bc96a76a015d193225_dedrone-rf-300-data-sheet-en.pdf (accessed on 13 January 2022).
113. Medaiyese, O.O.; Syed, A.; Lauf, A.P. Machine Learning Framework for RF-Based Drone Detection and Identication System. *arXiv* **2020**, arXiv:2003.02656. Available online: http://arxiv.org/abs/2003.02656 (accessed on 13 January 2022).
114. Robin Radar Systems. Available online: https://www.robinradar.com/press/blog/9-counter-drone-technologies-to-detect-and-stop-drones-today?fbclid=IwAR2Mnxiqbl1JLYmQJ5FXOe-UCKHfoi9jf8T6HbXW7b-LNzX4YkEphGqigEM (accessed on 13 January 2022).
115. Raytheon. Phaser High-Power Microwave System. 2020. Available online: https://www.raytheon.com/capabilities/products/phaser-highpower-microwave-system (accessed on 12 January 2022).
116. Zohuri, B. High-power microwave energy as weapon. In *Directed-Energy Beam Weapons*; Springer: Cham, Switzerland, 2019; pp. 269–308. [CrossRef]
117. Radasky, W.A.; Baum, C.E.; Wik, M.W. Introduction to the special issue on high-power electromagnetics (HPEM) and intentional electromagnetic interference (IEMI). *IEEE Trans. Electromagn. Compat.* **2004**, *46*, 314–321. [CrossRef]
118. Olivares, G.; Gomez, L.; de los Monteros, J.E.; Baldridge, R.J.; Zinzuwadia, C.; Aldag, T. *Volume II-UAS Airborne Collision Severity Evaluation-Quadcopter*; National Institute for Aviation Research: Wichita, KS, USA, 2017; Tech. Rep.; DOT/FAA/AR xx/xx.
119. RoboTiCan. Goshawk. 2020. Available online: https://robotican.net/goshawk/(2020) (accessed on 12 January 2022).
120. AerialX: DroneBullet. Available online: https://www.aerialx.com/defeat.shtml (accessed on 12 January 2022).
121. Anduril Industries. 2020. Available online: https://www.anduril.com/ (accessed on 12 January 2022).
122. Akhlou, M.A.; Arola, S.; Bonnet, A. Drones chasing drones: Reinforcement learning and deep search area proposal. *Drones* **2019**, *3*, 58. [CrossRef]
123. Ban Lethal. Slaugtherbots. 2020. Available online: https://autonomousweapons.org/ (accessed on 12 January 2022).
124. MBDA Missile Systems. Dragonfire Laser Turret Unveiled at DSEI. 2017. Available online: https://www.mbdasystems.com/press-releases/dragonre-laser-turret-unveiled-dsei-2017/ (accessed on 12 January 2022).
125. India Today. KALI: India'sWeapon to Destroy Any Uninvited Missiles and Aircrafts. 2015. Available online: https://www.indiatoday.in/education-today/gk-current-affairs/story/indias-top-secret-weapon-264111-2015-09-21 (accessed on 12 January 2022).
126. Daily Sabah. Turkey's Laser Weapon ARMOL Passes Acceptance Tests. 2019. Available online: https://www.dailysabah.com/defense/2019/09/30/turkeys-laser-weapon-armol-passesacceptance-tests (accessed on 12 January 2022).
127. Sudakov, D. Russia's Combat Laser Weapons Declassified. 2016. Available online: https://www.pravdareport.com/russia/135198-russia_laser_weapons/ (accessed on 12 January 2022).
128. Zeng, Y.; Lyu, J.; Zhang, R. Cellular-connected UAV: Potential, challenges, and promising technologies. *IEEE Wirel. Commun.* **2019**, *26*, 120–127. [CrossRef]
129. Lin, J.; Singer, P. Drones, Lasers, and Tanks: China Shows Off Its Latest Weapons. 2017. Available online: https://www.popsci.com/china-new-weapons-lasers-drones-tanks/ (accessed on 12 January 2022).
130. Josh Spires, Dubai-Made Magnetic Counter-Drone System to Launch Soon. 2021. Available online: https://dronedj.com/2021/01/04/dubai-made-magnetic-counter-drone-system-to-launch-soon/?fbclid=IwAR3e5Lk5TQZzxU_ovMRYqJ6rDm2XU4_T247rZSH9rJlNotDHMQQQkWQIByU (accessed on 12 January 2022).
131. Atherton, K.D. Trained Police Eagles Attack Drones on Command. 2016. Available online: https://www.popsci.com/eagles-attackdrones-at-police-command/ (accessed on 12 January 2022).
132. This Anti-Drone Net Gun Was Built from Scratch. 2017. Available online: https://www.popularmechanics.com/flight/drones/a27427/anti-drone-net-gun-diy/ (accessed on 12 January 2022).
133. Gettinger, D.; Michel, A.H. A Brief History of Hamas and Hezbollah's Drones. 2014. Available online: https://dronecenter.bard.edu/hezbollah-hamas-drones/ (accessed on 13 January 2022).
134. Taylor, H. Knight, Use of Water for Counter Unmanned Aerial Systems (C-UAS). Available online: https://dsiac.org/wp-content/uploads/2020/10/TI-Response-Report_Use-of-Water-for-C-UAS.pdf (accessed on 12 January 2022).
135. Guvenc, I.; Koohifar, F.; Singh, S.; Sichitiu, M.L.; Matolak, D. Detection, tracking, and interdiction for amateur drones. *IEEE Commun. Mag.* **2018**, *56*, 75–81. [CrossRef]
136. Multerer, T.; Ganis, A.; Prechtel, U.; Miralles, E.; Meusling, A.; Mietzner, J.; Vossiek, M.; Loghi, M.; Ziegler, V. Low-cost jamming system against small drones using a 3DMIMOradar based tracking. In Proceedings of the 14th European Radar Conference (EURAD), Nuremberg, Germany, 11–13 October 2017; pp. 299–302.
137. Li, A.; Wu, Q.; Zhang, R. UAV-enabled cooperative jamming for improving secrecy of ground wiretap channel. *IEEE Wirel. Commun. Lett.* **2019**, *8*, 181184. [CrossRef]

138. Curpen, R.; Balan, T.; Miclos, I.A.; Comanici, I. Assessment of signal jamming efficiency against LTE UAVs. In Proceedings of the International Conferenece on Communications (COMM), Bucharest, Romania, 14–16 June 2018; pp. 367–370.
139. Roh, Y.; Jung, S.; Kang, J. Cooperative UAV jammer for enhancing physical layer security: Robust design for jamming power and trajectory. In Proceedings of the IEEE Military Communications Conference (MILCOM), Norfolk, VA, USA, 12–14 November 2019; pp. 464–469.
140. Li, K.; Voicu, R.C.; Kanhere, S.S.; Ni, W.; Tovar, E. Energy efficient legitimate wireless surveillance of UAV communications. *IEEE Trans. Veh. Technol.* **2019**, *68*, 2283–2293. [CrossRef]
141. Noh, J.; Kwon, Y.; Son, Y.; Shin, H.; Kim, D.; Choi, J.; Kim, Y. Tractor beam: Safe-hijacking of consumer drones with adaptive GPS spoofing. *ACM Trans. Priv. Secur.* **2019**, *22*, 12:1–12:26. [CrossRef]
142. Moskvitch, K. Are Drones the Next Target for Hackers? 2014. Available online: https://www.bbc.com/future/article/20140206-candrones-be-hacked (accessed on 13 January 2022).
143. Hooper, M.; Tian, Y.; Zhou, R.; Cao, B.; Lauf, A.P.; Watkins, L.; Robinson, W.H.; Alexis, W. Securing commercial WiFi-based UAVs from common security attacks. In Proceedings of the IEEE Military Communications Conference (MILCOM), Baltimore, MD, USA, 1–3 November 2016; pp. 1213–1218.
144. Kerns, A.J.; Shepard, D.P.; Bhatti, J.A.; Humphreys, T.E. Unmanned aircraft capture and control via GPS spoofing. *J. Field Robot.* **2014**, *31*, 617–636. [CrossRef]
145. Summers, N. Icarus Machine Can Commandeer a Drone Mid-Flight. 2016. Available online: https://www.engadget.com/2016-10-28-icarus-hijack-dmsx-drones.html (accessed on 12 January 2022).
146. Directive 2014/53/EU of the European Parliament and of the Council of 16 April 2014. Available online: https://www.ancom.ro/en/uploads/links_files/Directive_RED_2014_53_UE.pdf (accessed on 12 January 2022).
147. Hotărâre Privind Punerea la Dispoziție pe Piață a Echipamentelor Radio. Available online: https://www.ancom.ro/uploads/articles/file/industrie/Echipamente%20radio/HG_740_2016_privind_punerea_la_dispozitie_pe_piata_a_echipamentelor_radio_in_vigoare_din_08_08_2019EN.pdf (accessed on 12 January 2022).
148. Taking Flight: The Future of Drones in the UK Government Response. Available online: https://assets.publishing.service.gov.uk/government/uploads/system/uploads/attachment_data/file/937275/future-of-drones-in-uk-consultation-response-web.pdf (accessed on 12 January 2022).
149. Drone Jammers: How They Work, Why They Exist, and Are They Legal? Available online: https://pilotinstitute.com/drone-jammers/ (accessed on 12 January 2022).
150. Drone Laws in Russia. Available online: https://drone-laws.com/drone-laws-in-russia/ (accessed on 12 January 2022).
151. Here's How China is Battling Drones. Available online: https://www.popsci.com/chinas-new-anti-drone-weapons-jammers-and-lasers/ (accessed on 12 January 2022).
152. Nie, W.; Han, Z.; Li, Y.; He, W.; Xie, L.; Yang, X.; Zhou, M. UAV Detection and Localization Based on Multi-dimensional Signal Features. *IEEE Sens. J.* **2021**. [CrossRef]
153. Basak, S.; Rajendran, S.; Pollin, S.; Scheers, B. Drone classification from RF fingerprints using deep residual nets. In Proceedings of the 2021 International Conference on COMmunication Systems & NETworkS (COMSNETS), Bengaluru, India, 5–9 January 2021; pp. 548–555. [CrossRef]
154. Ezuma, M.; Erden, F.; Anjinappa, C.K.; Ozdemir, O.; Guvenc, I. Detection and Classification of UAVs Using RF Fingerprints in the Presence of Wi-Fi and Bluetooth Interference. *IEEE Open J. Commun. Soc.* **2020**, *1*, 60–76. [CrossRef]
155. Xu, C.; Chen, B.; Liu, Y.; He, F.; Song, H. RF Fingerprint Measurement for Detecting Multiple Amateur Drones Based on STFT and Feature Reduction. In Proceedings of the 2020 Integrated Communications Navigation and Surveillance Conference (ICNS), Virtual Conference, 8–10 September 2020; pp. 4G1-1–4G1-7. [CrossRef]
156. Nemer, I.; Sheltami, T.; Ahmad, I.; Yasar, A.U.-H.; Abdeen, M.A.R. RF-Based UAV Detection and Identification Using Hierarchical Learning Approach. *Sensors* **2021**, *21*, 1947. [CrossRef]
157. Bisio, I.; Garibotto, C.; Lavagetto, F.; Sciarrone, A.; Zappatore, S. Blind Detection: Advanced Techniques for WiFi-Based Drone Surveillance. *IEEE Trans. Veh. Technol.* **2019**, *68*, 938–946. [CrossRef]
158. Flak, P. Drone Detection Sensor with Continuous 2.4 GHz ISM Band Coverage Based on Cost-Effective SDR Platform. *IEEE Access* **2021**, *9*, 114574–114586. [CrossRef]
159. Kaplan, B.; Kahraman, İ.; Görçin, A.; Çırpan, H.A.; Ekti, A.R. Measurement based FHSS–type Drone Controller Detection at 2.4GHz: An STFT Approach. In Proceedings of the 2020 IEEE 91st Vehicular Technology Conference (VTC2020-Spring), Online, 18 November–16 December 2020; pp. 1–6. [CrossRef]
160. IGelman, S.; Loftus, J.P.; Hassan, A.A. *Adversary UAV Localization with Software Defined Radio*; Worcester Polytechnic Institute: Worcester, MA, USA, 2019; Tech. Rep.; E-project-041719-144214.
161. Miranda, R.K.; Ando, D.A.; da Costa, J.P.C.L.; de Oliveira, M.T. Enhanced Direction of Arrival Estimation via Received Signal Strength of Directional Antennas. In Proceedings of the 2018 IEEE International Symposium on Signal Processing and Information Technology (ISSPIT), Louisville, KY, USA, 6–8 December 2018; pp. 162–167. [CrossRef]
162. Brito, A.; Sebastião, P.; Souto, N. Jamming for Unauthorized UAV Operations-Communications Link. In Proceedings of the 2019 International Young Engineers Forum (YEF-ECE), Costa da Caparica, Portugal, 10 May, 2019; pp. 94–98. [CrossRef]
163. Ferreira, R.; Gaspar, J.; Souto, N.; Sebastião, P. Effective GPS Jamming Techniques for UAVs Using Low-Cost SDR Platforms. In Proceedings of the 2018 Global Wireless Summit (GWS), Chiang Rai, Thailand, 25–28 November 2018; pp. 27–32. [CrossRef]

64. Pärlin, K.; Alam, M.M.; le Moullec, Y. Jamming of UAV remote control systems using software defined radio. In Proceedings of the 2018 International Conference on Military Communications and Information Systems (ICMCIS), Warsaw, Poland, 22–23 May 2018; pp. 1–6. [CrossRef]
65. Fang, L.; Wang, X.H.; Zhou, H.L.; Zhang, K. Design of Portable Jammer for UAV Based on SDR. In Proceedings of the 2018 International Conference on Microwave and Millimeter Wave Technology (ICMMT), Chengdu, China, 7–11 May 2018; pp. 1–3. [CrossRef]
66. Skorobogatov, G.; Barrado, C.; Salamí, E. Multiple UAV systems: A survey. *Unmanned Syst.* **2020**, *8*, 149–169. [CrossRef]
67. Yavariabdi, A.; Kusetogullari, H.; Celik, T.; Cicek, H. FastUAV-NET: A Multi-UAV Detection Algorithm for Embedded Platforms. *Electronics* **2021**, *10*, 724. [CrossRef]
68. Li, J.; Ye, D.H.; Chung, T.; Kolsch, M.; Wachs, J.; Bouman, C. Multi-target detection and tracking from a single camera in Unmanned Aerial Vehicles (UAVs). In Proceedings of the 2016 IEEE/RSJ International Conference on Intelligent Robots and Systems (IROS), Daejon, Korea, 9–14 October 2016; pp. 4992–4997. [CrossRef]
69. Sazdić-Jotić, B.; Pokrajac, I.; Bajčetić, J.; Bondžulić, B.; Obradović, D. Single and multiple drones detection and identification using RF based deep learning algorithm. *Expert Syst. Appl.* **2022**, *187*, 115928. [CrossRef]
70. The Most Promising Defense against Militarized Drone Swarms. Available online: https://mindmatters.ai/2021/06/the-most-promising-defense-against-militarized-drone-swarms/ (accessed on 13 January 2022).
71. Cyberwall. Available online: http://cyberwall.ro (accessed on 12 January 2022).
72. DronEnd Research Project. Available online: http://dronend.ro (accessed on 12 January 2022).
73. Martian, A.; Chiper, F.-L.; Craciunescu, R.; Vladeanu, C.; Fratu, O.; Marghescu, I. RF Based UAV Detection and Defense Systems: Survey and a Novel Solution. In Proceedings of the 2021 IEEE International Black Sea Conference on Communications and Networking (BlackSeaCom), Bucharest, Romania, 16–18 June 2021; pp. 1–4. [CrossRef]
74. Vladeanu, C.; Nastase, C.; Martian, A. Energy Detection Algorithm for Spectrum Sensing Using Three Consecutive Sensing Events. *IEEE Wirel. Commun. Lett.* **2016**, *5*, 284–287. [CrossRef]
75. Martian, A.; Al Sammarraie, M.J.A.; Vlădeanu, C.; Popescu, D.C. Three-Event Energy Detection with Adaptive Threshold for Spectrum Sensing in Cognitive Radio Systems. *Sensors* **2020**, *20*, 3614. [CrossRef] [PubMed]
76. Urkowitz, H. Energy Detection of Unknown Deterministic Signals. *Proc. IEEE* **1967**, *55*, 523–531. [CrossRef]
77. Ettus Research USRP X310. Available online: https://www.ettus.com/all-products/x310-kit/ (accessed on 12 January 2022).
78. Ettus Research Twin-RX RF Daughterboard. Available online: https://www.ettus.com/all-products/twinrx/ (accessed on 12 January 2022).
79. DJI Mavic Air Drone. Available online: https://www.dji.com/mavic-air (accessed on 12 January 2022).
80. DJI Phantom 4 Pro v2.0 Drone. Available online: https://store.dji.com/product/phantom-4-pro-v2/ (accessed on 12 January 2022).
81. DJI Mini 2 Drone. Available online: https://store.dji.com/product/mini-2 (accessed on 12 January 2022).
82. Mini-Circuits ZN4PD1-63HP-S+ 4 Ways DC Pass Power Splitter. Available online: https://www.minicircuits.com/WebStore/dashboard.html?model=ZN4PD1-63HP-S%2B (accessed on 12 January 2022).
83. Ettus Research VERT2450 Antenna. Available online: https://www.ettus.com/all-products/vert2450/ (accessed on 12 January 2022).
84. Ettus Research B200mini SDR Platform. Available online: https://www.ettus.com/all-products/usrp-b200mini/ (accessed on 12 January 2022).
85. Mini-Circuits ZHL-2W-63-S+ Power Amplifier. Available online: https://www.minicircuits.com/WebStore/dashboard.html?model=ZHL-2W-63-S%2B (accessed on 12 January 2022).
86. Ubiquiti UMA-D Antenna. Available online: https://dl.ubnt.com/datasheets/unifi/UMA-D_DS.pdf (accessed on 12 January 2022).

sensors

Review

A Survey of Blind Modulation Classification Techniques for OFDM Signals

Anand Kumar [1], Sudhan Majhi [2,*], Guan Gui [3], Hsiao-Chun Wu [4] and Chau Yuen [5]

1 Department of Electrical Engineering, Indian Institute of Technology Patna, Patna 801103, India; anand_1921ee15@iitp.ac.in
2 Department of Electrical Communication Engineering, Indian Institute of Science (IISc), Bangalore 560012, India
3 College of Telecommunications and Information Engineering, Nanjing University of Posts and Telecommunications, Nanjing 210003, China; guiguan@njupt.edu.cn
4 School of Electrical Engineering and Computer Science, Louisiana State University, Baton Rouge, LA 70803, USA; wu@ece.lsu.edu
5 Engineering Product Development (EPD) Pillar, Singapore University of Technology and Design, Singapore 487372, Singapore; yuenchau@sutd.edu.sg
* Correspondence: smajhi@iisc.ac.in

Abstract: Blind modulation classification (MC) is an integral part of designing an adaptive or intelligent transceiver for future wireless communications. Blind MC has several applications in the adaptive and automated systems of sixth generation (6G) communications to improve spectral efficiency and power efficiency, and reduce latency. It will become a integral part of intelligent software-defined radios (SDR) for future communication. In this paper, we provide various MC techniques for orthogonal frequency division multiplexing (OFDM) signals in a systematic way. We focus on the most widely used statistical and machine learning (ML) models and emphasize their advantages and limitations. The statistical-based blind MC includes likelihood-based (LB), maximum a posteriori (MAP) and feature-based methods (FB). The ML-based automated MC includes k-nearest neighbors (KNN), support vector machine (SVM), decision trees (DTs), convolutional neural networks (CNNs), recurrent neural networks (RNNs), and long short-term memory (LSTM) based MC methods. This survey will help the reader to understand the main characteristics of each technique, their advantages and disadvantages. We have also simulated some primary methods, i.e., statistical- and ML-based algorithms, under various constraints, which allows a fair comparison among different methodologies. The overall system performance in terms bit error rate (BER) in the presence of MC is also provided. We also provide a survey of some practical experiment works carried out through National Instrument hardware over an indoor propagation environment. In the end, open problems and possible directions for blind MC research are briefly discussed.

Keywords: blind modulation classification; orthogonal frequency division multiplexing; higher-order cumulant and cyclic cumulant; maximum-likelihood; maximum a posteriori; deep learning; convolutional neural networks; probability of correct classification; testbed implementation

Citation: Kumar, A.; Majhi, S.; Gui, G.; Wu, H.-C.; Yuen, C. A Survey of Blind Modulation Classification Techniques for OFDM Signals. *Sensors* **2022**, *22*, 1020. https://doi.org/10.3390/s22031020

Academic Editor: Peter Han Joo Chong

Received: 25 December 2021
Accepted: 24 January 2022
Published: 28 January 2022

1. Introduction

Blind modulation classification (MC) determines the modulation type of the received signal, ensuring proper demodulation and retrieval of the transmitted data [1–4]. Recently, MC has played a significant role in both military and civilian communications, such as cognitive radio, signal intelligence, link adaptation, signal control, and SDR [3–7]. With an intelligent receiver, blind parameter estimation and classification algorithms may be used, resulting in a substantial increase in spectral efficiency since no predefined training or pilot sequence is needed [8–10].

Over the years, various MC algorithms for single-carrier (SC) systems have been developed, which can be divided into likelihood-based (LB) and feature-based methods

(FB) [1,5,6,11–17]. Although the LB approaches are optimal in a Bayesian context, they have high computational complexity [11]. They often necessitate prior information about the signal parameters in order to distinguish modulation formats, which is typically undesirable in an intelligent or adaptive transceiver system. Furthermore, FB algorithms, which consist of features extraction and classifier construction, usually provide a sub-optimal solution. They are inherently simpler to implement, have less computational complexity, and may not necessitate prior information about the signal parameters and channel statistics. To identify the modulation schemes, existing FB approaches extract specific features, such as cumulants [12,13], cyclic statistics [5,6,14–16], and wavelet transform [17], and use threshold values to distinguish the extracted features. As a result, they are better fit for fading and additive white Gaussian noise (AWGN) channels. The algorithms [5,6,14–16,18–21] based on higher-order cyclic statistics are reliable and perform well in flat fading as well as in frequency-selective fading channels. They consider M-ary phase-shift keying (M-PSK) and M-ary quadrature amplitude modulation (M-QAM) modulation schemes by using non-zero cyclic frequencies of received signals. The combination of higher-order correlation-, cumulant-, cyclic cumulant-, and cyclostationarity-based MC algorithm for multiple-antenna systems is analyzed in [5]. The algorithm described in [6] is designed for single-antenna and single-carrier (SC) systems. It requires the combined features of cumulants and cyclic cumulants and performs well over flat fading channels. Furthermore, the algorithm proposed in [5,6] can also distinguish various quadrature PSK (QPSK) variants, such as offset QPSK (OQPSK), minimum-shift keying (MSK), and $\pi/4$-QPSK modulation types.

Recent advances in machine learning (ML) and data science have resulted in its extensive application in various fields. Artificial intelligence (AI) and other advanced ML approaches have significantly improved state-of-the-art outcomes in computer vision, speech recognition [22], drug discovery, genomics, and, most recently, physical layer communication [23]. MC algorithms [24–36] focused on various ML algorithms. In [24], the MC technique is evaluated using genetic programming (GP) and K-nearest neighbor (KNN). Cumulants are utilized by GP as input features to distinguish modulation types. In [25], extreme learning machine (ELM) and higher-order statistics-based MC algorithms for multiple antenna systems are presented. Convolutional neural networks (CNNs) are explored in [26] that can distinguish modulation schemes even at low signal-to-noise ratio (SNR) scenarios. Furthermore, CNN-based MC techniques are robust to prediction errors on carrier phase offset and SNR. The approach investigated in [27] extracts unique characteristics using higher-order cumulants (HOCs), and then the feed-forward neural network model is developed to distinguish modulation schemes. In comparison to typical centralized training, distributed learning-based MC (DistMC) based on several edge devices can achieve a faster training process and reduce communication costs through collaborative training [28]. Multi-task learning (MTL) based MC has a single trained model for all SNRs under carrier frequency offset (CFO) and phase offset (PO) conditions [29,30]. In [35], an adversarial transfer learning-based MC developed a framework for SC systems that combines transfer learning with adversarial networks to handle the problem of limited data in a realistic scenario. A complex-valued network [36] is presented to illustrate the enormous potential for MC and show the higher classification performance as compared to the real-valued network. The authors [37] studied a phoneme-based distribution regularization algorithm for speech enhancement by utilizing speech recognition information in the modulation domain. However, the approaches mentioned above [1,5,6,11–17,24–36] are only applicable to SC systems.

Orthogonal frequency division multiplexing (OFDM) is a well-known multicarrier modulation technology used in advanced wireless communications systems. OFDM is employed in the 4G Third-Generation Partnership Project (3GPP) Long Term Evolution-Advanced (LTE/LTE-A), Worldwide Interoperability for Microwave Access (WiMAX), and high-speed wireless local area network (WLAN) standards such as 802.11n [19]. It is also an integral part of 5G New Radio (NR) cellular. The key feature of OFDM is the ability to convert frequency-selective fading to flat fading channels. Due to the high spectrum utilization and strong anti-multipath interference ability, the OFDM modulation scheme has been employed as the main transmission approach for high data rate systems [38,39]. M-PSK and M-QAM are the two most popular modulation schemes that are used with OFDM. MC for OFDM signals is a critical research challenge for 5G and beyond wireless communication, where AI would be a fundamental aspect of the communication system [40–43].

Various MC algorithms for the OFDM systems were carried out in [44–97]. The algorithms for multiple-input multiple-output and OFDM (MIMO-OFDM) systems based on deep neural network (DNN) and Gibbs sampling are investigated in [44]. Moreover, these methods are restricted to known channel conditions and/or perfect synchronization. The likelihood-based MC algorithm for index modulation investigated in [47,71] is applicable to both known and unknown channel state information (CSI). However, both techniques require perfect synchronization classification of M-PSK/M-QAM modulation types. The likelihood and maximum a posteriori [50] based MC approach are employed when CSI is known. The MC approach based on the statistical features of the received OFDM signal is studied in [61]. This technique uses mean, skewness, and kurtosis as features to distinguish QPSK, 16-QAM, and 64-QAM modulation schemes. However, this technique does not perform well with timing and frequency synchronization errors. The MC algorithm based on amplitude moments is discussed in [62]. This method distinguishes between 16-QAM and 64-QAM modulation schemes by using the correlation between any two subcarriers. The non-parametric Kolmogorov–Smirnov (KS) based technique presented in [98,99] is used to classify M-PSK/M-QAM modulation schemes. It operates in the presence of known timing offset and unknown frequency and phase offsets, and the non-Gaussian noise channel. Most of the above MC algorithms for the OFDM signal are restricted to known CSI and/or perfect synchronization cases. Moreover, a discrete Fourier transform (DFT) and normalized higher-order cumulant [63] based blind MC is discussed to classify the lower-order digital modulation schemes for the OFDM system. However, the classification accuracy is unsatisfactory, subjected to channel degradation. In [96], the authors developed a high-performance deep residual network (ResNet) with a triple-skip residual stack (TRNN) based MC algorithm for real-time OFDM signal classification in dynamic fading channel conditions.

The objective of this paper is to present a comprehensive review of various MC techniques for OFDM signals. The statistical approach and the AI approach are two main classes of MC algorithms that will be discussed in detail. We concentrate on the most common statistical and ML models, emphasizing their benefits and drawbacks. The contributions of various research papers are summarized into compact forms. This will make it easier for the reader to recognize the important features of each approach. Furthermore, we also present results obtained by applying some statistical and ML algorithms with a testbed based on the National Instrument (NI) radio frequency (RF) hardware over an indoor transmission environment. Finally, challenges and potential research directions are briefly explored.

The remainder of the paper is organized as follows. The signal model of the received OFDM system is presented in Section 2. The statistical approach for MC is discussed in Section 3. We summarize the advantages and the limitations of AI models in MC in Section 4. Finally, challenges and future research directions involved in MC are discussed in Section 5. The organization of the paper is provided in Figure 1. The abbreviation used in the rest of the paper is listed in Table 1.

Table 1. List of abbreviations in alphabetical order.

Acronym	Explanation
AI	Artificial Intelligence
ALRT	Average Probability Ratio Test
AMAP	Approximated Maximum a Posteriori
ASB	Amplitude Spectrum of Bispectrum
AWGN	Additive White Gaussian Noise
BAT	Bit Allocation Table
BFSF	Bi-Fold Signal Fortification
BICM-ID	Bit-Interleaved Coded Modulation Iterative Decoding
CNN	Convolutional Neural Network
CSI	Channel State Information
DBN	Deep Belief network
DVB	Digital Video Broadcasting
FB	Feature Based
FCP	False Classification Probability
FFT	Fast Fourier Transform
FNSF	Frequency Non-Selective Fading Channel
FPGA	Field Programmable Gate Array
FSF	Frequency Selective Fading Channel
FSST	Fourier Synchrosqueezing Transformation
GLRT	Generalized Likelihood Ratio Test
HGWO	Hybrid Grey Wolf Optimization
HLRT	Hybrid Likelihood Ratio Test
HOC	Higher Order Cumulant
HOS	Higher Order Statistics
ICI	Inter-carrier Interference
IQ	In-phase and Quadrature
IQL	Improved Q-learning
KNN	K-Nearest Neighbors
KS	Kolmogorov–Smirnov
LLR	Log-likelihood ratio
MAP	Maximum a Posteriori
MC	Modulation Classification
MFCC	Mel Frequency Cepstral Coefficient
MDNCC	Multi-Distance-Based Nearest Centroid Classifier
NOMA	Non-Orthogonal Multiple Access
OFDM-IM	Orthogonal Frequency Division Multiplexing with Index Modulation
PCC	Percentage of Correct Classification
PDF	Probability Density Function
PER	Packet Error Ratio
PSO	Particle Swarm Optimization
SC	Single Carrier
SDR	Software-defined Radio
STFT	Short-Time Fourier Transform
TDD	Time Division Duplex
TF-HMS	Twin-Functioned Human Mental Search
UMP	Uniformly Most Powerful
VLC	Visible Light Communication
WOA	Whale Optimization Algorithm
WPS	Wavelet Packet Signals
WT	Wavelet Transform

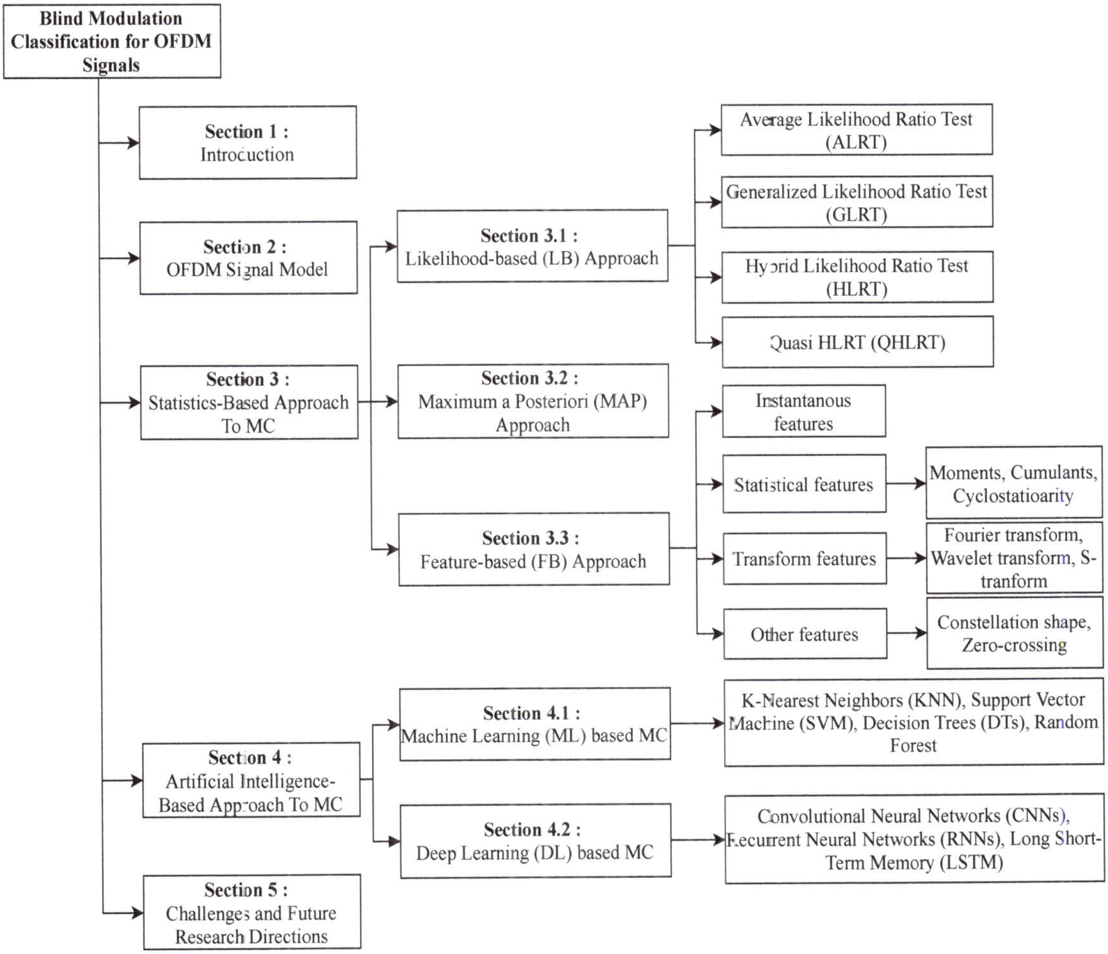

Figure 1. The organization of the paper.

2. OFDM Signal Model

The system model of MC for the OFDM system is shown in Figure 2. It consists of an adaptive OFDM transmitter, a receiver with statistics-based MC, ML-based MC, and DL-based MC. The transmitter can adjust its baseband modulation format and the number of subcarriers according to the requirement of the data rate and the available CSI. The signal is transmitted over a frequency-selective fading channel. This channel introduces all kinds of impairments into the transmitted signal, including timing, frequency, and phase offsets. The receiver consists of an MC system pre-processing block and selection of a proper MC algorithm. In the following subsections, we provide the mathematical framework of the OFDM signal for MC.

The discrete baseband OFDM samples $d_m[n]$ of the mth OFDM symbol, obtained by N-point inverse discrete Fourier transform (IDFT), which can be written as

$$d_m[n] = \sum_{k=0}^{N-1} D_m[k] e^{j2\pi kn/N}, \quad 0 \leq n \leq N-1, \tag{1}$$

where $N = \rho_s \times N_d$, ρ_s is the oversampling factor, N_d is the number of data subcarriers, and $D_m[k]$ is the baseband modulated oversampled data obtained by zero-padding the baseband modulated information, i.e., M-PSK/M-QAM and denoted by $\hat{D}_m[k]$. Thus, $D_m[k]$ is given by

$$D_m[k] = \begin{cases} \hat{D}_m[k] & 0 \le k \le N_d/2 - 1 \\ \hat{Z}_0 & N_d/2 \le k \le N_d(\rho_s - 1/2) - 1 \\ \hat{D}_m[k] & N_d(\rho_s - 1/2) \le k \le N - 1, \end{cases} \quad (2)$$

where $\hat{Z}_0$ is a vector of zeros of length $N_d(\rho_s - 1)$. To combat the effect of intersymbol interference (ISI), a cyclic-prefix (CP) of N_{cp} samples from the end of the OFDM symbol are added at the beginning of the OFDM symbol before the transmission. The transmitted baseband OFDM symbol $\bar{d}_m[n]$ of length $N + N_{cp}$, with CP is then given by

$$\bar{d}_m[n] = \begin{cases} d_m[n+N] & -N_{cp} \le n \le -1 \\ d_m[n] & 0 \le n \le N-1. \end{cases} \quad (3)$$

After passing through a frequency-selective fading channel with impulse response $g[l]$ of length L, the received baseband OFDM samples of the mth OFDM symbol are given by

$$x_m[n] = e^{(j2\pi\epsilon n/N+\phi)} \sum_{l=0}^{L-1} g[l]\bar{d}_m[n-l-\tau] + \omega[n], \quad 0 \le n \le N_s - 1 \quad (4)$$

where ϵ is the normalized carrier frequency offset (CFO), ϕ is the phase offset, τ is the symbol timing offset (STO), N_s length of the OFDM symbol with CP, $N_s = N + N_{cp}$ and $N_{cp} \ge L$, and $\omega[n]$ is the AWGN with zero mean and variance σ_ω^2.

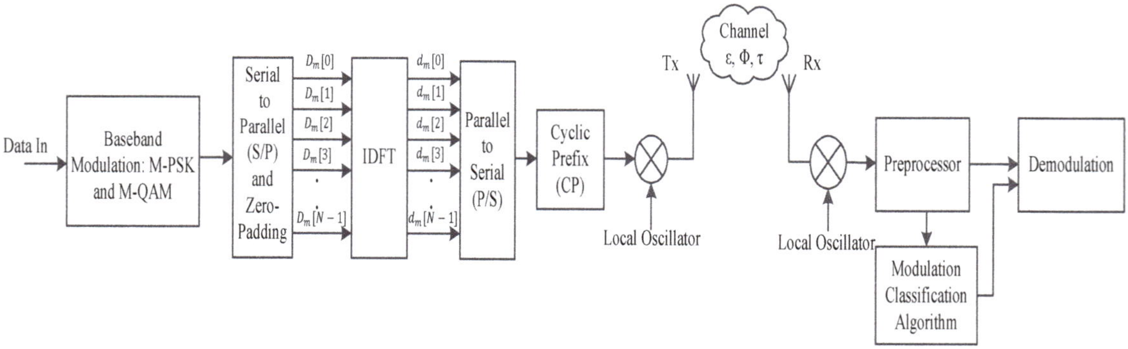

Figure 2. Block diagram of blind modulation classification for OFDM system.

3. Statistics-Based Approach to MC

3.1. LB Approach

In the LB system, MC refers to numerous composite hypothesis problems. The LB-MC is based on the assumption that the probability density function (PDF) of the analyzed waveform includes all classification information, depending on the embedded modulated signal. The average likelihood ratio test (ALRT) [47], generalized likelihood ratio test (GLRT) [46], and hybrid likelihood ratio test (HLRT) [47] are the main three LB-MC techniques studied in the literature, based on the model selected for the unknown parameter In some works in the literature, quasi-ALRT [46] and quasi-HLRT [45,46] are also described

ALRT: In this method, unknown parameters are considered random variables with specific PDFs. For the hypothesis H_j, which represents the jth modulation, $j = 1, 2, ..., M$, the likelihood function (LF) is as follows

$$\Lambda_{ALRT}^{j} = \sum_{v_j} \Lambda[x_m[n]|v_j, H_j] p(v_j|H_j), \tag{5}$$

where $\Lambda[x_m[n]|v_j, H_j]$ denotes the conditional LF of the received signal $x_m[n]$ associated with noise over H_j, conditioned on the undefined vector v_j under H_j. By integrating over v_j and using its known PDF, the problem is reduced to a basic hypothesis-testing problem. The conditional LF for a baseband complex AWGN is provided by

$$\Lambda[x_m[n]|v_j, H_j] = \frac{1}{\pi N_0} exp\left(-\frac{\frac{1}{\eta_0}\sum_{n=0}^{N-1}|x_m(n) - s_m[n]|^2}{N_0} \right) \tag{6}$$

where N_0 represents the power spectral density (PSD) of AWGN in W/Hz, with the autocorrelation $E\{\omega[n], \omega^*[n+\tau]\} = N_0\delta[n]$, with $E\{.\}$ denoting the expectation and * representing the complex conjugate. Furthermore, here $s_m[n] = e^{(j2\pi\epsilon n/N+\phi)} \sum_{l=0}^{L-1} g[l]\bar{d}_m[n-l-\tau]$. ALRT produces an optimal classifier in the Bayesian context when the chosen $p(v_j|H_j)$ is the same as the true PDF.

GLRT: This approach considers the unknown parameters to be unknown deterministic. The best result is obtained by carrying out the so-called uniformly most powerful (UMP) test [100]. If UMP test does not exist or is difficult to obtain, a rational technique is used to estimate the unknown parameters based on the assumption that H_j is true, and then utilize these estimations in ALRT as if they were accurate. When maximum likelihood is applied for estimations, the hypothesis test is known as GLRT. The unknown parameters of GLRT are, of course, considered deterministic unknowns, and LF under H_j is provided by

$$\Lambda_{GLRT}^{j}[x_m[n]] = \max_{v_j} \Lambda[x_m[n]|v_j, H_j]. \tag{7}$$

HLRT: This is the combined approaches of the above techniques, where the LF under H_j is defined by

$$\Lambda_{HLRT}^{j} = \max_{v_{j1}} \sum_{v_{j2}} \Lambda[x_m[n]|v_{j_1}, v_{j2}, H_j] p(v_{j2}|H_j) dv_{j2}, \tag{8}$$

where $v_j = \left[v_{j_1}^{\dagger} v_{j_2}^{\dagger}\right]^{\dagger}$ with $\dagger$ as the transpose and v_{j_1} and v_{j_2} are vectors of unknown parameters treated as unknown deterministic and random variables, respectively. Generally, v_{j_1} and v_{j_2} are made up of parameters and data symbols, respectively.

It is to be noted that ALRT necessitates multidimensional convergence, while GLRT necessitates multidimensional maximization. ALRT could be unrealistic due to the difficulties of performing multidimensional integration in the presence of a large number of unknown parameters and the requirement to know the PDFs. Furthermore, maximization over unknown parameters in GLRT results yield the same LF value for nested signal constellations, such as BPSK and QPSK, 16-QAM, and 64-QAM, resulting in inaccurate classification. However, with HLRT, averaging over unknown data symbols eliminates the GLRT problem of nested constellations. In the case of a two-hypothesis classification problem, a decision is made on the basis of

$$\Lambda_H^{(1)}[x_m[n]]/\Lambda_H^{(2)}[x_m[n]] \underset{H_2}{\overset{H_1}{\gtrless}} \eta_l, l = A(\text{ALRT}), G(\text{GLRT}), H(\text{HLRT}), \tag{9}$$

where η_l represents the threshold. The left-hand side of (9) represents the likelihood ratio and the test is referred to as the ALRT, GLRT, and the HLRT, respectively, depending on the approach used to estimate the LF. The extension of (8) to multiple classes is simple. Likewise, the log function can be extended to the two members of the inequality (9). Table 2 lists multiple LB-MC algorithms proposed in the literature, outlining the modulation types, uncertain parameters, and the channel employed.

Table 2. Summary of LB approaches for OFDM signals.

Author(s)	Classifier(s)	Modulation(s)	Parameter(s)	Channel	Average PCC at 20 dB SNR
T. Yucek [45]	Sub-optimum algorithm	BPSK, QPSK, 16-QAM and 64-QAM	Imperfect noise variance	AWGN	99.9%
J. Leinonen [46]	Quasi-log-likelihood Ratio Test based classifer	BPSK, QPSK, 16-QAM and 64-QAM	Known channel correlation between adjacent subchannels	AWGN	98.50%
J. Zheng [47]	ALRT, HLRT and Energy-based detector	BPSK, QPSK, 8-PSK and 16-QAM	Known CSI, Known noise variance and unknown CSI	Rayleigh	97.40%
T. Fang [48]	Expectation maximization block-quasi HLRT (EM-Block-QHLRT)	BPSK, QPSK, 8-PSK and 16-QAM	Unknown CSI and unknown noise power	Acoustic Rayleigh	100%
M. Marey [49]	Iterative EM-based MC algorithm, bit-interleaved coded modulation iterative decoding (BICM-ID) scheme	QPSK, 64-QAM, 1024-QAM and 8194-QAM	Presence of synchronization error and known and unknown CSI	Rayleigh	99%

An LB-MC for the OFDM system is studied in [45]. The aim of this work is limited to reliable blind MC schemes. A maximum likelihood that provides optimal performance in the presence of AWGN is introduced. A sub-optimal classifier is obtained based on the optimal maximum-likelihood classifier to minimize the computational complexity. The accuracy of such classifiers is evaluated through Monte Carlo simulations. In the simulation, an OFDM system with 64 subcarriers is considered. The subcarriers are divided into 4 bands, each of which has 16 subcarriers. In each sub-band, four distinct modulation formats, namely BPSK, QPSK, 16-QAM, and 64-QAM, are used to transmit the signal according to the channel conditions. Perfect CSI is considered for the simulation. It is observed that the proposed sub-optimal algorithm achieves near to optimal performance with significantly less complexity. As a result, it can be used rather than signaling in realistic systems to improve spectral efficiency.

In the proposed method [46], a modified quasi-log-likelihood ratio (QLLR) based MC for the OFDM system is studied. The ALRT- and GLRT-based classifiers need few symbols to achieve acceptable classification performance in the presence of appropriate channel estimation with relatively high SNR. To achieve acceptable performance, a modified QLLR based classifier needs high SNR and more symbols but their computational complexity remains lower compared to the ALRT- and GLRT-based classifiers. In order to classify QPSK, 16-QAM, and 64-QAM, the modified QLLR test is applied on received symbol sets. This method seems to be feasible if the operating point of SNR is comparatively high as compared to ALRT- and GLRT-based classifiers.

Another LB-MC for OFDM with index modulation (OFDM-IM) is analyzed in [47]. The modulation parameters in OFDM-IM often include the number of active subcarriers in addition to the constellation of signals, which distinguishes them from traditional modulations. Specifically, two MC cases are assumed. One is the MC with known CSI and another is the MC with unknown CSI. ALRT, HLRT-LLR, and HLRT-energy-based classifiers are considered for the case of known CSI. When compared to ALRT, both HLRT-LLR and HLRT-energy have lower computational complexity, but show degradation in classification performance. In the case of unknown CSI, the energy-based detector is first used to recognize the active subcarriers, then the expectation-maximization (EM) algorithm is employed to estimate the CSI for each hypothesis. The number of subcarrier $N = 128$

CP length $N_{cp} = 15$, number of channel tap $L = 5$ and Rayleigh channel are considered the simulation parameters. The simulation results revealed that with an increment in the observed data, the classification accuracy of MC with unknown CSI is near the MC with known CSI. Furthermore, a numerical analysis of MC for OFDM and OFDM-IM shows that OFDM-IM has less classification accuracy than the OFDM. It illustrates that OFDM-IM would have less MC efficiency than OFDM because of the identification of the additional parameter, i.e., the number of active subcarriers that would be necessary for OFDM-IM.

In [48], a MC for OFDM signal underwater acoustic multipath channel is studied. It works in the presence of unknown channel impulse response (CIR) and noise power. Channel is first estimated by the EM block. If the number of blocks in EM increases, the channel estimation increases accordingly. Then, the QHLRT method is used to classify the subcarrier modulations. The EM-block-QHLRT method is compared with the EM-QHLRT. The number of subcarrier $N = 1024$, CP length $N_{cp} = N/4$, sampling frequency 48 kHz and acoustic Rayleigh channel are considered the simulation parameters for this technique. It is observed that after 5 dB SNR, the classification rate achieved by EM-block-QHLRT is higher than 90%, which shows a higher accuracy compared to the EM-QHLRT-based classifier.

In [49], an iterative EM-based MC algorithm is used for OFDM-SDR systems. The soft information provided by the channel decoder of bit-interleaved coded modulation iterative decoding (BICM-ID) scheme is utilized as a priori information to the proposed classifier. Simulation is done for the perfect CSI and imperfect CSI for higher-order modulations over the Rayleigh fading channel. The results show a slight difference between the perfect CSI and imperfect CSI, which shows the robustness of the suggested method. The suggested method improves significantly with iterations and outperforms traditional uncoded algorithms. The suggested method obtained acceptable classification performance in the presence of synchronization error, i.e., timing, frequency, and phase offset with reduced processing time. Furthermore, as the constellation size increases, the identification performance degrades. This is because of the less reliable soft information provided by the channel decoder.

3.2. Maximum a Posteriori (MAP) Approach

MC is the process of determining the modulation format of received signals from a set of L modulation formats $\mathfrak{M} = \{M_j,\ j = 1, 2, ..., M\}$, based on a series of N received samples $\boldsymbol{x_m} = [x_m[0], x_m[1], ..., x_m[N_s - 1]]$. The maximum a posteriori (MAP) criteria can be used to find the optimal modulation classifier by using the Bayes decision principle [57]. For received signal x_m, the a posteriori probability of M_j is defined as $P(M_j|x_m)$, and the decision is made by

$$\hat{M}_j = \arg\max_{M_j \in \mathfrak{M}} P(M_j|x_m), \tag{10}$$

Another well-known classifier originating from the MAP criteria is the ML classifier. The a posteriori probability can be expressed using the Bayes' rule as

$$P(M_j|x_m) = \frac{P(x_m|M_j)P(M_j)}{P(x_m)}, \tag{11}$$

where $P(x_m|M_j)$ denotes the likelihood of the received samples x_m when the modulation format M_j is given, $P(M_j)$ is the prior likelihood of the modulation format M_j, and $P(x_m)$ is the marginal likelihood of the received samples x_m, which is independent of M_j. When all the candidate modulation formats are equiprobable, then the MAP classifier is identical to the ML classifier [51].

$$\hat{M}_j = \arg\max_{M_j \in \mathfrak{M}} P(x_m|M_j). \tag{12}$$

Table 3 lists multiple MAP-based MC algorithms studied in the literature, outlining the modulation types, uncertain parameters, and the channel employed.

Table 3. Summary of maximum a posteriori (MAP) based classifiers for OFDM signals.

Author(s)	Classifier(s)	Modulation(s)	Parameter(s)	Channel	Average PCC at 20 dB SNR
L. Häring [50]	MAP Algorithm, channel reciprocity in TDD systems	BPSK, 4-QAM, 16-QAM and 64-QAM	Perfect knowledge about data rate	Rayleigh	99%
L. Häring [51]	ML and MAP Algorithm	no modulation, BPSK, QPSK, 16-QAM and 64-QAM	Perfect synchronization and unknown CSI	Rayleigh	99%
L. Häring [52]	Simplified MAP algorithm that utilized frame structure, channel reciprocity, total number of transmitted data	no modulation, BPSK, QPSK,16-QAM and 64-QAM	Perfect knowledge about data rate	AWGN	100%
L. Häring [53]	Improved Approximated MAP Algorithm	QPSK, 16-QAM and 64-QAM	Perfect synchronization	-	79.5%
L. Häring [54]	Signalling-assisted modulation classifier	M-QAM	Known CSI, knowledge about total number of loaded bits and coding rate	AWGN	98.5%
L. Häring [55]	Jointly optimizes the bit loading algorithm	M-QAM	Perfect synchronization and knowledge about signalling	AWGN	99%
L. Häring [56]	Influence of imperfect reciprocity	IEEE 802.11a/n	Unknown CSI and knowledge about total number of loaded bits	Rayleigh	100%
C. Husmann [57]	MAP Algorithm	BPSK, QPSK, 16-QAM and 64-QAM	Perfect time and frequency synchronization	AWGN	97.5%
S. Bahrani [58]	Improved Approximated MAP Algorithm, channel prediction method	BPSK, QPSK, 16-QAM 64-QAM and no modulation	Perfect synchronization and unknown CSI	AWGN	98%
M. Karabacak [59]	Adaptive Pilot Based	BPSK, QPSK, 16-QAM and 64-QAM	Perfect synchronization and known CSI	AWGN	99.8%
S. bahrani [60]	Rate adaptive (RA) bit loading algorithm	BPSK, QPSK, 16-QAM and no modulation	Perfect synchronization and unknown CSI	Rayleigh	100%

A MAP-based MC algorithm in time division duplex (TDD) based OFDM systems with adaptive QAM modulation is studied in [50]. It takes advantage of the channel reciprocity in TDD systems and the data rate of transmission. Unlike the signaling-free adaptive modulation technique, MC and data detection are decoupled here, resulting in significantly decreased computational complexity. Moreover, this technique utilizes the fixed bit allocation table (BAT) for all transmission frames. As a result, more symbols of the same modulation scheme can be employed to make a decision. Compared to the traditional ML method, simulations have validated the superior classification performance of the modified MAP algorithm. This technique allows adaptive modulation to be applied in wireless OFDM systems without reducing the effective data rate due to the signaling of the BAT.

A novel efficient MC algorithm in wireless TDD-based OFDM systems with adaptive modulation is analyzed in [51]. The frequency-selective behavior of the channel is experienced by a finite impulse response (FIR) filter model with Rayleigh fading coefficients Jakes' spectrum with the Doppler frequency f_{dm} is used to model the time correlation of the different path coefficients. This adaptive modulation approach adapts modulation formats among BPSK, QPSK, 16-QAM, and 64-QAM to a group of two adjacent subcarriers The conventional maximum-likelihood method is modified to a MAP classifier that uses reciprocity of the channels in TDD systems. Moreover, a less computationally complex classifier based on the MAP criteria is developed and evaluated, which is desirable for real-time implementations. The feasibility of complexity reductions is validated by simulations The classification performance of the proposed technique is slightly reduced in terms of the packet error rate compared to perfectly known modulation schemes.

A framework of MAP algorithms for MC in OFDM-based communication systems with adaptive modulation is studied in [52]. This work extends the achievements in MAP-based MC [50,51] by adding a new constraint to the framework. In this paper, a metric approximation is used, whose accuracy increases with rising SNR; the reason behind using this is the high computational complexity of the optimal algorithm. The side information like the known frame structure, channel reciprocity, and the knowledge of total data transmission rate, which are typically available in wireless TDD systems are intensively utilized by the proposed classifiers. By utilizing this information, the proposed likelihood-based MC algorithms are highly effective for the short OFDM frames.

Another MAP-based MC for OFDM systems with adaptive coding and modulation (ACM) is carried out in [53]. The proposed classifier for QAM schemes utilizes the channel reciprocity in TDD systems that requires knowledge about the joint probabilities of the subcarrier-wise bit efficiencies at the transmitter and receiver sides. In contrast to prior heuristic approaches [52], these probabilities are calculated analytically if the transmitter and receiver apply the same bit loading (BL) algorithm on their erroneously estimated channel state information. Furthermore, the performance of the proposed MC algorithm employing analytical results is comparable to the simulated joint probabilities. However, it is still somewhat superior due to the subsidiary-independent technique's sub-optimal approach [50]. Analytical and simulation results outperform the heuristic approach [52], especially at higher SNRs.

Another modulation classification algorithm for wireless TDD-based OFDM systems with adaptive modulation and coding is analyzed in [54]. The proposed MAP-based classifiers use the distinct signaling bits that are transmitted along with the information symbol. Thus, these can be viewed as a hybrid of MC and a signaling-based transmission principle. According to the signal structure of the received data symbols, these classification algorithms are characterized as bit allocation tables, i.e., a list of modulation formats used on each subcarrier. These received bit allocation tables are explicitly transmitted auxiliary information. Numerical studies indicate that the reliability of the classifier can be significantly enhanced by the use of the specified auxiliary information in a standard indoor propagation environment. Moreover, the simulation results of effective spectral performance show that the proposed method can be a reliable alternative in pure signaling-based or MC schemes in adaptive OFDM transmission. It outperforms the non-adaptive OFDM transmission system. However, this algorithm works in the presence of known CSI, knowledge about the total number of loaded bits, and coding rate.

An adaptive transmission algorithm for TDD-based wireless OFDM systems is carried out in [55]. In this technique, at the transmitter side, the BL algorithm and at the receiver side modulation classification algorithm are jointly optimized. To increase the effective data rate, a MAP modulation classification algorithm is applied in place of signaling the complete BAT to the receiver. The classification reliability is increased while preserving the enhanced link quality and low signaling overhead with this optimization on the BL algorithm. The idea behind this contribution is to maximize the effective bandwidth efficiency by this joint optimization of the BL algorithm at the receiver side and the modulation classification algorithm at the receiver side. Thus, the data rate loss caused by the signaling overhead is reduced. The simulations are performed in a typical indoor propagation scenario using burst transmission. It shows the enhancement of bandwidth and reduces the signaling overhead compared to the conventional methods.

A reciprocity-based MC algorithm for adaptive OFDM transmission systems in TDD mode is studied in [56]. This proposed transmission technique used the BL algorithm at the transmitter and MC at the receiver. A MAP-based MC is proposed, which is already effective for short frames if channel reciprocity in TDD systems is assumed. In this contribution, the authors analyze the performance of an improved version of this algorithm in a more realistic scenario. Simulations are carried out to validate the accuracy of the MC algorithm in the presence of imperfections caused by channel time-variance, channel estimation errors, and non-reciprocal transceiver filters. Simulation setup investigations are focused on indoor propagation scenarios typical for WLAN. For calibrated transceivers, the simulations sho superior performance of the proposed adaptive transmission scheme with MC compared t a non-adaptive transmission in a typical indoor propagation scenario. It also has superio classification performance as compared to the signaling-based technique.

A simplified MAP-based MC is analyzed in [57]. An adaptive OFDM based on a IEEE 802.11a system is simulated. The system occupies a bandwidth of 20 MHz, which i split into $N = 64$ subcarriers. Among these subchannels, $N_d = 48$ subchannels are use for data transmission: 4 are reserved for channel tracking and synchronization purpose and the remaining 12 are unused. Throughout this paper, they have assumed perfec time and frequency synchronization, which they consider to be a typical indoor scenari The number of multipath components is assumed to be 16 such that the length of the guar interval is set to be 16 too. The maximum Doppler frequency is assumed to be f_d = 55 H corresponding to a speed of 3.33 ms, and the Doppler spectrum follows Jakes' mode The correlation is very strong in the considered system due to the quantized structure o the effective channel. The quantization is a result of adaptive power allocation. In th context of this paper, a MAP-based MC approach is investigated in wireless local are network (WLAN) based OFDM systems with adaptive modulation. The receiver has t estimate the channel, which is modeled by a slowly varying multipath Rayleigh fadin channel and AWGN. The performance of the MC algorithm is measured in terms of th end-to-end packet error rate (PER). Package errors occur due to data detection errors an MC classification errors. The PER of the proposed MC algorithm is almost identical to th PER of an error-free MC algorithm. This exemplifies the potential of MC applications i real-time scenarios.

Another MAP-based MC for the TDD-based OFDM system is studied in [58]. Thi paper proposes a channel prediction approach for improving the efficiency of the MC use in the adaptive OFDM scheme. To achieve an acceptable prediction performance, effectiv noise reduction and interpolation techniques are used. The channel is supposed to b frequency-selective, with Rayleigh fading coefficients and a power delay profile decide by the standard indoor environment for IEEE 802.11a models. For time correlation, Jake Doppler spectrum is presumed, with the maximum Doppler frequency f_d set to 20 Hz b default. Finally, simulations for the channel modeled with the Gaussian Doppler spectru are carried out to explore the robustness of the proposed approach to the channel mode The probability of incorrect classification for both Jakes' and Gaussian Doppler spectrum i compared, in the case f_d = 20 Hz. In this case, it can be shown that the proposed techniqu is sufficiently robust to the model of the channel's time variance.

The importance of adaptive modulation for effective usage of channel capacity in th OFDM system is shown in [59]. The need to transfer the information about modulation t the receiver is abolished by the MC algorithm, and thus, these algorithms are a very usefu method to increase the channel capacity. However, in practice, two different sets of pilo symbols are used for the identification of the modulation type and for the estimation of th channel impulse response. The author proposes only one set of pilot symbols to find th information about the modulation type as well as the channel in this paper. As the pilo symbols are related to the modulation type, so they are named "adaptive pilots". The ide tification of the modulation type is successfully done with the help of these adaptive pilot without affecting the performance of the channel estimation. By assigning unique pilots t every possible modulation type, the modulation information is embedded. BPSK, QPSK 16-QAM, and 64-QAM are the possible pilot patterns with corresponding modulation type It is shown by the simulation results that modulation types are successfully identifie by the proposed adaptive pilots, while no effect is introduced to the channel estimatio process. For the application of the proposed algorithm, pilots can be located at differen locations with different values. However, when more number modulation formats ar involved in the communication, more adaptive pilots may be required, which degrades th spectrum efficiency of the transmission.

The MC approach enables the estimation BAT technique in adaptive OFDM systems [60]. The authors analyze a less computationally complex MAP-based MC algorithm. They derive an estimation of the probability of classification error of a MAP-based classifier. Moreover, based on the derived estimation, a rate-adaptive (RA) BL algorithm is developed. The findings of the simulation reveal that the proposed RA algorithm greatly improves the accuracy of modulation classification. Furthermore, it is also shown that, in comparison to traditional RA methods, the proposed BL approach improves classification performance for SNR above 15 dB.

3.3. FB Approach

In the FB algorithm, the expert domain feature needs to be extracted first and then decisions are made for the classification. Some of the expert domain features are the variance of the normalized signal amplitude, phase, and frequency [101], the variance of the zero-crossing interval [102], moments, cumulants [63], cyclic cumulants [5], cyclostationarity [103], Fourier transform [63], wavelet transform (WT) [17], and constellation shape [104] of the received signal. The fuzzy logic [105], entropy [106], and constellation shape recovery technique also have been used for MC. Various decision-making approaches have been employed, including maximum-likelihood detector [63], Hellinger distance [107], Euclidean distance [108], and unsupervised clustering techniques [109].

MC with Higher Order Statistics (HOS)

Here, we provide a framework of the MC method with HOS [110]. The moment with the kth order and pth conjugations for x_m associated with $x_m[n]$ is defined as

$$M_{kp,x_m} = E\left[x_m^{k-p}(x_m^*)^p\right]. \tag{13}$$

where $()^*$ represents a complex conjugate. The corresponding cumulant with kth order and pth conjugations is defined as

$$C_{kp,x_m} = cum(\underbrace{x_m, x_m, ..., x_m}_{k-p}, \underbrace{x_m^*, x_m^*, ..., x_m^*}_{p}), \tag{14}$$

where $cum()$ represents the joint cumulant function. HOS provides an integrated technique as well as a nonlinear signal processing viewpoint. Nevertheless, the information in the power spectrum of the second-order statistics is only appropriate for describing Gaussian processes statistically. In MC applications [111], a general fourth-order statistics C_{42,x_m} is frequently used. According to (13) and the fourth-order cumulant formula for four random variables, X, Y, Z, and W can be expressed as

$$cum(X,Y,Z,W) = E[XYZW] - E[XY]E[ZW] - E[XZ]E[YW] - E[XW]E[YZ], \tag{15}$$

and

$$\begin{aligned} C_{42,x_m} &= cum(x_m, x_m, x_m^*, x_m^*) \\ &= E\left(|x_m|^4\right) - \left(\left|E(x_m^2)\right|\right)^2 - 2E^2\left(|x_m|^2\right), \end{aligned} \tag{16}$$

In a similar fashion, a typical second-order cumulant can be written as

$$C_{21,x_m} = E\left(|x_m|^2\right). \tag{17}$$

The normalized fourth-order cumulant [12] is typically used to calculate MC, defined as

$$\hat{C}_{42,x_m} = \frac{C_{42,x_m}}{C_{21,x_m}^2}. \tag{18}$$

The FB-MC approaches are listed in Table 4, which highlights selected features, modulation types, channels, and undefined parameters.

Table 4. Summary of FB approaches for OFDM signals.

Author(s)	Feature(s)	Modulation(s)	Parameter(s)	Channel	Decision-Making Approaches	Average PCC at 20 dB SNR
A. D. Pambudi [61]	Mean,Variance, Skewness, Kurtosis and Moment Order	QPSK,16-QAM and 64-QAM	-	Rayleigh	Threshold based technique	91%
D. Shimbo [62]	Amplitude, Moments and Correlation	16-QAM and 64-QAM	Prior knowledge about CFO	AWGN	Threshold based technique	89%
R. Gupta [63]	Using discrete Fourier transform (DFT) and normalized fourth-order cumulants	BPSK, QPSK, MSK, OQPSK, and 16-QAM	Unknown Signal Parameters, unknown CSI and imperfect synchronization	Rayleigh	Likelihood ratio test	97.5%
J. Zhang [64]	Wavelet transform (WT), Transient characteristics	4-FSK, QPSK, 16-QAM and OFDM	Unknown Signal Parameters	Rayleigh	-	100%
Y. Zhu [65]	Kurtosis coefficient, Power spectral parameter, Energy distribution parameter	2-ASK, 4-ASK, 2-FSK, 4-FSK and OFDM	Unknown symbol rate and carrier frequency	AWGN, FNSF, FSF and Rayleigh	Threshold based technique	97%
Y. Ma [66]	Constellation cluster, number of cluster center	QPSK, 8-QAM, 16-QAM, 32-QAM and 64-QAM	Rotation plane and angle	AWGN	Peak-density clustering algorithm	87.5%
Tomoya [67]	Identification estimation method, Modulation parameters of rotation planes and angles	OFDM, CDMA, a block of QAM and so on	-	AWGN	-	92.5%
J. Chen [68]	Inter-class identification, Higher order cumulants	OFDM, 2-FSK, 4-FSK, 8-FSK, BPSK, QPSK, 8-PSK, 16-QAM, 32-QAM and 64-QAM	Perfect CSI	Rayleigh	Threshold based technique	100%
H. Li [69]	Empirical Distribution Function-Based Gaussian Test	M-QAM	Unknown symbol duration, cyclic prefix duration and number of subcarriers	AWGN	-	95%
Y. Liu [70]	Latent Dirichlet Bayesian network, Gibbs sampling method	QPSK, 8-PSK and 16-QAM	Imperfect CSI and unknown SNR	Flat fading	-	97.5%
Y. Liu [71]	Optimal Bayesian Method, latent Dirichlet model, mean field variation inference	QPSK, 8-PSK, 16-QAM, and 16-PSK	Imperfect CSI and unknown SNR	Flat fading	-	97%
A.K. Pathy [72]	Using DFT and normalized fourth-order and sixth-order cumulants	BPSK, QPSK, MSK, OQPSK, and 16-QAM	Unknown Signal Parameters, unknown CSI and imperfect synchronization	Rayleigh	Likelihood ratio test	97%

Multicarrier modulation given by the OFDM signal generator using an IEEE 802.16e standard is studied in [61]. Based on the standard of IEEE 802.16e, three possible modulation formats can be used, such as QPSK, 16-QAM, and 64-QAM. The mean, variance, skewness, kurtosis index, and moment order of the received signal are all considered and compared in order to determine the modulation scheme non-line-of-sight (NLOS) with six multipath components. The dominant statistic features capable of separating the QPSK modulation scheme from 16-QAM and 64-QAM are skewness, kurtosis, and variance, as determined by the statistical properties of the received signal. Furthermore, the high order moment is one of the most important statistical features that distinguish the 16-QAM modulation scheme from the 64-QAM modulation scheme. However, in the context of timing and frequency synchronization issues, this approach does not perform well.

In another FB-MC [62], the amplitude moments and correlation properties are used to classify the modulation scheme for OFDM systems. This technique considers the presence of CFO, which is the cause of intercarrier interference (ICI) in the amplitude moments of the received signal. Therefore, the ICI component is estimated by using the correlation between the subcarriers. To determine the influence of ICI components in the amplitude moments, the authors derive the amplitude moment in the form of infinite series of elementary functions. It is observed that the amplitude moments increase as the frequency offset increases. Considering 4096 subcarriers in an OFDM symbol, at least 10 OFDM symbols are required to achieve the desired classification accuracy at 30 dB SNR. This approach outperforms the existing amplitude moment-based approach with the prior information about CFO. This is due to the estimation and elimination of the ICI components in the amplitude moments. However, this MC algorithm is restricted to known CSI and proper synchronization circumstances.

In [63], another FB blind MC approach is suggested and implemented on radio frequency (RF) testbed for OFDM signals. The authors use the combined features of DFT and the fourth-order cumulant, as shown in Figure 3. This algorithm does not need prior information about the signal parameters and CSI. It also works effectively when there are synchronization problems, such as timing, frequency, and phase errors. Before the feature extraction process, a random uniformly distributed timing offset is added in each OFDM symbol to reduce the influence of the timing offsets. The authors have listed BPSK, QPSK, OQPSK, MSK, and 16-QAM for the OFDM signal. The number of subcarrier $N = 1024$, CP length $N_{cp} = N/4$, channel tap $L = 4$, number of OFDM symbol 50, normalized CFO $-0.5 < \epsilon < 0.5$, symbol timing offset $[-N/2, N/2]$, sampling rate 50 Msamples/s, symbol rate 1 Msymbols/s, and Rayleigh channel are considered the simulation parameters for this technique. Classification is carried out in two stages. First, the received signal is transformed into the frequency domain by using the DFT operation, then the normalized fourth-order cumulant of the frequency domain signal is calculated. The modulation formats OQPSK, MSK, and 16-QAM can be distinguished by the normalized fourth-order cumulant, which is expressed as

$$\tilde{C}_{42_R} = \frac{1}{K}\sum_{m=1}^{K} \frac{\vec{C}_{42_{X_m}} - \left|\frac{1}{K}\sum_{v=0}^{K-1} e^{-j4\pi v/K(\tau+\theta_u)} X_m^2[v]\right|^2}{\frac{1}{K}\sum_{v=0}^{K-1}|X_m[v]|^2 - C_{21,W}}, \tag{19}$$

where $X_m[v]$ represents the DFT of the received signal $x_m[n]$, $C_{21,W} = \sigma_W^2$ represents the estimated variance of AWGN, and K is the total number of OFDM symbols.

The histogram of the above is given in Figure 4. The second stage performs the DFT of the square of the received signal then calculates the normalized fourth-order cumulant, which is expressed as

$$\tilde{C}_{42_U} = \frac{1}{K}\sum_{m=1}^{K} \frac{\vec{C}_{42_{U_m}} - \left|\frac{1}{K}\sum_{v=0}^{K-1} e^{-j4\pi v/K(\tau+\theta_u)} U_m^2[v]\right|^2}{\frac{1}{K}\sum_{v=0}^{K-1}|U_m[v]|^2 - C_{21,W}}, \tag{20}$$

where $U_m[v] = X_m[v] \circledast X_m[v]$, $\circledast$ denotes the linear convolution operator. For BPSK and QPSK modulation schemes, the above Equation (20) gives different values, as shown in Figure 5.

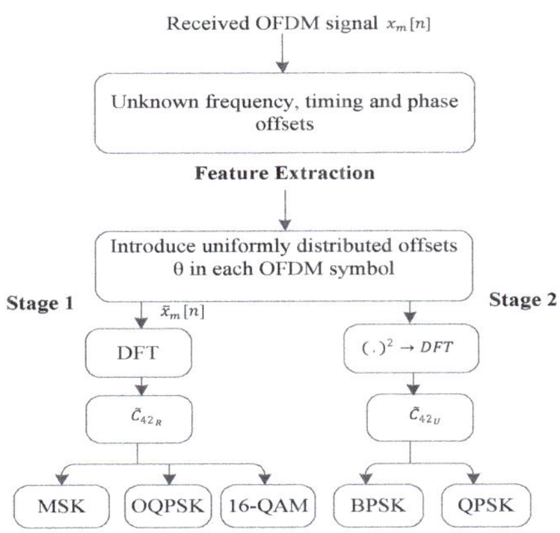

Figure 3. Schematic diagram of blind modulation classification studied in [63].

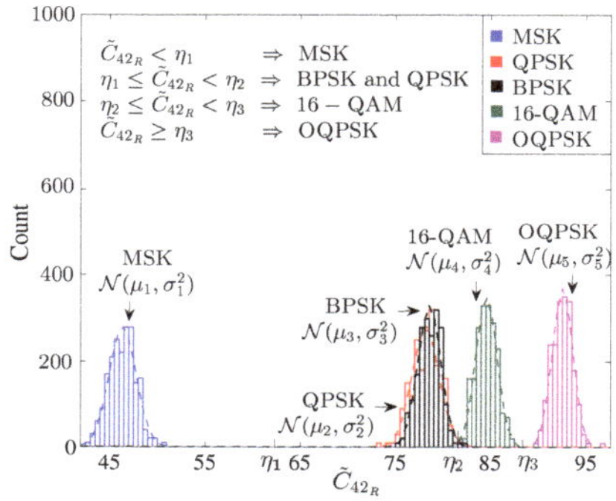

Figure 4. Histogram of $\tilde{C}_{42_R}$ for BPSK, QPSK, OQPSK, MSK, and 16-QAM. Adapted with permissio from Ref. [63]. Copyright 2021 IEEE.

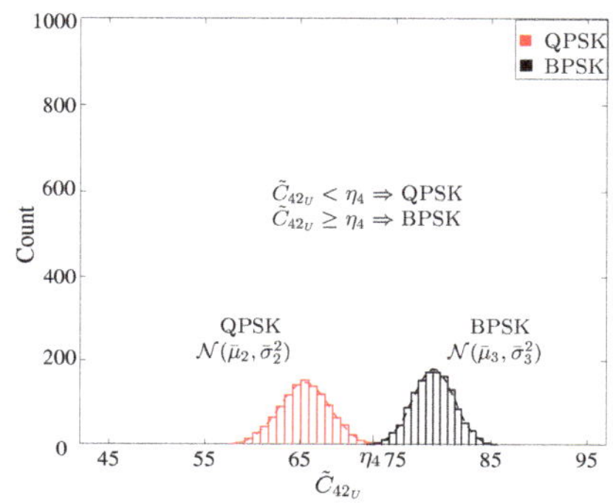

Figure 5. Histogram of $\tilde{C}_{42_U}$ for BPSK and QPSK. Adapted with permission from Ref. [63]. Copyrigh 2021 IEEE.

In the paper [64], the authors use a method for applying wavelet transform (WT) t OFDM and SC signals to extract their transient characteristics and then use the transien characteristics to identify the two types of signals. Good performance can be achieve in low SNR and multipath channel conditions. In addition, the effects of the sample rat and symbol rate on the identification algorithms are analyzed and simulated. The autho conducts a variety of simulation experiments to assess the performance of the propose identification algorithms and the effects of the sample rate and symbol rate on the ider tification algorithms. All results are based on 100 Monte Carlo trials. The percentage c correct classification (PCC) versus the SNR plot represents the average variance of signal versus SNR. Every 8000 data samples make up of a trial source. We notice that the averag variance is large in the OFDM curve and small in the SC curve. The result is well separate between OFDM and SC modulations. When SNR is 0 dB, the PCC between OFDM and S signals can reach 100% when the symbol rate is greater than 20 kHz.

Another FB-MC [65] is using spectrum analysis to classify the OFDM and SC. The au thors utilize the energy distribution parameter and the kurtosis of the power spectrun coefficient to classify OFDM and SC. This method does not need any prior informatio about the symbol rate, carrier frequency, etc. In simulation results, it is found that extracte spectrum parameters have better performance over AWGN as well as Rayleigh fadin channels. It has a classification rate of up to 97% with an SNR at 10 dB.

In [66], the peak-density clustering algorithm is used to investigate an MC techniqu for adaptive optical OFDM systems. The clustering technique is used to find the center of the signal constellation clusters. The number of cluster centers is calculated using the density and distance metrics of samples. The number of cluster centers is utilized to distinguish M-QAM. The OFDM signals are fed into an arbitrary waveform generator (AWG) with a sampling rate of 50 GSamples/s. The electrical OFDM signal is then converted into an optical signal using an external cavity laser (ECL) and an intensity modulator. The modulated optical signal is then routed through a variable optical attenuator (VOA) and an erbium-doped fiber amplifier (EDFA) to alter the signal SNR and mimic different transmission circumstances. They use a 50 GSamples/s real-time oscilloscope to capture data and another VOA to regulate the input power before the photodetector. Finally, OFDM MC and demodulation are conducted for a test sample of 8192 lengths in each optical SNR.

In the paper [67], the identification of these orthogonal modulations, i.e., OFDM, code division multiple access (CDMA), is studied. The classification method is based on general orthogonal modulations, whose modulation parameters should be estimated. The identification method applies to both adaptive modulation and increased security in radio communications. General orthogonal modulations are employed to identify modulations. First, the modulation parameters of rotation planes and angles are estimated. Orthonormal vectors are derived by received signal samples and rotated to hold orthogonality among time slots. Then, the inverse rotation corresponds to the modulation parameters to be estimated. The difference vector between the received signal vectors is used for this method. In computer simulations, OFDM, CDMA, a block of QAM, and so on are considered candidate modulations. The bit error probability of the estimated modulation is presented to compare the performance from the point of view of SNR and the number of samples. The proposed estimation performance is evaluated in the AWGN channel by computer simulations.

Based on the higher-order cumulants, an MC algorithm is carried out that discriminates the OFDM signals from SC signals [68]. First, OFDM signals are discriminated from SC signals based on distinct features parameters over the Rayleigh channel. In order to verify the effectiveness, the modulation set is assumed as OFDM, 2-FSK, 4-FSK, 8-FSK, BPSK, QPSK, 8-PSK, 16-QAM, 32-QAM, and 64-QAM. The combination of the second- and fourth-order cumulants is used as the feature to discriminate the OFDM signals from the SC signals. Simulation results show that the algorithm is stable with low computational complexity and high PCC in low SNR level.

In the paper [69], a classification technique is devised for identifying the OFDM signals from the SC. In addition to differentiating the OFDM signals from SC, some important parameters of OFDM signals are estimated for further processing. The estimated parameters include the number of subcarriers, the length of the OFDM symbol, and the CP length. Using these parameters, traditional modulation classification techniques may be used to identify the linear modulation format on each OFDM subcarrier. The analytical distribution function-based Gaussian test technique is shown to differentiate OFDM from SC modulations effectively, and the correlation test is shown to estimate the cyclic prefix length effectively. A fast Fourier transform (FFT) is used to effectively estimate the number of subcarriers. The simulation findings show that the proposed technique provides classification performance of more than 90% for SNR greater the 15 dB.

In the paper [70], a Bayesian inference-based MC technique for the MIMO-OFDM signal is used. This technique uses the Gibbs sampling convergence approach on a latent Dirichlet model as a baseline. However, the inference-based technique has a significant computational overhead, and it also needs perfect synchronization at the receiver.

In the paper [71], an MC algorithm for the MIMO-OFDM system is analyzed under the unknown frequency-selective fading channels and SNR. This work is an extension of the achievements in MAP-based MC [50,51] by adding a new constraint to the framework. The classification problem is presented as a Bayesian inference task, with solutions provided based on Gibbs sampling and mean-field variational inference. The Gibbs sampling method yields the best Bayesian result. It is shown that after multiple iterations, switching to the mean-field variational inference technique improves classification accuracy for the small length of the received signal. However, most of the existing MC consider channels as flat fading when the number of receiving antennas exceeds the number of transmitting antennas. However, under more general circumstances, the proposed algorithm works quite well. It is shown that the proposed Bayesian methods outperform existing non-Bayesian techniques based on independent component analysis (ICA). However, this inference-based technique is quite difficult, and it also necessitates perfect synchronization at the receiver.

A tree-based blind MC method for asynchronous MIMO-OFDM is developed in [72]. It extracts unique features for different modulation schemes using normalized fourth-order and sixth-order cumulants. It then performs a threshold-based classification using the likelihood ratio test to determine the modulation format of the received signal. The number of subcarrier $N = 128$, CP length $N_{cp} = N/4$, number of channel tap $L = 4$, number of OFDM symbol 50, normalized CFO $-0.5 < \epsilon < 0.5$, symbol timing offset $[-N/2, N/2]$, sampling sampling rate 50 Msamples/s, symbol rate 1 Msymbols/s, and Rayleigh channel are considered the simulation parameters for this technique. The classification performance of this algorithm is validated by using the RF testbed in a realistic scenario. The authors consider the higher number of transmitting and receiving antennas in the simulation process. However, the actual experimental systems in this paper only contain at most two transmitter antennas and two receiver antennas.

In the paper [112], signal parameter estimation, modulation classification, and synchronization are carried out for the OFDM signal. At the first stage, the cyclic cumulant is used to estimate the number of subcarriers, symbol length, useful symbol length, CP length, and oversampling factor. At the second stage, the elementary cumulant is used to classify the BPSK, QPSK, OQPSK, MSK, and 16-QAM modulation scheme over the Rayleigh fading channel. After that, a modified maximum likelihood technique is used to estimate the CFO and STO for the OFDM system jointly. After correction of the CFO and STO, recovery of the constellation diagram of modulation schemes and BER analysis is performed. The BER is found approximately 8.5×10^{-3} and 6.5×10^{-2} at 20 dB SNR for QPSK and 16-QAM modulation schemes, respectively. This technique is also validated over the NI RF testbed setup over an indoor propagation environment.

4. Artificial Intelligence-Based Approach to MC

AI is certainly the next big "game-changing" technology that includes both ML and DL. In MC, ML finds lots of significance in terms of decision trees, KNN, support vector machine (SVM), artificial neural network (ANN), and some hybrid algorithms. DL is a kind of a subsidiary of ML, which originates from the study of ANN. Neural networks are inspired by biology and try to mimic the neural structure of the human brain [113,114]. Recently, researchers in the field of wireless communication stated using DL extensively. It finds applications especially in the field of communication systems, such as non-orthogonal multiple access (NOMA) technology, MIMO technology, resource allocation scheme, and signal MC. Tables 5 and 6 lists a few of the ML- and DL-based MC algorithms studied in the literature, outlining the modulation types, uncertain parameters, and the channel employed.

4.1. ML-Based MC

In this paper [73], the authors extract the features by calculating higher-order cumulants, then the extracted features are applied to naïve Bayes classifier for MC. However, the authors assume proper equalized and perfectly synchronized signals received at the receiver. The features, combinations of fourth-order C_{42} and sixth-order cumulants C_{63} often produce better classification performance than using each of these features alone By using the same set of features, the naïve Bayes classifier is compared with the ML-based classifier and SVM-based classifier. It is observed that the naïve Bayes classifier outperforms the ML-based classifier and SVM-based classifier with less computational complexity.

This paper [74] introduces a technique for classifying OFDM signals using higher-order moments and cumulants with multiple types of classifiers and cluster techniques There are four considered methods of classification, namely, KNN, ML, SVM, and neural network (NN) classifiers. Fuzzy k-Means and fuzzy c-means are two cluster techniques that are used for the two classes of OFDM signals. One class is considered fixed WiMAX (IEEE 802.16d), which includes BPSK, QPSK, 16-QAM, and 64-QAM modulations. Another class is considered OFDM signals used in Wi-Fi (IEEE 802.11a), which includes and BPSK, QPSK, 16-QAM, and 64-QAM modulations. In the simulations, the input signals are normalized to have zero mean and unit variance after transmitting through the Rayleigh fading channel. The normalized output signal is then used for the feature extraction process. Higher-order moments and cumulants up to the 8th order are used to extract features. The extracted features are used as input to the different types of classifiers, such as SVM, KNN, ML, and NN classifiers, which use the fuzzy k-means and fuzzy c-means as clustering techniques. The performance of the SVM classifier with the fuzzy k-mean is better than all the combinations of classifiers and clustering algorithms for most of the SNR values.

Table 5. Summary of ML-based classifiers for OFDM signals.

Author(s)	Classifier(s)	Modulation(s)	Parameter(s)	Channel(s)	Average PCC at 20 dB SNR
M.L.D. Wong [73]	Optimize Shannon's channel capacity, Naive Bayes classifier	BPSK, QPSK, 16-QAM and 64-QAM	Perfect synchronization	AWGN	96.8%
S. E. El-Khamy [74]	Higher order moments and cumulants, Fuzzy K-Means and Fuzzy C-means	BPSK, QPSK, 16 QAM, and 64 QAM	-	Rayleigh	100%
X. Yuan [75]	Higher-order cumulants, random forest based MC algorithm	QPSK, 16-QAM and 64-QAM	Imperfect time synchronization	Frequency-selective	100%
W. Machid [76]	Least squares (LS) method and iterative closest point (ICP)	BPSK, QPSK, 16-QAM, and 64-QAM	Unknown noise variance and CSI	Flat fading	97.5%
Y. Zhang [77]	High order cumulants, Decision Tree classifier	BPSK, QPSK, GFSK, 16-QAM, 64-QAM and OFDM	Presence of timing offset	Flat fading	99.5%
B. Dehri [78]	Higher order statistics, pattern recognition methods, ANN or SVM, or RFC or KNN	QPSK and 16-QAM	Presence of CFO and Imperfect CSI	Rayleigh	100%
Y. Gu [79]	Peaks in the distribution of amplitude, the variance of the amplitude, the variance of the phase, and the variance of the spectrum, SVM classifier	BPSK, QPSK, 16-QAM, 64-QAM, 256-QAM and GMSK	Unknown CFO	AWGN	100%
J. He [80]	Clustering and Gaussian model	QPSK, 16-QAM, 64-QAM	-	AWGN	100%
L. Gaohui [81]	High order cumulants and bi-spectral envelope peaks, hierarchical iterative SVM classifier model	M-QAM, MFSK and MPSK	Perfect synchronization	Rayleigh	100%

Table 6. Summary of DL-based classifiers for OFDM signals.

Author(s)	Classifier(s)	Modulation(s)	Parameter(s)	Channel(s)	Average PCC at 20 dB SNR
R. M. Al-Makhlasawy [82]	Mel Frequency Cepstral Coefficients (MFCCs) and multi-layer feed -forward neural network	QPSK, 8-QAM, 16-QAM, 32-QAM, 64-QAM and 128-QAM	Perfect synchronization	AWGN	100%
Y. Li [83]	Bispectrum and CNN Alexnet model	BPSK, 2-ASK, 2-FSK, 4-FSK, 8-FSK, LFM, and OFDM	-	AWGN	97.5%
S. Hong [84]	CNN with dropout layer	BPSK, QPSK, 8-PSK, 16-QAM and 64-QAM	Perfect synchronization	Rician fading	99%
J. Shi [85]	CNN, ReLU and PReLu activation	BPSK, QPSK, 8-PSK, and 16-QAM	Presence of phase offset and imperfect CSI	AWGN	100%
S. Hong [86]	CNN	BPSK, QPSK, 4-PAM, 8-PSK, and 16-QAM	Perfect synchronization	Rician fading	97.5%
F. Meng [87]	CNN with two step training, Transfer learning	BPSK, QPSK, 8-PSK, 16-PSK, 16-QAM, 32-PSK and 64-QAM	Unknown CFO and unknown SNR	Time invariant and frequency non-selective	100%
D. H. AlNuaimi [88]	GaFP-Net, TF-HMS, MDNC, and IQL	QPSK, BPSK, DPSK, ASK, FSK, 16-QAM, 32-QAM, 64-QAM, and 128-QAM	Unknown CFO	AWGN	86%
Z. Zhang [89]	CNN-LSTM	BPSK, QPSK, 8-PSK, AM-DSB, AM-SSB, CPFSK, GFSK, WBFM, 4-PAM, 16-QAM, and 64-QAM	Presence of CFO and STO	Rayleigh	91%
M.C. Park [90]	IQ and FFT window bank (FWB), CNN-LSTM-based classifier	QPSK, 16-QAM, 32-QAM, and 64-QAM	-	Rayleigh	98.5%
Y. Zhang [91]	Mixed order moment, hybrid grey wolf optimization (HGWO) algorithm, DNN-based classifier	QPSK, 16-QAM, 32-QAM, and 64-QAM	Presence of CFO and STO	Rayleigh	100%
Z. Zhao [92]	AlexNet/GoogLeNet-TL-based classifier	BPSK, QPSK, 8-QAM, 16-QAM, 32-QAM and 64-QAM	-	AWGN	100%
J. Yin [93]	Lightweight CNN (LCNN)-based Shuffle MC, FFT, l_2 regularization	BPSK, QPSK, 8-PSK, 16-QAM	-	Rician fading	100%
G. Kong [94]	Fourier synchrosqueezing transformation (FSST), Independent component analysis (ICA), hierarchical CNN-based MC	16-QAM, 64-QAM, and 256-QAM	Perfect Synchronization	Rayleigh	90%
Q. Zheng [95]	Spectrum interference-based two-level data augmentation method, deep CNN	BPSK, QPSK, 8-PSK, 16-QAM, 64-QAM, GFSK, CPFSK, 4-PAM, WBFM, AM-SSB, and AM-DSB	-	Rayleigh	89.3%
T. Huynh-The [97]	CNN with integrated attention and residual connections	BPSK, QPSK, 8-PSK, and 16-QAM	Presence of CFO	Rayleigh	88%

The MC problem for MIMO systems employing OFDM under imperfect timing synchronization scenarios is studied in [75]. The proposed algorithm first uses the HOC of the received signal to extract the unique features, which show the robustness to STO. After that a random forest classifier is used as the decision criterion to perform the classification problem. The main benefits of random forests are their better classification performance and low exposure to noise. The number of subcarrier $N = 128$, CP length $N_{cp} = N/4$ channel tap $L = 5$, number of transmitting antennas 2, number of receiving antennas and frequency-selective channel are considered the simulation parameters. The simulation results show that the proposed classifier can work well in the presence of STO with satisfactory classification accuracy. In a realistic scenario, where perfect STO estimation is difficult to achieve, these algorithms can provide conceptual help.

In the paper [76], a modulation classifier without knowing noise variance is studied for the OFDM system. In order to estimate the amount of phase rotation caused by flat fading the authors investigate adopting the iterative closest point, which is a kind of template matching technique. Combining the least squares-based phase estimation, the classification performance of the proposed method can be improved significantly. The PCC at several SNRs when each correction is performed in flat fading, where four types of modulation schemes, i.e., BPSK, QPSK, 16-QAM, and 64-QAM, are used. From these results, it is found that, as compared with the method of using only the least-squares method, the method combining the least squares (LS) method with the iterative closest point (ICP) algorithm does not deteriorate the accuracy of the phase correction, even if the number of signal points decreases.

In this paper [77], the classification of OFDM, BPSK, QPSK, Gaussian frequency-shift keying (GFSK), 16-QAM, and 64-QAM is realized by MATLAB programming based on the characteristic of HOCs. A new feature parameter is proposed according to the second- and sixth-order cumulant. Simulations are conducted with classifiers, including KNN, SVM, decision theory, and back-propagation neural network (BPNN). It is found that the average classification rate is greater than 95%.

In the paper [78], the authors propose a blind MC algorithm for space-time block coding (STBC)-based MIMO-OFDM system, which works in the presence of CFO, channel estimation errors, and impulsive noise. Multiple signal classification (MUSIC) algorithms are used to estimate the CFO and channel statistics. The estimated CFO and channels are compensated and equalized, then features are extracted using higher-order moments (HOMs) and HOCs. Finally, the extracted features are applied to ANN, SVM, RF classifier (RFC), and KNN classifier. The simulation results show that the SVM and ANN classifiers have better classification performance, even at low SNR.

In [79], an SVM-based MC algorithm is studied for the OFDM system in the presence of frequency offset in which statistics-based features are used as input of the SVM classifier. The number of peaks in the distribution of amplitude, the variance of the amplitude, the variance of the phase, and the variance of the spectrum are extracted from the received signal. These extracted features are used to make a dataset. This dataset is applied to the SVM classifier to classify the modulation scheme of the received signal. The proposed method shows great accuracy in high SNR channels with over 80% accuracy. It also shows robustness against the frequency offset. However, when the signal is flooded by noise and extremely influenced by frequency offset, the proposed still has over 50% accuracy. The algorithm is tested experimentally on the SDR platform, which can realize a variety of communication systems by updating the software. Based on such a popular SDR hardware platform and using GNU Radio, the modulation formats are generated, transmitted, and classified.

In [80], the design and implementation of the MC algorithm for the OFDM visible light communication (OFDM-VLC) system are explored. Clustering and Gaussian model analysis are used to obtain the classification feature values. The modulation format is then classified using these classification feature values. The simulation results show that the suggested method can achieve 100% classification accuracy at 1 dB to 2 dB lower than that of the clustering scheme. Furthermore, the experimental findings show that the suggested MC technique is feasible in an OFDM-VLC system.

In [81], an MC algorithm base on a hierarchical iterative SVM classifier is studied for the OFDM signal. To extract characteristic values from OFDM signals, higher-order cumulants and bi-spectral envelope peaks are used, and the resulting characteristic values are then processed to create fresh training sample data. The feature extracted by using HOCs is used to distinguish multicarrier signals from the SC signals. The bi-spectral envelope peaks are used to distinguish the OFDM signal from the multicarrier signal. The training dataset obtained from the higher-order cumulants and bi-spectral envelope peaks of the received signal is applied to the input of a hierarchical iterative SVM classifier. The number of subcarrier $N = 128$, CP length $N_{cp} = N/4$, symbol rate 1024 bps, sampling rate 3000 kHz and Rayleigh channel are considered the simulation parameters for this technique. It is found the classification accuracy of the SVM-based classifier is improved when compared with the wavelet transform method and higher-order cumulant-based method.

4.2. DL-Based MC

A lot of focus has recently been drawn by DL due to its effective ability to integrate offline preparation and online deployment [115]. DL is a specialist in automated feature extraction from a huge amount of data instead of the complicated and challenging nature of man-made features [116,117].

In the paper [82], a cepstral algorithm for MC is proposed with adaptive modulation in OFDM systems. The expert domain features of the received signal are extracted using Mel frequency cepstral coefficients (MFCCs), and the modulation formats and their order are classified using a multi-layer feed-forward neural network. This classifier has the capability of recognizing the M-ary amplitude-shift keying (M-ASK), MSK, M-PSK, M-ary frequency-shift-keying (M-FSK), M-QAM signals and the order of the identified modulation. The classification performance of the proposed technique is evaluated using the false classification probability (FCP). The AWGN channel is taken into account when creating the mathematical model for most of the results. The simulation results reveal that the modulation format and order can identify by extracting cepstral features from the received signal and with the help of the transforms, such as discrete cosine transform (DCT), discrete sine transform (DST), and the discrete wavelet transform (DWT). These classify the distinct features using a robust back-propagation feed-forward neural network for different modulations, such as QPSK, 8-QAM, 16-QAM, 32-QAM, 64-QAM, and 128-QAM. The proposition to identify the modulation type and order is proven to be considered effective.

In the paper [83], the authors develop an MC algorithm that is based on the bispectrum and CNN AlexNet models. As we know, bispectrum is a high-order statistic that suppresses AWGN well and is frequently utilized in signal detection and nonlinear system characterization areas. Furthermore, AlexNet exhibits outstanding image classification performance despite having a very basic structure of eight layers. First, the authors compute the bispectrum of received signals, then take the amplitude spectrum of the bispectrum (ASB), which is used as input to the CNN network. After that, they fine-tune the chosen AlexNet, which automatically extracts the distinct features from ASB images. Finally, these features are passed into a softmax classifier, which classifies the modulation type. The simulations are performed under different noise environments for the dataset that includes BPSK, 2-ASK, 2-FSK, 4-FSK, 8-FSK, linear frequency modulation (LFM), and OFDM signals. It is observed that the bispectrum-AlexNet model has a classification accuracy greater than 97.7% when the SNR is greater than 5 dB.

The above MC techniques are developed by utilizing feature extraction-based machine learning. Moreover, the standard approaches face bottlenecks, where the PCC is very small and it is also impossible to incorporate them in realistic OFDM systems because it is difficult to extract distinct features from OFDM signals using conventional methods. In order to address this problem, the authors [84] suggest a CNN-based MC system for recognizing OFDM signals. In particular, a CNN is used to train in-phase (I) and quadrature (Q) samples for OFDM signals. The authors construct two datasets with separate modulations for the MC function. Dataset 1 contains BPSK, QPSK, 8-PSK, and 16-QAM modulations, while dataset 2 contains BPSK, QPSK, 8-PSK, 16-QAM, and 64-QAM modulations. These two datasets are utilized to test the robustness of the proposed system. Each modulation format considers 20,000 data samples for training and research. Since CNN can efficiently extract the distinct features of received OFDM signals, the simulation results indicate that CNN trained on I and Q samples achieves better classification performance than conventional machine learning based approaches.

In the paper [85], a CNN-based MC method by considering the phase offset effect is studied. As shown in Figure 6, CNN-based MC is first trained by the received I and Q samples in the presence of phase offsets at different values of SNR. As shown in Figure 7, the DL-based MC technique is mainly implemented by CNN. The PReLU is used as an activation function for all layers, except the last layer, and the softmax is used as an activation for the last layer to implement a multi-classification problem. The authors use two sets of data with different modulation modes for the MC issue, i.e., Dataset 1 and Dataset 2, to verify the robustness of the classification technique [85]. The number of subcarrier $N = 16$, CP length $N_{cp} = 2$, number of OFDM symbol 6 and AWGN channel are considered the simulation parameters for this technique. Comparative experiments show that its performance of classification is much higher than the conventional extraction methods. Moreover, the classification accuracy relatively reduces at the low SNRs due to the presence of phase offsets. By gradually increasing the SNRs, effective classification accuracy can be achieved eventually.

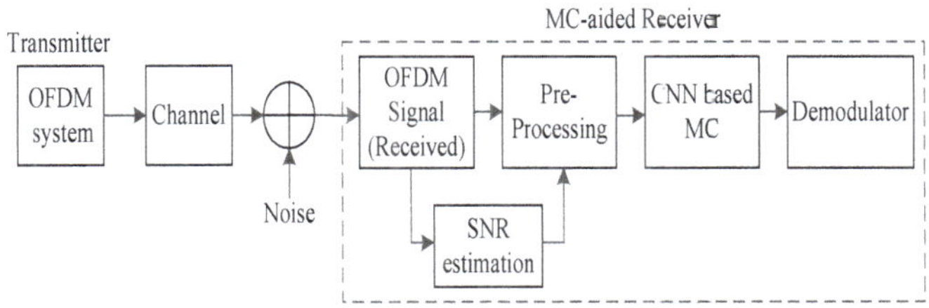

Figure 6. Framework of the proposed CNN-based MC system [85].

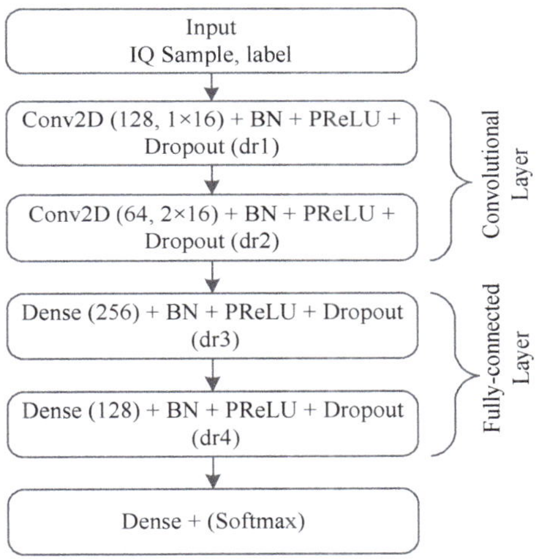

Figure 7. CNN structure design in the proposed CNN-based MC method [85].

An adaptive modulation model based on machine learning for a MIMO-OFDM system is carried out in [44]. The 5G new radio (NR) technology can be used in a wider range of internet of things (IoT) applications than traditional systems. The adaptive modulation technique, which changes data rate and latency based on channel conditions, can be efficiently employed in 5G digital NR technology. The traditional adaptive modulation technique is developed by assigning modulation formats based on the channel conditions since the rule-based MC is unable to analyze transmission efficiency based on channel correlations between antennas. The number of propagation modes is enhanced exponentially based on the available number of modulations and antennas. So, these are not appropriate for 5G NR systems. The proposed adaptive modulation technique is learned by the training data, which are generated by the feature extracted from the received signal. The DNN application for adaptive modulation is the primary method of the main component analysis, which improves the model efficiency. The simulation results on the optimal transmission mode classification for the MIMO-OFDM signal show that the proposed model supports adaptability according to the condition of the complex MIMO channel.

In [86], the authors here present a CNN-based MC approach for the identification of OFDM signals, which is linked to a CNN that is trained on I and Q samples. The suggested CNN-MC technique is made up of two parts: three convolutional layers and four fully connected layers. The number of subcarrier $N = 16$, CP length $N_{cp} = 2$, number of OFDM symbol 6 and Rician channel are considered the simulation parameters for this techniqu The suggested technique outperforms existing modulation classification algorithms i terms of accuracy and reliability. However, any parameters, such as number of subcarrier number of null subcarriers, STO, CFO, phase offset, and CP length, change these MC an do not provide accuracy of more than 50% for adaptive OFDM systems.

The other approach described in [87] focuses on two-step training to enhance th classification performance of CNN-based classifiers. Transfer learning is also introduced t increase the performance of the retraining. A wider range of modulation formats for th OFDM signal, such as BPSK, QPSK 8-PSK, 16-QAM, 32-QAM, and 64-QAM, is recognize by the suggested technique.

In [88], the MC algorithm for the OFDM signal is developed by using an intelligen pyramid model. This algorithm has four stages, i.e., pre-processing, feature extractio feature clustering, and classification. In the pre-processing step, the authors improve th received signal quality, which involves two steps quality evaluation and quality augmer tation, using the bi-fold signal fortification (BFSF) approach. The number of subcarrie $N = 2048$, CP length $N_{cp} = 3$, sampling frequency 5 MHz and AWGN channel are consic ered the simulation parameters for this technique. If the received signal quality is poor, the quality augmentation is performed, taking into account noise reduction, equalization, quar tization, and CFO compensation. Then the feature extraction process is performed by th gated feature pyramid network (GFP-Net). After that, the authors make the cluster from th extracted feature by using an intelligent twin-functioned human mental search (TF-HMS optimizer to minimize the classification complexity. Finally, they offer the multi-distanc based nearest centroid classifier (MDNCC) technique, as well as improved Q-learnin (IQL), to determine the correct modulation format for the received signal. However, thi technique only considers the CFO as the synchronization when performing the modulatio classification of the received signal.

In [89], a CNN long short-term memory (CNN-LSTM) based dual-stream structur for MC is developed. The first stream extracts local raw temporal characteristics fron raw signals, while the second stream learns knowledge from amplitude and phase dat To learn spatial and temporal information from each stream, CNN-LSTM is used, whic combines the spatial feature extraction ability of CNN and superior capacity of processin time-series data of LSTM. Furthermore, the features learned from two streams interact i pairs as a result of an effective operation, expanding the diversity of characteristics an therefore, improving the classification performance of the received signal.

In [90], a CNN-based MC is studied in order to classify SC and OFDM systems wit varying symbol lengths. The majority of older DL-based MC algorithms misinterprete OFDM-based signals with varying OFDM usable symbol lengths. To address this issu FFT window banks (FWB) are utilized as input to the CNN model to estimate the length o an OFDM symbol. After estimating the OFDM symbol length, a CNN-based MC techniqu is utilized to categorize the OFDM and SC modulation formats concurrently using FW and IQ samples as combined input. However, compared to the traditional DL-based MC this technique needed a longer received symbol to obtain the correct classification.

In [91], an OFDM signal identification technique based on a hybrid grey wolf opt mization (HGWO) algorithm to optimize with a deep neural network model is carrie out. This technique can distinguish the OFDM modulation signal from complex signal such as SC, OFDM signals, and wavelet packet signals (WPM) in a multipath channe In this technique, mixed order moment $u_{20} = M_{42}(x_m)/M_{20}^2(x_m)$, characteristics paramete $R = \sigma^2/\mu^2$, and $N = \frac{BW}{\Delta f} - 1$ are extracted from the received signal. Then, a dataset i prepared by using u_{20}, R, and N, which are the input of the classifier. Then the HGWC algorithm is used to optimize the weights and thresholds of the DNN. The experimenta findings demonstrate that the suggested method significantly speeds and improves the cor vergence speed of GWO. When compared to traditional methods, such as particle swarn optimization (PSO) and whale optimization algorithm (WOA), this technique outperform the two. However, the HGWO technique in this study is limited because it is only used to improve the weight and threshold of the DNN model, and the network structure must be chosen manually. The intelligent optimization technique applied to the structure of the deep learning network may enhance classification accuracy.

In [92], the MC technique for OFDM VLC systems based on transfer learning (TL) is developed. For virtually all SNR values, the suggested AlexNet/GoogLeNet-TL-based strategy outperforms previous approaches in which the AlexNet/GoogLeNet is trained from scratch (AlexNet/GoogLeNet-SC). In more practical, few-training-data circumstances, AlexNet/GoogLeNet-TL outperforms AlexNet/GoogLeNet-SC by a wide margin.

In [93], the authors design and implement lightweight CNN (LCNN) based MC methods, i.e., the ShuffleMC method for the IoT cyber–physical systems. The ShuffleMC technique requires considerably fewer parameters and is far less computationally complex than the CNN-based MC method but the classification performance of both is almost the same at high SNR. Furthermore, the authors introduce the FFT to pre-process the received OFDM signals for improved classification performance and training acceleration. In addition, l_2 regularization is used in the training procedure to minimize over-fitting and marginally enhance classification performance.

In [94], a hierarchical CNN-based MC is developed for the waveform and MC in radar communications systems. Using time-frequency representation of the received signal from the Fourier synchrosqueezing transformation (FSST) and deep CNN, the received signal is categorized as either SC radar signals or multicarrier radar signals. Then the cyclic prefix duration, the number of subcarriers, and subcarriers spacing are estimated for the received OFDM signal. After that, the independent component analysis (ICA) operation is used to make the I- and Q-components, which are fed into the CNN classifier for MC.

In [95], a spectrum interference-based two-level data augmentation method in CNN for MC is studied. The short-time Fourier transform (STFT) and IFFT are used to assist in the expansion of signals and the introduction of variations while maintaining the key characteristics. The frequency-domain data are provided to radio signals to improve modulation classification. Experimental results demonstrate that using a two-level data augmentation approach based on spectrum interference may considerably enhance the accuracy of the deep CNN for MC, especially when the SNR is low. This methodology obtains state-of-the-art classification accuracy when compared to a range of data augmentation approaches and leading modulation classification algorithms using the public dataset RadioML 2016.10a.

In [97], a CNN-based MC algorithm is designed, which used a novel data generation technique allowing deep networks to compute correlations between samples inside each OFDM symbol and between symbols. The authors construct a unique advanced processing block that integrates attention and residual connections to boost the learning efficiency of the model. This approach is tested on a synthetic OFDM signal dataset and shows improved classification performance under various channel circumstances.

The cross-talk between sub-carrier has been addressed in terms of CFO. The errors in CFO destroys the orthogonality among the subcarriers or subchannels, thereby introducing ICI. Therefore, classification performance for the MC algorithm may degrade due to ICI or the presence of CFO. Therefore, we need to estimate and compensate for the CFO before MC [63]. In the paper [62], the amplitude moments and correlation properties are used to classify the modulation scheme for OFDM systems. This technique considered the presence of CFO, which is the cause of ICI in the amplitude moments of the received signal. Therefore, the ICI component is estimated and eliminated by using the correlation between the subcarriers. This approach achieves the desired classification accuracy at 30 dB SNR for the normalized CFO for range $0.1 \leq \epsilon \leq 0.2$. In [63], the authors use the DFT and fourth-order cumulant to classify the modulation scheme in the presence of CFO. However, this technique has good classification accuracy for the normalized carrier frequency offset of range $-0.5 \leq \epsilon \leq 0.5$. In [90], a CNN-based MC is studied to classify modulation format for OFDM systems in the presence of CFO. FFT window banks (FWB) are utilized as input to the CNN model to estimate the length of an OFDM symbol. After estimating the OFDM symbol length, a CNN-based MC technique is utilized to categorize the OFDM and SC modulation formats concurrently, using FWB and IQ samples as combined input. The classification performance of this technique degrades to 87.3% in a minor CFO and 83.4% in a moderate CFO. However, it has a classification accuracy of 98.5% at high SNR in the absence of CFO.

5. Challenges and Future Research Directions

Based on an exhaustive literature review, this paper summarizes the two major MC approaches for the OFDM signal: statistics based and AI based, and also highlights their advantages and disadvantages. In the statistics-based approach, the LB approach provides optimal classification performance. As the number of unknown parameters increases, it becomes more computationally complex to find a desired analytical solution for the decision problem. If there is a closed-form solution made, it can be impractical because of its high computational complexity. A sub-optimal classifier is obtained from the optimal ML classifier to minimize computational complexity. In the FB algorithm, the expert domain feature needs to be extracted first, then decisions are made for the classification FB algorithms are easier to implement, despite being sub-optimal. Many of the ML- and DL-based MC first use the signal pre-processing step, which includes noise reduction, parameter estimation, and making the signal synchronized, which enhances the quality of the received signal. After that, proper selection of classification model that can reduce the signal processing steps, increase the modulations classification accuracy and provide more reliable and effective methods of modulation classification, compared to conventional modulation methods.

Nevertheless, several studies are mainly based on ideal hypotheses and rely on a large number of labeled signals. Most of the MC research is still focused on the simulation stage. The communication environment is more sophisticated, and signal frame lengths are varied in the realistic implementation scenario. However, with the increasing complexities of the communication environment and the increasing need for numerous particular tasks, it is difficult to make sure that a huge training data set is generated effectively for particular tasks. The development of semi-supervised algorithm systems is needed to solve this problem. Effective semi-supervised algorithms may be able to fulfill the increasing need for diverse signal processing demands by collecting a large amount of data, only a small fraction of which is labeled data. Another potential task is to figure out how to develop hardware platforms, implant applications, and evaluate algorithms employing measured data.

Another challenge in the future is how to incorporate a DL-based transmission signal modulation identifier for OFDM signals on a field-programmable gate array (FPGA), which would necessitate further research into data quantization, model compression, and other related studies. Finally, DL techniques have a wide range of applications and growth potential as a powerful method for processing data and extracting features. In various fields, combining the DL model with other intelligent algorithms will yield more efficient results. Furthermore, traditional DL-based MC is challenging to implement in OFDM-based narrow-band (NB)-IoT devices, as it requires high computational complexity and more power as well as memory resources. However, implementing light-weight DL-based blind MC for NB-IoT devices that need less computational, space, and power requirements might be a difficult task for future adaptive transceiver systems. Another challenge in the future is modulation classification for OQPSK, $\pi/4$-QPSK, and MSK. Higher-order modulation classification for OFDM, MIMO-OFDM system, and adaptive OFDM systems over a randomized environment using a hybrid model need to be proposed in future wireless communication. In addition, we have to extend to a large number of modulation formats that work for all types of systems. MC can be implemented for massive MIMO systems, such as intelligent reflective surfaces, to reduce the distortion due to the non-line-of-sight (NLOS) component of the signal in future wireless communication.

In OFDM-IM, the number of active subcarriers can be adjusted to achieve the desired spectral efficiency and BER performance. Thus, the MC algorithm for OFDM-IM needs to be explored. As compared to the OFDM system, the filter bank multicarrier (FBMC) system does not require a CP, so it makes the use of radio resources more efficient. Therefore, MC for FBMC can be a future problem. In NOMA, if a different user uses a different modulation format, then MC for NOMA can be a challenging task. As compared to the OFDM system, orthogonal time frequency space (OTFS) has significantly high error performance over delay-Doppler channels with a wide range of Doppler frequencies. MC for OTFS can be a future research problem for designing advanced wireless communication systems. Due to the high peak-to-average power ratio (PAPR), it is difficult to use OFDM on the uplink. To overcome this problem, single-carrier frequency division multiple access (SC-FDMA) is used on the uplink. Therefore, the MC algorithm for SC-FDMA needs to be developed. MC for multicarrier code-division multiple access (MC-CDMA) can also be a critical research problem for future wireless communication.

Author Contributions: Conceptualization, A.K. and S.M.; methodology, A.K.; validation, S.M.; formal analysis, A.K. and S.M.; investigation, A.K. and S.M.; writing, original draft preparation, A.K. and S.M.; writing, review and editing, A.K., S.M., G.G., H-C.W. and C.Y.; visualization, A.K., S.M., G.G., H-C.W. and C.Y.; supervision, S.M., G.G., H-C.W. and C.Y. All authors have read and agreed to the published version of the manuscript.

Funding: This work was partially supported by the Ministry of Electronics and Information Technology, Government of India, under Project 13 (2)/2020-CC&BT and in part by the Empowerment and Equity Opportunities for Excellence in Science schemes by the Science and Engineering Research Board through the Department of Science and Technology, Government of India, under Project EEQ/2018/000201.

Institutional Review Board Statement: Not applicable.

Informed Consent Statement: Not applicable.

Data Availability Statement: Not applicable.

Acknowledgments: The authors would like to thank Octavia A. Dobre, Memorial University, Canada for the precious help and suggestions.

Conflicts of Interest: The authors declare no conflict of interest.

References

1. Dobre, O.A.; Abdi, A.; Bar-Ness, Y.; Su, W. Survey of automatic modulation classification techniques: Classical approaches and new trends. *IET Commun.* **2007**, *1*, 137–156. [CrossRef]
2. Dobre, O.A. Signal identification for emerging intelligent radios: Classical problems and new challenges. *IEEE Instrum. Meas. Mag.* **2015**, *18*, 11–18. [CrossRef]
3. Eldemerdash, Y.A.; Dobre, O.A.; Öner, M. Signal identification for multiple-antenna wireless systems: Achievements and challenges. *IEEE Commun. Surv. Tutor.* **2016**, *18*, 1524–1551. [CrossRef]
4. Jiang, W.; Wu, X.; Wang, Y.; Chen, B.; Feng, W.; Jin, Y. Time–Frequency-Analysis-Based Blind Modulation Classification for Multiple-Antenna Systems. *Sensors* **2021**, *21*, 231. [CrossRef]
5. Gupta, R.; Majhi, S.; Dobre, O.A. Design and Implementation of a Tree-Based Blind Modulation Classification Algorithm for Multiple-Antenna Systems. *IEEE Trans. Instrum. Meas.* **2019**, *68*, 3020–3031. [CrossRef]
6. Majhi, S.; Gupta, R.; Xiang, W.; Glisic, S. Hierarchical Hypothesis and Feature-Based Blind Modulation Classification for Linearly Modulated Signals. *IEEE Trans. Veh. Technol.* **2017**, *66*, 11057–11069. [CrossRef]
7. Xu, J.L.; Su, W.; Zhou, M. Software-Defined Radio Equipped With Rapid Modulation Recognition. *IEEE Trans. Veh. Technol.* **2010**, *59*, 1659–1667. [CrossRef]
8. Majhi, S.; Kumar, M.; Xiang, W. Implementation and Measurement of Blind Wireless Receiver for Single Carrier Systems. *IEEE Trans. Instrum. Meas.* **2017**, *66*, 1965–1975. [CrossRef]
9. Majhi, S.; Ho, T.S. Blind Symbol-Rate Estimation and TestBed Implementation of Linearly Modulated Signals. *IEEE Trans. Veh. Technol.* **2015**, *64*, 954–963. [CrossRef]
10. Zhang, J.K.; Yuen, C.; Huang, F. Full diversity blind signal designs for unique identification of frequency selective channels. *IEEE Trans. Veh. Technol.* **2012**, *61*, 2172–2184. [CrossRef]

11. Wei, W.; Mendel, J.M. Maximum-likelihood classification for digital amplitude-phase modulations. *IEEE Trans. Commun.* **200** *48*, 189–193. [CrossRef]
12. Swami, A.; Sadler, B.M. Hierarchical digital modulation classification using cumulants. *IEEE Trans. Commun.* **2000**, *48*, 416–42 [CrossRef]
13. Wu, H.C.; Saquib, M.; Yun, Z. Novel Automatic Modulation Classification Using Cumulant Features for Communications vi Multipath Channels. *IEEE Wirel. Commun.* **2008**, *7*, 3098 –3105.
14. Oner, M.; Dobre, O.A. On the Second-Order Cyclic Statistics of Signals in the Presence of Receiver Impairments. *IEEE Tran Commun.* **2011**, *59*, 3278–3284. [CrossRef]
15. Majhi, S.; Gupta, R.; Xiang, W. Novel blind modulation classification of circular and linearly modulated signals using cycli cumulants. In Proceedings of the 2017 IEEE 28th Annual International Symposium on Personal, Indoor, and Mobile Radi Communications (PIMRC), Montreal, QC, Canada, 8–13 October 2017; pp. 1–5.
16. Gupta, R.; Majhi, S.; Dobre, O.A. Blind Modulation Classification of Different Variants of QPSK and 8-PSK for Multiple-Antenn Systems with Transmission Impairments. In Proceedings of the 2018 IEEE 88th Vehicular Technology Conference (VTC-Fall Chicago, IL, USA, 27–30 August 2018; pp. 1–5.
17. Daponte, P.; Mercurio, G.; Rapuano, S. A wavelet networks-based method for the digital telecommunication system monitorin *IEEE Trans. Instrum. Meas.* **2001**, *50*, 1773–1780. [CrossRef]
18. Jerjawi, W.A.; Eldemerdash, Y.A.; Dobre, O.A. Second-order cyclostationarity-based detection of LTE SC-FDMA signals fo cognitive radio systems. *IEEE Trans. Instrum. Meas.* **2014**, *64*, 823–833. [CrossRef]
19. Dobre, O.A.; Venkatesan, R.; Popescu, D.C. Second-order cyclostationarity of mobile WiMAX and LTE OFDM signals an application to spectrum awareness in cognitive radio systems. *IEEE J. Sel. Top. Signal Process.* **2011**, *6*, 26–42.
20. Karami, E.; Dobre, O.A. Identification of SM-OFDM and AL-OFDM signals based on their second-order cyclostationarity. *IEE Trans. Veh. Technol.* **2014**, *64*, 942–953. [CrossRef]
21. Punchihewa, A.; Zhang, Q.; Dobre, O.A.; Spooner, C.; Rajan, S.; Inkol, R. On the cyclostationarity of OFDM and single carrie linearly digitally modulated signals in time dispersive channels: Theoretical developments and application. *IEEE Trans. Wire Commun.* **2010**, *9*, 2588–2599. [CrossRef]
22. Santhanavijayan, A.; Kumar, D.N.; Deepak, G. A semantic-aware strategy for automatic speech recognition incorporating dee learning models. In *Intelligent System Design*; Springer: Singapore, 2021; pp. 247–254.
23. LeCun, Y.; Bengio, Y.; Hinton, G. Deep learning. *Nature* **2015**, *521*, 436–444. [CrossRef]
24. Aslam, M.W.; Zhu, Z.; Nandi, A.K. Automatic Modulation Classification Using Combination of Genetic Programming and KN *IEEE Trans. Wirel. Commun.* **2012**, *11*, 2742–2750.
25. Liu, X.; Zhao, C.; Wang, P.; Zhang, Y.; Yang, T. Blind modulation classification algorithm based on machine learning for spatiall correlated MIMO system. *IET Commun.* **2017**, *11*, 1000–1007. [CrossRef]
26. Wang, Y.; Liu, M.; Yang, J.; Gui, G. Data-Driven Deep Learning for Automatic Modulation Recognition in Cognitive Radios. *IEE Trans. Veh. Technol.* **2019**, *68*, 4074–4077. [CrossRef]
27. Xie, W.; Hu, S.; Yu, C.; Zhu, P.; Peng, X.; Ouyang, J. Deep Learning in Digital Modulation Recognition Using High Orde Cumulants. *IEEE Access* **2019**, *7*, 63760–63766. [CrossRef]
28. Wang, Y.; Guo, L.; Zhao, Y.; Yang, J.; Adebisi, B.; Gacanin, H.; Gui, G. Distributed learning for automatic modulation classificatio in edge devices. *IEEE Wirel. Commun. Lett.* **2020**, *9*, 2177–2181. [CrossRef]
29. Wang, Y.; Gui, G.; Ohtsuki, T.; Adachi, F. Multi-task learning for generalized automatic modulation classification unde non-Gaussian noise with varying SNR conditions. *IEEE Trans. Wirel. Commun.* **2021**, *20*, 3587–3596. [CrossRef]
30. Chang, S.; Huang, S.; Zhang, R.; Feng, Z.; Liu, L. Multi-Task Learning Based Deep Neural Network for Automatic Modulatio Classification. *IEEE Internet Things J.* **2022**, *9*, 2192–2206. [CrossRef]
31. Huang, S.; Jiang, Y.; Gao, Y.; Feng, Z.; Zhang, P. Automatic modulation classification using contrastive fully convolutiona network. *IEEE Wirel. Commun. Lett.* **2019**, *8*, 1044–1047. [CrossRef]
32. Huang, J.; Huang, S.; Zeng, Y.; Chen, H.; Chang, S.; Zhang, Y. Hierarchical Digital Modulation Classification Using Cascade Convolutional Neural Network. *J. Commun. Inf. Netw.* **2021**, *6*, 72–81.
33. Qi, P.; Zhou, X.; Zheng, S.; Li, Z. Automatic modulation classification based on deep residual networks with multimoda information. *IEEE Trans. Cogn. Commun. Netw.* **2020**, *7*, 21–33. [CrossRef]
34. Han, H.; Ren, Z.; Li, L.; Zhu, Z. Automatic Modulation Classification Based on Deep Feature Fusion for High Noise Level an Large Dynamic Input. *Sensors* **2021**, *21*, 2117. [CrossRef] [PubMed]
35. Bu, K.; He, Y.; Jing, X.; Han, J. Adversarial transfer learning for deep learning based automatic modulation classification. *IEE Signal Process. Lett.* **2020**, *27*, 880–884. [CrossRef]
36. Tu, Y.; Lin, Y.; Hou, C.; Mao, S. Complex-valued networks for automatic modulation classification. *IEEE Trans. Veh. Technol.* **202** *69*, 10085–10089. [CrossRef]
37. Liu, Y.; Peng, X.; Xiong, Z.; Lu, Y. Phoneme-Based Distribution Regularization for Speech Enhancement. In Proceedings of th ICASSP 2021-2021 IEEE International Conference on Acoustics, Speech and Signal Processing (ICASSP), Toronto, ON, Canad 6–11 June 2021; pp. 726–730.
38. Hwang, T.; Yang, C.; Wu, G.; Li, S.; Li, G.Y. OFDM and Its Wireless Applications: A Survey. *IEEE Trans. Veh. Technol.* **200** *58*, 1673–1694. [CrossRef]

39. Lorincz, J.; Ramljak, I.; Begušić, D. A Survey on the Energy Detection of OFDM Signals with Dynamic Threshold Adaptation: Open Issues and Future Challenges. *Sensors* **2021**, *21*, 3080. [CrossRef]
40. Qin, Z.; Ye, H.; Li, G.Y.; Juang, B.F. Deep Learning in Physical Layer Communications. *IEEE Wirel. Commun.* **2019**, *26*, 93–99. [CrossRef]
41. Gui, G.; Huang, H.; Song, Y.; Sari, H. Deep Learning for an Effective Nonorthogonal Multiple Access Scheme. *IEEE Trans. Veh. Technol.* **2018**, *67*, 8440–8450. [CrossRef]
42. Chen, S.; Zhang, Y.; He, Z.; Nie, J.; Zhang, W. A Novel Attention Cooperative Framework for Automatic Modulation Recognition. *IEEE Access* **2020**, *8*, 15673–15686. [CrossRef]
43. Nie, J.; Zhang, Y.; He, Z.; Chen, S.; Gong, S.; Zhang, W. Deep Hierarchical Network for Automatic Modulation Classification. *IEEE Access* **2019**, *7*, 94604–94613. [CrossRef]
44. Ha, C.; You, Y.; Song, H. Machine Learning Model for Adaptive Modulation of Multi-Stream in MIMO-OFDM System. *IEEE Access* **2019**, *7*, 5141–5152. [CrossRef]
45. Yucek, T.; Arslan, H. A novel sub-optimum maximum-likelihood modulation classification algorithm for adaptive OFDM systems. In Proceedings of the IEEE Wireless Communications and Networking Conference, Atlanta, GA, USA, 21–25 March 2004; Volume 2, pp. 739–744.
46. Leinonen, J.; Juntti, M. Modulation classification in adaptive OFDM systems. In Proceedings of the IEEE 59th Vehicular Technology Conference. VTC 2004-Spring, Milan, Italy, 17–19 May 2004; Volume 3, pp. 1554–1558.
47. Zheng, J.; Lv, Y. Likelihood-based automatic modulation classification in OFDM with index modulation. *IEEE Trans. Veh. Technol.* **2018**, *67*, 8192–8204. [CrossRef]
48. Fang, T.; Liu, S.; Ma, L.; Zhang, L.; Khan, I.U. Subcarrier modulation identification of underwater acoustic OFDM based on block expectation maximization and likelihood. *Appl. Acoust.* **2021**, *173*, 107654. [CrossRef]
49. Marey, M.; Mostafa, H. Turbo modulation identification algorithm for OFDM software-defined radios. *IEEE Commun. Lett.* **2021**, *25*, 1707–1711. [CrossRef]
50. Häring, L.; Chen, Y.; Czylwik, A. Automatic modulation classification methods for wireless OFDM systems in TDD mode. *IEEE Trans. Commun.* **2010**, *58*, 2480–2485. [CrossRef]
51. Häring, L.; Chen, Y.; Czylwik, A. Efficient modulation classification for adaptive wireless OFDM systems in TDD mode. In Proceedings of the IEEE Wireless Communication and Networking Conference, Sydney, Australia, 18–21 April 2010; pp. 1–6.
52. Haring, L.; Chen, Y.; Czylwik, A. Utilizing side information in modulation classification for wireless OFDM systems with adaptive modulation. In Proceedings of the 2011 IEEE Vehicular Technology Conference (VTC Fall), San Francisco, CA, USA, 5–8 September 2011; pp. 1–5.
53. Häring, L.; Kisters, C. MAP-based automatic modulation classification for wireless adaptive OFDM systems. In Proceedings of the IEEE International Conference on Acoustics, Speech and Signal Processing, Vancouver, BC, Canada, 26–31 May 2013; pp. 5204–5208.
54. Haering, L.; Kisters, C. Signalling-assisted modulation classification in wireless OFDM systems with adaptive modulation and coding. In Proceedings of the International Conference on Communications (ICC), Budapest, Hungary, 9–13 June 2013; pp. 5037–5041.
55. Häring, L.; Kisters, C. Joint optimization of bit loading and modulation classification in wireless OFDM systems. In Proceedings of the IEEE 9th International Conference on Wireless and Mobile Computing, Networking and Communications (WiMob), Lyon, France, 7–9 October 2013; pp. 402–407.
56. Haering, L.; Kisters, C. Influence of imperfect reciprocity on MAP-based automatic modulation classification for adaptive OFDM systems in TDD mode. In Proceedings of the 18th International OFDM Workshop 2014 (InOWo'14), Essen, Germany, 27–28 August 2014; pp. 1–6.
57. Husmann, C.; Chen, Y. Modulation classification for adaptive mobile OFDM systems. In Proceedings of the 18th International OFDM Workshop (InOWo'14), Essen, Germany, 27–28 August 2014; pp. 1–8.
58. Bahrani, S.; Derakhtian, M.; Zolghadrasli, A. Effect of channel prediction on automatic modulation classification for adaptive OFDM Systems. In Proceedings of the 20th Iranian Conference on Electrical Engineering (ICEE2012), Tehran, Iran, 15–17 May 2012; pp. 1280–1285.
59. Karabacak, M.; Cırpan, H.A.; Arslan, H. Adaptive pilot based modulation identification and channel estimation for OFDM systems. In Proceedings of the 21st Annual IEEE International Symposium on Personal, Indoor and Mobile Radio Communications, Istanbul, Turkey, 26–30 September 2010; pp. 730–735.
60. Bahrani, S.; Derakhtian, M.; Zolghadrasli, A. Performance analysis of a low-complexity MAP algorithm for automatic modulation classification in adaptive OFDM systems. *IET Commun.* **2016**, *10*, 2363–2371. [CrossRef]
61. Pambudi, A.D.; Tjondronegoro, S.; Wijanto, H. Statistical properties proposed for blind classification OFDM modulation scheme. In Proceedings of the IEEE International Conference on Aerospace Electronics and Remote Sensing Technology, Yogyakarta, Indonesia, 13–14 November 2014; pp. 89–93.
62. Shimbo, D.; Oka, I. A modulation classification using amplitude moments in OFDM systems. In Proceedings of the International Symposium On Information Theory & Its Applications, Taichung, Taiwan, 17–20 October 2010; pp. 288–293.
63. Gupta, R.; Kumar, S.; Majhi, S. Blind Modulation Classification for Asynchronous OFDM Systems Over Unknown Signal Parameters and Channel Statistics. *IEEE Trans. Veh. Technol.* **2020**, *69*, 5281–5292 [CrossRef]

64. Zhang, J.; Li, B. A new modulation identification scheme for OFDM in multipath rayleigh fading channel. In Proceedings of the International Symposium on Computer Science and Computational Technology, Shanghai, China, 20–22 December 2008; Volume 1, pp. 793–796.
65. Zhu, Y.; Tian, B.; Ma, R.; Sun, Y.; An, J.; Yi, K.; Ren, Y. An OFDM modulation recognition algorithm based on spectrum analysis In Proceedings of the 10th International Conference on Signal Processing Proceedings, Beijing, China, 24–28 October 2010; pp. 1557–1560.
66. Ma, Y.; Gao, M.; Ye, Y.; Chen, W.; Wang, L.; Sha, Y.; Yan, Y. Modulation Format Identification for Adaptive Optical OFDM System In Proceedings of the 24th OptoElectronics and Communications Conference (OECC) and International Conference on Photonics in Switching and Computing (PSC), Fukuoka, Japan, 7–11 July 2019; pp. 1–3.
67. Katayama, T.; Oka, I.; Ata, S. Modulation identification by general orthogonal modulations. In Proceedings of the International Conference on Advanced Technologies for Communications, Hanoi, Vietnam, 6–9 October 2008; pp. 12–15.
68. Chen, J.; Kuo, Y.; Liu, X. Modulation identification for MIMO-OFDM signals. In Proceedings of the 2007 IET Conference on Wireless, Mobile and Sensor Networks (CCWMSN07), Shanghai, China, 12–14 December 2007.
69. Li, H.; Bar-Ness, Y.; Abdi, A.; Somekh, O.S.; Su, W. OFDM modulation classification and parameters extraction. In Proceedings of the 1st International Conference on Cognitive Radio Oriented Wireless Networks and Communications, Mykonos, Greece, 8–10 June 2006; pp. 1–6.
70. Liu, Y.; Simeone, O.; Haimovich, A.M.; Su, W. Modulation classification for MIMO-OFDM signals via Gibbs sampling. In Proceedings of the IEEE 49th Annual Conference on Information Sciences and Systems (CISS), Baltimore, MD, USA, 18–20 March 2015; pp. 1–6.
71. Liu, Y.; Simeone, O.; Haimovich, A.M.; Su, W. Modulation classification for MIMO-OFDM signals via approximate Bayesian inference. *IEEE Trans. Veh. Technol.* **2016**, *66*, 268–281. [CrossRef]
72. Pathy, A.K.; Kumar, A.; Gupta, R.; Kumar, S.; Majhi, S. Design and Implementation of Blind Modulation Classification for Asynchronous MIMO-OFDM System. *IEEE Trans. Instrum. Meas.* **2021**, *70*, 5504011. [CrossRef]
73. Wong, M.D.; Ting, S.K.; Nandi, A.K. Naive Bayes classification of adaptive broadband wireless modulation schemes with higher order cumulants. In Proceedings of the 2nd International Conference on Signal Processing and Communication Systems, Gold Coast, Australia, 15–17 December 2008; pp. 1–5.
74. El-Khamy, S.E.; Elsayed, H.A.; Rizk, M.M. C45. Classification of OFDM signals using higher order statistics and clustering techniques. In Proceedings of the 29th National Radio Science Conference (NRSC), Cairo, Egypt, 10–12 April 2012; pp. 541–549
75. Yuan, X.; Li, Y.; Gao, M.; Li, T.; Zhang, H. Automatic modulation classification for MIMO-OFDM systems with imperfect timing synchronization. In Proceedings of the IEEE 86th Vehicular Technology Conference (VTC-Fall), Toronto, ON, Canada, 24–27 September 2017; pp. 1–5.
76. Machida, W.; Ichijo, K.; Sugiura, Y.; Shimamura, T. Phase Correction for Automatic Modulation Classification Using Iterative Closest Point. In Proceedings of the International Symposium on Intelligent Signal Processing and Communication Systems (ISPACS), Taipei, Taiwan, 3–6 December 2019; pp. 1–2.
77. Zhang, Y.; Wu, G.; Wang, J.; Tang, Q. Wireless Signal Classification Based on High-Order Cumulants and Machine Learning. In Proceedings of the International Conference on Computer Technology, Electronics and Communication (ICCTEC), Dalian, China, 19–21 December 2017; pp. 559–564.
78. Dehri, B.; Besseghier, M.; Djebbar, A.B.; Dayoub, I. Blind digital modulation classification for STBC-OFDM system in presence of CFO and channels estimation errors. *IET Commun.* **2019**, *13*, 2827–2833. [CrossRef]
79. Gu, Y.; Xu, S.; Zhou, J. Automatic Modulation Format Classification of USRP Transmitted Signals Based on SVM. In Proceedings of the 2020 Chinese Automation Congress (CAC), Shanghai, China, 6–8 November 2020; pp. 3712–3717.
80. He, J.; Zhou, Y.; Shi, J.; Tang, Q. Modulation classification method based on clustering and gaussian model analysis for vlc system. *IEEE Photonics Technol. Lett.* **2020**, *32*, 651–654. [CrossRef]
81. Gaohui, L.; Jiakun, C. Research on Modulation Recognition of OFDM Signal Based on Hierarchical Iterative Support Vector Machine. In Proceedings of the 2020 International Conference on Communications, Information System and Computer Engineering (CISCE), Kuala Lumpur, Malaysia, 3–5 July 2020; pp. 38–44.
82. Al-Makhlasawy, R.M.; Elnaby, M.M.A.; El-Khobby, H.A.; El-Samie, F.E.A. Automatic modulation recognition in OFDM systems using cepstral analysis and a fuzzy logic interface. In Proceedings of the 8th International Conference on Informatics and Systems (INFOS), Giza, Egypt, 14–16 May 2012; pp. CC-56–CC-62.
83. Li, Y.; Shao, G.; Wang, B. Automatic Modulation Classification Based on Bispectrum and CNN. In Proceedings of the IEEE 8th Joint International Information Technology and Artificial Intelligence Conference (ITAIC), Chongqing, China, 24–26 May 2019; pp. 311–316.
84. Hong, S.; Zhang, Y.; Wang, Y.; Gu, H.; Gui, G.; Sari, H. Deep Learning-Based Signal Modulation Identification in OFDM Systems. *IEEE Access* **2019**, *7*, 114631–114638. [CrossRef]
85. Shi, J.; Hong, S.; Cai, C.; Wang, Y.; Huang, H.; Gui, G. Deep Learning-Based Automatic Modulation Recognition Method in the Presence of Phase Offset. *IEEE Access* **2020**, *8*, 42841–42847. [CrossRef]
86. Hong, S.; Wang, Y.; Pan, Y.; Gu, H.; Liu, M.; Yang, J.; Gui, G. Convolutional neural network aided signal modulation recognition in OFDM systems. In Proceedings of the IEEE 91st Vehicular Technology Conference (VTC2020-Spring), Antwerp, Belgium, 25–28 May 2020, pp. 1–5.

7. Meng, F.; Chen, P.; Wu, L.; Wang, X. Automatic modulation classification: A deep learning enabled approach. *IEEE Trans. Veh. Technol.* **2018**, *67*, 10760–10772. [CrossRef]
8. AlNuaimi, D.H. AMC2-Pyramid: Intelligent Pyramidal Feature Engineering and Multi-Distance Decision Making for Automatic Multi-Carrier Modulation Classification. *IEEE Access* **2021**, *9*, 137560–137583. [CrossRef]
9. Zhang, Z.; Luo, H.; Wang, C.; Gan, C.; Xiang, Y. Automatic modulation classification using CNN-LSTM based dual-stream structure. *IEEE Trans. Veh. Technol.* **2020**, *69*, 13521–13531. [CrossRef]
0. Park, M.C.; Han, D.S. Deep Learning-Based Automatic Modulation Classification With Blind OFDM Parameter Estimation. *IEEE Access* **2021**, *9*, 108305–108317. [CrossRef]
1. Zhang, Y.; Liu, D.; Liu, J.; Xian, Y.; Wang, X. Improved deep neural network for OFDM signal recognition using hybrid grey wolf optimization. *IEEE Access* **2020**, *8*, 133622–133632. [CrossRef]
2. Zhao, Z.; Wei, Z.; Wang, Z.; Zhang, Y.; Li, M.; Khan, F.N.; Fu, H. Modulation Format Recognition based on Transfer Learning for Visible Light Communication Systems. In Proceedings of the Optoelectronics and Communications Conference, Hong Kong, China, 3–7 July 2021; pp. JS2B.12.
3. Yin, J.; Guo, L.; Jiang, W.; Hong, S.; Yang, J. ShuffleNet-inspired lightweight neural network design for automatic modulation classification methods in ubiquitous IoT cyber-physical systems. *Comput. Commun.* **2021**, *176*, 249–257. [CrossRef]
4. Kong, G.; Jung, M.; Koivunen, V. Waveform Classification in Radar-Communications Coexistence Scenarios. In Proceedings of the GLOBECOM 2020-2020 IEEE Global Communications Conference, Taipei, Taiwan, 7–11 Decemebr 2020; pp. 1–6.
5. Zheng, Q.; Zhao, P.; Li, Y.; Wang, H.; Yang, Y. Spectrum interference-based two-level data augmentation method in deep learning for automatic modulation classification. *Neural Comput. Appl.* **2021**, *33*, 7723–7745. [CrossRef]
6. Zhang, L.; Lin, C.; Yan, W.; Ling, Q.; Wang, Y. Real-Time OFDM Signal Modulation Classification Based on Deep Learning and Software-Defined Radio. *IEEE Commun. Lett.* **2021**, *25*, 2988–2992. [CrossRef]
7. Huynh-The, T.; Pham, Q.V.; Nguyen, T.V.; Pham, X.Q.; Kim, D.S. Deep Learning-based Automatic Modulation Classification for Wireless OFDM Communications. In Proceedings of the 2021 International Conference on Information and Communication Technology Convergence (ICTC), Jeju Island, Korea, 20–22 October 2021; pp. 47–49.
8. Wang, F.; Wang, X. Fast and Robust Modulation Classification via Kolmogorov-Smirnov Test. *IEEE Trans. Commun.* **2010**, *58*, 2324–2332. [CrossRef]
9. Mohammadkarimi, M.; Dobre, O.A. Blind identification of spatial multiplexing and Alamouti space-time block code via Kolmogorov-Smirnov (KS) test. *IEEE Commun. Lett.* **2014**, *18*, 1711–1714. [CrossRef]
00. Trees, V.; Harry, L. *Detection, Estimation, and Modulation Theory-Part L-Detection, Estimation, and Linear Modulation Theory*; John Wiley & Sons: New York, NY, USA, 2001.
01. Azzouz, E.; Nandi, A.K. *Automatic Modulation Recognition of Communication Signals*; Springer: New York, NY, USA, 2013.
02. Hsue, S.Z.; Soliman, S.S. Automatic modulation classification using zero crossing. *IEE Proc. (Radar Signal Process.)* **1990**, *137*, 459–464. IET Digital Library. [CrossRef]
03. Ramkumar, B. Automatic modulation classification for cognitive radios using cyclic feature detection. *IEEE Circuits Syst. Mag.* **2009**, *9*, 27–45. [CrossRef]
04. Mobasseri, B.G. Digital modulation classification using constellation shape. *Signal Process.* **2000**, *80*, 251–277. [CrossRef]
05. Lopatka, J.; Pedzisz, M. Automatic modulation classification using statistical moments and a fuzzy classifier. In Proceedings of the WCC 2000-ICSP 2000, 2000 5th International Conference on Signal Processing Proceedings, 16th World Computer Congress 2000, Beijing, China, 21–25 August 2000; Volume 3, pp. 1500–1506.
06. Paris, B.P.; Orsak, G.C.; Chen, H.; Warke, N. Modulation classification in unknown dispersive environments. In Proceedings of the 1997 IEEE International Conference on Acoustics, Speech, and Signal Processing, Munich, Germany, 21–24 April 1997; Volume 5, pp. 3853–3856.
07. Huo, X.; Donoho, D. A simple and robust modulation classification method via counting. In Proceedings of the 1998 IEEE International Conference on Acoustics, Speech and Signal Processing, ICASSP'98 (Cat. No. 98CH36181), Seattle, WA, USA, 15 May 1998; Volume 6, pp. 3289–3292.
08. Dobre, O.A.; Bar-Ness, Y.; Su, W. Robust QAM modulation classification algorithm using cyclic cumulants. In Proceedings of the 2004 IEEE Wireless Communications and Networking Conference (IEEE Cat. No. 04TH8733), Atlanta, GA, USA, 21–25 March 2004; Volume 2, pp. 745–748.
09. Ali, A.; Yangyu, F. Unsupervised feature learning and automatic modulation classification using deep learning model. *Phys. Commun.* **2017**, *25*, 75–84. [CrossRef]
10. Muhlhaus, M.S.; oner, M.; Dobre, O.A.; Jkel, H.U.; Jondral, F.K. Automatic modulation classification for MIMO systems using fourth-order cumulants. In Proceedings of the 2012 IEEE Vehicular Technology Conference (VTC Fall), Quebec City, QC, Canada, 3–6 September 2012; pp. 1–5.
11. Li, T.; Li, Y.; Dobre, O.A. Modulation Classification Based on Fourth-Order Cumulants of Superposed Signal in NOMA Systems. *IEEE Trans. Inf. Forensics Secur.* **2021**, *16*, 2885–2897. [CrossRef]
12. Chaudhari, M.S.; Kumar, S.; Gupta, R.; Kumar, M.; Majhi, S. Design and Testbed Implementation of Blind Parameter Estimated OFDM Receiver. *IEEE Trans. Instrum. Meas.* **2021**. [CrossRef]
13. Wang, C.X.; Di Renzo, M.; Stanczak, S.; Wang, S.; Larsson, E.G. Artificial intelligence enabled wireless networking for 5G and beyond: Recent advances and future challenges. *IEEE Wirel. Commun.* **2020**, *27*, 16–23. [CrossRef]

114. Zha, X.; Peng, H.; Qin, X.; Li, G.; Yang, S. A deep learning framework for signal detection and modulation classification. *Sensor* **2019**, *19*, 4042. [CrossRef]
115. Hinton, G.E.; Salakhutdinov, R.R. Reducing the dimensionality of data with neural networks. *Am. Assoc. Adv. Sci.* **200** *313*, 504–507. [CrossRef]
116. Sun, J.; Shi, W.; Yang, Z.; Yang, J.; Gui, G. Behavioral modeling and linearization of wideband RF power amplifiers using BiLSTM networks for 5G wireless systems. *IEEE Trans. Veh. Technol.* **2019**, *68*, 10348–10356. [CrossRef]
117. Gui, G.; Liu, F.; Sun, J.; Yang, J.; Zhou, Z.; Zhao, D. Flight delay prediction based on aviation big data and machine learning. *IEE Trans. Veh. Technol.* **2019**, *69*, 140–150. [CrossRef]

MDPI
St. Alban-Anlage 66
4052 Basel
Switzerland
Tel. +41 61 683 77 34
Fax +41 61 302 89 18
www.mdpi.com

Sensors Editorial Office
E-mail: sensors@mdpi.com
www.mdpi.com/journal/sensors

www.ingramcontent.com/pod-product-compliance
Lightning Source LLC
Chambersburg PA
CBHW061456220926
51850CB00004B/104
9783036571201